Lecture Notes in Networks and Systems

Volume 353

The series "Lecture Notes in Networks and Systems" publishes the latest developments in Networks and Systems—quickly, informally and with high quality. Original research reported in proceedings and post-proceedings represents the core of LNNS.

Volumes published in LNNS embrace all aspects and subfields of, as well as new challenges in, Networks and Systems.

The series contains proceedings and edited volumes in systems and networks, spanning the areas of Cyber-Physical Systems, Autonomous Systems, Sensor Networks, Control Systems, Energy Systems, Automotive Systems, Biological Systems, Vehicular Networking and Connected Vehicles, Aerospace Systems, Automation, Manufacturing, Smart Grids, Nonlinear Systems, Power Systems, Robotics, Social Systems, Economic Systems and other. Of particular value to both the contributors and the readership are the short publication timeframe and the world-wide distribution and exposure which enable both a wide and rapid dissemination of research output.

The series covers the theory, applications, and perspectives on the state of the art and future developments relevant to systems and networks, decision making, control, complex processes and related areas, as embedded in the fields of interdisciplinary and applied sciences, engineering, computer science, physics, economics, social, and life sciences, as well as the paradigms and methodologies behind them.

Indexed by SCOPUS, INSPEC, WTI Frankfurt eG, zbMATH, SCImago.

All books published in the series are submitted for consideration in Web of Science.

More information about this series at http://www.springer.com/series/15179

Aleksei Muratov · Svetlana Ignateva
Editors

Fundamental and Applied Scientific Research in the Development of Agriculture in the Far East (AFE-2021)

Agricultural Innovation Systems, Volume 1

Springer

Editors
Aleksei Muratov
Far Eastern State Agrarian University
Blagoveshchensk, Russia

Svetlana Ignateva
ITMO University
St. Petersburg, Russia

ISSN 2367-3370 ISSN 2367-3389 (electronic)
Lecture Notes in Networks and Systems
ISBN 978-3-030-91401-1 ISBN 978-3-030-91402-8 (eBook)
https://doi.org/10.1007/978-3-030-91402-8

This Springer imprint is published by the registered company Springer Nature Switzerland AG
The registered company address is: Gewerbestrasse 11, 6330 Cham, Switzerland

Contents

Screening of Potato Cultivars (*Solanum Tuberosum* L.) and Identification of Markers of Fesistance Genes to PVX, PVY, *Globodera Pallida* and *Globodera Rostochiensis* 1
Petr Fisenko, Olga Sobko, Irina Kim, Nathalia Matsishina, and Dmitriy Volkov

Intra-callus Genetic Variability of Rice *Oryza Sativa* L. Doubled Haploids Regenerated Throuth Androgenesis *in vitro* 9
Marina Ilyushko, Marina Romashova, and Svetlana Guchenko

Depth of Fertilizer Application to Justify Subsoiler Body in Soybean Cultivation 19
Viktor Epifantsev, Alexander Panasiuk, Yakov Osipov, and Yuriy Vaitekhovich

Studies on Correction of the Trajectory of a Tractor Train 28
Alexey Kushnarev, Alexandr Shuravin, Evgeny Kuznetsov, Sergey Shchitov, Aleksey Goncharuk, and Olesya Mitrokhina

Improving the Efficiency of Wheel-Wheeling Machine-Tractor Units in Areas with Special Natural and Climatic Conditions 36
Alexander Slepenkov, Evgeny Kuznetsov, Sergey Shchitov, Tatiana Sharipova, Olga Shchegorets, and Andrey Burmaga

Characteristic of Spring Vetch Breeding Lines in the Conditions if the Steppe Zone of Primorsky Krai 44
Olga Telichko, Oksana Mokhan, and Alexey Emelyanov

The Problem of Attracting Private Investment in Agriculture 54
Sergey Yekimov, Viktoriia Nianko, Natalia Sirenko, Alla Burkovska, and Olena Bodnar

The Impact of State Regulation on the Sustainable Development of Rural Areas in Ukraine 62
Sergey Yekimov, Viktoriia Nianko, Olga Kuhareva, Tetiana Mishustina, and Olena Lysa

Distribution of Ascariosis and Asophagostomosis of Pigs in Central Yakutia 70
Alexander Reshetnikov, Lyubov Kozlova, and Anastasia Barashkova

Effect of Salicylic Acid and Copper Sulfate on the Synthesis of Rutin in *Fagopyrum Esculentum* Moench Plants *in vitro* 80
Elena Barsukova, Alexey Klykov, and Elena Chaikina

Promising Middle-Early Potato Varieties for Baikal Region 88
Nadezhda Bolsheshapova, Sergey Burlov, Irina Abramova, and Evgeny Boyarkin

Technology for Protecting Cattle from Blood-Sucking Dipteric Insects and Myiasis Gadflys 96
Anastasia Barashkova

Features of the New Mid-season Soybean Variety Zolushka 105
Evgeniya Fokina and Galina Belyaeva

Study of Anthocyanins in Tubers of Potato Hybrids (Solanum Tuberosum L.) of Primorsky Krai Selection 113
Irina Kim, Valentina Vozniuk, Dmitriy Volkov, and Aleksey Klykov

Identification of Bioactive Compounds in Clove (*Syzygium Aromaticum L.*) 122
Fadi Darwish, Tatiana Kalenik, Tamara Senotrusova, Elena Motkina, Maya Razgonova, and Natalia Li

Efficiency of Applying Sowing Machines for Planting Spring Wheat in the Irkutsk Region 131
Vladimir Solodun, Evgeny Boyarkin, Alexsander Zaitsev, and Tatyana Amakova

Development a Way of Protection for Domestic Reindeer from the Attack of Mosquitoes 138
Alexander Reshetnikov

Influence of Submerged Cultivation Conditions on the Titer of Active Cells of Rhizobia 147
Mariya Yakimenko, Arina Sorokina, and Stepan Begun

The Influence of Mineral Fertilizers on the Productivity of Spring Triticale in the Conditions of the Southern Zone of the Amur Oblast 156
Alexey Muratov, Pavel Tikhonchuk, and Eugenia Tuaeva

Agrotechnical Assessment of the Work of the Furrow Opener During the Cultivation of Cereals Using Intensive Technology 164
Maxim Chekusov, Andrey Schmidt, and Aleksander Kem

Adaptability of Determinate and Indeterminate Varieties of *Fagopyrum Esculentum* Monch in a Monsoon 174
Alexey Klykov, Galina Murugova, Oksana Timoshinova, Elena Barsukova, and Elena Chaikina

Resource Provision of Innovations in Agricultural Enterprises in the South of the Russian Far East 185
Tatiana Potenko, Irina Zhupley, Oksana Mamai, and Ludmila Kvashko

Improving the Methods of Preserving the Genetic Identity of the Cucumber Variety Population Naslednik During Reproduction 193
Galina Kuzmitskaya and Nadezhda Kuliakina

Measures to Increase Water Supply to Agroecosystems in the Subarctic Zone of Yakutia 202
Mikhail Loskin, Alexandr Reshetnikov, and Alexander Filatov

Evaluation of Technical and Chemical Parameters of Fruits of Apricot, Plum, and Apple Varieties of Far Eastern Selection 211
Oleg Mikhailichenko, Evgeniya Tikhomirova, Olga Tokareva, and Natalia Yudaeva

Comparative Assessment of Promising Soybean Varieties at the Final Selection Stage 219
Ekaterina Butovets, Ludmila Lukyanchuk, and Evgenia Vasina

Checking the Validity of Soy Threshing and Separation Models Using Experimental Data 230
Irina Prisyazhnaya, Serafima Prisyazhnaya, Valentina Sinyegovskaya, and Mikhail Sinegovskiy

Adaptive Potential Assessment of the Potato Varieties in the Conditions of the Amur Region 238
Sergey Rafalsky, Olga Rafalskaya, and Tatiana Melnikova

Changes in the Yield and Quality Grain of Soybean When Using Liquid Fertilizers 246
Sergey Fokin and Olga Piletskaya

Cost Management of Soybean Production in the Amur Region, Russia 255
Anastasiia Boiarskaia, Boris Boiarskii, Mikhail Sinegovskii, and Hideo Hasegawa

Peopleware of the Region's AIC Industry 264
Lyudmila Chausova, Nikolai Chausov, and Natalia Ogorodnikova

Assessment of the Adaptive Ability of Corn Hybrids in Primorsky Krai 275
Irina Danilenko, Natalia Kraskovskaya, and Ekaterina Butovets

Problems of State Regulation of Intra-industry Integration of Business Structures of the Primorsky Region 285
Alexander Latkin, Olga Fedoreeva, Dina Mukhina, and Ekaterina Saveleva

Phytosanitary Monitoring of Grain Crops and Soybeans in the Khabarovsk Territory 293
Anna Tishkova and Olga Imtosimi

Influence of Alfalfa Changeable (Medicago Varia Mar.) on the Fertility of Meadow-Brown Bleached Soil of the Primorsky Krai 303
Elena Ivanova and Lyubov Yayuk

Fixed Capital Investments in the Russian Federation: Territorial Differences 311
Natalia Ivanova and Lubov Pashina

Findings of the Study of Potato Varieties of Different Geographical Origin in the Kamchatka Territory 321
Anna Ivashchenko, Tamara Sherstyukova, and Galina Vlasenko

Application of Bio-Mix Organic Fertilizer in Potato Cultivation in the Kamchatka Territory 329
Vera Gainatulina, Olga Khasbiullina, and Roman Khasbiullin

Study of Breeding Lines of Soft Spring Wheat in a Competitive Variety Trial in Irkutsk Region 339
Natalia Klimenko, Irina Abramova, and Elena Kuznetsova

Aspects of the Productive Longevity of Cows in the Dairy Farms of the Middle Priamurye 347
Natalia Klyuchnikova, Michael Klyuchnikov, and Elena Klyuchnikova

Prospects of Using Whole Grain Flour from Recognized Selection Wheat Varieties of the Far Eastern State Agrarian University in Food Technologies 357
Elena Gartovannaya and Anna Ermolaeva

Results of Spring Barley Hybridization in the Amur Region 366
Alexandra Kuznetsova

Actual Problems of State Support of Entrepreneurship in the Agro-Industrial Complex of the Far East 376
Alexander Latkin, Dina Mukhina, Olga Fedoreeva, and Ekaterina Saveleva

Ecological Plasticity of Varieties *Triticum Aestivum L., Triticum Durum Desf.* And *Hordeum Vulgare L.* Of Various Origins in the Conditions of Primorsky Krai 386
Galina Murugova, Alexey Klykov, Polina Bogdan, Inna Konovalova, and Nadezhda Kryuchkova

Telescopic Seeder Coulter Suspension 397
Sergey Inshakov, Anna Redkokashina, Aleksander Redkokashin, and Viktor Balabanov

The Research of Genotoxic Potential of Water Samples Taken from Agricultural Areas Based on Test-System of the Soybean 404
Sergey Shishlov, Natalia Bykovskaya, Marina Shishlova, and Natalia Chugaeva

Improving the Efficiency of Combine Harvesters 415
Irina Lontseva and Vyacheslav Sennikov

Management of Demographic Processes in the Countryside of the Far East of Russia 424
Yurij Shmidt, Irina Zhupley, Yuliya Schmidt, and Lyubov Soldatova

Phytoviruses in the Potato Field Tripartite Agroecosystem 434
Olga Sobko, Nathalia Matsishina, Petr Fisenko, Irina Kim, and Natalia Boginskaya

Comparative Evaluation of Different Potato Varieties for Their Suitability for Starch Processing 443
Dmitry Volkov, Irina Kim, Aleksey Klykov, and Natalia Matsishina

The Effect of Tillage Minimization on the Weed Infestation of Soybean and Grain Crops in the Conditions of the Amur Region in Russia 451
Elena Zakharova and Aleksej Nemykin

Reproduction of Bee and Drone Brood in Bee Colonies of the Far Eastern Honey Bee (*Apis Mellifera* Linnaeus, 1758) in the Conditions of Primorsky Krai 460
Maxim Sharov

The Dynamics of the Main Fertility Elements Content in a Long-Term Agrochemical Stationary Experiment 469
Roman Timoshinov, Elena Kushaeva, Lyudmila Marchuk, Alexandr Dubkov, and Alexey Klykov

Automating the Calculation of Damage Caused by Illegal Logging 479
Alexander Baranov and Victoria Sergeeva

Blueberry Ordinary as an Environmentally Friendly Product in the Amur Region 487
Irina Berkal, Natalia Yust, and Elena Osipenko

The State of the Population of the Common Lynx (Lynx Lynx Stroganovi), Inhabiting the Territory of the Amur Region 495
Roman Chikachev

Forestry and Economic Justification of Commercial Logging in the Amur Region 505
Olga Dyadchenko, Natalya Ivanova, Alexander Baranov, and Natalya Timchenko

Restoration Prospects of Abies Holophylla Forests in the Southern Russian Far East 514
Alyona Kiselyova, Ilona Rodnikova, Olga Ukhvatkina, Kirill Ganzei, and Nina Pshenichnikova

The Current State of the Northern Coniferous-Deciduous Forests (on the Example of the State Nature Reserve "BASTAK") 522
Ekaterina Lonkina and Tamara Rubtsova

Problems of Effective Forestry Fund Land Management in the Amur Region 532
Marina Makannikova and Lyudmila Lapshakova

Forest Burnability Analysis and Preliminary Fire Forecasting Dangers on the Territory of the Amur Region 542
Tatiana Molchanova, Elena Grebenshchikova, Natalia Shelkovkina, Natalia Gorbacheva, and Evgeniy Khovanets

Assessment of the Condition of Coniferous Trees in Urban and Suburban Forests of the Krasnoyarsk City 553
Nina Pakharkova, Olga Shabalina, Galina Sorokina, Irina Bezkorovainaia, and Ekaterina Tegnerenko

Study of Crocosmia × Crocosmiiflora (Lemoine) N.E.Br. in the Agrometeorological Conditions of the Southern Zone of the Amur Region for the Prospects of Use in Landscaping 563
Irina Kurkova and Svetlana Stokoz

Climate Warming and Changes in the Reproductive Capacity of Woody Plants 573
Kirill Tkachenko, Gennady Firsov, and Alexandra Volchanskaya

The History of the Introduction of Species of the Genus *Rosa* to St. Petersburg, Russia 581
Kirill Tkachenko and Alla Kapelian

Comparative Analysis of Dendroflora of Parks in Blagoveschensk 589
Natalia Yust, Natalia Timchenko, and Olesia Shcherbakova

Fortified Sour-Milk Beverages with the Use of the Far Eastern Region's Wild Berries 602
Elena Zakipnaya and Svetlana Parfyonova

Use of Pine Nut Resources in Food Technology as One of the Steps of Sustainable Forestry 611
Yulia Derzhapolskaya, Ekaterina Reshetnik, and Svetlana Gribanova

Development Technology of Functional Drink 620
Anna Ermolaeva and Elena Gartovannaya

The Use of the Far East Wild Plants in the Production of Functional Alimentary Products 628
Elena Osipenko and Svetlana Kostrykina

Efficiency of the Application of Growth Stimulants Ribav-Extra and Zircon in the Germination of Manchurian Fir (*Abies Holophylla* Maxim.) Seeds 637
Valentina Ostroshenko and Lyudmila Ostroshenko

Changes in the Areas of the Predominant Tree-And-Shrub Species of the Transbaikal Territory 648
Larisa Pak

Ecological and Chemical Assessment of the Habitats of Cranes in the Khingan State Nature Reserve, Russia 658
Antonina Pakusina, Tatyana Platonova, Tatyana Parilova, Mikhail Parilov, Nikita Malinovsky, and Irina Balan

Distribution of Hippophae Rhamnoides L. In Primorsky Krai and Its Invasive Potential 667
Nina Kolyada, Tatyana Moskaliuk, and Aleksandr Kolyada

Control Over the State of Specially Protected Natural Areas and Objects of Regional Significance of the Amur Region 676
Tatiana Zolotukhina and Galina Stekolnikova

Author Index 687

Screening of Potato Cultivars (*Solanum Tuberosum* L.) and Identification of Markers of Fesistance Genes to PVX, PVY, *Globodera Pallida* and *Globodera Rostochiensis*

Petr Fisenko, Olga Sobko, Irina Kim(✉), Nathalia Matsishina, and Dmitriy Volkov

Federal Scientific Center of Agricultural Biotechnology of the Far East named after A.K. Chaiki, Volozhenina street, 30, Settlement Timiryazevsky, 692539 Ussuriysk, Primorsky Krai, Russia

Abstract. Currently, the control of many pathogens and pests is carried out by repeated chemical treatments. However, this method of protection requires large material and technical costs, is not environmentally safe and is not effective for all types of pathogens. Therefore, the main direction in the control of numerous potato diseases and pests is to identify and create varieties with complex resistance that can provide the required level of plant protection from phytopathogens. Selective sampling with different types of resistance is important. In the Potato and Vegetable Growing Department of Federal State Budget Scientific Institution "Federal Scientific Center of Agricultural Biotechnology of the Far East named after A. K. Chaika" work is underway to identify donor varieties with resistance genes to PVX and PVY viruses, pests-the pale potato nematode Globodera pallida and the golden cyst-forming potato nematode Globodera rostochiensis. As a result of the conducted studies, a number of donor varieties with complex resistance to pathogens and pests were identified: Smak, Bellarosa, Red Scarlett, Sante (Rx1, H1, Gpa2); Augustine, Kazachok, Yantar (Rx1, Gpa2); Gulliver, Queen Anne, Fresco, Gala, Impala, Red Lady, Secura (H1, Gpa2); Dachny (Rx1, H1-N195). The selected varieties are included in targeted crosses, hybrid combinations are obtained.

Keywords: *Solatium tuberosum* · Potato varieties · MAS · DNA markers of R-genes · *Globodera pallida* · *Globodera rostochiensis* · Potato virus X · Potato virus Y

1 Introduction

Potato is one of the most affected by viral diseases and pests crops. Currently, there are about 30 of the most common potato diseases [1]. A constant factor that reduces the potato yield is the development of viral diseases that lead to its degeneration. About 40 viruses affecting potatoes have been detected in various countries and regions with different natural climatic conditions [2, 3]. Potato as a vegetatively propagated crop accumulates

A. Muratov and S. Ignateva (Eds.): AFE 2021, LNNS 353, pp. 1–8, 2022.
https://doi.org/10.1007/978-3-030-91402-8_1

viruses that can cause 10–80% of crop losses. The creation of potato varieties resistant to phytopathogens is the most reliable and economical way to reduce yield losses [4]. The greatest harm is caused by the potato Y virus (PVY), the infestation of which can lead to complete degeneration of plants, especially in combination with X (PVX) and M (PVM) viruses. Currently, 9 groups of strains are detected and more new PVY isolates are being detected [5]. The molecular markers YES3-3A and YES3-3B of the Ry_{sto} gene,RYSC3of the Ry_{adg}gene, Ry186of the Ry_{chc}gene, and PVX of the *Rx1* gene have been widely used in virus resistance breeding [6–8].

Potato cyst-forming nematodes of the genus *Globodera* belong to quarantine facilities in many countries, including Russia. Potato nematodes include two species of *Globodera rostochiensis* (Woll.) Behrens and *Globodera pallida* (Stone) Behrens. The populations of these pests are heterogeneous, there are eight pathotypes (Ro1, Ro2, Ro3, Ro4, Ro5) in *Globodera rostochiensis* and three in *Globodera pallida* (Pa1, Pa2, Pa3). Both types of potato production can cause significant yield losses, up to 70–90%, in addition, nematodes can spread with the seed material, which leads to the spread of the pest to new regions, or infection with more aggressive pathotypes. The cultivation of resistant potato varieties not affected by nematodes is an important indicator of these quarantine pests spread and a priority when creating new varieties. Of the eight pathotypes of *Globodera rostochiensis*, only Ro1 is widely spread on the territory of the Russian Federation [9]. Resistance to Ro1 is an important and necessary feature of new potato varieties. Phenotyping for resistance to potato cyst-forming nematodes is laborious and requires a large amount of time. Marker-assisted selection is currently an alternative and complementary method for achieving resistance to potato cyst nematodes. Mapping of resistance to this pathogen genes in potato served as the basis for the development of molecular markers for their diagnosis. Most of the R genes (*Gro1.2, Gro1.3, Gro1.4, Gpa, Grp1, Gpa5, Gpa6, GpaM1, GpaM2, GpaM3, RGp5-vrn HC,* $GpaXI^{l}_{tar}$, $GpaIV^{s}_{adg}$) provide only partial potato protection, while four *H1, GroV1, Gro1* and *Gpa2* cause specific resistance to nematodes [10].

In the Potato and Vegetable Growing Department of Federal State Budget Scientific Institution "Federal Scientific Center of Agricultural Biotechnology of the Far East named after A. K. Chaika" is conducting research work on the study of the potato bioresource collection. Screening of varietal samples for economically valuable characteristics, including resistance to pathogens and pests, is carried out annually [11]. In this regard, the purpose of this work is to identify donor varieties with resistance genes to PVX and PVY viruses, pests-pale potato nematode *Globodera pallida* and golden cyst - forming potato nematode *Globodera rostochiensis.*

2 Materials and Methods

The research was carried out in the breeding and seed nurseries of the Federal State Budget Scientific Institution "Federal Scientific Center of Agricultural Biotechnology of the Far East named after A. K. Chaika" In the field, the bioresource collection was evaluated in the amount of 300 varieties according to ecological and geographical characteristics, as a result of which a group of varieties showing economically valuable signs in the conditions of the region was distinguished. The object of molecular genetics research was 28 potato varieties of various genetic origin: Augustine, Bryanskij delikates, Vector, Gulliver, Dachny, Kazachok, Queen Anne, Meteor, Pamyati Rogacheva, Smak, Udacha, Fioletovyj, Yubilya, Yantar, Adretta, Gala, Bellarosa, Belmonda, Impala, Fresco, Labella, Laperla, Latona, Lilly, Red Lady, Red Scarlett, Sante, Secura. The selection of molecular markers for detecting resistance genes to potato PVX virus (*Rx1*), PVY (Ry_{sto}), pale potato nematode *Globodera pallida*(*Gpa2*), golden cyst-forming potato nematode *Globodera rostochiensis* (*H1*), as well as primers for them, was carried out according to literature sources [12, 13]. The markers PVX of the *Rx1* gene, PVY of the Ry_{sto}gene, Gpa2–2 of the *Gpa2* gene,N195 and 57R of the *H1* gene (Table 1) were selected.

Table 1. DNA markers used to identify potato resistance genes to pathogens and pests.

Gene	Marker	Primers	Fragment	$MgCl_2$ (mM)
DNA markers of resistance to the PVY virus				
Ry_{sto}	YES3-3A (STS)	TAACTCAAGCGGAATAACCC AATTCACCTGTTTACATGCTTCTTGTG	341	0.8
DNA marker of resistance to the PVX virus				
Rx1	PVX (STS)	ATCTTGGTTTGAATACATGG CACAATATTGGAAGGATTCA	1230	2.5
DNA markers of resistance to *Globodera rostochiensis*				
H1	N195(SCAR)	TGGAAATGGCACCCACTA CATCATGGTTTCACTTGTCAC	337	0.8
	57 R (SCAR)	TGCCTGCCTCTCCGATTTCT GGTTCAGCAAAAGCAAGGACGTG	450	0,7
DNA markers of resistance to *Globodera pallida* (Wollenweber) Behrens				
Gpa2	Gpa2–2 (STS)	GCACTTAGAGACTCATTCCA ACAGATTGTTGGCAGCGAAA	452	2

PCR optimization was carried out by selecting the ratio of the components of the reaction mixture using positive controls - the DNA of potato varieties having the desired genes in their genome according to literary sources. To test the method, the varieties Meteor (Ry_{sto}, *Rx1*, *Sen1*, *Gpa2*, *H1*), Vector (*Rx1*, *Gpa2*), Yubilyar (*Gpa2*), Zhukovskij rannij (*Gpa2*, *Rx1*) were selected [13]. Total DNA was isolated by the salt method [14] with an additional step of purification of the extract with a mixture of chloroform/isoamyl alcohol (24/1) and/or chloroform/phenol (1/1). DNA samples of varieties - positive controls were isolated individually from several plants, after PCR testing, a total sample was prepared by mixing aliquots of chemicals. The DNA of the varieties for screening was isolated from the leaf tissues of a total field sample of 10 plants. Only total DNA samples were used directly for screening markers of the desired genes of the studied varieties. PCR was performed in 10 μl of a 2X reaction mixture of BioMaster HS-Taq PCR-Color (Biolabmix) in a thermal cycler T100 (Bio-Rad) in two or three repetitions. 10–50 ng was used in the reaction. DNA nanomatrix The temperature protocol of the multiplex reaction according to Sinaykov A. B. et al., 2018 was used [15], the reactions were set individually. The PCR products were separated by electrophoresis in a 2% agarose gel stained with ethidium bromide. DNA fragments were visualized by UV irradiation using the Gel-Doc XR + (Bio-Rad) gel-documenting system. Fragment length markers were used to determine the length of fragments Step 100 and Step 100 Long(Biolabmix). The presence or absence of the desired marker was determined by comparison with a positive control having a fragment of a certain length.

3 Results and Discussion

At the present stage, breeders should have databases that include the results of phenotyping of the breeding material and information about the presence of DNA markers associated with QTL genes that determine economically valuable traits. Over the past two decades, numerous DNA markers have been developed for potato that allow marker-associated selection (MAS) of promising genotypes and are widely used in molecular screening by researchers from different countries [15, 18, 19].

Many selectively valuable traits of potato, including resistance to pathogens and pests – potato PVX virus (Potaro Virus X), PVY (Potato Virus Y), PVS (Potaro Virus S), Potato Leaf Roll Virus (Potato Leaf Roll Virus, PLRV), cyst-forming nematodes, etc. - have a monogenic nature. Molecular markers linked to genes Ry_{sto}andRy_{adg} controlling immune to PVY, markers, gene *Rx1*immune to PVX genes *H1, Gro 1–4* resistance to the Golden nematode *Globodera rostochiensis* (Wollenweber) Behrens and gene *Gpa2* resistance to pale potato cyst nematode *Globodera pallida* (Stone) Behrens, can be an effective tool to intense breeding work [4].

In 2019, the identification of markers 57R, N195 and YES3A of the PCR resistance genes to the golden nematode *Globodera rostochiensis* (Wollenweber) Behrens (*H1*) and PVY virus (Ry_{sto}), respectively, was tested. In 2020, the identification of markers of resistance genes to the pale potato nematode and potato X virus was worked out. Using the mastered markers, a screening of 28 potato varieties was carried out, which were distinguished by the results of ecological and geographical testing and used in breeding and seed production of the Federal Research Center.

As a result of working out the method, the optimal concentrations of magnesium chloride were selected - 2 mM for *Gpa2* and 2.5 mM for *Rx1* (Table 1). When choosing the reaction conditions, we were guided by the fact that during PCR, a marker fragment should be synthesized in most samples of the variety - positive control (regardless of the quality of the DNA used), while non-specific reaction products are absent, or are present in the smallest amount and intensity of luminescence, in the absence of a product of the desired length in the negative control. Such criteria were chosen to level out possible heterogeneity, as well as the influence of the DNA quality of the total sample of field samples of each studied variety used during screening. This approach was chosen to reduce the consumption of reagents and accelerate the study as a whole.

It should be noted that the Zhukovskij rannij variety was distinguished by the greatest heterogeneity and dependence on the PCR conditions. From four plants, depending on the concentration of magnesium chloride, the marker fragment was detected from 1 to three plants for the Gpa2–2 marker, and in two plants for the PVX marker. This can be either the result of poor DNA quality, or the heterogeneity of the variety caused by contamination.

As positive controls, we used the DNA of potato varieties having, according to the literature data, markers of the desired genes - Meteor, Zhukovskij rannij, Yubilyar, and Vector. Based on the data obtained, 28 potato varieties were screened, including the positives used (Table 2).

Table 2. Results of screening of potato varieties to identify genes of resistance to pathogens and pests.

Variety	The resistance gene				
	Ry_{sto}	*Rx1*	*H1*		*Gpa2*
	YES3A	PVX	57R	N195	Gpa 2–2
Vector	–	+	–	–	+
Zhukovskij rannij	–	+	+	+	+
Meteor	+	+	+	+	+
Yubilyar	–	+	+	+	+
Bellarosa	–	+	+	+	+
Augustine	–	+	–	–	+
Yantar	–	+	–	–	+
Gala	–	–	+	+	+
Sante	–	+	+	+	+
Latona	–	–	–	–	+
Bryanskij delikates	–	–	–	–	–
Dachny	–	+	–	+	+
Gulliver	–	–	+	+	+
Secura	–	–	+	+	+
Udacha	–	–	–	–	+
Pamyati Rogacheva	–	–	+	+	–
Smack	–	+	+	+	+
Impala	–	–	+	+	+
Fioletovyj	–	–	–	–	–
Fresco	–	–	+	+	+
Kazachok	–	+	–	–	+
Adretta	–	–	–	–	+
Red Scarlett	–	+	+	+	+
Laperla	–	+	+	+	+
Lilly	–	–	+	+	+

(*continued*)

Table 2. (*continued*)

Variety	The resistance gene				
	Ry_{sto}	*Rx1*	*H1*		*Gpa2*
	YES3A	PVX	57R	N195	Gpa 2–2
Queen Anne	–	–	+	+	+
Red Lady	–	–	+	+	+
Labella	–	–	+	+	–
Belmonda	–	–	+	+	–

The patterns of identification and distribution of marker fragments obtained during the development of the method for identifying selected genes are confirmed by the literature data for the varieties selected as standards, which indicates the reliability of the results of screening the sample in our study. The marker of the *Rx1* gene was detected in the varieties -- Augustine, Vector, Dachny, Zhukovskij rannij, Kazachok, Meteor, Smak, Yubilyar, Yantar, Bellarosa, Sante, Red Scarlett, and Laperla. The marker of the *Gpa2* gene was detected in the varieties - Augustine, Vector, Gulliver, Dachny, Zhukovskij rannij, Kazachok, Queen Anne, Meteor, Smak, Udacha, Yubilyar, Yantar, Adretta, Bellarosa, Fresco, Gala, Impala, Laperla, Latona, Lilly, Red Scarlett, Sante, Secura, and Red Lady. In addition, based on the data obtained in 2019, a sample was also screened for the presence of resistance genes to the PVY virus and *Globodera rostochiensis* (Wollenweber) Behrens (Table 2). No marker YES3A of the *Ry_{sto}*gene was found among the studied cultivars, except for the Meteor variety, which acted as a positive control for this marker. Markers 57R and N195of the *H1* - gene were detected in varieties - Gulliver, Zhukovskij rannij, Queen Anne, Meteor, Pamyati Rogacheva, Smak, Yubilyar, Bellarosa, Belmonda, Gala, Impala, Fresco, Labella, Laperla, Lilly, Red Lady, Red Scarlett, Secura. In the Dacha variety, only the marker N195 is consistently detected, while 57R is absent.

4 Conclusion

As a result of the conducted studies, a number of donor varieties with complex resistance to pathogens and pests were identified: Smak, Bellarosa, Red Scarlett, Sante (*Rx1, H1, Gpa2);*Augustine, Kazachok, Yantar (*Rx1, Gpa2);* Gulliver, Queen Anne, Fresco, Gala, Impala, Red Lady, Secura (*H1, Gpa2*); Dachny (*Rx1, H1*-N195). The selected varieties are included in targeted crosses, hybrid combinations are obtained. The most valuable populations will be subjected to a similar screening to detect the presence of genes of resistance to pathogens and pests.

The support of the breeding process using molecular genetic markers allows to reduce the volume of the analyzed material in generations and to conduct a more targeted selection, which allows to reduce the breeding load and speed up the process of breeding new varieties.

Acknowledgements. The research was funded by a government grant as a subsidy from the federal budget with the purpose of creating and implementing in the agricultural sector of modern technology based on their own scientific research and educational institutions in the framework of the implementation of the decree of the President of the Russian Federation of July 21, 2016, No. 350; "On measures for the implementation of the state scientific and technical policy in the development of agriculture," including with the participation of genomic research centers of the world level.

The authors are grateful R. Romanciuc (Alecu Russo State University, Faculty of Philology, Ocnita, Moldova) for translating this manuscript from Russian to English.

References

1. Edward, B.R., William, D.H., Rafael, E.C.: Integrated Pest Management: Concepts, Tactics, Strategies and Case Studies, p. 529. Cambridge University Press, Cambridge (2009)
2. Salazar, L.: Potato viruses after the XX centry: effects, dissemination and their control. Mater. In: Participants in Pyongyang Intern. Scientific Simpos. on Potato, Peohangan, pp. 35–42. DPRK ((2003))
3. Milošević, D., Milenković, S., Perić, P., Stamenković, S.: The effects of monitoring the abundance and species composition of aphids as virus vectors on seed potato production in Serbia. Pestic. Phytomed. (Belgrade) **29**(1), 9–19 (2014). https://doi.org/10.2298/PIF1401009M
4. Ramakrishnan, A.P., Ritland, C.E., Blas Sevillano, R.H., Riseman, A.: Review of potato molecular markers to enhance trait selection. Am. J. Potato Res. **92**, 455–472 (2015)
5. Chrzanowska, M.: New isolates of the necrotic strain of potato virus Y (PVY^N) found recently in Poland. Potato Res. **34**, 179–182 (1991). https://doi.org/10.1007/BF02358039
6. Song, Y.S., Hepting, L., Schweizer, G., Hartl, L., Wenzel, A.S.: Mapping of extreme resistantce to PVY (Ry_{sto}) on chromosome XII using anther-culture-derived primary dihaploid potato lines. Theor. Appl. Genet. **111**, 879–887 (2005)
7. Kasai, K., Morikawa, Y., Sorri, V.A., Valkonen, J.P.T., Gebhardt, C., Watanabe, K.N.: Development of SCAR markers to the PVY resistance gene Ry_{adg} based on a common feature of plant disease resistance genes. Genome **43**, 879–887 (2000)
8. Mori, K., Sakamoto, Y., Mukojima, N., et al.: Development of a multiplex PCR method for simultaneous detection of diagnostic DNA markers of five disease and pest resistance genes in potato. Euphytica **180**, 347–355 (2011). https://doi.org/10.1007/s10681-011-0381-6
9. Khiutti, A.V., Antonova, O., Mironenko, N.V., Gavrilenko, T.A., Afanasenko, O.S.: Potato resistance to quarantine diseases. Vavilovskii Zhural Genetiki i Selektsii=Vavilov J. Genet. Breed. **21**(1), 51–61 (2017). https://doi.org/10.18699/VJ17.223
10. Dalamu, B.V.: Potato cyst nematode (PCN) resistance: genes, genotypes and markers - an update. SABRAO J. Breed. Genet. **44**(2), 202–228 (2012)
11. Kim, I.V., Volkov, D.I., Klykov, A.G.: The study of bioresource collection of potato (*Solanum Tuberosum* l.) in the conditions of Primorsky Krai. IOP Conf. Ser. Earth Environ. Sci. **547**, 012013 (2020). https://doi.org/10.1088/1755-1315/547/1/012013
12. Klimenko, N.S., Gavrilenko, T.A., Kostina, L.I., Mamadbokirova, F.T., Antonova, O.: Search for resistance sources to Globodera pallida and potato virus X in the collection of potato varieties using molecular markers. Plant Biotechnol. Breed. **2**(1), 42–48 (2019). https://doi.org/10.30901/2658-6266-2019-12 48.

13. Biryukova, V.A., Shmyglya, I.V., Mityushkin, A.V., et al.: Molecular screening of samples of breeding and genetic collections of State scientific institution All-Russian research institute of potato farming by Lorh AG. Theoretical foundations and applied research in potato breeding and seed production: abstracts of reports of the scientific conference (Russian Academy of Sciences, Siberian Branch, Feder. research. center for Cytology and Genetics Institute), Novosibirsk (2018). https://doi.org/10.18699/Potato-2018-04
14. Aljanabi, S.M., Martinez, I.: Universal and rapid salt-extracion of high qualiny genomic DNA for PCR-based techniques. Nucl. Acid Res. **25**(22), 4692–4693 (1997)
15. Saynakova, A.B., et al.: Testing potato collection samples for the presence of genes for resistance to phytopathogens by means of DNA markers. Vavilovskii Zhurnal Genetiki i Selektsii=Vavilov J. Genet. Breed. **22**(1), 18–24 (2018). https://doi.org/10.18699/VJ18.326

Intra-callus Genetic Variability of Rice *Oryza Sativa* L. Doubled Haploids Regenerated Throuth Androgenesis *in vitro*

Marina Ilyushko(✉), Marina Romashova, and Svetlana Guchenko

Federal Research Center of Agricultural Biotechnology of the Far East named after A.K. Chaik, Volozhenina Street, 30, stl. Timiryazevsky, 692539 Ussuriysk, Russia

Abstract. The intracallus variability frequency of *Oryza sativa* L. doubled haploids, obtained *in vitro* androgenesis was studied. Molecular genetic analysis of 855 plants (56 callus lines) for the presence of resistance or susceptibility alleles to five rice resistance genes to *Pyricularia oryzae* Cav. was performed. Among the doubled haploids, one–three genes were detected depending on their presence in the donor hybrids. Callus lines with identical regenerants on one of the genes, as well as possessing two resistance alleles of the studied genes, were revealed. An analysis of doubled haploids obtained from hybrids, which possess the resistance alleles of three genes, revealed no more than two combinations of alleles among the doubled haploids on one callus line out of a possible eight. With the same combination of resistance alleles of two genes, up to 28 plants were formed, and with the same combination of three genes, up to 18 plants were formed on the alone callus line. Thus, low intracallus genetic variation of rice doubled haploids due to gametoclonal variability was revealed. A quarter (26%) of the callus lines show variation among doubled haploids. There is veritable cloning of doubled haploids within the callus lines in androgenesis *in vitro*.

Keywords: Doubled haploids · Intra-callus gene variability · *Pi-b* · *Pi-ta*^{2} · *Pi-z* · *Pi-1* · *Pi-2*

1 Introduction

Cells and tissues culture *in vitro* of agricultural crops for their intended purpose can be conditionally divided into two groups: for obtaining genetically modified initial breeding material and for mass cloning of already existing forms and varieties. Anthers culture (androgenesis *in vitro*) allows switching the microspore development program from the gametophytic path to the sporophytic pathway, as a result of spontaneous doubling, doubled haploids are formed in diploid species, or dihaploids (polyhaploids) are fixed in tetraploid species, which are widely used in plant breeding [1]. Variability among plants regenerated from gametes is commonly referred to as «gametoclonal variability» [2] Formally speaking, these regenerants are not clones of the same donor plant genotype,

A. Muratov and S. Ignateva (Eds.): AFE 2021, LNNS 353, pp. 9–18, 2022.
https://doi.org/10.1007/978-3-030-91402-8_2

through the lack of complete genetic identity with it, due to recombination during the formation of microspores. In fact, each microspore is a new genotype, albeit with the same combination of genes. In addition, variability can be induced by the culture *in vitro* [3].

The plants variability obtained in the anthers or microspores culture *in vitro* from alone donor plant has been studied to a greater extent at the genomic and chromosomal level, since researchers are primarily interested in spontaneous chromosome duplication and, as a consequence, in completely homozygous fertile offspring. The following main mechanisms of genomic duplication have been suggested: endoreduplication, nuclear fusion, endomitoses, and c-mitosis [4]. D.E. Daghma et al. [5] found exclusively nuclear fusion in barley anther culture using time-lapse imaging technology. Indirect methods have shown the possibility of the regenerants emergence from unreduced 2n-gametes. The unicellular origin of wheat haploids has been proven [6]. In direct embryogenesis from a microspore all plants a priory differ genotypically. As a result among androgenic regenerants of alone donor plant multiple and non-multiple changes in the chromosome numbers were revealed, molecular genetic variability, variability of selectively significant traits and resistance to pathogens [7, 8].

In anther culture *in vitro*, rice *Oryza sativa* L. goes through an additional stage of callus formation, from which regenerant plants are formed [9]. Upon callus formation initiation, a highly vacuolated cereal microspore undergoes symmetric mitotic division with the formation of two equal cells which is realized in different pathways of morphogenesis [2]. In this case, the cells and regenerants variability additionally arises, associated with the phenomenon of somaclonal variability [11]. Rice calli contain haploid, diploid, tetraploid, hexaploid and octaploid cells. Prolonged passages reduced the number of haploid and highlyploid cells. Prolonged callus cultivation is manifested in an increase of the proportion of diploid and highlyploid regenerants, as well as in a deviation from the normal distribution of economically valuable rice traits. There is a deviation from the expected segregation for molecular markers [7]. And if in barley cultivated through microspores, up to 50 regenerants from the anther are obtained [2], then in rice that forms callus, in some cases several hundred haploid plants or more than 120 doubled haploids are formed. There are calli (17%) with a small number of tetraploids (up to 18 pcs.) or seedless non-haploid regenerants. Intracallus morphological plants variability under *ex vitro* conditions obtained from calli with multiple regeneration was studied on variety Kaskad. Significant haploids variability and doubled haploids monomorphism in morphological indices were revealed. However, on the F_1 rice hybrid plant, several callus lines were obtained, where among the doubled haploids of alone callus line, variability in awnness and anthocyanin pigmentation of the stem was demonstrated, molecular genetic variability in the *Pi-ta* and *Pi-ta*^{2} rice blast resistance genes to *Pyricularia oryzae* Cav. [*Magnaporthe grisea* (Hebert Barr.)] was detected [11]. In theory, a callus line can be formed by one or more microspores. Rice has more than 1000 pollen grains in the anther [12]. The frequency of intracallus variability occurrence can reveal the number of callus founders microspores. The study aim was to determine the degree of doubled haploids genetic homogeneity or heterogeneity in alone anther.

2 Materials and Methods

The studies were carried out among the doubled haploids (DH_1) obtained in androgenesis *in vitro* of F_1 hybrids of rice *Oryza sativa* L. subspecies *japonica* Kato, the following combinations: Lugovoy × Maratteli 5A – L × 5A; Rassvet × (Oxy 2x × Darius 23) – R × O × 23; Almaz × [(Maratteli 5A × Boyarin) × Maratteli 5A] – 4P; 242–01 × Rassvet – 242 × R; Almaz × Magnat – A × M; Dolinny × Magnat – D × M; Dubrava × Viola – Db × V; Khankaisky 52 × 242–01 – K × 242; Dolinny × Don 4237 – D × D4237 and F_2 hybrid Rassvet × Oxy 2x – R × O. The hybrids donor plants were grown on a vegetative plot in vessels until the period of panicle collection. The method of anthers cold pretreatment, anthers, calli, and regenerants culture *in vitro* is published earlier [13].

The term «callus line» was understood as all callus aggregates formed on alone anther. Callus aggregates (calli) 2–5 mm in size were transplanted from the induction nutrient medium to the regeneration medium with an interval of seven days and were assigned serial numbers.

R_0 regenerants with a developed root system were planted in a pot and continued to grow in chamber until seeds were formed. After that, the green leaves were used for DNA extraction. The DNA extraction technique and modes of DNA amplification were described earlier [11]. The sequence, the primers annealing temperature, and the target product size for the studied genes are shown in Table 1. Varieties with known genes were used as control (Table 1). Amplification was performed in triplicate.

Table 1. Molecular markers specification for identification of blast resistance genes to *Pyricularia oryzae* Cav. in *Oryza sativa* L. doubled haploids regenerated through androgenesis *in vitro.*

Gene	Marker name	Nucleotide sequence (5'-3')→		Size, bp*	T, °C**	References	Control varieties
Pi-1	Rm224	F	atcgatcgatcttcacgagg	158	56	[14]	Magnat
		R	tgctataaaaggcattcggg				
Pi-2	Rm527	F	ggctcgatctagaaaatccg	233	56	[14]	Magnat
		R	ttgcacaggttgcgatagag				
Pi-b	*Pi-b*	F4	catcaacgaagtccagctca	490	60	[14]	Oxy 2x
		R5	ccgcgctatcttgtacattc				
		R6	ctcagcatatgtggcagctc				
Pi-ta^2	*Pi-ta*^{2}	F	cagcgaactccttcgcatac gca	515	60	[15]	CD8 Pi №4
		R	cgaaaggtgtatgcactata gtatcc				
Pi-z	*Pi-z*	F	ggacccgcgttttccacgtg taa	292	60	[15]	Maratteli 5A
		R	aggaatctattgctaagcac gac				

Remark. * Indicates predicted size of the PCR product (bp).
** Indicates annealing temperature of PCR.

3 Results

56 callus lines with three or more doubled haploids were formed on the anthers of ten hybrid plants. Eight callus lines formed two–four callus aggregates with green regenerants. Data analysis of the callus line 415.2 L × 5A hybrid showed the following: the resistance alleles of the two genes *Pi-z* and *Pi-ta*^{2} are present in all doubled haploids of callus aggregate 415.2.1, and on callus aggregate 415.2.2 only the resistance allele of the *Pi-z* gene is present (Table 2). This means that callus aggregates are initiated by various immature rice microspores, i.e. the phenomenon of alone anther gametoclonal variability is presented. Earlier, differences in morphological indices between doubled haploids of different callus aggregate on the alone anther were confirmed. Later, each callus aggregate was considered as a separate callus line.

The work carried out molecular genetic analysis of 855 plants. The number of doubled haploids from each callus line with resistance and susceptibility alleles of genes for resistance to rice blast disease is presented in Table 2 and in the Fig. 1.

Table 2. Resistance/susceptibility alleles of blast resistance genes to *Pyricularia oryzae* Cav. existence in *Oryza sativa* L. doubled haploids regenerated through androgenesis *in vitro* F_1 hybrids.

Hybrid	Callus line	Gene			Hybrid	Callus line	Gene
		Pi-ta^2	*Pi-1*	*Pi-2*			*Pi-2*
242 × R	454.1.1	–	0/3	3/0	4P	402.1.2	23/0
	554.1.1		2/3	2/3		402.1.4	3/18
	583.2.1		3/28	3/28		404.1.1	3/0
	582.2.1		0/18	18/0		404.1.2	5/0
D × M	405.1.1	14/0	14/0	0/14		411.1.1	0/6
	405.1.3	2/2	4/0	0/4		413.1.2	0/10
A × M	520.2.1	–	0/8	0/8		417.1.3	24/0
	563.1.1		4/0	0/4		417.1.4	3/0
	564.2.1		5/0	3/2		418.2.2	14/0
Db × V	476.2.1	2/1	2/1	2/1		431.1.3	0/5
	610.2.1	1/6	7/0	6/1		446.1.1	13/0
K × 242	574.1.1	–	0/21	21/0		447.1.1	3/8
	574.1.2		0/12	12/0		621.1.1	0/26
		Pi-z		*Pi-ta2*			*Pi-ta2*
L × 5A	415.2.1	4/0		4/0	R × O × 23	463.2.1	0/2
	415.2.2	16/0		0/16		509.1.1	4/10

(continued)

Table 2. (*continued*)

Hybrid	Callus line	Gene			Hybrid	Callus line	Gene
		Pi-ta^2	*Pi-1*	*Pi-2*			*Pi-2*
	416.1.2	–		10/0		527.2.1	1/6
	433.1.1	–		0/4	D × D4237	1.2.1	12/0
	435.2.1	–		2/1		4.1.2	9/0
	465.1.3	–		0/5		7.2.2	11/0
	467.1.1	–		5/3		8.2.1	36/8
Remark. DH_1 number with resistance allele is over the line; DH_1 number with susceptibility allele is under the line, pieces. «–» – not detected						21.2.1	15/0
						36.2.3	15/0

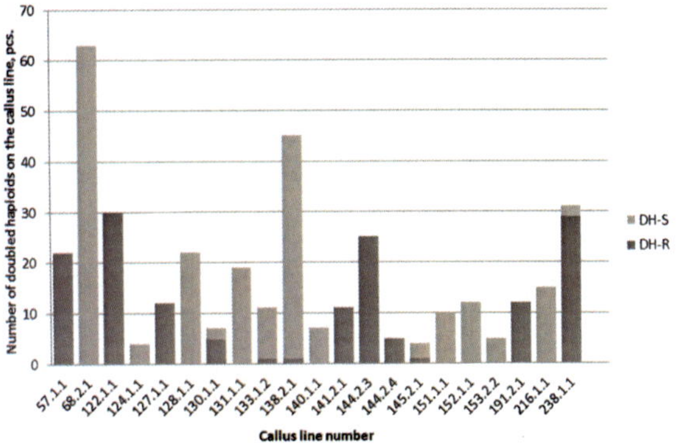

Fig. 1. Resistance/susceptibility alleles of blast resistance genes *Pi-b* to *Pyricularia oryzae* Cav. existence in *Oryza sativa* L. doubled haploids regenerated through androgenesis *in vitro* F_2 hybrid Rassvet × Oxy 2x. DH-R – doubled haploids with resistance allele, DH-S – doubled haploids with susceptible allele.

Among them, there are both callus aggregates with allele identical regenerants and with two alleles of the studied genes. A quarter (26%) of the callus lines show variation among doubled haploids (Table 3).

Table 3. The proportion of *Oryza sativa* L. callus lines with intracallus genetic variability among doubled haploids regenerated through androgenesis *in vitro*.

Hybrid	Gene	Callus line number (pcs.)		
		Only resistance allele	Only susceptibility allele	Variable
L × 5A	*Pi-ta2*	2	3	2
	Pi-z	2	0	0
R × O × 23	*Pi-ta2*	0	1	2
4P	*Pi-2*	7	4	2
242 × R	*Pi-1*	0	2	2
	Pi-2	2	0	2
	Pi-z	0	1	0
D × M	*Pi-ta*^{2}	1	0	1
	Pi-1	2	0	0
	Pi-2	0	2	0
A × M	*Pi-1*	2	1	0
	Pi-2	0	2	1
Db × V	*Pi-ta*^{2}	0	0	2
	Pi-1	1	0	1
	Pi-2	0	0	2
K × 242	*Pi-1*	0	2	0
	Pi-2	2	0	0
D × D4237	*Pi-ta*^{2}	5	0	1
R × O	*Pi-b*	7	10	4
Callus line proportion (%)	–	40	34	26

Analysis of callus lines obtained from hybrids D × M and Db × V, which have three genes resistance alleles, revealed no more than two combinations of alleles among the doubled haploids of alone callus line out of possible eight. So, for example, on the callus line 610.2.1 seven doubled haploids were formed. In six plants *Pi-1* and *Pi-2* resistance alleles, and in one regenerant the *Pi-1* and *Pi-ta*^{2} resistance alleles were identified (Table 4). With the same combination of two genes resistance alleles up to 28 plants, and with the same combination of three genes up to 14 plants were formed (Table 2).

4 Discussion

It is important to obtain the greatest doubled haploids diversity per plant in haploid breeding. This can be achieved due to high callus formation values and induced immature microspores abundance. For species capable of direct embryogenesis, microspore culture provides up to 50 regenerants from the anther [2]. *O. sativa* is a callusing in androgenesis *in vitro* [9], and inoculation of microspores on culture media is ineffective [16]. The mature rice anther opens on two terminal sides – basal and apical [17]. We have repeatedly observed the onset of callus formation *in vitro* at opposite anther flanks (Fig. 2).

Table 4. Combinations of resistance genes in *Oryza sativa* L. doubled haploids studied on three blast resistance genes to *Pyricularia oryzae* Cav.

Hybrid	Callus line	Doubled haploids number (pcs.)	Gene		
			Pi-ta^2	*Pi-1*	*Pi-2*
D × M	405.1.1	14	+	+	–
	405.1.3	2	–	+	–
		2	+	+	–
Db × V	476.2.1	1	–	–	–
		2	+	+	+
	610.2.1	1	+	+	–
		6	–	+	+

Remark. « +» – Resistance allele of blast resistance genes;
«–» –Susceptibility allele of blast resistance genes.

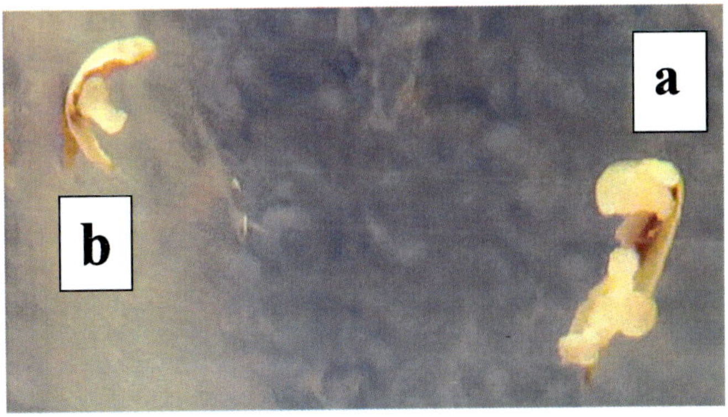

Fig. 2. Formation of calli in androgenesis in vitro *Oryza sativa* L.: a – two callus aggregates on basal and apical terminal sides of anther; b – one callus aggregate at the middle of anther.

Subsequently, these callus aggregates grew to a suitable size for transplantation onto a regeneration nutrient medium. The presence of only two combinations of blast resistance alleles for three genes (*Pi-1*, *Pi-2* and *Pi-ta*^{2} – D × M and Db × V hybrids doubled haploids) and single combination (*Pi-z*, *Pi-1* and *Pi-2* – 242–01 × P hybrid doubled haploids) of the possible eight ones confirms the visible observations. It is possible that the callus formation induction on alone microspore inhibits the neighboring microspores development, but does not interfere progress at the terminal flanks in at least 26% of cases.

The second and subsequent callus aggregates of the alone anther are often, but not always, identical in the genes combination among doubled haploids (404.1.1 and 404.1.2, 417.1.3 and 417.1.4 – 4P hybrid; 574.1.1 and 574.1.2 – X × 242 hybrid; Table 2). This is evidence in favor of veritable doubled haploids cloning both within the same callus aggregate and in different callus aggregates of the alone anther. Indirectly, cloning is supported by the number of regenerants of different ploidy on the callus aggregate. Up to 18 tetraploids on callus are formed, doubled haploids – up to 125 pcs., and haploids up to 349 pcs. The haploid cells mitotic index exceeds this indicator of doubled haploids and tetraploids [18], which contributes to the rapid haploids cloning on callus. In addition, the preceding cell fusion process or endomitoses reduces the number of cells involved in the formation of doubled haploids. Cell fusion can occur under certain research protocols at the initial stage during the anthers and microspores pretreatment [2], but it cannot be ruled out that this also occurs immediately before the formation of regenerants with a high ploidy level (doubled haploids, triploids, tetraploids, etc.). It is known that in wheat anther callus in vitro regenerants (hemmorizogenic embryos) are formed from a group of cells [6].

Rice callus formation in anther culture in vitro contributes to the large number production of doubled haploids, but in a limited genotypes number. An increase in the genetic variability of doubled haploids can be facilitated by the somaclonal variability inherent in rice anther culture, as in any plant cells and tissues culture *in vitro* [11]. This is rather an undesirable phenomenon in androgenesis *in vitro*, since again it is not clear whether somaclonal variants of cells arise before their fusion with subsequent mitotic division and regeneration, or after this process. In the second case, i.e. upon regeneration after fusion, a different development of events is possible. The cell, which emerged as a somaclonal variant, multiplied and assisted, along with other cells groups, in the formation of doubled haploids. Hence, plant mixoploidy occurring in androgenesis *in vitro*, and even heterozygous doubled haploids are possible. It contradicts the idea of haploid breeding, where the goal of obtaining homozygous seed offspring is pursued. Although it is generally accepted that heterozygosity in androgenesis *in vitro* is a sign of the anther walls somatic tissues proliferation [2], the absence of this phenomenon has been proven in rice, i.e. callus is always formed from an immature microspore [10]. The propensity for somaclonal variability in diploid cells of different cultures is genetically determined [3]. Apparently, this pattern also applies to haploid cells. And since each microspore in anther and plant is a separate genotype, the occurrence of the somaclonal variability phenomenon in calli initiated by various anthers may differs.

To ensure the protectability of breeding achievements, arbitration criteria are required, therefore according to the UPOV method (Union for the Protection of Plant varieties) tests are carried out for distinction, uniformity and stability of varieties [19]. However, uniformity in the crop population is not always achievable and necessary. It is known that a polymorphic variety, due to its buffering composition, can be more plastic. Therefore, rice breeders often follow the combining genotypes path of different origins with target traits in a single selection sample. As a result of such combination, the average of the plant population quantitative indicators improve, and resistance to pathogens increases, in particular due to the heterozygous effect with a favorable genes combination [20]. Obviously, complete uniformity for increased yields is not always a desirable characteristic. The veritable cloning phenomenon of rice doubled haploids will speed up the breeding process. After a preliminary molecular genetic analysis for the presence of target genes and an initial assessment of morphological indices, the seeds of one callus line must be combined into one selection sample, which will allow us to quickly switch to yield accounting. Occurring intracallus polymorphism for neutral traits, for example, the stem anthocyanin pigmentation [11] is acceptable to ensure the variety buffering. The heterozygosity at some loci, even in strict self-pollinators as rice, is not critical for their commercial use for a long time [19]. Moreover, in the case of a crops marker assisted selection for three genes, it is recommended to recruit several hundred doubled haploids to maintain genetic variability and eliminate gene drift in the population [10]. Molecular markers can serve as reliable arbitration criteria for distinctness, uniformity and stability of crop varieties [19], especially if there is two or three target genes combination.

5 Summary

Low intracallus genetic variability of doubled rice haploids in androgenesis *in vitro*, caused by gametoclonal variability, was revealed. Proportion of variable callus lines is a quarter (26%) of all studied calli. Alone callus line contains one or two combinations of alleles of rice blast resistance genes. There is a veritable cloning of doubled rice haploids within the callus line in androgenesis *in vitro*, which should be used in the breeding acceleration.

Acknowledgements. The authors express their gratitude to P.I. Kostylev (Federal Research Center Agrarian Center «Donskoy», Russian Federation) and V.N. Morokhovetz (Far Eastern Scientific Research Institute of Plant Protection, Russian Federation) for the provided seeds of control varieties.

References

1. Niazian, M., Shariatpanahi, M.E.: In vitro-based doubled haploid production: recent improvements. Euphytica **216**(5), 1–21 (2020). https://doi.org/10.1007/s10681-020-02609-7
2. Germana, M.: Gametic embryogenesis and haploid technology as valuable support to plant breeding. Plant Cell Rep. **30**, 839–857 (2011). https://doi.org/10.1007/s00299-011-1061-7
3. Kuznetsova, O., Ash, O., Gostimskij, S.: The effect of the duration of callus culture on the accumulation of genetic alterations in pea *Pisum sativum* L. Russ. J. Genet. **42**(5), 555–562 (2006). https://doi.org/10.1134/s1022795406050139
4. Segui-Simarro, J., Nuez, F.: Pathways to doubled haploidy: chromosome doubling during androgenesis. Cytogenet. Genome Res. **120**, 358–369 (2008). https://doi.org/10.1159/000121085

5. Daghma, D., Hensel, G., Rutten, T., Melzer, M., Kumlehn, J.: Cellular dynamics during early barley pollen embryogenesis revealed by time-lapse imaging. Front. Plant Sci. **5**, 1–14 (2014). https://doi.org/10.3389/flps.2014.00675
6. Seldimirova, O., Titova, G., Kruglova, N.: A complex morpho-histological approach to the in vitro study of morphogenic structures in a wheat anther culture. Izvestiia akademii nauk Seria bilogicheskaia **2**, 155–161 (2016). https://doi.org/10.7868/S0002332916020089
7. Goncharova, J.: Selective elimination of alleles in rice anther cultures. Russ. J. Genet. **49**(2), 170–177 (2013). https://doi.org/10.1134/S102279541210002X
8. Yi, G., Lee, H.-S., Kim, K.-M.: Improved marker-assisted selection efficiency of multi-resistance in doubled haploid rice plants. Euphytica **203**(2), 421–428 (2014). https://doi.org/10.1007/s10681-014-1303-1
9. Tripathy, S., et al.: Exploring factors affecting anther culture in rice (Oryza sativa L.). J. Appl. Biol. Biotechnol. **7**(02), 87–92 (2019). https://doi.org/10.7324/JABB.2019.70216
10. Sarao, N., Gosal, S.: In vitro androgenesis for accelerated breeding in rice. In: Gosal, Satbir Singh, Wani, Shabir Hussain (eds.) Biotechnologies of Crop Improvement, Volume 1, pp. 407–435. Springer, Cham (2018). https://doi.org/10.1007/978-3-319-78283-6_12
11. Ilyushko, M., et al.: Intra-callus variability of rice doubled haploids generated through *in vitro* androgenesis. Agricul. Biol. **55**(3), 533–54 (2020). https://doi.org/10.15389/agrobiology.2020.3.533eng
12. Farrell, T., Fox, K., Williams, R., Fukai, S.: Genotypic variation for cold tolerance during reproductive development in rice: screening with cold air and cold water. Field Crop Res **98**, 178–194 (2006). https://doi.org/10.1016/j.fcr.2006.01.003
13. Ilyushko, M.: Effect of growing conditions of rice donor plants on anther culture *in vitro*. J. Agric. Sci. Technol. **5**, 686–694 (2015). https://doi.org/10.17265/2161-6256/2015.08.007
14. Dubina, E.V., Kostylev, P.I., Garkusha, S.V., Ruban, M.G.: Development of blast-resistant rice varieties based on application of DNA technologies. Euphytica **216**(10), 1–12 (2020). https://doi.org/10.1007/s10681-020-02698-4
15. Wang, J., Correll, J., Jia, Y.: Characterization of rice blast resistance genes in rice germplasm with monogenic lines and pathogenicity assays. Crop Prot. **72**, 132–138 (2015). https://doi.org/10.1016/j.cropro.2015.03.014
16. Ferrie, A., Caswell, K.: Isolated microspore culture techniques and recent progress for haploid and doubled haploid plant production. Plant Cell, Tissue Organ Cult. **104**, 301–309 (2011). https://doi.org/10.1007/s11240-010-9800-y
17. Win, A., Tanaka, T., Matsui, T.: Panicle inclination influences pollination stability of rice (Oryza sativa L.). Plant Prod. Sci. **23**(1), 60–68 (2020). https://doi.org/10.1080/1343943X.2019.1698971
18. Sahoo, S.A., Jha, Z., Verulkar, S.B., Srivastava, A.K., Suprasanna, P.: High-throughput cell analysis based protocol for ploidy determination in anther-derived rice callus. Plant Cell Tissue Organ Cult. (PCTOC) **137**(1), 187–192 (2019). https://doi.org/10.1007/s11240-019-01561-2
19. Zhang, S., Li, B., Chen, Y., Shaibi, A., Zheng, H., Sun, J.: Molecular-assisted distinctness and uniformity testing using SLAF-sequencing approach in soybean. Genes **11**, 175 (2020). https://doi.org/10.3390/genes11020175
20. Wu, Y., et al.: Comprehensive evaluation of resistance effects of pyramiding lines with different broad-spectrum resistance genes against *Magnaporthe oryzae* in rice (*Oryza sativa* L.). Rice **12**(1), 1–13 (2019). https://doi.org/10.1186/s12284-019-0264-3

Depth of Fertilizer Application to Justify Subsoiler Body in Soybean Cultivation

Viktor Epifantsev[1], Alexander Panasiuk[2](✉), Yakov Osipov[1], and Yuriy Vaitekhovich[1]

[1] Federal Research Center «All-Russian Research Institute of Soybeans», Ignat'yevskoye shosse, 19, Blagoveshchensk 575027, Russia
[2] Federal State Budgetary Educational Institution of Higher Education "Far Eastern State Agrarian University", Polytechnic, 86, Blagoveshchensk 575005, Russia
alex28rus@list.ru

Abstract. The article presents data on the effect of the type and depth of fertilizer application on the development of the root system and stem growth of soybeans. Organic farming provides a set of measures aimed at preserving the fertility and ecological well-being of the area. Experiment included factors and variants: A - 1. Control - without fertilizers; 2. Sapropel; 3. Wood ash; 4. Chicken manure. Factor B - depth of fertilizer application below seed sowing: 1. Control - did not apply; 2. 60 mm; 3. 80 mm; 4. 100 mm. The research showed that the soybean plants grow better and reach a height of 259–411 mm, form a branched root system going deep into the soil at the depth of 305–311 mm, with the root volume 51 ml and weight 39 g when the granulated wood ash is put in the soil at the depth of 100 mm below putting the seeds. When designing machines for simultaneous deepening, seeding and local fertilizer application, the design of working tools for granulated wood ash application should be oriented.

Keywords: Type · Fertilizer · Depth of application · Working body · Soil dredger · Soybean

1 Introduction

When growing soybeans using organic technology, technogenic influences should ensure the reduction of soil compaction by machines and mechanisms, the abandonment of monoculture and the introduction of multispecies crop rotations with sideral or mulching crops, the transition to the use of allowed agromeliorants [1]. Biological preparations and new methods of soil treatment optimize the conditions for the growth of roots and above-ground parts of plants, preserve fertility and increase crop productivity, and minimize the development of water and wind erosion [2–6].

Minimum tillage and No-till technology (no-till), allow to achieve preservation and increase of soil fertility, preservation of moisture in it in drought periods. Cultivation of soybeans on No-till, compared with the traditional technology in the rotation reduces the number of perennial weeds by 41%, in the permanent crops - by 30%. It increases

A. Muratov and S. Ignateva (Eds.): AFE 2021, LNNS 353, pp. 19–27, 2022.
https://doi.org/10.1007/978-3-030-91402-8_3

the humus content in the rotation by 0.27%, in no-till crops - by 0.31%. No-till reduces production costs by 22.5%, decreases production costs by 10.0%, increases economic efficiency of soybean growing by 10.6% and profitability by 32%. The soybean yield is high in the rotation after corn - 2.52 t/ha, low in the monoculture - 1.14 t/ha. Soil density in crop rotation is 1.13 g/cm^3, and in monoculture 1.20–1.25 g/cm^3 [7]. The effect of No-till technology is not manifested immediately, but after 4–10 years. In 50% of cases the soil remains compacted after No-till application. Negative results of No-till are associated with suboptimal parameters of organic mulch, because its weight depends on the crop. No-till development is associated with risks, and the key to success is the implementation of optimal mulch parameters for specific growing conditions [8–10]. No-till crops in Kursk and Belgorod regions were thinned. They were observed severe weed infestation of crops, mass damage by mice. As a result, we got a decrease in grain quality, grain yields and other crops. No-till system, promoted by scientists, cannot be ideal. It can be used for grain crops if crop rotations, fertilizers, and herbicides are followed. The spread of No-till in Russia, can lead to serious environmental consequences, which will be difficult to eliminate [10].

In recent years, agricultural production in Russia begins to focus on ecologization and biologization with a decrease in the chemical load on the soil and even a complete rejection of the use of chemicals. An alternative to No-till can be the Mow-till method, in which a strip with crops is cultivated, and 2/3 of the field is left uncultivated, agromeliorants are applied to two depths simultaneously with soil loosening. The use of navigational devices, allows to accurately conduct soil preparation, sowing and subsequent cultivation. The sowing strip is shifted to the uncultivated inter-row annually, providing crop rotation. In Mow-till, it is better to take 20 cm as the basic width of the soybean planting strip [11]. Mowing in the inter-row strips of soybean, where rape is sown according to Mow-till, allows to refuse from herbicides, and provides an ecological production. The deep loosening of the arable layer is positive in the ecological aspect on large and small fields. Deep loosening can prevent erosion processes and, in the second half of summer, excessive overwatering and compaction from the work of tillage tools and vehicles [12]. The creation of a well-cultivated topsoil is one of the urgent tasks of farming.

This paper presents research on the effect of fertilizers and the depth of their application on stem growth and the root system of soybeans carried out to justify the design of a combined subsoiler.

2 Materials and Methods

Experimental work was carried out in 2020 in the laboratory of FSBSI FERIMEA. Rationale for the experiment scheme. To support the development of organic food programs and conservation of the environment in the Russian Federation adopted GOST 33980-2016, which authorizes the use of dry poultry manure, natural phosphorites, wood ash and other soil-improving substances in crop production. More than 60% of the Amur region is forested. Natural agromeliorants are extracted in the Arkharinsky, Bureisky, and Ivanovsky districts. Two large poultry farms operate here. There are no problems with finding a base for the production of organic fertilizers and soil improvers for growing soybeans in the region. The type of fertilizer and the depth of its application affect the growth and development of the root system, vegetative organs and their marketable yields.

A two-factor vegetative experiment to justify the depth of fertilizer application included the following factors and options. Schematic of the experiment:

Factor A - type of fertilizer: 1. Control - no fertilizer; 2. Pelletized sapropel - N - 1.2%; P 0.18%; Ca - 34%; Mg - 0.8; 3. Pelletized wood ash - K - 3%; P - 7%; Ca - 40%; 4. Pelletized chicken manure - N:P:K - 4:3:3.

Factor B - the depth of fertilizer application below the level of the soybean seedbed: 1. Control - did not apply; 2. Application of fertilizer to a depth of 60 mm; 3. Application to a depth of 80 mm; 4. Application to a depth of 100 mm. Plastic vessel area - 6358.5 mm^2. Volume - 11763225 mm^3. Repetition of variants in the experiment - 3 times. Placement of variants - systematic. Repetition is 2 times.

To exclude accidental influences on the growth and development of plants, they were grown in identical plastic vessels of 90 mm × 1850 mm size (the first number corresponds to the diameter of the vessel, the second - its height). Six round holes with a diameter of 3 mm were made in the bottom of the vessel. Plastic vessels are easy to wash, their walls are not corroded by soil solution, through the plastic we can see how stuffed the vessel, the soil structure and grow the roots of plants. Soybean plants were grown on the southeast side of the building at a constant temperature of 20–22 °C. Under conditions of insufficient light with a short day in January-March and on cloudy days, the plants were supplemented with 14 h of light. To achieve a total power of 300 W per 1 m^2 installed four lamps type LB-40 with a light flux of 3000 lumens and four lamps type LD-80 with a light flux of 4070 lm per 1 m^2 area.

Freshly plowed typical meadow-chernozem-like soil was harvested for the experiment in autumn 2019 from a plot of fallow land located near the place of field experiments to the depth of the arable layer. The soil contains NO_3 - 7.7 mg/kg, P_2O_5 - 62.0, K_2O - 164 mg/kg and pH - 5.0 (according to GOST R 54650-2011). It was thoroughly mixed on the prepared site. Then it was passed through a sieve with 3 mm holes, removing stones, roots, debris and dry plant debris. Knowing the percentage of hygroscopic water and the amount of air-dry soil in aluminum cylinders, we calculated its moisture capacity. After that we calculated the amount of water for moistening the soil in the vessels up to 60% of full moisture capacity. The selected by weight vessels were checked by diameter, height and weight. They were weighed beforehand with the accuracy of 0.1 g, and on their walls the borders of agromeliorants application and seeds sowing were marked with a felt-tip pen. 100 g of broken glass was poured on the bottom of the vessels at an angle of 300, occupying 2/3 of its part. Then the soil, evenly compacted and leveled to the bottom border of the depth of fertilizer application. The bottom layer of soil in the vessel, 3–4 cm thick was placed more densely than the rest of the soil. Then a layer of fertilizer was also distributed. They were applied in the form of granules. Doses of fertilizers were calculated according to the recommendations of reference books. For one vessel according to the variants: granulated wood ash - 5 g, granulated chicken manure - 22 g, sapropel - 55 g. After that filled up the soil to the border of the seeding, according to the variants of experiment, the filled vessels were brought to the same weight. Seeding rate - 6 seeds per 1 vessel, slightly more than the calculated number of 3 pcs. Preliminary laboratory experiment in 4-fold repetition showed that the germination of seeds of Azure variety was 62%. The seeds were sown to a depth of 5 cm, evenly pouring and compacting the soil over them. Sowing dates for soybeans were January 22 and February 28. Soil weight in all vessels was 610 g. Subsequently, soil was evenly compacted and its surface was leveled; about 1.5–2 cm of free space was left to the top edge of the vessel. After the vessels were filled, soil humidity was maintained at the same level by watering during soybean plants cultivation. Watering was performed 2–3 times a week, the water consumption per vessel was 15–30 ml. Tubes were 170 mm in diameter and 2350 mm long. For the convenience of irrigation, the vessels were calibrated, i.e. brought to the same weight. When sprouts appeared, soil of 10 mm thickness was added to the vessels again, and unnecessary plants were carefully removed. Care of the crops included watering, loosening the soil, and removing weeds. Similar conditions were in the second repetition.

Phenological observations - seed germination, seedlings (cotyledonous and primordial leaves), first true triple leaves, subsequent leaves (single entering the phase of 10% of plants, mass - 80%) were carried out in the experiment. Biometric records - counting the number of plants, leaves; measuring stem height and depth of root penetration into the soil, stem thickness at the level of the seedling knee. Root length was measured after careful extraction from each vessel. Weight was determined by weighing, and plant volume by the amount of water displaced from the vessel. Provision with nutrients - visually (by signs of starvation and excess). The growth and development of soybean plants were recorded on March 3 and April 10.

3 Results and Discussion

As can be seen from Table 1, the deeper the fertilizers were put into the soil, the higher the soybean stems grew. The highest stems were in the variant of applying wood ash to a depth of 100 mm below the seedbed and reached 259 mm. When sapropel was applied 60 and 80 mm below the seedbed, and chicken manure at a similar depth, the stems were 1 to 59 mm shorter than the control. The excess of soybean stalks height in the variants of sapropel, chicken manure and wood ash incorporated at the depth of 100 mm below the seeds over the control was 1 mm, 34 and 42 mm, respectively.

Table 1. Influence of fertilizer application depth on the soybean stalk growth dynamics, mm.

Fertilizer	Depth under seed, mm	Date							
		29.01	30.01	31.01	3.02	5.02	7.02	10.02	13.02
Control without fertilizers	Didn't contribute	0	20	38	78	98	142	174	217
Sapropel	80	0	14	44	80	118	160	182	202
	60	17	48	63	122	138	152	170	188
	100	0	48	73	82	98	153	183	218
Woodash	80	0	38	58	106	176	179	198	242
	60	0	32	57	89	132	166	200	234
	100	0	36	73	108	132	166	188	259

(continued)

Table 1. (*continued*)

Fertilizer	Depth under seed, mm	Date							
		29.01	30.01	31.01	3.02	5.02	7.02	10.02	13.02
Chicken manure	80	0	0	03	34	94	108	198	216
	60	0	0	0	50	84	92	131	158
	100	27	67	94	108	148	158	189	251

In the repeated experiment, the highest stem height was in the variant with wood ash 100 mm (Table 2).

Table 2. Effect of fertilizer and depth of its application on the dynamics of soybean stalk growth, mm.

Fertilizer	Depth under seed, mm	Date							
		4.03	5.03	6.03	10.03	13.03	18.03	24.03	10.04
Control without fertilizers	Didn't contribute	10	30	48	138	154	202	235	258
Sapropel	80	0	30	54	138	154	226	250	318
	60	5	40	48	147	180	228	240	303
	100	0	25	50	148	197	234	298	374
Woodash	80	5	30	50	172	200	218	264	394
	60	0	05	38	128	168	224	273	373
	100	3	30	40	100	200	243	300	411
Chicken manure	80	0	0	12	70	142	191	245	386
	60	0	0	41	116	160	218	225	330
	100	0	05	51	70	138	237	269	407

The experiments revealed a regular excess of soybean plants stem height in the variants with fertilizer application in comparison with the control variant. As well as the efficiency of wood ash in comparison with chicken manure and sapropel. With increasing the depth of all fertilizers there was an increase in stem growth. In the variant of sapropel 80 and 100 mm the growth was 15 and 71 mm more in comparison with the depth of 60 mm. In the 80 and 100 mm chicken manure variant, growth was 56 and 77 mm and the 80 and 100 mm wood ash variant, growth was 21 and 38 mm greater than the 60 mm depth. The excess of soybean stalk height in the variants of application of sapropel, chicken manure and wood ash applied at a depth of 100 mm over the control reached 116 mm, 149 and 382 mm, respectively. Growing soybean plants in the experiment had the following appearance and are shown in Fig. 1.

Fig. 1. Growing soybean plants in the experiment.

In the primary experiment all soybean plants in all variants reached the phase of three triple leaves of the beginning of bud formation by the time of accounting. The root system penetrated most deeply into the soil up to 305 mm in the variant with application of wood ash to a depth of 100 mm below the level of seed sowing. Weight and volume of roots depended on the depth of their penetration into the soil of the type of fertilizer used and the depth of its application. Wood ash provides quite good branching of soybean roots and their penetration into deep layers. When wood ash is applied to a depth of 100 mm below the seed sowing the root system of soybean actively branches. The results of the secondary experiment, confirmed the same patterns and dependencies that were noted initially.

Table 3. Effect of fertilizers on soybean root system indicators (average for 2 replications).

Fertilizer	Depth under seeds, mm	Root placement depth, mm	Root weight, g	Main root length, mm	Number of plants with roots reached the bottom of the vessel, pcs.	Root volume, ml
Control without fertilizers	Didn't contribute	230	28.5	64	2.0	40.0
Sapropel	80	238	32.0	131	2.5	35.0
	60	225	30.0	91	2.3	30.0
	100	269	36.0	142	2.8	37.5
Woodash	80	263	36.0	142	3.0	45.0
	60	247	32.0	138	3.0	50.0
	100	308	39.0	159	3.0	51.0

(continued)

Table 3. (*continued*)

Fertilizer	Depth under seeds, mm	Root placement depth, mm	Root weight, g	Main root length, mm	Number of plants with roots reached the bottom of the vessel, pcs.	Root volume, ml
Chicken manure	80	210	26.5	45	1.0	28.0
	60	180	19.0	37	1.0	25.0
	100	292	30.0	69	2.8	31.5
$SSD_{0.5}$ in the primary experiment By factor A By factor B AB interaction		16 13 21	2.6 2.2 3.4	13.8 7.5 19.0	0.1 0.1 0.2	3.1 2.5 4.5
$SSD_{0.5}$ in the secondary experiment For factor A Byfactor B AB interaction		19 14 32	2.7 2.3 3.5	14.5 8.1 21.0	0.1 0.1 0.2	4.8 1.5 5.3

As can be seen from Table 3, in two experiments sapropel, applied at different depths of 60 mm, 80 and 100 mm below the sowing of soybean seeds limits root penetration to the deep layers on clay soil, respectively, they reach a depth of 91 mm, 131 and 142 mm. It does not decompose well, and after irrigation does not provide fast water filtration. The length of the main root depended on the type of fertilizer used.

The shortest main roots were when chicken litter was applied 60 and 80 mm below the seed sowing depth, they reached 37 and 45 mm. In these variants, the roots branched poorly, and rotting of root tips was noted in some plants. The greatest number of soybean plants whose roots reached the bottom of the vessel were in the variants with application of wood ash below the seed sowing at 60 mm, 80, 100 mm.

In the experiment $F_f > F_{0,5}$, Tukey's criterion proved that H_0:d = 0, the null hypothesis is rejected. The absolute error of the difference in the average depth of soybean plant roots in the primary experiment S_d = 10.3 mm, in the repeated experiment S_d = 15.7 mm. With the value of $t_{0.5}$ = 2.04, the relative value of $SSD_{0.5}$ in the primary experiment is equal to 8.5%, in the secondary 12.9%. According to the weight of the roots in the primary experiment S_d = 1.8 g, in the repeated experiment S_d = 1.7 g, by the length of the main root respectively S_d = 9.3 mm and S_d = 10.3 mm. According to the number of plants with roots reaching the bottom of the vessel in both cases S_d = 0.098 pc. According to the volume of roots, the error of the difference of averages in the earlier experiment S_d = 2.2 ml, in the later one S_d = 2.6 ml, and the relative value of the $SSD_{0.5}$, respectively, was 16.5% and 19.4% for this indicator.

According to all indicators of soybean root system, wood ash is significantly superior to the control. When sapropel is applied, the root indicators are at the level of control, and chicken manure is inferior to it. The regular dependence of the increase in the soybean root system indicators with increasing depth of fertilizer application was established. In order to carry out soil deepening and simultaneous application of agromeliorant in the field conditions there was designed a technical unit with working bodies, which is shown in Fig. 2.

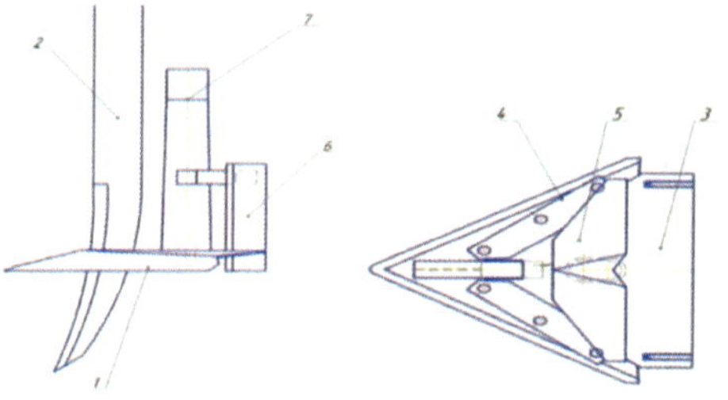

Side view Bottom view

Fig. 2. Scheme of the working body for the strip preparation for sowing. Where: 1 - Cultivator foot; 2 - Slitter stand; 3 - plate; 4 - mounting plate; 5 - Spreader; 6 - Knives; 7 - tube-seed pipe.

Technical unit for preparing a strip for soybean crops simultaneously performs operations: soil deepening (slitting), cutting a 200 mm wide strip for soybean crops from the main mass of stubble, undercutting mulching crop roots in the soybean cropping strip, and applying fertilizer doses in the horizon of undercut mulching crop roots taking into account the results of the experiments.

Invention patent № 2717298 of 19.03.2020 was received for this technical unit.

4 Conclusion

After the researches on justification of the depth of agromeliorants application with a subsoiler for ecological soybean cultivation it can be concluded that soybean plants grow better and reach a height of 259–411 mm when wood ash is applied at a depth of 100 mm. They form a branched root system going deep into the soil at the depth of 305–311 mm with the root volume 51 ml and their mass 39 g when applying granulated wood ash at the depth of 100 mm below the level of sowing seeds. When designing machines for subsoiler, simultaneous seeding and local fertilizer application, it should be oriented on the development of working tools for granulated wood ash.

Acknowledgments. The work was carried out on the topic of R&D № 0818-2019-0005 by the FSBSI FERIMEA.

References

1. Kudeyarov, V.N.: Soil-biogeochemical aspects jf arable farning in the Russian Federation. Pochvovedeniye **1**, 109–121 (2019). https://doi.org/10.1134/s0032180x1901009
2. Zhang, Y., Li, C., Wang, M.: National scale of meta-analysis of soil carbon and nitrogen accumulation potential in the "'Grain for Green"' program in China. Eur. Soil Sci. **53**(7), 892–901 (2020). https://doi.org/10.1134/S1064229320070170
3. Litvin, L.F., Kiryukhina, Z.P., Krasnov, S.F., et al.: Dynamics of Agricultural Soil Eiberia and far East. Pochvovedeniye **1**, 136–148 (2021). https://doi.org/10.31857/S0032180X2101007X
4. Olson, K.R., Gennadiev, A.N.: Dynamics of accumulation and erosion of organic carbon in the soil as a result of land use change (Illinois, USA). Eur. Soil Sci. **53**(4), 436–445 (2020). https://doi.org/10.1134/S1064229320040122
5. Shen, H.O., Wang, D.L., Wen, L.L., et al.: Soil erosion and typical measures for the protection of soil and water on hillsides in the Chinese region. Eur. Soil Sci. **53**(10), 1509–1519 (2020). https://doi.org/10.1134/S1064229320100178
6. Mrabet, R., Moussadek, R., Fadlaoui, A., Van Ranst, E.: Conservation agriculture in dry areas of Morocco. Field Crop Res. **132**, 84–94 (2012)
7. Nikul'chev, K.A., Banetskaya, Y.: Influence of crop rotation cultures on microbiological activity, soil agrophysical properties and soybean yield. Zemledeliye **1**, 11–14 (2020). https://doi.org/10.24411/0044-3913-2020-10103
8. Polyakov, D.G., Bakirov, F.G.: Organic mulch and No-till in agriculture: a review of international experience. Zemledeliye **1**, 3 (2020). https://doi.org/10.24411/0044-3913-2020-10101
9. Jordan, A., Zavala, L.M., Gil, J.: Effects of mulching on soil physical properties and runoff under semi-arid conditions in southern Spain. CATENA **81**(1), 77–85 (2010)
10. Chenu, C., Le Bissonnais, Y., Arrouays, D.: Organic matter influence on clay wettability and soil aggregate stability. Soil Sci. Soc. Am. J. **64**(4), 1479–1486 (2000). https://doi.org/10.2136/sssaj2000.6441479x
11. Epifantsev, V.V., Panasyuk, A.N., YaA, O., Vaytekhovich, Y.: Optimum strip width at Mow-till for organic accumulation and soybean yield. Siberian Herald Agric. Sci. **50**(4), 13–22 (2020). https://doi.org/10.26898/0370-8799-2020-4-2
12. Epifantsev, V.V., Panasyuk, A.N., YaA, O., Vaytekhovich, Y.: The effect of soil deepening on soybean yield by using various sowing machines. Siberian Herald Agric. Sci. **50**(1), 13–22 (2020). https://doi.org/10.26898/0370-8799-2020-1-2

Studies on Correction of the Trajectory of a Tractor Train

Alexey Kushnarev, Alexandr Shuravin, Evgeny Kuznetsov(✉), Sergey Shchitov, Aleksey Goncharuk, and Olesya Mitrokhina

Far Eastern State Agrarian University, Politehnicheskaja St. 86, 675005 Blagoveshchensk, Russia

Abstract. The simplest way to increase the efficiency of the available energy resources is the most complete implementation of the traction and coupling properties inherent in them, observing all the requirements for the transportation of agricultural goods. The conducted research has established. that an important limiting factor in the technical implementation of solutions devoted to this problem is the condition and width of the roads along which the multi-link tractor-transport trains (MTTT) move. At the same time, one of the main reasons hindering the operation of the MTTT is the width of the transport corridor when making turns. The article proposes a solution to the above problem through the use of devices that have a design and scientific novelty, and are designed to optimize the width of the transport corridor of the MTTT in turns. Conducted theoretical, experimental studies and the obtained experimental data in terms of comparing related indicators in previous studies on this issue allow us to conclude that in the modern period of development of agricultural science, the presented solution is the most rational, which undoubtedly distinguishes the proposed methodological and technical solutions.

Keywords: Tractor and transport unit · Device · Trailer · Width of the transport corridor · Turn · Traction-coupling properties · Efficiency

1 Introduction

The efficiency of the use of energy resources in most regions is determined by the possibility of realizing their traction and coupling properties, which is confirmed by the results of studies presented in the works [1, 2].

It was found that the efficiency of the use of energy resources is greatly influenced by natural production conditions of the region of operation. This factor is especially significant in areas with a sharply continental climate, where a sharp change in temperature conditions during the day significantly changes the operating conditions.

Based on this, its own system of technologies and machines is being developed for each production region, which provides for the influence of natural production conditions on the technology of agricultural work [3, 4].

An important factor is also the influence of the road surfaces on which the goods are transported on the operation of tractor-transport units. In the conditions of the Amur

A. Muratov and S. Ignateva (Eds.): AFE 2021, LNNS 353, pp. 28–35, 2022.
https://doi.org/10.1007/978-3-030-91402-8_4

Region, the largest percentage of roads are on field and unpaved roads, for which the width of the carriageway does not always reach even the minimum permissible values [5]. All of the above conditions impose their own requirements when using tractor trains to transport goods, especially from the fields.

Thus, in the conditions of the Amur Region, in order to increase the efficiency of the use of tractor-transport trains, it is necessary to follow the path of increasing the traction-coupling qualities of power facilities with a simultaneous optimization of the width of the transport corridor.

On the basis of previous studies [6, 7] on the use of tractor-transport trains for the transportation of goods [8, 9] it is noted that it is possible to increase the efficiency of the MTTT operation in such conditions by using special devices and devices that can eliminate the above problems.

Based on the above, we can conclude that the purpose of the work is to install devices in the tractor-transport unit that ensure their reliable and safe operation by optimizing the width of the transport corridor.

2 Material and Methods

As an example of the use of devices for shifting the connection points of the components of a tractor-transport unit, original solutions are proposed that allow you to optimize the width of the transport corridor when making turns on roads with a limited width of the carriageway.

As a result of the studies, it was found that to ensure the optimal width of the transport corridor, it is necessary to install two devices for the MTTT running gear. The first device is installed on the tractor and provides a displacement of the connection point of the power unit and the first trailer. The second device is installed on the first trailer and provides a displacement of the connection point of the first and second trailer.

Let us dwell in more detail on the second device according to RF patent No. 2739635 [10], which provides a displacement of the trailer point of the first and second trailers. The main feature of this design is the ability to ensure the displacement of the connection point of two trailers automatically, without operator intervention while driving on turns.

Fig. 1. Tracking device mounted on the turntable of the first trailer.

For this purpose, on the turntable of the first trailer, tracking devices are installed on both sides (Fig. 1), which are connected to the towing device itself by means of a flexible cable connection (Fig. 2). To ensure constant tension of the flexible cable connection, spring cable tensioners are installed (Fig. 3). The installed cable connection provides the connection of the follower (Fig. 1) and the towing hitch at the connection point of the first and second trailers (Fig. 4 and 5).

Fig. 2. Flexible cable power connection of the device.

Fig. 3. Automatic spring cable tensioner.

When performing a maneuver, the turntable of the first trailer makes a pivot movement (Fig. 6) and the follower (Fig. 1) installed on it automatically moves the connection point through a flexible cable connection depending on the direction of rotation (Fig. 7 and 8).

Fig. 4. General view of a trailer with a tracking device.

Fig. 5. Sliding drawbar.

Fig. 6. Fragment of rotation during device operation.

A multi-link tractor train consisting of an energy unit (a wheeled tractor of class 1.4) and two agricultural trailers (2TTU-4) was taken as the object of the study of the

Fig. 7. Example of displacement (trailer hitch is shifted to the right side).

Fig. 8. Location of the trailer hitch when the trailer is moving in a straight line.

operation of the MTTT. The studies were carried out according to well-known general and specific methods using specialized programs for mathematical calculations, experimental modeling and methods of regression analysis [11].

At the same time, the following parameters were measured: the turning radius of the follower, the angle of rotation of the trailer guide wheels, the change in the attachment point of the trailer to the tractor, the length of the hydraulic cylinder rod outlet. During the research, specialized instruments and equipment were used to measure the listed parameters. The processing of experimental data obtained in the course of the research was carried out by the well-known methods of mathematical statistics using information technologies.

3 Results

Theoretical and experimental studies on the use of the device to optimize the width of the corridor, carried out on the basis of the Federal State Budgetary Educational Institution "Far Eastern GAU", the Amur Region, the city of Blagoveshchensk, the Russian Federation and the peasant farms of the region, showed the validity and correctness of the chosen method of increasing the traction and coupling properties of the tractor.

In the course of the carried out theoretical and experimental studies, it was found that the total displacement of the MTTT connection points consists of two components:

$$L_0 = L_1 + L_2 \quad (1)$$

where L1 and L2 accordingly, the length of movement of the points of connection of the MTTT, m

In general, the total length of movement of the connection points can be determined by the formula:

$$L_0 = \mathrm{Rarccos}\frac{ab + \sqrt{(R^2 - b^2)\cdot(R^2 - \alpha^2)}}{R^2} + 2R(0_2 - 0_1)\cos\alpha \quad (2)$$

where R- turning radius of the device, m, φ_1 and φ_2 - device rotation angles, degrees.

The analysis of formula (2) showed that when using the proposed devices, the width of the transport corridor is reduced in comparison with the serial MTTT.

To confirm the obtained theoretical studies, experimental studies were carried out directly in real operating conditions, the processed test data are presented in Fig. 9.

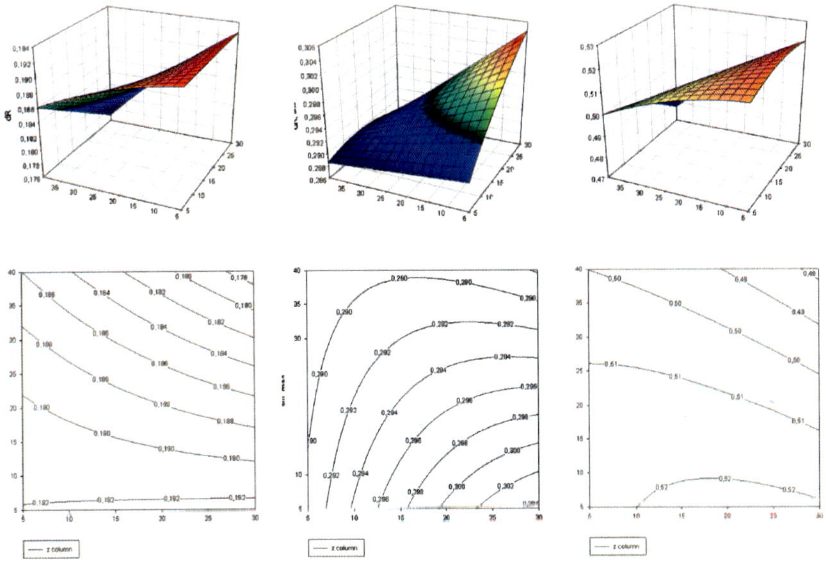

Fig. 9. Surface and cross-section of the response surface dR depending on changes in the parameters of the tug-loading device.

When constructing the response surfaces, using the SigmaPlot v.11.0 program, two factors were varied - the angle of change in the direction of movement and the length of the displacement of the TCD.

As a result of analysis of variance (Fig. 9), it was found that the use of the proposed devices makes it possible to adjust the width of the transport corridor due to:

displacement of the connection point of the power plant and the first trailer;
moving the connection point of the first and second trailers.

The data obtained allow us to conclude that the proposed device allows to optimize the width of the transport corridor up to 14… 18% in comparison with the serial MTTT.

The introduction of the results of research and development work into the technology of transport and technological support at "Soyuz "of the Ivanovo district made it possible to obtain energy savings from 8… to 10 MJ/tkm, which in ruble terms amounted to 840 thousand rubles per one unit per year.

The data obtained as a result of the conducted theoretical and experimental studies are within the confidence interval, which indicates their reliability.

4 Conclusion

Based on the foregoing, it can be concluded that the proposed device is a highly efficient design that implements original ideas and has a design novelty, designed to maintain the optimal width of the transport corridor. The research materials are used in the technology of transport and technological support used by" Soyuz"in the Oktyabrsky District and a number of other peasant farms in the Amur Region.

The implementation of the results obtained in production has made it possible to increase the efficiency when using class 1.4 wheeled tractors in transport operations and to obtain significant economic benefits.

Conducted theoretical, experimental studies and the obtained experimental data in terms of comparing related indicators in previous studies [12–15] on this issue allow us to conclude that in the modern period of development of agricultural science, the presented solution is the most rational, that undoubtedly highlights the proposed methodological and technical solutions, forming a new conceptual line in the environment of applied science, in particular, dedicated to the study of interaction in the field of knowledge "man-machine-nature" in agricultural production and carrying out transport operations.

Based on the results of the work, the team of authors responsibly declares that there is no conflict of interest, information data and results were obtained as a result of the activities of the entire research group. At the same time, the material resources spent are not attracted funds and belong personally to the co-authors.

References

1. Rehnberg, A., Drugge, L.: Snaking stability of articulated frame steer vehicles with axle suspension. Int. J. Heavy Veh. Syst. **17**(2), 119–138 (2010). https://doi.org/10.1504/IJHVS.2010.033178

2. Azad, N., McPhee, J.: Off-road laterial stability analysis of an articulated steer vehicles with a rear-mounted load. Int. J. Veh. Syst. Model. Test. **1**(3), 106–130 (2005)
3. Nsari, M., Esmailzadeh, E., Younesian, D.: Longitudinal dynamics of freight trains. Int. J. Heavy Veh. Syst. **16**(2), 102–131 (2009). https://doi.org/10.1504/IJHVS.2009.023857
4. Kropáč, O., Múčka, P.: Effect of obstacles on roads with different waviness values on the vehicle response. Veh. Syst. Dyn. **46**(3), 155–178 (2008). https://doi.org/10.1080/00423110701211548
5. Shchitov, S., Tikhonchuk, P., Bumbar, I.: Research in traction and coupling properties of wheeled tractors class 1.4 equipped with a trailing weight distribution device. Revista ESPACIOS **38**(48), 35 (2017)
6. Shchitov, S., Kuznetsov, E.: Improvement of efficiency of use of wheeled transport vehicles in the agro-industrial complex. J. Adv. Res. Dyn. Control Syst. (JARDCS) **13**, 707–714 (2018)
7. Shchitov, S.V., Krivutsa, Z.F., Kuznetsov, E.: Experimental studies of the effectiveness of the design for the cross-axle redistribution of the weight load of the car. Int. J. Appl. Eng. Res. (IJAER) **14**(24), 16747–16752 (2018)
8. Michałek, M., Pazderski, D.: Forward tracking of complex trajectories with non-Standard N-Trailers of non-minimum-phase kinematics avoiding a jackknife effect. Int. J. Control **92**(11), 2547–2560 (2018). https://doi.org/10.1080/00207179.2018.1448117
9. Martinod, R., Betancur, G., Osorio, J., Castañeda, L.: A study of the effect of the transition curve in the coupling elements between the carbody and the bogie. Int. J. Heavy Veh. Syst. **21**(4), 351 (2014). https://doi.org/10.1504/IJHVS.2014.068113
10. Shchitov, S.V., Kuznetsov, E.E.: Patent for invention No. 2728162 (RU), 2018114603. Accessed 28 July 2020 (2020)
11. Bojic, I., Nymoen, K: Survey on synchronization mechanisms in machine-to-machine systems. Eng. Appl. Artif. Intell. **45**, 361–375 (2015)
12. Bojic, I., Nymoen, K.: Survey on synchronization mechanisms in machine-to-machine systems. Eng. Appl. Artif. Intell. **45**, 361–375 (2015). https://doi.org/10.1016/j.engappai.2015.07.007
13. Vantsevich, V.: Vehicle systems: coupled and interactive dynamics analysis. Veh. Syst. Dyn. **52**(11), 1489–1516 (2014). https://doi.org/10.1080/00423114.2014.944869
14. Thomas, G., Vantsevich, V.: Wheel-terrain-obstacle interaction in vehicle mobility analysis. Veh. Syst. Dyn. **48**(1), 139–156 (2010). https://doi.org/10.1080/00423111003690496
15. Wan, C., Markine, V.L., Shevtsov, I.Y.: Improvement of vehicle-turnout interaction by optimising the shape of crossing nose. Veh. Syst. Dyn. **52**(11), 1517–1540 (2014). https://doi.org/10.1080/00423114.2014.944870

Improving the Efficiency of Wheel-Wheeling Machine-Tractor Units in Areas with Special Natural and Climatic Conditions

Alexander Slepenkov, Evgeny Kuznetsov(✉), Sergey Shchitov, Tatiana Sharipova, Olga Shchegorets, and Andrey Burmaga

Far Eastern State Agrarian University, Politehnicheskaja Street 86, 675005 Blagoveshchensk, Russia

Abstract. The harrowing effect is the leveling of the surface, which makes it possible to optimize the distribution of the incoming atmospheric moisture over the field. The processes of high-quality tillage are relevant for the conditions of the Amur Region, as an area of risky agriculture with special natural and climatic conditions. So in the fall, a large amount of precipitation falls in the region, which leads to a violation of the water-air state of the soil and the formation of a soil crust on the surface of the field. To eliminate the above mentioned effect, it is necessary to carry out such a field operation as harrowing using heavy disc harrows due to the fact that the land resources of the Amur Region are represented mainly by heavy loamy soils. The article presents the results of the research and optimization parameters that confirm the effectiveness of the modernization of disc harrows by adjusting the vertical load on the working body of the harrow due to the redistribution of the adhesion weight in the scheme of a wheeled harrow machine-tractor unit.

Keywords: Wheeled tractor · Weight · Load optimization · Disc harrow · Soil · Density · Efficiency

1 Introduction

Creation of favorable conditions for the growth of agricultural crops is the main criterion for obtaining a high yield and gross product. In this case, the main role is assigned to pre-sowing soil cultivation, which is implemented during a number of technological operations - harrowing, cultivation, disking, rolling (before and after sowing) [1, 2]. The list of the above-mentioned operations for different natural and climatic zones of countries is also not the same [3, 4].

For the Amur region, it is characteristic of its attribution to the zone of "risky agriculture", in which naturally a sharp change in natural and climatic conditions during the year, due to the possibility of the arrival of unfavorable natural and abnormal phenomena, which regulates the cultivation and cultivation of crops [5, 6]. Also, a feature of the region is the cultivation of such an agricultural legume crop as soybeans on a large area

A. Muratov and S. Ignateva (Eds.): AFE 2021, LNNS 353, pp. 36–43, 2022.
https://doi.org/10.1007/978-3-030-91402-8_5

of land, which makes the region one of the most significant in the Russian Federation in terms of its collection.

It is necessary to take into account that the harvesting of soybeans ends at the onset of freezing temperatures, usually at the end of October-November. In addition, during this period of time, a large amount of precipitation falls in the form of rain and snow, which does not allow the autumn post-harvest tillage due to the weak bearing properties of the upper soil layer due to high humidity.

This all imposes certain necessary conditions on the timing and duration of spring field work, since it must be borne in mind that they begin only when the surface thawing of the upper layer of the soil allows it to be processed, but most often in the field, pre-sowing soil cultivation is carried out simultaneously with sowing.

Along with this, the timing of spring field work in the region is strictly limited, since spring precipitation contributes to the rapid melting of the permafrost layer, which reduces the bearing capacity of the soil. Small peasant farms (PF) are most susceptible to the influence of factors of natural and climatic conditions, the number of mechanization means and tractors in which is quite limited, and where the main energy unit is wheeled tractors of classes 1, 4 with a wheel formula 4K2 [7, 8]. Moreover, in the total volume of agricultural production in the region, peasant farms account for, as shown by the analysis of statistical data, up to 25% of the total gross harvest [9]. It is in these farms that harrowing with the use of heavy disc harrows is provided for one of the operations of pre-sowing soil cultivation by a simplified production technology.

At the same time, in the conditions of insufficient traction and coupling properties of tractors with wheel formula 4K2, their use is limited, which reduces the efficiency of using the fleet of mobile energy facilities. Taking into account that the land potential of the Amur Region is represented mainly by heavy loamy soils of various textures, for high-quality processing of fields it is required to load the working bodies of the harrow with ballast.

To solve the above-mentioned problems, on the basis of earlier studies [8, 9], a device-regulator of the adhesion weight of the harrow unit was proposed, for which the RF patent for intellectual property No. 196181 [10], shown in Fig. 1, was obtained, allowing, depending on the conditions of the road surface and the required quality of work, due to the redistribution of the adhesion weight in the scheme of the wheeled

Fig. 1. General view of the adhesion weight regulator of the harrow unit.

harrow machine-tractor unit, which allows you to regulate the vertical load on the harrow working body without using ballast.

Thus, the purpose of the research is to check the efficiency of using the proposed device in real production conditions and to assess its impact on the technological parameters of the harrow unit.

2 Material and Methods

The research was based on the standard methods set forth in the state standards of the Russian Federation GOST 7057–2001 "Agricultural tractors. Test methods", GOST 30745–2001 "Agricultural tractors. Methods for determining traction indicators", GOST 24055–2016, GOST 24059–2017 "Agricultural machinery. Operational and Technological Assessment Methods". On the basis of the above requirements, private author's methodological techniques were developed that allow assessing the effect of the proposed device on the main operational and design-technological parameters of the unit.

In the course of experimental tests related to the redistribution of the adhesion weight, we used platform electronic scales MVSK (V) for weighing vehicles MVSK-10–3-A, weighing limit 10 t and portable underlay industrial scales RW-10 (15) P, intended for weighing any power equipment with up to three axles and a maximum axle load of 20 tons (Fig. 2).

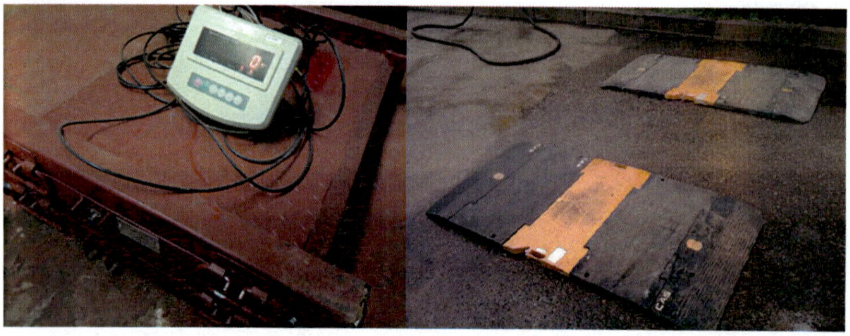

Fig. 2. Platform electronic scales MVSK (V) and scales RW-10 (15) P.

The experimental data were recorded by the instrumentation complex (Fig. 3).

Fig. 3. Instrument-measuring complex and a fragment of field tests.

The processing of the data obtained during the experiment was carried out by the well-known methods of mathematical statistics using information technologies and the use of specialized programs Mathcad and SigmaPlot v.11.0.

3 Results

High-quality soil preparation for sowing is difficult to carry out without the help of agricultural equipment - a disc harrow. It is simultaneously able to level the soil surface, mix it, and also loosen its surface without turning the layer. Using a harrow reduces the time and cost of cultivating the land. With these quality standards for harrows, certain requirements are imposed on them, among which the main ones are:

- clear correspondence of the depth of soil cultivation to the specified parameters;
- uneven depth should be no more than 0.01 m.

Disc harrows available today are mainly designed for light and medium soils. When working on heavy clay and loamy soils, they cannot fully meet the requirement for them - the depth of processing, due to insufficient vertical load on the working body (disk). Most often, to eliminate this drawback, additional weights (ballasting) are used, installed on the harrow frame. This ultimately increases the mass of the harrow itself and, as a result, increases the energy consumption of labor and reduces the safety of movement of the unit, especially when moving.

Experiments aimed at industrial verification of the technical solution were carried out in field conditions, the soil background of which, in its characteristics, corresponded to the agrotechnical requirements. According to its properties (mechanical composition), the soil was a heavy loam. In order to find acceptable values of the factors, it was necessary to solve a compromise problem for the four obtained regression equations. At the same time, the tillage depth, the width of the disc harrow, the speed of the unit and the lift angle of the tractor hitch are represented by independent functions. The search for compromise values was carried out using the programs "Sigma Plot 11.0" and "Compass 3D V16 LT" (Fig. 4).

In a graphical form, the effect of the lift angle of the tractor's hitch on the working depth is shown in Fig. 5.

The regression equation took the form:

$$a = 0.73h + 0.0356h^2 \tag{1}$$

Analysis of the data obtained (Fig. 5) allows us to conclude that with an increase in the lifting angle of the traverse of the tractor hitch, the depth of soil cultivation increases. So with an increase in the lifting angle of the linkage and adjusting the device from 5° to 30°, the processing depth increased from 0.05 m to 0.02 m. Thus, by adjusting the angle of rise, you can change the depth of cultivation of the soil layer.

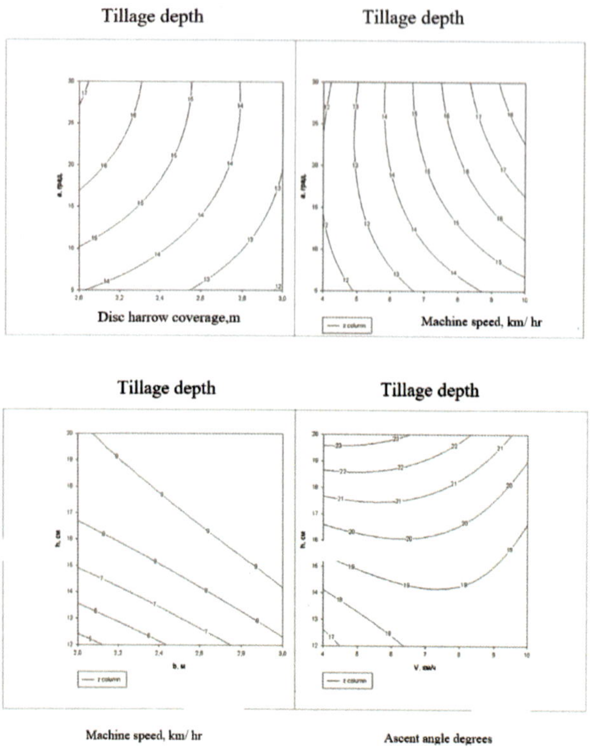

Fig. 4. Cross-section of the response surface (3D models) of the dependence of the working depth and width of the disc harrow on the speed of the disc harrow and the angle of the traverse of the tractor hitch.

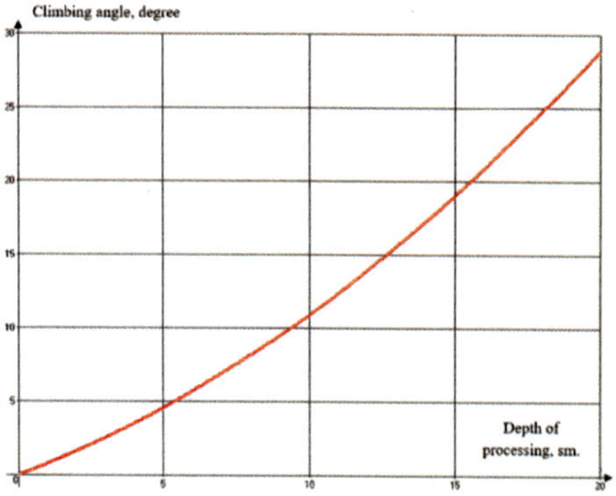

Fig. 5. Dependence of the depth of tillage on the angle of the device (V = 7 km/h, b = 2.5 m).

It should be noted that the lifting angle of the hitch and the adjustment of the device also affects the speed of the machine-tractor unit (MTA), since the traction resistance of the harrow changes. The research results are shown in Fig. 6.

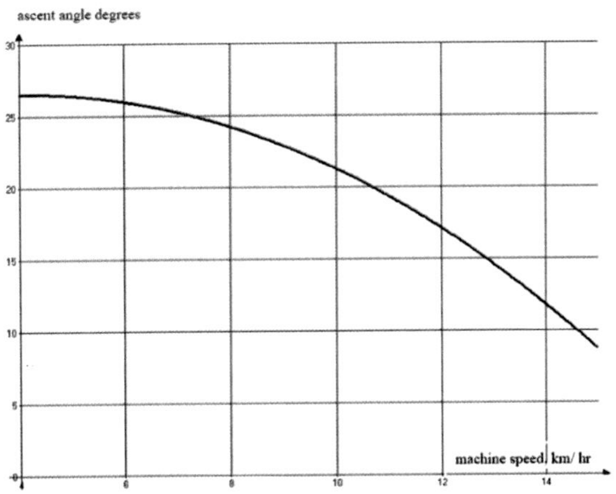

Fig. 6. Dependences of the speed of the unit on the lifting angle (b = 2.5 м, h = 16 cm).

The process under consideration can be described by the following regression equation:

$$a = 24 + 1.23V - 0.087V^2 \quad (2)$$

It was found that with a decrease in the angle of ascent of the device, the speed of the MTA increases. So with a decrease in the angle of elevation of the device from 25° to 8° , the speed of movement increased from 7.3 km/h to 13.4 km/h.

The experimental data obtained within the confidence interval correspond to the results of theoretical studies, which indicates their reliability.

As a result of solving the compromise problem, the following acceptable values are justified, all other things being equal: with an average depth of soil cultivation, the angle of the tractor's hitch is 26–30° from the transport position of the harrow, the width of the disc harrow is 2.2–2.8 m, the travel speed of unit - 7–10 km/h. At the maximum lifting angle of the hitch, the processing depth is 19–20 cm, the speed of the unit is 4–7 km/h, the width of the disc harrow is 2.9–3.0 m. At the optimal value (6–8 km/h) of the unit's speed, the tillage depth is 13.5–17 cm, the width of the disc harrow is 2.4–3.0 m, the lifting angle is 23–30°. With the maximum working width of the harrow, the lifting angle is 26–30°, the tillage depth is 13–17 cm, the speed of the unit is 11–12 km/h.

4 Discussion

The carried out production check made it possible to determine the optimal design and technological parameters of the experimental MTA with the proposed device.

The research materials are used in the agricultural production technology used by "Soyuz" in the Oktyabrsky district and a number of other peasant farms in the Amur Region. The implementation of the results of research and development work has shown the high efficiency of the proposed solution and confirmed empirically that the use of a machine-tractor unit with a harrow unit adhesion weight regulator allows improving the quality of pre-sowing soil cultivation by optimizing the weight, while reducing fuel and energy MTA costs by 15… 17%.

The experimental data obtained in terms of comparing related indicators in earlier studies [7, 9, 11, 12] on this issue allow us to conclude that in the modern period of development of agricultural science, the proposed solution is the most rational, that undoubtedly highlights the proposed methodological and technical solutions, forming a new conceptual line in the environment of applied science, in particular devoted to the study of interaction in the field of knowledge "man-machine-nature" in agricultural production in the conditions of risky farming areas, devoted to studies of increasing the efficiency of soil preparation for sowing.

5 Conclusion

Based on the foregoing, it can be concluded that the proposed device-regulator of the grip weight of the harrow unit is a highly efficient design that implements original technical ideas, which has a design novelty and high industrial applicability.

The implementation of the results obtained in production has made it possible to increase the efficiency when using wheeled tractors of class 1.4, aggregated with a heavy disc harrow, in the technology of presowing soil cultivation.

Acknowledgements. Based on the results of the work, the team of authors responsibly declares that there is no conflict of interest, information data and results were obtained as a result of the activities of the entire research group. At the same time, the spent material resources are not attracted funds and belong personally to the co-authors.

References

1. Gholkar, M., Salokhe, V., Keen, A.: The Effect of axle load and tyre inflation pressure on the tractive performance of a two wheel drive tractor on soft clay paddy field. In: An ASABE Meeting Presentation, vol. 1–12, p. 096606 (2009)
2. Avarvarei, B.: The impact of working speed, soil and plough type on tractors power force at normal tillage. Medunarodni Simhjzij agronoma zbornikradova. Ostjek **2**, 892–896 (2009)
3. Troyanovskaya, I.: Ploughing tractor lateral withdrawal model. Procedia Eng. **206**, 1540–1545 (2017). https://doi.org/10.1016/j.proeng.2017.10.674
4. Belyaev, V.I., Fruhauf, M., Mainel, T.: Ecological consequences of conversion of steppe to arable land in Western Siberia. Europa Reg. **1**(4), 13–21 (2004)
5. Gopichandran, R.: Mitigation-adaptation nexusfor sustainability: some importantcrosscutting and emerging considerations. Vulnerabil. Agric. Water Fisher. Clim. Change Toward Sustain. Adapt. Strat. (2014). https://doi.org/10.1007/978-94-017-8962-2_1
6. Crisan, V.: Evalution of small seeds distribution uniformity. Actual Tasks Agric. Eng. **40**, 243–251 (2012)
7. Shchitov, S., Tikhonchuk, P., Bumbar, I.: Research in traction and coupling properties of wheeled tractors class 1.4 equipped with a trailing weight distribution device. Revista ESPACIOS **38**(48), 35 (2017)
8. Shchitov, S., Kuznetsov, E.: Improvement of efficiency of use of wheeled transport vehicles in the agro-industrial complex. J. Adv. Res. Dyn. Control Syst. **13**, 707–714 (2018)

9. Shchitov, S.V., Krivutsa, Z.F., Kuznetsov, E.: Experimental studies of the effectiveness of the design for the cross-axle redistribution of the weight load of the car. Int. J. Appl. Eng. Res. (IJAER) **14**(24), 16747–16752 (2018)
10. Slepenkov, A.E., Polikutina, E.S., Shchitov, S.V., Kuznetsov, E.E., Krivutsa, Z.F.: Increasing the efficiency of use of wheeled harrow units in regions of risk farming. E3S Web Conf. **262**, 01003 (2021). https://doi.org/10.1051/e3sconf/202126201003
11. Li, H., He, J., Gao, H., Zhang, Z., Chen, Y.: The effect of conservation tillage on crop yield in china. Front. Agric. Sci. Eng. **2**(2), 179–185 (2015). https://doi.org/10.15302/J-FASE-2015058
12. Wulfsohn, D., Way, T.: Factors that influence tractive performance of wheels, tracks and vehicles. Adv. Soil Dyn. **3**, 209–252 (2009)

Characteristic of Spring Vetch Breeding Lines in the Conditions if the Steppe Zone of Primorsky Krai

Olga Telichko(✉), Oksana Mokhan, and Alexey Emelyanov

FSC of Agricultural Biotechnology of the Far East named after A.K. Chaika, 30, Volozhenin Street, Timiryazevsky Stl., 692539 Ussuryisk, Russia

Abstract. Breeding lines of spring vetch, one of the promising annual crops, when cultivated in the conditions of Primorsky Krai, is characterized. Spring vetch has high fodder advantages (yield, nutritional value). The purpose of the work is to create and study new breeding material for spring vetch in order to obtain new variety. The results of studies of the main economically valuable traits of seed productivity (plant height, number of beans, seed productivity per plant, 1000 seed weight) are presented. In terms of the complex of economic parameters of seed productivity, promising breeding lines have been identified. Breeding line No. 72 (Lugovskaya 85 × Uzunovskaya 34) is characterized by the best economically valuable traits: seed productivity - 20.39 g/plant, number of beans - 45.1 pcs/plant, 1000 seed weight - 50.75 g. The most productive in terms of green mass are: No. 40, 42, 49, 63 and 69 (8.0–9.4 t/ha) are the most productive in terms of green mass among the studied samples; they exceed the standard by 1.9–3.3 t/ha. The content of raw protein in the studied samples was at the level of 19.0% (No. 63) - 23.9 (No. 66), fiber - 20.8% (No. 69) - 27.6 (No. 47).

Keywords: Vetch · Breeding lines · Seed productivity · Yield

1 Introduction

The agro-industrial complex of the Russian Federation is currently working on the establishment of territorial and industrial production of livestock products. An increase in the number of animals requires an improvement in fodder, and the lack of vegetable protein in fodder is the most pressing problem.

Spring vetch (*Vicia sativa L.),* which has high forage characteristics, is of great interest among annual forage grasses. As one of the most common forage crops in the production of annual legumes it is used in all agricultural regions of Russia. It has a high content of digestible protein, carotene, carbohydrates, fats, various vitamins and minerals. Spring vetch is superior in total nutritional value and crude protein content to peas and broad beans. Grass mixtures using spring vetch can be used as intermediate, stubble, postcut and undersown crops. The inclusion of spring vetch annual grasses in the composition of mixed crops makes it possible to obtain fodder of high lifestock

A. Muratov and S. Ignateva (Eds.): AFE 2021, LNNS 353, pp. 44–53, 2022.
https://doi.org/10.1007/978-3-030-91402-8_6

level. This crop is successfully used for making hay, haylage and silage. The use of spring vetch as a legume component in a grass mixture allows to obtain fodder with the highest nutritional characteristics and maximum protein digestibility. The most common mixtures of spring vetch are with oats (*Avena sativa L.*), wheat (*Triticum aesturum L.*), and barley (*Hordeum vulgare L.*) [1–8].

Spring vetch seeds, as well as their processed products, are valuable fodder for pigs and poultry and one of the unused reserves of highly concentrated protein fodder [9, 10]. Leguminous crops do not only provide valuable forage products, but also have a beneficial effect on soil fertility. These crops are capable of fixing nitrogen from the atmosphere and accumulating it in the soil. The use of vetch grown on green manure as a precursor can significantly increase the biological activity of the soil [11].

In the conditions of Primorsky Krai, spring vetch is characterized by a high potential for seed productivity. This crop is capable of meeting the needs of farms in green mass, suitable for grain forage purposes, however, due to the lack of stable seed production, spring vetch was not widespread in the region. As well as other crops, spring vetch is realized through varieties. The varieties approved for use in the Far East region have many advantages, however, according to individual economic parameters, they are not ideal. Therefore, further improvement aimed at increasing the productive potential of newly created varieties, giving them plasticity, adaptability, and resistance to stress factors of the environment should be carried out.

Spring Vetch breeding at the Federal State Budgetary Scientific Institution "Federal Research Center of Agricultural Biotechnology of the Far East named after A.K. Chaika" is aimed not only at improving the forage productivity of green mass, as well as at increasing the yield of seeds [12].

The objective of the current study was to create and study new breeding material for spring vetch to obtain a new variety.

Tasks:

- to study and select the most promising breeding lines of spring vetch;
- to identify correlations between economically valuable traits.

The scientific novelty lies in the fact that for the first time in the conditions of the Primorsky Krai of the Russian Federation, suitable for further breeding promising breeding samples (lines) of spring vetch with a complex of economically valuable traits were created.

2 Materials and Methods

The study of spring vetch was carried out in the areas of breeding crop rotation of the laboratory of field and grassland forage production of the Federal State Budgetary Scientific Institution "Federal Research Center of Agricultural Biotechnology of the Far East named after A.K. Chaika" in 2018–2020.

The meteorological conditions of the growing season (2018–2020) were characterized by significant differences in the distribution of precipitation and temperature regime. The sum of positive temperatures was 2625 °C in 2018, 2746 °C in 2019, and 2684 °C in

2020; precipitation total - 694.7, 437.1, 510.8 mm, respectively. To establish the influence of the amount of precipitation and heat on the duration of interphase periods, the hydrothermal coefficient (HTC) introduced by Selyaninov. It is calculated as the sum of precipitation divided by the sum of temperatures, reduced by a factor of 10. In 2018, the HTC was 2.72; in 2019 - 1.72; in 2020 - 2.2 (the obtained coefficients characterize 2018 and 2020 years of research as excessively wet, and 2019 – as wet).

The arable layer thickness was 22–25 cm, the humus content was 53 ± 0.53% (low), the pH of the salt extract was 5.0 ± 0.1 (moderately acidic), the content of mobile phosphorus was 56 ± 11 mg/kg (average), exchangeable potassium - 60 ± 12 mg/kg (average), nitrogen - 62 ± 7 mg/kg (average).

The following breeding of spring vetch were studied: No. 32 - Lugovskaya 85 × Ugolyok; No. 36 - Lugovskaya 85 × Lyudmila; No. 40 Lugovskaya 85 × no name (36133); No. 42, 53 - Lugovskaya 85 × Elena (36089); No. 47, 49, 55 - Lugovskaya 85 × no name (35798); No. 57, 59 - Lugovskaya 85 × no name (36105); No. 63, 64 - Lugovskaya 85 × no name (35257); No. 66 - Lugovskaya 85 × VIR 356 (36331); No. 69, 70, 72 - Lugovskaya 85 × Uzunovskaya 34). The variety Lugovskaya 85, bred in the V.R. Williams Federal Research Center of Forage Production and Agroecology ", was used as control.

The plot included two rows (ratio of spring vetch to supporting culture - 1:1, 40 seeds each), area - 2 m^2. One row was collected for green mass and dry matter count, the other one - for seed productivity count. Oats was used as maintaining culture. The control was located 10 numbers away. The breeding lines, inferior to the control in terms of economically valuable traits, were rejected.

Soil was cultivated by underwinter plowing, early spring harrowing, simultaneous harrowing and marking. Sowing was performed during the third decade of April, manually. The breeding line nursery was weeded as needed.

Chemical analyzes of plant samples of spring vetch and soil were carried out in the agrochemical laboratory of the Federal State Budgetary Scientific Institution "Federal Research Center of Agricultural Biotechnology of the Far East named after A.K. Chaika". While counting the accumulation of nutrients in the green mass, the following was determined: the content of crude protein according to the Kjeldahl method, crude fiber - according to Henneberg and Stohmann, crude fat - according to the Rushkovsky method.

Nitrate nitrogen, mobile phosphorus, mobile potassium, $pH_{sol.}$, and organic matter were identified in soil samples. The counts and analysis of hybrid samples were carried out according to the methodological instructions of the VIR [13]. The energy nutritional value of spring vetch was calculated in accordance with the methodology "Calculation of metabolic energy in fodder based on the content of raw nutrients". Correlation analysis was carried out according to B.A. Dospekhov.

3 Results and Discussion

72 breeding lines of spring vetch were studied from 2018 to 2020; each of them was obtained with the participation of the Lugovskaya 85 variety, recommended for cultivation in the Far East region (bred in the FWRC FPA). The growing season was 81–96 days (90 days for the standard). The hybrid samples No. 66, 69, 70, 72 (95–96 days) had a longer growing season. The earliest ripening numbers are 57 and 59 (81 days) (Table 1).

Table 1. Spring vetch breeding lines characteristics in terms of economically valuable traits (2018–2020).

Line no.	Vegetation period. 24 h	Biological yield. t/ha	Plant height. cm	Number of beans per plant. ea	Number of seeds per bean. ea	Bean length. cm
No. 32	88	1.84	116.2	16.1	7–9	5.6–6.5
No. 36	85	1.99	117.5	20.3	7–9	5.6–6.2
No. 40	85	1.91	118.8	22.0	5–9	4.6–6.5
No. 42	85	2.14	132.1	24.5	5–9	4.0–6.1
No. 47	85	2.94	113.5	27.1	6–9	5.3–6.4
No. 49	88	2.62	104.6	21.2	7–9	5.6–6.9
No. 53	84	2.57	99.3	25.1	6–9	5.1–7.0
No. 55	87	2.65	109.6	37.4	7–10	5.7–6.6
No. 57	81	3.00	94.0	41.1	6–9	5.0–6.9
No. 59	81	2.77	105.1	35.0	6–9	5.0–7.0
No. 63	84	3.78	114.7	40.4	5–10	4.5–6.5
No. 64	84	2.81	102.0	36.3	6–9	4.9–6.9
No. 66	96	2.37	135.0	23.1	7–10	5.7–6.3
No. 69	95	2.84	111.4	24.3	7–9	6.0–6.3
No. 70	95	2.72	110.3	25.0	6–10	5.7–6.4
No. 72	95	4.46	113.5	45.1	6–10	5.6–6.5
Lugovskaya 85 (standard)	90	1.19	104.7	14.1	7–9	5.6–6.1

Plant height is an important indicator of forage productivity. It varied from 94.0 to 135.0 cm depending on the sample. The tallest samples are No 42 and 66; they exceeded the standard by 27.4–30.3 cm.

The assessed breeding lines differed in the number of beans per plant, which ranged from 16.1 to 45.1 g/plant. Samples No. 55, 57, 59, 63, 64, 72 (35.0–45.1 pieces/plant) had the largest number of beans. According to this indicator, all studied lines exceeded the standard by 1.2–3.2 times.

When cultivating spring vetch for seeds, the biological yield of the studied samples was 1.84–4.46 t/ha. The lines No. 47, 57, 63, 72, exceeded the standard by 2.5–3.7 times and were the most productive.

The productivity of seeds from a plant plays a significant role in breeding spring vetch for seed productivity. It was 7.28–20.39 g/plant in selection samples. Sample No. 72 had the highest productivity of seeds per plant - 20.39 g, exceeding the standard by 3.6 times (Fig. 1).

1000 seed weight is affected by the nature of the variety as well as the growing conditions. Diseases and lack of moisture in the soil result in a decrease in this indicator. Due to the lack of moisture during the vegetation period of 2019, 1000 seed weight was 1.4–1.7 times lower in comparison with other vegetation periods. There was enough rainfall for the formation of vetch seeds in other years. In the studied samples, this indicator varied from 36.4 (Lugovskaya 85 × no name (35257), No. 64) to 62.5 g (Lugovskaya 85 × no name (35798), No. 49). The largest-seeded lines under study are Nos. 32, 47, 49, 69.

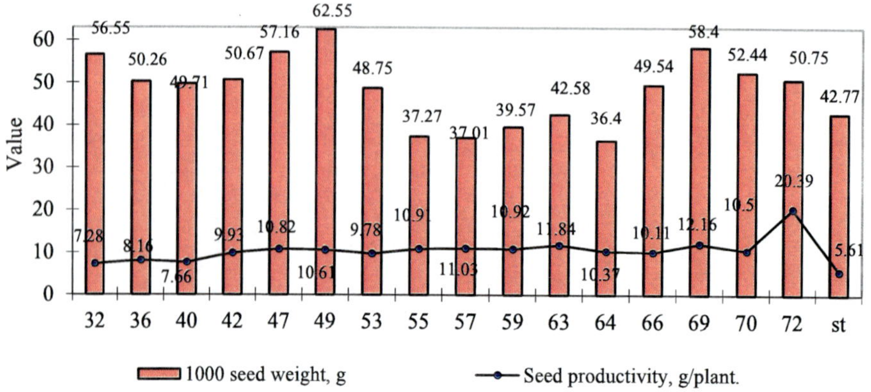

Fig. 1. Spring vetch seed productivity indicators (2018–2020).

The correlation analysis revealed a weak negative relationship between the plant height and the number of beans per plant (r = −0.29), seed productivity (r = −0.03); 1000 seed weight and the number of seeds in a pod (r = −0.15). During the research, it was found that the 1000 seed weight has an average negative relationship with the number of beans (r = −0.56) and a weak positive relationship with the productivity of seeds per plant (r = 0.04), as well as with the length of the bean (r = 0.03). Plant height was positively correlated with the weight of 1000 seeds (r = 0.33). Seed productivity strongly depended on the number of beans per plant (r = 0.74) and weakly on the number of grains (r = 0.17). The data obtained indicate that the 1000 seeds weight has the least effect on the productivity of seeds (r = 0.04).

One of the indicators of the spring vetch fodder value is the green mass yield and the collection of dry matter. Green mass is the main fodder for farm animals in the summer. When cultivating spring vetch for fodder purposes, it is possible to obtain from 5.0 to 9.4 t/ha of highly nutritious green plant mass (Fig. 2). The most productive among the studied samples are No. 40, 42, 49, 63 and 69, which exceed the standard sample by 1.9–3.3 t/ha. For collection of dry matter (DM) from 1 ha, lines No. 40, 42 and 49 have an advantage, exceeding the standard by 0.69–1.09 t/ha.

As a result of the research, it was found that the collection of dry matter is highly dependent on the yield of green mass (r = 0.88). The “plant height” index has little effect on dry matter collection (r = 0.26) and green mass yield (r = 0.25).

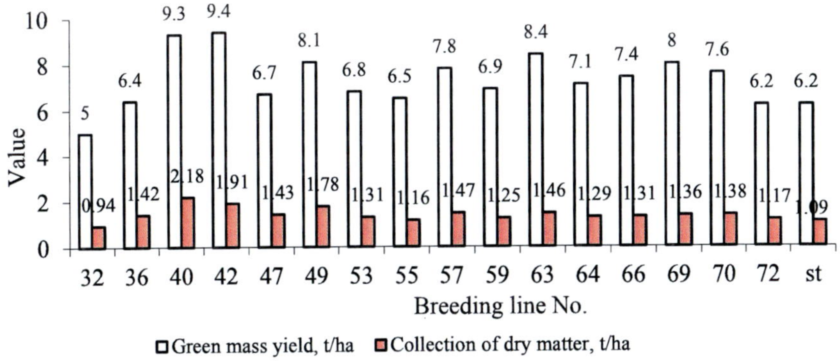

Fig. 2. Green mass yield and dry matter collection of spring vetch per 1 m^2 (2018–2020).

In order to identify the nutritional value of the green mass, a biochemical analysis of spring vetch was carried out. Protein plays an important role in the nutrition of animals. All vital processes of animals are associated with protein metabolism in the body. Protein deficiency leads to increased production costs of livestock products. Samples 57, 66 and 69 stand out with the highest content of crude protein (Table 2). Compared to the standard, the content of crude protein in them was 12.1–21.3% higher comparing to the standard.

Crude fiber is a necessary factor that improves digestion, although, overly high fiber protein content in fodder leads to insufficient palatability by animals. Fiber content in the studied samples was 20.8 (No. 69) – 27.6 (No. 47). Samples 66, 69, 70, 72 had less fiber.

The fat content in fodder plays a huge role in animal nutrition; it is part of the structural material of the protoplasm of all cells. Fat is essential for the proper functioning of the digestive glands, it is the body's main energy store and an important source of heat. Depending on the sample, the crude fat content varied from 2.3 to 5.7%. Nos 55, 66, 69 and 70 differ in the maximum content of crude fat.

Table 2. Biochemical indicators of the spring vetch green mass, % (2018–2020).

Breeding line No	Crude protein	Crude fiber	Crude fats	Absolute dry matter
No 32	21.3	26.7	2.3	18.8
No 36	20.2	27.3	4.8	22.2
No 40	20.6	26.3	4.4	23.5
No 42	19.4	27.2	2.6	20.3
No 47	20.6	27.6	4.2	21.3
No 49	21.1	26.9	4.1	22.0
No 53	21.5	27.5	4.3	19.3
No 55	20.4	27.0	5.3	18.0

(*continued*)

Table 2. (*continued*)

Breeding line No	Crude protein	Crude fiber	Crude fats	Absolute dry matter
No 57	22.2	26.8	4.8	18.8
No 59	19.7	26.6	3.4	18.1
No 63	19.0	27.1	3.1	17.4
No 64	19.4	26.8	3.2	18.2
No 66	23.9	21.4	5.5	17.7
No 69	22.1	20.8	5.7	17.0
No 70	21.9	21.4	5.2	18.2
No 72	20.9	21.0	5.3	18.9
Lugovskaya 85 (Standard)	19.7	25.3	4.8	17.8

The cultivation of spring vetch for forage purposes makes it possible to increase the nutritional value of forage. According to zootechnical standards, 1 fodder unit (f. unit) must contain at least 110 g of DP (digestible protein - the portion of the protein that is used in the digestion process). In terms of the provision with 1 fodder unit of digestible protein, breeding lines No. 32, 53, 57 (180.3–181.8 g) stand out (Fig. 3). 1 kg of dry matter of the studied samples contains from 142.5 (No. 63) to 179 g (No. 66) of DP. The DP content in the standard sample is 161.2 g and 147.8 g, respectively. Contents of fodder units in 1 kg of dry matter corresponds to 0.87 (No. 42, 63) - 1.02 (No. 66) (Table 3). The collection of digestible protein from 1 ha, depending on the sample, varied from 0.15 to 0.34 tons. No. 40 has the highest content (an excess of 0.24 tons over the standard).

In general, all studied samples of spring vetch, including the standard, comply with zootechnical standards.

An important indicator of the energy nutritional value of fodder is fodder unit. Fodder rates for farm animals are calculated on its basis. The nutritional value of 1 kg of dry oats is taken as a unit. 1 kg of dry matter in the studied samples contains from 0.87 to 1.02 fodder units. The maximum content was noted in No. 66 and 69, in No. 70 this indicator corresponds to 1.00. The collection of fodder units from 1 ha was within the range of 0.83 tons (No. 32) - 1.97 tons (No. 40). Almost all samples (with the exception of No. 32) exceeded the Lugovskaya 85 variety by 6.1–39.0%.

The scientifically grounded criteria for the energy assessment of the nutritional value of fodder is metabolic energy, which is capable of replenishing resources of the animal's body spent on production. The provision of farm animals with energy is one of the important factors that determine the level of their productivity. The content of exchange energy in 1 kg of dry matter, depending on the sample, ranged from 10.33 to 11.23 MJ. Nos. 66, 69, 70 and 72 are distinguished by the increased energy content. Samples Nos. 40, 42, 49 are characterized by the maximum yield of metabolic energy: the excess over the variety Lugovskaya 85 is 1.6–2.0 times.

Table 3. Productivity and nutritional value of spring vetch samples (2018–2020).

Breeding line no.	Fodder units content in 1 kg of dry matter	Collection from 1 ha. t		Metabolic energy content in 1 kg of dry matter. kg	Metabolic energy output. GJ/ha
		Fodder unit	Digestible protein		
No. 32	0.89	0.83	0.15	10.46	9.83
No. 36	0.89	1.27	0.22	10.49	14.89
No. 40	0.91	1.97	0.34	10.57	23.05
No. 42	0.87	1.65	0.28	10.33	19.74
No. 47	0.88	1.26	0.22	10.45	14.94
No. 49	0.90	1.60	0.28	10.53	18.74
No. 53	0.89	1.17	0.21	10.51	13.76
No. 55	0.90	1.05	0.18	10.55	12.24
No. 57	0.92	1.35	0.25	10.63	15.63
No. 59	0.88	1.11	0.19	10.45	13.06
No. 63	0.87	1.27	0.21	10.35	15.11
No. 64	0.88	1.13	0.19	10.40	13.42
No. 66	1.02	1.34	0.24	11.23	14.71
No. 69	1.02	1.38	0.23	11.20	15.23
No. 70	1.00	1.38	0.23	11.11	15.33
No. 72	0.99	1.17	0.18	11.10	12.99
Lugovskaya 85 (standard)	0.92	0.99	0.16	10.64	11.59

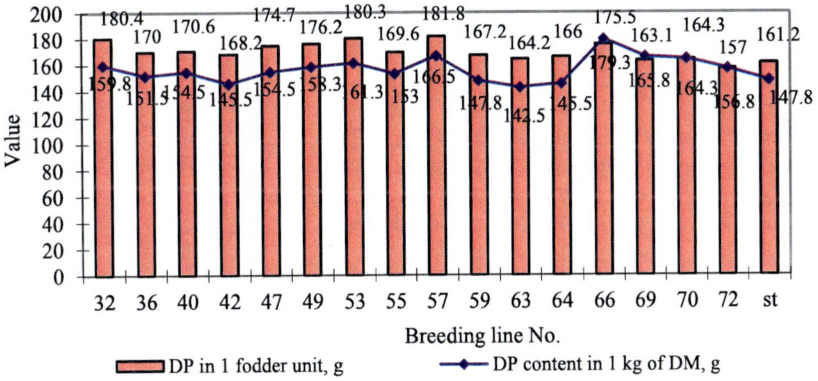

Fig. 3. Digestible protein content in the spring vetch samples (2018–2020).

In the course of research, it was found that the collection of digestible protein per hectare significantly depends on the yield of green mass (r = 0.86) and weakly on the content of digestible protein in 1 kg of DM (r = 0.17). The collection of exchangeable energy from 1 ha has a strong correlative relationship with the yield of green mass (r = 0.89), and weak negative relationship with the content of digestible protein and fodder units in 1 kg of DM (r = −0.02 and r = −0.05, respectively). It was found that the height of plants has little effect on the content of DP in 1 fodder unit. (r = 0.12) and in 1 kg of DM (r = 0.13).

As a result of the research, it can be concluded that No. 40 and 42 are characterized by a complex of economically valuable traits and are the most suitable for further breeding for fodder productivity.

4 Conclusion

As a result of the study of breeding lines of spring vetch, the best samples that significantly exceed the standard in plant height, 1000 seed weight, seed productivity per plant, number of beans per plant, biological seed yield, forage productivity and nutritional value were identified.

Nos. 40, 42, 72 are characterized by the complex of economically valuable traits. They will be further studied according to the breeding process scheme for the subsequent transfer of the variety to the State Variety Testing.

References

1. Buyakin, N., Krasnoperov, A.: Efficiency of bean and cereal mixes in different terms of sowing. Legumes Roat Crops **3**(35), 44–50 (2020). https://doi.org/10.24411/2309-348X-2020-11184
2. Donskaya, M., Velkova, N., Naumkin, V.: Leguminous crops (Lathyrus, Vetch, Pea) in mixed plantings with white mustard. Zemledelie **4**, 25–28 (2019). https://doi.org/10.24411/0044-3913-2019-10406.
3. Zenkova, N., Kovaleva, I., Shloma, T., Moiseeva, M.: Legumes Cereals **4**(36), 99–105 (2020). https://doi.org/10.24411/2309-348X-2020-11211
4. Andreeva, O., Pilipenko, N., Sidorova, L., Kharchenko, N.: Productivity and nutritional value of sudan grass in single-species and multi-species sowings. Forage Prod. **7**, 28–31 (2020). https://doi.org/10.25685/KRM.2020.7.2020.004
5. Mao, Z., Fu, H., Nan, Z., Wang, J., Wan, C.: Fatty acid content of common vetch (Vicia sativa L.) in different regions of Northwest China. Biochem. Syst. Ecol. **44**, 347–351 (2012). https://doi.org/10.1016/j.bse.2012.06.021
6. Enrico, J.M., Piccinetti, C.F., Barraco, M.R., Agosti, M.B., Eclesia, R.P., Salvagiotti, F.: Biological nitrogen fixation in field pea and vetch: response to inoculation and residual effect on maize in the Pampean region. Eur. J. Agron. **115**, 126016 (2020). https://doi.org/10.1016/j.eja.2020.126016
7. Deng, J., et al.: Replacing summer fallow with annual forage improves crude protein productivity and water use efficiency of the summer fallow-winter wheat cropping system. Agric. Water Manag. **230**, 105980 (2020). https://doi.org/10.1016/j.agwat.2019.105980

8. García-Garijo, A., Palma, F., Lluch, C., Tejera, N.A.: Physiological and biochemical responses of common vetch to the imazamox accumulation. Plant Physiol. Biochem. **73**, 321–325 (2013). https://doi.org/10.1016/j.plaphy.2013.10.013
9. Kononchuk, V., Timoshenko, S., Shtyrkhunov, V., et al.: Productivity of vetch-cereal mixtures for haylage and grain depending on the composition and fertilizers under different weather conditions in the central non-chernozem region. Legumes Groat Crops **3**(35), 78–83 (2020). https://doi.org/10.24411/2309-348X-2019-11189
10. Tabeidian, S.A.: Effect of processing on the nutritional value of common vetch (Vicia sativa) seed as a feed ingredient for broilers. J. Appl. Poultry Res. **20**(4), 498–505 (2011). https://doi.org/10.3382/japr.2010-00306
11. Gurin, A., Chadaev, I.: The role of legume predecessors in the activation of biological activity of gray forest soil. Legumes Groat Crops **1**(29), 21–25 (2019). https://doi.org/10.24411/2309-348X-2019-11068
12. Telichko, O., Mokhan, O.: Forage advantages of vetch sowing hybrids in the steppe zone of primorsky krai. Agrarian Russia **5**, 7–11 (2019). https://doi.org/10.30906/1999-5636-2019-5-7-11
13. Repiev, S.: Spring Vetch Selection: Methods and Directions. L, VIR (1991)

The Problem of Attracting Private Investment in Agriculture

Sergey Yekimov[1](✉), Viktoriia Nianko[2], Natalia Sirenko[3], Alla Burkovska[3], and Olena Bodnar[3]

[1] Publishing House "Education and Science" s.r.o., Olstynska 607/1, Praha 8, 18100 Troja, Czech Republic

[2] Dniprovsk State Technical University, Dnipropetrovsk Region, Kamianske, Ukraine

[3] Mykolayiv National Agrarian University, Mykolayiv, Ukraine

Abstract. Improving mortgage lending, increasing government spending on financing scientific research in the field of agriculture, improving the system of agricultural risk insurance, and providing tax incentives to agricultural enterprises can help increase the investment attractiveness of agricultural businesses. In order to attract external investors, agricultural enterprises should, in our opinion, offer their investment projects more widely at various forums dedicated to investment activities, and make efforts to form their own good investment image. One of the reasons for the high interest rates for lending to agricultural enterprises is the size of the Central Bank's discount rate. In banking, the Central Bank applies only a single discount rate, but in many industrialized countries, its size can be zero or even negative. The authors proposed to apply not one, but several discount rates of the Central Bank, depending on what specific purposes the commercial bank will spend the financial resources received from the Central Bank. In our opinion, for lending to agricultural enterprises, a commercial bank should be able to receive funds from the Central Bank at a discount rate close to zero, and in our opinion it will make it possible to reduce the interest rate when granting loans to agricultural enterprises.

Keywords: Agriculture · Investments · Investments in agriculture · State regulation of agriculture · Agricultural sector of the economy

1 Introduction

In agricultural production, the main tool of production is the land. It is not subject to depreciation, but nevertheless, it can have an impact on the cost of production if it is an agricultural enterprise that leases it. The yield of agricultural land is heterogeneous and largely depends on the influence of climatic and natural factors. Some of the livestock products remain in the use of the enterprise in the further production process, and the rest is sold to consumers. Animal husbandry and crop production are interrelated activities of an agricultural enterprise. Crop production consumes organic fertilizers produced by animal husbandry, and animal husbandry consumes feed produced by crop production.

A. Muratov and S. Ignateva (Eds.): AFE 2021, LNNS 353, pp. 54–61, 2022.
https://doi.org/10.1007/978-3-030-91402-8_7

The production cycle in the production of agricultural products can last for several months.

The need for the use of borrowed funds by an agricultural enterprise is due to the following reasons: the need to replenish working capital, the need to update agricultural machinery, as well as to cover losses in lean years, as a result of natural disasters and mass destruction of agricultural plantings by pests. The specifics of agricultural production is also that the final result of the economic and financial activities of the enterprise can be estimated only by the results of the work for the year. The evaluation of the efficiency of the functioning of an agricultural enterprise can be characterized by economic and production indicators in physical or monetary terms. A specific feature of most types of agricultural machinery is its narrow specialization for performing work only with certain types of agricultural crops. This often requires the presence of many varieties of agricultural machinery in the arsenal of an agricultural enterprise. Mainly, agricultural production is represented by a large number of small rural farms and large agricultural holdings.

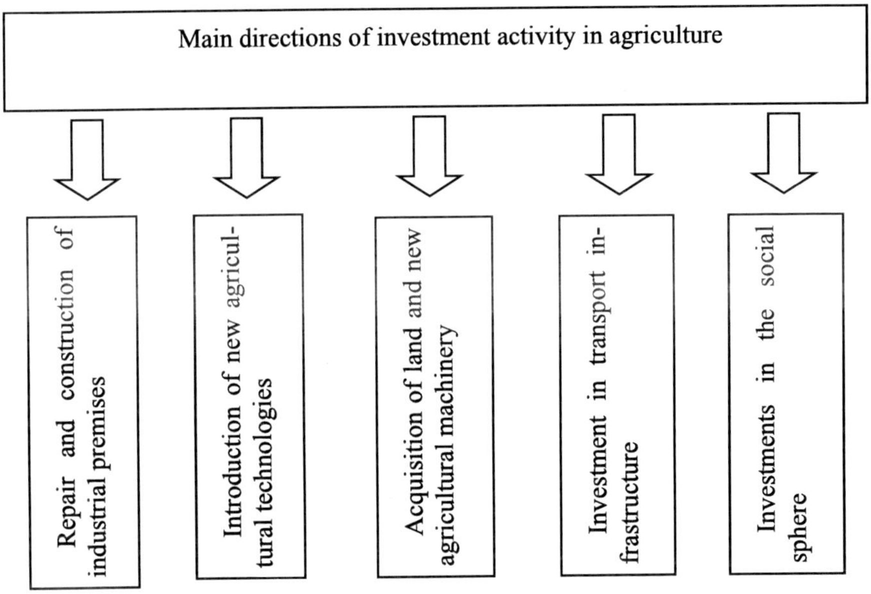

Fig. 1. Main directions of investment activity in agriculture.

According to [1], the need for agricultural enterprises to attract external financing is due to the following factors: the development of new agricultural machinery and new agricultural technologies, seasonal replenishment of working capital and the need to modernize the material and technical base.

According to [2], the lack of investment resources in the agro-industrial complex of Ukraine has become one of the main reasons for the deterioration of the financial condition of agricultural enterprises, the deterioration of their material and technical base and the weak development of market infrastructure.

According to [3], the technical equipment of agricultural production largely depends on the structure, volume and direction of investment (Fig. 1).

According to [4], the peculiarity of the innovation process in the agricultural sector of the economy is that the process of expanded production in it depends on the most important factors that need to be taken into account when implementing innovation and investment activities.

The authors [5] divide the factors influencing the innovation and investment activity of an agricultural enterprise into external (positive and negative) and internal (unfavorable and favorable.

According to [6], one of the important factors influencing the economic development of the state is the successful and stable functioning of the agricultural sector of the economy.

According to [7], for the effective development of the agro-industrial complex, there is a need for the introduction of new innovative agricultural technologies, improving relations between consumers and producers of agricultural products.

According to [8], agricultural enterprises form investment resources at the expense of their own funds and funds attracted from outside.

Self-financing of agricultural enterprises, in our opinion, is generally available. large agricultural holdings due to the limited financial resources of small farms.

According to [9], with a lack of own financial resources, it is advisable to solve the problem of financing agricultural enterprises by attracting external investment.

According to [10], the problem of attracting investment in agriculture requires a comprehensive approach that involves the use of all tools to increase the investment attractiveness of the enterprise, reduce investment risks and stimulate the emergence of its own investment resources.

According to [11], the increase in state support for agricultural producers, accelerated depreciation of fixed assets of agricultural enterprises, the use of long-term bank lending to agricultural enterprises, and the attraction of foreign investment in the agricultural sector of the economy should help to increase the investment attractiveness of agriculture. the development of various forms of cooperation in rural areas.

According to [12], investment activities in rural areas should be aimed at the development of production and transport infrastructure in rural areas.

According to [13], the main problems with attracting investment in agriculture are:

1) Availability of credit resources for agricultural enterprises.
2) Implementation of effective control over the intended use of credit funds.
3) Protection of creditors' rights.

In our opinion, the state investment policy in the CIS countries should provide for the introduction of preferential tax rates for agricultural enterprises.

According to [14, 15], the investment attractiveness of agriculture is reduced due to: low elasticity of demand for products produced by agricultural enterprises, low mobility of resources used in agriculture, as well as the dependence of the results of financial and economic activities on natural and climatic conditions.

2 Methods

When implementing this scientific research, the authors used an analytical method, which made it possible to consider the problems studied in the work in their development and unity. Taking into account the purpose and task of the study, we used the functional-structural method of scientific cognition. As a result we were able to study a number of problems related to attracting private investment in agriculture.

3 Results

The basis for effective socio–economic development of the state is the progress in the development of agriculture. Successful development of agriculture requires constant development and implementation of new innovative agricultural technologies. The development of innovative processes in the agricultural sector of the economy of most industrialized countries largely depends on state support. In our opinion, innovation policy in the agricultural sector of the economy should be focused on creating favorable conditions for investment and innovation processes. In our opinion, the state innovation policy in the field of agriculture should include:

1. Creating conditions for the maximum realization of agricultural enterprises of their production potential.
2. Creating equal conditions for enterprises of all forms of ownership to use their innovative potential.
3. Ensuring competitive conditions and cooperation in the implementation of agricultural enterprises of their innovative potential

In our opinion, the organizational and economic mechanism of innovation management in the agricultural sector of the economy is a system that ensures the interaction of individual components of innovation activity within the boundaries of the selected management structure. planning of innovation activities and optimal attraction and allocation of resources for innovation activities.

Effective implementation of innovative activities is possible only if there is organizational and financial support for the process of innovative development.

In our opinion, the state innovation policy in agriculture should be focused on creating favorable conditions for innovation and investment processes.

We count. that the problem of attracting investment in agricultural production should be solved comprehensively and using all the tools to form the investment attractiveness of agriculture. We count. that the main sources in the agricultural sector of the economy should be enterprises of all forms of ownership that carry out their production and economic activities in rural areas.

The effective operation of agricultural enterprises in the medium and long term and ensuring their competitiveness largely depends on the provision of agricultural enterprises with investment resources the degree of availability of investment resources for them.

In our opinion, the main problems hindering the attraction of private investment in agriculture are (Fig. 2):

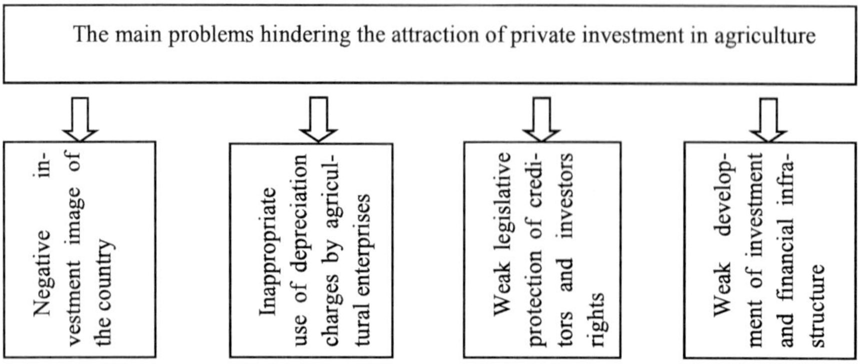

Fig. 2. The main problems hindering the attraction of private investment in agriculture.

In our opinion, improving the investment activity of agricultural enterprises in Ukraine can be facilitated by improving the system of mortgage lending, expanding budget funding for research in the field of agricultural sciences, improving the system of risk insurance in the agricultural business and providing tax incentives to agricultural enterprises.

In order to attract external investment, in our opinion, agricultural enterprises should offer their investment projects more widely at various investment forums, and make every effort to create a good investment image.

We adhere to the point of view that the state should improve the investment attractiveness of the agricultural sector of the economy for private investors:

1. Allocate subsidies for the creation of rural transport infrastructure and environmental protection.
2. Provide legal protection for investors and creditors.
3. Ensure equal access to state support for enterprises of all forms of ownership.
4. Create preferential tax conditions for agricultural enterprises that export their products abroad.
5. Use public-private partnership in the creation of transport infrastructure in rural areas.
6. Subsidize the production of agricultural products that are important for ensuring the food security of the state.
7. Expand state financing of objects of social significance for citizens living in rural areas
8. To encourage the use of own funds of agricultural enterprises for the implementation of their investment activities.

Commercial banks are one of the main investors in the agricultural sector of the economy. They provide agricultural enterprises with financial resources for the implementation of specific projects.

Among the disadvantages of bank lending to agricultural enterprises is an important drawback, which is the high interest rates on loans issued. This, in our opinion, significantly hinders the effective development of the agricultural sector of the economy.

One of the reasons for high credit rates is the high discount rate of the Central Bank. In banking practice, only one Central Bank discount rate is used, but in many countries this rate can be zero or even take negative values.

The authors suggest using not one, but several discount rates of the Central Bank, depending on the purposes for which commercial banks will spend the funds received on a returnable basis from the Central Bank. In our opinion, for lending to agricultural enterprises, commercial banks should be able to receive funds from the Central Bank at a discount rate close to zero, this will allow them to reduce the interest rate when lending to agricultural businesses.

In our opinion, the use of special discount rates of the Central Bank will be more effective than conventional government subsidies, since a commercial bank will be interested in making a profit and will not allow the irrational use of borrowed funds. At the same time, state subsidies can encourage agricultural enterprises to search for new subsidies, rather than to increase the level of efficiency of their production.

4 Discussion

Investment is an important component of financing extended reproduction. which determines the economic growth of an agricultural enterprise in the medium and long term. They. As a rule, they have a significant impact on the dynamics of the development of agricultural production, their widespread use allows for the modernization and technical re-equipment of the production process, the introduction of modern resource-saving and high-performance technologies, as well as to strengthen the market position of the enterprise.

The activation of the state investment policy in the agricultural sector of the economy, in our opinion, should primarily be aimed at creating conditions for attracting private capital to this industry.

Within the framework of the state investment policy in the agricultural sector, the state should effectively use all the opportunities at its disposal: reducing tax rates and providing tax benefits, increasing the labor potential of rural residents, developing transport infrastructure and the social sphere in rural areas.

Investing in the agricultural sector of the economy largely depends on the payback period of the investment. In our opinion, the achievement of a quick payback period for investments can be achieved with the provision of high-quality agricultural products, as well as reducing the level of its cost. In addition, the profit received from the production of products can also be invested in agricultural production.

The use of leasing, in our opinion, can be considered as one of the forms of attracting investment in agriculture. The form of financing using leasing requires less diversion of own financial resources due to the fact that the interest rate on leasing is usually lower than even when using a bank loan.

An additional bonus for the company to use leasing can be considered depreciation deductions, which allow reducing the tax base of the agricultural enterprise.

In addition, obtaining financing through leasing does not require collateral, which has a great advantage over bank lending.

Agricultural production is a risky type of economic activity due to the fact that its profitability largely depends on many objective and subjective factors, the impact of which is difficult to predict and control. Therefore, the use of property and crop insurance for agricultural enterprises, in our opinion, will further contribute to attracting private investment in agriculture.

5 Conclusions

Investments in agriculture will expand the production of agricultural products, ensure the food security of the state and saturate the domestic market with food. However, without government support, this is difficult to implement. Environmental education should contribute to the formation of the young generation of ecological culture, ecological thinking and environmental awareness, as well as promote the promotion of a healthy lifestyle.

Environmental education can create a sense of responsibility for the preservation of the natural environment among today's young people.

The main task of innovation and investment activities in the agricultural sector of the economy, in our opinion, should be to contribute to the creation of favorable conditions for the activation and development of the application of the innovative potential of the agricultural sector on the basis of attracting private investment.

In our opinion, the priority is the development of innovative and investment activities of agricultural enterprises in order to ensure the food security of the state.

The main principles of innovation and investment activity in the agricultural enterprise, in our opinion, are adaptability, efficiency, multi-variant, as well as economic, social and economic security.

The problem of attracting private investment in agriculture requires an integrated approach with a low use of tools to increase the level of its investment attractiveness, reduce risks, attract investment resources, as well as use the company's own financial resources for investing.

References

1. Kurmanova, G., Sukhanberdina, B., Kim, A., Urazova, B.: Investments in agriculture in the region. Bull. Natl. Eng. Acad. Republic Kazakhstan **1**, 183–189 (2021). https://doi.org/10.47533/2020.1606-146X.77
2. Borodin, K., Gataulina, E.: Investing in agriculture. Probl Econ. Trans. **60**, 437–447 (2018). https://doi.org/10.1080/10611991.2018.1547597
3. Bosanquet, C.: Investment in agriculture. J. Agric. Econ. **19**, 3–22 (2008). https://doi.org/10.1111/j.1477-9552.1968.tb01309.x
4. Khoruzhy, L., Tryastsina, N.: Analysis of real investments in agricultural enterprises. Account. Agric. 36–47 (2020). https://doi.org/10.33920/sel-11-2007-03
5. Judd, M., Boyce, J., Evenson, R.: Investment in agricultural research and extension (2019). https://doi.org/10.1201/9780429301940-2
6. Orshanskaya, J., Kukin, I.: Attraction of investments in agriculture of Russia. Econ. Labor Manag. Agric. 121–126 (2019). https://doi.org/10.33938/198-121

7. Hrytsaienko, H., Hrytsaienko, I., Bondar, A., Zhuravel, D.: Mechanism for the maintenance of investment in agriculture. In: Nadykto, V. (ed.) Modern Development Paths of Agricultural Production, pp. 29–40. Springer, Cham (2019). https://doi.org/10.1007/978-3-030-14918-5_4
8. Oborin, M.: Investments in agriculture in the Volga Federal District. AIC Econ. Manag. 30–39 (2020). https://doi.org/10.33305/209-30
9. Dudareva, A.B.: Improvement of state support of investment activities in agriculture. Bull. Agrarian Sci. **2**, 108–114 (2020). https://doi.org/10.17238/issn2587-666X.2020.2.108
10. Redbond, M.: Investing in agricultural innovation. **56**, 304–307 (2014)
11. Latysheva, L.A.: Investment potential and sources of nvestments in agriculture in the Stavropol Kray. Econ. Agric. Russia 13–17 (2020). https://doi.org/10.32651/204-13
12. Cheremukhin, A.D., Igoshin, A.N., Fedotov, A.V., Ivanova, O.P.: Proceedings of International Scientific and Practical Conference "Russia 2020 – A New Reality: Economy And Society". Constructing a Multi-Production Function for Investment in Agricultural Organizations (2021). https://doi.org/10.2991/aebmr.k.210222.091
13. Tyu, L.: Improvement of state support for investments in agriculture. AIC Econ. Manag. 23–30 (2019). https://doi.org/10.33305/1911-23
14. Harding, A., Jiang, L., Anseeuw, W., Alden, C.: The drive for Chinese investments in agriculture: comparing South Africa to the continent. In: Alden, C., Wu, Y.-S. (eds.) South Africa–China Relations, pp. 179–198. Springer, Cham (2021). https://doi.org/10.1007/978-3-030-54768-4_9
15. Yekimov, S., et al.: IOP Conf. Ser. Earth Environ. Sci. **677**, 022075 (2021)

The Impact of State Regulation on the Sustainable Development of Rural Areas in Ukraine

Sergey Yekimov[1](✉), Viktoriia Nianko[2], Olga Kuhareva[3], Tetiana Mishustina[4], and Olena Lysa[3]

[1] Publishing House "Education and Science" s.r.o., Olstynska 607/1, Praha 8, 18100 Troja, Czech Republic
[2] Dniprovsk State Technical University, Dnipropetrovsk Region, Dnepropetrovskaya 2, Kamenskoe 51918, Ukraine
[3] Oles Honchar Dnipro National University, Dnipropetrovsk Region, Gagarina Avenue, 72, Dnipro 49000, Ukraine
[4] Alfred Nobel University, Dnipropetrovsk Region, Sicheslavskaya Embankment, 18, Dnipro 49000, Ukraine

Abstract. The implementation of long-term development of rural areas necessitates the definition of certain benchmarks. The factors that have a significant impact on the level of development of rural areas can, in our opinion, include the integration of the region into world and interregional trade, the general state of the economy and the development of scientific and technical potential in the state, the degree of development of civil society, the level of corruption, the quality of life of citizens, the availability of natural resources and highly qualified labor resources in the region, natural and climatic conditions. One of the main problems of rural development is the lack of fair employment standards and social standards. Rural areas are characterized by high poverty of the population and an unfair distribution of the results of labor activity. The main tools for the development of rural areas, in our opinion, should be the creation of new jobs, the implementation of housing programs, education and socio-cultural sphere in rural areas, the provision of targeted material assistance for the least socially protected segments of the population.

Keywords: Agriculture state regulation of agriculture · Development of agriculture · Agriculture of Ukraine · Development of rural areas

1 Introduction

Territorial development of rural areas should take into account all components of their socio-economic potential. Along with the agricultural sector, tourism, the food industry, the production of energy from renewable sources and much more can develop in rural areas.

According to [1], rural areas cannot be considered only as a base for agricultural production, since such an approach will contribute to the most effective development.

A. Muratov and S. Ignateva (Eds.): AFE 2021, LNNS 353, pp. 62–69, 2022.
https://doi.org/10.1007/978-3-030-91402-8_8

According to [2, 3], the main problems that hinder the development of rural regions in Ukraine are:

1. Due to the ill-conceived state agrarian policy, during the transition to market-based forms of management, there was a sharp decline in the production potential of agricultural enterprises and the outflow of qualified labor resources to urban areas.
2. Agricultural enterprises are experiencing difficulties in financial resources and, for this reason, have difficulties with increasing the volume of production, and this has a negative impact on the amount of tax revenues to local rural budgets.
3. As a result of the migration of the city's labor force, the demographic situation in rural areas deteriorated.
4. The increase in imports of food products, including those traditionally produced in Ukraine, displaces domestic agricultural producers from the domestic agricultural market.
5. Most of the state funds are allocated for the development of agriculture, and not for the development of rural areas.

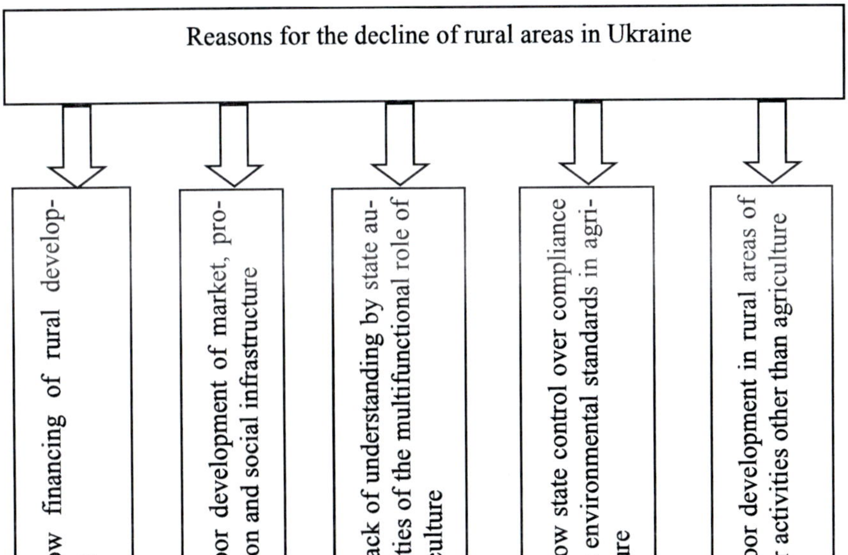

Fig. 1. Reasons for the decline of rural areas in Ukraine.

According to [3–6], achieving social progress in rural areas is possible only if economic growth is achieved. However, economic growth is not possible without social development.

Currently, due to the lack of financial resources in rural budgets, many rural residents cannot receive the necessary assistance from the state. Rural residents 'incomes are at a low level, their main source of income is the labor available on household land plots.

There is practically no renewal of the social infrastructure, and there are difficulties with obtaining medical and educational services.

The authors [7] explain the decline of rural areas in Ukraine by the following reasons (Fig. 1). The state budget of Ukraine has very low social payments in rural areas. Many social programs are curtailed, because the cost of providing a unit of services is significantly higher than in the city.

The state standards of social security of the population living in rural areas do not meet their needs. In our opinion, it is possible to maximize the production and labor potential of a rural region by using all its competitive advantages.

According to [8], the competitive advantages of the region include:

1. The existence of a balanced financial and budgetary system;
2. The availability of natural resources in the region. which would be of interest to investors.
3. Availability of a convenient geographical location.
4. Good condition of the transport infrastructure.
5. The presence of stable demand for products produced in the region.
6. Guarantees of protection of the rights of investors and creditors.
7. Stability of the tax system.
8. Availability of clear licensing principles
9. Close distance from economically highly developed regions.
10. Availability of qualified labor resources in the region.
11. Competitive quality of products produced in the region.

In our opinion, the competitive advantages of a rural region determine (Fig. 2):

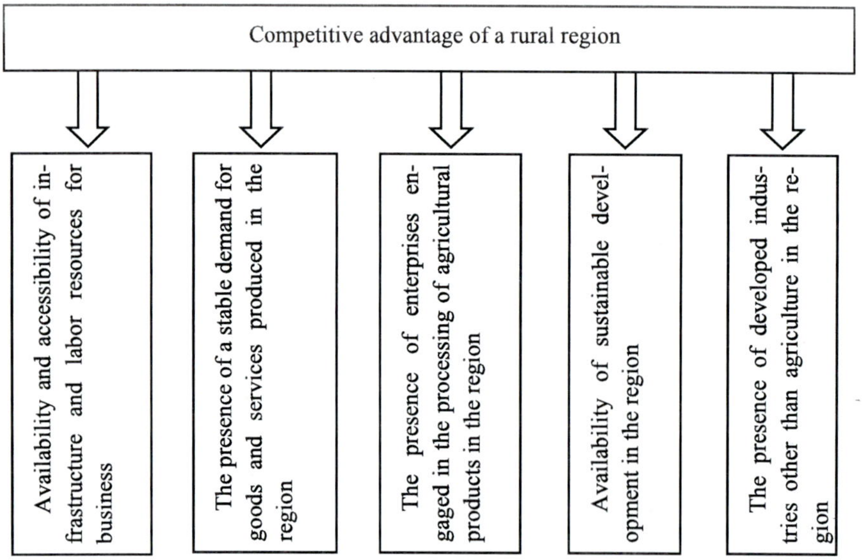

Fig. 2. Competitive advantage of a rural region.

2 Methods

In carrying out this study, the authors used an analytical method, through which the problems studied in it were considered with their development and unity. Taking into account the objectives of this study, the structural and functional method of scientific research was used. This ultimately allowed us to study some problems related to the impact of state regulation on the sustainable development of rural areas in Ukraine.

3 Results

State regulation of the socio-economic development of rural areas in Ukraine is implemented at three levels: local, regional and state. At the state level, the regulation of socio-economic development of rural areas is carried out on the basis of budget and tax instruments, which are based on standards that take into account the interests of regional and local development. At the regional level, social and economic development of the territories is monitored and analyzed, as well as targeted assistance to rural territories in the form of grants and transfers.

At the local level, various programs are being developed and implemented aimed at the socio-economic development of rural areas based on the interests of their residents.

According to [9–12], sustainable development is the ability to develop even in the presence of adverse external factors.

The authors [13–15] distinguish potential, absolute, actual, and normative sustainable development.

In our opinion, the most sustainable development of the region is possible with the maximum use of the production and labor potential of the region.

We believe that the real sustainable development of the rural region should provide for the optimal efficiency of the existing production facilities in the region, in compliance with the conditions of competition.

Sustainable development of an agricultural region should provide for the most effective use of the material and technical, human and intellectual, resource and natural, as well as cultural and social potentials of this region, which tend to steadily increase.

We hold the view that the main economic and social components of the stable evolution of rural areas are:

1. The production sector, which includes agricultural and other enterprises operating in the region.
2. Socio-cultural sphere, the task of which is to preserve folk culture, spiritual and moral values.
3. The social sphere, it is connected with the formation of the labor potential of the region.
4. Environmental activities aimed at preserving land resources and increasing their fertility.

We believe that the stable development of rural areas should become one of the most important priorities of the state economic policy, which should be based on efforts to develop rural infrastructure, increase the use of labor and natural resources available in the region, develop rural tourism and other non-agricultural activities in rural areas, improve the quality of medical services for rural residents, as well as improve cultural and educational activities in rural areas.

The stable development of rural areas in Ukraine requires the improvement of spiritual, cultural and educational institutions of rural settlements, as well as the common interests of people living in rural settlements.

Solving the problem of rural development, in our opinion, requires a comprehensive approach from the state, focused on increasing the income of rural residents, improving the efficiency of agricultural production, improving the social and transport infrastructure in rural areas, increasing the investment and environmental attractiveness of the region.

At the same time, it is very important that state regulation is carried out with minimal interference in the work of market mechanisms.

The main driving force for the stable development of rural regions, in our opinion, should be the activities of enterprises in these regions.

Among the non-agricultural activities in rural areas, the development of tourism and the production of energy from renewable natural sources deserves attention.

We believe that due to the fact that agriculture provides citizens with food, the state should pay increased attention to the development of agriculture when addressing the issue of ensuring sustainable development of rural areas, since ensuring the country's food security depends on it.

The lack of financial and material resources, the worn-out material and technical base, and the low motivation of labor determine the need for state support for the agricultural sector.

In our opinion, the main areas of state regulation of agriculture are:

1. Elimination of the price disparity for industrial and agricultural products.
2. Creation of transport and market infrastructure facilities in rural areas.
3. Introduction of tax incentives for domestic agricultural producers
4. Training of personnel for agricultural enterprises.
5. Providing conditions for granting preferential loans to agricultural enterprises.
6. The use of customs duties to protect domestic agricultural enterprises that produce products that are important for ensuring the country's food security.
7. Stimulating the development of agricultural exports.

In our opinion, the effective development of rural areas is not possible without an appropriate state policy in the field of innovation and science, which provides for the creation of an infrastructure consisting of educational and scientific organizations, venture investment funds, and technology distribution centers.

This, in our opinion, will allow us to attract the personnel potential of leading scientific schools, as well as foreign investors and partners for the implementation of regional development programs to address issues of rural development.

Agriculture is organically connected with the territory and the labor resources that live in this territory. Agricultural enterprises are the main taxpayers of local rural budgets and the efficiency of the functioning of municipal institutions of consumer services and public utilities depends on their successful functioning.

In this regard, for the development of rural areas, in our opinion, it is also advisable to create favorable conditions for agricultural production.

In our opinion, there is a need to develop a scientifically based methodology for assessing the ratio of funds transferred to the state budget and financial resources received from higher budgets in the form of subventions and subsidies for a specific rural area. This problem is also of political importance along with the economic one. This will make it possible to make adjustments to the state economic and social policy and determine the most attractive rural areas in terms of investment.

4 Discussion

One of the main problems of the deterioration of the standard of living in rural areas is the inefficient use of the human, material, natural and other types of resources available in the region.

The strategic development of rural areas requires defining some long-term benchmarks and adapting to changes. The factors influencing the development of rural regions include the integration of the region into interregional and world trade, the general state of the economy and the level of development of scientific and technological progress in the state, the development of civil society, the quality of life of people, the level of corruption in the state, the perfection of the legislative framework, the availability of qualified labor and natural resources in the region, natural and climatic conditions.

The formation of financial resources of local budgets to support the cultural and social infrastructure in rural areas is the main problem of the functioning of rural social infrastructure. Its solution depends on the financial capabilities of taxpayers, as well as subsidies and subventions received by local budgets from the state and regional budgets. The destruction of the production potential of agricultural enterprises in Ukraine, their technical and technological backwardness not only does not allow them to produce competitive agricultural products, but also has a negative impact on the competitiveness of the country, region, and rural territory in which they are located.

The implementation of the cluster policy in the development of small agro-industrial enterprises in the region allows, in our opinion, to use the competitive advantage of rural regions in which these agricultural enterprises operate.

5 Conclusions

One of the main problems of rural development is the lack of adequate social standards and employment standards for rural residents. The main reason for this phenomenon is the low efficiency of the state's management of rural areas.

For rural areas, important problems are the uneven and unfair distribution of labor results, high poverty of the population.

These and other problems do not allow us to maximize the production and labor potential of rural areas. The main tools aimed at the development of rural areas, in our opinion, can be the promotion of the creation of new jobs, the implementation of housing programs, support for the development of human capital, housing and communal services, targeted financial assistance to the least socially protected segments of the population.

In modern conditions, the standard of living of the rural population in Ukraine remains at an extremely low level, and many rural areas are in a depressed state, the possibility of diversifying the rural economy may be important. One of the options for such activities can be the development of green tourism in rural areas. Although it is difficult to expect large tax revenues to the budget, green tourism can be considered as a form of employment for the rural population and can provide an additional impetus for the development of some rural areas, reducing labor migration from rural areas to cities by creating new jobs. Green tourism can also contribute to reducing the level of social tension in rural areas and can be an alternative sector of the rural economy to agriculture.

The lack of transport infrastructure, the imperfection of the legal framework, the lack of the necessary number of qualified personnel, as well as the lack of state support for this type of activity. Taking into account the current state of the ecological situation in the industrial centers of Ukraine and the increased demand for treatment and recreation in a clean natural environment, the manifestation of interest in folk culture and folk art is important in our opinion should have the support of the state. In our opinion, this support should include the creation of preferences for the rural population in this field of activity, including tax benefits.

References

1. Dubovitskii, A.A., Klimentova, E.A., Pechurkin, A.S.: Unemployment in the context of sustainable rural development. Reg. Econ. Theory Pract. **18**, 2334–2353 (2020). https://doi.org/10.24891/re.18.12.2334
2. Shpykuliak, O., Sakovska, O.: Agricultural cooperation as an innovation for rural development. Baltic J. Econ. Stud. **6**, 183–189 (2020). https://doi.org/10.30525/2256-0742/2020-6-3-183-189
3. Kostyaev, A.: On the scientific basis for developing rural development strategies. Agric. Sci. Euro-North-East **21**, 462–474 (2020). https://doi.org/10.30766/2072-9081.2020.21.4.462-474
4. Lazhentsev, V.N., Ivanov, V.A.: Rural development strategy of the Northern Region. Econ. Reg. **16**, 696–711 (2020). https://doi.org/10.17059/ekon.reg.2020-3-2
5. Tsvettsykh, A.V., Shevtsova, N.V.: Methodology for assessing sustainable rural development. Azimuth Sci. Res. Econ. Adm. **9** (2020). https://doi.org/10.26140/anie-2020-0903-0094
6. Bukhtiyarova, T., Demyanov, D., Tenetko, A., Nemykina, Y.: Ensuring sustainable rural development. IOP Conf. Ser. Earth Environ. Sci. **421**, 022002 (2020). https://doi.org/10.1088/1755-1315/421/2/022002
7. Sayabaev, K., Zholmukhanova, A., Nukesheva, A., et al.: Sustainable rural development through tourism. J. Environ. Manag. Tour. **11**, 1550 (2020). https://doi.org/10.14505/jemt.v11.6(46).25
8. Alimkulova, E., Aitmukhanbetova, D.: State regulation of agricultural sector. Probl. AgriMarket **4**, 47–53 (2020). https://doi.org/10.46666/2020-4-2708-9991.05
9. Yekimov, S., et al.: IOP Conf. Ser. Earth Environ. Sci. **677**, 022075 (2021)
10. Kenduh, E., Capova, O., Demyanov, V.: State regulation of agricultural economy. Sci. Person Humanitarian Res. **14**, 147–152 (2020). https://doi.org/10.17238/issn1998-5320.2020.14.2.25

11. Poperechnyi, S., Tarnavska, O.: Problems of state regulation of Ukraine. Vet. Med. Biotechnol. **21**, 42–46 (2019). https://doi.org/10.32718/nvlvet-e9309
12. Barabash, L., Malovanyi, M., Osipova, A.: Social aspects of state regulation of agricultural production. SHS Web Conf. **100**, 04001 (2021). https://doi.org/10.1051/shsconf/202110004001
13. Derbeneva, E., Polushkina, P.: Foreign experience of state regulation of agriculture. Russ. J. Resour. Conserv. Recycl. **4** (2017). https://doi.org/10.15862/05RRO417
14. Nikiforova, Y.: Methodology for estimating the efficiency of state regulation of agriculture. In: Proceedings of the Kuban State Agrarian University, vol. 1, pp. 38–44 (2019). https://doi.org/10.21515/1999-1703-79-38-44
15. Baimisheva, T., Kurmaeva, I., Gazizyanova, Y., et al.: State regulation systems of agriculture insurance. IOP Conf. Ser. Earth Environ. Sci. **315**, 022090 (2020). https://doi.org/10.1088/1755-1315/315/2/022090

Distribution of Ascariosis and Asophagostomosis of Pigs in Central Yakutia

Alexander Reshetnikov, Lyubov Kozlova(✉), and Anastasia Barashkova

FRC YaSC SB RAS M. G. Safronov Yakut Scientific Research Institute of Agriculture, 23/1/ Bestuzheva-Marlinskogo Str., 677001 Yakutsk, Russia

Abstract. Purpose of research: To study the spread of ascariosis and esophagostomosis of pigs in industrial pig breeding and in farms of various types in Central Yakutia. Materials and methods: The study of the species composition of pathogens and the spread of intestinal nematodes was carried out in 2001–2004. In industrial pig farm and subsidiary plots pigs were tested for infection helminthological diseases. For this, a saturated solution of sodium chloride was prepared (400–420 g of sodium chloride per 1 L of boiling water. 3 g of faeces were weighed according to a sample, which were placed in a glass with a capacity of 50 ml, were poured with a small amount of a saturated solution of sodium chloride and stirred. 3 drops were removed from the film surface with a metal loop, transferred to a glass slide and examined under a microscope with low magnification. A total of 1236 samples of pig feces were examined, including 262 young animals from 2 to 4 months and 974 adult pigs. Results and discussion: The pigs of Yakutia are parasitized by 2 types of nematodes: Ascaris suum, Oesophagostomum dentatum.

Keywords: Intestinal nematodes · Ascariosis · Esophagostomosis · Yakutia of Russia

1 Introduction

Pig breeding is the second most important branch in animal husbandry after cattle breeding. Pig breeding is a rapidly developing branch of animal husbandry in the Republic of Sakha (Yakutia). Industrial pig breeding began to develop in the republic in 1987, with the commissioning of the Khatas pig-breeding complex with a capacity of 12 thousand pigs. The total number of pigs was brought to more than 24 thousand animals in 2016. The main area for the development of pig breeding is Central Yakutia, which includes 9 large livestock and agricultural regions, in which 47% of the republic's pigs are concentrated [1].

Domestic pigs are susceptible to various infectious, invasive and non-infectious diseases, often causing significant damage to industrial pig breeding. In some countries, pig production is strategically important in meat production. One of the most ferocious pig diseases is African swine fever. The African swine fever outbreak has caused severe economic losses in Asia since 2018. As African swine fever is a new and emerging disease, many farmers are hesitant to breed pigs before biosafety procedures have been

A. Muratov and S. Ignateva (Eds.): AFE 2021, LNNS 353, pp. 70–79, 2022.
https://doi.org/10.1007/978-3-030-91402-8_9

assessed as effective. A comprehensive procedure called the Quadruple Protection Procedure (QPP) has been tested on 35 small farms that have been confirmed with African swine fever virus to support small farms in the resumption of pig production. By the end of the study, a total of 18,730 pigs and 3006 sows were healthy, with a percentage of 100 and 98.8, respectively, indicating that QPP may protect small-farm pigs from pathogens in China. In this study, an effective system of protective procedures was developed for small-scale pig farms at risk of an outbreak of African swine fever [2].

Aujeszky's disease (AD), also called pseudorabies, is an infectious viral disease caused by the alpha herpes virus and a natural host of domestic and feral pigs. The movement of infected animals is considered to be the most likely route for the Aujeszky virus (ADV) to enter Spain. For this reason, the Spanish AD control and eradication program has instituted measures to improve animal safety in an attempt to protect areas with free or low prevalence status. A strategy for testing pigs in fattening farms 15 days prior to shipment and using a sample size sufficient to detect a prevalence of 5% with 95% confidence. Testing can reduce the likelihood of introducing ADV-infected animals by 91% at no additional cost [3].

Blastocystis sp. is a unicellular microorganism found in the gastrointestinal tract of humans and various animals and is distributed throughout the world. Blastocystis displays an extensive genetic diversity of 28 subtypes (STs) based on the small subunit of ribosomal RNA (SSU rRNA) gene. In this study, the genetic diversity and zoonotic potential of Blastocystis were assessed using fecal samples of pigs from two farms in Slovakia. Blastocystis spp. have been found in pigs intended for sale and slaughter. The ST 5 subtype was identified in all positive samples and age categories with a prevalence of 12%. However, the prevalence on one of the farms was up to 28.6%. This is the first pig blastocystis study conducted in Slovakia. It may have zoonotic potential and therefore may be a risk factor due to close contact between humans and pigs in breeding farms [4].

Respiratory diseases are one of the main causes of losses in the pig industry worldwide. Pig production is the largest component of the livestock sector in the Philippines. This study included a total of 260 pigs from Batangas and 300 pigs from Albay, belonging to commercial or private farm types. Pigs affected by respiratory diseases often have multiple chest lesions, especially pneumonia and pleurisy. However, studies reporting such lesions tend to report individual prevalence of lesions and do not examine patterns of joint occurrence of lesions [5].

Thailand is an agricultural country for which animal husbandry is a key component. The private household with a small number of livelihood pigs plays a critical role in the livelihoods of many households. Over the past two decades, Asia has seen several outbreaks of major swine diseases, one of which is porcine reproductive and respiratory syndrome virus (PRRSV). An outbreak of PRRSV has been detected in domestic pigs in many countries, especially in Southeast Asia such as Cambodia, Vietnam and Myanmar. In particular, influenza A virus (IAV) is suspected to occur in private households because the closeness between humans, pigs and poultry is often practiced in the backyard. Classical Swine Fever Virus (CSFV) is a highly contagious transboundary disease. As for CSFV, pigs are usually infected with oronasally, and spread occurs both directly through horizontal and vertical transmission and indirectly through contaminated fomites and pig products. CSFV has been found among domestic pigs in Vietnam, Indonesia. For porcine circovirus type 2 (PCV2), as immunosuppressive in nature, it has been found to continuously coinfect with other viruses such as CSFV and PRRSV, resulting in an outbreak and mortality of pigs. It has been found in domestic pigs in Indonesia and Vietnam [6].

In recent years, the Brazilian National Veterinary Service (BNVS) has intensified its efforts to surveillance, control and eradicate diseases, especially foot and mouth disease and classical swine fever. Therefore, it is imperative to understand the patterns and risks associated with animal movement in order to better understand the potential risk of spreading pathogens along the production chain. This is especially true for national efforts to move from a country-wide disease-free to vaccine-free FMD status, which is expected to be achieved by 2023, as infected animals can move from one region to another with disease-free and/or geographically distant status. In this context, contact network analysis is a powerful tool for describing, predicting and evaluating the role of trade in the spread of disease, producing important data for understanding, preventing and mitigating possible outbreaks in the region [7].

Canine tapeworm Echinococcus multilocularis causes alveolar echinococcosis (AE) in humans and other intermediate hosts. Depending on the intermediate host, the larval form of E. multilocularis (metacestode) can be either fertile, for example, in rodents, and thus support the life cycle of the parasite, or sterile, for example, in pigs, and thus interrupt the life cycle. Routine examination of the liver for slaughter served as the basis for a large-scale surveillance study of E. multilocularis infection in pigs. The aim of this combined crossover and case-control study was to estimate the minimum preva-lence of E. multilocularis in pigs in Switzerland, find factors associated with infection, and evaluate potential regional clusters of infection. The study provided additional evidence for widespread environmental pollution by E. multilocularis eggs in Switzerland. In addition, the lack of protoscolecules in any of the lesions supported the concept that pigs only act as dead-end hosts and thus do not contribute to the parasite's life cycle. Factors associated with E. multilocularis infection have been consistent with the biology of the parasite, and many of them can be eliminated by improving hygiene and management standards [8].

F18 enterotoxigenic and verotoxigenic Escherichia coli (E. coli) are widespread and cause diarrhea after weaning of piglets from sows. There is no commercial vaccine. Thus, the serum prevalence of F18 in 178 randomly selected pig farms in northern Belgium was estimated by determining the titers of F18-specific IgG antibodies in pooled sera of sows on the breeding farm. This review presents the first data on the serological prevalence of F18 in the herd and shows that the prevalence of F18 + E. coli is common in northern Belgium. Significantly more open pig farms were F18 seropositive (96.4%) compared to closed ones (88.8%). In addition, no clear relationship was found between the serology of the F18 herd and the number of sows per breeding farm (rsp = 0.19) or the number of fattening pigs per breeding farm (rsp = 0.17) [9].

Swine Reproductive Respiratory Syndrome was first diagnosed in the United States in the mid-1980s and quickly spread to the commercial pig population in the country. Currently, swine reproductive respiratory syndrome is endemic in the United States and causes significant economic losses in pig production every year. The syndrome is characterized by fever, anorexia, late abortion, respiratory symptoms, increased mortality before and after weaning, and poor growth and feeding efficiency. The virus can be transmitted to pig farms through infected breeding stock and semen, exposure to contaminated vehicles, tools and through the air from neighboring infected farms. An epidemiological study showed a significant decrease in the incidence of reproductive-respiratory syndrome after the introduction of an air filtration system compared to the incidence observed on the same farm before the introduction of air filtration. The same study estimates that 80% of cases of reproductive-respiratory syndrome in unfiltered farms located with high density pigs can be associated with airborne transmission and, therefore, can be prevented by air filtration [10].

Pork meat is one of the most consumed in the world, not only because of its taste, but also because of its excellent source of protein, energy and micronutrients. In re-cent decades, its consumption has grown exponentially along with the demand for lean meat, which has led to the emergence of new pig genetic lines for obtaining meat in less time, which led to the emergence of pigs with a higher ratio of muscle mass to fat. However, this genetic breeding carries undesirable traits, as in the case of porcine stress syndrome (PSS) or malignant hyperthermia (MH).

Pigs carrying a single deleterious allele show a transition from C to T at nucleotide 1843 (c.1843C > T), causing an arginine to cysteine substitution at position 615 of the amino acid sequence (p.Arg615Cys), thus producing an abnormal protein. In skeletal muscle, RYR1 encodes the calcium channel of the sarcoplasmic reticulum, which, in recessive homozygosity, causes hyperactivation, leading to several pathologies in pigs such as muscle stiffness, hyperthermia, cardiac arrhythmias and sudden death. In addition, Pig Stress Syndrome affects not only animal welfare, but also meat quality, creating a pale, soft and exudative (PSE) product. The result is low quality meat for the current market. Thus, the presence of pigs with pig stress syndrome results in huge economic losses and is a major concern in pig production worldwide [11].

Haemophilus parasuis, a gram-negative bacterium that infects the upper respiratory tract of weaned pigs, is one of the most important bacterial pathogens affecting pig populations. It is the causative agent of Glesser's disease, which causes systemic infection and polyserositis, meningitis and arthritis. H. parasuis infection can lead to high mortality and morbidity with significant economic losses for pig producers. The overall prevalence of H. parasuis among pigs in China was 27.8%, with the highest prevalence occurring between 2011 and 2015 (41.0%). With regard to the age of the pigs, the point estimate of the prevalence of H. parasuis was higher in suckling piglets (29.2%) compared with pigs of other ages. The prevalence in the serum subgroup (29.8%) was higher than in the nasal swab subgroup (12.5%). The analysis showed that H. parasuis infection was common in pig populations in China. Consequently, effective control measures are needed to reduce this threat to the pig herd [12].

Mycoplasmahyopneumoniae is the main causative agent of enzootic pneumonia and one of the most common pathogens in pigs worldwide, causing severe economic losses in pig production. Mycoplasmahyopneumoniae is also considered one of the main agents involved in the Porcine Respiratory Disease Complex (PRDC), predisposing animals to secondary infections with other respiratory pathogens, including bacteria, parasites and viruses. Pigs infected with Mycoplasmahyopneumoniae develop a chronic unproductive cough that appears 10–16 days after infection. The duration of the infection is relatively long and is detected in samples from the respiratory tract of infected pigs for 214 days after experimental inoculation. The usual route of transmission is nose-to-nose contact, and infection in herds appears to be maintained by transmission from gilts and from low litter sows to their litters. The impact on productivity caused by Mycoplasmahyopneumoniae infection is usually associated with decreased average daily weight gain (ADG), increased mortality, decreased feed efficiency, and increased use of antibiotics to combat PRDC diseases. Consequently, it takes longer to sell pigs infected with Mycoplasmahyopneumoniae, there is a decrease in the number of animals sold per female per year and the potentially low efficacy of some antibiotics due to the emergence of antimicrobial resistant strains [13].

Toxoplasma gondii infections are common in humans and animals around the world. Toxoplasma gondii infection in pigs continues to be a public health concern. Pigs are important to the economics of many countries, in particular the United States, China and European countries. Among the many food animals, pigs are considered the most important for the transmission of T. gondii in the United States and China, as viable parasites are rarely isolated from beef or domestic chickens. In addition to public health concerns, T. gondii is causing outbreaks of clinical toxoplasmosis in pigs in China, associated with the unique genotype T. gondii (ToxoDB genotype №9 or Chinese 1), which is rare elsewhere. The safety of ready-to-eat pork products against T. gondii infection is a matter of debate [14].

Salmonella is a worldwide pathogen that affects both humans and animals and is commonly associated with intensive livestock production such as poultry and pigs. The purpose of this work was to assess the relationship between the presence of genes associated with virulence and the phenotype of antimicrobial resistance. The study found that isolates from intensive farms were resistant to more antimicrobial agents than isolates from subsidiary farms [15].

Based on the foregoing, taking into account the state importance of industrial pig breeding, we set the goal to study the spread of ascariosis and esophagostomosis of pigs in industrial pig breeding and in farms of various types in Central Yakutia.

2 Materials and Methods

The study of the species composition of pathogens and the spread of intestinal nematodes was carried out in 2001–2004 on industrial, farm, subsidiary and backyard farms using helminthic studies using of Fulleborn method. For this, a saturated solution of sodium chloride was prepared (400–420 g of sodium chloride per 1 L of boiling water, specific gravity 1.18–1.9). Weighed 3 g of feces, placed in a 50 ml glass, poured in a small amount of saturated sodium chloride solution and stirred. 3 drops were removed from the film surface with a metal loop, transferred to a glass slide and examined under a microscope with low magnification. A total of 1236 samples of pig feces were examined, including 262 young animals from 2 to 4 months, 775 for fattening livestock, 185 for sows and 14 for boars. Of these, 425 samples were examined from the Khatassky pig complex by helminthic ovoscopy, and from farms with. Tabaga, "Siberia", "Uchur", from the farm of the boarding school for the disabled s. Kapitonovka, and from private farmsteads of the villages of Khatassa, Vladimirovka and Poultry factories - 811 samples. The number of helminth eggs was counted in three drops and their average number in one drop was deduced. To calculate the number of eggs and larvae in 1 g of feces, we used the counting chamber of the All-Russian Research Institute of Fundamental and Applied Parasitology of Animals and Plants named after K.I. Scriabin, developed by L.D. Migacheva, G.A. Kotelnikov. The intestines were separated from the animals killed for meat, which was placed in a separate vessel and cut along the entire length, washed off with water and set to settle for 10–15 min. The resulting matrix was examined for the presence of helminths by scanning with a magnifying glass. Determination of the types of helminths was carried out in laboratory conditions according to K.Ya. Scriabin.

3 Results

According to our research by the methods of incomplete parasitological autopsy of animals killed for meat, as well as ovoscopic studies on Fulleborn, carried out in the Khatas pig complex, in the farms of the village of Khatassa, the subsidiary farm of the colony of the village of Novaya Tabaga and the private sector of the village. Vladimirovka of sick pigs with ascariasis, esophagostomosis and trichocephalosis was not found.

Pigs of the Tretyakov farm s. Staraya Tabaga, subsidiary farm “Uchur” are infected with helminthiases by 100%; in the subsidiary farm of the boarding school in the village of Kapitonovka, the number of sick animals is 87.5%. In the private sector the poultry farm - 66.6% and in the collective farm “Siberia” - 50%. In these farms, on average, from 4.58 ± 0.56 to 14 ± 2 helminths were found per animal (Table 1).

Table 1. Infection of pigs with helminths according to the results of helminthological studies of those killed for meat (data for 2001–2004).

Farms	Studied animals by farms	Number of sick animals		Percentage of sick animals		The number of helminths per animal	
		I	II	I	II	I	II
Tretyakov farm of Staraya Tabaga	8	8	–	100	–	7.12 ± 1.2	–
Subsidiary farm “Uchur”	12	12	12	100	100	4.58 ± 0.56	10.75 ± 1.64
Collective farm of “Siberia”	4	2	–	50	–	2.5 ± 0.5	–
Subsidiary farm of the boarding school in the village of Kapitonovka	8	7	3	87.5	37.5	5.12 ± 1.17	6.33 ± 0.87

(*continued*)

Table 1. (*continued*)

Farms	Studied animals by farms	Number of sick animals		Percentage of sick animals		The number of helminths per animal	
		I	II	I	II	I	II
The private sector of village Vladimirovka	15	0	0	0	0	0	0
Subsidiary farm of the colony. settlement of Novaya Tabaga	12	0	0	0	0	0	0
In the private sector the poultry farm	3	2	–	66.6	–	14 ± 2	–
Khatas industrial pig-breeding complex	81	0	0	0	0	0	0
Household farm of Khatassi	14	0	0	0	0	0	0
Total	76	31	15	40.78	19.74	5.83 ± 0.54	9.8 ± 0.9

* Note: I - Ascariasis, II - Oesophagostomosis.

Along with helminthological autopsy, we carried out systematic helminthic ovoscopy studies of pigs of different ages in farms of all types using a VIGIS counting chamber. The results of the studies have shown that the egg production of Ascaris remains at the same level over the seasons. The average content of eggs of various helminths in 1 g of pig feces on average in Yakutia is characterized by the following indicators: Ascarissuum - 9.78 ± 2.83, Trichocephalus suis - 11.29 ± 4.08 and Oesophagostomum dentatum 6.3 ± 1.06 specimens. In the Khatas pig complex, sows and boars often have balantidia, and piglets from 3 to 20 days of age have eimeriosis (Table 2).

Table 2. Infection of pigs according to the results of oovoscopic examinations.

Farms	Average number of helminth eggs in 1 g of feces		
	Ascaris suum	Oesophagostomum dentatum	Trichocephalus trichiurus
1. Subsidiary farm of the colony. settlement of Novaya Tabaga	83.6 ± 9.9	0	127.6 ± 10.8
2. Subsidiary farm "Uchur"	72.9 ± 11.4	45.6 ± 5.37	0

(*continued*)

Table 2. (*continued*)

Farms	Average number of helminth eggs in 1 g of feces		
	Ascaris suum	Oesophagostomum dentatum	Trichocephalus trichiurus
3. Collective farm of "Siberia"	50.1 ± 5.16	0	0
4. Subsidiary farm of the boarding school in the village of Kapitonovka	114 ± 6.79	34.9 ± 5.15	0
5. In the private sector the poultry farm	158 ± 9.78	0	0
6. Khatas industrial pig-breeding complex	0	0	0
7. Household farm of Khatassi	0	0	0

4 Discussion

Our study of the spread of ascariosis and esophagostomosis of pigs in industrial pig breeding and on farms of various types in Central Yakutia revealed the following pattern: in industrial pig breeding with a closed production cycle, pigs are not infested with helminthic diseases such as ascariosis, trichocephalosis and others. The enterprise maintains a staff of veterinary specialists who routinely carry out all veterinary measures such as vaccinations, deworming, and the development of an appropriate diet for the age and sex groups of pigs.

Sick pigs with ascariosis, esophagostomosis and trichocephalosis was not found in the farms such as the farmsteads of the village Khatassy, the subsidiary farms of the colony of the village Novaya Tabaga and the private sector of the village Vladimirovka. When researching these farms, it was found that they were buying healthy young pigs not infected with helminths from the Khatas pig complex for rearing.

All pigs are sick with helminthic diseases in the Tretyakov's farms of the village Staraya Tabaga, in the subsidiary farm "Uchur", in the collective farm "Siberia", in the subsidiary farm of the boarding school in the village Kapitonovka and in the private sector of the Poultry Farm. The infection of pigs with helminths in these farms is 100%.

5 Conclusions

Pigs do not suffer from helminthic diseases in industrial pig breeding of a closed cycle. Due to the daily cleaning of manure, the planned disinfection of the buildings with the help of hydraulic flushing and well-established veterinary work, the appearance of helminthiasis is prevented in a timely manner.

Pigs in private households, which are not provided with veterinary care, suffer from parasitic diseases. The pigs of Yakutia are parasitized by 2 types of nematodes: *Ascaris suum, Oesophagostomum dentatum.*

Acknowledgement. The authors are deeply grateful to the director N.V. Klarov and the staff of the Khatass pig farm for their assistance in collecting the data.

References

1. Official information portal of the Republic of Sakha (Yakutia), The livestock of horses and pigs is growing in Yakutia. (2016). https://www.sakha.gov.ru/news/front/view/id/2688156. Accessed 24 Mar 2021
2. Lang, T., et al.: A quadruple protection procedure for resuming pig production in small-scale ASFV-positive farms in China. Curr. Res. Microbial Sci. **2**, 100014 (2021). https://doi.org/10.1016/j.crmicr.2020.100014
3. Martínez-López, B., Carpenter, T.E., Sánchez-Vizcaíno, J.M.: Risk assessment and cost-effectiveness analysis of Aujeszky's disease virus introduction through breeding and fattening pig movements into Spain. Prev. Vet. Med. **90**(1–2), 10–16 (2009). https://doi.org/10.1016/j.prevetmed.2009.03.004
4. Danisova, O., Valencaková, A.: First detection of Blastocystis sp. in pigs in Slovakia and in Europe. Parasitol. Int. **81**, 102235 (2021). https://doi.org/10.1016/j.parint.2020.102235
5. Barnes, T.S., Lajarca, A., Blackall, P.J.: Latent class analysis identifies multimorbidity patterns in pigs with respiratory disease. Prev. Vet. Med. **186**, 105209 (2021). https://doi.org/10.1016/j.prevetmed.2020.105209
6. Sutiwat, C., Kannika, N.L., LertrakS, K.P.: Seroprevalence of the viral pig diseases among backyard pigs in Chiang Mai, Thailand. Prev. Vet. Med. **190**, 105330 (2021). https://doi.org/10.1016/j.prevetmed.2021.105330
7. Cespedes, C.N., Kimberly, V.W., Pereira, V.F., Onell, G.A.J., Marcos, A., Grisi-Filho José, H.H.: Spatio-temporal network analysis of pig trade to inform the design of risk-based disease surveillance. Prev. Vet. Med. **189**, 105314 (2021). https://doi.org/10.1016/j.prevetmed.2021.105314
8. Meyer, A., et al.: Combined cross-sectional and case-control study on *Echinococcus multilocularis* infection in pigs in Switzerland. Vet. Parasitol. **4**, 100031 (2020). https://doi.org/10.1016/j.vpoa.2020.100031
9. Verdonck, F., Cox, E., Ampe, B., Goddeeris, B.M.: Open status of pig-breeding farms is associated with slightly higher seroprevalence of F18^{+}*Escherichia coli* in northern Belgium. Prev. Vet. Med. **60**(2), 133–141 (2003). https://doi.org/10.1016/S0167-5877(03)00121-1
10. Sanhueza, J.M., Stevenson, M.A., Vilalta, C., Kikuti, M., Corzo, C.A.: Spatial relative risk and factors associated with porcine reproductive and respiratory syndrome outbreaks in United States breeding herds. Prev. Vet. Med. **183**, 105128 (2020). https://doi.org/10.1016/j.prevetmed.2020.105128
11. Acosta, D.B., et al.: Wild pigs (*Sus scrofa*) population as reservoirs for deleterious mutations in the RYR1 gene associated with Porcine Stress Syndrome. Vet. Animal Sci. **11**, 100160 (2021). https://doi.org/10.1016/j.vas.2020.100160
12. Hong-Bo, N., Qing-Long, G., Quan, Z., Xiao-Yue, L., Xiao-Xuan, Z.: Prevalence of *Haemophilus parasuis "Glaesserella parasuis"* in pigs in China: a systematic review and meta-analysis. Prev. Vet. Med. **182**, 105083 (2020). https://doi.org/10.1016/j.prevetmed.2020.105083

13. Silva, G.S., Yeske, P., Morrison, R.B., Linhares, D.C.L.: Benefit-cost analysis to estimate the payback time and the economic value of two *Mycoplasma hyopneumoniae* elimination methods in breeding herds. Prev. Vet. Med. **168**, 95–102 (2019). https://doi.org/10.1016/j.prevetmed.2019.04.008
14. Dubey, J.P., et al.: All about *Toxoplasma gondii* infections in pigs: 2009–2020. Vet. Parasitol. **288**, 109185 (2020). https://doi.org/10.1016/j.vetpar.2020.109185
15. Joaquim, P., Herrera, M., Dupuis, A., Chacana, P.: Virulence genes and antimicrobial susceptibilityin *Salmonella enterica* serotypes isolated from swine production in Argentina. Revista Argentna de microbiologia (2020). https://doi.org/10.1016/j.ram.2020.10.001. Accessed 9 Jan 2021

Effect of Salicylic Acid and Copper Sulfate on the Synthesis of Rutin in *Fagopyrum Esculentum* Moench Plants *in vitro*

Elena Barsukova[1], Alexey Klykov[1](✉), and Elena Chaikina[2]

[1] Federal Scientific Center of Agricultural Biotechnology of the Far East named after A.K. Chaiki, 30 Volozhenina Street, Timiryazevsky Stl., 692539 Ussuriysk, Primorsky Krai, Russia
[2] Pacific Institute of Bioorganic Chemistry G.B. Elyakov FEB RAS, 159/2 Centennial Avenue, 690022 Vladivostok, Russia

Abstract. Breeding buckwheat varieties with high lavonoids content is one of the priorities at the Federal Research Center of Agricultural Biotechnology of the Far East named after A.K. Chaika. The aim of this study was to find ways of increasing the accumulation of rutin in *F. esculentum* plants *in vitro* culture using salicylic acid (SA) and copper sulfate. The effective concentrations and exposures of elicitors, which ensure the maximum content of rutin in the plant tissues of buckwheat microplants, were established as a result of the study. The microplants of the Izumrud × Inzerskaya hybrid synthesized the largest content of rutin (3.27%) under prolonged exposure (24 days) to 6.9 mg L^{-1} of SA concentrate. Microplants of the Izumrud variety synthesized 2.8% of rutin when exposed to the same SA concentrtion for 2 days. Using SA allowed to increase the accumulation of rutin in *F. esculentum* plants by 1.7–1.9 times, copper sulfate - by 1.2 times compared with the control. The effect of such treatment, which persists for a long time, makes obtaining buckwheat genotypes with a high level of flavonoid synthesis possible.

Keywords: *F. esculentum* · *in vitro* · Rutin · Microplants · Salicylic acid (SA) · Copper sulphate

1 Introduction

Synthesis of phenolic compounds is a typical feature of the secondary metabolism of Tatar buckwheat (*F. tataricum* (L.) Gaertn and common buckwheat (*F. esculentum* Moench) [1]. Buckwheat contains an increased content of flavonoids, such as: rutin, orientin, vitexin, quercetin, isovitexin, and isoorientin [2]. Breeding buckwheat varieties with high content of bioflavonoids (rutin in particular) is currently a developing direction in Russian and global buckwheat breeding [3, 4]. Flavonoids contained in buckwheat products play an important therapeutic and prophylactic role, since they have vaso-strengthening, anti-inflammatory, hypotensive, anti-edematous, anti-diabetic, and anti-allergic effects, anti-tumor activity, and reduce blood cholesterol levels [5].

A. Muratov and S. Ignateva (Eds.): AFE 2021, LNNS 353, pp. 80–87, 2022.
https://doi.org/10.1007/978-3-030-91402-8_10

The ability to accumulate phenolic compounds as important components of the antioxidant system of plant protection can serve as a criterion for high stress factor resistance. In plants adaptive to stress factors a significant increase in the content of phenolic compounds in cells comparing to forms with low viability and adaptive response was revealed [6]. Identifying buckwheat forms with high flavonoid content contributes to the development of adaptive varieties that are resistant to various stress factors of the external environment.

The intensification of the process of biosynthesis of flavonoids in various parts of the plant can be influenced by a number of factors: pathogenic fungi, bacteria, viruses, temperature changes, mechanical damage, bright light, ultraviolet radiation, imbalance of mineral components in the soil, drought, salinity, imbalance of mineral components in the soil, the effect of ozone, herbicides, heavy metal salts. The conditions for cell cultivation in vitro, as well as additional stress effects (salinity, heavy metal ions, etc.) are capable of providing significant changes in the secondary metabolism of cultured cells [7–9]. It has been established that secondary metabolites such as isoflavonoids, terpenes, phytosterols, and phenylpropanoids can be induced by salicylic acid and methyl jasmonate [10]. However, S. Hou et al. [10] provides information on the effect of exogenous elicitors on an increase in the accumulation of flavonoids in buckwheat and Tartar buckwheat immediately after treatment; there is no information on the persistence of this effect in further plant generations.

The purpose of our experiments was to find ways to increase the accumulation of rutin in buckwheat plants by using SA and copper sulfate *in vitro* culture.

2 Materials and Methods

The experiments were carried out in the laboratory of agricultural biotechnology of the Federal State Budgetary Scientific Institution «Federal Research Center of Agricultural Biotechnology of the Far East named after A.K. Chaiki». Buckwheat (*Fagopyrum esculentum* Moench): Izumrud variety and Izumrud hybrid × Inzerskaya was used as an object of the research. The disinfection was carried out by immersing mature seeds in concentrated sulfuric acid (H_2SO_4) for 2 min, then rinsed three times with sterile distilled water in a laminar box, and the pericarp was removed. The disinfected seeds without pericarp were passaged on to a nutrient medium with the Murashige and Skoog (MS) mineral base containing (mg L^{-1}): thiamine - 2, pyridoxine - 1, casein hydrolyzate - 1000, as well as 20 g L^{-1} of sucrose and 6 g L^{-1} of agar, pH 5.8–6.0. The culture medium was autoclaved for 20 min at a temperature of 121 °C.

Aseptic seeds and plants were cultivated at a temperature of 22 ± 2 °C, light day of 16 h, illumination of 4.5 klx. In buckwheat seedlings, a part of the stem of 1.0–1.5 cm long without cotyledon leaves was removed and placed in test tubes on MS medium. The stem of microplants was cut into microscopic cuttings (segments with one axillary bud 1.0 cm long) and passaged on selective MS media supplemented with salicylic acid ($C_7H_6O_3$) at a concentration of 6.9; 13.8; 20.7 mg L^{-1} and copper sulfate ($CuSO_4 \times 5\ H_2O$) - 23, 46, 69, 161, 184 mg L^{-1}. Cultivation of buckwheat explants on selective media with SA was carried out for 2 and 24 days, and 25 days on media with copper sulfate. Control plants were grown on MS medium without salicylic acid and with a standard copper sulfate content in MS medium (0.025 mg L^{-1}). Further micropropagation of plants was also carried out on MS medium, the composition of which is given above.

The rutin content was determined in microplants four months after treatment with chemical agents and cultivation in a control medium in vitro. Phenolic compounds were extracted from dry plant tissues with 70% ethanol. The extraction was carried out in a flask connected to a reflux condenser, heated in boiling water bath for 1 h. After extraction, the extract was cooled and filtered through filter paper. The filtered liquid was used for spectrophotometric studies of phenolic compounds. Determined the content of the sum of flavonoids with 2% $AlCl_3$ solution at a wavelength of 413 nm. Spectrophotometric studies were carried out using a Shimadzu UV mini-1240 spectrophotometer (Japan). Calibration curves were plotted according to the routine (Chemapol, Czech Republic), expressing the content of these substances in percentage [11].

Statistical data processing was carried out using Statistica 8. Mean values ± standard error, Student's coefficient at a confidence level of 95% were calculated. Each experiment was carried out in three biological replicates.

3 Results

Short-term exposure to salicylic acid (for 2 days) did effect plant regeneration negatively. Cultivation of microcuttings for 24 days on a nutrient medium with SA at a concentration of 13.8 mg L^{-1} led to death of 25% of explants of the Izumrud × Inzerskaya hybrid, the viability of microshoots of the Izumrud variety was 100%. With an increase in the SA content in the medium to 20.7 mg L^{-1}, the survival rate of microscopes decreased to 33.3% in both genotypes. Long-term cultivation of explants (24 days) at a low SA concentration (6.9 mg L^{-1}) in a nutrient medium did not have negative effect on viability; all microshoots retained the ability to regenerate (Fig. 1). Salicylic acid promoted the acceleration of the process of setting flower buds in regenerant plants, which impeded further micropropagation.

a b

Fig. 1. Buckwheat plants *in vitro*: a - microshoots, b – microplants.

During the experiments, a different response of buckwheat genotypes to the SA exposure was observed. In the control plants of the Izumrud variety, the content of rutin was 1.46%, the value of this indicator in microplants after SA treatment varied from 1.23 to 2.8% (Fig. 2). The maximum content of rutin (2.8%) was observed in plants of the Izumrud variety, the microshoots of which were cultivated for 2 days on a medium containing 6.9 mg L^{-1} salicylic acid. With an increase in the SA concentration to 13.8 mg L^{-1}, the accumulation of rutin in plants *in vitro* decreased and amounted to 2.47%. Long-term treatment for 24 days and a high concentration of SA in the nutrient medium (20.7 mg L^{-1}) negatively affected the biosynthesis of rutin. It was observed to decrease in plant tissues to 1.23%, that is, its content was lower than in control plants.

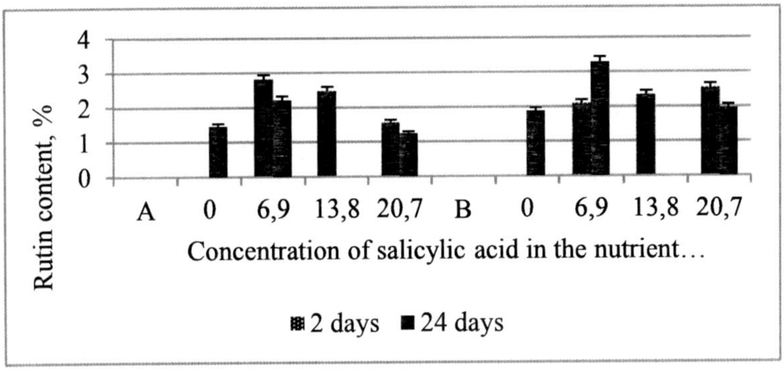

Fig. 2. The effect of salicylic acid on the accumulation of rutin in buckwheat plants *in vitro* culture: A - Izumrud variety, B - Izumrud × Inzerskaya $P < 0.05$.

Microplants of the Izumrud × Inzerskaya hybrid without SA treatment (control) were characterized by higher rutin content (1.87%) compared to microplants of the Izumrud variety (1.46%). The variation of this indicator among the Izumrud × Inzerskaya plants in the experiment ranged from 1.96 to 3.27%. For the hybrid, long-term treatment for 24 days with low concentration of SA (6.9 mg L^{-1}), when the accumulation of rutin was maximum, 3.27%, turned out to be the most effective. High content of rutin (2.53%) was also noted during short-term treatment (2 days) at a high SA concentration of 20.7 mg L^{-1}. Cultivation of regenerant plants on a medium with SA (20.7 mg L^{-1}) for 24 days led to a decrease in the flavonoid content in plants to the control level.

The results of the experiment on the effect of copper sulfate on the biosynthesis of flavonoids in plants *in vitro* showed that the studied buckwheat genotypes reacted to copper salt differently. Microshoots of buckwheat variety Izumrud were characterized by greater resistance to high concentrations of copper sulfate than explants of the Izumrud × Inzerskaya hybrid. Microplants of the hybrid showed low tolerance to copper salts. When cultivating microshoots of the hybrid on a nutrient medium with high concentrations of copper salt in the medium of 161 and 184 mg L^{-1}, regeneration of the shoots was absent, and their death occurred.

Chemical analysis of the quantitative content of rutin in the green mass of buckwheat microplants after cultivation on nutrient media with copper sulfate showed direct relationship between the content of rutin and the concentration of copper sulfate in the medium. With an increase in the concentration of copper ions in the medium, the amount of rutin in plants increased (Fig. 3). With copper salt content in the selective medium of 69 mg L^{-1}, microplants of the Izumrud variety accumulated 2.55%, and the Izumrud × Inzerskaya hybrid - 2.73% rutin, the excess compared to the control (2.15%), respectively, was 12.8 and 26.9%. With an increase in the concentration of copper salt in the selective medium to 161 mg L^{-1}, the content of rutin in the plants of the Izumrud variety increased by 20.35% (2.72%), and with a concentration of copper sulfate in the medium of 184 mg L^{-1} - by 22.57% (2.77%) compared with control microplants (2.26%).

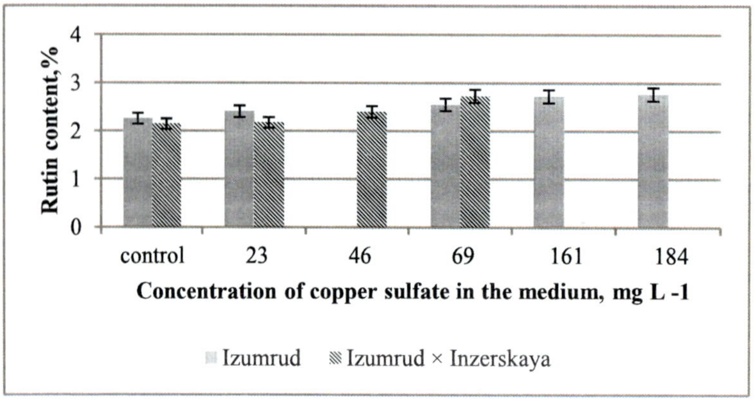

Fig. 3. The effect of copper sulfate on the accumulation of rutin in buckwheat plants *in vitro* P < 0,05.

4 Discussion

Salicylic acid is a phytohormone of a phenolic nature, which together with other phytohormones has a regulatory effect on many physiological processes and the formation of plant defense reactions in response to the attack of pathogens and the action of unfavorable environmental factors [12]. SA *in vitro* treatment of buckwheat plants *F. esculentum* and *F. tataricum* was used in the experiments of S. Hou et al. [10]. According to the data of Chinese scientists, treatment with SA for 12–48 h significantly increased the level of rutin only in Tatar buckwheat; after 24 h of treatment with SA, the highest content of rutin was observed - 3 times more than in the control. In buckwheat, the sowing effect of SA (within 12 and 48 h) led to an increase in the content of rutin only 1.7 times as compared with the control. In our experiment, the action of SA after treatment continued for a long time. After four months of cultivation *in vitro*, buckwheat microplants of the Izumrud variety, which were treated with SA for 2 days at a concentration of 6.9 mg L^{-1}, accumulated 1.9 times more rutin compared to the control. In plants of the Izumrud × Inzerskaya hybrid, the synthesis of rutin increased by 1.7 times after prolonged

(24 days) treatment with SA at a concentration of 20.7 mg L^{-1}. The data of our experiment are consistent with the results of S. Hou et al. [10]: treatment with SA increases the accumulation of rutin to a lesser extent than in Tatar buckwheat. In our study, the same trend persists - 1.7–1.9 times. The authors also note the fact that the content of rutin in *F. esculentum* plants in their experiment decreased after treatment with SA for 24 h to 60% of the control, which indicates a negative relationship between the accumulation of rutin and SA. It has been suggested that SA stimulates the accumulation of rutin, but, apparently, inhibits the continuous synthesis of rutin above a certain level until the plant adapts to the effects of SA, when the content of rutin can be restored [13]. This assumption is also confirmed in our results, when the content of rutin decreased in plant tissues upon exposure to high SA concentrations for a long time (24 days). Four months after the treatment of plants, the synthesis of rutin in microplants of the Izumrud × Inzerskaya hybrid recovered to the control level, while in the plants of the Izumrud variety it was lower than the control.

Heavy metals are one of the abiotic stressors in plants. The use of copper sulfate in a selective medium during the cultivation of *F. esculentum* regenerated plants promoted a more intense (1.2 times) accumulation of flavonoids, compared with the control. This is probably due to the plant response to ionic stress. Many genes for flavonoid biosynthesis are also induced under stress. Therefore, the biosynthesis of flavonoids is often stimulated to a greater extent in stress-sensitive species than in stress-resistant species [14]. This statement by G. Agati et al. is consistent with the results of our experiment. The hybrid Izumrud × Inzerskaya showed a lower tolerance to copper; at concentrations of copper salt of 161 and 184 mg L^{-1}, the death of microshoots occurred. However, the plants adapted to the content of copper sulfate in the medium of 69 mg L^{-1}, at which the maximum increase in the amount of rutin was observed by 26.9% more than in the control. Enhancing the synthesis of phenols in plants is a nonspecific response to stress and plays an important role in the protection, adaptation of cells and the ability to survive under stress conditions. In buckwheat regenerated plants that survived cultivation on media with increased concentrations of copper salt, along with an increase in copper tolerance, there were changes in the biosynthesis of rutin. Flavonoids are among the most biologically active secondary metabolites in plants and represent a secondary antioxidant system that is activated as a result of depletion of the activity of antioxidant enzymes [15]. The active accumulation of flavonoids in the cells of assimilatory organs can increase the efficiency of the antioxidant system in the processes of neutralizing the products of oxidative stress and contribute to an increase in plant resistance to the action of toxicants.

It is generally known that *F. tataricum* accumulates more flavonoids than *F. esculentum*. For this reason, *F. tataricum* is the object of studies of the biosynthesis of rutin more often. Our results on the regulation of the accumulation of rutin in *F. esculentum* confirm the possibility of an increase in flavonoids due to the use of exogenous substances *in vitro*. It is necessary to conduct further experiments to study various chemical agents affecting the biosynthesis of rutin in buckwheat. The manifestation of different reactions of *F. esculentum* genotypes to the action of the applied elicitors also prompts the search for genotypes that are responsive to their action.

5 Conclusion

The experiments carried out using salicylic acid and copper sulfate in the composition of the nutrient medium show the possibility of improving *F. esculentum* plants in in vitro culture. The preservation of the positive effect of the studied chemical agents for a long time makes it possible to count on obtaining *F. esculentum* genotypes with increased content of rutin. The exogenous use of salicylic acid in the cultivation of *F. esculentum* plants in vitro made it possible to increase the accumulation of rutin by 1.7–1.9 times, copper sulfate - by 1.27 times compared with the control.

References

1. Gumerova, E., Akulov, A., Rumyantseva, N.: Effect of methyl jasmonate on growth characteristics and accumulation of phenolic compounds in suspension culture of tartary buckwheat. Rus. J. Plant Physiol. **62**(2), 195–203 (2015). https://doi.org/10.1134/S1021443715020077
2. Zhang, Z., et al.: Bioactive compounds in functional buckwheat food. Food Res. Int. **49**, 389–395 (2012). https://doi.org/10.1016/j.foodres.2012.07.035
3. Klykov, A., Chaikina, E., Anisimov, M., Borovaya, S., Barsukova, E.: Rutin content in buckwheat (*Fagopyrum esculentum* Moech, *F. tataricum* (L.) Gaertn and *F. cymosum* Meissn.) grow in the Far East of Russia. Folia biologica et geologica **61**(1), 61–68 (2020). https://doi.org/10.3986/fbg0068
4. Suzuki, T., Noda, T., Morishita, T., Ishiguro, K., Otsuka, S., Brunori, A.: Present status and future perspectives of breeding for buckwheat quality. Breed Sci. **70**(1), 48–66 (2020). https://doi.org/10.1270/jsbbs.19018
5. Kreft, I., Fabjan, N., Yasumoto, K.: Rutin content in buckwheat (*Fagopyrum esculentum* Moench). Food Chem. **98**(3), 508–512 (2006). https://doi.org/10.1016/j.foodchem.2005.05.081
6. Kazantseva, V., Goncharuk, E., Fesenko, A., Shirokova, A., Zagoskina, N.: Features of the phenolics' formation in seedlings of different varieties of buckwheat (*Fagopyrum esculentum moench*). Agric. Biol. **50**(5), 611–619 (2015). https://doi.org/10.15389/agrobiology.2015.5.611rus
7. Ramakrishna, A., Ravishankar, G.: Influence of abiotic stress signals on secondary metabolites in plants. Plant Signal Behav. **6**(11), 1720–1731 (2011). https://doi.org/10.4161/psb.6.11.17613
8. Wasternack, C., Hause, B.: Jasmonates: biosynthesis, perception, signal transduction and action in plant stress response, growth and development. Ann. Bot. **111**(6), 1021–1058 (2013). https://doi.org/10.1093/aob/mct067
9. Barsukova, E., Klykov, A., Fisenko, P., Borovaya, S., Chaikina, E.: The usage of the method of biotechnology in the selection of buckwheat plants in the Far East. Vestnik Far East Branch **4**, 58–66 (2020). https://doi.org/10.37102/08697698.2020.212.4.010
10. Hou, S., Sun, Z., Linghu, B., Wang, Y., Huang, K., Xu, D., Han, Y.: Regeneration of buckwheat plantlets from hypocotyls and the influence of exogenous hormones on rutin content and rutin biosynthetic gene expression in vitro. Plant Cell Tiss. Org. Cult. **120**, 1159–1167 (2015). https://doi.org/10.1007/s11240-014-0671-5
11. Kreft, S., Strukelj, B., Gaberscik, A., Kreft, I.: Rutin in buckwheat herbs grown at different UV-B radiation levels: comparison of two UV spectrophotometric and an HPLC method. J. Exp. Bot. **53**(375), 1801–1804 (2002). https://doi.org/10.1093/jxb/erf032

12. Shugaev, A., Butsanets, P., Andreev, I., Shugaeva, N.: Effect of salicylic acid on the metabolic activity of plant mitochondria. Rus. J. Plant Physiol. **61**(4), 520–528 (2014). https://doi.org/10.1134/S1021443714040189
13. Murthy, H.N., Lee, E.-J., Paek, K.-Y.: Production of secondary metabolites from cell and organ cultures: strategies and approaches for biomass improvement and metabolite accumulation. Plant Cell Tissue Organ. Cult. (PCTOC) **118**(1), 1–16 (2014). https://doi.org/10.1007/s11240-014-0467-7
14. Agati, G., Azzarello, E., Pollastri, S., Tattini, M.: Flavonoids as antioxidants in plants: location and functional significance. Plant Sci. **196**, 67–76 (2012). https://doi.org/10.1016/j.plantsci.2012.07.014
15. Gondor, O., et al.: Salicylic acid induction of flavonoid biosynthesis pathways in wheat varies by treatment. Front. Plant Sci. **7**, 1447 (2016). https://doi.org/10.3389/fpls.2016.01447

Promising Middle-Early Potato Varieties for Baikal Region

Nadezhda Bolsheshapova(✉), Sergey Burlov, Irina Abramova, and Evgeny Boyarkin

Irkutsk State Agrarian University named after A. A. Ezhevsky, Molodezhny 1/1, Irkutsk District, 664038 Irkutsk, Irkutsk Region, Russia

Abstract. Potato breeding stock is of major importance in the potato growing industry. In the Irkutsk region, potatoes are one of the main food products. For successful growing, some specific characteristics are required in potatoes. A middle-early potato variety shall be high-yielding, have good taste and cooking qualities, as well as be resistant to pathogens. This contributes to minimizing crop losses caused by viral, fungal diseases, and pests challenging potato growing in our region. The article provides an overview of the research conducted at the Irkutsk State Agrarian University named after A.A. Ezhevsky on yields and quality characteristics of middle-early potato varieties. Same tillage and cultivation technologies were generally accepted for zonal conditions. Sarma variety, recognized in the Irkutsk region, was selected as a reference. The study covered 15 middle-early potato varieties. The highest yields were shown by the following varieties: Dar, Krinitsa, Sappho, Ladozhsky, 22,009, Ryabinushka, Mars, Briz, and Ketsky. High starch content was observed in Krinitsa, Garnet, and Ladozhky varieties. The average starch content was registered in the following varieties: Sarma, Sappho, Khoziayushka, Ketsky, Mustang, Briz, Mars. A number of potato varieties were selected for official testing.

Keywords: Potato · Hybrid · Variety · Quality · Starch · Yield

1 Introduction

In the Irkutsk region, potatoes are one of the main food products. For successful growing, some specific characteristics are required in potatoes. A middle-early potato variety shall be high-yielding, have good taste and cooking qualities, as well as be resistant to pathogens. This contributes to minimizing crop losses caused by viral, fungal diseases, and pests challenging potato growing in our region [1].

Eastern Siberia is one of the largest regions of Russia with a population of about nine million. The potato cultivation area counts more than 150 ha, mostly in the forest-steppe zone. The potato is primarily grown for food purposes, with about 236 kg of tubers produced per each resident of the Irkutsk region, or 180% of the demand [2–6]. The region is considered to be one of the most favourable in Siberia in terms of conditions for organic potato cultivation.

A. Muratov and S. Ignateva (Eds.): AFE 2021, LNNS 353, pp. 88–95, 2022.
https://doi.org/10.1007/978-3-030-91402-8_11

The potato planting area in the region amounts to 38.5 thousand hectares, the yield remains low—14–15 t/ha, croppage – 567.6 thousand tons. 236 kg of potatoes, or + 180% to the standard required value, are produced per each resident in the region. In general, it is worth noting that potato growing as an industry is suboptimal here. Seed breeding is poorly developed, especially for domestic varieties [4, 5].

According to the State Commission of the Russian Federation for Selection Achievements Test and Protection, potato varietal resources are represented by 318 varieties of different maturation periods, including early-season—28.6%, middle-early—30.2%, mid-season—24.8%, middle-late—12.9%, late-season—3.5%, of which 52.5% are bred domestically (167 varieties), and 47.5% are bread abroad (151 varieties). 163 potato varieties are protected selection achievements [4, 5].

The Irkutsk region is the area of risk farming, while the natural environment of the primary agricultural sites located in the forest-steppe and steppificated areas, allow for achieving high and sustainable yields of crops for this region, including small grains and potatoes. Thus, growing high-performance and high-quality crops, as well as providing these products to the population, is a relevant task for the farmers of the region [2].

The purpose of the research is to study and evaluate promising middle-early potato varieties in the Baikal region in terms of yield and quality of potato tubers.

Research objectives: to assess potato varieties as per the economically valuable characteristics, to determine the yield and quality of tubers of middle-early potato varieties, to identify the best varieties for further breeding.

Soil and climate conditions and the location of the experiments.

According to the agroclimatological zoning, the territory of Molodezhny research and production site of the Irkutsk State Agrarian University, where the experimental field is located, is part of the Irkutsko-Cheremkhovskaya plain of the Predsayansky foredeep. This plain contains a strip of low-level swampy areas. Terrain is flat; slope steepness is 2–5°. The site is located in a normal flat-topped dry land [6, 7].

The climate of the region is sharp continental with long dry severe winters and relatively warm summers. The considerable latitudinal extent and complexity of the terrain result in a highly variable air temperature distribution. The average annual air temperature is generally below zero: from - 9 °C in the north and up to - 1–2 °C in the west, south and south-east of the region [3, 7–9].

During the years of research, the meteorological conditions were not uniform and were relatively favorable. In general, the analysis of meteorological conditions showed that they differed from the long-term average [4, 5].

2 Results and Discussion

The studies were carried out on the experimental field of the Irkutsk State Agrarian University named after A.A. Ezhevsky and the potato laboratory of the Department of Agriculture and Crop Production in 2013–2015. The experiments were arranged as two-field crop rotation: weed-free fallow – potatoes. The soil of the experimental site was gray forest soil. The work was conducted within the framework of the comprehensive scientific and technical project Selection and Seed Breeding of New High-Yielding, Disease-Resistant Potato Varieties. The Development of Propagation Methods for Obtaining

High-Quality Original Seed Tubers of Potato Varieties Bred by the Irkutsk State Agrarian University for Resolving the Issue of Import Substitution in Potato Growing in the Irkutsk Region."

15 middle-early potato varieties breeded in Russia and abroad, and being of interest for cultivation in the Irkutsk region, were taken as the source material. Sarma, a recognized middle-early potato variety in the Irkutsk region, was used as a reference.

The area of the plot was 8 m^2, with a triple tier. Plots were located along the slope, and their arrangement was randomized.

In our experiments, the planting was carried out on May 20–24, and full potato seedlings emerged 17–23 days after planting. Between June 10 and 12, Briz, Garnet, and Skarb varieties showed early emergence. The bud phase in Briz, Garnet, and Dar varieties was observed from June 28 to July 9, and their first flowering occurred from July 9 to July 11. Table 1 provides the yield and quantitative indicators of middle-early varieties.

Table 1. Yield and quantitative indicators of middle-early potato varieties

Variety	Yield t/ha	Number of tubers in the clump, pcs.	Mean tuber weight, g	Marketability, %
Dar	33.5	11.1	96.0	84.6
Krinitsa	31.3	13.7	86.3	77.4
Sappho	31.1	7.9	122.3	89.9
Ladozhsky	30.2	7.4	115.0	95.6
22,009	30.2	10.3	98.7	90.4
Ryabinushka	29.6	7.3	118.3	92.0
Mars	29.6	10.9	88.3	88.5
Sarma (ref.)	29.2	11.3	103.7	92.9
Briz	28.8	8.6	101.3	91.4
Ketsky	28.1	9.1	99.7	86.4
Khoziayushka	26.3	7.3	110.7	93.7
Skarb	22.5	8.1	87.7	93.5
Garnet	21.5	9.7	80.3	80.1
Mustang	20.0	5.7	102.3	85.7
HCP_{05}	1.3			

The studied potato varieties differed in their plant height and the number of stems per plant. So, in the group of middle-early potato varieties, the number varied from 2.3 pcs./clump (Khoziayushka) to 6.8 pcs./clump (hybrid 22,009). The variation in middle-early and mid-season potato varieties was 26.45%, which is considered significant.

In terms of highest performance, varieties Dar, Krinitsa and Sappho were identified, having provided a yield of 33.5, 33.3 and 33.1 t/ha, respectively, over 3 years, while the yield of Sarma variety was 29.2 t/ha. Varieties Ryabinushka, Mars had the standard yield level. The other varieties were 0.5–9.2 t/ha inferior to the standard reference.

When selecting varieties for crossbreeding, marketability is considered of no less importance, and breeders take it into account. The harvest marketability of Ladozhsky, Khoziayushka, and Skarb varieties was higher than that of the Sarma reference variety. The main advantage of the Krinitsa variety is the large number of tubers in the loculus. Mars, Sarma, Dar, and hybrid 22,009 showed the highest number of total tubers in the clump among the middle-early varieties. The average number of tubers in the group was in the range of 5.7 pcs./clump up to 13.7 pcs./clump. Sappho, Ryabinushka, Sarma, Ladozhsky, Briz, Khoziayushka, and Mustang varieties showed excellent tuber weight.

According to Z.I. Usanova (2013), the dry matter content in potato tubers depends more on the variety, the heat and moisture availability of the plants, and to a lesser extent—on the level of mineral nutrition. In Eastern Siberia, potatoes are primarily used as food. Therefore, tuber quality characteristics are as important as yield characteristics. In terms of starch content, they exceeded the standard by 2 and 3.7%, Garnet and Krinitsa, also exceeded the standard reference by 2–3.8% in terms of dry matter content. In terms of taste, all varieties were not inferior to the standard reference (Table 2).

Table 2. Quality indicators of middle-early potato varieties (average)

Variety	Dry matter content, %	Starch content %	Taste, score	Late blight resistance %
Krinitsa	26.9	19.6	9	20
Garnet	25.1	17.9	7	24
Ladozhsky	24.5	17.2	9	22
Zekura	23.9	16.7	9	25
22,009	24.1	16.6	9	32
Sarma (ref.)	23.1	15.9	9	37
Sappho	23.1	15.9	5	26
Khoziayushka	22.7	15.5	5	24
Ketsky	22.5	15.2	9	32
Mustang	22.2	14.9	9	48
Briz	21.3	14.2	9	26
Mars	21.4	14.2	7	22
Skarb	21.1	13.8	7	22
Ryabinushka	19.8	12.6	9	30
Dar	18.9	11.7	9	22

Increased starch content improves the taste of tubers and enables using the varieties for technical purposes.

Notably, all the tested varieties in 2013–2015 were highly resistant to late blight.

The biochemical analysis of protein, vitamin C and sugar content was conducted in the testing laboratory of the Federal State Budgetary Institution of the State Center of Agrochemical Service Krasnoyarsky, the city of Krasnoyarsk. Khoziayushka variety was singled out by the protein content in potato tubers (2.06%), Krinitsa and Ladozhsky varieties had the protein content at the level of Sarma reference variety (1.94%). In other potato varieties, the protein content was within the range of 1.59 to 1.74%. The highest sugar content was observed in Briz variety (0.51%). The distinguished varieties Krinitsa, Mars, Sappho, Ladozhsky, Khoziayushka, hybrid 22,009, Sarma, had a low content of reducing sugars under 0.5%. The mentioned varieties meet the requirement for food purposes in terms of reducing sugar content (Gusev, 2014) [8]. Following the studies, we have identified the following varieties with the highest content of vitamin C in potato tubers: Sarma, Sappho, Garnet, and hybrid 22,009, —these varieties can be used in breeding programs [10–13].

Variety assessment by ecological flexibility is of interest both for theoretical research and for breeding practice. Of great importance to the national economy are the recognized varieties maintaining highly stable yields under various environmental conditions. Thus, the stability shall be assessed by two indicators – the regression coefficient (flexibility) bi and the mean square deviation (stability variance) S2d from the regression line [4, 5, 14].

The analysis of potato varieties shown in Table 3 allowed us to identify samples that have advanced adaptive properties in terms of yield. On average, for the collection of potato varieties, the variation of the regression coefficient (flexibility) bi in terms of yield in varieties ranged from –0.18 to 2.30. The stability variance (S^2d) ranged from 0.71 to 118.51.

The middle-early varieties Mars (bi = 1.12), Sappho (bi = 1.21), Sarma (bi = 1.49), Skarb (bi = 1.52), Mustang (bi = 1.98), belong to the intensive type, these varieties have a regression coefficient (flexibility) bi significantly higher than 1. Varieties Mars, Sappho, Sarma, Skarb, Mustang, respond well to the improvement in cultivation conditions. In unfavorable weather conditions, as well as at a low agricultural background, the yield of these varieties decreases sharply. The most flexible varieties are Ryabinushka (bi = 0.66), Ketsky (bi = 0.72), Garnet (bi = 0.75), Ladozhsky (bi = 0.92). In these varieties, the regression coefficient (flexibility) bi is closer to 1, i.e. these varieties are well adapted to various environmental conditions. The regression coefficient (flexibility) bi below 1 was identified in the potato varieties Briz (bi = 0.09), Khoziayushka (bi = 0.22), hybrid 22,009 (bi = 0.31)—these varieties belong to the neutral type with low ecological flexibility. They provide little response to changes in environmental factors under heavy cropping conditions. Krinitsa and Dar varieties with a regression coefficient (flexibility) bi = −0.18 and −0.60 did not provide significant yield response to the improvement in environmental conditions.

The most stable varieties in terms of yield were Sarma ($S^2d = 1.62$), Ladozhsky ($S^2d = 2.87$), Ketsky ($S^2d = 2.97$), Skarb ($S^2d = 3.18$), and Briz ($S^2d = 9.02$)—they made the best use of favorable conditions for crop formation. The remaining varieties exhibited relatively average stability. According to the S^2d stability index, the least stable under various environmental conditions were: the early-season variety Rosara ($S^2d = 65.94$), and the middle-early varieties Dar ($S^2d = 103.21$), and Ryabinushka ($S^2d = 118.51$).

The best combination of flexibility and yield stability was registered in the middle-late Mustang variety with a regression coefficient bi = 1.98, which refers the variety to the intensive type, and it responds well to improved cultivation, as the variety maintains high stability ($S^2d = 0.71$).

The potato hybrid 22,009 is a neutral type with low ecological flexibility, the regression coefficient amounted to bi = 0.31, however, this hybrid can be considered stable with $S^2d = 0.79$.

Table 3. Parameters of potato varieties adaptability by yield

Variety	Yield, t/ha		Regression coefficient (flexibility) (b_i)	Stability variance (S^2d)
	$X_{avg.}$	Lim-opt		
Dar	33.5	27.2–42.0	−0.60	103.21
Krinitsa	31.3	26.9–36.7	−0.18	50.29
Sappho	31.1	23.4–39.2	1.21	16.52
Ladozhsky	30.2	25.3–36.7	0.92	2.87
22,009	30.2	28.6–32.9	0.31	0.79
Ryabinushka	29.6	19.5–35.7	0.66	118.51
Mars	29.6	22.2–37.1	1.12	17.08
Sarma (ref.)	29.2	23.7–39.6	1.49	1.62
Briz	28.8	26.1–30.4	0.09	9.02
Ketsky	28.1	24.8–33.5	0.72	2.97
Khoziayushka	26.3	21.8–29.3	0.22	26.18
Skarb	22.5	15.3–32.7	1.52	3.18
Garnet	21.5	14.7–26.1	0.75	26.44
Mustang	20.0	11.8–33.1	1.98	0.71

Thus, following the analysis of yield adaptability, Mars, Sappho, Sarma, Skarb, and Mustang are considered intensive (highly flexible) varieties. The yield-flexible varieties include Ryabinushka, Ketsky, Garnet, and Ladozhsky.

Mustang variety and hybrid 22,009 showed high yield stability over the years. The most stable varieties are Sarma, Ladozhsky, Ketsky, Skarb, and Briz.

Results of potato breeding in Irkutsk State Agrarian University. In a comparative test, promising varieties of potatoes are studied:

Malibu (Siren × Victoria), middle-early variety, tubers and pulp are red. High yield, large tubers, cancer resistance.

Molodezhny (Sarma × Victoria). Early-season variety, food grade. Average marketable yield: 210 dt/ha. Maximum yield: 279 dt/ha. Tuber shape: oval. Marketable tuber weight: 78–144 g. Taste: Good to Excellent. Marketability: 91% Tuber preservation capacity: 92%. The variety is resistant to cancer, golden nematode.

Babr (Sante × Pushkinets). Middle-early variety, food grade. Average yield: 234 dt/ha (36 dt/ha over the standard). Maximum yield: 387 dt/ha. Tuber shape: round to oval. Marketable tuber weight: 94–176 g. Taste: Good to Excellent. Tuber marketability: 92%. Tuber preservation capacity: 92%. The variety is resistant to cancer, golden nematode.

22,009 (Peresvet × Nevsky). Middle-early variety. Average marketable yield: 262 dt/ha (42 dt/ha over the standard). Maximum yield: 367 dt/ha. Tuber shape: round to oval. Marketable tuber weight: 102–186 g. Taste: Good to Excellent. Tuber marketability: high. Tuber preservation capacity: good. The variety is resistant to late blight, golden nematode.

3 Conclusion

The results obtained allowed for the following conclusions: Dar, Krinitsa, Sappho, Ladozhsky, 22,009, Ryabinushka, Mars, Briz, Ketsky, which give a yield at the standard level or higher, were recognized as the highest-yielding varieties.

The highest starch content was observed in Krinitsa, Garnet, and Ladozhsky varieties (17.2–19.6%). The average starch content was registered in the following varieties: Sarma (ref.), Sappho, Khoziayushka, Ketsky, Mustang, Briz, Mars (14.2–15.9%).

Khoziayushka variety was determined as having the highest protein content; Sarma, Sappho, Garnet and hybrid 22,009—highest vitamin C content; Garnet and Krinitsa—highest tuber dry matter content (17.9% and 19.6% respectively).

Varieties of interest in terms of taste and cooking properties are Mars, Sarma, Sappho, Garnet, Khoziayushka, Mustang, Dar, and 22,009.

Krinitsa, Skarb, Mars, Ladozhsky, Garnet, Sappho, and Khoziayushka varieties are characterized by high resistance to potato late blight, black speck, and scab.

An analysis of the ecological flexibility and stability of potato varieties in terms of yield enabled identification of highly flexible varieties: Sappho, Mars, Skarb, Sarma, Mustang; flexible varieties: Ladozhsky, Ryabinushka, Ketsky, Garnet. The most stable varieties in terms of yield were: Sarma, Ketsky, Skarb, Briz.

A number of potato varieties were selected for official testing.

References

1. Burlov, S.P., Bol'sheshapova, N.I.: Vestnik IRGSKHA **92**, 7–16 (2019)
2. Abramova, I.N., Bol'sheshapova, N.I., Burlov, S.P., Boyarkin, E.V.: Materialy Vserossijskoj nauchnoj konferencii s mezhdunarodnym uchastiem. M., Izd-vo RGAU – MSKHA (2020)
3. Bol'sheshapova, N.I., Burlov, S.P.: Otsenka perspektivnykh gibridov kartofelya dlya usloviy Irkutskoy oblasti. Kartofel' i ovoshch **12**, 36–37 (2019). https://doi.org/10.25630/PAV.2019.33.53.010
4. Bol'sheshapova, N.I.: Ocenka sortov i gibridov kartofelya na ekologicheskuyu plastichnost' i stabil'nost' urozhajnosti, kachestva klubnej v lesostepi Irkutskoj oblasti: dis… kand. s.-h. nauk. Tyumen' (2019)

5. Solodun, V.I., Zajcev, A.M., Filippov, A.S., Takalandze, G.O.: Nauchnye osnovy adaptivno-landshaftnyh sistem zemledeliya Predbajkal'ya Uchebnoe posobie. Irkutsk, Izd-vo IrGSKHA (2012)
6. Wang, C., Shi, X., Chu, Q.: Interdecadal variation of potato climate suitability in China. Agric. Ecosyst. Environ. **6**, 310 (2021). https://doi.org/10.1016/j.agee.2020.107293
7. Mal'cev, V.T., Sultanov, F.S., Ostanin, V.A., Solodun, V.I.: Osnovy resursosberegayushchego zemledeliya. Irkutsk, Vost.-Sib. Izdatel'skaya kompaniya (2001)
8. Gusev, A.N.: Sovremennye tekhnologii produktov pitaniya. Sbornik nauchnyh statej mezhdunarodnoj nauchno-prakticheskoj konferencii, Kursk (2014)
9. Alaoui, K., Chafik, Z., Kharmach, E.Z.: In vitro antifungal activity of Lactobacillus against potato Late blight. Phytophthora Infestans **45**(8), 7725–7733 (2021). https://doi.org/10.1016/j.matpr.2021.03.338
10. Zamalieva, F.F., Stasevski, Z., Safiulina, G.F.: 17-th Conference of European Association for Potato Research, pp. 320–323 (2008)
11. Marion, L.: Acta agron ovariensis. Mosonmagyarovar **42**(1), 81–93 (2000)
12. Torikov, V.E., Bogomaz, O.A.: Bull. Bryansk State Agric. Acad. **4**, 64 (2008)
13. Yi, L., Ivanyo, Ya.M., Solodun, V.I., Verkhoturov, V.V., Burlov, S.P.: Advances in engineering research Comparison of potato chlorophyll fluorescences variables and analysis of clusterization. Agro-SMART 453–456 (2018). https://doi.org/10.2991/agrosmart-18.2018.85
14. Muhametshin, I., Potorochina, N.: Glavnyj agronom **6**, 46–49 (2014)

Technology for Protecting Cattle from Blood-Sucking Dipteric Insects and Myiasis Gadflys

Anastasia Barashkova(✉)

FRC YaSC SB RAS M. G. Safronov Yakut Scientific Research Institute of Agriculture, 23/1 Bestuzheva-Marlinskogo Street, 677001 Yakutsk, Russia

Abstract. Objective: To develop a technology for protecting cattle from blood-sucking insects and myiasis gadfly of cattle in Yakutia. Methods: The experiments were carried out in farms "Tumul" and "Khorobut" Yakutia. The protective effectiveness of spraying aerosol mist over a herd of cows with 0.05% aqueous emulsion of deltamethrin was calculated using a method based on the determination of the coefficient of protective action (CPA), expressed as a percentage. The number of attacking insects was determined by direct counting. Results. In Yakutia, harmful blood-sucking and non-blood-sucking insects are widespread. To protect cattle from them, the technology of spraying with an aerosol mist has been developed, which sharply increases the effectiveness of animal protection, environmental friendliness, insecticidal, repellent and anti-stress treatment of cattle. Discussion: The effectiveness of the technology is based on the dispersion of liquid substances. Due to this, which causes a sharp increase in the activity of the active substance, accelerating the physicochemical processes. Conclusions: A technology has been developed to protect cattle from diptera blood-sucking insects and adult gadfly, patent for the invention of RUS No. 2724462.

Keywords: Cattle protection · Mosquitoes · Horseflys · Midges · Flies · Blood-sucking dipteric insects · Myiasis gadfly of cattle

1 Introduction

Ensuring food security in the dairy market is one of the strategic objectives of the Russian Federation [1]. Modern data from nutritional science state that cow's milk is an irreplaceable food product at any age, with the exception of certain diseases such as lactose intolerance or milk protein allergy. Milk consumption, especially in infancy, introduces all the necessary micro and macro elements into the diet, when the growth of bone mass is in a critical phase. When milk is consumed by adults, there is a protective effect against overweight, obesity, diabetes and cardiovascular disease [2].

Recently, alternative dairy products such as soy, almond, oat and goat are making their way onto store shelves. However, comparative data on the nutritional and energy values of cow's milk and alternative dairy products are difficult to obtain. In a comparative analysis of the nutritional value of natural cow's milk with alternative dairy products, it

A. Muratov and S. Ignateva (Eds.): AFE 2021, LNNS 353, pp. 96–104, 2022.
https://doi.org/10.1007/978-3-030-91402-8_12

was found that cow's milk in terms of the content of proteins, fats and microelements and the balance of indicators turned out to be the best against alternative dairy products such as soy, almond, oat and goat milk. Primary care pediatricians should inform parents when making nutritional decisions for children [3].

In many countries, a blessed summer with green pasture fodder is darkened for cattle by a massive attack on them by blood-sucking dipteric insects: horsefly (Tabanidae), mosquitoes (Culicidae), midges (Simuliidae), biting midges (Ceratopogonidae) and myiasis flies (Calliphoridae). Horsefly pricks are very painful and cause great distress to the animals. Diptera insects all over the world are of great medical and veterinary importance as dangerous pathogens of myiasis and vectors of infectious and invasive diseases. In India, cutaneous myiasis from flies attacks is causing significant damage to the wool industry. Animal health and safety costs are AUD $ 4 per sheep. Infectious diseases of livestock such as foot and mouth disease, mastitis and avian influenza, carried by zoophilic flies and horseflies, are economically important diseases that lead to large economic losses [4].

In many countries, scientific research is underway to protect domestic animals from harmful insects, vectors of dangerous animal and human diseases. To find effective methods and means of protecting cattle from tsetse flies in areas in Kenya, a new technology was tested using the repellent 2-methoxy 4-methylphenol to combat trypanosomiasis. A repellent dispenser was attached to a collar that was worn around the animal's neck. The trials began in April 2005. 12 cattle herds were recruited in each district. As a result, a 50% reduction in the incidence of trypanosomiasis in animals was obtained in comparison with the control. Thus, this technology turned out to be insufficiently effective [5].

Horny buffalo flies (Haematobia irritans and H. exigua) are small (3–5 mm long) dark gray hematophagous ectoparasites closely associated with livestock (cattle and buffalo) and sometimes with horses grazing in open pastures. The flies Haematobia irritans and H. exigua constantly attack and feed on the blood of cattle and buffaloes, resulting in significant dairy losses. Horn flies are not only annoying bloodsuckers, but also mechanical carriers of bacteria that cause mastitis in cattle and intermediate hosts of the nematodes Stephanofilaria and Parabronema. The main problem in protecting cattle and buffaloes from the horny buffalo flies Haematobia irritans and H. exigua is their insecticide resistance [6].

Mosquitoes (Phlebotominae) are of great medical and veterinary importance. They are common throughout the Old World and are carriers of leishmaniasis, vesicular stomatitis, bartonellosis, and sand fly fever [7].

Arthropods include many species of medical or veterinary importance. Among them, some are responsible for the transmission of certain dangerous diseases to humans, such as malaria, Chagas disease, leishmaniasis, African trypanosomiasis, lymphatic filariasis, onchocerciasis, dengue, yellow fever, and Zika. These diseases have a huge socioeconomic impact in the countries where they occur. Human-induced environmental changes, with a particular focus on urbanization and global warming, are contributing to their emergence or re-emergence in areas where they have not previously been recorded. Under these conditions, the control of arthropod vectors of especially dangerous diseases requires taking into account the knowledge of their genetic, behavioral and ecological characteristics and an integrated approach to their control [8].

Cattle breeding plays an important role in the production of milk and meat. One of the economically important regular endemic diseases of livestock in the northern hemisphere of the globe is hypodermosis. A map of the spatial distribution of this disease in different countries of the world has been compiled to identify endemic areas. The information will be useful for predicting areas of increased risk of disease in livestock with hypodermosis [9].

Currently, environmental changes continue. The likelihood of a link between the environment and parasitic zoonoses will be important in protecting global health and well-being in those regions of the world where these effects are most severe [10].

According to the forecast of FAS USDA analysts, milk production in Russia will grow to 31.8 million tons in 2021. Russia's largest milk producing companies make the main contribution to the positive dynamics of the industry by investing in the best genetics and equipment. Labor productivity is gradually increasing along with the expansion of leading companies into new regions [11]. Cattle breeding is successfully developing in the regions of Russia. The vast territory of Yakutia is a region of unique traditional cattle breeding, currently there are more than 190 thousand heads of cattle [12].

Yakutia is characterized by a high seasonality of milk production. During the autumn and long winter, cows are kept in tied stalls for a long time, when their milk production is sharply reduced. To this is added the two-month period required for cows before calving. Only in the spring after the mass calving of cows, and in the summer when the herd is driven to pastures, there is a sharp increase in milk production. In this case, a long daylight hours in Central Yakutia, which lasts almost round the clock in June to mid-July, play a positive role. The harm caused to animal husbandry by harmful dipterans is very great, which prompts scientists to look for effective measures to protect animals from them [13].

Research by S. N. Rzhanikov found that by spraying with 0.001% of the active ingredient, an aqueous emulsion of deltsid per cow, 500 ml, and 250 ml for young cattle, using a stationary horizontal spray installation, were treated against mosquitoes (Culicidae) and horseflies (Tabanidae) in one season thousand cows. At the same time, the economic effect per 1 ruble of expenses was 80.44–76.73 rubles [14].

The order Diptera is found in all inland waters from ephemeral ditches to Great Lakes and rivers to rivers, and they are also very common in the open oceans. Unlike all other orders of invertebrates, Diptera contain many species that are harmful or at least annoying to humans in adulthood. Many adult Diptera transmit parasites or diseases that can be debilitating or fatal to humans, such as malaria and West Nile virus. Other flies appear in such large numbers (some midges) that they can cause allergic reactions in humans or clog air conditioners. On the other hand, Diptera larvae are extremely important in aquatic food webs leading to the emergence of commercial fish and some groups are raised in hatcheries as fish feed, and many adult Diptera are important plant pollinators or food for dragonflies and birds such as swallows, swifts. The important medical and veterinary importance of the order of Diptera insects requires constant research to develop preventive measures to protect against them [15].

Tabanids, commonly referred to as horse flies, deer flies or clegs, feed on many vertebrate hosts, including humans. Their medical and veterinary importance resides mainly in the painful bite and their role of biological and mechanical vectors of various human and animal disease agents as viruses, bacteria, and parasites. Tabanids mostly occur in warm areas with suitable moist locations for breeding, occupying a wide range of habitats. They are large, robust and sturdy flies, strong fliers, with a striking appearance, variously colored from black through brown to greens and yellows, bold stripes, and large brilliant colored eyes. Due to their very complex and long-time life cycle, and other various biological features, no tabanid species has been successfully colonized. The effective tabanid population management is challenging due to several factors related mostly to the biological features of these insects and include mainly insecticides, trapping and grazing management [16].

Culicidae mosquitoes are main vectors of arboviruses that cause arboviral diseases in humans. Studies on fauna, ecology, biology, resting behaviors of Culicidae mosquitoes are important and greatly impacts the control of arboviral diseases that are transmitted by vectors. The aim of the present study was to determine fauna of mosquitoes (Diptera: Culicidae) based on morphological and molecular (genomic) identification and their habitats in Lorestan province, Western Iran. Culicidae mosquitoes were caught in 2015 and 2016 from human homes, animal dwellings, storehouses and pit shelters in Lorestan province, Western Iran. Among all of the trapped mosquitoes (4211), 94.68% were from genus Culex mosquitoes (3987), which indicate that this genus is the dominant in Lorestan province, Western Iran. Anopheles comprised of 201 individuals out of the total catch. Arboviruses were not detected in these samples [17].

Considering the above, the goal of our work was to develop a technology for protecting cattle from blood-sucking dipteric insects and myiasis gadfly in Yakutia with an ultra-small amount of a drug with high environmental friendliness, insecticidal, repellent and anti-stress treatment of animals.

2 Materials and Methods

The work was carried out in 2018–2020. according to the program of fundamental scientific research of the state academies of sciences for 2013–2020 0821-2019-0003 "To develop methods for creating new biological products, technologies for the control and prevention of infectious, parasitic and non-infectious diseases of animals and birds, taking into account the climatic conditions and the peculiarities of conducting livestock industries in The Far North". The patent search was carried out in accordance with the task and the search regulations to determine the achieved level and development trends on the research topic. Enough literary sources and titles of protection were found.

The test of the effectiveness of the protection of cattle from blood-sucking dipteric insects and myiasis gadfly in Yakutia was carried out in the agricultural production complex "Tumul" and livestock enterprise "Khorobut" of the Megino-Kangalassky region of the Republic of Sakha (Yakutia). The numbers of attacking mosquitoes, horseflys, and myiasis gadfly's of cattle were counted by capturing them from a bait cow with an entomological net with removable bags. One count consisted of 10 strokes ("figure eight") in 10 repetitions. Every day, during the entire flight period, insects were recorded 3 times a day (at 7, 13, and 19 h's local time) [18].

Every day, during the entire period of flight and attacks of blood-sucking dipteric insects on animals, meteorological data were recorded 3 times a day (at 7, 13 and 19 h local time). Air temperature and humidity were measured with an aspiration psychrometer, wind speed - with an ASO-3 anemometer, atmospheric pressure - with an aneroid barometer, illumination - with a Yu-116 luxmeter, cloudiness - visually on a 10-point scale, precipitation - with a rain gauge. Cameral processing of arthropod collections was carried out in the laboratory of arachnoentomology of the Yakutsk Scientific Research Institute named after M. G. Safronov. When identifying species, the identification keys of the monographs by N. G. Olsufyev, I. A. Violovich, G. Ya. Bei-Bienko and I. A. Rubtsov were used.

For aerosol spraying of cattle herds, we chose an aerosol generator with a capacity of 850 m^3/h with a dispersion of 70 μ for aqueous solutions with an air flow rate of 125 mps, with an engine of 72.3 cm^3/4.1 horses/power. The volume of the solution container was 14 L. The experiment tested the effectiveness of the aqueous emulsion deltamethrin. The deltamethrin concentration in the aqueous emulsion was 0.05%. The herd of cows was protected from blood-sucking and non-blood-sucking dipteric insects by spraying with an aerosol during the maximum number of insects attacking the cows [19].

The protective efficiency of an ultra-low volume of spraying a herd of cows with 0.05% aqueous emulsion of deltamethrin was calculated using a method based on the determination of the coefficient of protective action (CPA), expressed as a percentage using the formula:

$$CPA\frac{(B-A)}{B}\cdot 100 \tag{1}$$

where:

CPA - coefficient of protective action in percent.

A - is the number of insects on the treated animal.

B - the number of insects on the control animal.

3 Results

As a result of the studies carried out on the pastures of South and Central Yakutia, it was found that 23 species and 1 subspecies of horsefly (Tabanidae) were parasitized on animals. On horse pastures in Central Yakutia (Ust-Aldansky, Churapchinsky and Tattinsky regions), horseflys flies from the end of the first decade of June to the middle of the first decade of August, which averaged 58 days. The period of mass flight, when one animal is attacked by 83 to 150 horseflys in a 15-min count, lasts about a month from the third decade of June to the second decade of July. The daily activity of horseflys fits into a one-peak curve and lasts from 8 to 20 h, with the maximum number observed at 12–16 h. Mosquitoes (Culicidae) in Yakutia are the most numerous family in terms of numbers. The mosquitoes fly season begins in the first ten days of May and lasts until September. The number of mosquitoes in Central Yakutia is more than 400 individuals per count. Throughout the territory of Yakutia, in the summer, in addition to horsefly (Tabanidae) and mosquitoes (Culicidae), animals are greatly disturbed by zoophilic flies

(Calliphoridae), midges (Simuliidae), biting midges (Ceratopogonidae) and imago of gadfly (Hypoderma bovis).

The regularities of the dynamics of the species composition and structure of the fauna of horseflies (Tabanidae), mosquitoes (Culicidae), midges (Simuliidae) and their quantitative aspects in the central, western, southern, northeastern and arctic tundra natural and climatic zones of the Republic of Sakha (Yakutia) were revealed. In the Primorsky tundra (The Anabarsky district is located 96°2′-97°9′N–111°3′-113°2′E.), previously not found species of horseflys were identified: Hybomitra montana montana and H. nigricomis. For plain pastures - Chiysops divaricatus, Chr. suavis and Hybomitra aequetincta. In the Kolyma River basin (collection point 67°27′N 153°42′E), a previously unseen species of midges Simulium posticatuni was recorded. In agrocenoses, the number of species is less than in natural ecosystems. For the first time for the Republic of Sakha (Yakutia), a zoogeographic analysis of the species composition and structure of the horsefly fauna has been carried out. Six faunistic assemblages of three types of fauna have been identified.

When performing a patent search, prototype 1 was found - patent RU 2558970, a universal installation for spraying animals, installed in the gate for pasture of animals, consisting of two branches, each of which has five nozzles with two modes for spraying liquid. For cows, the first regime is applied medium volume spraying: where 500 ml of insecticide emulsion is consumed per animal. For young animals, a second low-volume spraying mode is used: where 100–50 ml of insecticide emulsion is consumed per young animal.

The disadvantage of prototype 1 are: first - the use of a very large amount of solution (emulsion) of the insecticide, by abundant shower of cows, which creates a stressful situation for dairy cattle, has a negative effect on milk and meat productivity; second - on the routes of summer migrations of herds of domestic reindeer along the tundra, there are no fenced areas with gates necessary for the shower system of prototype 1.

Prototype 2. Known protection of cattle and reindeer from blood-sucking dipteric insects and gadflies in the Tyumen region: Methodological recommendations. Tyumen, 2010. Pp. 1–59. Ultra-low-volume sprayed on cattle and domesticated reindeer are carried out using a diesel-fueled solution.

The disadvantage of prototype 2 is the use of diesel fuel to reduce aerosol particles to the required dispersion. Breathing diesel fuel into the lungs of an animal can be harmful.

Prototype 3. There is a known method of protecting domesticated reindeer from attacks by mosquitoes, patent RU 2595831, by ultra-low-volume spraying over a herd of deer with a concentration of 0.05% based on 7 ml of an aqueous emulsion of insecticide per one deer.

Disadvantage of the prototype 3. The method is used only for domesticated reindeer in the seaside tundra from the mosquitoes and is not used for cattle. Cattle and reindeer differ in live weight, therefore, the doses of the medicinal emulsion are also very different.

The technical objective of our claimed invention for the development of technology for the protection of cattle from blood-sucking dipteric insects and imago of gadflies is to achieve high environmental friendliness of animal protection from midges, improve environmental friendliness, insecticidal, repellent, and anti-stress providing 6-h protection of cows from attacks of harmful insects.

According to the act on the acceptance of experiments by the commission of the Scientific and Methodological Council for the study of the ecology of the components of blood-sucking and non-blood-sucking dipteric insects in cattle pastures and testing the optimal scheme for the use of ultra-low-volume spraying of cattle from blood-sucking dipteric insects in the Agricultural production enterprise "Tumul" on July 23, 2020, the following data were received: During the mass flight in the morning and evening peaks, the number of mosquitoes for a 5-min count ranged from 136 to 220 (on average 178) specimens. Mass flight of horseflys was observed from late June to mid-July, up to 51–83 (on average 67) horseflys attacked the bait cow in a 15-min count. After the treatment of cows with ultra-low-volume spraying, the protection of cows from the attack of blood-sucking dipteric insects and myiasis gadfly was maintained for 6 h, starting from the first hour of treatment (Table 1).

Table 1. Evaluation of the economic efficiency of cow anti-cow protection.

No.	Indicators	Unit of measurement	Per herd of cows
1	Damage from reduced productivity of cows during acute entomosis, from attacks of harmful insects	₽	1533168
2	Cost of processing a herd of cows in 2020 prices	₽	46578
3	Duration of protective action against insect attacks	Hour	6
4	Estimated annual economic benefit	₽	1486590
5	Economic effect for 1 ₽ costs	₽	32.92

A similar experience in studying the ecology of the harmful insects components and testing the effectiveness of the use of ultra-low-volume spraying of cattle from blood-sucking dipteric insects and myiasis gadfly attack was carried out in parallel at "Khorobut" farm on August 01, 2020. The following data were obtained: in the morning and evening peaks, the number of mosquitoes for a 5-min count ranged from 63 to 172 (on average 117) specimens. Mass flight of horseflys was observed from late June to mid-July. The bait cow was attacked by up to 57–62 (on average 59) horseflys in a 15-min count. After treatment of cows with ultra-low-volume spraying, the protection of cows of blood-sucking dipteric insects was maintained for 6 h.

On the basis of the experimental studies, the authors have developed a technology for protecting cattle from blood-sucking dipteric insects and adult gadflies, protected by RF patent No. 2724462. An assessment of the economic efficiency of the technology for protecting cattle from blood-sucking insects in Yakutia is presented in Table. The economic effect per 1 ruble of costs is 32.92 rubles.

4 Discussion

Blood-sucking and non-blood-sucking dipteric insects are widespread in the Arctic, Subarctic, Central and Southern climatic zones of the Republic of Sakha (Yakutia). On the basis of the experimental studies, the authors have developed a technology for protecting cattle from blood-sucking dipteric insects and adult gadflies, protected by RF patent No. 2724462. The economic effect per 1 ruble of costs is 32.92 rubles.

The technical result of solving the problem is solved by the fact that the liquid substance is transferred to an aerodispersed state, which leads to a sharp increase in the specific surface area. Due to this, the molecules of the dispersed substance are located in the outer layer of the micro-aerosol particle, which causes a sharp increase in the activity of substances, accelerating their physical and chemical processes.

5 Conclusions

A technology has been developed for the protection of cattle from blood-sucking dipteric insects and myiasis gadfly of cattle in Yakutia with an ultra-small amount of a preparation with high environmental friendliness, efficiency, insecticidal, repellent and anti-stress treatment of animals. The developed technology for the protection of cattle from dipterans and myiasis gadfly as an object of industrial property is protected by RF patent No. 2724462.

References

1. Decree of the President of the Russian Federation dated 21 January 2020. No. 20, On the approval of the Food Security Doctrine of the Russian Federation. http://kremlin.ru/acts/bank/45106/page/1
2. Kalyn, M., Collard, M.D., David, P., McCormick, M.D.: A nutritional comparison of cow's milk and alternative milk products. Acad. Pediatr. **21**(6), 1067–1069 (2020). https://doi.org/10.1016/j.acap.2020.12.007
3. Marangoni, F., Pellegrino, L., Verduci, E., Ghiselli, A., et al.: Cow's milk consumption and health: a health professional's guide. J. Am. Coll. Nutr. **38**(3), 197–208 (2018). https://doi.org/10.1080/07315724.2018.1491016
4. Aruna, P., Chakravarty, A.K.: Disease-resistant livestock populations-their development and impact on farm economics. In: Genetics and Breeding for Disease Resistance of Livestock, vol. 22, pp. 351–358 (2020). https://doi.org/10.1016/B978-0-12-816406-8.00022-X
5. Bett, B., Randolph, T.F., Irungu, P., Nyamwaro, S.O., et al.: Field trial of a synthetic tsetse-repellent technology developed for the control of bovine trypanosomosis in Kenya. Prev. Vet. Med. **97**(3–4), 220–227 (2010). https://doi.org/10.1016/j.prevetmed.2010.09.001
6. Votýpka, J., Mihalca, A.D.: Horn flies (Haematobia and Haematobosca). In: Reference Module in Biomedical Science (2020). https://doi.org/10.1016/B978-0-12-818731-9.00021-5
7. Munstermann, L.E.: Phlebotomine sand flies and moth flies (psychodidae). In: Medical and Veterinary Entomology, 3rd edn., vol. 12, pp. 191–211 (2019). https://doi.org/10.1016/B978-0-12-814043-7.00012-1
8. Benelli, G., Canale, A., Lucchi, A., Di Giovanni, F.: Insects and mites of medical and veterinary importance: a broad overview. In: Reference Module in Biomedical Science (2020). https://doi.org/10.1016/B978-0-12-818731-9.00040-9

9. Haroon, A., Afzal, M.S., Mobeen, M., Simsek, S.: An overview on different aspects of hypodermosis: current status and future prospects. Acta Trop. **162**, 35–45 (2016). https://doi.org/10.1016/j.actatropica.2016.05.016
10. Polley, L., Kutz, S.J., Hoberg, E.P.: Parasite zoonoses. In: Encyclopedia of Environmental Health, 2nd edn., pp. 9–26 (2011). https://doi.org/10.1016/B978-0-444-63951-6.00677-X
11. DairyNews.ru. Forecast: Milk production in Russia will grow to 31.8 million tons in 2021 (2021). https://www.dairynews.ru/news/prognoz-proizvodstvo-moloka-v-rossii-vyrastet-v-20.html
12. Yakutia 24 Online Edition. In Yakutia, there are more than 190 thousand head of cattle and over 181 thousand horses (2021). http://yakutia24.ru/news/obshchestvo/54604-v-yakutii-naschityvaetsya-bolee-190-tysyachi-golov-krupnogo-rogatogo-skota-i-svyshe-181-tysyachi-loshadej
13. Barashkova, A.I., Reshetnikov, A.D.: Research for the practice of means and methods of protecting farm animals from the attack of diptera blood-sucking insects in Yakutia. Vestnik Buryat St. Acad. **3**(36), 7–13 (2014)
14. Rzhanikov, S.N.: Efficiency and safety of deltsid in the treatment of cattle against blood-sucking insects: Abstract cand thesis in vet sci Tyumen 1–23 (2005)
15. Thorp, J.H., Rogers, D.C.: Midges, mosquitoes, blackflies, and other true flies: insect order diptera. Fiel. Guid. Freshwat. Inverteb. Nor. Am. **27**, 247–260 (2011). https://doi.org/10.1016/B978-0-12-381426-5.00027-2
16. Mihalca, A.D., Votýpka, J.: Horse flies (Diptera: Tabanidae). In: Reference Module in Biomedical Science (2020). https://doi.org/10.1016/B978-0-12-818731-9.00007-0
17. Kayedi, M.H., Sepahvand, F., Mostafavi, E., Chinikar, S., et al.: Morphological and molecular identification of Culicidae mosquitoes (Diptera: Culicidae) in Lorestan province, Western Iran. Heliyon **6**(8), e04480 (2020). https://doi.org/10.1016/j.heliyon.2020.e04480
18. Detinova, T.S., Rasnitsyn, S.P., Markovich, N.Ya., et al.: Unification of the methods for counting the number of blood-sucking dipteric insects. Med. Parasitol. Parasitic Dis. **47**(5), 84–92 (1978)
19. Ministry of Agriculture of the Russian Federation: State catalog of pesticides and agrochemicals approved for use on the territory of the Russian Federation. Part 1: Pesticides. 1–795 (2021). https://mcx.gov.ru/ministry/departments/departament-rastenievodstva-mekhanizatsii-khimizatsii-i-zashchity-rasteniy/industry-information/info-gosudarstvennaya-usluga-po-gosudarstvennoy-registratsii-pestitsidov-i-agrokhimikatov/

Features of the New Mid-season Soybean Variety Zolushka

Evgeniya Fokina(✉) and Galina Belyaeva

Federal State Budget Scientific Institution Federal Research Center All-Russian Scientific Research Institute of Soybean, Ignat'yevskoe shosse, 19, 675000, 675027 Blagoveshchensk, Amur Region, Russia

fem@vniisoi.ru

Abstract. The needs of agricultural production determine the direction of breeding research. The goal is to create a new competitive soybean variety adapted to local growing conditions, combining high potential productivity, resistance to major diseases, pests and low temperatures during germination. The work on creating a new variety was carried out in 2004–2016 in the selection crop rotation in the All-Russian Research Institute of Soybeans, using the classical scheme of the selection process for self-pollinating crops. As a result of long-term selection work in the conditions of the Amur Region, a new high-yielding mid-season soybean variety Zolushka with a potential yield of 3.52 t/ha was created. The new variety is characterized by resistance to low positive temperatures during the germination period (+4… 7 °C), at the optimal sowing time it exceeds the Dauria standard by 0.28 t/ha, at early dates on May 7 and 14 by 0.53 and 0.48 t/ha, respectively. It has complex resistance to pathogens and pests, provides a high yield of crude protein and fat from a unit area, exceeding the standard by 11.2%. The Zolushka variety is suitable for use for food purposes, well adapted to the difficult natural and climatic conditions of the Amur Region, its industrial use would increase the profitability of the crop cultivation in the Far East region, which would provide potential commercial attractiveness for agricultural producers and soybean processors.

Keywords: Soybeans · New variety · Standard · Growing season · Hybridization · Yield · Cold resistance · Selection

1 Introduction

The main goal of agriculture is to increase grain production. The key way to achieve it is to increase yields through the use of new zoned varieties with improved qualities and properties. The variety is the genetic basis of yield, the main product of selection, which is the principal means of producing cheap products. The increase and stabilization of productivity in a particular region of cultivation depends on the set of varieties, since the biological characteristics of a variety, as a rule, always correspond to the ecological resources in which it was created [1–3].

At the present stage, the selection of domestic competitive varieties is of particular relevance, taking into account the goals of large-scale import substitution in the modern

A. Muratov and S. Ignateva (Eds.): AFE 2021, LNNS 353, pp. 105–112, 2022.
https://doi.org/10.1007/978-3-030-91402-8_13

conditions of the existence of an expanded variety of foreign agricultural crops on the Russian market. At the same time, increasing the efficiency of agricultural production is inextricably linked with the further systematic improvement of the domestic assortment - one of the leading elements of modern innovative technologies [4–6].

The varietal component is of fundamental importance for increasing the efficiency of the soybean industry in the Russian Federation. Due to the increasing requirements for new varieties on the part of production, breeding tasks become more complicated and the requirements for the degree of study of the source material increase [1, 7, 8]. The potential productivity of varieties as well as their resistance and tolerance to pests and to extreme environmental factors is gaining in importance [9–11]. In recent years, special attention has been paid to the development of soybean varieties with high adaptive potential in combination with increased cold resistance. In this connection, the breeding material is tested for the possibility of germination under conditions of low positive temperatures of soil and air (4… 7 °C) at early terms of field sowing [12–14]. At the same time, to create varieties as parental elements, as a rule, local forms, well adapted to the specific soil and climatic conditions of a particular region, as well as soybean samples from the VIR genetic collection are used [9, 15].

As is known, the Amur Region is a zone of risky agriculture with an unstable hydrothermal regime, early autumn frosts, which negatively affect the quality of seeds of mid-ripening and late-ripening soybean varieties. In this regard, in order to carry out early sowing of seeds and obtain full maturation of plants before frost it becomes necessary to create varieties resistant to low positive temperatures during germination.

The goal is to create a new competitive soybean variety adapted to local growing conditions, combining high potential productivity, resistance to major diseases, pests and low temperatures during germination.

2 Conditions, Materials and Methods

The work on creating a new variety was carried out in 2004–2016 in the selection crop rotation in the All-Russian Research Institute of Soybeans, Sadovoe vil., Tambov region, using the classical scheme of the selection process for self-pollinating crops. The cultivation of soybeans was carried out according to the technology developed for the southern agricultural zone of the Amur region*. The soil of the experimental plot is meadow chernozem-like, medium-thick. Tillage included autumn plowing. Before sowing, disking, the application of mineral fertilizers (ammophos N12P52) - 0.10 t/ha, followed by cultivation, the introduction of soil herbicide Frontier Optima (1.2 l/ha) were carried out. Selection in hybrid nurseries was carried out according to the pedigree method with tracing the pedigree of the offspring. At the final stage of the breeding process, the material was studied in nurseries for preliminary and competitive variety testing (PVT, CVT). Sowing of CVT was carried out in 4-fold repetition, by the method of randomized repetitions (blocks), standard varieties were sown in each block using the SN-P-16 seeder. Mid-season soybean samples were compared with the Dauria variety. The area of the CVT plots was 40.5 m^2. Phenological observations, assessments, counts during the entire growing season were carried out according to the methodology of the State variety testing of agricultural crops. In the period of flowering and ripening, roguing was performed. The harvesting of the plots was carried out by continuous threshing. The yield of the varieties was determined in terms of the standard moisture content of seeds (14%) according to GOST 12041–182, the weight of 1000 seeds - according to GOST 12042–80. Mathematical processing of the data was carried out according to the method of B.A. Dospekho, description of varieties - according to the methodology of the SVT and VIR. The content of protein and oil in seeds was determined using an IR analyzer "FOSS NIRSystem 5000". Evaluation of the damage to varieties and variety samples by diseases during the period of mass flowering was carried out against a natural infectious background. To test samples of CVT and soybean varieties for resistance to low positive temperatures at different sowing dates - May 7, 14 and 21 - the studies were carried out in 2014... 2018. Plot area was 1.35 m^2, replication was 4-fold. Sowing seeds was carried out by hand.

3 Results of the Study

A new variety of Zolushka soybeans, breeding number Am.2345, created as a result of intraspecific hybridization. The samples of soybeans from the VIR collection and the lines and varieties obtained on their basis adapted to the natural and climatic conditions of the Amur Region were used as the initial material (see Figure). At the final stage of crossing, the local soybean variety ♀MK-100 was used as the maternal form, and the variety of Chinese origin ♂Heihe 14 was used as the paternal form (Fig. 1).

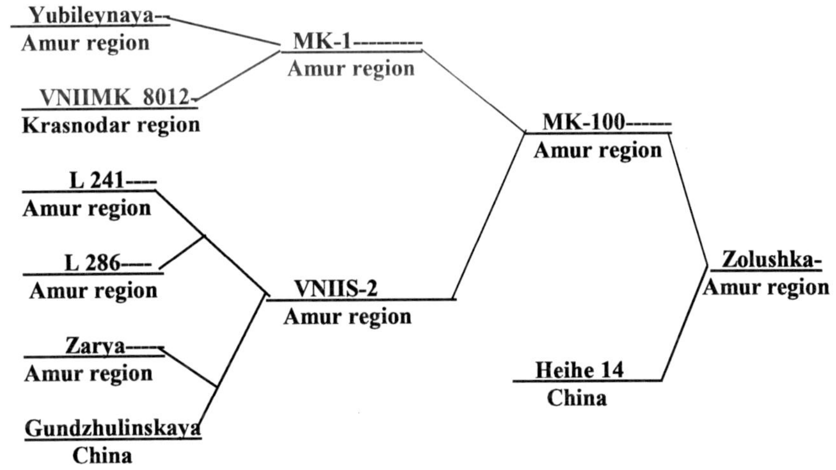

Fig. 1. Zolushka soybean pedigree.

The Zolushka soybean variety was created for 12 years. The crossing of parental components was carried out in 2004, followed by multiple individual selection and tracing the pedigree by offspring (pedigree method). In 2005–2010 hybrid material was studied in F1... F6 generations nurseries. The constant form was isolated in the breeding nursery in 2011. The study of this form in the control nursery was carried out in 2012.

The study of the variety sample accession in the PVT - 2013, in the CVT - 2014… 2016. The Zolushka variety, according to the production classification adopted in the Amur Region, belongs to the group of mid-season varieties, the growing season is 112… 115 days, on average 114, and is intended for a zone with a sum of active temperatures of 2200… 2600 °C (see Table 1).

Table 1. Characteristics of the Zolushka soybean variety based on the results of competitive variety testing for commercially valuable traits, 2014–2016.

Indicators	Zolushka soybean variety				Dauria soybean variety (standard)			
	Year				Year			
	2014	2015	2016	3-year average	2014	2015	2016	3-year average
Seed yield at standard moisture content, t/ha (LSD_{05} - 0.16 t/ha)	2.50	3.06	3.28	2.95	2.34	2.94	2.73	2.67
Vegetation period, days	112	115	115	114	103	106	110	106
Plant height, cm	76	71	81	76	67	68	71	69
Height of attachment of the lower beans, cm	13	11	17	14	11	9	9	10
Weight of 1000 seeds, g	150.9	177.2	159.6	162.6	154.9	180.2	167.6	167.6
Protein content in seeds,%	37.1	37.6	39.3	38.0	37.9	36.1	38.8	37.6
Gross crude protein yield, t/ha	0.80	0.99	1.11	0.96	0.76	0.91	0.91	0.86
Fat content in seeds,%	18.6	17.8	18.3	18.2	20.4	19.3	19.0	19.6
Gross fat yield, t/ha	0.40	0.47	0.52	0.46	0.41	0.49	0.45	0.45

The Zolushka variety is recommended for cultivation in the Far East (12) region. Over the years of study in the nursery of competitive variety testing, the yield of seeds of the Zolushka (Cinderella) variety was 2.50… 3.28 t/ha (average 2.95 t/ha), exceeding the standard Dauria variety by 0.28 (0.12… 0.55) t/ha. The grain productivity potential of the Zolushka variety in 2018 at the demonstration plot reached 3.52 t/ha. The variety is characterized by a semi-determinant type of growth, the shape of the bush is erect (compact), does not lodge. The stem is straight, it mainly forms one branch with a completed apex. The height of the plants is 76 (71… 81) cm. The height of attachment of the lower beans is 14 cm, and in different years it varied from 11 to 17 cm. The leaves are trifoliate, medium-sized, lanceolate. Inflorescence is a raceme, the number of flowers in a raceme is 13… 15, in the middle part of the stem there are 2 racemes (sessile and on a short stalk), the apical raceme contains up to 16 purple flowers. Beans are slightly curved, almost straight, gray in color. The pubescence of plants is gray, of medium density. Seeds are yellow, slightly shiny, spherical-flattened, the surface of the seeds is smooth. The hilum is short, oval, seed-colored (yellow). The mass of 1000 seeds is 150.9… 177.2 g, the average is 162.6 g. The content of protein in seeds is 38.0% (37.1… 39.3%), content of fat is 18.2% (17.8… 18.6%). Plants of this variety accumulate an increased percentage of protein in the seeds (0.4% more than the standard). At the same time, the gross yield of both protein and fat is 0.1 t/ha, or 11.2% higher than the yield of the Dauria variety. Laboratory analysis showed that light-colored vegetable oil and soy milk distinguished by good taste were obtained from Zolushka grains. In the process of creating a new variety, studies were carried out to determine resistance to low temperatures during germination (see Table 2). Sowing of soybeans was performed in the first, second and third decades of May, on May 7, 14 and 21, respectively.

Table 2. Productivity of soybean varieties at different sowing dates, 2014… 2018.

Variety	Yield, t/ha (average)				Vegetation period, days (average)			
	May 7	May 14	May 21	Average on sowing periods	May 7	May 14	May 21	Average on sowing periods
Dauria - st	2.34	2.65	2.89	2.62	116	112	111	113
Zolushka	2.87	3.13	3.22	3.07	119	116	115	117
Deviation from standard, ± Δ	+0.53	+ 0.48	+ 0.33	+0.44	+3	+4	+4	+3.6
%	22.6	18.1	11.4	17.3	–	–	–	–
LSD_{05}	0.17	0.21	0.15					

The optimal sowing time in the Amur region for mid-season varieties is the period from 20 to 25 May. Conducted in 2014… 2018 studies have shown that early sowing dates in the first and second decades of May were favorable for the Zolushka variety. With the sowing date on May 7, the excess over the standard averaged 0.53 t/ha, with the sowing on May 14–0.48 t/ha. Vegetation period at the first sowing date on May 7 for varieties increased by 3… 5 days in comparison with the second and third sowing dates. This is a slight increase as the sowing was performed 14 and 7 days earlier and harvesting started earlier, respectively.

According to the results of the immunological assessment, it was found that the new Zolushka soybean variety is characterized by complex resistance to the main disease-causing pathogens widespread in this region (Table 3).

Testing for resistance to the most harmful, leaf and root forms of diseases has shown that Zolushka variety is characterized by high resistance to bacteriosis (Bacterium glycineum Coerper), cercosporosis (Cercospora kikuhii), phylostictosis (Fhyllosticta soyaecola Macc), root rot (Fusarium); resistant to septoria (Septoria glycines Hemmi).

Table 3. Characteristics of the Zolushka soybean variety for resistance to pathogens common in the Amur Region, 2014–2016.

Name of pathogens	Degree of defeat affection							
	Zolushka soybean variety				Dauria soybean variety (standard)			
	Year				Year			
	2014	2015	2016	Average over 3 years	2014	2015	2016	Average over 3 years
Ascochytosis *(Ascochyta sojaecola Abramov)*	0	0	0	0	0	0	0	0
Cercosporosis *(Cercospora kikuhii)*	0	0	1.9	0.6	0	0.8	7.5	2.8
Root rot *(Fusarium)*	5.2	12.6	9.2	9.0	15.0	20.8	18.9	18.2
Septoria *(Septoria glycines Hemmi)*	21.0	14.2	13.2	16.1	21.5	17.3	22.1	20.3
Phylostictosis (olive spot) *(Phyllosticta soyaecola Macc)*	14.0	5.0	6.2	8.4	9.0	3.3	10.4	7.6
Bacteriosis *(Bacterium glycineum Coerper)*	8.0	5.0	6.3	6.4	11.0	10.8	24.0	15.3
Peronosporosis or downy mildew *(Perenospora manchurika N. Naumov)*	0	0	0.3	0.1	6.0	0	7.2	4.4
Soybean moth *Leguminivora glycinivorella Mats*	1.2	1.3	2.9	1.8	3.5	2.7	6.4	4.2

During the years of study, the affection of this variety by peronosporosis (Perenospora manchurika) was insignificant, and scochitosis (Ascochyta sojaecola Abramov) was absent. Damage to seeds by the soybean moth (Leguminivora glycinivorella Mats) is 2.3 times lower than that of the Dauria standard.

The Zolushka soybean variety is suitable for mechanized cultivation at a seeding rate of 500 thousand viable seeds per 1 hectare, row spacing 15, 30, 45 cm, sowing period from May 7 to 25. The variety is resistant to waterlogging and lodging. In 2017–2018 the Zolushka soybean variety passed the State variety testing at 7 state seed-trial grounds (SSTG) of the Far East region - Tambov, Svobodnensky, Mazanovsky (Amur Region), Vyazemsky (Khabarovsk Krai)), Amur (Jewish Autonomous Region), Kirovsky, Ussuriysky (Primorsky Krai). Following the results of the SVT, it was included in the State Register of Breeding Achievements for use in production (patent No. 10251 dated April 26, 2019). Since 2020, the seeds of a new variety have been multiplied in primary seed nurseries for its introduction into production.

4 Conclusion

A new high-yielding mid-season soybean variety Zolushka has been created with a vegetation period of 112… 115 days, a potential yield of 3.52 t/ha, which meets the technological requirements for plant height (71… 81 cm), attachment of lower beans (11… 17 cm), uniformity of ripening, resistance to lodging, and it is suitable for mechanized cultivation. The new variety is characterized by resistance to low positive temperatures during the germination period, at the optimal sowing time it exceeds the Dauria standard by 0.28 t/ha, at early dates on 7 and 14 May by 0.53 and 0.48 t/ha, respectively. In addition, it has a complex resistance to diseases and pests common in the region, is distinguished by an increased protein content in seeds (+0.4% to st) compared to the standard, and provides a high yield of crude protein and fat per unit area (+11.2% to st). The Zolushka variety is suitable for use for food purposes, well adapted to the difficult conditions of the Amur Region. Based on the results of the State Variety Testing, it was included in the State Register of Breeding Achievements for use in the Far Eastern Region (patent No. 10251 dated April 26, 2019). The commercial use of the Zolushka variety would increase the profitability of crop cultivation, which would provide potential commercial attractiveness for agricultural producers and soybean processors. In the laboratory of soybean breeding and genetics of the Federal State Budgetary Scientific Institution of the Federal Research Center of All-Russian Research Institute of soybeans, the Zolushka variety is included in the scheme of future crosses as a source of cold resistance, high yield, complex immunity to disease-causing pathogens.

References

1. Akpertey, A., Singh, R.J., Diers, B.W., et al.: Genetic introgression from Glycine tomentella to soybean to increase seed yield. Crop Sci **58**(3), 1277–1291 (2018). https://doi.org/10.2135/cropsci2017.07.0445
2. Tolokonnikov, V.V., Koshkarova, T.S.: New soya variety Volgogradka. Nauchnyi zhurnal Rossiiskogo NII problem melioratsii **11**(1), 14–23 (2021). https://doi.org/10.31774/2222-1816-2021-11-1-14-23

3. La, T., Large, E., Taliercio, E.W., et al.: Characteristics of selected wild soybean samples in the USDA germplasm collection for seed composition and agronomic traits. Crop Sci **59**(1), 233–251 (2019). https://doi.org/10.2135/cropsci2017.08.0514
4. Sinegovskaya, V.T., Fokina, E.M.: Soybean Breeding as a Tool for Solving Import Substitution Problems in the Far Eastern Federal District. Trudy Kubanskogo gosudarstvennogo agrarnogo universiteta **72**, 328–331 (2018). https://doi.org/10.21515/1999-1703-72-328-331
5. Kashevarov, N.I., Polyudina, R.I., Potapov, D.A.: Soybean breeding at the Siberian Research Institute of Forages. Dostizheniya nauki i tekhniki APK **34**(8), 28–32 (2020). https://doi.org/10.24411/0235-2451-2020-10804
6. Abugalieva, A.I., Didorenko, S.V.: Genetic diversity of soybean cultivars belonging to different ripeness groups with regard to performance and quality. Vavilov J of Gen and Breed **20**(3), 303–310 (2016). https://doi.org/10.18699/VJ16.168
7. Ru, S., Bernardo, R.: Predicted genetic gains from introgressing chromosome segments from exotic germplasm into an elite soybean cultivar. Theor Appl Genet **133**, 605–614 (2020). https://doi.org/10.1007/s00122-019-03490-2
8. Butovets, E.S., Vasina, E.A., Luk'yanchuk, L.M.: Screening of soybean germplasm under conditions of the Primorsky Territory. Dostizheniya nauki i tekhniki APK **34**(8), 23–27 (2020). https://doi.org/10.24411/0235-2451-2020-10803
9. Omelyanyuk, L.V., Asanov, A.M., Yusova, O.A., Churakov, A.A.: New early maturing soybean cultivar Zaryanitsa for conditions of Siberia. Maslichnye kul'tury **3**(183), 161–168 (2020). https://doi.org/10.25230/2412-608X-2020-3-183-161-168
10. Pryanishnikov AI, Savchenko IV, Mazurov VN (2018) Adaptive breeding: theory and practice of selection for productivity. Vestnik RASKHN 3:29–32. doi: https://doi.org/10.30850/vrsn/2018/3/29-32
11. Haffke, S., Wilde, P., Schmiedchen, B.: Toward a selection of broadly adapted germplasm for yield stability of hybrid rye under normal and managed drought stress conditions. Crop Sci **5**(3), 1026–1034 (2015). https://doi.org/10.2135/cropsci2014.08.0532
12. Zelentsov, S.V., Moshnenko, E.V., Bubnova, L.A., et al.: A cold-resistant soybean cultivar of the northern ecotype Barguzin. Maslichnye kul'tury **1**(181), 132–139 (2020). https://doi.org/10.25230/2412-608X-2020-1-181-132-139
13. Li, M., Liu, Y., Wang, C., et al.: Identification of Traits Contributing to High and Stable Yields in Different Soybean Varieties Across Three Chinese Latitudes. Front Plant Sci **10**, 1642 (2020). https://doi.org/10.3389/fpls.2019.01642
14. Zelentsov SV, Moshnenko EV, Bubnova LA (2019) Justification of the criteria of soybean selection for increased frost resistance based on the phenomenon of cytocolloids cryogenic sedimentation. Maslichnye kul'tury 1(177):128–143. https://doi.org/10.25230/2412-608X-2019-2-178-128-143
15. Fokina, E.M., Belyaeva, G.N., Razantsvei, D.R.: Characteristic collection of soybeans as a basis for creating new generation varieties. Vestnik DVO RAN **4**, 86–92 (2020). https://doi.org/10.37102/08697698.2020.212.4.014

Study of Anthocyanins in Tubers of Potato Hybrids (Solanum Tuberosum L.) of Primorsky Krai Selection

Irina Kim(✉), Valentina Vozniuk, Dmitriy Volkov, and Aleksey Klykov

Federal Scientific Center of Agricultural Biotechnology of the Far East named after A.K Chaika, 30 Volozhenin Str., Timityazevskiy Stl., 692539 Ussuriysk, Primorsky Krai, Russia

Abstract. In the presented study, using liquid chromatography and second-order mass spectrometry methods, anthocyanins were identified and their content was determined in promising potato hybrids bred by the FSBSI "*FSC of Agricultural Biotechnology of the Far East named after A.K Chaika*". This work aimed to determine the qualitative composition and quantitative content of anthocyanins in tubers of potato hybrids for cultivation and selection in the Russian Far East. Anthocyanin profiles were analyzed by the method of double identification: by the retention time on the reverse phase column C18 with detection in the wavelength range of 400–700 nm and by the method of second-order mass spectrometry in the positive ionization mode with subsequent fragmentation of ions. As a result of second-order mass spectrometry in tubers of potato hybrids in the determination of anthocyanins, it was found that the most common were petunidin-3-arabinoside and pelargonidin-3-glucoside: they were identified in all studied samples. The highest content of anthocyanins was noted in the varieties with purple peel and yellow pulp with pigment spots: Pri-15-12-14 and Pri-14-52-2. Samples with pink skin of tubers were distinguished by a high content of pelargonidine-3-glucoside (98.2–190.4 mg/kg).

Keywords: Solanum tuberosum L. · Potatoes · Anthocyanins · Delphinidin · Petunidin · Mal-vidin · Cyanidin · Pelargonidin · Mass spectrometry · HPLC

1 Introduction

Potatoes are an important food crop valuable as a food product containing antioxidant substances such as anthocyanins. Due to its high prevalence, pigmented potatoes can serve as a source of natural anthocyanins [1]. Anthocyanins (from Greek anthos - color, and kyanos - blue) or anthocyans are an extensive group of water-soluble plant pigments that determine red, blue and purple colors of fruits, flowers, leaves and other parts of plants. Anthocyanins belong to flavonoids and are glycosides of flavilia cations - anthocyanidins [2].

Anthocyanidins are rarely found in free form in nature. In flowers, fruits and other parts of plants, they are in the form of glycosides - anthocyanins. According to literature data, more than 600 natural anthocyanins have been identified [3]. The glycoside

A. Muratov and S. Ignateva (Eds.): AFE 2021, LNNS 353, pp. 113–121, 2022.
https://doi.org/10.1007/978-3-030-91402-8_14

part contains mono-, di- or trisaccharides. Monosaccharides are represented by hexoses (galactose, glucose, rhamnose, etc.) and pentoses (xylose, arabinose, etc.). Di- and trisaccharides are formed mainly from monosaccharides.

Within the limits of the genetic diversity of potatoes, it is possible to choose from selection samples rich in red, violet and blue pigments that provide color and act as antioxidants [4]. Red and purple tuber pulp contains the anthocyanins pelargonidin, malvidin, petunidin, peonidin, and delphinidin, which provide tissue color [5]. According to literature data, the content of anthocyanins in potatoes varies from 5.5 to 35 mg/100 g of dry mass [6].

Anthocyanins are involved in the protection of the photosynthetic apparatus, neutralize free radicals, increase the efficiency of assimilation of phosphorus and nitrogen, enhance the osmoregulatory function, and reduce the freezing point of cell contents. These properties allow plants to adapt to unfavorable conditions - excessive UV radiation, drought, extreme temperatures, soil salinity, deficiency of phosphorus and nitrogen, toxic effects of herbicides, and heavy metal ions [7, 8]. Anthocyanins are known to have antimicrobial activity. For example, they contribute to potato plants resistance to wet rot: on the sections of tubers with purple pulp, the lesion area was, on average, 28.6% less than that of potatoes with yellow pulp [9].

Anthocyanin pigments are used in the food and pharmaceutical industries. They belong to water-soluble natural dyes - antioxidants that can color food in different shades of red. According to numerous literature data, it is known that anthocyanins have capillary-strengthening, antioxidant and anticarcinogenic properties and are used in medicine for the treatment and prevention of type II diabetes mellitus and some forms of tumors [10].

The development of potato varieties with antioxidant properties is of great social importance, since the availability of potatoes makes it possible to provide a valuable dietary product to the general population. Breeding work on the breeding of table dietary varieties has been carried out since the beginning of the new millennium, and varieties with a high content of carotenoids and anthocyanins, distinguished by red and purple pulp of tubers, have already been created [11].

Scientists of the FSBSI "FSC of Agrobiotechnology of the Far East named after A.K. Chaika" are working on creating hybrids with colored pulp [12]. Samples with purple flesh, pink and purple rind have been created. In this regard, the purpose of this work is to determine the composition and quantitative content of anthocyanins in tubers of potato hybrids for cultivation and breeding in the Russian Far East.

2 Materials and Methods

Vegetation and field studies were carried out in 2019–2021 at the breeding plots of the FSBSI "FSC of Agrobiotechnology of the Far East named after A.K. Chaika". Five hybrids of Primorsky Krai selection - Pri-14-52-2 Lomonosovsky × Purple potato, Pri-15-12-14 Purple potato × Manifest, Pri-12-35-4 Debryansk × Mustang, Pri-15-41-8 Russkaya krasavitsa × Irbitsky, At-15-7-16 Irbitsky × Avrora – were the object of the reseearch.

The study of tuberous reproductions of seedlings and hybrids was carried out in the field following the selection process plan. During the growing season of plants, phenological observations were carried out, the degree of viral damage by fungal and viral diseases was determined. The yield was counted by weighing tubers from the plot. In laboratory conditions, a tasting assessment, biochemical analysis of tubers (content of dry matter, starch, protein, vitamin C, reducing sugars) were carried out. During the storage period of the hybrids (October-June), the tuber storage capacity was assessed.

Testing was based on the methods of the Federal Research Center N.I. Vavilov All-Russian Institute of Plant genetic Resources [13] and Lorch Potato Research Institute [14].

The composition and accumulation of anthocyanins were analyzed according to C.E. Lewis et al. [15]. To determine the content of anthocyanins, we used pulp tissue and tuber skin. The obtained sample material was stored until analysis in a cool place without direct sunlight. The samples were rinsed in cold water, weighed, crushed and filled with a solution (40% ethyl alcohol and 1% formic acid (5 g of crushed mass + 25 ml of the resulting solution). The mass was subjected to freezing-thawing, as well as ultrasonic disintegration to destroy the walls and membranes of cells and organelles Anthocyanins were extracted for 90 min at 40 °C in a closed vessel to prevent the access of atmospheric oxygen. The extract was centrifuged (CM-6M, Elmi, Latvia) at 3500 g for 30 min, the supernatant was filtered through syringe filters (size pore 0.45 μm). The extract was stored in a freezer at a temperature of 20 °C for one week.

Anthocyanins were separated by HPLC on a liquid chromatograph with an LC-20AD high-pressure gradient pump and a CTO-20A column thermostat (Shimadzu, Japan). Chromatography was carried out on a Shodex C18-4E reverse phase column (250½4.6 mm), the sorbent grain diameter was 5 μm (Shodex, Japan), the temperature was 50 °C, the mobile phase rate was 0.58 ml/min. Eluent A - acetonitrile (AppliChem GmbH, Germany), eluent B - 1% formic acid solution (Sigma-Aldrich, USA). Gradient: 0.00–5.00 min - B concentration varied from 100% to 92%; 5.00–45.00 min - the concentration of B varied from 92% to 80%; 45.00–45.01 min - the concentration of B varied from 80% to 10%. Detection was carried out in the wavelength range of 300–600 nm with UV/VIS spectrophotometric detector SPD-20A (Shimadzu, Japan). For analysis, 5 μl of the extract was injected. The analysis was carried out in 15 biological and 3 analytical replicates.

Anthocyanins were identified by second-order mass spectrometry using an amaZon SL trap (Bruker, Germany) equipped with an electrospray ionization source. For this, anthocyanins isolated earlier by HPLC were analyzed by direct injection. Detection was carried out in the mode of positive and negative ions. The mass scanning range is from 150 to 2200 amu, the maximum scanning speed is 32000 Da/s, the voltage on the atomizing capillary is 4500 V, the pressure on the nebulizer is 29 psi, the dry gas flow is 10 l/s, the capillary temperature is 180 °C. The fragmentation of ions was carried out with an electron beam with 1.5 eV of energy. The analysis was carried out in 15 biological and 3 analytical replicates.

To prove the presence of significant differences between the means, one-way analysis of variance was used, followed by multiple comparisons of Fisher means (LSD method).

Table 1. Morphological and economic characteristics of potato hybrids.

Characteristic	Pri-14-52-2 Lomonosovsky × Purple potato	Pri-15-12-14 Purple potato × manifest	Pri-12-35-4 Debryansk × Mustang	Pri-15-41-8 Russkaya krasavitsa × Irbitsky	Pri-15-7-Irbitsky × Avrora
Vegetation period. days[1]	100	120	109	108	113
Productivity. t/ha	34.9±1.3	54.7±3.1	30.8±0.9	42.8±2.8	64.4±4.0
Marketable tuber mass. g	103.1±3.0	131.5±4.1	103.3±3.0	134.6±4.2	148.6±4.5
Tuber form	oval-elongated	oval-elongated	rounded oval	rounded oval	rounded oval
Eyes on tuber (depth)	deep	small	medium	medium	medium
Peel color	purple	purple	pink	pink	pink
Pulp color	yellow with purple pigmentation	yellow with purple pigmentation	yellow	white	yellow
Content of:					
Starch. %	16.1±0.1	17.8±0.1	14.2±0.1	17.5±0.1	15.3±0.1
ascorbic acid. mg/100g	39.3±0.4	13.4±0.1	18.9±0.2	24.9±0.3	18.3±0.2
Taste. points[2]	7.5	7.0	9.0	9.0	8.3
Resistance to:[3]					
Late blight. point	5.7	7.0	7.0	6.0	6.5
Early blight. point	7.5	8.0	6.3	7.0	8.0
Brown patch. point	9.0	9.0	9.0	9.0	9.0

1- Days between planting and collecting: >130 – very late ripening. 121-130 – late ripening. 91-120 – middle-late. 71-90 – mid-season-ripening. 61-70 –middle-early. 51-60 – early-ripening. <50 – very early ripening.

2 – Taste characteristics: 9 points – excellent. 7 points – good. 5 points – acceptable. 3 points – unpalatable. 1 point – poor.

3 – Late blight. early blight. and brown patch resistance : 9 points – extremely high. 8 points – high. 7 points – relatively high. 5 points – medium. 3 points – low. 1 point – extremely low.

3 Results and Discussion

The results of evaluating hybrid populations allowed identifying promising varieties from the nurseries of the main and competitive trials (Table 1).

Five isolated hybrids were characterized by increased productivity (30.8–64.4 t/ha), resistance to fungal pathogens (5.7–9.0 points), colored skin, good and excellent taste (7.0–9.0 points). The highest yield was observed in the sample Pri-15-7-16–64.4 t/ha. According to the number of days of physiological ripeness, the hybrids are assigned to the group of medium late (100–120 days from planting to wilting of the ground part of the plants). Hybrid Pri-15-7-16 was characterized by high mass of marketable tuber - 148.6 g. The studied samples had tubers of a round-oval and oval-elongated shape. The results of biochemical analysis showed the starch content in the tubers of the studied samples in the range of 14.2–17.8%. The starchiest hybrids were Pri-15-12-14 and Pri-15-41-8 (starch content 17.5 and 17.8%, respectively). The amount of ascorbic acid (vitamin C) in tubers varied from 13.4 to 39.3 mg/100 g. High content of vitamin C was found in hybrids Pri-14-52-2 - 39.3 mg/100 g and Pri-15- 41-8–24.9 mg/100 g.

Pigmented parts of plants are an important trait that can be effectively used in breeding for an increased content of anthocyanins [16, 17]. The studied hybrids differed in the morphological characteristics of the tubers. Their skin was pink and purple, pulp - white, yellow, and yellow with purple pigmentation. Two hybrids (At-14-52-2 and At-15-12-14) were characterized by purple skin and yellow pulp with purple pigmentation. As a result of the determination of the anthocyanin profile in tubers, five different anthocyanins were identified. Second-order mass spectrometry identified the isolated anthocyanins as delphinidin, petunidin, malvidin, cyanidin, and pelargonidin (Fig. 1; Table 2).

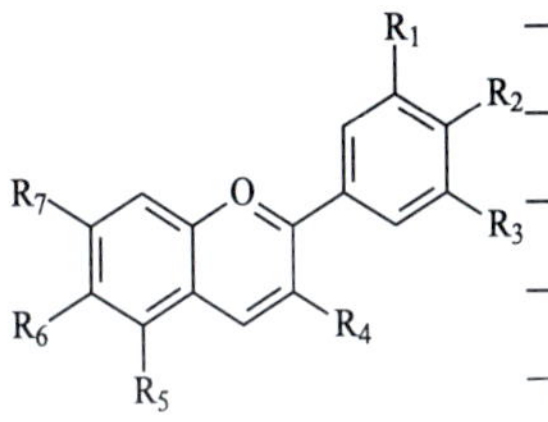

Anthocyanin	R_1	R_2	R_3	R_4	R_5	R_6	R_7
Delphinidin-3-glucoside	-OH	-OH	-Glc	-OH	-OH	-H	-OH
Petunidin 3-arabinoside	-OH	-OH	-Ara	-OH	-OH	-H	-OH
Malvidin-3-glucoside	-OCH3	-OH	-O-Glc	-OH	-OH	-H	-OH
Cyanidin-3-glucoside	-OH	-OH	-Glc	-OH	-OH	-H	-OH
Pelargonidin-3-glucoside	-H	-OH	-Glc	-OH	-OH	-H	-OH

Fig. 1. Structural formula of anthocyanins.

Petunidin-3-arabinoside and pelargonidin-3-glucoside were the most common anthocyanins: they were identified in all studied samples. According to literature data, petunidin is contained in parts of plants that have a purple color, and pelargonidine - red-orange [18]. As a result of the study, the presence of pelargonidin-3-glucoside and petunidin-3-glucoside was revealed in tubers with pink and purpleskin, but in different quantitative ratios (Table 3).

Violet skinned hybrids Pri-15-12-14 and Pri-14-52-2 were the highest in anthocyanins content. All five anthocyanins were found in the same samples. The noted samples were allocated a significant amount of petunidin-3-arabinoside - 310.0 and 360.0 mg/kg, respectively. Cyanidin-3-glucoside in the amount of 149.8 mg/kg and 54.2 mg/kg of delphinidin-3-glucoside were identified in skin of the Pri-15-12-14. Pelargonidin-3-glucoside (69.1 mg/kg) was detected in tubers of Pri-14-52-2. Samples with pink skin of tubers were distinguished by a high content of pelargonidine-3-glucoside (98.2–190.4 mg/kg). Pri-15-7-16 had the maximum amount of this anthocyanin, which was also characterized by an increased amount of petunidin 3-arabinoside (38.4 mg/kg) and cyanidin-3-glucoside (74,3 mg/kg).

Table 2. Anthocyanin profile of tuber skin in promising potato hybrids (*Solanum tuberosum* L.), 2019–2021.

Hybrid	Molecular ion [M + H]+	HPLC time, min	Anthocyanin
Pri-12-35-4	465,3; 303,2	12,5–13,0	Delphinidin-3-glucoside
Pri-15-41-8; Pri-12-35-4	493,3; 331,3	22,5–23,0	Malvidin-3-glucoside
Pri-15-7-16; Pri-15-12-14; Pri-14–52-2; Pri-15-41-8; Pri-12-35-4	433,3; 271,1	13,5–14,9	Pelargonidin-3-glucoside
Pri-15-7-16; Pri-15-12-14; Pri-14-52-2; Pri-15-41-8; Pri-12-35-4	479,3; 317,2	15,5–16,0	Petunidin 3-arabinoside
Pri-15-12-14	449,2; 287,2	10,1–10,8	Cyanidin-3-glucoside

Table 3. Content of anthocyanins (mg/kg) in promising potato hybrids (*Solanum tuberosum L.)* skin, 2019–2021.

Hybrid	Delphinidin-3-glucoside	Malvidin-3-glucoside	Pelargonidin-3-glucoside	Petunidin 3-arabinoside	Cyanidin-3-glucoside
Pink-skinned tubers					
Pri-15-7-16	1.2 ± 0.1	6.7 ± 0.1	190.4 ± 0.4	38.4 ± 0.1	74.3 ± 0.2
Pri-15-41-8	5.9 ± 0.1	3.9 ± 0.1	123.0 ± 0.3	14.8 ± 0.1	3.6 ± 0.1
Pri-12-35-4	2.7 ± 0.1	< 0.5	98.2 ± 0.2	5.0 ± 0.1	<0.5
Purple-skinned tubers					
Pri-15-12-14	54.2 ± 0.1	8.0 ± 0.1	3.9 ± 0.1	310.0 ± 0.5	149.8 ± 0.4
Pri-14-52-2	7.6 ± 0.1	1.3 ± 0.1	69.1 ± 0.1	360.0 ± 0.5	1.2 ± 0.1

When studying the pulp of tubers, petunidin and cyanidin were found in hybrids with purple pigmentation - Pri-15-12-14 and Pri-14-52-2. In samples with yellow and white colored pulp, anthocyanin substances were detected either in insignificant amounts, or their traces were found (Table 4).

Pri-15-12-14 and Pri-14-52-2 had 39.8 and 81.2 mg/kg of cyanidin-3-glucoside and 91.1 and 143.0 mg/kg of petunidin-3-arabinoside respectively, which is the highest of the studied samples. The same anthocyanins are found in the isolated hybrids skin. The tuber pulp tissue contained almost no phenolic substances in the three other hybrids.

Petunidin-3-arabinoside turned out to be the main component in the tubers of Pri-14-52-2. It is found in large quantities in skin and pulp of the sample (Fig. 2 and Fig. 3).

Table 4. Content of anthocyanins (mg/kg) in promising potato hybrids (*Solanum tuberosum L.)* pulp, 2019–2021.

Hybrid	Delphinidin-3-glucoside	Malvidin-3-glucoside	Pelargonidin-3-glucoside	Petunidin 3-arabinoside	Cyanidin-3-glucoside
White pulp					
Pri-15-41-8	<0.5	<0.5	<0.5	<0.5	<0.5
Yellow pulp					
Pri-15-7-16	<0.5	<0.5	<0.5	1.4 ± 0.1	2.3 ± 0.1
Pri-12-35-4	1.2 ± 0.1	<0.5	<0.5	<0.5	<0.5
Yellow pulp with purple pigmentation					
Pri-15-12-14	1.7 ± 0.1	1.0 ± 0.1	< 0.5	91.1 ± 0.3	39.8 ± 0.4
Pri-14-52-2	1.5 ± 0.1	1.1 ± 0.1	1.2 ± 0.1	143.0 ± 0.4	81.2 ± 0.3

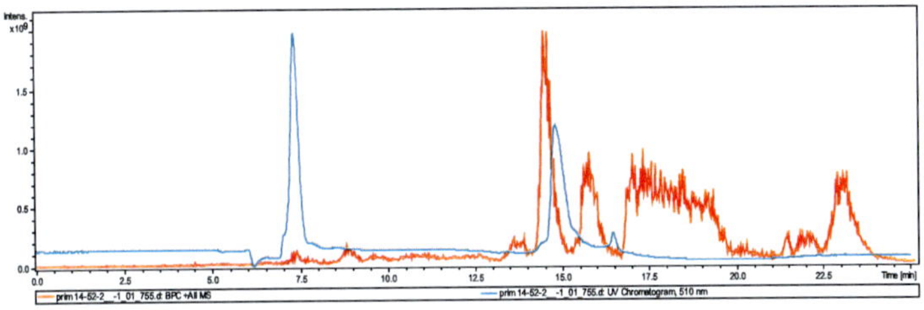

Fig. 2. Elution profile of anthocyanins found in the skin of potato (*Solanum tuberosum* L.) hybrid Pri-14-52-2.

During fixation of the detection, an intense signal was detected at a wavelength of 510 nm (retention time 15.5 min), which corresponds to the main anthocyanin petunidin-3-arabinoside of Pri-14-52-2, in addition, pelargonidin-3- glucoside was in significant amounts (peak at 14.9 min).

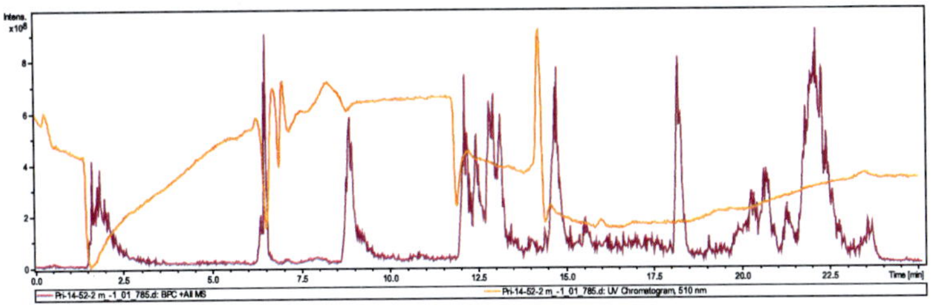

Fig. 3. Elution profile of anthocyanins found in the pulp of potato (*Solanum tuberosum* L.) hybrid Pri-14-52-2.

When analyzing the pulp of Pri-14-52-2, a peak with an intense signal was recorded at 14.3 min, which corresponds to petunidin-3-arabinoside. There was also a signal at 10.5 min, where the anthocyanin cyanidin-3-glucoside was identified. The elution profiles revealed connection between the content of specific anthocyanins and the color of tubers, as well as their belonging to a certain variety. Some publications confirm that the qualitative composition of anthocyanins depends on the characteristics of the genotype, which determine the activity of the corresponding enzymes that promote the synthesis of certain components of the anthocyanin complex [19].

4 Conclusion

As a result of second-order mass spectrometry, delphinidin, petunidin, malvidin, cyanidin, and pelargonidin were identified in the tubers of potato. Petunidin-3-arabinoside and pelargonidin-3-glucoside were the most common: they were identified in all studied samples. The highest content of anthocyanins was noted in the varieties with purple skin and yellow pulp with purple pigments Pri-15-12-14 and Pri-14-52-2. Pri-14-52-2 and Pri-15-12-14 were characterized by the fact that they had the same anthocyanins in skin and pulp. Samples with pink skin were distinguished by a high content of pelargonidine-3-glucoside - 98.2–190.4 mg/kg. The maximum amount of this anthocyanin was noted in Pri-15-7-16, as well as an increased content of petunidin-3-arabinoside (38.4 mg/kg) and cyanidin-3-glucoside (74.3 mg/kg). The selected potato hybrids are characterized by increased yield (30.8–64.4 t/ha), resistance to fungal pathogens (5.7–9.0 points), colored skin and pulp of tubers, good and excellent taste (7.0–9.0 points).

The research results showed that the peel and pulp color can be used as a trait in the selection of promising hybrids with an increased content of anthocyanins.

References

1. Jansen, G., Flamme, W.: Antioxidant activity of potatoes (Solanum tuberosum L.) and anthocyanin content, its biosynthesis and physiological role. Gen. Res. Crop. Evol. **53**, 1321–1331 (2006). https://doi.org/10.18619/2072-9146-2019-6-84-90
2. Tamimi, R.M., Lagiou, P., Adami, H.-O.: Prospects for chemoprevention of cancer. Trichopoulos J. Inter. Med. **251**, 286–300 (2002). https://doi.org/10.1046/j.1365-2796.2002.00969.x
3. Velioglu, Y.S., Mazza, G., Gao, L., Oomah, B.D.: Antioxidant activity and total phenolics in selected fruits, vegetables, and grain products. J. Agr. F Chem. **46**, 4113–4117 (1998). https://doi.org/10.1021/jf9801973
4. Brown, C.R., Durst, R.W., Wrolstad, R., De Long, W.: Variability of phytonutrient content of potato in relation to growing location and cooking method. Pot Res. **51**, 259–270 (2008).https://doi.org/10.1007/s11540-008-9115-0
5. Chalker-Scott, L.: Photochem. Invited review environmental significance of anthocyanins in plant stress responses. Photobiol **70**(1), 1–9 (1999). https://doi.org/10.1111/j.1751-1097.1999.tb01944.x
6. Hale, K.L., Tufan, H.A., Pickering, I.J., et al.: Anthocyanins facilitate tungsten accumulation in Brassica. Phys. Plant **116**(3), 351–358 (2002). https://doi.org/10.1034/j.1399-3054.2002.1160310.x
7. Khlestkina, E.: The adaptive role of flavonoids: emphasis on cereals. Cereal Res. Commun. **41**(2), 185–198 (2013). https://doi.org/10.1556/CRC.2013.0004
8. Kita, A., Bakowska-Barczak, A., Hamouz, K., Kułakowska, K., Lisińska, G.: The effect of frying on anthocyanin stability and antioxidant activity of crisps from red- and purple-fleshed potatoes (Solanum tuberosum L.). J. F. Comp. Anal. **32**(2), 169–175 (2013). https://doi.org/10.1016/j.jfca.2013.09.006

9. Cisowska, A., Wojnicz, D., Hendrich, A.B.: Anthocyanins as antimicrobial agents of natural plant origin. Nat. Prod. Commun. **6**(1), 149–156 (2011). PMID: 21366068
10. Sancho, R.A.S., Pastore, G.M.: Evaluation of the effects of anthocyanins in type 2 diabetes. Food Res. Inter. **46**(1), 378–386 (2012). ISSN: 0963–9969
11. Ezekiel, R., Singh, N., Sharma, S., Kaur, A.: Beneficial phytochemicals in potato—a review. Food Res. Int. **50**, 487–496 (2013). https://doi.org/10.1016/j.foodres.2011.04.025
12. Kim, I.V., Klykov, A.G.: Vestnik of FEB RAS **3**, 12–15 (2018)
13. Kiru, S.D., Kostina, L.I., Truskinov, E.V., Zoteeva, N.M., et al.: Methodological giudance directory for studying and maintaining VIR collection of potato. VIR, St. Petersburg (2010)
14. Pshechenkov, K.A., Davydenkova, A.N., Sedova, V.I., et al.: Guidelines for the assessment of potato varieties for suitability for processing and storage. M., Lorch Potato Research Institute (2008)
15. Hellmann, H., Goyer, A., Navarre, D.A.: Antioxidants in potatoes: a functional view on one of the major food crops worldwide. Molecules **26**, 2446 (2021). https://doi.org/10.3390/molecules26092446
16. Simakov, E.A., Anisimov, B.V., Mityushkin, A.V., et al.: Potato varieties of the Lorch Potato Research Institute breeding center. Consumer and dining qualities, culinary type (2016)
17. Kim, I.V., Volkov, D.I., Zakharenko, V.M., et al.: Agric Biol **55**(5), 995–1003 (2020)
18. Wang, H., Race, E.J., Shrikhande, A.J.: Anthocyanin transformation in cabernet sauvignon wine during aging. J. Agric. Food Chem. **51**(27), 7989–7994 (2003). https://doi.org/10.1021/jf034501q
19. Kalita, D., Jayanty, S.S.: Comparison of polyphenol content and antioxidant capacity of colored potato tubers, pomegranate and blueberries. J. Food Proc. Tech. **5**, 358 (2014). https://doi.org/10.4172/2157-7110.1000358

Identification of Bioactive Compounds in Clove (*Syzygium Aromaticum L.*)

Fadi Darwish(✉), Tatiana Kalenik, Tamara Senotrusova, Elena Motkina, Maya Razgonova, and Natalia Li

School of Biomedicine, Far Eastern Federal University, 10, Ajax, Russky Island, 690922 Vladivostok, Russia

Abstract. Clove (*Syzygium aromaticum L.*) is an aromatic plant that contains a rich composition of biological active substances. In this study, *Syzygium aromaticum L.* was extracted by supercritical CO_2 extraction (SC-CO_2). The identification of biologically active substances of SC-CO_2 extract of *Syzygium aromaticum L.*was carried out by tandem mass spectrometry using an amaZon SL mass spectrometer equipped with an electrospray ionization source (ESI) in negative ion mode for MS/MSn spectra analysis. Forty-one different compounds were identified in the SC-CO_2 extract of *Syzygium aromaticum L.* Among these compounds, 16 new compounds were first time identified in clove extracts. The results of the study showed that the Syrian clove extract contains many biologically active compounds, mainly flavonoids, therefore, SC-CO_2 extract of *Syzygium aromaticum L.* can be recommended as a good source of natural antioxidants for use in the food industry. The supercritical carbon dioxide extraction method (SC-CO_2) is considered one of the best methods to obtain a clove extract with a high content of biological active substances.

Keywords: Clove (Syzygium aromaticum L.) · Supercritical fluid extraction · HPLC-MS/MS · Flavanoids

1 Introduction

Aromatic plants known as herbs have been used as food preservatives and medicinal plants for centuries. Currently, the demand for these herbs has increased because they are natural and safe products [1]. Herbal extracts are complex mixtures, they contain many biologically active compounds, mainly polyphenolics, responsible for their antimicrobial, antioxidant, and anti-inflammatory properties. Clove is a well-known aromatic plant that belongs to the *Myrtaceae* family. It is one of the most popular herbs that have been used in beverages or foods, because of the therapeutic properties they introduce to human health. Clove is known to be a rich source in phenolic compounds, especially in phenolic acids (gallic, ferulic, coffee, salicylic and elagic acids) and flavonoids (quercetin, kempferol and its derivatives). The main antioxidant activity in clove is due to the existence of polyphenols that include the following compounds: phenolic volatile

A. Muratov and S. Ignateva (Eds.): AFE 2021, LNNS 353, pp. 122–130, 2022.
https://doi.org/10.1007/978-3-030-91402-8_15

oils namely eugenol acetyl and eugenol, phenolic acids like gallic acid and flavonol glucosides [2].

Pérez-Jiménez, J., Neveu, V., Vos, F., Scalbert, A. (2010) studied the content of polyphenols in some plants, and found that the largest amount of polyphenols was found in cloves (15,000 mg/100 g in cloves, 1018 mg/100 g in rosemary, 322 mg/100 g in basil, 878 mg/100 g in thyme) [3]. In this regard, it is promising to obtain a clove extract, which will be rich in the above biologically active components, as well as to study its composition.

Various extraction methods were used for the extraction of bioactive phytochemicals from many natural products, with supercritical liquid extraction (SC-CO2), currently gaining an increasing research interest, comparing to conventional extraction methods such as reduced volume of organic co-solvent and the possibility of more selective extraction, higher diffusion coefficient and density, lower viscosity of supercritical fluids compared to organic solvents [4].

The aim of the study was to identify the composition of biologically active substances present in the CO2-extract of *Syzygium aromaticum L.*

2 Materials and Methods

2.1 Materials

Clove flower buds (*Syzygium aromaticum* L.) was collected from the area near Tartous, Syria and dried inside an oven (Binder oven M115 PM, Germany) at 50 °C for 300 min. Drying was stopped once the material reached a moisture content of $8 \pm 1\%$.

2.2 Chemicals and Reagents

HPLC-grade acetonitrile was purchased from Fisher Scientific (Southborough, UK), MS-grade formic acid was from Sigma-Aldrich (Steinheim, Germany) and ultra-pure water was prepared from a SIEMENS ULTRA clear (SIEMENS water technologies, Germany). The solvent used for supercritical carbon dioxide extraction (SC-CO_2) was CO_2 (PLAMYA, Russia) with 99.9% purity. All other solvents and chemicals used were of analytical grade.

2.3 Supercritical Carbon Dioxide Extraction (SC-CO_2)

Supercritical carbon dioxide extraction method (SC-CO_2) was implemented using the Thar SCF-500 automated supercritical fluid extraction apparatus (Thar SCF Waters, Milford, USA). The system was also equipped with a co-solvent pump for extracting polar samples (Thar Waters P-50 High Pressure Pump) and CO_2 flow meter (Siemens, Germany) and multiple extraction vessels. Flow rate was 10–25 mL/min for liquid CO_2 and 1.00 mL/min for ethanol (EtOH). For extraction, sample of 10 g *Syzygium aromaticum L.*was used. The extraction time was counted after reaching the working pressure and equilibrium flow; and it was 60–90 min.

2.4 Liquid Chromatography - Mass Spectrometry

Chemical analysis was performed using a Shimadzu LC-20 Prominence HPLC (Shimadzu, Japan) equipped with an UV-sensor (SPD-20A, Shimadzu, Japan) and a HPLC reverse phase column (Shodex ODP-40 4E), in order to perform the separation of multicomponent mixtures and coupled to MS ion trap (amaZon SL, BRUKER DALTONIKS, Berlin, Germany), equipped with an electrospray ionization source (ESI) in negative ion mode for MS/MSn spectra analysis.

HPLC gradient elution program was as follows: 0.01–4 min, 100% A; 4–60 min, 100–25% A; 60–75 min, 25–0% A; control washing 75–120 min 0% A. Sample injection volume was 20 μL, column temperature was set at 17 °C and UV-VIS detector was operating at 230 and 330 nm wavelengths. Mass spectrometry (MS) analysis was performed with the ion trap ESI source in negative ion mode and the optimized parameters used were: ionization source temperature: 70 °C, gas flow: 4 L/min, nebulizer gas (atomizer): 7.3 psi, capillary voltage: 4500 V, end plate bend voltage: 1500V, fragmentary: 280 V, collision energy: 60 eV. The MS An ion trap was used in the scan range: m/z 100–1.700 for MS and MS/MS, with a capture rate: one spectrum for MS and two spectrums for MS/MS. Data collection were controlled by CromQuest 4.2 (Thermo Electron Corporation). All experiments were repeated three times.

3 Results and Discussion

The extract of *Syzygium aromaticum L.*was obtained using supercritical fluid extraction. The identification of biologically active substances of the CO2 extract of *Syzygium aromaticum L.*was carried out by tandem mass spectrometry using an amaZon SL mass spectrometer equipped with an ESI source in negative mode.

Sample evaluation was performed with the analysis of over 300 mass spectra for each analytical replicate (Fig. 1) and resulted to the identification of 41 compounds.

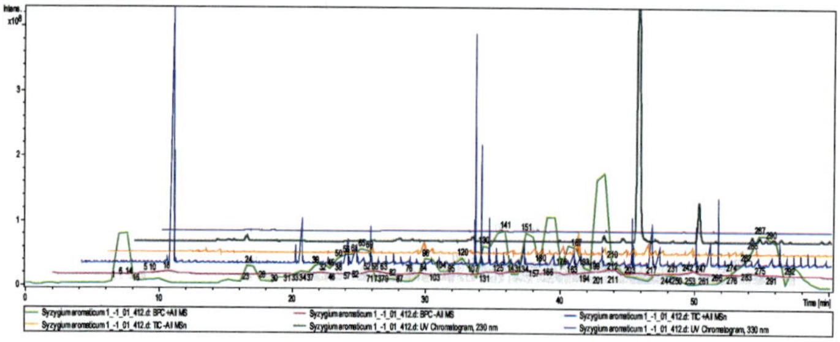

Fig. 1. Chemical profiles of *Syzygium aromaticum L.*, ion chromatogram from SC-CO_2 extract.

Various chemical compound categories (29 flavonoids, 2 fatty acids, 1 hydroxydicarboxylic acid, 3 phenolic acids,1 phenolic compound, 2 tannins, 2 diterpenes and 1 tocopherol) were identified in the SC-CO_2 extract of *Syzygium aromaticum L.* by comparing the data with the literature [4–10].

The biological active compounds identified in the Syrian clove extract were as follows:

3.1 Flavonoids from SC-CO2 Extract of *Syzygium Aromaticum L.*

Compound (1) showed $[M–H]^-$ ion at m/z 313.12 produced 3 fragment ion with m/z 298.02, m/z 283 and m/z 255.01 was identified as 4', 7'-dimethoxyluteolin ($C_{17}H_{14}O_6$). The CID spectrum in negative ion modes of 4',7'-dimethoxyluteolin from *Syzygium aromaticum L.*is shown in Fig. 2.

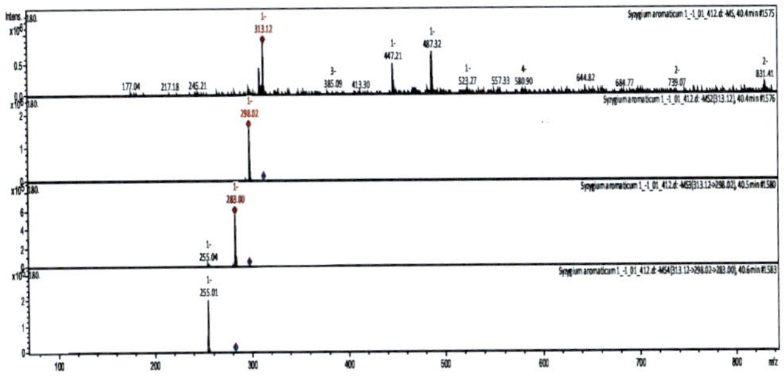

Fig. 2. CID spectrum of 4', 7'-dimethoxyluteolin from *Syzygium aromaticum L.*, m/z 313.12.

Compound (2) exhibited $[M–H]^-$ ion at m/z 315.09 produced 10 fragment ion with m/z 300.05, m/z 282.97, m/z 271.04, m/z 227.11, m/z 151.07, m/z 164.04, m/z 270.86, m/z 243.96, m/z 227.03 and m/z 137.12 was identified as isorhamnetin ($C_{16}H_{12}O_7$) [6]. The CID spectrum in negative ion modes of isorhamnetin from *Syzygium aromaticum L.*is shown in Fig. 3.

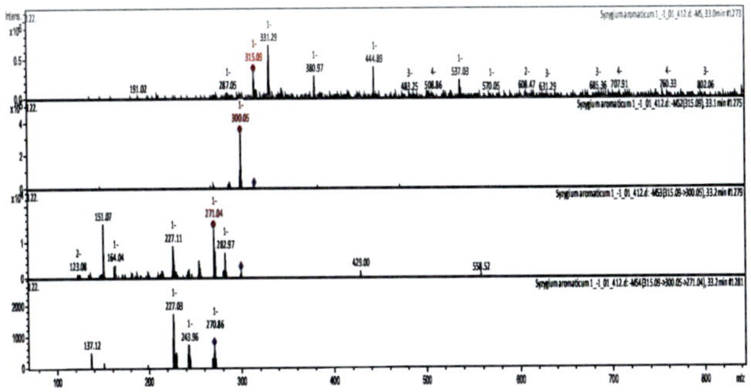

Fig. 3. CID spectrum of isorhamnetin from *Syzygium aromaticum L.*, m/z 315.09.

Compound (3) generated $[M–H]^-$ ion at m/z 271.05 produced three fragment ion with m/z 177.11, m/z 151.03 and m/z 119.19 was identified as naringenin ($C_{15}H_{12}O_5$) [6].
Compound (4) gave $[M + H]^+$ ion at m/z 271.13 produced two fragment ion with m/z 253.14 and m/z 186.04 was identified as apigenin ($C_{15}H_{10}O_5$).
Compound (5) exhibited $[M–H]^-$ ion at m/z 285.24 produced three fragment ion with m/z 239.02, m/z 221 and m/z 211 was identified as kaempferol ($C_{15}H_{10}O_6$) [6].

Compound (6) showed $[M + H]^+$ ion at m/z 285.09 produced 2 fragment ion with m/z 239.02 and m/z 132.99 was identified as 6,7-dihydroxy-4'-Methoxyisoflavone ($C_{16}H_{12}O_5$) [7].
Compound (7) exhibited $[M + H]^+$ ion at m/z 291.04 produced two fragment ion with m/z 231.06 and m/z 230.21 was identified as catechin ($C_{15}H_{14}O_6$) [10].
Compound (8) gave $[M + H]^+$ ion at m/z 291.07 produced one fragment ion with m/z 231.10 was identified as epicatechin ($C_{15}H_{14}O_6$) [10].
Compound (9) showed $[M + H]^+$ ion at m/z 301.02 produced three fragment ion with m/z 285.98, m/z 285 and m/z 257 was identified as chrysoeriol ($C_{16}H_{12}O_6$) [7].
Compound (10) exhibited [M + H] + ion at m/z 301.02 produced two fragment ion with m/z 285.97 and m/z 284.98 was identified as hispidulin ($C_{16}H_{12}O_6$) [7].
Compound (11) showed $[M + H]^+$ ion at m/z 301.01 produced 2 fragment ion with m/z 285.97 and m/z 257.93was identified as kaempferide ($C_{16}H_{12}O_6$).
Compound (12) generated $[M–H]^-$ ion at m/z 301.11 produced two fragment ion with m/z 273.04 and m/z 245.05 was identified as herbacetin ($C_{15}H_{10}O_7$).
Compound (13) gave $[M–H]^-$ ion at m/z 301.07 produced 4 fragment ion with m/z 273.04, m/z 245.05, m/z 179 and m/z 151 was identified as quercetin ($C_{15}H_{10}O_7$) [6].
Compound (14) showed $[M + H]^+$ ion at m/z 301.04 produced two fragment ion with m/z 285.97 and m/z 258.06 was identified as diosmetin ($C_{16}H_{12}O_6$) [7].
Compound (15) exhibited $[M–H]^-$ ion at m/z 311.18 produced 2 fragment ion with m/z 311.14 and m/z 183.06 was identified as 8-Demethyleucalyptin ($C_{18}H_{16}O_5$) [4].
Compound (16) generated $[M–H]^-$ ion at m/z 315.05 produced 4 fragment ion with m/z 300.01, m/z 193.04, m/z 165.04 and m/z 121.11 was identified as rhamnetin I ($C_{16}H_{12}O_7$).
Compound (17) showed $[M–H]^-$ ion at m/z 359.08 produced two fragment ion with m/z 343.99 and m/z 328.96 was identified as jaceidin isomer 1 ($C_{18}H_{16}O_8$) [8].
Compound (18) exhibited $[M–H]^-$ ion at m/z 329.07 produced 2 fragment ion with m/z 314 and m/z 298.95 was identified as Quercetin dimethyl ether ($C_{17}H_{14}O_7$) [8].
Compound (19) gave $[M–H]^-$ ion at m/z 329.09 produced two fragment ion with m/z 314.05 and m/z 298.97 was identified as jaceosidin ($C_{17}H_{14}O_7$) [7].
Compound (20) showed $[M + H]^+$ ion at m/z 344.98 produced 2 fragment ion with m/z 329.98 and m/z 314.96 was identified as xanthomicrol ($C_{18}H_{16}O_7$) [7].
Compound (21) exhibited $[M + H]^+$ ion at m/z 345.03 produced two fragment ion with m/z 330 and m/z 314.98 was identified as 5,6-dihydroxy-7,3',4'-trimethoxy ($C_{18}H_{16}O_7$) [7].
Compound (22) gave [M + H] + ion at m/z 345.04 produced 3 fragment ion with m/z 330.01, m/z 301.02 and m/z 300.95 was identified as nevadensin ($C_{18}H_{16}O_7$) [7].
Compound (23) showed $[M + H]^+$ ion at m/z 355.02 produced 3 fragment ion with m/z 337, m/z 288.99 and m/z 205.01 was identified as hinokinin ($C_{20}H_{18}O_6$).
Compound (24) exhibited $[M–H]^-$ ion at m/z 358.90 produced two fragment ion with m/z 343.97 and m/z 328.96 was identified as thymonin ($C_{18}H_{17}O_8$) [7].
Compound (25) generated $[M–H]^-$ ion at m/z 383.25 produced 2 fragment ion with m/z 365.14 and m/z 215.06 was identified as fraxetin-7-O-beta-glucuronide ($C_{16}H_{16}O_{11}$) [4].
Compound (26) gave $[M–H]^-$ ion at m/z 447.22 produced two fragment ion with m/z 301.20 and m/z 283.20 was identified as quercetin-3-O-rhamnoside ($C_{21}H_{20}O_{11}$).

Compound (27) exhibited $[M–H]^-$ ion at m/z 447.22 produced 2 fragment ion with m/z 301.17 and m/z 283.20 was identified as quercitrin ($C_{21}H_{20}O_{11}$) [10].
Compound (28) showed $[M–H]^-$ ion at m/z 447.21 produced two fragment ion with m/z 301.17 and m/z 283.20 was identified as rhodiolatuntoside ($C_{21}H_{20}O_{11}$) [10].
Compound (29) exhibited $[M–H]^-$ ion at m/z 461.09 produced 2 fragment ion with m/z 299.03 and m/z 284.04 was identified as chrysoeriol-7-O-glucoside ($C_{22}H_{22}O_{11}$).

3.2 Phenolic Acids from SC-CO_2 Extract of *Syzygium Aromaticum L.*

Compound (1) showed $[M + H]^+$ ion at m/z 195.09 produced 5 fragment ion with m/z 174.98, m/z 149.09, m/z 137.06,m/z 147.08 and m/z 122.14 was identified as ferulic acid ($C_{10}H_{10}O_4$) [10].

The CID spectrum in positive ion modes of ferulic acid from *Syzygium aromaticum L.*is shown in Fig. 4.

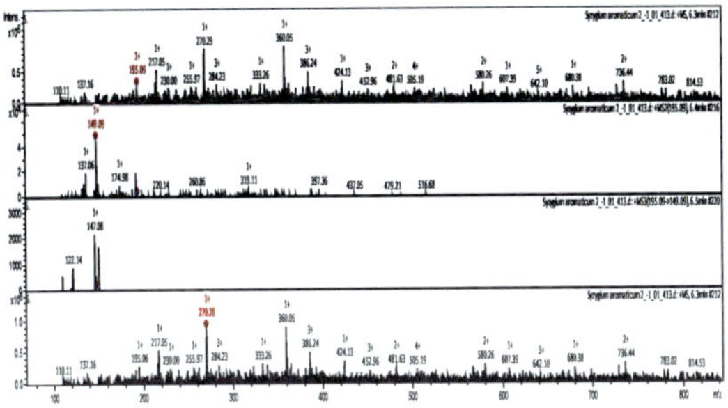

Fig. 4. CID spectrum of ferulic acid from *Syzygium aromaticum L.*, m/z 195.09.

Compound (2) exhibited [M–H]- ion at m/z 169.10 produced two fragment ion with m/z 125.09 and m/z 125.09 was identified as gallic acid (C7H6O5) [6].

Compound (3) generated [M–H]- ion at m/z 301.11 produced 6 fragment ion with m/z 273.04, m/z 257, m/z 245.05, m/z 201, m/z 173 and m/z 151was identified as ellagic acid (C14H6O8) [6].

3.3 Diterpenes from SC-CO_2 Extract of *Syzygium Aromaticum L.*

Compound (1) showed [M + H] + ion at m/z 297.98 produced three fragment ion with m/z 279, m/z 208.69 and m/z 198.45 was identified as cryptotanshinone (C19H20O3).

Compound (2) showed $[M–H]^-$ ion at m/z 315.07 produced two fragment ion with m/z 287.03 and m/z 259.05 was identified as 20-deoxocarnosol ($C_{20}H_{28}O_3$) [9].

3.4 Fatty Acids from SC-CO_2 Extract of *Syzygium Aromaticum L.*

Compound (1) showed [M + H] + ion at m/z 277.13 produced 5 fragment ion with m/z 207.06, m/z 259.08, m/z 233.02, m/z 147.02 and m/z 124.06 was identified as stearidonic

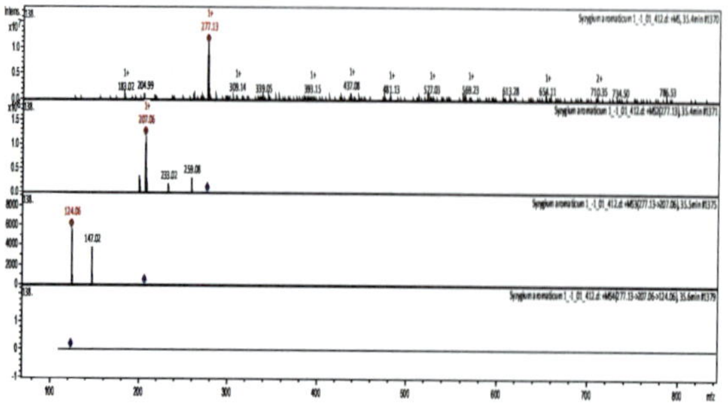

Fig. 5. CID spectrum of stearidonic acid from *Syzygium aromaticum L.*, m/z 277.13.

acid (C18H28O2) [9]. The CID spectrum in positive ion modes of stearidonic acid from *Syzygium aromaticum L.*is shown in Fig. 5.

Compound (2) exhibited [M + H] + ion at m/z 369.16 produced 2 fragment ion with m/z 351.18 and m/z 331.08 was identified as lignoceric acid (C24H48O2)[4].

3.5 Hydroxydicarboxylic Acid from SC-CO_2 Extract of *Syzygium Aromaticum L.*

Compound (1) showed [M–H]- ion at m/z 133.10 produced two fragment ion with m/z 115.07 and m/z 115.11 was identified as L-malic acid (C4H6O5) [5].

3.6 Tannins from SC-CO_2 Extract of *Syzygium Aromaticum L.*

Compound (1) gave [M–H]- ion at m/z 331.28 produced 7 fragment ion with m/z 313.23, m/z 295.21, m/z 277.20, m/z 171.15, m/z 276.21, m/z 180.83 and m/z 217.36 was identified as galloyl glucose (C13H16O10). The CID spectrum in negative ion modes of galloyl glucose from *Syzygium aromaticum L.*is shown in Fig. 6.

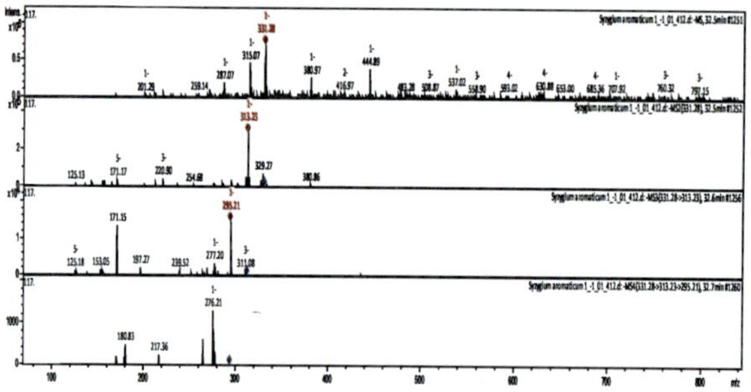

Fig. 6. CID spectrum of galloyl glucose from *Syzygium aromaticum L.*, m/z 331.28.

Compound (2) exhibited [M–H]- ion at m/z 331.28 produced 2 fragment ion with m/z 313.25 and m/z 295.21 was identified as monogalloyl glucose (C13H16O10) [6].

3.7 Tocopherol from SC-CO_2 Extract of *Syzygium Aromaticum L.*

Compound (1) exhibited [M + H] + ion at m/z 593.18 produced 2 fragment ion with m/z 533.10 and m/z 461.05 was identified as alpha.-Tocopherol-Beta-d-Mannoside (C35H60O7) [4].

3.8 Phenolic Compound from SC-CO_2 Extract of *Syzygium Aromaticum L.*

Compound (1) showed [M + H] + ion at m/z 185.01 produced 2 fragment ion with m/z 168.08 and m/z 138.74 was identified as methyl gallic acid (C8H8O5).

From the results, it can be concluded that the Syrian clove extract is characterized by a high content of biological active substances, especially flavonoids, so it can be used as a natural antioxidant in the food industry.

Compared with the results of previous studies [6] of the clove content of biological active substances, this study reveals the presence of 16 new compounds (alpha.-Tocopherol-Beta-d-Mannoside, lignoceric acid, xanthomicrol, cryptotanshinone, hinokinin, 5,6-dihydroxy-7,3, 4'-trimethoxyflavone, herbacetin, fraxetin-7-O-beta-glucuronide, diosmetin, chrysoeriol-7-O-glucoside, 20-deoxocarnosol, stearidonic acid, hispidulin, 8-demethyleucalyptin, chrysoeriol, 6,7-dihydroxy-4'-methoxyisoflavone) that have been identified for the first time.

By comparing the results of the analysis of the biological active substances in the supercritical clove extract with the clove extract obtained using conventional organic solvents in the study conducted by Fathoni and others (2017), it was found that the use of supercritical extraction allows obtaining a clove extract with a higher content of the biological active substances.

4 Conclusion

In the present study, the composition of SC-CO2 extract of *Syzygium aromaticum L.* was analyzed by HPLC with MS/MS identification. As a result, 41 compounds were identified. 16 new compounds were discovered for the first time compared to the results of other scientists' work. Among extraction methods, supercritical carbon dioxide (SC-CO2) was shown to be the method of choice for obtaining a clove extract with a high content of biologically active substances. The Syrian clove extract can be used as an important source of natural antioxidants in the food industry.

References

1. Christaki, E., Bonos, E., Giannenas, I., Florou-Paneri, P.: Aromatic plants as a source of bioactive compounds. Agriculture **2**(3), 228–243 (2012). https://doi.org/10.3390/agriculture2030228
2. Cortés-Rojas, D.F., de Souza, C.R.F., Oliveira, W.P.: Clove (Syzygium aromaticum): a precious spice. Asian Pacific J. Trop. Biomed. **4**(2), 90–96 (2014). https://doi.org/10.1016/S2221-1691(14)60215-X

3. Pérez-Jiménez, J., Neveu, V., Vos, F., Scalbert, A.: Identification of the 100 richest dietary sources of polyphenols: an application of the Phenol-Explorer database. Euro. J. Clin. Nutrit. **64**(3), S112–S120 (2010). https://doi.org/10.1038/ejcn.2010.221
4. Razgonova, M., Zakharenko, A., Ercisli, S., Grudev, V., Golokhvast, K.: Comparative Analysis of Far East Sikhotinsky Rhododendron (Rh. sichotense) and East Siberian Rhododendron (Rh. adamsii) Using Supercritical CO2-Extraction and HPLC-ESI-MS/MS Spectrometry. Molecules **25**(17), 3774 (2020). https://doi.org/10.3390/molecules25173774
5. Cirlini, M., Mena, P., Tassotti, M., Herrlinger, K.A., Nieman, K.M., Dall'Asta, C., Del Rio, D.: Phenolic and volatile composition of a dry spearmint (Mentha spicata L.) extract. Molecules **21**(8), 1007 (2016). https://doi.org/10.3390/molecules21081007
6. Fathoni, A., Saepudin, E., Cahyana, A.H., Rahayu, D.U.C., Haib, J.: Identification of non-volatile compounds in clove (Syzygium aromaticum) from Manado. In: AIP Conference Proceedings, vol. 1862(1), p. 030079 (2017). https://doi.org/10.1063/1.4991183
7. Xu, L.L., et al.: Analysis of non-volatile chemical constituents of Menthae Haplocalycis herba by ultra-high performance liquid chromatography-high resolution mass spectrometry. Molecules **22**(10), 1756 (2017). https://doi.org/10.3390/molecules22101756
8. Taamalli, A., Arráez-Román, D., Abaza, L., et al.: LC-MS-based metabolite profiling of methanolic extracts from the medicinal and aromatic species Mentha pulegium and Origanum majorana. Phytochem. Anal. **26**(5), 320–330 (2015). https://doi.org/10.1002/pca.2566
9. Yang, S.T., Wu, X., Rui, W., Guo, J., Feng, Y.F.: UPLC/Q-TOF-MS analysis for identification of hydrophilic phenolics and lipophilic diterpenoids from Radix Salviae Miltiorrhizae. Acta Chromatographica **27**(4), 711–728 (2015). https://doi.org/10.1556/AChrom.27.2015.4.9
10. Han, F., et al.: A rapid and sensitive UHPLC-FT-ICR MS/MS method for identification of chemical constituents in Rhodiola crenulata extract, rat plasma and rat brain after oral administration. Talanta **160**, 183–193 (2016). https://doi.org/10.1016/j.talanta.2016.07.014

Efficiency of Applying Sowing Machines for Planting Spring Wheat in the Irkutsk Region

Vladimir Solodun[1,2], Evgeny Boyarkin[1,2], Alexsander Zaitsev[1,2], and Tatyana Amakova[1](✉)

[1] Irkutsk State Agrarian University named after A.A. Ezhvevskiy, Molodezhny 1/1, Irkutsk District, 664038 Irkutsk Region, Russia

[2] Irkutsk Scientific Research Institute of Agriculture, Pivovarikha, Dachnaya Street, 14, Irkutsk District, 664511 Irkutsk Region, Russia

Abstract. In recent years, a wide range of sowing machinery is actively used for seeding in the Irkutsk region. Currently, the list of sowing and till-plant units applied includes 19 equipment brands. Up to 50% of the total grain crop area is sown with state-of-the-art multifunctional sowing units. They are applied not only for direct sowing, but also for fallow sowing, plowed or non-plowed autumn tillage, as well as for shallow spring tillage. No comparative studies on the efficiency of these machines with traditional sowing methods were conducted in the region prior to our work, resulting in absence of clear prospects for improving the process of combining tillage with sowing, and application of a large variety of sowing unit makes. The conducted research revealed that advanced sowing and till-plant units manufactured in Russia and abroad with tine coulters and broadcast-strip planting show little promise for sowing grain crops in the Irkutsk region. The utilization of multifunctional machines Ob-4, Kuzbass, Concord, Morris provide for optimal seeding depth and plant alimentation area, increase field seed germination by 6–8%, and spring wheat yield by 0.28–0.35 t/ha.

Keywords: Seeding complexes · Spring cereals · Tillage machines · Sowing · Tillage

1 Introduction

Today, in the Irkutsk region, sowing units performing multiple process operations (tillage, sowing, mineral fertilization, leveling and packing) during a single pass are increasingly used for sowing grain crops.

According to the source [1], this is driven by the gradual abandonment of the traditional plowing system applying single-operation tillage tools, and the transition to multifunctional, resource-saving techniques combining tillage and sowing. As noted by Polyakov [2], the range of such sowing and till-plant units utilized in the region includes 19 equipment brands of both domestic and foreign production. As the end of 2020, the following sowing units were registered in the agro-industrial complex of the region: Omichka – 11 units, Horsh – 2 units, Salford-4 units, Kuzbass-8.5 – 64 units, John

A. Muratov and S. Ignateva (Eds.): AFE 2021, LNNS 353, pp. 131–137, 2022.
https://doi.org/10.1007/978-3-030-91402-8_16

Deere – 3 units, DMS – 13 units, DKT – 1 unit, Agromaster – 11 units, Concord – 1 unit, Tom-8/10 – 8 units, Morris – 4 units, Atlas – 2 units, SKP-2,1 – 64 units, Ob-4 – 48 units, SZS-2,1 – 10 units.

As per the data available [3, 4], up to 50% of the total grain crop area is processed by these sowing units. Therewith, these sowing units are used for direct sowing (using the No-Till method), for soils fine-processed by the disc harrow in autumn and spring tillage, as well as for plowed fallow and autumn-tilled soil sowing, as indicated by Solodun [5].

A major part of the tilled area is represented by clay-loamy grey forest and sod-carbonate soils, which are still processed as per the traditional ploughing method, and the sowing is performed by seed drills with disc coulters SZ-3,6 and SZP-3,6 with 15 cm row spacing and gutter planting.

In the region, the largest number of seeders are of these brands (more than 1,500 pieces according to the source) and they are used for sowing half of the entire grain and grain-legume crop area.

As specified in the special sources [6–8], in contrast to the gutter sowing with these seed drills, the sowing units are equipped mainly with tine and hoe coulters for strip, broadcast-strip or band sowing.

The conditions of the Irkutsk region, according to regional sources, are characterized by the highest rates of grain crops (wheat, barley, oats) seeding in Siberia and the Russian Federation ranging from 6.5 to 9 million germinating grains per hectare, conditioned upon the low field germination rate (40–50% in some years). Therefore, following these specifics, broadcast-strip sowing methods tend to be the most promising for the region, since they provide a more optimal seed placement over the alimentation area, eliminate excess seed density in rows and mutual plant competition, as indicated by the researchers [9–13].

Prior to our previous studies [14], no special research was conducted on the comparison of traditional drill row and broadcast-strip seeding methods.

The purpose of this study is to assess the impact of sowing machines manufactured in Russia and abroad on the distribution of seeds in the alimentation area and depth, as well as on the spring wheat yield.

2 Materials and Methods of Research

The research was performed in 2016–2018 on typical gray forest clay-loamy soils in the primary farm of Okinsky agricultural production co-operative, well equipped with advanced sowing equipment. In a stationary field experiment No. 1, 6 sowing machines were studied: traditional disk seeders SZ-3,6, sowing units Ob-4, Kuzbass-8,5, Concord-2000, Morris Concept-2000, Tom-10,6. Wheat sowing was performed over the preceding crop – wheat planted in bare fallow at 2 depths: 4–5 and 5–6 cm. Prior to sowing with SZ-3,6 and Tom-10,6 seed drills, pre-sowing tillage was performed by KPE-3,8 up to the depth of 4–6 cm. Prior to sowing with soil-tilling and planting units, no pre-sowing tillage was performed. Upon sowing with SZ-3,6, packing by KZK-6–01 was performed.

Tiers – 3, cultivar – Iren, sowing period – the second decade of May, seeding rate – 7 million germinating grains per 1 ha. Field discount area – 250 m^2. Observations, records, analyses, and statistical data processing were conducted according to Dospekhov.

In field experiment No. 2, the use of different seeding machines for clean fallow, autumn plowing and spring tillage was studied. In the experiment, 3 sowing machines were studied: seeders SZ-3,6, SZM-400 and a sowing complex Kuzbass-8,5. In the experiment, wheat Tulunskaya 12 was sown. The areas of the plots, replicates and studies corresponded to experiment No. 1.

Weather conditions in the years of research were characterized by noticeable deviations in the amount of precipitation falling over the months of the growing season and the average daily air temperatures.

Table 1 shows that the most critical period for spring wheat (June) was characterized by a two-fold (or higher) underprecipitation, and in May, the lowest precipitation was registered in 2018. In general, the amount of precipitation in May-June was lower than the long-run annual average, which gives grounds to consider the years of research as moisture-deficient, and in terms of average daily temperatures, all months were significantly warmer than the long-run annual average.

Table 1. Hydrothermal conditions for the growing season months in the years of research (weather station of the Irkutsk Research Institute of Agriculture).

Year	Month					May–September Total
	May	June	July	August	September	
Precipitation, mm						
2016	28	37	86	177	99	427
2017	67	13	105	55	37	277
2018	16	27	77	100	56	276
Long-run annual average	30	62	110	95	47	344
Average daily air temperature, °C						
2016	10.0	17.9	21.0	14.6	10.7	14.8
2017	10.5	18.5	19.7	16.9	9.5	15.0
2018	10.6	19.8	18.3	18.4	9.5	15.3
Long-run annual average	9.0	14.7	16.5	14.6	7.6	12.7

3 Results

The yield data given in Table 2 showed considerable variations in the spring wheat yield associated with the use of different sowing machines. Sowing machines with tine coulters and broadcast-strip sowing: Ob-4, Kuzbass-8,5, Concord-2000, Morris Concept-2000 provided for a considerably higher yield with both shallower and deeper seeding compared to SZ-3,6 seed drill and Tom-10,6 sowing unit with disc coulters.

Table 2. Yield of spring wheat machine-sown at different depths, t/ha.

Sowing machine brand	Coulter type. seeding method. row spacing width	Prescribed seeding depth. cm	Year			Average in 2016–2018	± against reference
			2016	2017	2018		
SZ-3.6 – control	Double-disk	4–5	1.42	1.81	1.76	1.66	–
	Drill. 15 cm	5–6	1.54	1.93	1.81	1.76	–
Ob-4	Tine	4–5	1.81	2.20	1.96	1.99	+0.33
	Broadcast-strip	5–6	1.84	2.24	2.05	2.04	+0.28
Kuzbass-8.6	Tine	4–5	1.86	2.19	2.00	2.00	+0.34
	Broadcast-strip	5–6	1.90	2.22	2.08	2.06	+0.30
Concord-2000	Tine	4–5	1.81	2.19	2.00	2.00	+0.34
	Broadcast-strip	5–6	1.87	2.21	2.10	2.06	+0.30
Morris Concept-2000	Tine	4–5	1.85	2.20	2.10	2.05	+0.39
	Broadcast-strip	5–6	1.87	2.23	2.12	2.07	+0.31
Tom-10.6	Double-disk	4–5	1.50	1.84	1.91	1.75	+0.09
	Drill. 19 cm	5–6	1.61	1.98	1.93	1.84	+0.8
Least significant difference $_{05.\ t/ha}$			0.17	0.22	0.20		

The average increase provided by the sowing and till-plant units amounted to 0.33–0.39 t/ha at the prescribed seeding depth of 4–5 cm and 0.28–0.31 t/ha at the seeding depth of 5–6 cm.

To identify the causes of the resulting yield variance, the actual seeding depth was studied after sowing, as well as the field germination rate and the number of yielding stems for harvesting were determined.

4 Discussions

The results obtained, shown in Table 3, allowed for identifying a number of yield generation specifics, depending on the seeding depth and sowing methods.

Thus, when sowing with disc-coulter seeders SZ-3,6 and Tom-10,6, field germination was more dependent on the methods of sowing than on the seeding depth. With an increase in the seed placement depth by disc seeders, the field germination rate raised by 6–8%, and when sowing with tine coulters, the highest raise amounted to 3%. The average increase in field germination from broadcast-strip sowing amounted to 15–23%.

This difference is attributable to the fact that the actual depth of seeding with tine coulters was closer to the optimal value (5–6 cm), and the seeds fell into a part of the seeding layer with higher moisture content during seeding. Disc coulters, unlike tine coulters, provide for shallower seeding, vibrate during sowing, and the majority of seeds are distributed in the regularly drying upper part of the seeding layer.

Table 3. Seeding depth, field germination rate, and number of yielding stems depending on different sowing machines application (average in 2016–2018).

Sowing machine brand	Sowing method and coulter type	Prescribed seeding depth					
		4–5	5–6	4–5	5–6	4–5	5–6
		Actual seeding depth. cm		Field germination. %		Number of yielding stems for harvesting. ea./m2	
SZ-3.6 – control	Drill. Double-disk	2–6	3–7	47	53	377	395
Ob-4	Tine. Broadcast-strip	3–6	4–8	64	67	421	446
Kuzbass-8.6	Tine. Broadcast-strip	3–7	3–8	62	65	420	441
Concord-2000	Tine. Broadcast-strip	3–7	3–8	63	66	423	443
Morris Concept-2000	Tine. Broadcast-strip	4–6	4–8	64	67	422	444
Tom-10.6	Drill. Double-disk	2–5	3–6	44	52	369	396

These field germination data correlate with the number of yielding stems. At the seeding rate of 7 million germinating grains per 1 ha accepted for the region, 359–396 yielding stems survived until harvesting with field germination of 44–53% when sown with disc seeders SZ-3,6 and Tom-10,6, and with broadcast-strip planting by sowing units this number amounted to 441–446 stems with field germination of 62–67%, depending on the seeding depth.

The results of research on experiment No. 2 showed (Table 4), that different sowing machines with working bodies of different designs differently affected the yield of wheat on the background of the main tillage. When sowing on pure steam, the best results were obtained when sowing with a single-disc planter SZM-400 with a significant increase in yield of 0.43 t/ha, slightly lower when sowing Kuzbass-8.5. In the autumn cold, the Kuzbass-8.5 and SPM-400 PCs showed approximately equal productivity.

This indicates that, given deeply cultivated backgrounds, these seeding units should replace the traditionally and massively used SZ-3,6 seeders and their modifications with double-disc openers and 15 cm row spacing.

When sowing on stubble (zero basic tillage), the Kuzbass-8.5 sowing complex gave a clearly higher yield (an increase of 0.51 t/ha). The SZ-3.6 seeder was also more efficient than the SZM-400 seeder with preliminary cultivating stubble cultivation.

Table 4. Yield of spring wheat Tulunskaya 12, depending on the background of the main tillage and sowing machines (average for 2016–2018).

Brand of sowing machine	Type of coulter and method of sowing	Yield. t/ha	± t/ha to control
Background A – Pure steam			
SZ-3.6 – control	Drill. Double-disk. 15 cm	18.3	–
SZM-400	Ordinary. Single-disk 14 cm	22.6	+0.43
Kuzbass-8.6	Tine. Broadcast-strip	21.4	+0.31
Least significant difference05		2.5	
Background B – Autumn plowing to a depth of 20–22 cm			
SZ-3.6 – control	Drill. Double-disk. 15 cm	16.4	–
SZM-400	Ordinary. Single-disk 14 cm	19.3	+0.29
Kuzbass-8.6	Tine. Broadcast-strip	18.3	+0.22
Least significant difference05		2.0	
Background V – Spring tillage and sowing			
Processing of KPE-3.8 to a depth of 6–7 cm. seeding of SZ-3.6-control	Drill. Double-disk. 15 cm	14.3	–
Processing of KPE-3.8 to a depth of 6–7 cm. seeding of SZM-400	Ordinary. Single-disk 14 cm	16.6	+0.23
Direct seeding Kuzbass-8.5	Tine. Broadcast-strip	19.4	+0.51
Least significant difference05	Background		
	A	0.17	
	B	0.17	
	AB	0.24	

5 Conclusions

On clay-loamy grey forest soils of the forest-steppe agrarian landscape of the Irkutsk region, on fields non-tilled since autumn, when sowing spring wheat, the use of traditional seeders with disc coulters and single-line sowing with pre-sowing tillage by cultivators or disc harrows at the accepted high seeding rates is considered to be an ineffective method. When sowing on dump-treated clean pairs, the SPM-400 seeding machines with single-disc coulters and the Kuzbass-8.5 seeding complexes provide approximately equal yields.

The most promising method of sowing combined with tillage, is direct sowing with sowing and till-plant machines (units) of domestic and foreign production (Ob-4,

Kuzbass-8,5, Concord-2000, Morris Concept-2000) with tine coulters and broadcast-strip seeding method These units provide for optimal seed depth and plant alimentation area, increase field seed germination by 6–8%, and spring wheat yield by 0.28–0.35 t/ha.

References

1. Solodun, V.I., Zaytsev, A.M., Filippov, A.S.: Nauchnyye osnovy sistem zemledeliya Predbaykal'ya. Irkutsk, IrGSHA (2012)
2. Polyakov, G.N., Solodun, V.I., Shukhanov, S.N.: Sostav i izmeneniye struktury sel'skokhozyaystvennykh mashin dlya pochvoobrabotki v Irkutskoy oblasti. Izvestiya-maao **47**, 28–31 (2019)
3. Polyakov, G.N., Perevalov, V.M., Brichagina, A.A.: Sel'skokhozyaystvennyye mashiny i orudiya dlya vozdelyvaniya zernovykh kul'tur po resursosberegayushchim tekhnologiyam v Irkutskoy oblasti. Irkutsk, IrGSHA (2012)
4. Solodun, V.I.: Mekhanicheskaya obrabotka pochvy i yeyo nauchnoye obosnovaniye v Predbaykal'ye: monografiya. Irkutsk, IrGSHA (2009)
5. Gol'tyapin, V.Y.: Innovatsionnyye tekhnologii pryamogo poseva zernovykh kul'tur: nauch.analit.obzor: monografiya, Moscow, FGBNU "Rosinformagrotech" (2019)
6. Ivan'o, Y.M., Dmitriyev, N.N., Adushinov, D.S., et al.: Sistema vedeniya sel'skogo khozyaystva Irkutskoy oblasti: monografiya. Irkutsk, Megaprint (2019)
7. Alforova, P.A.: Semennoye zerno Vostochnogo Zabaykal'ya: nauch.monografiya. Chita: Chita mountain.type (2013)
8. Zaitsev, A.M., Solodun, V.I., Gorbunova, M.S.: Sravnitel'naya otsenka sposobov poseva yarovoy pshenitsy pri ispol'zovanii raznykh tipov soshnikov. In: IOP Conference Series: Earth and Environmental Science. Conference proceedings. Krasnoyarsk Science and Technology City Hall of the Russian Union of Scientific and Engineering Associations, p 62017. https://doi.org/10.1088/1755-1315/421/6/062017
9. Solodun, V.I.: Otsenka sposobov poseva zernovykh kul'tur pri primenenii raznykh tipov soshnikov. Vestnik KrasGAU **4**, 37–40 (2015)
10. Solodun, V.I., Gabdrakhimov, O.B., Zaitsev, A.M., Kusnetsova, E.N.: Vliyaniye urovney khimizatsii na urozhaynost' i kachestvo zerna rayonirovannykh sortov yarovoy pshenitsy Pribaykal'ya. IOP Conference Series: Earth and Environmental Science. III International Scientific Conference: AGRITECH-III-2020: Agribusiness, Environmental Engineering and Biotechnologies. Krasnoyarsk Science and Technology City Hall of the Russian Union of Scientific and Engineering Associations, p 22082 (2020). https://doi.org/10.1088/1755-1315/548/2/022082
11. Zhang, X., Xina, X., Yang, W., et al.: Soil and Tillage Research Soil respiration and net carbon flux response to long-term reduced/no-tillage with and without residues in a wheat-maize cropping system. Soil Tillage Res. **214**, 105182 (2021). https://doi.org/10.1016/j.still.2021.105182
12. Resursosberegayushchiye tekhnologii v zemledelii (2009) Rekomendatsii
13. Känkänen, H., Alakukku, L., Salo, Y., Pitkänen, T.: Growth and yield of spring cereals during transition to zero tillage on clay soils. Eur. J. Agron. **34**(1), 35–45 (2011). https://doi.org/10.1016/j.eja.2010.10.002
14. Kiryushin, B.D., Usmanov, R.R., Vasil'yev, I.P.: Osnovy nauchnykh issledovaniy v agronomii. Colossus, Moscow (2009)

Development a Way of Protection for Domestic Reindeer from the Attack of Mosquitoes

Alexander Reshetnikov(✉)

FRC YaSC SB RAS M. G. Safronov Yakut Scientific Research Institute of Agriculture, 23/1, Bestuzheva-Marlinskogo Str., 677001 Yakutsk, Russia

Abstract. Objective: To develop a method for protecting domestic reindeer from mosquitoes and to assess its economic efficiency in the Arctic of Yakutia. Methods: The experiments were carried out in reindeer herd in the Anabar region of Yakutia. The protective efficiency of aerosol mist spraying over a herd of reindeer with deltamethrin emulsion was calculated by a method based by S.D. Pavlov (1982). Results: In summer, in the Anabar tundra, mosquitoes attack reindeer around the clock. One deer is attacked from 3328 to 6080 specimens of the genus Aedes. Then the reindeer died in large numbers from acute entomosis. To protect reindeer from them, an ultra-low-volume spraying technology has been developed. The economic effect of technology in Yakutia per 1 ruble of costs is 28.7 rubles. Discussion: The effectiveness of the technology is based on the dispersion of liquid substances. Due to this, a huge number of molecules of the dispersed substance is located in the outer layer of the aerosol particle, which causes a sharp increase in the activity of the active substance.

Keywords: Reindeer · Protection · Mosquitoes · Tundra · Blood-sucking dipteric insects · Biotopes

1 Introduction

Reindeer husbandry is the only branch of the northern economy in which only representatives of the indigenous peoples of the Far North are employed. According to the results of the All-Russian Agricultural Census of 2016, the number of domesticated reindeer in the Republic of Sakha (Yakutia) was 172.8 thousand animals and ranked third in the Russian Federation after Chukotka (182 thousand heads) and the Yamalo-Nenets Autonomous Okrug (886.85 thousand heads) [1]. A decrease in the number of reindeer in Yakutia can greatly affect the quality of life of the indigenous population of the republic: Dolgans, Evens, Yukagirs, Evenks and Chukchi. A decrease in their consumption of traditional food products such as deer and fish meat can lead to a deficiency in protein, amino acid, vitamin and mineral nutrition.

In 2007, the United Nations adopted the Declaration of the Rights of Indigenous Peoples as a universal basis for the standards of their survival, dignity, well-being and rights of indigenous peoples and promotes their full and effective participation in all

A. Muratov and S. Ignateva (Eds.): AFE 2021, LNNS 353, pp. 138–146, 2022.
https://doi.org/10.1007/978-3-030-91402-8_17

issues that concern them. It also guarantees their right to remain self-reliant and pursue their own priorities in economic, social and cultural development. The Declaration explicitly encourages harmonious cooperative relations between states and indigenous peoples [2].

Despite the importance of indigenous people in the Arctic, there is no accurate estimate of their size and distribution. We defined indigenous people as those groups represented by the "permanent participants" of the Arctic Council. The census in Canada, Russia and the United States records status as an indigenous person. In Greenland, a proxy measure is place of birth supplemented by other information. For the Nordic countries we utilized a variety of sources including registered voters' lists of the various Sami parliaments and research studies that established Sami cohorts. Overall, we estimated that there were about 1.13 million indigenous people in the northern regions of the 8 Member States of the Arctic Council. There were 8,100 Aleuts in Alaska and the Russian North; 32,400 Athabaskans in Alaska and northern Canada; 145,900 Inuit in Alaska, northern Canada and Greenland; 76,300 Sami in northern Norway, Sweden, Finland and Russia; and 866,400 people in northern Russia belonging to other indigenous groups [3].

Observations of Leung L. in Canada have established that among the Aboriginal population of Canada (which includes the First Nations, Inuit and Metis), diabetes mellitus contribute significantly to their higher morbidity and increased health disparity when compared to the non-Aboriginal Canadians. In view of this, the Federal Government of Canada had launched the Aboriginal Diabetes Initiative (ADI) in 1999 as part of the bigger Canadian Diabetes Strategy to provide a better framework for surveillance, public education and community-based management of diabetes [4].

Indigenous people in Canada are disproportionately affected by an increasing rate of chronic diet-related diseases and are undergoing rapid lifestyle changes that affect diet. Two different versions of "healthy eating" have been identified: one related to traditional food and cooking methods; the other reflects medical reports of diseases and diagnosed conditions. These findings expand the current understanding of the context of nutrition and healthy eating in Iyou Eastchi Canada, highlighting the country's current and historical experience. Ignoring the interdependence of traditional lifestyles and nutrition, respect for individuality, does not contribute to a deep understanding of the shaping experiences of indigenous peoples in the field of food and health [5].

The One Health paradigm recognizes the interdependence among the health and well-being of people, animals and the environment. As such, the framework aligns well with many Indigenous world views. This proactive, interdisciplinary, constructivist, and collaborative approach promise earlier detection of risks and threats, as well as more effective responses, in part by engaging community level stakeholders in all stages of the process. In the far North, humans, especially indigenous peoples, continue to live closely connected to their environment, in settings that exert significant impacts on health [6].

Population growth, socio-cultural and economic changes as well as technological progress have an immediate impact on the environment and human health in particular. Our steadily rising needs of resources increase the pressure on the environment and narrow down untainted habitats for plants and wild animals. Balance and resilience of ecosystems are further threatened by climate change, as temperature and seasonal shifts increase the pressure for all species to find successful survival strategies. Arctic and subarctic regions are especially vulnerable to climate change, as thawing of permafrost significantly transforms soil structures, vegetation and habitats. With rising temperature, the risk of zoonotic diseases in the Republic of Sakha (Yakutia) has also increased. As vegetation periods prolong and habitats broaden, zoonotic pathogens and their vectors find more favourable living conditions. Moreover, permafrost degradation may expose historic burial grounds and allow for reviving the vectors of deadly infections from the past. To assess the current state of knowledge and emerging risks in the light of the "One Health" concept, a German-Russian Symposium took place on 13 August 2018 in Yakutsk, Russian Federation. This symposium report presents the main findings generated from presentations and discussions [7].

When the wind dies down and the endless summer sun, burns the ponds, covers the frozen tundra, some of the most ferocious predators of the Arctic appear and form formidable blizzards from millions of mosquitoes, darken the horizon - and the mood of everyone. Large, blood-sucking mosquitoes already are the bane of people, caribou, reindeer, and other mammals eking out a living in the frozen north. Large, blood-sucking mosquitoes already are the bane of people, caribou, reindeer, and other mammals eking out a living in the frozen north. When the water temperature becomes warm, mosquitoes over the Arctic Circle appear en masse, grow faster, turning into adult winged pests in very large numbers [8].

Currently, environmental changes continue. The likelihood of a link between the environment and parasitic zoonoses will be important in protecting global health and well-being in those regions of the world where these effects are most severe. The possible effects of environmental change on life support systems, most importantly food, water, and shelter, are also discussed, particularly as they affect parasite transmission to people [9]. Diptera insects all over the world are of great medical and veterinary importance as dangerous pathogens of myiasis and vectors of infectious and invasive diseases. In India, cutaneous myiasis from flies attacks is causing significant damage to the wool industry. Animal health and safety costs are AUD $ 4 per sheep. Infectious diseases of livestock such as foot and mouth disease, mastitis and avian influenza, carried by zoophilic flies and horseflies, are economically important diseases that lead to large economic losses [10].

In many countries, scientific research is underway to protect domestic animals from harmful insects, vectors of dangerous animal and human diseases. To find effective methods and means of protecting cattle from tsetse flies in areas in Kenya, a new technology was tested using the repellent 2-methoxy 4-methylphenol to combat trypanosomiasis. A repellent dispenser was attached to a collar that was worn around the animal's neck. The trials began in April 2005. 12 cattle herds were recruited in each district. As a result, a 50% reduction in the incidence of trypanosomiasis in animals was obtained in comparison with the control. Thus, this technology turned out to be insufficiently effective [11].

Horny buffalo flies (Haematobia irritans and H. exigua) are small (3–5 mm long) dark gray hematophagous ectoparasites closely associated with livestock (cattle and buffalo) and sometimes with horses grazing in open pastures. The flies Haematobia irritans and H. exigua constantly attack and feed on the blood of cattle and buffaloes, resulting in significant dairy losses. Horn flies are not only annoying bloodsuckers, but also mechanical carriers of bacteria that cause mastitis in cattle and intermediate hosts of the nematodes Stephanofilaria and Parabronema. The main problem in protecting cattle and buffaloes from the horny buffalo flies Haematobia irritans and H. exigua is their insecticide resistance [12].

Mosquitoes (Phlebotominae) are of great medical and veterinary importance. They are common throughout the Old World and are carriers of leishmaniasis, vesicular stomatitis, bartonellosis, and sand fly fever [13]. Arthropods include many species of medical or veterinary importance. Among them, some are responsible for the transmission of certain dangerous diseases to humans, such as malaria, Chagas disease, leishmaniasis, African trypanosomiasis, lymphatic filariasis, onchocerciasis, dengue, yellow fever, and Zika. These diseases have a huge socioeconomic impact in the countries where they occur. Human-induced environmental changes, with a particular focus on urbanization and global warming, are contributing to their emergence or re-emergence in areas where they have not previously been recorded. Under these conditions, the control of arthropod vectors of especially dangerous diseases requires taking into account the knowledge of their genetic, behavioral and ecological characteristics and an integrated approach to their control [14]. Research by S. N. Rzhanikov found that by spraying with 0.001% of the active ingredient, an aqueous emulsion of deltsid per cow, 500 ml, and 250 ml for young cattle, using a stationary horizontal spray installation, were treated against mosquitoes (Culicidae) and horseflies (Tabanidae) in one season thousand cows. At the same time, the economic effect per 1 ruble of expenses was 80.44–76.73 rubles [15].

The purpose of our research was to develop a method for protecting domestic reindeer from blood-sucking dipterans insects and to assess its economic efficiency in the Arctic Yakutia.

2 Materials and Methods

The work was carried out in 2012–2020 under the program of the state academies of sciences 0821–2019-0003 "To develop methods for creating new biological products, technologies for the control and prevention of infectious, parasitic and non-infectious diseases of animals and birds, taking into account the climatic conditions and the peculiarities of the livestock industries in the Far North ". The patent search was carried out in accordance with the task and search regulations to determine the achieved level and development trends on the research topic. Enough literary sources and titles of protection were found.

The test of the effectiveness of protection was carried out in reindeer herd No. 7 of the Municipal Unitary Enterprise named after Hero of Labor Ilya Spiridonov, Anabar region of the Republic of Sakha Yakutia. The Anabarsky district is located 96°2′-97°9′N и 111°3′-113°2′E.

The counts of the number of attacking mosquitoes, horseflies and gadflies were carried out by catching them from a lure deer with an entomological net with removable bags. One count consisted of 10 strokes ("figure eight") in 10 repetitions. Every day, during the entire flight period, insects were recorded 3 times a day (at 7, 13, and 19 h local time).

In summer, every day during the entire period of activity and attacks of blood-sucking insects on reindeer, meteorological data were measured 3 times a day at 7, 13 and 19 h local time. Air temperature and humidity were measured with an aspiration psychrometer, wind speed - with an ASO-3 anemometer, atmospheric pressure - with an aneroid barometer, illumination - with a Yu-116 luxmeter, cloudiness - visually on a 10-point scale, precipitation - with a rain gauge. Cameral processing of arthropod collections was carried out in the arachnoentomology laboratory of the Yakutsk Research Institute. Identifying keys of the monographs by N. G. Olsufiev, I. A. Violovich, G. Ya. Bei-Bienko and I. A. Rubtsov were used in the identification of species. For ultra-low-volume spraying of a herd of domestic reindeer, an aerosol generator with a capacity of 850 m3/h with a dispersion of 70 microns was chosen for aqueous solutions with an air flow rate of 125 mps, with an engine of 4.1 L/sec. The volume of the container for the aqueous solution was 14 L.

In the experiment, the effectiveness of an aqueous emulsion of 0.05% deltamethrin was tested. Protective spraying of a herd of domesticated reindeer was carried out by spraying aerosol fog over the herd. The reindeer herd consisted of 2000 animals. When spraying 14 L of water emulsion over a herd of reindeer, which consisted of 2,000 animals one deer comes 7 ml of emulsion in the form of aerosols.

The protective efficiency of an ultra-low volume of spraying a herd of cows with 0.05% aqueous emulsion of deltamethrin was calculated using a method based on the determination of the coefficient of protective action (CPA), expressed as a percentage using the formula:

$$CPA = \frac{(B - A)}{B} \cdot 100 \tag{1}$$

where: CPA - coefficient of protective action in percent; A - is the number of insects on the treated animal; B - the number of insects on the control animal.

The number of attacking insects was determined by direct counting of mosquitoes that sat on the animal's skin, of the animal for 5 min.

3 Results

In the Anabar region, the Primorskaya tundra stretches in a discontinuous strip and occupies vast areas of grassy boggy lowlands; in elevated places, there are shrub tundra with forbs, moss, dwarf birches and willows. Tundra swamps are a continuous biotope for the mass development of mosquito larvae. In 2012–2013 the density of mosquito larvae in them ranged from 3000 to 7000 specimens per m^2. In the second decade of June, with warming, the flight of single mosquitoes was noted. Mass flight of adults and their attack on reindeer is observed from 12 to 28 July. On these days, daytime temperatures constantly rose to 19.5–29.8 °C, and nocturnal ones did not fall below 12.0–15 °C, which contributed to the mass activity of adults and attacks on animals.

In the context of the polar day, the diurnal variation of mosquito activity from 12 to 28 July was observed around the clock without a decline in numbers. During daytime counts, the number of mosquitoes ranged from 3328 to 6080 individuals for one entomological registration. The species composition was represented by mosquitoes of the genus Aedes.

With such a massive attack of blood-sucking mosquitoes on a herd of domesticated reindeer, their massive mortality from acute entomosis begins. At the same time, animals become uncontrollable, scatter across the endless tundra, or endless herd whirling in one place begins until the complete loss of strength. And starting with calves and young deer, exhausted animals are trampled in the cycle of the herd. In a few days, up to a thousand or more animals die at the same time.

Based on the foregoing, the task was set to develop a method to protect domesticated reindeer from mosquito attacks. When performing a patent search, the prototype 1 - RU No. 1656713 was found.

Prototype 1. A method of protecting reindeer from the attack of blood-sucking insects, in which the reindeer are treated by spraying with an aqueous emulsion of stomozan in an amount of 0.04–0.12%. Aqueous emulsions of stomozan provided a protective effect against mosquitoes for 1.0–2.8 h. The disadvantages of this method were: 1) insufficient protective effect of the stomozan repellent in protecting reindeer from the attack of blood-sucking dipteric insects. The technical objective of the claimed invention is the use of safe insecticides approved for veterinary medicine to protect domestic reindeer from the attack of gnat, while reducing labor intensity and increasing the efficiency of treatment. The technical result is solved by the fact that an ultra-small amount of the drug is used with the method of ultra-low-volume aerosol treatment in the form of fog of a herd of domestic reindeer, which ensured high environmental friendliness, insecticidal, repellent and anti-stress treatment of animals. The claimed method of protecting domesticated reindeer from a midge attack differs from the prototype in that:

1. a drug from the group of synthetic pyrethroids is used - deltamethrin, approved for use in Russia;
2. instead of fine-droplet spraying, ultra-low-volume aerosol treatment of a herd of domestic reindeer is used with an insecticide concentration of 0.036–0.05% with a consumption rate per animal of 7 ml of emulsion;
3. the complexity of processing a herd of deer decreases, against 120 L of emulsion solution in the prototype in the claimed method, it is enough to prepare 14 L;
4. the duration of the protective action is almost doubled - up to 6 h in comparison with the analogue.

When determining the economic efficiency of the developed method of protecting domestic reindeer from the attack of blood-sucking mosquitoes by ultra-low-volume spraying on the example of herd No. 7 with the number of animals more than 2000 heads in the seaside tundra of the Municipal Unitary Enterprise of the Anabar region, the following data were used: the number of animals by sex and age groups, died from the attack mosquitoes, economic damage from mortality. At the same time, the economic damage from mortality due to the attack of blood-sucking mosquitoes in 2011 amounted to 1,236 rubles, in 2012 it amounted to 1,411.92 rubles, in 2014 it was 3,174.6 rubles, an average of 1940.84 rubles. The cost of processing is 65352 rubles.

When the developed method of protecting domestic reindeer from blood-sucking dipteric insects by ultra-low-volume spraying was applied in 2013, no mortality of domestic reindeer was observed in herd No. 7. The annual economic effect of the developed technology for protecting domestic reindeer in herd No. 7 with a population of 2 thousand animals amounted to 1875490 rubles. The economic effect per 1 ruble of costs amounted to 28.7 rubles in Table 1.

Table 1. Evaluation of economic efficiency ways to protect domesticated reindeer from the attack of mosquitoes in the Arctic tundra of Yakutia.

No. of line	Indicators	Unit of measurement	Reindeer herd with 2,000 animals
1	The amount of damage to reindeer herd No. 7 on average over 3 years	₽	1940840
2	The cost of processing the herd, in prices of one year (2014)	₽	65350
3	The duration of the protective action (with a CPD of more than 75%)	hour	6
4	Estimated annual economic effect (1 line + 2 line) = 4 line	₽	1875490
5	Economic effect per 1 ruble of costs (1 line / 2 line) = 5 line	₽	28.7

4 Discussion

The Anabar region is located in the Russian Arctic. The vast tundra stretches in a discontinuous strip along the coast of the Laptev Sea, where large herds of domesticated reindeer are bred. Herds of deer roam the tundra in the summer to the sea, and in the winter to the forest-tundra. In summer, the tundra is occupied by vast areas of grassy lowlands; in high places, there are shrub tundra with forbs, moss, birches and willows with dwarfs.

Tundra swamps are a continuous biotope for the mass development of mosquito larvae. In 2012–2013 the density of mosquito larvae in them ranged from 3 to 7 thousand specimens per m2. In the second decade of June, with warming, the flight of single mosquitoes was noted. Mass flight of adult mosquitoes and attacks on reindeer were observed from 12 to 28 July. On these days, daytime temperatures constantly rose to 19.5–29.8 °C and nocturnal ones did not fall below 12.0–15 °C, which contributed to the mass emergence of adults.

In the context of the polar day, the daily variation of mosquito activity from 12 to 28 July was observed around the clock without a decline in numbers. During daytime counts, the number of mosquitoes ranged from 3328 to 6080 individuals for one entomological registration. The species composition is represented by mosquitoes of the genus Aedes. With such a massive attack of blood-sucking mosquitoes on a herd of domesticated reindeer, their massive mortality from acute entomosis began. The loss of reindeer for reindeer herders means the loss of work, wages and, most importantly, the traditional way of life of Dolgan reindeer herders, the indigenous peoples of Yakutia. In order to help them, the goal was to develop a method for protecting domestic reindeer from blood-sucking dipteric insects and the economic efficiency of the method in the Anabar region of Yakutia was determined.

As a result of a patent search, a laborious, ineffective analogue was found. Instead, a modern method has been developed to protect domesticated reindeer from being attacked by blood-sucking mosquitoes. The developed method of protecting domestic reindeer from harmful insects differs from the prototype in its low labor intensity in use, convenience in the constant movement of the herd from the forest-tundra to the seashore and back. The high efficiency of the protective action on reindeer from the attack of blood-sucking mosquitoes is based on the use of aerosol technology, in which the liquid is dispersed to an ultra-low-volume state. Due to this, a huge number of molecules of the dispersed substance are located in the outer layer of the aerosol particle, which causes a sharp increase in the activity of the active substance, accelerating the physical and chemical processes. The dispersion of particles of 70 microns and above ensures the achievement of an effective insecticidal effect in ultra-low doses, thereby ensuring environmental friendliness, safety and repellency. Anti-stress treatment of domesticated reindeer is ensured by the fact that aerosol mist is sprayed over the herd from a distance of 30 m. The economic effect of the method of protecting domesticated reindeer from mosquitoes in Yakutia per 1 ruble of costs is 28.7 rubles.

5 Conclusions

Reindeer husbandry is the only branch of the northern economy in which only representatives of the indigenous peoples of the Far North are employed. A decrease in the number of deer in Yakutia can greatly affect the quality of life of the indigenous small population. The preservation of reindeer husbandry contributes to the preservation of the traditional way of life.

A method has been developed to protect domestic reindeer from mosquitoes with an ultra-small amount of the preparation with high environmental friendliness, efficiency, insecticidal, repellent and anti-stress treatment of animals. The high efficiency of the protective action on reindeer from the attack of blood-sucking mosquitoes is based on the use of aerosol technology, in which the liquid is dispersed to an ultra-low-volume state. The economic effect of the method of protecting domesticated reindeer from mosquitoes in Yakutia per 1 ruble of costs is 28.7 rubles. The developed technology for the protection of cattle from blood-sucking dipterans and adult gadflies as an object of industrial property is protected by RF patent No. 2595831.

Acknowledgments. The authors are deeply grateful to Roman Dmitrievich Tuprin and Ekaterina Tuprina in the Anabar District for their assistance in collecting data.

References

1. Total of the All-Russian Agricultural Census of 2016: The number of reindeer in farms of all categories (2018) Federal State Statistics Service, Information and Publishing Center "Statistics of Russia". https://rosstat.gov.ru/storage/mediabank/VSHP%202016_T5_k1_web.pdf
2. United Nations Declaration on the Rights of Indigenous Peoples. Manual Natio. Hum. Rights Institu (2007) Office of the High Commissioner for Human Rights. https://www.ohchr.org/EN/Issues/IPeoples/Pages/Declaration.aspx
3. Young, T.K., Bjerregaard, P.: Towards estimating the indigenous population in circumpolar regions. Int. J. Circum. Health **78**, 1653749 (2019). https://doi.org/10.1080/22423982.2019.1653749
4. Leung, L.: Diabetes mellitus and the aboriginal diabetic initiative in Canada: an update review. JFMPC **5**, 259–265 (2016). https://doi.org/10.4103/2249-4863.192362
5. Goettke, E., Reynolds, J.: "It's all interconnected… like a spider web": a qualitative study of the meanings of food and healthy eating in an Indigenous community. Int. J. Circum. Health **78**, 1648969 (2019). https://doi.org/10.1080/22423982.2019.1648969
6. Hueffer, K., Ehrlander, M., Etz, K., Reynolds, A.: One health in the circumpolar North. Int. J. Circum. Health **87**, 1607502 (2019). https://doi.org/10.1080/22423982.2019.1607502
7. Huber, I., Potapova, K., Reshetnikov, A.D., et al.: Symposium report: emerging threats for human health – impact of socioeconomic and climate change on zooanthroponosis in the Republic of Sakha (Yakutia) Russia. Int. J Circum. Health **79**, 1715698 (2020). https://doi.org/10.1080/22423982.2020.1715698
8. Welch, C.: Why the Arctic's mosquito problem is getting bigger, badder. The trend could spell disaster for caribou (and more nasty bites for humans). Nation Geog (2015)
9. Polley, L., Kutz, S.J., Hoberg, E.P.: Parasite Zoonoses. Encycl Env Health (2011).https://doi.org/10.1016/B978-0-444-63951-6.00677-X
10. Aruna, P., Chakravarty, A.K.: Disease-resistant livestock populations-their development and impact on farm economics. Genet. Breed Dis. Resist. Livest. **22**, 351–358 (2020). https://doi.org/10.1016/B978-0-12-816406-8.00022-X
11. Bett, B., Randolph, T.F., Irungu, P., et al.: Field trial of a synthetic tsetse-repellent technology developed for the control of bovine trypanosomosis in Kenya. Prev. Vet. Med. **97**, 220–227 (2010). https://doi.org/10.1016/j.prevetmed.2010.09.001
12. Votýpka, J., Mihalca, A.D.: Horn Flies (Haematobia and Haematobosca). Refer Mod. Biomed. Sci. (2020). https://doi.org/10.1016/B978-0-12-818731-9.00021-5
13. Munstermann, L.E.: Med. Vet. Entom. (2019). https://doi.org/10.1016/B978-0-12-814043-7.00012-1
14. Benelli, G., Canale, A., Lucchi, A., Di Giovanni, F.: Insects and mites of medical and veterinary importance: a broad overview. Refer Module Biomed. Sci. (2020). https://doi.org/10.1016/B978-0-12-818731-9.00040-9
15. Rzhanikov, S.N.: Efficiency and safety of deltsid in the treatment of cattle against blood-sucking insects: abstract cand thesis in vet sci, Tyumen (2005). https://search.rsl.ru/ru/record/01002751816

Influence of Submerged Cultivation Conditions on the Titer of Active Cells of Rhizobia

Mariya Yakimenko(✉), Arina Sorokina, and Stepan Begun

All-Russian Research Institute of Soybeans, 19, Ignat'evskoe shosse, Amur Region, 675027 Blagoveshchensk, Russia

Abstract. The studies were carried out in 2019–2020 in order to select the optimal conditions for submerged cultivation of collection strains of rhizobia, selection of All-Russia Scientific Research Institute of Soybean, providing a high titer CFU/ml. The objects of research were CM-42 and 648a strains of *Bradyrhizobium japonicum* (Jordan, 1982) and BB-49, KB-11 *Sinorhizobium fredii* (Scholla, Elkan, 1984), for which copyright certificates and patents were obtained. The effectiveness of the CM-42 and 648a strains of *B. japonicum* and BB-49 and KB-11 strains of *S. fredii* selected for research was tested in 2014–2016 on meadow chernozem soils of the Amur Region in field experiments on soybeans of the Harmonia, Evgeniya and MK 100 varieties, selection of All-Russia Scientific Research Institute of Soybean. Under the conditions of 72-h submerged cultivation in a laboratory fermenter on nutrient medium No. 1, where sucrose was used as a carbon source, at a cultivation temperature of +28 °C, stirrer rotation 70 rpm and pH of the medium 6.73–7.1, the maximum titer of active cell of the strains of *S. fredii* BB-49 were obtained, amounting to $17 \cdot 10^9$ CFU/ml.

Keywords: Rhizobia · Strain · B. japonicum · S. fredii · Submerged cultivation · Titer · Carbon sources · Nutrient medium

1 Introduction

In modern economic conditions, the main factor in increasing soil fertility and obtaining high yields without environmental damage is the biologization of agriculture, aimed at partial or complete replacement of agrochemicals with biological preparations [1–3]. For example, the experience of using preparations based on nodule bacteria of the Rhizobiaceae family, in symbiosis with which leguminous plants acquire the ability to assimilate and use atmospheric nitrogen for their nutrition, shows that this method for increasing productivity is an effective, economical and completely safe for humans, animals and other components of biocenoses, and makes it possible to abandon the use of mineral nitrogen fertilizers [4–6].

Obtaining a high-quality rhizobial preparation directly depends on the producer strain, which should have not only such important properties as virulence, competitiveness, efficiency, but also the ability to accumulate a sufficient number of active bacterial cells in the culture medium [7–9].

A. Muratov and S. Ignateva (Eds.): AFE 2021, LNNS 353, pp. 147–155, 2022.
https://doi.org/10.1007/978-3-030-91402-8_18

This is due to the peculiarities of the technology for the production of rhizobial preparations, where the central stage is the cultivation of microorganisms on production nutrient media in order to obtain a high titer of active cells (at least 10^7 CFU/ml) – the main criterion of their effectiveness [10–12]. For the successful cultivation of microorganisms in industrial conditions, the composition of the medium, duration, temperature, pH, and other parameters are very important [13–15].

The purpose of this work is to select a carbon source and cultivation time that provide a high titer of bacterial cells of the patented *B. japonicum* and *S. fredii* strains isolated from Far Eastern natural populations by submerged cultivation.

2 Objects and Methods

Experimental studies were carried out at the All-Russia Scientific Research Institute of Soybean (Blagoveshchensk). The objects of research were the strains of *Bradyrhizobium japonicum* (CM-42, 648a) and *Sinorhizobium fredii* (BB-49, KB-11), from the collection of the Federal State Budgetary Scientific Institution FSC VNII of Soybeans [16], possessing stable efficacy, for which copyright certificates [17] and patents [18–20] were obtained.

Cultivation is the main stage of the biotechnological process. There are two main ways of cultivating microorganisms - surface and submerged. Submerged cultivation is carried out on liquid nutrient media, in which microorganisms develop in the entire volume of the nutrient medium. Submerged cultivation is carried out in special apparatus - fermenters (bioreactors), equipped with stirrers and a sterile air supply system to ensure the growth of aerobic microorganisms. The submerged method is more profitable for the industry, as it allows for complete mechanization and automation of the process, and makes it possible to avoid infection of the technological process by extraneous microflora [21–23].

As production media used nutrient media No. 1 and No. 2 of the following composition, g/l: K_2HPO_4 – 0.5; KH_2PO_4 – 0.5; $MgSO_4$ – 0.1; NaCl – 0.2; $CaSO_4$ – 0.1; agar-agar - 1.0; where sucrose (nutrient medium No. 1) and glucose (nutrient medium No. 2) were used as a carbon source.

The selection of the optimal carbon source was carried out according to the indicator of the maximum number of cells of the studied strains for the minimum period of growth in the process of submerged cultivation. The adjustment of the submerged cultivation parameters was conducted on special-purpose equipment - a laboratory TN-50L fermenter (maximum stirrer rotation 70 rpm), volume 50.0 l, air compressor power 1.1 kW (maximum pressure 0.7 MPa, volume of produced gas 0.11 m^3) at a temperature of 28 $\pm$ 1 °C, pH = 6.40. For inoculation of the fermenter, a 5–7-day-old inoculum was used in an amount of 10% of the volume of the nutrient medium. The titer in CFU/ml and the pH of the culture liquid were determined after 24, 48, 72 h of cultivation. The inoculum was prepared by growing pure cultures on a laboratory SHO-1D shaker (rotational speed 180–200 rpm) in 250 ml Erlenmeyer flasks with 100 ml of liquid culture medium in a thermostat at a temperature of +27… 28 °C [9, 13, 24]. The number of cells of the studied strains in the inoculum was determined by the method of serial dilutions and their subculture by the Koch method on a solid nutrient medium [25]. For this, prepared tubes with 9 ml of sterile saline solution were numbered, and 1.0 inoculum was added to the first tube. Then, after mixing, in the same volume, each time with a new sterile pipette, the inoculum was transferred to the next tube. From dilutions 10^7–10^9, deep sowing was done in Petri dishes using agar mineral-plant nutrient medium (MRS) of the following composition, in g/l: soy flour – 10; mannitol – 20; agar-agar – 15–20; K_2HPO_4 – 0.5; KH_2PO_4 – 0.5; $MgSO_4$ – 0.1; NaCl – 0.2; $CaSO_4$ – 0.1; molybdenum salt - traces. The inoculated Petri dishes were incubated at a temperature of +27… 28°C. After 3–4 days for *S. fredii* strains and after 8–11 days for soybean rhizobia strains of *B. japonicum*, the number of colonies in Petri dishes was counted and the titer of these bacteria was determined per 1 ml of culture liquid.

Field studies were carried out on the experimental field of the All-Russia Scientific Research Institute of Soybean (Amur Region, Tambov District, Sadovoe village). The registration plot area was 2.7 sq. m (length – 3.0 m, width – 0.9 m.). The repetition of the experiment was 4 times Soybean seeds were sown by hand, 2 seeds per point at intervals of 10 cm to a depth of 4–5 cm. Wide-row sowing with 45 cm row spacing. The predecessor was wheat. On the day of sowing, soybean seeds were treated with the tested strains at the rate of 1 million bacterial cells per 1 seed. The agro-engineering preparation of the experimental plot included autumn plowing and harrowing, spring disking, pre-sowing application of soil herbicide, harrowing, rolling. Hand weedings were carried out during the growing season. The soil of the experimental plot is meadow chernozem, medium-thick (humus layer (A + AB) is 20–30 cm), characterized by the following parameters - humus content (GOST 26 213–84) 4.3%; mobile phosphorus P_2O_5 and potassium K_2O (according to Kirsanov) – respectively 37 mg/kg and 211 mg/kg of soil; mineral nitrogen $NO_3 + NH_4$ 18.6 mg/kg; the reaction of the soil solution is slightly acidic (pH_{KCl} 5.2 units); hydrolytic acidity (Hr) – 2.52 mg-Eq per 100 g of soil.

The establishment, the conducting of field experiments, the accounting for the harvest, statistical processing of harvest data were carried out according to the method of B.A. Dospekhov [26].

3 Results and Discussion

The effectiveness of the strains *B. japonicum* CM-42, 648a and *S. fredii* BB-49, KB-11 selected for research was tested in 2014–2016 on meadow chernozem soils of the Amur Region in field experiments on soybeans of the varieties Harmonia, Evgeniya and MK 100, selection of All-Russia Scientific Research Institute of Soybean Over the years of study, the tested strains of rhizobia increased the yield of the air-dry mass of soybean plants by an average of 0.57 t/ha, seeds – by 0.27 t/ha. The highest increases in the yield of biomass (0.52 t/ha) and soybean seeds (0.24 t/ha) of the Harmony variety relative to the control were obtained in the variant with seed treatment with strain KB-11, biomass of MK 100 varieties (0.79 t/ha) and Evgeniya (0.98 t/ha) - in the variant with the strain CM-42, soybean seeds of the MK 100 variety (0.58 t/ha) – in the variant with the strain BB-49, soybean seeds of the Evgeniya (0, 27 t/ha) – in the variant with strain KB-11 (Table 1).

Table 1. Influence of rhizobia strains on the productivity of soybean varieties Harmonia, MK 100, Evgeniya.

Strain	Soybean variety Harmony. 2014		Soybean variety MK 100. 2015		Soybean variety Evgeniya. 2016		Increase in yield from strains on average. t/ha	
	Biological productivity. t/ha							
	Aboveground mass	Seeds	Aboveground mass	Seeds	Aboveground mass	Seeds	Aboveground mass	Seeds
Control	5.37	2.49	6.62	3.07	5.84	2.54	–	–
648a	5.49	2.62	7.11	3.38	6.42	2.71	0.4	0.2
BB-49	5.71	2.73	7.31	3.65	6.78	2.77	0.66	0.35
KB-11	5.89	2.73	6.95	3.26	6.33	2.81	0.45	0.23
CM-42	5.74	2.68	7.41	3.53	6.82	2.75	0.72	0.29
LSD_{05}						0.1	0.08	

For the production of microbiological fertilizers based on rhizobia, special attention is paid to the main parameters as sources of carbon nutrition in the nutrient medium, as well as the cultivation time until the end product is released, since the temperature ($t = \pm 28$ °C) and pH (6.4–7.5) are optimal for the development of these microorganisms and are supported automatically during the entire cultivation period.

Table 2. Growth dynamics of active cells of *B. japonicum* and *S. fredii* strains during 72-h cultivation in a laboratory fermenter on nutrient medium No. 1 (carbon source suc rose).

Strain	Fermentation period, hours	pH of the medium	Titer CFU$\cdot 10^7$/ml	Titer CFU$\cdot 10^9$/ml
CM-42	24	7.12	5.0	–
	48	7.05	4.0	–
	72	7.17	–	–
648a	24	7.44	4	–
	48	7.37	12	–
	72	7.10	61	3
KB-11	24	6.77	61	–
	48	6.62	36	–
	72	6.46	8	–
BB-49	24	6.74	–	–
	48	6.55	4	2
	72	6.73	40	17

During 72-h cultivation in a laboratory fermenter of the studied strains, on nutrient medium No. 1, the maximum titer of bacterial cells was obtained in strains 648a ($3{\cdot}10^9$ CFU/ml at a pH of the medium of 7.1) and BB-49 ($17{\cdot}10^9$ CFU/ml at pH of the medium 6.73). Observations of the growth of strains CM-42 and KB-11 during submerged cultivation showed that after 24-h fermentation the titer of microbial units of these strains was high and amounted to $5{\cdot}10^7$ CFU/ml at a pH of the medium 7.12 and $61{\cdot}10^7$ CFU/ml at a pH of the medium 6.77, respectively. An increase in fermentation time led to a decrease in the titer of CFU $\cdot$ 10^7/ml of CM-42 and KB-11 strains, the growth of microbial units of these strains in a dilution of 10^9 was absent throughout the entire fermentation period (Table 2).

Under the conditions of submerged 72-h cultivation on nutrient medium No. 2, where glucose was used as a carbon source, all studied strains (CM-42, 648a, BB-49, and KB-11) showed a titer of CFU$\cdot10^9$/ml at different fermentation time (Table 3). Fluctuations in the pH of the nutrient medium during fermentation were noted within the normal range from pHof the medium 6.41 to pH of the medium 7.76. The maximum titers of bacterial cells were shown by strains CM-42 ($8{\cdot}10^9$ CFU/ml at pH of the medium 6.77) and KB-11 ($2{\cdot}10^9$ CFU/ml at pH of the medium 6.65) after 72 and 48 h of cultivation, respectively.

Table 3. Growth dynamics of active cells of *B. japonicum* and *S. fredii* strains during 72-h cultivation in a laboratory fermenter on nutrient medium No. 2 (carbon source glucose).

Strain	Fermentation period, hours	pH of the medium	Titer CFU$\cdot10^7$/ml	Titer CFU$\cdot10^9$/ml
CM-42	24	6.83	1	1
	48	6.89	13	2
	72	6.77	37	8
648a	24	7.76	7	1
	48	7.35	1	–
	72	6.89	1	–
KB-11	24	7.03	1	1
	48	6.65	2	2
	72	6.52	1	1
BB-49	24	7.00	13	1
	48	6.86	3	1
	72	6.41	1	–

Note: "–" no growth

The titer CFU/ml of strains BB-49 and 648a, under conditions of 72-h fermentation on nutrient medium No. 2, was lower than during cultivation on nutrient medium No. 1, where sucrose was used as a carbon source. The maximum titer of microbial units of strains BB-49 and 648a as a result of using nutrient medium No. 2 was obtained during

24-h cultivation (pH $_{\text{of the medium}}$ 6.77–7.76) and amounted to $1 \cdot 10^9$ CFU/ml. A further increase in the fermentation time led to the growth termination of the titer of bacterial cells of BB-49 and 648a strains, so after 48 h of cultivation of strain 648a and 72 h of cultivation of strain BB-49 there was no growth of microbial units in a dilution of 10^9.

In the studied strains, a high degree of dependence of the titer of microbial cells on their cultivation time in the fermenter was observed (Fig. 1).

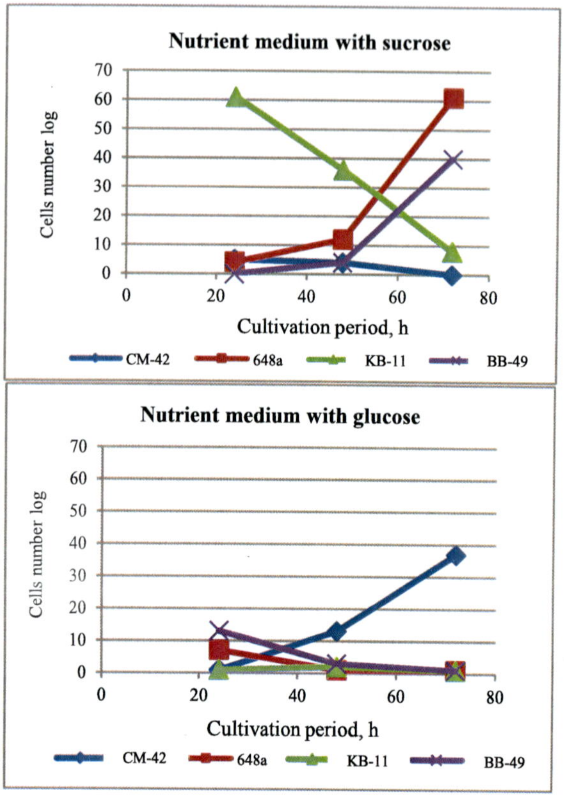

Fig. 1. Dynamics of cell growth of the patented strains *B. japonicum* and *S. fredii* during submerged cultivation in a fermenter.

Strains 648a and BB-49 (r = 0.92 and r = 0.91) showed a direct dependence of the CFU titer on the cultivation time on a nutrient medium with sucrose, while an inverse relationship was noted on a nutrient medium with glucose (r = –0.87 and r = –0.93, respectively).

The strain CM-42 at 72-h cultivation showed a high inverse dependence of the CFU titer on the cultivation time when using sucrose as a nutrition source (r = –0.94), and a high direct dependence when using glucose as a carbon source (r = 0.98). The dependence of the CFU titer of KB-11 strain on the cultivation time on a nutrient medium with glucose had not been established, while on a nutrient medium with sucrose, the dependence of the CFU titer on the cultivation time was very high. It can be assumed

Table 4. Correlation dependence of CFU titer on fermentation time on media with different carbon sources, *r*.

Nutrient medium with sucrose			Nutrient medium with glucose		
Strain	CFU·10^7/ml	CFU·10^9/ml	Strain	CFU·10^7/ml	CFU·10^9/ml
CM-42	−0.94	–	CM-42	0.98	0.93
648a	0.92	0.87	648a	−0.87	−0.87
KB-11	−0.99	–	KB-11	–	–
BB-49	0.91	0.92	BB-49	- 0.93	- 0.87

that the CFU titer of the KB-11 strain more depends on the carbon source than on the cultivation time (Table 4).

4 Conclusion

Studies of strains B. japonicum CM-42, 648a and S. fredii BB-49, KB-11, selected by the All-Russian Scientific Research Institute of Soybeans, have shown that they have a stable efficacy. Over the years of study, the tested strains of rhizobia increased the yield of the air-dry mass of soybean plants by an average of 0.57 t/ha, seeds – by 0.27 t/ha compared with the control. During submerged cultivation of the studied strains in a laboratory fermenter of the TN-50L series, it was found that the most technologically advanced in terms of their ability to accumulate a CFU·10^9/ml titer on nutrient media with various sources of carbon nutrition are the BB-49 strain of S. fredii and 648a strain of B. japonicum. The maximum cell titers of S. fredii BB-49 and B. japonicum 648a strains, which amounted to 17·10^9 CFU/ml and 3·10^9 CFU/ml, respectively, were obtained under conditions of 72-h submerged cultivation in a laboratory TN-50L fermenter on nutrient medium No. 1, where sucrose was used as a carbon source, at a cultivation temperature of +28 °C, stirrer rotation of 70 rpm and pH of the medium 6.73–7.1. The KB-11 and CM-42 strains showed the maximum titers of bacterial cells, amounting to 2·10^9 CFU/ml after 48 h and 8·10^9 CFU/ml, respectively, after 72-h submerged cultivation in a laboratory TN-50L fermenter on a nutrient medium No. 2, where glucose was used as a carbon source, at a cultivation temperature of +28°C, a stirrer rotation of 70 rpm and a pH of the medium 6.65–6.77.

The results obtained should be used in the production of inoculants to optimize the conditions for the cultivation of rhizobia.

References

1. Tikhonovich, I., Provorov, N.: Microbiology is the basis of sustainable agriculture: an opinion. Ann. Appl. Biol. **159**(2), 155–168 (2011). https://doi.org/10.1111/j.1744-7348.2011.00489.x
2. Nosheen, S., Ajmal, I., Song, Y.: Microbes as biofertilizers, a potential approach for sustainable crop production. Sustainability **13**, 1868 (2021). https://doi.org/10.3390/su13041868

3. Itelima, J., et al.: Bio-fertilizers as key player in enhancing soil fertility and crop productivity: a review. J. Agric. Food Sci. **6**(3), 73–83 (2018). https://doi.org/10.26765/DRJAFS.2018.4815
4. Minnebaev, L., Kuzina, E., Rafikova, G., Chanyshev, I., Loginov, O.: Productivity of legume-rhizobia complex under the influence of growth-stimulating microorganisms. Agric. Biol. **54**(3), 481–493 (2019). https://doi.org/10.15389/agrobiology.2019.3.481rus
5. Glyan'ko, A., Ishchenko, A., Filinova, N.: Effect of nodule and pathogenic bacteria on levels of nitric oxide and cyclic adenosine monophosphate in pea roots at initial stages of interaction. Vestnik Khar'kovskogo natsional'nogo agrarnogo universiteta. Biol. Ser. **3**(42), 6–22. (2017). https://doi.org/10.21285/2227-2925-2020-10-2-294-302
6. Jesus, E., Leite, R., Bastos, R., Aragão, O., Araújo, A.P.: Co-inoculation of Bradyrhizobium stimulates the symbiosis efficiency of Rhizobium with common bean. Plant Soil **425**(1–2), 201–215 (2018). https://doi.org/10.1007/s11104-017-3541-1
7. Gavrilova, N., Sadanov, A., Dadonova, T., Ratnikova, I.: Criteria of selection of strains of nodule bacteria in structure of biological nitrogen and increase of bean cultures. Izvestiya Natsional'noi akademii nauk Respubliki Kazakhstan. Biol. Med. Ser. **1**, 78–83 (2015). https://doi.org/10.32014/2020.2224-526
8. Teresa, M., Goma-Tchimbakala, J., Eckzechel, N.S.A., Aimé, L.A.: Isolation and characterization of native rhizobium strains nodulating some legumes species in South Brazzaville in Republic of Congo. Adv. Biosci. Biotech. **12**, 10–30 (2021). https://doi.org/10.4236/abb.2021.121002
9. Okafor, N.: Modern Industrial Microbiology and Biotechnology. Science Publishers, Enfield, NH, USA. Printed in India (2007)
10. Bhutani, N., Maheshwari, R., Kumar, P., Suneja, P.: Bioprospecting of endophytic bacteria from nodules and roots of Vigna radiata, Vigna unguiculata and Cajanus cajan for their potential use as bioinoculants. Plant Gene (2021). https://doi.org/10.1016/j.plgene.2021.100326
11. Laktionov, Y., Popova, T., Andreev, O., Ibatullina, R., Kozhemyakov, A.: Creating new forms of growth-stimulating microbial preparations and their performance on various cultures. Agric. Biol. **3**, 116–118 (2011)
12. Buntić, A., Stajković-Srbinović, O., Knežević, M., Kuzmanović, D., Rasulić, N., Delić, D.: Development of liquid rhizobial inoculants and pre-inoculation of alfalfa seeds. Arch. Biol. Sci. **71**(2), 379–387 (2019). https://doi.org/10.2298/ABS181008062B
13. Garipova, S., Irgalina, R., Dmitrieva, D., Kutueva, A.: Evaluation of new strains of endophytic bacilli and rhizobia when inoculated of common bean Ufimskaya variety under South Ural. Vestn. Orenburgskogo un-ta **10**, 437–439. (2009). 10/25198/1814-6457
14. Kozhemyakov, A., Laktionov, Y., Popova, T., et al.: The scientific basis for the creation of new forms of microbial biochemicals. Agric. Biol. **50**(3), 369–376 (2015). https://doi.org/10.15389/agrobiology.2015.3.369eng
15. Pajcin, I., Vlajkov, V., Dodic, J., Jokic, A., Grahovac, J.: Biotechnological production of plant inoculants based on nitrogen-fixing bacteria. J. Process Energy Agric. **2**, 25 (2021). https://doi.org/10.5937/jpea25-31071
16. Yakimenko, M., Begun, S.: Distinctive Features of Rhizobia Species *Sinorhizobium fredii* and *Bradyrhizobium japonicum* inhabiting. Soils Russ. Far East, Russ. Agric. Sci. **47**(1), 58–62 (2021). https://doi.org/10.3103/S1068367421010201
17. Til'ba, V., Begun, S.: R. japonicum 648a nodule bacteria strain - producer of rhizotorphin for soybeans: a. s. 1221865 SSSR. № 3785868/30-15/applied. Bul. **20**, 5 (1984)
18. Til'ba, V., Begun, S., Yakimenko, M.: Bradyrhizobium japonicum BKM B-2455D nodule bacterial strain for manufacturing of bacterial soya fertilizer. Patent 2415925. № 2009139609. Bul. **10**(CM-42). 26 Sept 2009

19. Til'ba, V., Begun, S., Yakimenko, M.: Strain of nodule bacteria of the species S. fredii KB-11 for the manufacture of bacterial fertilizer for soybeans. Patent 2413707. № 2009112119. Bul. **7**, 1 Apr 2009
20. Til'ba, V., Begun, S., Yakimenko, M.: Strain of nodule bacteria S. fredia VKM V-2458D for the production of bacterial fertilizer for soybeans. Patent 2415924. № 2009139614. Bul. **10**(BB-49), 26 Oct 2009
21. Shevelukha, V., Kalashnikova, E., Kochieva, E., et al.: Sel'skokhozyaistvennaya biotekhnologiya. Vyssh shk (2008)
22. Chechina, O.: General Biotechnology. Yurayt (2019)
23. Lukanin, A.: Engineering biotechnology: fundamentals of microbiological production. INFRA-M (2017)
24. Kartyzhova, L., Semenova, I., Korolenok, N., Aleshchenkova, Z., Romanova, L.: Efficient strain of slow-growing nodulating bacteria *Bradyrhizhobium jponicum* 84KL as a basis of biofertilizer Soyarhiz. Vestsi Natsyyanal'nai Akadehmii Navuk Belarusi. Ser. Biyalagichnykh Navuk **2**, 107–112 (2014). https://doi.org/10.29235/1029-8940
25. Lavrenchuk, L.: Praktikum po mikrobiologii. Ural University Publishing House (2019)
26. Dospekhov, B.: Metodika polevogo opyta (s osnovami statisticheskoi obrabotki rezul'tatov issledovanii. Book on Demand (2012)

The Influence of Mineral Fertilizers on the Productivity of Spring Triticale in the Conditions of the Southern Zone of the Amur Oblast

Alexey Muratov(✉), Pavel Tikhonchuk, and Eugenia Tuaeva

Far Eastern State Agrarian University, 86, Politeknicheskaya Street, Blagoveschensk 675005, Russia

Abstract. High yields can be achieved by using modern agricultural techniques, including the optimization of mineral nutrition. In this regard, the aim was to study the possibility of increasing the photosynthetic activity and plant productivity of spring triticale based on creating the most optimal nutrient regime of the soil as a result of different doses of mineral fertilizers during sowing. The research was conducted in 2017–2019 in the of the Amur region. As a result of studies, it was found that the application of nitrogen fertilizer N_{30} contributes to the highest level of field germination to 14% relative to the control. The size of leaf area also depends on the level of mineral nutrition and it was already significant in the initial period of growth, to the phase of earing the level of leaf surface area reached the maximum in the variant with the fertilizer rate $N_{60}P_{30}$ (20.5 thousand m^2/ha), while the difference with the variant without fertilizer was 23%. The highest yield (3.38 t/ha) was noted in the variant when the highest dose of $N_{60}P_{60}$ was applied: the increase in this variant compared to the control was 0.70 t/ha or 26%, it shows the positive effect of mineral fertilizers on the productivity of spring triticale.

Keywords: Spring triticale · Mineral fertilizers · Yield

1 Introduction

The priority direction in the agricultural sector of the Amur Oblast today is the development of crop production, so there is a question of obtaining high yields of agricultural products. A partial solution to this issue is facilitated by the expansion of cultivated areas under a relatively new grain crop - triticale, especially since the share of soybean crops [1] is too high; therefore, it is necessary to introduce new alternative crops in the region's dropping.

The range of spring triticale in Russia is gradually expanding, occupying a certain niche in the structure of dropping, expanding the list of biodiversity of cultivated crops and contributing to the strengthening of the fodder base of animal husbandry. After all, triticale combines the best features and properties of both parental forms: high productivity, resistance to viral and fungal diseases, high protein and lysine content in grain and herbage, and low demand for soil fertility [2, 3].

A. Muratov and S. Ignatova (Eds.): AFE 2021, LNNS 353, pp. 156–163, 2022.
https://doi.org/10.1007/978-3-030-91402-8_19

The production of spring triticale in the Amur Oblast is very promising in connection with the set task of the development of animal husbandry. This crop has already proved itself as a fodder-grain crop [4, 5]. At the same time, high yields of spring triticale grain are based on the study of the peculiarities of its production process, methods of its regulation, including optimization of mineral nutrition.

Fertilizer application is one of the effective elements of increasing productivity processes because it determines the level of provision of crops with nutrients, primarily nitrogen and phosphorus, the availability of which in the soils of the region is at a low level. Therefore, one of the tasks of modern farming is developing optimal cultivation technologies for arable crops.

Studies on the effect of mineral fertilizer application on spring triticale in other regions and countries show that this crop is responsive to fertilizer application, which can significantly increase both grain yield and its quality. However, the payback on mineral fertilisers is observed only at medium application rates [6–8].

The experiments on the comparative study of the effect of different doses of mineral fertilizers on photosynthetic activity of spring triticale crops were not conducted, however, these studies may contribute to the optimization of the productive process of the crop.

The research aims to study the possibility of increasing the photosynthetic activity and productivity of plants of spring triticale cultivar Karmen based on the creation of the most optimal nutrient regime of the soil as a result of different doses of mineral fertilizers.

2 Materials and Methods

The study area is located in the southern agricultural zone of the Amur Oblast, characterized as a sharply continental climate with the sum of active temperatures 2160–2300 °C, the average annual precipitation of 462 mm. The hydrothermal coefficient (HTC) for the growing season was 1.5 in 2017. - 1.5, it is characterized as optimal moistening, in 2018. - 2.0, in 2019 - 1.9, it indicates excessive moisture.

The field studies were carried out on the experimental field of the Far Eastern GAU in the village of Gribskoye. The precursor was a typical crop for the region - soybean. The soil of the experimental plot is meadow-chernozemic. Humus content is 3.7 - 3.9%. According to agrochemical properties characterized by medium reaction of the environment, hydrochemical acidity is low [9].

The object of research was spring triticale - Karmen. The setting of experiments was carried out according to the "Field-plot technique".

As part of this study, the field study was laid out according to the following scheme: 1. Control (without fertilizer application); 2.N30; 3.N30P30; 4.N60P30; 5.N60P60. Fertilizer was applied in spring under pre-sowing cultivation. The total plot area was 32 m^2, counting area 24 m^2, repetition four times, placement of plots was systematic.

3 Results

Yield formation is characterised by field germination and survival to harvest (Fig. 1). As shown by the research, the lowest field germination by year is observed in the first year of the trial and ranges from 288 to 327 plants/m^2. In the next two years of testing, field germination is higher and reaches 433 plants per m^2.

These differences seem to have been influenced by the amount of rainfall at the beginning of the growing season. If we consider the doses of mineral fertilizers, we can confidently assert that with an increase in their concentration, there is a decrease in the density of seedlings. As shown in Picture 1, regardless of the year of research, the application of nitrogen fertilizers at a dose of 30 kg. a./ha contributes to the highest level of field germination and in 2017 it is 303 pcs./m^2, in 2018 - 430 pcs./m^2 and in 2019 - 433 pcs./m^2, which is higher than the control variant by 4.7%, 2.6% and 14.3%. However, application of the maximum dose of N60P60 resulted in a 7.7% decrease in field germination in 2018 and 2.3% in 2019, respectively, which generally suggests the need to increase the seeding rate if higher doses of fertilizer are applied.

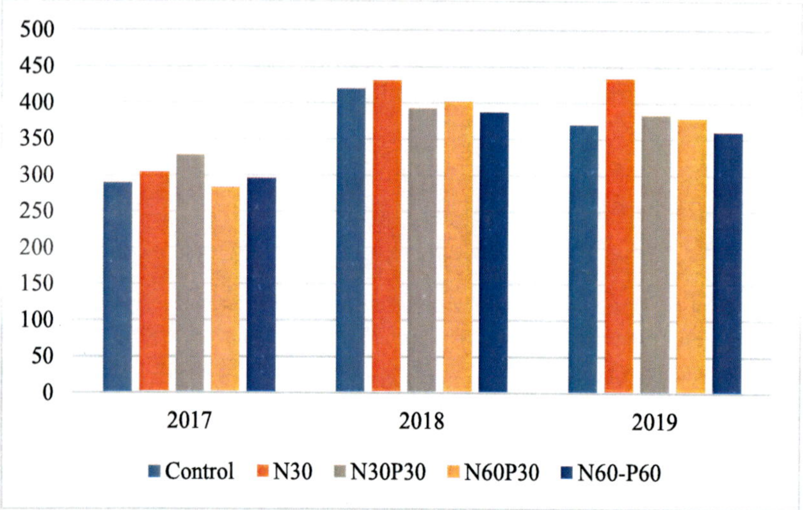

Fig. 1. Effect of mineral fertilizers on the dynamics of germination of spring triticale plants, pcs/m^2.

In general, the level of field germination allows predicting future yields, but considering the fact that the Amur Oblast is a region considered to be a risky farming area, it is most important to pay attention to the number of plants that have survived before harvesting (Fig. 2).

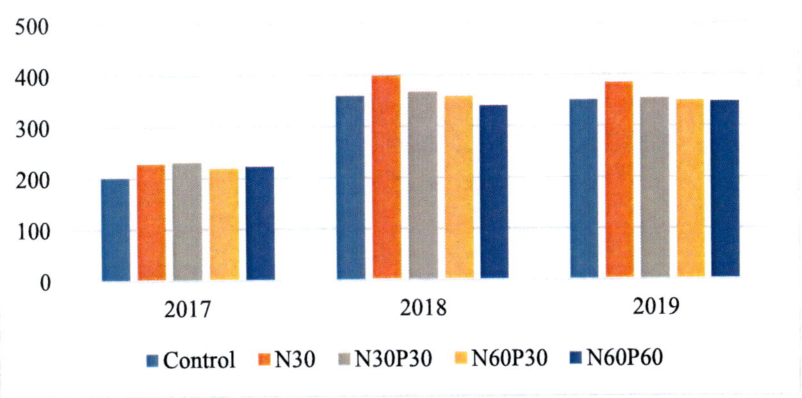

Fig. 2. The effect of mineral fertilizers on the dynamics of spring triticale survival rates, pcs/m^2.

As the survey data showed, 2017 as a whole was characterised as a drier year according to a hydrothermal index and by harvest time the number of plants had fallen to almost half the number sown, being at 69–77% of field germination. In 2018 and 2019, In 2018 and 2019, with higher water availability, the hydrothermal index reached 1.9, and not only field germination was higher, but also plant survival to harvest was 83–93% in 2018 and 88–96% in 2019. If we consider the effect of mineral fertilizers on the number of plants at harvest, the trend noted in the assessment of germination persists: there is the highest density degree in the variant with the application of nitrogen at a dose of 30 kg/day/ha relative to not only the control but also other variants of the experiment. In general, the number of plants harvesting relative to field germination did not change significantly, so in assessing the effect of mineral fertilizers on field germination and viability of plants in the crops of spring triticale we can say that mineral fertilizers have a greater impact on germination than on viability.

The number of plants per unit area does not give a complete picture when assessing the productivity of a given crop. The most accurate prediction of future yield can be made by assessing its photosynthetic activity.

Analysis of the research results revealed that the size of leaf area depends on the rate of mineral fertilizer application at sowing. As we can see from picture 3 the influence of fertilizers is already significant in the initial period of growth and development of plants of spring triticale, the highest indicator of 2.82 thousand m^2/ha is observed in the variant N30P30 (Fig. 3).

In the phase of booting the difference between the variants of the experiment increased, exceeding the control variant by 15–31%. The highest indicator - 18.24 thousand m^2/ha was observed at a dose of N30P30, which is 7–19% higher than other variants with fertilizers. This tendency is also saved in the earing phase when the maximum level of leaf size is reached, but it is in the variant with N60P30 fertilizer (20,5), the difference with the control variant (without fertilizer) is 23%. The analysis of photosynthetic apparatus in the phase of milky ripeness in variants with fertilizers, on average for three years of research, shows that it is at the level of 1,91–2,51 thousand m^2/ha, which is higher than the control by 5–27%.

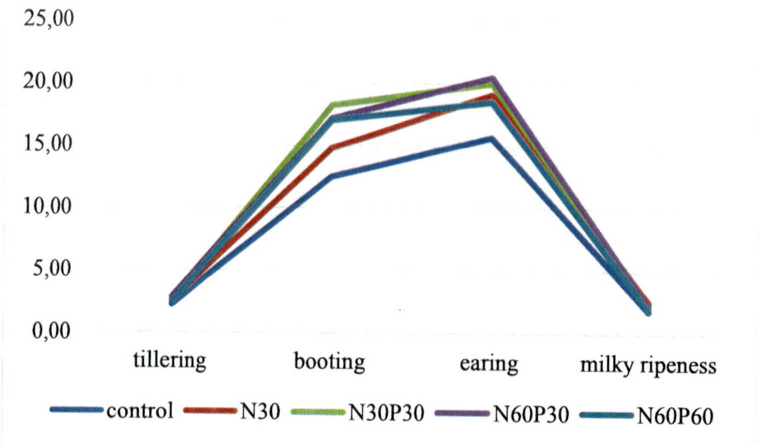

Fig. 3. The influence of mineral fertilizers on the leaf surface size of spring triticale plants, thousand m^2/ha, (average for 2017–2019).

The highest leaf size values were in 2018. (up to 26.25 thousand m^2/ha). The weather conditions during the growing season - warm weather with plenty of rainfall - contributed to this. In 2017, due to the lack of moisture, there was reduced field germination of spring triticale plants in the spring, which was further reflected in the total leaf surface size of the crops (reaching a maximum level in the earing phase in the variant N30P30 - 17.21 thousand m2/ha). In 2019, the leaf size was lower than in 2018 but higher than in 2017. Apparently, in 2019 the amount of rainfall was at a higher level, but the intensity of the heat was lower than in 2018.

But whatever the indicator of the intensity of photosynthetic apparatus activity is, the target indicator is the accumulation of dry matters by plants as a whole.

In 2017, dry matters accumulation by triticale plants in the tillering phase was most intense in the control, exceeding the variants with fertilizer application by 10–20%. In subsequent phases, in the fertilized variants, dry matters accumulation intensity was higher than the control variant by 18% at $N_{60}P_{60}$ and by 31% at $N_{60}P_{30}$, respectively. In the $N_{30}P_{30}$ and N_{30} variants the level of dry matters was at the control level of 1.05–1.22 t/ha. During the subsequent phases of growth and development of triticale plants, the difference in dry matter accumulation between the variants was 3–10%, reaching the highest level in the phase of milky ripeness in the control variant - 4.65 t/ha.

In 2018 and 2019, with sufficient moisture during the growing season, the amount of dry matter in the control variant exceeded the variants with fertilizer application only in the initial period in the tillering phase, and, amounted to 1.45 t/ha, which is higher by 7–20% respectively. In subsequent phases of observation, the fertilised variants exceeded the control by up to 41%, with the N60P60 dose in 2018 to 6.83 t/ha and 2019 to 3.91 t/ha respectively.

As shown in Fig. 4, on average over the three years of research, starting from the phase of booting, the intensity of dry matter growth of plants of spring triticale cultivar Karmen was higher in the variants of experiment with the application of mineral fertilizers at sowing.

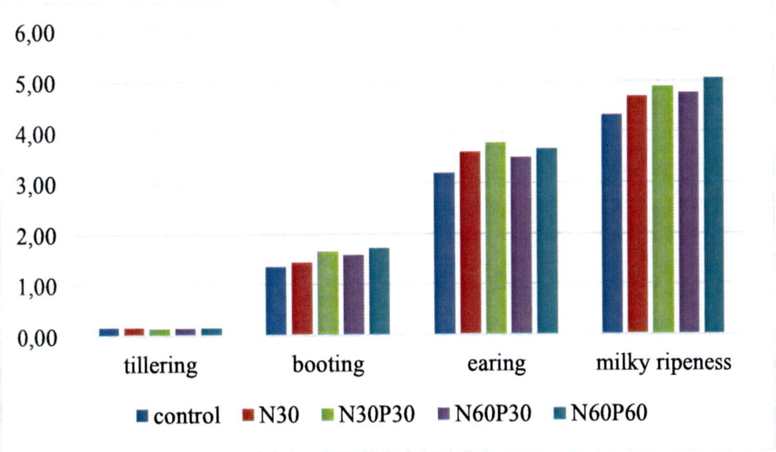

Fig. 4. Dry matter accumulation of spring triticale plants, t/ha, (averaged over 2017–2019).

It was noted that in the phase of tillering the level of accumulation of dry matter in the control variant was by 3–14% higher than in variants with the application of fertilizers, but to the phase of tillering, the amount of dry matter in variants with the application of fertilizers increased by 6–27% relative to control at the rate of N60P60 was 1.72 t/ha. In the following phases, the difference between the control and the variants with the application of fertilizers was not less than 10%, which indicates the positive effect of the application of mineral fertilizers at the sowing. At the same time, the difference between the variants of the experiment with the application of fertilizers averaged only about 5%.

The target result of photosynthetic activity, first of all, is yield.

Table 1. Effect of mineral fertilisers on spring triticale yields, t/ha.

Variant	2017	2018	2019	Average value
Control	2.88	2.64	2.53	2.68
N_{30}	3.06	3.12	2.87	3.02
$N_{30}P_{30}$	3.13	3.09	2.95	3.06
$N_{60}P_{30}$	3.25	3.57	3.05	3.29
$N_{60}P_{60}$	3.42	3.60	3.11	3.38

As shown in Table 1, according to the results in 2017, under satisfactory moisture conditions, the yield of spring triticale in the experimental variants ranged from 2.88 to

3.42 t/ha. At the same time, the application of fertilizer at sowing time promotes a yield increase from 6% (N30) to 19% (N60P60) relative to the variant without fertilizer, and the overall level of increase with increasing fertilizer rate varies from 3 to 6% between the experimental variants.

2018 was marked by the most favourable ratio of precipitation and temperature during the growing season, which contributed to the highest grain yield in the 3 years of observations. The highest yield was in the $N_{60}P_{60}$ variant and was 3.60 t/ha, which was 0.96 t/ha higher than the control, but also in the $N_{60}P_{30}$ variant, the yield was only 0.03 t/ha lower relative to the highest dose of fertiliser application, which is within the error of experiment. This pattern is also observed in the variants N_{30} and $N_{30}P_{30}$, as a result, we can conclude that the application of nitrogen fertilizers to a greater extent contributes to increasing the yield of the grain of spring triticale than the application of phosphorus fertilizers.

In 2019, the indicator of moisture intensity of the growing season is at a high level, however, the uniformity of their fallout had a contrasting character and all this led to the lowest level of grain yield in all years of observation. At the same time, in 2019, as in the previous years of research, the application of mineral fertilizers at sowing contributed to an increase in grain yield by 13–23% relative to the control, increasing the rate of fertilizers between the variants of experience contributed to an increase in yield by only 2–3%.

The average yield of spring triticale cultivar Karmen during three years of research, when different doses of mineral fertilizers were applied, was 3.19 t/ha, and in the control variant - 2.68 t/ha. The highest crop capacity (3.38 t/ha) was noted in the variant where the highest dose of N60P60 was applied: the increase in this variant compared to the control was 0.70 t/ha or 26%. At the same time, the significant increase of the crop capacity on the average for the three years of research was observed in the variants with the increase of nitrogen fertilizer rate N30 and N60P30 by 13 and 23% respectively. Application of phosphorous fertilizers increased grain yield, but only by 1 and 3% relative to the variants N30P30 and N60P60, respectively.

4 Conclusions

A comparative evaluation of the use of different doses of fertilizers in growing promising grain crop spring triticale in the Amur Oblast, it may be noted that the use of the minimum dose of N30 at sowing enhances both field germination and viability of plants before harvesting, but at higher doses of fertilizers N60P60 there is the opposite trend - reducing the field germination, and as a consequence, reducing the viability of plants. The size of the leaf also depends on the level of fertilizer application during sowing. If during the initial period it was practically indistinguishable by variants of the experiment, starting from the phase of booting intensity of the leaf size growth increased relative to control by 15–31%, reaching its maximum value at the earing stage in the variant N60P30 (20.5 thousand m^2/ha). At the same time, the difference between the control variant (without fertilizer) was 23%. As a consequence, the amount of dry matter in the experiment variants with fertilizer application also increased, reaching the highest level by the end of the growing season of more than 5 t/ha at the highest dose of fertilizer. As a result, the grain yield in the experiment with the use of fertilizer was significantly higher than in the version without fertilizer. However, at sowing, the plants of spring triticale responded more positively to the application of nitrogen fertilizers than phosphorus fertilizers. Therefore, when choosing a particular type of fertiliser, nitrogen fertilisers are preferable, as they give the greatest effect at the lowest cost.

References

1. Selikhova, O.A., Tikhonchuk, P.V.: Problems of rational varietal placement of soybean in the amur region. In: Conference Series: Earth and Environmental Science, p. 012033 (2020). https://doi.org/10.1088/1755-1315/547/1/012033
2. Rahemi-Karizaki, A., Khaliliaghdam, N., Biabani, A.: Predicting time trend of dry matter accumulation and leaf area index of winter cereals under nitrogen limitation by non-linear models. Plant Physiol. Rep. **26**(3), 443–456 (2021). https://doi.org/10.1007/s40502-021-00597-x
3. Gamayunova, V.V., Sydiakina, O.V., Dvoretskyi, V.F., et al.: Productivity of spring triticale under conditions of the southern steppe of Ukraine. Ecol. Eng. Environ. Tech. **22**(2), 104–112 (2021). https://doi.org/10.12912/27197050/133456
4. Muratov, A.A., Nizkii, S.E.: The dependence of spring triticale yield and its structure on harvesting time and methods. In: Conference Series: Earth and Environmental Science, p. 012023 (2020). https://doi.org/10.1088/1755-1315/547/1/012023
5. Muratov, A.: Growth and development of triticale culture in the amur region (Russia). In: E3S Web of Conferences, p. 02007 (2020). https://doi.org/10.1051/e3sconf/202020302007
6. Selle, P.H., Moss, A.F., Khoddami, A., Chrystal, P.V., Liu, S.Y.: Starch digestion rates in multiple samples of commonly used feed grains in diets for broiler chickens. Anim. Nutr. **7**(2), 450–459 (2021). https://doi.org/10.1016/j.aninu.2020.12.006
7. Saglam, N., Ustunalp, G.: The effect of different sowing densities and nitrogen doses on yield and yield components in triticale (X Triticosecale Wittmack). APCBEE Proc. **8**, 354–358 (2014). https://doi.org/10.1016/j.apcbee.2014.03.053
8. Kádár, I., Ragályi, P.: Fertilizer effects in a triticale monoculture on sandy soil. Agrokemia es Talajtan **63**(2), 283–294 (2014). https://doi.org/10.1556/Agrokem.63.2014.2.8
9. Kalashnikov, N.P., Tikhonchuk, P.V., Fokin, S.A.: The influence of micronutrients on the productivity of corn during cultivation on green mass in the southern zone of Amur region. In: Conference Series: Earth and Environmental Science, p. 012043 (2020). https://doi.org/10.1088/1755-1315/547/1/012043

Agrotechnical Assessment of the Work of the Furrow Opener During the Cultivation of Cereals Using Intensive Technology

Maxim Chekusov, Andrey Schmidt, and Aleksander Kem(✉)

Omsk Agrarian Research Center FSBSI, Koroleva Avenue, 26, 644012 Omsk, Russia
kem@anc55.ru

Abstract. The results of laboratory and field comparative studies carried out in 2019–2020 on the effect of the rate and method of applying mineral fertilizers using serial spear point and combined furrow openers on the yield of spring wheat Omskaya-36 are presented. The design of a combined opener is proposed, which makes it possible to apply mineral fertilizers below the sowing of seeds of grain crops with a soil layer of 3–4 cm between seeds and fertilizers. Field comparative experiments were carried out in the fields of the FSBSI Omsk ASC with the determination of such indicators as: field germination of seeds (%); number of stems (pcs/m^2), yield accounting (t/ha). According to meteorological indicators, in 2020 there was 62 mm less precipitation than in 2019, and the air temperature turned out to be 2 °C higher, which directly influenced the research results. In 2019, the highest field germination values of 75% were observed when sowing with a combined opener with a mineral fertilizer application rate of 150 kg/ha, and in 2020, the highest field germination rate was 92% with the same fertilizer application rate. The number of stems, 337 and 412 pcs/m^2 in 2019 and 2020, respectively, was on plots with the application of mineral fertilizers (150 kg/ha) with a combined opener. In 2019, the grain yield for all plots when sowing with the SKP-2.1M seeder with combined openers is on average 14% higher than when seeding with the SKP-2.1 serial seeder with spear point openers. In 2020, the yield growth was 12.5%.

Keywords: Grain seeder opener · Mineral fertilizers · Seeds · Method of fertilizing · Opener · Grain yield

1 Introduction

Cereals are one of the bases of food security in the world, accounting for about half of all the cultivated area of world agriculture. Grain is valuable as a food product for humans, feed for farm animals and raw material for the processing industry. For comparison, the production of a unit of energy contained in meat requires 6–7 times more agricultural land than for the production of the same amount of energy in grain products [1, 2]. Cereals are cultivated in all regions of the country, but their largest areas are concentrated in Western Siberia, the Urals, the Volga region, the North Caucasus, in the central black

A. Muratov and S. Ignateva (Eds.): AFE 2021, LNNS 353, pp. 164–173, 2022.
https://doi.org/10.1007/978-3-030-91402-8_20

earth region, whose farms produce the bulk of marketable grain. The volume and quality of products of cultivated crops largely depend on the breeding of new varieties, the culture of agriculture, the introduction of scientifically grounded, intensive technologies, the rational use of natural resources, and the optimization of production conditions. “Programmed” crop cultivation is gaining more and more popularity, i.e., digital control of technological processes based on an integrated collection of information that directly affects the harvest [3, 4].

Over the past decades, the rate of technical and technological re-equipment in the agricultural sector of the country has sharply decreased. Often, the production of certain types of equipment has completely stopped. Along with this, the flow of organic and mineral fertilizers has decreased. Effective use of fertilizers is possible under the condition of a high farming standards [5].

The effectiveness of the introduction of intensive and resource-moisture-saving technologies for the cultivation of grain crops, which is the basis of agricultural production in most developed countries, largely depends on agrotechnical indicators and the timing of sowing. To increase yields in the process of agricultural work in the world, about 120 million tons of mineral fertilizers are used annually. Nowadays, the share of mineral fertilizers from the amount of nutrients applied to the soil accounts for about 60%. The interaction of fertilizers with soil can be both positive and negative for plant nutrition. The positive effect of the systematic application of mineral fertilizers is manifested in the change in the physicochemical properties of soils. The negative character from the long-term use of mineral fertilizers is associated with acidification of the reaction of the soil solution due to the displacement of hydrogen and aluminum ions from the absorbing complex, the use of physiologically acidic nitrogen and potassium fertilizers. The negative effect is often the result of misuse of agrochemicals. Up to 30% of the applied mineral fertilizers are lost due to violation of the justified technology of application, storage and transportation [6–8].

With the formation of stable yields, it becomes necessary to provide plants with accessible forms of nitrogen nutrition. A distinctive feature of ammonium nitrate is its ability to have a beneficial effect on the soil that is not yet sufficiently warmed up by the sun. Nitrogen fertilizers are able to quickly release the required amount of nitrogen for further germination of crops and at the same time retain beneficial properties due to the interaction with oxygen in the air [9].

Increasing the yield of agricultural crops and providing the processing enterprises with raw materials inevitably lead to the need to improve sowing equipment, its constructive refinement for a variety of physical and mechanical properties of seeds, schemes and methods of sowing in various soil and climatic conditions. For this method of application, the Duet double furrow opener is used, which provides broadband sowing at 18–20 cm with a sowing depth of up to 7 cm. The Duet furrow opener is capable of sowing, applying fertilizers and simultaneously working the soil and effectively removing crop residues from the sowing horizon in one working pass [10, 11]. A distinctive feature of this opener is that it allows applying a system of simultaneous fertilization into the soil exactly under the sowing strip to a depth of 4–5 cm below its horizon. This eliminates the likelihood of a chemical burn of the seeds. However, the disadvantages of this type of opener are that it does not perform continuous tillage and weed cutting, and also does not combine the operations of soil cultivation and sowing. As a result, the number of technological operations increases, and consequently, energy consumption for soil cultivation increases. Improving operational and technological indicators has always been and remains an urgent task.

When creating new sowing machines and implements, it is necessary to take into account the zonal conditions, the requirements of technology, and the characteristics of the seed. Scientists create universal working bodies, including sowing machines, for working with chemicals (granular mineral fertilizers) applied below the level of sowing seeds of grain crops [12–14].

Shovel-type furrow openers are used for sowing cereal seeds on untreated soil. Such an opener can perform several operations at the same time: loosen the soil, cut weeds, sow seeds of agricultural crops, apply mineral fertilizers. Most of the opener groups of sowing machines are designed in such a way that fertilization is carried out together with the seeds on one sowing bed. In this case, it is recommended to apply small doses of fertilizers to ensure a powerful start for the plants. For the entire growing season, their action is not enough, therefore, additional fertilizing is used in the future. This inevitably leads to additional production costs. If we apply a deliberately large dose of mineral fertilizers together with seeds in order to ensure plant nutrition for the entire growing season, then the negative effect of a high concentration of nutrients (in the area of young roots), which can negatively affect the growth and development of plants and germination of seeds, is inevitable. In this case, fertilizers are used ineffectively, since they are located above the root system of plants and are not used as starting ones, which negatively affects plant development and productivity. In addition, openers that apply fertilizers below the level of seeding are used, but these working bodies are very expensive and not adapted to local soil conditions [15–17].

The most rational is the multilevel fertilization together with sowing. Scientific research suggests that it is better to place mineral fertilizers at a depth twice the depth of sowing seeds of agricultural crops. The soil layer of 3–4 cm between seeds and fertilizers reduces the chemical effects of highly concentrated nitrogen fertilizers. In the early days of growth, the plant's root system is too weak to take all the necessary nutrients from the soil on its own. The main task of mineral fertilizers is primarily to provide seedlings with the necessary nutrition in the first days after sowing [13].

Intensive technology for the cultivation of grain crops includes such systems as: agrobiological (ABS), technical (TS), agrotechnical (ATS) systems. The agrobiological system (ABS) includes three main objects directly interacting with each other - soil, seed, mineral fertilizers. Accordingly, the system can be affected by a number of external factors $F_{ABC}(t)$, which include: natural and climatic conditions; state, properties and type of soil; biological characteristics of cultivated crops; seed quality, rate and method of sowing; type of mineral fertilizers, rate and method of their application. External factors $F_{TS}(t)$ acting on the technical system (TS) directly depend on the quality of preparation of seed material and fertilizers, compliance and implementation of agronomic requirements. Agrotechnical system (ATS) manages basic agrotechnical (ATR) and agrochemical (ACR) requirements. With regard to the opener group, the structural model of the combined opener for different-depth sowing and mineral fertilization can be represented by the following diagram (Fig. 1).

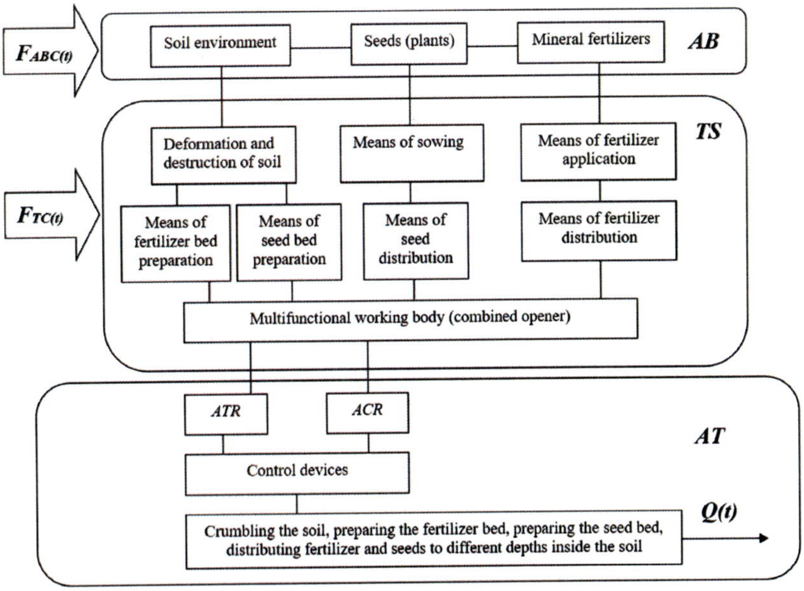

Fig. 1. Diagram of the structural model of the opener group.

Considering the structural model of the opener group, it can be said that in order to increase the yield of grain crops, obtain stable products, and control the production processes of plants, it is necessary to improve the opener group of a stubble seeder. Mineral fertilizers play an important role, since for the formation of high and stable crop yields, it is necessary to provide plants with accessible forms of nitrogen, phosphorus and potassium nutrition, including microelements, in sufficient quantities.

Moreover, the process of interaction of the components of the biological (ABS) and technical systems (TS) when fertilizing the soil is considered as one system, the task of which is to realize the potential of the cultivated crop to obtain the maximum yield. The feedback of the biological system (ABS) is expressed in the fact that mineral fertilizers, through the plant, affect the initial state of the soil, which, in turn, leads to an increase in the biological mass and root residues of plants, which themselves subsequently become fertilizers.

Studies have established that the increase in the yield of grain crops due to the use of chemicals can reach 64.3%. The annual removal of nutrients from the soil leads to its depletion, there is a gradual decrease in the yield of agricultural crops. The main factor in increasing the productivity of lands, and as a result, the yield of grain crops, remains the provision of soils with elements of mineral nutrition, compositions, doses, terms, and methods of their application. That is why the transition to multifunctional combined openers, which perform the preparation of the soil for sowing, the application of fertilizers with their closure using a soil layer, the creation of a compacted bed for seeds and sowing in one pass, is inevitable [13, 16].

Analyzing the patent and technical literature, the process of cultivation of cereals using intensive technology, scientists and engineers of the Omsk ASC have patented the design of a new combined opener, which allows sowing with fertilization below the seed level [13].

The novelty of the proposed seeder with combined openers for different-depth sowing and the application of mineral fertilizers is that the implementation of separate seed placement. The basis for the development of a seed-fertilizer stubble seeder with separate application of seeds and fertilizers is based on the scheme of SZS-type seeders-cultivators tested in various soil and climatic zones, namely, SKP-2.1. At the same time, the combined opener loosens the top layer of the soil, cuts weeds, makes pre-sowing application of mineral fertilizers, creates a compacted bed, and sows seeds above fertilizers. The principle of operation is shown in Fig. 2.

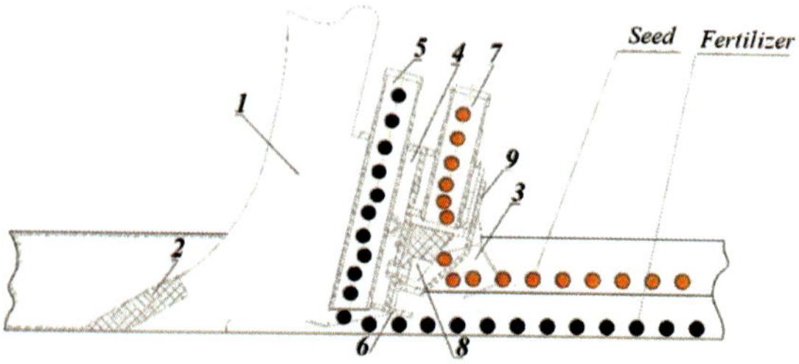

Fig. 2. Sectional view of the combined opener.

The working process of the combined opener is as follows: when moving in the soil, the tine (1) forms a groove using a plate (2) made of carbide metal. At the same time, cultivator blades (3) installed on both sides of the receiver-distributor (4) loosen the surface layer of the soil and cut weeds. Fertilizers are supplied to the bottom of the furrow via the fertilizer line (5). When the tine (1) moves, the sealing plate (6) rolls over and compacts the soil. The formed soil layer excludes contact between fertilizers and seeds. The seeds, passing through the seed guide (7), enter the distributor (8) and, depending on the presence of a removable plate (9), form a strip or two-line sowing method.

The purpose of the work is to conduct comparative studies of the effect of the rate and method of applying fertilizers by serial and combined openers on the yield of spring wheat.

2 Research Methodology

The field comparative experiment was conducted at the fields of the Omsk ASC in 2019–2020. Sowing was carried out with an SKP-2.1 seeder with two types of openers: 1) a standard opener with a duckfoot shovel; 2) an experimental combined opener for multilevel sowing of seeds and fertilizers (SKP-2.1M).

Fig. 3. MTZ-82 tractor combined with SKP-2.1M seeder.

Sowing of spring wheat Omskaya-36 without the use of fertilizers was taken as control. Several sowing options were compared: with a serial opener without fertilization and with application of starting doses of mineral fertilizers together with seeds into one soil horizon; with a combined opener without fertilization and with application below the sowing of seeds. Ammonium nitrate was used as a fertilizer at a rate of 100, 150, and 200 kg/ha. The moisture content of fertilizers is 7.2%. Spring wheat variety - Omskaya-36. The seed purity is 99%, the germination rate is 93%, the moisture content is 12.5%, the mass of 1000 seeds is 43 and 40 g in 2019 and 2020, respectively. The seeding depth of seeds is 5 cm, the depth of fertilization with the combined opener is 9 cm. The soil type is leached medium-thick chernozem, the field topography is even, the soil structure is fine crumbly, the soil moisture in the sowing layer is 16.2% and 15.4% in 2019 and 2020, respectively. The seeding rate was fixed at 4.5 million units of germinating grains per hectare. Sowing of the plots was carried out with an MTZ-82 tractor combined with a seeder. The entire amount of work was completed in one working shift.

The weather conditions of the Omsk region during the growing season (May-August) of 2019 were as follows: the average air temperature is 16.5 °C, which is 0.2 °C lower than the average long-term value; the amount of precipitation is 13 mm less and amounted to 193 mm. The meteorological indicators of the growing season in 2020 were marked by an increase in the average air temperature by 1.8 °C and a small amount of precipitation not exceeding 131 mm, which is significantly below the norm.

During the field comparative experiment, the following indicators were determined: field germination of seeds (%); number of stems (pcs/m^2), yield accounting (t/ha). The observations were carried out in accordance with the Russian State Standards STO AIST 10 5.6–2003 and GOST 31345–2007.

In the design and manufacture of the main components of the experimental sample, the combined openers were based on the works of the following scientists: Bosoy E.S., Buzenkov G.M., Gribanovsky A.P., Gyachev L.V., Karpenko A.N., Kardashevsky S.V., Listopad G.E., Semenov A.N. and others.

3 Research Results and Discussion

An important condition for obtaining a high yield is the timely appearance of full, even and well-developed sprouts after sowing. Field germination is influenced by environmental conditions: soil temperature at the depth of sowing seeds, air temperature, soil moisture, the presence of soil pests, soil crust. Field germination is an integral indicator of the quality of seeds and the level of agricultural technology. The sowing was carried out at the optimal time, the air temperature on the day of sowing was 240 C. The density of the plant stand depends on the seeding rate, field germination, bushiness and plant survival. Field germination and plant productivity are closely related. With low field germination, a significant part of the grain is wasted.

Field germination and the number of stems of spring wheat Omskaya-36, depending on the method and rate of application of mineral fertilizers, are shown in Fig. 4 and 5.

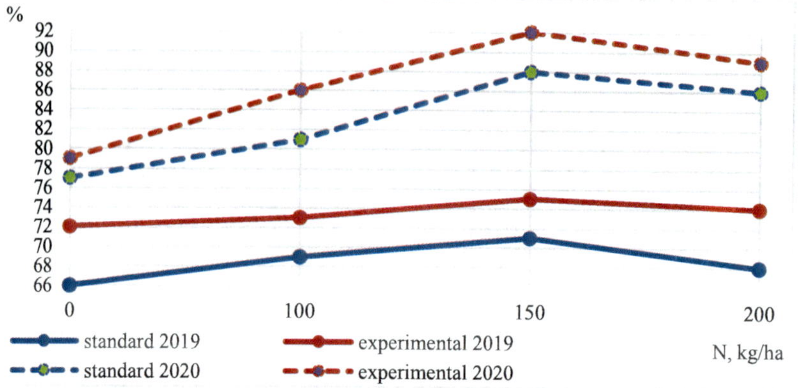

Fig. 4. Field germination of spring wheat Omskaya-36, depending on the method and rate of application of mineral fertilizers in 2019–2020.

The graph in Fig. 4 shows that the field germination rate on the experimental plots in 2019 is significantly lower than in 2020, which is explained by the untimely precipitation in the initial phase of plant development. The maximum value of 75% was observed in 2019 when sowing with a combined opener with a mineral fertilization rate of 150 kg/ha. Field experience in 2020 showed the highest field germination value of 92% when sowing with a combined opener with a mineral fertilizer application rate of 150 kg/ha. For all sowing options, there is a gradual increase in field germination to a value of N = 150 kg/ha, after which a slight decrease is observed, which may be caused by a stronger chemical effect of ammonium nitrate on the seed.

The graph of the number of stems in Fig. 5 has an identical appearance with the graph of field germination shown in Fig. 5. A larger number of stems were noted on the plot when applying N = 150 kg/ha with a combined opener and amounted to 337 and 412 pcs/m^2 in 2019 and 2020, respectively.

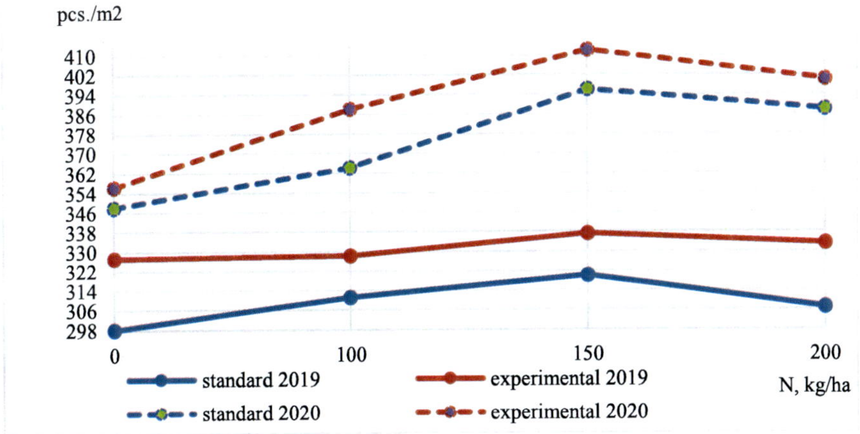

Fig. 5. The number of stems of spring wheat Omskaya-36, depending on the method and rate of application of mineral fertilizers in 2019–2020.

The experimental plots were harvested using a Hege-125 plot combine. Figure 6 shows a graph that shows how the yield of soft spring wheat Omskaya-36 changed depending on the rate of fertilization by the SKP-2.1 seeder with serial spear point openers and the SKP-2.1M seeder with combined openers for multilevel sowing and application of mineral fertilizers during the period 2019–2020.

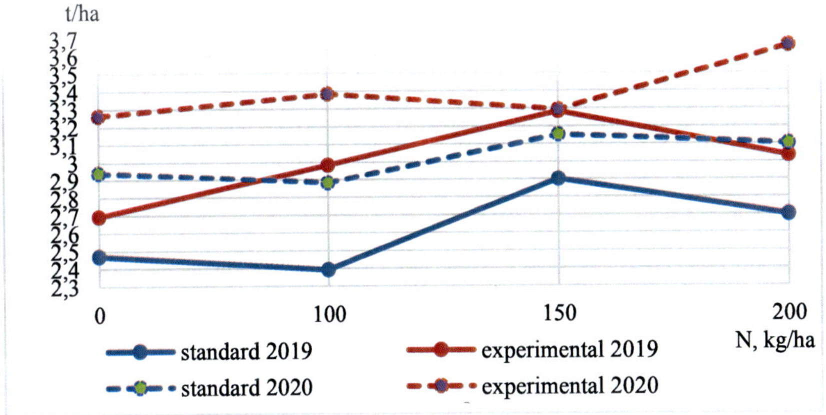

Fig. 6. Grain yield of spring wheat Omskaya-36, depending on the method and rate of application of mineral fertilizers in 2019–2020.

As can be seen from Fig. 3, the grain yield for all plots in 2019 when sown with the SKP-2.1M seeder is on average 14% higher. The maximum yield was observed on a plot where sowing was carried out with an SKP-2.1M seeder with a fertilizer application rate of 150 kg/ha and amounted to 3.28 t/ha, which is 13% higher than that for the SKP-2.1

seeder (control) at the same fertilization rate. The difference between the lowest and highest yields in 2019 was 0.89 t/ha, or 37%. In 2020, the experiment showed that on plots where sowing was carried out with an SKP-2.1M seeder equipped with combined openers, the grain yield is on average 12.5% higher. The highest yield of 3.65 t/ha was obtained on plots where sowing was carried out with an SKP-2.1M seeder with a fertilizer application rate of 200 kg/ha.

4 Conclusions

The existing technological processes of sowing seeds and mineral fertilizers require improvement in the area of their adaptation to the needs of cultivated crops in the methods and rates of their application. The scheme of the structural model of the opener group is proposed, including agrobiological (ABS), technical (TS), and agrotechnological (ATS) systems.

The data obtained in the course of a comparative field experiment on studying the effect of the rate and method of applying mineral fertilizers by serial (control) and combined openers on the yield of spring wheat showed that the field germination rate on the experimental plots in 2019 is significantly lower than in 2020 and amounted to 75% when sowing with a combined opener with a mineral fertilization rate of 150 kg/ha. In 2020, field germination on the control when sowing without the use of mineral fertilizers with serial spear point openers of the SKP-2.1 seeder was 77% versus 79% when sowing with combined openers. The highest field germination is achieved by both methods of application at a rate of 150 kg/ha of ammonium nitrate: using a seeder with serial spear point openers - 88%, and using a seeder with combined openers - 92%, respectively.

The largest number of stems, 337 and 412 pcs/m^2, were obtained on plots at a fertilization rate of 150 kg/ha with a combined opener in 2019 and 2020, respectively. The grain yield for all plots in 2019 when sowing with the SKP-2.1M seeder is on average 14% higher, and in 2020 - an increase of 12.5% compared to the SKP-2.1 serial seeder.

References

1. Rodríguez, J., Rahman, H., Thushar, S., Singh, R.: Healthy and resilient cereals and pseudo-cereals for marginal agriculture: molecular advances for improving nutrient bioavailability. Front. Genet. **11**, 49 (2020). https://doi.org/10.3389/fgene.2020.00049
2. Rendov, N., Gladkikh, A., Nekrasova, E.: The influence of the cultivation technology elements on the economic performance of the bare barley grain production. In IOP Conference Series: Earth and Environmental Science, International Conference on Sustainable Development of Cross-Border Regions SDCBR, vol. 395, p. 012011 (2019). https://doi.org/10.1088/1755-1315/395/1/012011
3. Visser, O., Spoor, M., Mamonova, N.: Is Russia the emerging global 'Breadbasket"? Re-cultivation, agroholdings and grain production.' Eur. Asia Stud. **66**, 1589–1610 (2014). https://doi.org/10.1080/09668136.2014.967569
4. Gostev, A., Pykhtin, A.: Normative-reference database structure for agricultural manufacturers support system and rational choice of cost-effective adaptive tichnologies for grain crops cultivation. In: 18th International Multidisciplinary Scientific GeoConferences SGEM, pp. 329–334 (2018). https://doi.org/10.5593/sgem2018/3.2/S13.043

5. Matveenko, D., Komarov, A., Nugis, E., Lekomtsev, P.: Optimization of options for differential application of nitrogen fertilizers in the precision farming system. J. Agric. Sci. **31**(1), 41–46 (2020). https://doi.org/10.15159/jas.20.06
6. Osipova, V., Buyankin, N., Krasnoperov, A., Avdeenko, A., Kondratenko, L.: The impact of soil microflora on the crop yield of cocultivated cereals and legumes. Eur. Asian J. BioSci. **14**, 1933–1940 (2020)
7. Kursacova, V., Stupina, L.: The influence of nitrogen-fixing bacteria preparations on morphogenetic indices of spring barley. In: IOP Conference Series: Earth and Environmental Science SDCBR, p. 012018 (2019). https://doi.org/10.1088/1755-1315/395/1/012018
8. Verma, S., Arora, K., Srivastava, A.: Monitoring of genotoxic risks of nitrogen fertilizers by Allium cepa L. mitosis bioassay. Caryologia **69**, 343–350 (2016). https://doi.org/10.1080/00087114.2016.1226540
9. Gostev, A., Pykhtin, A.: Structure of costs and expenditures in agro technologies of different intensity levels. J. Appl. Eng. Sci. **15**, 463–466 (2017). https://doi.org/10.5937/jaes15-15455
10. Aduov, M., Kapov, S., Nukusheva, S., Rakhimzhanov, M.: Components of coulter tractive resistance for subsoil throwing about seeds planting. Life Sci. J. **11**(5s), 67–71 (2014)
11. Shumaev, V., Kulikova, J., Orehov, A., Polikanov, A.: Investigation of the grain seeder opener operation for environmentally friendly technologies of crops production. Sci Pap. Ser. A-Agron. **63**, 527–532 (2020)
12. Andreev, K., Terentyev, V., Shemyakin, A.: Technological process of application of mineral fertilizers by self-loading machine IOP Conference Series: Earth and Environmental Science. In: 12th International Scientific Conference on Agricultural Machinery Industry, p. 012180 (2019). https://doi.org/10.1088/1755-1315/403/1/012180
13. Kem, A., Chekusov, M., Schmidt, A.: The effect of the combined coulter on the quality and yield of grain. Tract. Agr. Mach. **6**, 72–77 (2020). https://doi.org/10.31992/0321-4443-2020-6-72-77
14. Singh, K., Agrawal, K., Jat, D., et al.: Design, development and evaluation of furrow opener for differential depth fertilizer application. Ind. J. Agric. Sci. **86**, 250–255 (2016)
15. Aduov, M., Nukusheva, S., Kaspakov, E., et al.: The influence of random technological and control impacts on the process of seed sowing and mineral fertilizers. Ecol. Environ. Conserv. **23**, 267–277 (2017)
16. Shemyakin, A., Borychev, S., Uspenskiy, I., Andreev, K., Terentyev, V.: Improvement of the technological process of surface application of mineral fertilizers. In: BIO Web of Conferences. International Scientific-Practical Conference "Agriculture and Food Security: Technology, Innovation, Markets, Human Resources" (FIES 2019), p 00192 (2020). https://doi.org/10.1051/bioconf/20201700192
17. Aduov, M., Nukusheva, S., Esenalikaspakov, E., Kazbekisenov, K., Volodya, K.: Analysing the results field tests of an experimental seeder with separate introduction of seeds and fertilizers. Int. J. Mech. Prod. Eng. Res. Dev. **9**, 589–598 (2019). https://doi.org/10.24247/ijmperdaug201958

Adaptability of Determinate and Indeterminate Varieties of *Fagopyrum Esculentum* Monch in a Monsoon

Alexey Klykov[1](✉), Galina Murugova[1], Oksana Timoshinova[1], Elena Barsukova[1], and Elena Chaikina[2]

[1] Federal Scientific Center of Agricultural Biotechnology of the Far East Named After A.K. Chaika, 30, Volozhenin Street, Timiryazevsky Stl, 692539 Ussuriysk, Primorsky Krai, Russia
[2] Pacific Institute of Bioorganic Chemistry. G.B. Elyakov FEB RAS, 159/2, Centennial Avenue, 690022 Vladivostok, Russia

Abstract. The article presents the results of the study of determinate and indeterminate varieties of *Fagopyrum esculentum* Moench. The study was carried out at the FSBSI "FSC of Agricultural Biotechnology of the Far East named after A.K. Chaika" in 2015–2020. The objects of research were Buckwheat varieties of various ecological and geographical origins. The zoned variety Izumrud was taken as standard. Among the indeterminate varieties, Molva variety was distinguished in terms of short stem (41.9 cm), Kayemchataya variety – in terms of the first internode length (4.18 cm), first internode thickness (0.42 cm) - Nikolskaya and Kitavase, in terms of the content of rutin in the fruits of the first internode - Bashkirskaya Krasnostebelnaya (15 mg/100 g). The Ussurochka variety showed the highest yield (1.5 t/ha). According to the adaptability coefficient, varieties of the indeterminate type of growth were identified - Izumrud, Pri 7, Ussurochka, Nikolskaya, Zemlyachka, Inzerskaya, Kitavase, Mankan, and varieties of determinate growth type - Dialog, Devyatka, Dikul. The selected buckwheat varieties are recommended for breeding in the conditions of Primorsky Krai for high yield and grain quality.

Keywords: Buckwheat · *Fagopyrum esculentum* · Adaptability parameters · Plasticity · Stability · Yield · Rutin · Seed quality

1 Introduction

Fagopyrum esculentum Moench is a valuable agricultural crop [1, 2]. Its main producers are Russia and China [3]. Buckwheat is widely used in human nutrition as high-calorie, dietary, rich in easily digestible proteins, fats [4], carbohydrates, organic acids, minerals [5], vitamins E, B and P [6].

Buckwheat is characterized by high concentration of various flavonoid compounds, especially the dominant flavonoid rutin, which is a glycoside of the flavonoid quercetin and is found in the leaves, stems, flowers, roots and fruits of buckwheat. Flavonoids have antioxidant, hypotensive, and anti-inflammatory activity [7]. Buds and flowers

A. Muratov and S. Ignateva (Eds.): AFE 2021, LNNS 353, pp. 174–184, 2022.
https://doi.org/10.1007/978-3-030-91402-8_21

of Japanese Sophora (Sophora japonica L.) are the main source of rutin. There is no industrial raw material base for this plant in the Russian Federation, and the need for routine is met by imports from China. In a number of countries (Russia, China, Japan), special varieties of F. esculentum Moench with an increased content of rutin have been obtained for its production. Increased content of rutin in buckwheat products has an important therapeutic and prophylactic value and helps to reduce the occurrence of many diseases [7–9].

It should be noted that *F. esculentum* as a source of flavonoids is of interest not only for the food industry, but also for medicine as a possible medicinal raw plant material, especially since it is widely cultivated in Russia and the volume of the obtained harvest can provide raw materials necessary for pharmaceuticals.

Fagopyrum esculentum is grown in various environmental conditions of the Far East, therefore varieties adapted to abiotic and biotic environmental factors are required [10]. Over the past two decades, breeders have created new varieties of buckwheat, which differ in their morphology, growing season, seed productivity and grain quality [11].

Despite the long history of the study of *Fagopyrum esculentum*, studies aimed at creating adapted varieties of this culture with a high ability to accumulate rutin are still relevant [12].

The aim of this research was to study buckwheat varieties of various origins by economically valuable traits and parameters of adaptability in a monsoon climate.

2 Materials and Methods

The work was carried out in the grain and cereal Scientific Center of Agricultural Biotechnology of the Far East named after A.K. Chaika" in 2015–2020. The Izumrud variety, zoned for the Far Eastern region, was taken as standard. Varieties of various ecological and geographical origin were studied during the ecological test: three varieties - Izumrud, Pri 7, Ussurochka, and two breeding lines 24 and 27 obtained in the laboratory of agricultural biotechnology laboratory of the FSC of Agricultural Biotechnology of the Far East named after A.K. Chaika (Primorsky Krai), Bashkirskaya Krasnostebelnaya, Ufimskaya, Agidel, Ilishevskaya, Zemlyachka, Saulyk, Batyr, Cheremshanka, Inzerskaya, Svetlana (Republic of Bashkortostan), Strelka (Nizhny Novgorod region), Chatyr Tau, Nikolskaya (Republic of Tatarstan), Dizain, Devyatka, Dikul, Dialog, Yesen, Temp, Molva (Oryol region), Amurskaya Mestnaya (Amur Region), Natasha and Irmenka (Novosibirsk Region), Kitavase and Mankan (Japan).

Phenological observations and records were carried out according to the methodology of the State variety testing of agricultural crops and the methodological instructions of the VIR. The 1000 grain weigh and hull content were determined according to GOST. Rutin content was determined by the chromatospectrometric method at the G.B. Elyakov Pacific Institute of Bioorganic Chemistry. To identify rutin, NMR 1H spectra were used, which were recorded on a Bruker AC-250 spectrometer (250.13 MHz for 1H) (Germany) in CDCl3 (deuterochloroform) and acetone-d6, and compared with individually pure rutin (Chemopol, Czech Republic). Mass spectra were obtained on an LKB-9000S instrument (Sweden) with direct input at ionizing electron energies of 18 and 70 eV.

Statistical processing of crop data was carried out according to the B.A. Dospekhov. Environmental plasticity was calculated using the S.A. Eberhart and W.A. Russell method.

3 Results

The studied Buckwheat varieties were represented by determinate (8 cultivars) and indeterminate (21) types of stem growth. The most important indicator of the development of Fagopyrum esculentum is plant height. The length of the stem affects the amount of seed yield: the number of lateral branches, flowers and fruits increases [12]. Analysis of the obtained data showed that the varieties with the smallest plant height are the most resistant to lodging. The variety of the indeterminate type Molva stood out for its short stems - 41.9 cm.

The stem of the Fagopyrum esculentum plant consists of branching and fruiting zones. A more reliable system is a plant with a large number of developed branches. Branching gives the final structure of the vegetative sphere of the plant and the productivity potential that is a characteristic of a particular species, variety or plant under given conditions [10]. The studied varieties differed in the number of the first order branches. On these grounds, varieties of the indeterminate type (Kitavase) and the determinate type (Esen) were distinguished.

Plants with a shortened first internode and a thickened stem are more resistant to lodging and, as a rule, are more productive. The shortest first internode (4.18 cm) was noted in the variety of indeterminate type Kayemchataya and determinate type Yesen (4.30 cm). The thickness of the first internode varied among the indeterminate type from 0.27 cm (Strelka) to 0.42 cm (Nikolskaya and Kitavase). In determinate varieties, this trait varied from 0.30 cm (Irmenka) to 0.39 cm (Dizain).

The number of lateral branches of the first order was from 0.8 pcs. (Mankan) to 2.5 pcs. (Amurskaya Mestnaya) among the indeterminate varieties, and from 1.2 pcs. (Irmenka) to 2.6 pcs. (Dikul) among determinate varieties. In terms of the resistance to lodging (4 points), varieties of the indeterminate type were distinguished: Emerald, At 7, Zemlyachka, Saulyk, Kitavase, Mankan (Table 1).

The biological value of Fagopyrum esculentum proteins, the digestibility of which is 78%, includes essential amino acids (lysine, leucine, phenylanine, etc.). The Saulyk variety of indeterminate growth type was distinguished for 14.13% of protein; the indeterminate varieties Cheremshanka and Kitavase were distinguished for fat (2.38%), as well as the Temp variety among the determinate varieties (2.20%).

In terms of harvest among the indeterminate growth type varieties, Pri 7 stood out (79.7%); among the determinate type varieties, the variety Temp was distinguished (80%). Buckwheat grain is thin-filmed if its shell content is up to 20%, medium-filmed – up to 20–22%, thick-filmed over 22% [13]. Indeterminate varieties Pri 7, Amurskaya Mestnaya, Nikolskaya, Molva, Bashkirskaya Krasnostebelnaya, Batyr, and Kitavase are thin-filmed, as well as Yesen and Devyatka among determinate varieties (Table 2).

Table 1. Morphological characteristics of determinate and indeterminate buckwheat varieties *Fagopyrum esculentum*.

Variety	Plant height. cm	Plant components height. cm		Internodes. cm		Number of lateral branches of the 1st order. pcs	Lodging resistance. point
		branching	fruiting	1st internode length	1st internode thickness		
Indeterminate growth type							
Izumrud (st)	84.8	35.4	28.7	5.85	0.35	2.0	4
Pri 7	86.2	19.3	30.6	6.61	0.33	1.7	4
Ussurochka	77.0	22.9	31.5	4.80	0.37	2.6	4
Chatyr Tau	61.8	35.4	27.0	5.86	0.31	1.6	3
Amurskaya Mestnaya	67.1	27.9	37.3	4.74	0.28	2.5	3
Nikolskaya	74.1	27.9	35.0	6.95	0.42	1.6	3
Kayemchataya	61.9	16.8	25.1	4.18	0.28	1.6	3
Strelka	69.7	23.3	34.2	6.08	0.27	1.8	3
Svetlana	71.3	26.5	36.3	5.22	0.40	1.8	3
Molva	41.9	26.3	35.8	4.99	0.29	1.6	3
Zemlyachka	67.9	25.7	33.3	5.56	0.36	1.7	4
Saulyk	59.6	22.3	30.0	5.70	0.34	1.3	4
Bashkirskaya Krasnostebelnaya	70.4	23.4	37.6	5.67	0.35	1.8	3
Batyr	74.0	25.2	38.1	5.01	0.36	1.9	2
Cheremshanka	72.1	31.6	13.9	7.94	0.35	1.7	3
Natasha	71.7	29.4	28.4	5.13	0.38	1.5	3
Ilishevskaya	64.9	24.2	31.6	6.45	0.32	1.4	3
Agidel	72.4	28.4	34.1	5.56	0.37	2.0	3
Inzerskaya	71.4	25.2	28.5	5.68	0.34	1.6	3
Ufimskaya	64.3	19.7	27.7	4.46	0.31	1.7	3
Kitavase	103.0	41.9	50.1	6.27	0.42	1.8	4
Mankan	90.4	39.3	34.3	7.43	0.37	0.8	4
Determinate growth type							
Irmenka	75.4	22.9	30.2	6.43	0.30	1.2	3
Dizain	67.3	25.9	25.1	5.47	0.39	2.3	3
Dialog	68.3	27.4	31.9	6.22	0.35	2.0	3

(*continued*)

Table 1. (*continued*)

Variety	Plant height. cm	Plant components height. cm		Internodes. cm		Number of lateral branches of the 1st order. pcs	Lodging resistance. point
		branching	fruiting	1st internode length	1st internode thickness		
Devyatka	72.0	30.6	32.1	5.82	0.36	2.5	3
Esen	76.5	36.0	37.3	4.30	0.37	2.1	3
Druzhina	71.9	19.8	28.2	5.91	0.37	2.0	3
Dikul	64.1	27.1	29.8	5.00	0.37	2.6	3
Temp	62.1	26.5	32.1	5.49	0.35	2.2	3
LSD $_{05}$	7.4	1.0	4.5	0.50	0.06	0.2	–

The level of productivity of the genotype depends on the 1000 grain weight - the most important economically valuable trait. On this basis, the variety of the indeterminate type Mankan (33.1 g) was distinguished, and from the determinate type - the variety Irmenka (33.5 g).

Rutin is important substance that imparts special properties to the fruits of Fagopyrum esculentum. Therefore, the obtained buckwheat varieties with high rutin content are one of the important features of improving the quality of fruits. The highest ability to accumulate rutin in fruits (15.0 mg/100 g) is the characteristic of the indeterminate variety Bashkirskaya Krasnostebelnaya, which was obtained from a hybrid population with the participation of the red-flowered mutant of the Rubra variety (Kamenets-Podolsk Agrarian Technical University, Ukraine) and Ufimskaya and Chishminskaya (Bashkir Research Institute of Agriculture, Russia). In the rest of the indeterminate varieties, this indicator was in the range of 4.0–10.0 mg/100 g. The highest content of rutin in determinate varieties was noted in Druzhin - 9.0 mg/100 g and Dikul - 8.7 mg/100 g.

Table 2. Characteristics of determinate and indeterminate varieties of *Fagopyrum esculentum* by technological and biochemical grain qualities.

Variety	Crop harvest %	Hull content. %	1000 seed weight. gr	Fat. %	Protein. %	Rutin mg/100 g
Indeterminate growth type						
Izumrud (st)	75.4	24.6	33.0	1.80	12.83	7.5
Pri 7	79.7	20.3	26.1	1.93	13.25	5.0
Ussurochka	73.5	26.4	23.4	2.20	13.20	9.7

(*continued*)

Table 2. (*continued*)

Variety	Crop harvest %	Hull content. %	1000 seed weight. gr	Fat. %	Protein. %	Rutin mg/100 g
Chatyr Tau	74.0	26.0	30.1	1.62	12.87	7.7
Amurskaya Mestnaya	79.1	20.9	24.2	1.90	12.74	4.3
Nikolskaya	78.1	21.9	28.2	1.92	12.40	4.0
Kayemchataya	74.8	25.2	31.6	2.26	13.52	10.0
Strelka	76.1	23.9	24.8	2.07	12.89	4.0
Molva	79.3	20.7	24.8	1.99	12.66	8.0
Zemlyachka	76.9	23.1	27.1	1.77	12.76	6.0
Saulyk	75.5	24.5	28.6	1.83	14.13	7.3
Bashkirskaya Krasnostebelnaya	78.5	21.5	24.1	1.89	13.59	15.0
Batyr	78.8	21.2	28.3	1.92	13.10	4.7
Cheremshanka	73.7	26.3	26.1	2.38	12.02	7.7
Natasha	73.1	26.9	35.2	1.91	12.65	8.3
Ilishevskaya	74.9	25.1	23.5	1.96	12.66	7.1
Agidel	76.1	23.9	26.8	1.96	13.23	8.3
Inzerskaya	75.0	25.0	26.3	2.21	13.12	9.7
Ufimskaya	76.6	23.4	26.2	1.96	13.18	5.7
Kitavase	78.7	21.3	28.9	2.38	13.25	6.3
Mankan	76.8	23.2	33.1	2.15	12.15	5.7
Determinate growth type						
Irmenka	72.4	27.6	33.5	1.80	12.38	7.3
Dizain	77.1	22.9	29.4	1.72	12.82	4.0
Dialog	76.1	23.9	28.1	2.10	12.77	7.0
Devyatka	79.0	21.0	27.2	1.96	12.20	8.3
Esen	79.1	20.9	26.8	1.94	13.20	4.3
Druzhina	77.0	23.0	29.5	1.93	12.35	9.0
Dikul	77.7	22.3	25.7	1.92	12.77	8.7
Temp	80.2	19.8	24.7	2.20	13.65	6.3
LSD_{05}	2.8	2.8	4.6	0.30	0.84	0.4

Rutin content in flowers was higher compared to fruits (Fig. 1). Nine varieties of Fagopyrum esculentum were studied for rutin content in flowers. Rutin content depends on the genotype [14]. The amount of routine in flowers varied from 5.6% (Dizain) to 9.9% (Bashkirskaya Krasnostebelnaya). The maximum content of rutin in reproductive organs (flowers) is associated with an increase in the synthesis of phenolic compounds in comparison with other plant organs and indicates the possible participation of rutin in the processes of plant growth and development.

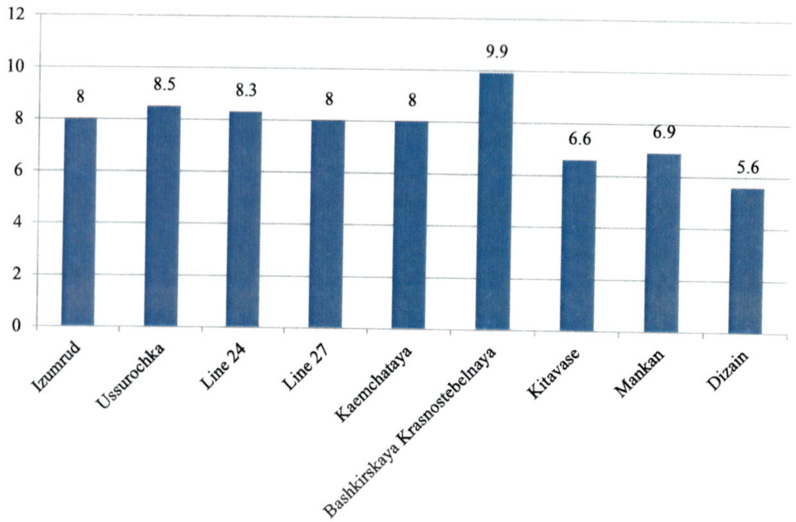

Fig. 1. Rutin content in flowers of *Fagopyrum esculentum* varieties of various origin, %.

Varieties that combine high biological, economic and technological properties and at the same time possess sufficient environmental stability and plasticity are of great value. The creation of environmentally sustainable varieties, that is, forms of medium intensity, capable of giving a stable yield in any conditions remains one of the important problems both in crop production and in breeding.

It is known that the higher the regression coefficient, the more significant the response of the variety to improvements in growing condition is, and the lower the stability is, the more stable the variety is against the background of varying environmental conditions.

Varieties, in which $b_i > 1$, and the S^2d indicator is unstable, should be considered as the most valuable in breeding. Such indicators were found in Batyr and Mankan (indeterminate type of growth). These varieties are best grown on an intensive background with a high level of agricultural technology. Forms with high $b_i > 1$ and S^2d values are highly responsive to growing conditions. Such varieties of the indeterminate type of growth include Izumrud, Amurskaya Mestnaya, Strelka, Molva, Zemlyachka, Bashkirskaya Krasnostebelnaya, Ilishevskaya, Agidel, Inzerskaya, Ufimskaya, Kitavase, and from determinate varieties - Irmenka, Dizain, Dialog, Devyatka, Esen, Dikul, Temp. These varieties are less responsive to improving environmental conditions and are more suitable to an extensive background. The varieties in which $b_i < 1$, regardless of the S^2d

value, should be considered as unpromising, since they lack such an important biological feature as an adequate response to an improvement in growing conditions. A similar reaction is characteristic of varieties of the indeterminate type of growth Pri 7, Chatyr Tau, Nikolskaya, Kayemchataya, Cheremshanka, Natasha, and the determinate type of growth variety Druzhina. Varieties with $b_i = 1$ and $S^2d = 0$ are of great importance, as such parameters indicate plasticity and stability. Ussurochka is one of these varieties.

Varieties with adaptability coefficient over 100% are adaptive. Therefore, the indicator was distinguished by varieties of indeterminate growth type - Izumrud, Pri 7, Ussurochka, Nikolskaya, Zemlyachka, Inzerskaya, Kitavase, Mankan, and varieties of determinate growth type Dialog, Devyatka, Dikul (Table 3).

Table 3. Parameters of plasticity and stability of determinate and indeterminate varieties of *Fagopyrum esculentum.*

Variety	Yield. t/ha		b_i	S^2d	Adaptability coefficient %
	Lim	Medium			
Indeterminate growth type					
Izumrud (st)	0.9–1.7	1.2	1.4	0.1	117.6
Pri 7	0.8–1.3	1.1	0.3	0.0	126.9
Ussurochka	1.2–1.8	1.5	1.0	0.1	150.0
Chatyr Tau	0.8–1.4	1.0	0.2	0.2	92.8
Amurskaya Mestnaya	0.6–1.3	0.9	1.9	0.3	75.7
Nikolskaya	1.0–1.2	1.1	0.9	0.1	110.7
Kaemchataya	0.4–0.7	0.6	0.2	0.0	62.7
Strelka	0.8–1.3	1.2	3.1	0.6	98.9
Molva	0.6–1.3	0.9	2.4	2.1	78.6
Zemlyachka	1.1–1.4	1.1	1.9	0.4	103.7
Saulyk	0.7–0.8	0.7	0.5	0.0	75.0
Bashkirskaya Krasnostebelnaya	0.6–1.1	0.9	1.9	0.6	76.4
Batyr	0.6–1.5	0.9	1.4	0.0	76.1
Cheremshanka	0.6–1.1	0.9	0.3	0.0	97.8
Natasha	0.6–0.7	0.6	0.0	0.0	69.0
Ilishevskaya	0.2–1.7	1.0	1.9	1.1	91.5
Agidel	0.9–1.2	0.9	1.9	1.6	84.7
Inzerskaya	0.3–1.3	1.1	1.3	0.8	100.3
Ufimskaya	0.3–1.1	0.9	1.5	0.7	83.7
Kitavase	0.9–1.2	1.1	1.5	1.4	119.5
Mankan	0.7–1.6	1.2	1.9	0.1	117.8

(*continued*)

Table 3. (*continued*)

Variety	Yield. t/ha		b_i	S^2d	Adaptability coefficient %
	Lim	Medium			
Determinate growth type					
Irmenka	0.8–1.0	0.9	3.0	0.4	96.7
Dizain	0.7–1.9	1.1	3.0	1.2	87.2
Dialog	0.7–1.4	1.2	2.9	0.8	102.2
Devyatka	0.5–1.3	1.1	5.6	2.2	133.9
Esen	0.7–0.9	0.8	1.4	0.1	87.5
Druzhina	0.6–1.0	0.9	0.2	0.0	91.2
Dikul	0.9–1.4	1.2	2.5	0.9	107.3
Temp	0.6–1.3	1.1	2.7	0.7	95.9

4 Discussion

Fagopyrum esculentum plants are subdivided into determinate and indeterminate, depending on the type of growth. In determinate buckwheat varieties, the apical inflorescence is represented not by a shield or a half-umbrella, but by a simple brush. Our research showed that the varieties of the indeterminate type have the highest morphological characteristics, and our data were confirmed by other researchers. Determinate varieties were distinguished by the number of lateral branches, which is consistent with the results obtained by Amelina - varieties of the determinate type of growth have a greater number of lateral branches than indeterminate ones.

There is a lot of literature on the role of rutin in the process of individual plant development [15]. The greatest amount of rutin in the studied varieties was found in flowers (5.6–9.5%). This is confirmed by the literature data of domestic and foreign studies [28–30]. The zoned varieties Izumrud and Pri 7, widespread in the Primorsky Krai, had a content of rutin in fruits of 7.5 and 5.0 mg/100 g, respectively. In the conditions of the Oryol Region, the maximum accumulation of flavonoids was found in the determinate varieties Dizain and Dialog. Dietrych-Szostak D, Oleszek W., Zhang H Z, Guo YR, Niu L. noted high content of rutin in the reproductive organs of plants in their studies [14].

Varieties of indeterminate type of growth were distinguished in terms of biochemical parameters, and determinate varieties - according to technological grain qualities. When studying the parameters of adaptability, the dependence of buckwheat varieties on weather conditions was revealed. Varieties of the indeterminate growth type from the Republic of Tatarstan, the Republic of Bashkotorstan and Japan, were adapted to the conditions of the monsoon climate; varieties from the Oryol Region turned out to be adapted to the conditions of the determinate type of growth.

5 Conclusion

To increase the efficiency of buckwheat breeding in a monsoon climate, it is recmmended to include sources with the following valuable traits in the hybridization as parental forms:

- Short stem: indeterminate variety Molva (41.9 cm);
- First internode length: indeterminate variety Kaemchataya (4,18 cm) and determinate variety Esen (4,30 cm);
- First internode thickness: indeterminate varieties Nikolskaya and Kitavase (0,42 cm);
- 1000 seed weight: indeterminate variety Mankan (33.1 g) and determinate variety Irmenka (33.5 g);
- Yield: indeterminate variety Ussurochka (1.5 t/ha);
- Rutin content in fruits and flowers: indeterminate variety Baskirskaya Krasnostebelnaya (15 mg/100 g and 9.9%, accordingly).

References

1. Cai, Y.Z., Corke, H., Li, W.D.: Reference modele in food science. Buckwheat (2016). https://doi.org/10.1016/B978-0-08-100596-5.00034-2
2. Mendler-Drienyovski, N., Cal, A.J., Dobrancski, J.: Progress and prospects for interspecific hybridization in buckwheat and the genus Fagopirum. Biotech. Adv. **31**, 1768–1775 (2013). https://doi.org/10.1016/j.biotechadv.2013.09.004
3. Amelin, A., Fesenko, A., Chekalin, E., Fesenko, I., Zaikin, V.: Higher yielding varieties of common buckwheat (Fagopyrum esculentum Moench) with determinate growth habit (singl mutation det) manifest higher photosynthesis rate at stage of grain filing. Acta Agric. Slov. Med. **1**(115), 59–65 (2020). https://doi.org/10.14720/aaa2020.115.1.1316
4. Klocek, J., Szwed, M., Koczkodaj, D., Mitrus, J., Saniewski, M., Horbowicz, M.: Methyl jasmonate vapors affect the composition and peroxidation of major fatty acids in common buckwheat seedlings. Acta Agrobotanica Med. **4**(69), 1–8 (2016)
5. Ikeda, S., Yamashita, Y., Tomura, K., Kreft, I.: Nutrritional cjmparison in mineral characteristics between buckwheat and cereals. Fagopyrum Med. **23**, 61–65 (2006)
6. Gasanzade, S.: Optimization of technological methods of buckwheat cultivation in the conditions of the Ganja-Kazak zone of Azerbaijan. Agr. Sci. Med. **11–12**, 45–48 (2018). https://doi.org/10.32634/0869-8155-2018-320-11-45-48
7. Wieslander, G., Vombergar, B., Kreft, I.: Bioactive flavonoids in buckwheat grain and green parts. Mol. Breed. Nutr. Aspects Buckwheat. (2016). https://doi.org/10.1016/B978-0-12-803692-1.00012-2
8. Kreft, I., Fabjan, K., Yasumoto, K.: Rutin contentin buckwheat (Fagopyrum esculentum Moench) food materials and products. Food Chem. Med. **3**(98), 508–512 (2006)
9. Brunori, A., Nobill, C., Procacci, S.: Molecular Breeding and Nutritional Aspects of Buckwhtat. Elsevier, London (2016)
10. Klykov, A., Murugova, G., Bogdan, P., Konovalova, I., Timoshinova, O., Kryuchkova, N.: Use of genetic resources in white straw and cereal crops breeding in the monsoon climate. In: IOP Conference Series: Earth and Environmental Science, vol. 548, no. 6, p. 062014 (2020). https://doi.org/10.1088/1755-1315/548/6/062014

11. Kodochilova, N.: Economic and biological assessment of buckwheat varieties in competitive variety testing. Agric. Sci. Med. **10**, 62–64 (2019). https://doi.org/10.32634/0869-8155-2019-332-9-62-64
12. Klykov, A., Moiseenko, L.: Influence of buckwheat cultivation conditions upon rutin content. In: Advances in Buckwheat Research Proceedings of 11th International Symptoms Buckwheat, pp. 475–483 (2010)
13. Kreft, I.: Rutin content in common and bitter buckwheat food meterials and products. In: III International Symposium on Buckwheat and the Dietary Culture, 8–12 August Xichang, China. Xichang, pp. 38–39 (2005)
14. Zhang, H., Guo, Y., Niu, L.: Study on dynamics of flavonoids content in different growth phase of buckwheat. J. Food Sci. Technol. **27**(8), 71–73 (2012)
15. Chitarrini, G., Nobili, C, Pinzari, F., et al.: Buckwheat achenes antioxidant profile moderates aspergillus flavus growth and aflatoxin production. Int. J. Food Microbiol. **189**, 1–10 (2014). https://doi.org/10.1016/j.ijfoodmicro2014.07.029

Resource Provision of Innovations in Agricultural Enterprises in the South of the Russian Far East

Tatiana Potenko[1] (✉), Irina Zhupley[2], Oksana Mamai[3], and Ludmila Kvashko[4]

[1] Federal Scientific Center of Agricultural Biotechnology of the Far East named after A.K. Chaika, 30 Volozhenin Street, Ussuriysk 692539, Russia

[2] Primorskaya State Academy of Agriculture, 44, Blucher Avenue, Ussuriysk 692510, Russia

[3] Samara State Agrarian University, 2, Uchebnaya Street, Kinel, Urban-type settlement Ust-Kinelsky, Samara Region 446442, Russia

[4] Primorsky Institute of Railway Transport, a Branch of Far Eastern, State Transport University in Ussuriysk, 166, Pushkin Street, 692522 Ussuriysk, Russia

Abstract. What resources can help innovate sustainable development? Does resource scarcity limit innovation towards sustainability? This study explores the resource-based innovation capabilities of soybean producers, based on empirical data from large and medium-sized enterprises in the southern Russian Far East. The findings suggest that businesses can innovate with very different combinations of resources. The limiting factors are "machinery and equipment" and "number of employees" (insufficient use from the standpoint of allocative efficiency), resources such as "sown area" and "material costs" are used excessively. In farms capable of innovative development, the value of revenue, as a rule, exceeds the standard value and is 2–3 times higher than the average revenue for the entire set of enterprises, which indicates the efficiency of resource use and the ability to innovative development.

Keywords: Innovation · Agricultural production · Cluster · Economic efficiency · Development · Tangible resources · Soybean

1 Introduction

Recent studies show that the main problem of agricultural production is low level of innovation absorption and insufficient investment in the development of innovations [1–4]. Due to the lack of own circulating assets, low liquidity of collateral and large volumes of accounts payable, the agricultural sector remains unattractive for investors [5]. There is an excess of the growth rate of the volume of credit investments over the growth rate of the volume of production. This may limit the ability of enterprises to innovate.

The innovations assimilation allows ensuring a continuous renewal of the technological, technical, organizational, and economic base of agricultural production and, on this basis, to increase the economic efficiency and competitiveness of agriculture.

A. Muratov and S. Ignateva (Eds.): AFE 2021, LNNS 353, pp. 185–192, 2022.
https://doi.org/10.1007/978-3-030-91402-8_22

Despite the implementation of important state programs that determine the prospects for the development of the Russian Far East in recent years, there are a number of problems that hinder the development of innovations in agriculture:

- Reducing the number of research and development;
- Insufficient financing of science and production related to innovations;
- Shortage of qualified personnel;
- Lack of information field about innovations in agriculture;
- Economic inability of agricultural enterprises to use innovations due to the crisis in the economy.

Availability of the necessary mechanisms for introducing scientific developments into production and the economic ability of enterprises to innovate that requires additional resources is the key to success.

If the enterprise has necessary resources, it can follow the path of the leader's strategy, developing and implementing new or basic innovations. If innovative opportunities are limited, then it is advisable to choose a follower strategy, i.e. implement improving technologies.

The above suggests that the resources behind innovation deserve careful scrutiny in order to better understand the potential of such innovation on the part of agricultural enterprises.

A wide range of resource classification may be found in the literature. They are considered as consisting of tangible or intangible assets, the former consist of financial and physical resources, and the latter - of human, social, organizational and technological resources and reputation [5–7]. The focus of this article is on tangible resources, which include (a) financial assets and (b) physical assets.

According to recent studies [1, 3], public policies that stimulate innovation among agricultural enterprises may be a more fruitful approach to improving the environmental and social impact of this sector. The government can help tackle funding gaps that are often cited as an obstacle to innovation. There are successful examples of financing innovation in Europe, achieved through the convergence of administrative and legal mechanisms [8–10].

In the study, we ask the following research questions and examine them with the help of three subtasks: With the help of what material resources can agricultural enterprises achieve progress in innovative development? What method can be used to assess the degree and possibilities of innovative development of agricultural enterprises? What are the differences between enterprises with scarce and abundant resources?

The work continues as follows: first, we will discuss the methodology for assessing the resource potential of agricultural enterprises; then we outline the resource categories used in this study and refer to the empirical results.

As this analysis shows that some resources are overused and some are underutilized. We will examine the reasons for this situation. Finally, we discuss the empirical findings as well as the methodological input.

2 Materials and Methods

Our methodology for assessing the innovative potential of agricultural enterprises is based on cluster approach, which allows singling out homogeneous groups-clusters from a set of farms according to the degree and possibilities of innovative development. The purpose of grouping is to identify homogeneous objects that are similar to each other. In 2009, 35 farms were studied, in 2019 - 37 farms specializing in the production of soybeans, with an area of crops of at least 500 hectares. This choice is justified by the fact that soybeans are an economically significant crop in the south of the Russian Far East and are in demand both on the domestic and foreign markets.

To simplify the comparative analysis, we studied only eight indicators that underlie the resource provision of innovations. This may mean that some potentially relevant resources were not included in the analysis.

The following indicators were used to classify enterprises by the number of resources:

Y1 – soybean productivity, t per 1 ha;
Y2 – expenditures per 1 ha of soybeans, rub000's;
Y3 – soybean production profitability, %;
Y4 – farm profitability, %;
Y5 – depreciation per 1 hectare of soybean sown area, rub000's;
Y6 – cost of fertilizers and plant protection products per 1 ha of soybeans, rub000's;
Y7 – cost of repair and maintenance of fixed assets per 1 ha of soybeans, rub000's;
Y8 – revenue per 1 ruble of salary for soybean, rub

The selected indicators have the same direction of action. Therefore, the greater their importance is, the higher their innovative development potential is. The quantitative value of the selected indicators reflects the regularity of the degree changes and possibilities of innovative development of enterprises.

Economic conditions of reproduction are reflected by indicators Y_3, Y_4; the technical level of production is shown by the Y_5 indicator; Y_7 - the effect of saving live labor due to a higher provision of machines and equipment is shown by indicator Y_8.

The proposed methodology is not supposed to use enterprises with a significant difference in values of indicators due to the peculiarities of production technology. For example, if a farm has incomparably higher costs per hectare in comparison with other enterprises, this does not mean that they are more innovative.

The standardization of the initial indicators was carried out according to the following

$$t_{ij} = (x_{ij} - \overline{x}) / \sigma_i \quad (1)$$

where tij – standardized value of the i-th indicator for the j-th object,

x_{ij} – initial value of the *i*-th indicator for the *j*-th object,
$\overline{x}$ – the average value of the *i*-th indicator,
σ_i – standard deviation of the *i*-th indicator.

Indicators transformed this way have an average variance equal to zero or equal to one. It is important that the standardization procedure does not change the structure of interaction between indicators.

The calculations were performed using STATISTICA 6.0. The method of means clustering was used, which made it possible to divide the initial set of agricultural enterprises (objects) into homogeneous groups (clusters) for all indicators. This minimizes the variability of objects within clusters and maximizes the variability between clusters. In the method of means, the number of clusters is set by the researcher.

3 Results

Soybeans are an economically significant crop for the south of the Russian Far East. The Amur Region and the Primorsky Krai are the leaders in soybean production in the region. Soybeans have great potential for expanding soybean production and increasing exports to China, given its proximity, which reduces transport costs [11]. Therefore, cluster analysis is applied to a set of large and medium-sized enterprises cultivating this valuable oilseed and high-protein crop.

The study identified three clusters. The first cluster includes enterprises with abundant resources, hence a high potential for innovative development. The second cluster is the average provision of resources (under certain conditions, innovative development is possible), the third cluster is the lowest resources (low resource availability and the absence of economic conditions for innovative development).

Table 1. Descriptive statistics of selected variables for assessing innovation potential (2009).

Variables	Means		
	Cluster I	Cluster II	Cluster III
Number of farms (no.)	11	14	10
Soybean productivity (ton per 1 ha)	1.4	1.1	0.9
Expenditures per 1 ha of soybeans. rub000's	12.61	8.38	6.09
Soybean production profitability. %	54.0	1.4	62.06
Farm profitability. %	30.2	−0.5	26.1
Cost of fertilizers and plant protection products per 1 ha of soybeans. rub000's;	2.12	0.88	0.4
Depreciation per 1 ha of soybean sown area. rub000's	2.02	1.63	0.89
Cost of repair and maintenance of fixed assets per 1 ha of soybeans. rub000's	2.38	0.78	1.09
Revenue per 1 ruble of salary for soybean	5.49	7.33	13.04

Table 1 show that the farms of the first cluster have higher costs per hectare. These farms have high productivity - 1.4 t/ha and farm profitability - 30.2%.

Resources are ineffectively used in the farms of the second cluster, their yield is lower - 1.1 t/ha and the provision of agricultural equipment. The production is generally unprofitable for the farm.

The farms of the third cluster have the lowest yield - 0.9 t/ha, but nevertheless they provide profitable production.

Figure 1 shows the results of the performed multivariate clustering.

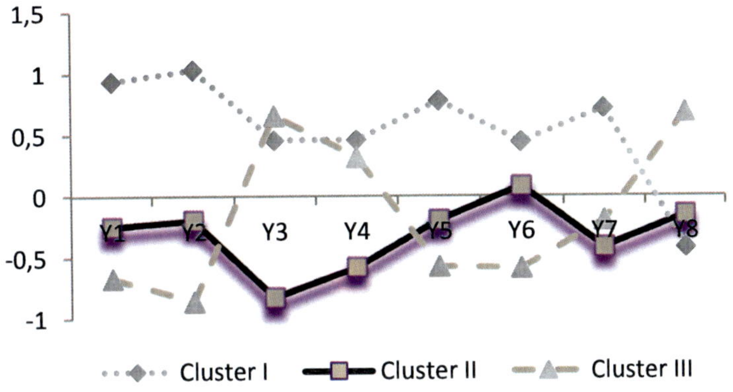

Fig. 1. Graphical display of clusters by indicators.

Figure 1 shows that there are stable differences between the clusters in terms of the totality of the indicators under consideration, while economically justified patterns are observed in the ratio of indicators for the clusters.

Similar calculations were performed for the clustering of agricultural enterprises according to 2019 data.

According to the Table 2, the enterprises included in the first cluster, have the greatest resource potential for the introduction of innovations, they have the highest yield and equipment availability. With soybean yield of 1.7 t/ha, they showed the best economic results - the profitability of soybean production was 38.22%.

For the second cluster enterprises, yields and equipment are lower, they occupy an intermediate position in terms of economic efficiency.

The enterprises of the third cluster are characterized by low yields, low technical availability, fertilizers are applied below the recommended rates. However, their activities provide a break-even production.

Table 2. Descriptive statistics of selected variables for assessing innovation potential (2019).

Variables	Means		
	Cluster I	Cluster II	Cluster III
Number of farms (no.)	8	16	13
Soybean productivity (ton per 1 ha)	1.7	1.3	1.0
Expenditures per 1 ha of soybeans. rub000's	27.8	25.7	22.5
Soybean production profitability. %	38.22	19.14	7.31
Farm profitability. %	22.14	13.31	2.45
Cost of fertilizers and plant protection products per 1 ha of soybeans. rub000's;	6.12	5.22	1.76
Depreciation per 1 hectare of soybean sown area. rub000's	3.17	1.88	2.25
Cost of repair and maintenance of fixed assets per 1 ha of soybeans. rub000's	2.22	1.85	1.07
Revenue per 1 ruble of salary for soybean	8.6	9.4	14.65

4 Discussion

The profitability of soybean production is largely determined by the level of the achieved yield, the growth of which is largely determined by using high-yielding varieties, adherence to the cultivation technology using high-tech energy-saving machines.

Soy is an economically important crop. The sowing area of soybeans increased by 2.2 times in 2019 compared to 2009. As can be seen from the comparative analysis, the yield in the best farms of the first cluster has also increased by 20%, and the average farms in the second cluster - by 18%. Average yield in 2009 was 1.0 t/ha, in 2019 - 1.3 t/ha.

Variety renewal contributed to the yield growth. Over the past 10 years, new varieties have appeared; their productivity at the genetic level is higher in comparison with previous ones. Protection of plants from pests, diseases and weeds, also play an important role. In terms of agroclimatic conditions, the southern regions of the Far East are most favorable for cultivation of ultra-early and early-maturing varieties, as well as some mid- and late-ripening varieties. These varieties have the same potential yield as the world average, and if all agronomic measures are followed, they can provide a yield of 3.9–4.3 tons per hectare. However, the region realizes only about 50 percent of this production potential. Further genetic improvement is required for the successful implementation of the soybean development program.

The obtained results indicate that the best farms (capable of innovative development) use resources most efficiently. Equity capital, machinery and equipment, the number of employees, the cultivated area are necessary resources for the development of innovation, but is not a sufficient condition for the introduction of innovations into production. To be effective, the innovations must be connected to R&D cooperation. Only five of the total number of the studied enterprises, including the poorest ones, had both financial capital in the form of equity capital and capital in the form of R&D cooperation. Despite limited resources, these enterprises showed signs of active cooperation in the field of R&D, thereby compensating for the lack of resources by cooperation with external partners.

This observation suggests that, during the technological innovation development phase, collaboration with relevant stakeholder groups enables low-income enterprises to acquire new ideas concerning their production.

5 Conclusion

Some economically developed countries have been able to achieve significant progress in innovative development. Therefore, overcoming the accumulated problems in the development of agriculture, such as technological backwardness and low profitability indicators, is possible on the basis of the widespread use of modern achievements of the scientific and technological process.

The study shows that although a lack of resources is a frequently encountered obstacle to innovation, a limited number of enterprises compensate for the lack of resources through R&D collaboration with other relevant stakeholders.

Our research also has a methodological contribution. We applied a cluster approach to the study of resources, which allows us to assess the degree and possibilities of innovative development of agricultural enterprises. Using "distances" to centers to identify in each cluster the typical enterprises that are located at a shorter distance from the center of the cluster, as well as the worst and best resource enterprises that are more distant from the center of the cluster.

The results of cluster analysis indicate that, in the overwhelming majority of agricultural enterprises, the effective use of resources and production opportunities are limited by the availability of agricultural machinery, working capital and largely depend on economic conditions. In the long term, the innovative activity of enterprises depends on the policy that ensures access of producers to credit resources, the development of public-private partnerships, international cooperation and the popularization of the rural way of life. However, joint programs must go beyond removing obvious obstacles. They also need to stimulate the development of processing facilities in the Far Eastern Federal District, which will make the regional economy fully integrated.

With regard to constraints and future research, we studied the use of material resources of the enterprises. In the control group we did not identify those enterprises that sought to create technological, environmental or social innovation and did/did not succeed. This study design may be useful in future studies that explore how relationships with external R&D partners are important for large and medium-sized agricultural enterprises.

In the midst of the economic recession and sanctions, which led to a reduction in world trade and consumption and capital outflow, hopes are being pinned on the contribution of agricultural enterprises to ensuring food security and solving problems of socio-economic development. Evaluation of the use of resource provision of enterprises allows obtaining information on the results of the economic activity of the enterprises, identifying reserves, assessing ways to improve the efficiency of the enterprises and move on to forecasting, as well as the implementation of innovations.

References

1. Moreddu, C., Tongeren, F.V.: Improving agricultural productivity sustainably at global level: the role of agricultural innovation policies. EuroChoices **12**(1), 8–14 (2013)
2. Moreddu, C., Contini, E., Ávila, F.: Challenges for the Brazilian agricultural innovation system. EuroChoices **16**(1), 26–31 (2013)
3. Ogundari, K., Bolarinwa, O.D.: Impact of agricultural innovation adoption: a meta-analysis. Aust. J. Agric. Res. Econ. **62**(2), 217–236 (2018)
4. Alston, J.M., Pardey, P.G.: Antipodean agricultural and resource economics at 60: agricultural innovation. Aust. J. Agric. Res. Econ. **60**(4), 554–568 (2016)
5. Gaisford, J.D., Hobbs, J.E., Kerr, W.A.: Making markets work: modeling agri-food systems in transition. Can. J. Agric. Econ/Revue Can. d'agroeconomie **49**(2), 181–202 (2005)
6. Harris, M., Fraser, I.: Natural resource accounting in theory and practice: A critical assessment. Aust. J. Agric. Res. Econ. **46**(2), 139–192 (2002)
7. Rounds, J., Segner, R.: Construction Supervision. Hoboken, New Jersey (2011)
8. Brunori, G., Barjolle, D., Dockes, A.-C., et al.: CAP reform and innovation: the role of learning and innovation networks. EuroChoices **12**(2), 27–33 (2013)
9. Cioloş, D., Geoghegan-Quinn, M.: Promoting EU research and innovation in agri-food. EuroChoices **12**(1), 4–7 (2013)
10. Zilgalvis, P.: The Need for an innovation principle in regulatory impact assessment: the case of finance and innovation in Europe. Policy Internet **6**(4), 377–392 (2014)
11. Yi, F., Gudaj, R.T., Arefieva, V., Yanbykh, R., et al.: Sino-Russian cooperation on soybean development in the Russian far east. Am. J. Econ. Soc. **79**(5), 1553–1586 (2020)

Improving the Methods of Preserving the Genetic Identity of the Cucumber Variety Population Naslednik During Reproduction

Galina Kuzmitskaya(✉) and Nadezhda Kuliakina

Far Eastern Research Institute of Agriculture, Khabarovsk Federal Scientific Center, Far Eastern branch, Russian Academy of Sciences, Klubnaya Street, 13, v. Vostochnoe, Khabarovsk District 680521, Russia

Abstract. An important task of seed production is to preserve the genotypic identity of varieties in order to maintain the genetic structure of heterogeneity. To accomplish this task, an in-depth study of the population structure is necessary for the sake of establishing the number of biotypes as well as their productivity and ecological stability. The shape and length of the seed fruit served as markers in the study of the structure of the variety population. In the course of studies carried out in 2016–2020, it was found that the variety population of Naslednik cucumber is heterogeneous in terms of the approbation characteristics of the seed fruit, and differs in the shape and length of biologically ripe fruits. The basis of the variety population is formed by ovoid and cylindrical testes having a length of 14–18 cm. However, in years with extreme weather conditions in the variety population, biotypes with short seed fruits (10–13 cm) of both forms, as well as with long fruits (21 and > cm), can be split off in elongated ovoid and elongated cylindrical shapes. The established structure of the variety population should be taken into account in the primary seed production of this variety.

Keywords: Cucumber (Cucumis sativus L.) · Population · Biotype · Genetic identity · Variety Naslednik · Fruit shape · Fruit length

1 Introduction

In the modern world, vegetable growing is one of the most dynamically developing branches of agriculture, and currently its production reaches 1 billion tons; this increases every year by 60–80 tons. Due to natural and anthropogenic factors, vegetable growing is regionally species-specific [1]. The natural and climatic potential of the Far Eastern region, in particular the Khabarovsk Territory, is quite favourable for the cultivation of heat-loving vegetable crops, including cucumber.

Cucumber (Cucumis sativus L.) is an economically important vegetable crop throughout the world [2–6]. It is a widely cultivated plant in the cucurbitaceae family. It belongs to the genus Cucumis L. and, according to various authors, includes 30 to 40 species. They originate mainly from India, China, Nepal and tropical regions of Africa, which differ sharply in growing conditions, from arid to excessively humid.

A. Muratov and S. Ignateva (Eds.): AFE 2021, LNNS 353, pp. 193–201, 2022.
https://doi.org/10.1007/978-3-030-91402-8_23

Among them, Cucumis sativus L. is of the greatest economic importance. Cucumber is grown in open ground and protected buildings [3, 7–10].

In terms of calorie content, cucumber is inferior to most vegetables, since it consists of 97% water, but it does possess high taste and dietary properties. In addition, cucumber is rich in sugar, vitamins, antioxidants and mineral components, and is easily used both for fresh consumption and for processing [11–16].

World experience shows that breeding and seed production play an important role in increasing the efficiency of agricultural production. The variety is the most centralized economically and environmentally effective means of increasing the quality of the crop, as well as of ensuring resource and energy efficiency, environmental sustainability, environmental protection and profitability of agricultural production [17]. An important task of seed production is to preserve the genotypic identity of varieties in order to maintain the genetic structure of heterogeneity. To accomplish this task, an in-depth study of the population structure is necessary for the sake of establishing the number of biotypes, their productivity and ecological stability.

All the sorts of vegetable cultures common in the country are the populations, the genetic composition of which is regulated by natural and artificial selection. It is necessary to have a good and correctly developed seed production for the implementation the new sorts in productions [18].

Currently, in the open ground vegetable growing context in Russia, populations as well as multi-line and synthetic varieties are still widespread. For their reproduction, seed growers need scientific advice based on knowledge of the genetic structure of the variety.

In Russia, seed production of varieties-populations is an important problem both in the production aspect and in terms of preserving the variety population as a manifold heterogeneous complex for maintaining the biological diversity of the plant gene pool.

An important role in this is assigned to primary seed production, the task of which is not only to introduce new varieties, but also to preserve their genetic identity from year to year or from generation to generation. During the multiplication process, the variety must have the same genetically determined uniformity that it had when it was created and registered. To correctly keep the main economic and biological properties of the sort population, characterized it originally, it is necessary to have the clear characteristic of sort population and the selection method for its maintenance. All that is necessary in reproduction. For analysis and assessment of the sort population the least changeable signs are necessary [18].

It is believed that the analysis of a population should begin with the characteristics of the phenotypic characteristics of the individuals included in it, with the establishment of the parameters of phenotypic variability in terms of quantitative traits.

The most stable part of plants is the generative organs; they change very little, since they are protected by the entire system of genetic resistance. The stable traits in cucumber include the fruit shape index, fruit length and width, and internode length.

Fruit size and shape are complex traits influenced by numerous factors, including genotype and environment, such as temperature, light, precipitation, cultivation conditions [19].

The aim of the research was to study the phenotypic spectrum of the cucumber cultivar Naslednik, specifically to determine the number and ratio of biotypes that differ in the length and shape of the seed fruit that make up this cultivar population.

2 Materials and Methods

The object of research was the Naslednik cucumber variety of the Far East Agricultural Research Institute, included in the State Register of Breeding Achievements of the Russian Federation in 2018.

Experimental work was carried out in the period 2016–2020. Studies examining the variety population of the Heir were carried out at the seed-growing plot of the Department of Vegetable Crops and Potatoes of the Far East Agricultural Research Institute of Agriculture. The area of the experimental site was 0.5 hectares. Analysis of variety populations in terms of composition and ratio of biotypes was carried out against a stabilizing background, which contributes to the preservation of the original phenotypic composition. This background usually corresponds to the conditions in which the variety was bred. The shape and length of the seed fruit served as markers in the study of the structure of the variety population. The analysis included 300 plants each year (800–1,000 seed fruits). It was found that such a number of plants is optimal to ensure the presence of all biotypes of the variety population in the studied sample [18]. The selection of plants was carried out, the length was measured, and the shape of the seed fruits was described.

The years of research were marked by a climatic factor's characteristic of the Amur region, and significantly differed among themselves (Fig. 1, 2).

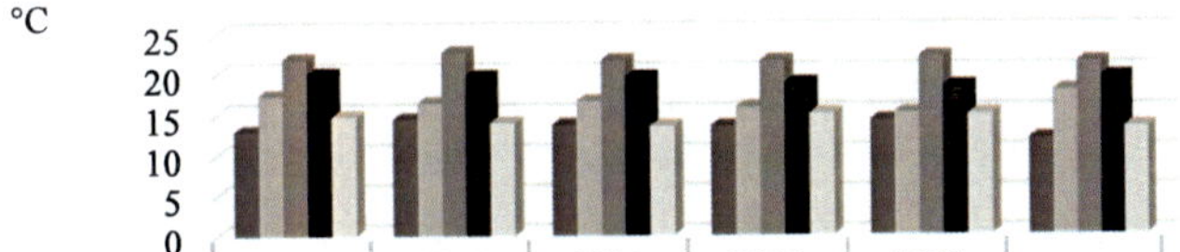

	2016	2017	2018	2019	2020	average annual
May	13	14.5	13.8	13.6	14.3	12
June	17.4	16.5	16.7	15.9	15.2	17.9
July	21.9	22.7	21.7	21.6	22.1	21.4
August	20.1	19.8	19.7	18.9	18.5	19.6
September	14.8	14	13.6	15.1	15	13.4

Fig. 1. Temperature during the years of research.

In general, for the period from May to September, the hydrothermal coefficient was: 2016 – HTC = 2.0; 2017 – HTC = 1.5; 2018 – HTC = 1.9; 2019 – HTC = 2.5; 2020 – HTC = 2.9 (Fig. 3). The hydrothermal coefficient (HTC) is the ratio of the amount of precipitation to the sum of temperatures, reduced by a factor of 10, and is used as an indicator of a plant's need for moisture.

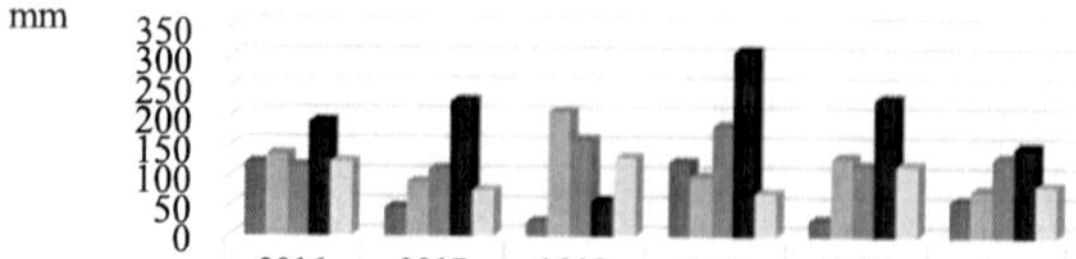

	2016	2017	2018	2019	2020	average annual
May	122	47	24	124	27	60
June	136	90	208	99	131	78
July	116	112	161	184	118	132
August	192	226.8	59	311	231	151
September	124	76	131	72	119	86

Fig. 2. Precipitation during the years of research.

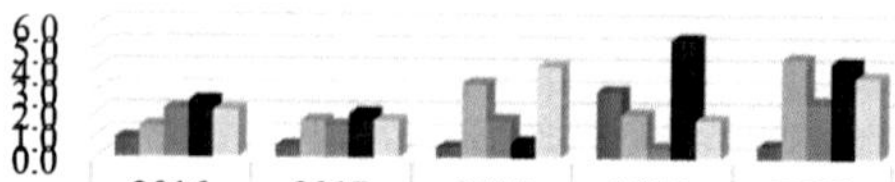

	2016	2017	2018	2019	2020
May	0.9	0.5	0.5	3.0	0.6
June	1.4	1.7	3.3	2.0	4.5
July	2.2	1.4	1.7	0.5	2.5
August	2.6	2.0	0.7	5.4	4.4
September	2.2	1.7	4.1	1.7	3.7

Fig. 3. Hydrothermal coefficient.

This made it possible to give an objective assessment of the material under study in accordance with the prevailing specific agroclimatic conditions. It was found that each year of research was characterized by excessive moisture, since the indicators of the hydrothermal coefficient annually exceeded 1.3 during the growing season of the cucumber. Nevertheless, in some periods of plant development, there were months of dry (HTC = 0.7–1.0) and dry farming (HTC = 0.5–0.7).

3 Results and Discussion

The Naslednik variety population turned out to be heterogeneous and differed in the length and shape of the seed fruits, depending on the hydrothermal conditions of a particular growing season. Over the entire period of research, all biologically ripe fruits were combined into groups according to shape and length.

It has been established that 4 forms (biotypes) of testes can be distinguished in the variety population: ovoid, elongated-ovate, cylindrical and elongated-cylindrical.

It was determined that the ovoid form, which averaged 50–71% over the years, was an exception in 2016, and the cylindrical form was 29–44%, which were the main biotypes (Table 1); elongated-ovoid and elongated-cylindrical forms were associated biotypes.

Table 1. Determination of the proportion of fruit shape in the structure of the Naslednik cucumber variety population.

Fruit forms, cm	Frequency of occurrence of the trait, f				
	2016	2017	2018	2019	2020
Cylindrical	90	87	101	111	83
$x \pm t_{05}s_x$	0.33 ± 0.027	0.29 ± 0.051	0.34 ± 0.053	0.44 ± 0.014	0.33 ± 0.059
Elongated cylindrical	106	–	–	8	–
x ± t05sx	0.39 ± 0.027			0.03 ± 0.014	
Ovoid	44	213	199	125	167
x ± t05sx	0.16 ± 0.027	0.71 ± 0.051	0.66 ± 0.053	0.50 ± 0.014	0.67 ± 0.059
Elongated ovoid	35	–	–	6	–
x ± t05sx	0.13 ± 0.027			0.02 ± 0.014	
n	275	300	300	250	250
s, %	22.7	45.4	47.4	10.7	47.1
Vp, %	90.8	90.8	94.8	42.8	94.2
sp	1.4	2.6	2.7	0.7	3.0

Table note: n – the number of measurements, s – the standard deviation of the proportion, Vp – coefficient of variation, sp – share error, x ± t05sx – confidence interval for 5% value level

The results of selective observation suggest that the general share of ovoid fruits in the studied population varied over the years in the following intervals: 13.3 ÷ 18.7% (2016), 65.9 ÷ 76.1% (2017), 60.7 ÷ 71.3% (2018), 48.6 ÷ 51.4% (2019), 61.1 ÷ 72.9%, (2020), and the proportion of cylindrical fruits is 30.3 ÷ 35.7% (2016), 23.9 ÷ 34.1% (2017), 28.7 ÷ 39.3% (2018), 42.6 ÷ 45.4% (2019) and 27.1 ÷ 38.9% (2020). Fruits of an elongated-cylindrical shape with qualitative variability were in the following ranges: 36.3 ÷ 41.7% (2016), 1.6 ÷ 4.4% (2019); fruits of an elongated-ovoid shape were in the following intervals: 10.3 ÷ 15.7% (2016), 0.6 ÷ 3.4% (2019). The significance level of this conclusion is 5%. Elongated-cylindrical and elongated-ovoid biotypes were not identified in 2017, 2018 or 2020.

The relative abundance (frequency of occurrence) of seed fruits of the same form in 2016, 2017, 2018 and 2020 was approximately the same and approached the maximum possible variability with these gradations of the trait, which is confirmed by the coefficients of variation (V = 90.8–94.8) tending to 100%, and only 2019 was an exception. An indicator of the coefficient of variation of 100% means that the ratio of qualitative features relative to each other is the same. In our studies, with four gradations of the 'shape of the fetus' trait, the share of each trait tends to 25%, and with two gradations, to 50%.

The length of biologically ripe fruits in the years of research varied within 10–27 cm (Table 2).

Table 2. The main statistical indicators of the sample for determining the average length of the fruit in the cucumber variety population Naslednik.

Fruit length. cm	Frequency of occurrence of the trait. f				
	2016	2017	2018	2019	2020
10	–	–	–	–	13
11	–	–	–	–	30
12	–	–	9	5	38
13	**1**	**7**	**16**	**29**	**51**
14	**12**	**14**	**50**	**35**	**42**
15	**19**	**34**	**83**	**46**	**34**
16	**24**	**74**	**82**	**44**	**26**
17	**39**	**83**	**45**	**39**	**13**
18	**42**	**53**	**14**	**33**	**2**
19	**32**	**26**	**1**	**10**	**1**
20	47	8	–	9	–
21	22	1	–	–	–
22	21	–	–	–	–
23	10	–	–	–	–
24	2	–	–	–	–
25	2	–	–	–	–
26	1	–	–	–	–
27	1	–	–	–	–
n	275	300	300	250	250
Average fruit length. cm	18.6	16.7	15.4	15.8	13.5
s^2	6.5	2.3	2.0	3.7	3.9
s. %	2.5	1.5	1.4	1.9	2.0
V. %	13.4	9.0	9.4	12.1	15.3
s_x	0.2	0.1	0.1	0.1	0.1
$s_{x\%}$. %	0.8	0.5	0.5	0.8	1.0
$x \pm t_{05}s_x$	19 ± 0.39	17 ± 0.20	15 ± 0.20	16 ± 0.20	13 ± 0.20

Table note: n – the number of measurements. s^2 – variance. s – standard deviation. V – coefficient of variation. s_x – sample mean error. $s_x\%$ – relative error of the sample mean. x ± t05sx – confidence interval for mean value

The average length of the fruit in the variety population during the research years was in the range of 13–19 cm, which is confirmed by the confidence intervals. The smallest value of the trait was noted in 2020 in the range of 12.8 to 13.2 cm, with an average variation of 15.3%. The highest indicators were noted in 2016, also with an average variation of the trait of 13.4%, and ranged from 18.6 to 19.3 cm. It should be noted that the variability of the average length of the fetus in 2017 and 2018 was insignificant, which is confirmed by the values the coefficient of variation being 9.0% and 9.4%, respectively. The accuracy of determining the average length during the years of research is high. The relative error of the sample mean did not exceed 1%. As a result, 6 biotypes were identified according to the length of seed fruits: 13–14 cm, 15–16 cm and 17–18 cm, which were the main biotypes and which amounted in total to more than 2/3 of the variety population. Lengths 10–12 cm, 19–20 cm and 21 and greater cm were the associated biotypes.

4 Conclusion

During the years 2016–2020, investigations obtained experimental data for the development of a methodology for maintaining the genetic structure of a new variety of cucumber Naslednik during reproduction, while maintaining a complex of economically valuable traits that were inherent in it initially when creating and increasing the productivity of seed crops of this variety of cucumber. It has been established that the variety population of the cucumber Heir is heterogeneous in terms of the approbation characteristics of the seed fruit, and differs in the shape and length of biologically ripe fruits. The conducted research allowed to identify 4 forms (biotypes) of seed fruits: ovoid and cylindrical – the main biotype, which constitutes 16.0–71.0% and 29.0–44.0% of the total population respectively; elongated-ovoid and elongated-cylindrical – related biotypes, which on average for 2016–2020 accounted for 2.0–39.0%. The length of the seed fruits varied within 10–22 cm. The 7 biotypes were revealed on this basis: 10–12 cm; 13–14 cm; 15–16 cm, 17–18 cm; 19–20 cm; 21–22 cm and > 22 cm. The basis of the variety population is formed by ovoid and cylindrical testes having a length of 14–18 cm. However, in years with extreme weather conditions, short seed fruits (10–13 cm) of both forms, as well as long fruits (21 and > cm), elongated-ovoid, can be split off in the variety population and elongated cylindrical shape. The established structure of the variety population should be taken into account in the primary seed production of this variety. The length of the fruits was less stable than the shape of the fruit, and is directly dependent on the sum of active temperatures and number of fallen during the growing period precipitation. Seeds of higher reproductions must be selected from seed fruits of the main biotypes in years with hydrothermal conditions characteristic of a given region.

References

1. Spiridonov, A.M., Danilova, T.A., Adrickaya, N.A.: The state of vegetable growing in the Northwestern Federal district of the Russian federation and the main directions of its development. Izvestiya Sankt-Peterburgskogo gosudarstvennogo agrarnogo universiteta 3(56), 43–49 (2019). https://doi.org/10.24411/2078-1318-2019-13043

2. Mengnan X, Shenhao W, Shu Z, Qingzhi C, Dongli G, Huiming C, Sanwen H (2015) A New Gene Conferring the Glabrous Trait in Cucumber Identified Using MutMap. Horticultural Plant J 1(1):29–34. http://dx.doi.org/https://doi.org/10.16420/j.issn.2095-9885.2015-0003
3. Mirzabe, A.H., Kakolaki, M.B., Abouali, B., Sadin, R.: Evaluation of some engineering properties of cucumber (Cucumis sativus L.) seeds and kernels based on image processing. Inf. Process. Agric. **4**, 300–315 (2017). https://doi.org/10.1016/j.inpa.2017.07.001
4. Zhao, L., et al.: The MIXTA-LIKE transcription factor CsMYB6 regulates fruit spine and tubercule formation in cucumber. Plant. Sci. **300**, 110636 (2020). https://doi.org/10.1016/j.plantsci.2020.110636
5. Huang, L.-Q., Niu, Y.-C., Su, L., Deng, H., Lyu, H.: The potential of endophytic fungi isolated from cucurbit plants for biocontrol of soilborne fungal diseases of cucumber. Microbiol Res **231**, 126369 (2020). https://doi.org/10.1016/j.micres.2019.126369
6. Sun, C., Song, X., Zheng, J., Li, X., Feng, Z., Yan, L.: Comparative proteomic profiles of resistant/susceptible cucumber leaves in response to downy mildew infection. Hortic. Plant. J. **7**(4), 327–340 (2021). https://doi.org/10.1016/j.hpj.2021.01.008
7. Mukherjee, P.K., Nema, N.K., Maity, N.I., Sarkar, B.K.: Phytochemical and therapeutic potential of cucumber. Fitoterapia **84**, 227–236 (2013). https://doi.org/10.1016/j.fitote.2012.10.003
8. Burenin, V.I., Piskunova, T.M., Gashkova, I.V.: To the problem of interspecific hybridization in the genus of Cucumis L. Vegetable Crops Russ. **1**(39), 28–31 (2018). https://doi.org/10.18619/2072-9146-2018-1-28-31
9. Pavlenko, V.N., Zvonkova, I., Pavlenko, V.I.: Scientific basis of modern technologies of cultivation of cucumbers in south regions of Russia. Prirodoobustrojstvo **1**, 89–93 (2018). https://doi.org/10.26897/1997-6011/2018-1-89-93
10. Mokryanskaya, T.I.: The selection of a bee pollinating cucumber hybrids of gherkin type. Vegetable Crops Russ. **1**(45), 16–19 (2019). https://doi.org/10.18619/2072-9146-2019-1-16-19
11. Steinert, T.V., Aliluev, A.V., Avdeenko, L.M., Kudryashov, A.V.: Breeding of cucumber in Sibniirs (siberian research institute of plant industry and breeding) – history, results, prospects. Vegetable Crops Russ. **1**(39), 37–42 (2018). https://doi.org/10.18619/2072-9146-2018-1-37-42
12. Kiełkiewicz, M., Gajc-Wolska, J., Slusarz, S., Szwacka, M.: Growth, development and yield of transgenic 35S-thaumatin II-expressing cucumber plants – open field evaluation. Sci. Hortic. **143**, 82–91 (2012). https://doi.org/10.1016/j.scienta.2012.06.007
13. Ramírez-Pérez, L.J., Morales-Díaz, A.B., Benavides-Mendoza, A., De-Alba-Romenus, K., González-Morales, S., Juárez-Maldonado, A.: Dynamic modeling of cucumber crop growth and uptake of N, P and K under greenhouse conditions. Sci. Hortic. **234**, 250–260 (2018). https://doi.org/10.1016/j.scienta.2018.02.068
14. Qi, X., et al.: Identification of genes related to Mesocarp development in cucumber. Hortic. Plant J. **6**(5), 293–300 (2020). https://doi.org/10.1016/j.hpj.2020.08.001
15. Feng, X., et al.: Dynamics and risk assessment of pesticides in cucumber through field experiments and model simulation. Sci. Total Env. **773**, 145615 (2021). https://doi.org/10.1016/j.scitotenv.2021.145615
16. Valverde-Miranda, D., Díaz-Pérez, M., Gómez-Galán, M., Callejón-Ferre, Á.-J.: Total soluble solids and dry matter of cucumber as indicators of shelf life. Postharvest Biol. Technol. **180**, 111603 (2021). https://doi.org/10.1016/j.postharvbio.2021.111603
17. Kulyakina, N.V., Yurechko, T.K.: Promising cucumber genotypes for open ground adapted to the conditions of the far eastern region. Dostizheniya nauki i tekhniki APK **34**(6), 70–74 (2020). https://doi.org/10.24411/0235-2451-2020-10613

18. Kuzmitskaya, G.A., Shestopalova, G.E.: Influence of hydrothermic factors on phenotypic composition of cucumber variety population Naslednik. Vestnik Far East Branch **4**, 28–33 (2020). https://doi.org/10.37102/08697698.2020.212.4.005
19. Gao, Z., Zhang, H., Cao, C., Han, J., Li, H., Ren, Z.: QTL mapping for cucumber fruit size and shape with populations from long and round fruited inbred lines. Horticultural Plant J. **6**(3), 132–144 (2020). https://doi.org/10.1016/j.hpj.2020.04.004

Measures to Increase Water Supply to Agroecosystems in the Subarctic Zone of Yakutia

Mikhail Loskin[1](✉), Alexandr Reshetnikov[1], and Alexander Filatov[2]

[1] FRS YaSC SB RAS M. G. Safronov Yakut Scientific Research Institute of Agriculture, 23/1, Bestuzhev-Marlinsky Street, 677001 Yakutsk, Russia

[2] Arctic State Agrotechnological University (ASAU), 3 km. house 3, Sergelyakhskoye Sh., Rep. Sakha (Yakutia), 677007 Yakutsk, Russia

Abstract. To develop measures to increase the water availability of the agroecosystem in the subarctic zone of Yakutia. Methods: Research was carried out in 2020 at a typical hydroelectric complex of the Maralayi reservoir in the Churapchinsky district of the Republic of Sakha (Yakutia). The reservoir area is 35 hectares. The total volume of water is 1 million m^3. Useful volume 650 thousand m^3. Results: The destruction of the dam and the banks occurs due to the thawing of the permafrost soils of the dam foundation. To protect the dam, a method is proposed that promotes deep freezing of the soils of the dam base. Discussion: The method consists in lowering the water level in the reservoir by removing it with a water outlet in a volume of 250 m^3 for lowering the water level by 0.71 m. An air temperature of minus 40–50 °C increases freezing of the foundations of structures, banks and reduces the area of thermal effect of water. Conclusion: To increase the water supply of typical agroecosystems located in the subarctic zone of Yakutia, it is proposed to protect dams through deep freezing of their foundations.

Keywords: Subarctic territory of Yakutia · Water supply to agro ecosystems · Hydraulic structures · Permafrost · Climate change · Pre-winter depression

1 Introduction

The territory of the Republic of Sakha (Yakutia) is a huge country located in the Arctic and Subarctic zones of the Russian Federation. The permafrost zone extends across the Arctic and Subarctic, where the republic with a population of 1 million is located. The most acute problem in the Arctic and Subarctic zones of the Republic of Sakha (Yakutia) is the problem of providing agroecosystems with high-quality household and drinking water, which directly affects not only agricultural production, but also the standard of living in rural areas as a whole. Water supply problems in rural areas are exacerbated by climate change in recent years.

The UN special report "Global warming by 1.5 °C" states that the consequences of climate change on the world's population are in the overwhelming majority of cases negative. An acceleration of the process of degradation of permafrost was noted. The

A. Muratov and S. Ignateva (Eds.): AFE 2021, LNNS 353, pp. 202–210, 2022.
https://doi.org/10.1007/978-3-030-91402-8_24

observed climate changes already have a noticeable impact on the environment and human economic activity [1].

The data of numerous studies on climate change in the world indicate an almost universal increase in atmospheric temperature in the next 20–30 years. Climate change is a major concern for food security and agricultural production. The article provides an overview of the main strategies for mitigating the effects of warming and an effective response to these global problems [2].

In recent years, serious drought-related disasters in China have received increased attention due to their significant impacts. The results of our research can help in understanding and describing drought for drought risk management in the context of global warming [3].

In Mediterranean climate and other semi-arid areas where the scarcity of water is a major problem. Our results show that crops´ water requirements increased by 21%. [4].

In Northern Germany, water resources will become more limited due to climate change. At the same time, the demand for irrigation is likely to increase due to the intensification of the cultivation of agricultural crops [5].

In the NW of Poland draining into Lake Miedwie, on which a drinking water source for the city of Szczecin is located. Despite more than 20 years since first programmes of measures towards protection of water quality have been introduced into the catch-ment, these have not been produced the expected results. The highest inputs from agriculture have been identified in the most upper sub-catchment and this area have been pointed out as most suitable for implementation of agricultural best management practices towards protection of water quality at a local level [6].

Small reservoir-based climate adaptation in African drylands carries multidimensional challenges. The planned interventions to enhance adaptation and build resilience of predominantly rainfed and vulnerable smallholder production systems to climate change in African drylands include small-scale irrigation using rainwater harvesting reservoirs. However, the required technological improvement, participatory breakthroughs, institutional settings, farm risk management, and investment justification are rarely established comprehensively and integrated into adaptive planning [7].

Understanding the impact of climate change on reservoirs in relation to the passage of time is an important component of water resource management and the maintenance of a stable water supply. The irrigation vulnerability in South Korea was estimated to evaluate the performance of water supplies in agricultural reservoirs; it is a recommended preparatory adaptive measure for developing a future water supply strategy in terms of using climate change scenarios reflecting different future conditions [8].

The water-climate-agriculture nexus quantifies synergies, trade-offs, advantages and disadvantages that arise between water management, food production, and climate change consequences taking into account the environmental impacts, economic conditions, and population growth. This work evaluates the water-climate-agriculture nexus in the Gharanghu basin, Iran, seeking to achieve sustainable management of its water system. An optimization model solves for water-release adaptive strategies corresponding to five nexus scenarios to cope with reduced water resources under climate change [9].

Baseflow is an important component of river runoff and an important ecological indicator of environmental health. A quantitative analysis of the impact of climate change and human activities on baseflow based on a scientific and reasonable baseflow separation scheme is helpful for the maintenance of river ecological health and the rational allocation of water resources, especially in the karst basins of Southwest China. The methodology was successfully implemented using a 55-year time series (1963–2017) of meteorological data and flow data recorded by the stations in the Chengbi River Karst Basin (Southwest China). The results show that annual rainfall and annual potential evapotranspiration display an upward and downward trend, respectively, while the average annual temperature is relatively stable [10].

Over 90% of Swedish agriculture is rainfed, and thus future climate change can pose a risk to agricultural production in coming decades. An overall increase in annual precipitation is predicted for northern Europe, but Sweden could still face an increasing need for irrigation, as shown by the drought summer of 2018. The results showed an increasing need for irrigation of cereal crops during the early season (May–June), for two main reasons: 1) A shift to an earlier start of the cropping period, leading to an earlier need for irrigation; and 2) a higher probability of dry spring weather, substantially increasing the irrigation requirement in dry years. However, the calculation approach developed for this study may have underestimated the need for irrigation, which could be higher than reported here [11].

Analysis of climate change impacts as well as conscious decision-making and long-term planning in complex water resources systems require use of innovative approaches under conditions of deep climate uncertainty. This research aims to design and evaluate robust adaptable plans under deep climate uncertainties in the agricultural sector. For this purpose, a combination of the Adaptation Pathways (AP) approach in conjunction with the Soil and Water Assessment Tool (SWAT) model is used to evaluate the robustness of adaptation actions and to design robust adaptation pathways under future climate uncertainties in the Hablehroud River Basin, Iran [12].

Wheat is one of the main crops cultivated in nearly all regions of the world, from the equator to temperate lands. Since the early 1960s, grain yield has increased almost linearly. However, in several European countries, a stagnation is observed since the mid-1990s. Several studies have been carried out to analyse the possible causes of this stagnation, particularly in Western Europe and France. Abnormally warm temperatures in late autumn, wet conditions and low solar radiations in spring explained this particularly low grain yield. Extreme events cause huge challenges to breeding that would require either adequate experimental conditions, trait assisted methods or genomic selection models that would enable to select for climatic scenarios that are very rare [13].

Wheat is one of the main crops cultivated in nearly all regions of the world, from the equator to temperate lands. Since the early 1960s, grain yield has increased almost linearly. However, in several European countries, a stagnation is observed since the mid-1990s. Several studies have been carried out to analyse the possible causes of this stagnation, particularly in Western Europe and France. Abnormally warm temperatures in late autumn, wet conditions and low solar radiations in spring explained this particularly low grain yield. Extreme events cause huge challenges to breeding that would require either adequate experimental conditions, trait assisted methods or genomic selection models that would enable to select for climatic scenarios that are very rare [14].

Using the example of a typical arid agroecosystem of the subarctic territory of Yakutia, the goal was to determine the possibility of improving measures to increase the water availability of agroecosystems in a changing climate, taking into account the volume of water consumption and the impact on the microclimate.

2 Materials and Methods

The studies were carried out in 2020 at a typical hydroelectric complex of the Maralaya reservoir in the Maralay village of the Churapchinsky district of the Republic of Sakha (Yakutia). The Maralaya reservoir was created in 1967 to provide household drinking water and create a favorable microclimate around the village Maralay. The reservoir area is 35 hectares. The total volume of water is 1 million m3. Useful volume is 650 thousand m3. The level of the normal retaining level is 190.75 m. 882 people live in the Maralay village. Cattle - 1177 heads, horses - 1290 heads, 24 peasant farms. There are 9 cowsheds in total, including: - 5 barns for 25 head of cattle; - 1 barn for 40 head of cattle; - 2 barns for 50 head of cattle; ¬ 1 barn for 100 head of cattle. 8 social and industrial institutions. The area of hayfields is 2575 hectares; the sown area is 80 hectares.

The Maralay hydroelectric complex was built on the principle of design and construction in permafrost conditions with the preservation of the permafrost state of soils, both during the construction process and throughout the entire period of operation, the structures include: - earth dam, 250 m long, 8 m wide at the top, 5 m high. Slopes 1:2. Siphon outlet from steel pipes in 2 lines, diameter 300 mm; - automatic open-type spillway. From gabions and Reno mattresses. Flow rate 10.7 m3/sec; - tubular sluice-regulator made of steel pipe with a diameter of 1100 mm. Flow rate 5.68 m3/sec.

It is located in the Lena-Amga interfluve of Central Yakutia in the area of continuous permafrost distribution. It is a dissected plain with a uniform relief structure and an absolute height of 400 m in the south and 300 m in the north, where it smoothly passes into the Tatta-Amga erosion-accumulative plain. The incision depth of rivers is 80–150 m, less often up to 200 m. The watersheds are flat, wide, rather abruptly turning into river valleys. The latter are wide, sharply expressed, with numerous lakes. Central Yakutia is a country of small lakes, mainly of thermokarst origin. They have a variety of sizes and shapes, depending on the stage of development and the degree of drying out. Abundant lakes are developed on the floodplain of large rivers. In addition to the floodplain, all lakes are mainly fed by snow and rain. In the feeding of thermokarst lakes, underground ice takes part only in the initial period of their development, when the under alas talik is being formed.

This territory has a sharply continental climate. Winters are cold and little snow, summers are short, mostly dry with relatively high temperatures. The lowest temperatures are observed in January in depressions, hollows or in narrow valleys with a difficult outflow of cold air. The average annual air temperature varies from -7 to -12 °C, which has a positive value for 5 months a year. A stable transition of the average daily air temperature over 0 °C is noted in early May, and in autumn - in early October. The annual amplitude of fluctuations in average monthly air temperatures reaches 100 °C.

Average annual precipitation ranges from 200 to 250 mm. The least amount of precipitation falls in winter, which is due to the anticyclonic state of the weather. The greatest amount of precipitation falls in the summer. The warm period (April-October) accounts for 75–8 5% of the annual amount. During the summer months (June-August), 115–130 mm falls.

The study area is located in a zone of continuous permafrost distribution. Permafrost thickness varies from 100 to 450 m. Soil temperature is minus 0.8–3.5 °C. The layer of seasonal thawing, depending on the relief and other conditions, ranges from 1 to 3 m. The temperature regime of permafrost to a depth of 15–20 m varies greatly depending on geological and geomorphological conditions. The increase and local uneven change in temperature is associated with differences in the vegetation cover, the moisture regime of rocks, as well as with the circulation of shallow groundwater and the presence of icing associated with them. In connection with the warming of the climate in the territory of Central Yakutia, there is a sharp activation of thermodenudation processes, which shows the degradation of the permafrost from above. An additional factor to this is the thermal effect of reservoirs on hydraulic structures, which negatively affect the stability of the foundations of structures. The process of the formation of the thermal and humidity regime of structures stretches for many years. At the same time, the scheme of the static operation of the structure and its foundation changes, the filtration regime and other parameters change radically.

For a detailed study, there was carried out an analysis of the passport of the settlement, statistical data, literature sources on climate change, its impact on engineering structures and agroecosystems in the permafrost zone. On-site technical inspection of facilities was carried out. The official documentation on monitoring the condition of a hydraulic structure has been studied and analyzed. There were applied the provisions of the current regulatory and legal documents on the safe operation and monitoring of the state of hydraulic structures and water bodies when analyzing the state of hydraulic structures and agroecosystems.

The methodology for determining the possibility of improving measures to increase the water availability of the agroecosystem, taking into account the volume of water consumption and the impact of the reservoir on the microclimate of the territory, is as follows:

I. The required volume of water consumption for the needs of consumers is calculated in accordance with the current water supply standards, including: - water consumption of the population; - water consumption for irrigation of plantings on personal plots of vegetable crops; - water consumption of cattle and horses; - water consumption for the processing of dairy products; - the need for water for fire extinguishing; - water consumption of social facilities; - the volume of ice in winter, excluded from water consumption.

II. Taking into account unforeseen situations (10–15%), the required minimum useful volume of the reservoir is determined.

III. Subtracting the available volume of the reservoir, the volume of water discharged in winter is determined and, depending on the area of the reservoir, the level of the pre-winter decrease is formed.

3 Results

At the time of the reconnaissance survey, the structures of the hydroelectric complex of this reservoir are in an unsatisfactory condition, besides for the tubular sluice-regulator. The main problematic issue for the water supply of the settlement is the destruction in recent years of the slopes of the dam and the banks of the reservoir (thermal abrasion), as well as the gradual expansion of its water area. Typical destructions are thermal abrasion of the slopes of the dam and the banks of the reservoir, as a result of which there is a threat of destruction of the highway and social facilities. Permafrost soils of the foundations of the dam and banks contain ice wedges. Thermokarst processes in the form of subsidence and troughs became more active along the perimeter of the reservoir (Table 1).

Table 1. Aggregated calculation of water consumption in Maralay village.

	Quantity	Specific consumption	Required volume of water, m3/year
Water consumption of the population, Qpop	882 people	0.2 m3/day	64 386
Water consumption for irrigation of plantings on personal plots of vegetable crops, Q_{irr}	–	0.05 m^3/day	5 292
Water consumption of cattle and horses, Q_{cattle}	Cattle–1 177 heads, Horses–1 290 heads	0.06–0.08 m^3/day	62 619
Water consumption for dairy products processing, Q_{proc}	1 749 tons	50 m^3	87 450
The need for fire extinguishing, Q_{fire}	–	10 l/s	39 420
Total water consumption, Q_{total}	–	–	259 167
Water consumption of social facilities, Q_{soc}	–	12% from Q_{total}	31 100
Ice volume in winter, excluded from water consumption, Q_{ice}	35 ha	–	78 750
Minimum useful reservoir volume	–	–	369 017

(*continued*)

Table 1. (*continued*)

	Quantity	Specific consumption	Required volume of water, m3/year
The required useful volume of the reservoir, taking into account unforeseen situations 10–15%, $W_{useful\ min}$	–	–	**400 000**

Note: the costs of evaporation from the water surface are conventionally taken equal to the amount of precipitation in the calculations

Survey results show that dam and bank destruction occurs as a result of wave erosion and thawing of permafrost underlying the foundations of the dam and banks. A significant role in the thawing of permafrost is played by the thermal effect of the reservoir in winter.

In order to determine the possibility of improving measures to increase water availability, we analyzed the water consumption of a typical settlement from a multipurpose reservoir.

Aggregated calculation of water consumption of Maralay village is determined in accordance with the standards of the current codes of practice for the water supply of settlements, fire protection systems, agricultural water supply standards and on the basis of data from state statistical reporting and is given in the table.

4 Discussion

The results of the reconnaissance survey show the unsatisfactory condition of the earth dam and banks due to thermal abrasion, as well as the expansion of the water area of this reservoir in recent years. These negative factors threaten to destroy the social infrastructure of the village. At the same time, the obtained results of water consumption from the multipurpose reservoir show that the available usable volume of water in the reservoir 650 thousand m3/year exceeds the needs of water consumers 400 m3/year by 1.6 times. Consequently, in order to reduce (eliminate) the processes of thermal abrasion, thermokarst of the reservoir banks, and preserve the operational characteristics of the hydroelectric complex, it is necessary to consider the possibility of improving measures to increase water availability.

We propose one of the ways to solve this issue, which consists in the fact that the preventive measures of this hydroelectric complex include a method of pre-winter lowering of the water level in the reservoir to increase the layer of freezing of soils of the foundations of structures and the banks of the reservoir [15]. The method consists in the fact that a pre-winter lowering of the water level in the reservoir is carried out to increase the contact area of the shores and bodies of hydraulic structures with the ambient air temperature of minus 40–50 °C to increase the freezing layer of the foundations of the banks and structures while reducing the area of the thermal effect of the reservoir (Fig. 1). At the Maralay hydroelectric complex, it is possible to remove excess water

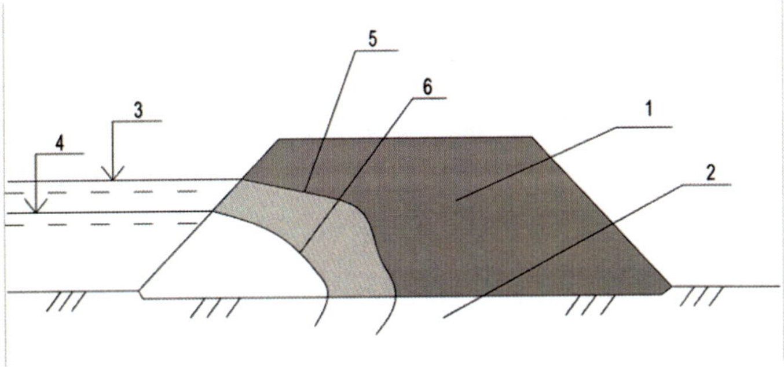

Fig. 1. Method of pre-winter lowering of the water level in the reservoir: 1 – the body of the dam; 2 – the base of the dam; 3 – water level before lowering; 4 – water level after lowering; 5 – the depth of freezing to the pre-winter lowering of the water level; 6 – the depth of freezing after the pre-winter lowering of the water level.

in the pre-winter period of September-October with a siphon outlet in the volume of 250 m3 by lowering the reservoir level by 0.71 m to a mark of 190.04 m.

5 Conclusion

Thus, in order to increase the water supply of typical agroecosystems located in the subarctic zone of Yakutia, it is necessary to include in the preventive operational measures the pre-winter lowering of the water level, taking into account the needs of water consumers remaining for the specified period and the impact of the reservoir on the microclimate of the territory. According to the research results, a patent for invention of the Russian Federation 2685476 was obtained [15].

References

1. Special report Global Warming of 1.5 °C. https://www.ipcc.ch/sr15/
2. Gupta, P., Singh, J., Verma, S., Chandel, A.S., Bhatla, R.: Impact of climate change and water quality degradation on food security and agriculture. In: Water Conservation in the Era of Global Climate Change, vol. 1, pp. 1–22 (2021). https://doi.org/10.1016/B978-0-12-820200-5.00006-3
3. Xu, Y., Zhang, X., Hao, Z., Hao, F., Li, C.: Projections of future meteorological droughts in China under CMIP6 from a three-dimensional perspective. Agric. Water Manage. **252**, 106849 (2021). https://doi.org/10.1016/j.agwat.2021.106849
4. Vila-Traver, J., Aguilera, E., Infante-Amate, J., González de Molina, M.: Climate change and industrialization as the main drivers of Spanish agriculture water stress. Sci. Total Environ. **760**, 143399 (2021). https://doi.org/10.1016/j.scitotenv.2020.143399
5. Egerer, S., Cotera, R.V., Celliers, L., Costa, M.M.: A leverage points analysis of a qualitative system dynamics model for climate change adaptation in agriculture. Agric. Syst. **189**, 103052 (2021). https://doi.org/10.1016/j.agsy.2021.103052
6. Kuczyńska, A., Jarnuszewski, G., Nowakowska, M., et al.: Identifying causes of poor water quality in a Polish agricultural catchment for designing effective and targeted mitigation measures. Sci. Total Environ. **765**, 144125 (2021). https://doi.org/10.1016/j.scitotenv.2020.144125

7. Koide, J., Yokoyama, S., Hirouchi, S., et al.: Exploring climate-resilient and risk-efficient cropping strategies using a new pond irrigation system: an experimental study in northern Ghana. Agric. Syst. **191**, 103149 (2021). https://doi.org/10.1016/j.agsy.2021.103149
8. Nam, W.-H., Choi, J.-Y., Hong, E.-M.: Irrigation vulnerability assessment on agricultural water supply risk for adaptive management of climate change in South Korea. Agric. Water Manage. **152**, 173–187 (2015). https://doi.org/10.1016/j.agwat.2015.01.012
9. Golfam, P., Ashofteh, P.-S., Loáiciga, H.-A.: Modeling adaptation policies to increase the synergies of the water-climate-agriculture nexus under climate change. Environ. Dev. **37**, 100612 (2021). https://doi.org/10.1016/j.envdev.2021.100612
10. Mo, C., Ruan, Y., Xiao, X., Lan, H., Jin, J.: Impact of climate change and human activities on the baseflow in a typical karst basin. Southwest Chin. Ecol. Indic. **126**, 107628 (2021). https://doi.org/10.1016/j.ecolind.2021.107628
11. Grusson, Y., Wesström, I., Joel, A.: Impact of climate change on Swedish agriculture: growing season rain deficit and irrigation need. Agric. Water Manage **251**, 106858 (2021). https://doi.org/10.1016/j.agwat.2021.106858
12. Babaeian, F., Delavar, M., Morid, S., Srinivasan, R.: Robust climate change adaptation pathways in agricultural water management. Agric. Water Manage. **252**, 106904 (2021). https://doi.org/10.1016/j.agwat.2021.106904
13. Gouis, J.L., Oury, F.-X., Charmet, G.: How changes in climate and agricultural practices influenced wheat production in western Europe. J. Cerea. Sci. **93**, 102960 (2020). https://doi.org/10.1016/j.jcs.2020.102960
14. Liu, Z., Herman, J.D., Huang, G., Kadir, T., Dahlke, H.E.: Identifying climate change impacts on surface water supply in the southern central valley. California. Sci. Total Environ. **759**, 143429 (2021). https://doi.org/10.1016/j.scitotenv.2020.143429
15. Loskin, M., Knysh, A., Reshetnikov, A.D.: Patent 2685476 RF, E02B 7/06 (2006.01) Method of preventive protection of hydraulic structures by pre-winter lowering of the water level in the reservoir. Application date: 23 Apr 2018, Published: 18.04.2019 Bul. No. 11 (2019)

Evaluation of Technical and Chemical Parameters of Fruits of Apricot, Plum, and Apple Varieties of Far Eastern Selection

Oleg Mikhailichenko(✉), Evgeniya Tikhomirova, Olga Tokareva, and Natalia Yudaeva

Khabarovsk Federal Research Center of the Far Eastern Branch of the Russian Academy of Sciences, Separate Division of the Far Eastern Branch of the Agricultural Research Institute, 107, Karl Marx street, 680031 Khabarovsk, Russia

Abstract. The article presents the evaluation of technical and biochemical parameters of apricot, plum, and apple varieties and variety samples cultivated in the Far East. These varieties are adapted to the local conditions of the fruit plants and have good taste qualities of fruit, used fresh and canned. Varieties of fruit crops cultivated in the Far East are varied in taste, chemical composition, juiciness, and flavor, as well as in the way of storage and transportation. At present, a qualitatively new stage of development is observed in horticulture in the Far East. New specialized fruit plots, farms of various forms of ownership are being established. The cultivation of fruit crops in favorable areas of the Far Eastern region will increase the consumption of fresh and processed fruits by the region residents, which has remained at a low level for a long time.

Keywords: Fruits · Technological properties · Apricot · Plum · Apple tree · Dry substances · Titratable acids

1 Introduction

The horticulture industry occupies a particular place in providing the population of the Russian Federation with products containing the required complex of vitamins, have preventive and curative properties, and their consumption determines the life quality of citizens of the country. Residents of the Far East receive 70–90% less high-vitamin medicinal and dietary fruits and berries, which are so necessary for a fruitful, everyday human life [1].

At the same time, the staff of the Far Eastern Research Institute of Agriculture has done a lot of research and selection work to create varieties of fruit crops with high commercial and productive qualities, adapted to the harsh conditions of the Far East, in relation to fruit growing areas [2].

Northeast Asia is the center of origin and one of the oldest centers of apricot formation. Here, as a result of the joint growth and natural hybridization of the common apricot (*Prunus armeniaca*) with northern species – Manchurian (*Armeniaca mandschurica*),

A. Muratov and S. Ignateva (Eds.): AFE 2021, LNNS 353, pp. 211–218, 2022.
https://doi.org/10.1007/978-3-030-91402-8_25

Siberian (*Armeniaca sibirica*), David (*Armeniaca davidiana*) – the most winter-hardy forms emerged, which served as the starting material for the creation of cultivated varieties of apricot in the Far East [3]. The most important indicators of fruit quality in apricot varieties are the size, appearance, and taste advantages. Apricot fruits contain a significant amount of biologically active substances that positively affect health and the human body. Apricot fruits are used for food in fresh form and as various processed products (compotes, jams, juices) [4].

Plum has been one of the most promising fruit crops in the Far East. The value of plums is determined, first of all, by a successful combination of such important economic and biological qualities as high winter hardiness of plants, resistance to sunburn damage, high and annual yield, and good taste of fruits [5].

The high taste and dietary advantages of plum fruits are determined by the successful combination of sugars and acids. In addition, they contain vitamins and biologically active substances, increasing their importance in human nutrition. Dried fruits, juices, compotes, jams, and other products of technical processing are produced from the fruits [6].

Far Eastern varieties of plums are the most rapid-fruiting among the world assortment of this breed. Plum trees are relatively long-lived. Under optimal conditions of cultivation and the organization of careful control of pests and diseases, their productive age is 15–20 years. The yield of some varieties in optimal years reaches up to 80 kg per tree [7].

Apple tree (*Malus domestica Borkh.*) is one of the most important fruit crops of the Far East. Wild-growing species are the progenitors of all existing varieties of domestic apple trees [8]. In recent years, non-regional varieties of apples have become widespread in the gardens of the Far East. The consumer's requirements for the commercial and specific qualities of fruits are diverse. They are caused by both the size of the fruit, the color of the peel, and the chemical composition that determines their nutritional value and medicinal properties [9]. Apple fruits are a source of sugars, vitamins, mineral salts, trace elements, and pectin substances. Apples contain up to 10 different vitamins; the main ones are ascorbic acid (vitamin C) and P-active substances (catechins, leucoanthocyanins, flavonic glycosides, etc.). Apple fruits also contain chlorogenic and ursolic acids. Apples are also able to remove heavy metal ions from the body, including radioactive elements and carcinogenic substances [10].

2 Materials and Methods

Research work on fruit crops was carried out on the territory of the garden of the Far Eastern Agricultural Research Institute (FE ARI), which is located in the city of Khabarovsk. Geographical coordinates: latitude – 48°28′57″ N, longitude – 135°05′01″ E, elevation differences in this area – from 63 to 74 m above sea level.

The object of the study was the technological characteristics of fruits of area-specific and promising varieties of apricot, plum, and apple trees grown in the conditions of the south of the Far East. The description of the fruits was carried out during the ripening period according to the "Program and methodology of variety study of fruit, berry and nut crops" (Orel, 1999). The chemical analysis of the fruits was performed in the FE ARI biochemistry laboratory following the accepted methods: dry substances as per GOST-29030-91; total sugar per GOST 8756-13.87; titrated acidity per GOST 25555.0-82, vitamin C (ascorbic acid) following the method of A. I. Ermakov.

3 Results and Discussion

3.1 Apricot

The total consumption period of apricot fruits is about 45 days (from July 20 to September 5). However, depending on the thermal resources of the year, the ripening time may vary.

According to the average long-term dates of ripening dates (July 19–25), Ranniy Marusicha and Khabarovskiy varieties can be attributed to early-maturing varieties: Akademik, Amur, and Titan - to average-maturing varieties; Gritikaz and Kazmar to relatively late varieties.

The most important indicator of technical characteristics is the size of the fruit, which largely depends on the variety, and within the same variety - on the place and conditions of cultivation, a load of trees with the harvest.

The studied varieties can be divided into two groups according to the size of the fruits: with medium-sized fruits - Amur, Gritikaz, Kazmar, Ranniy Marusicha and relatively large-fruited - Akademik, Titan, and Khabarovskiy. As a rule, large-fruited varieties have the best marketable appearance (Table 1).

Table 1. Technical indicators of fruits of apricot varieties selected by FE ARI.

Variety	Ripening period	Maximum fruit weight. g	Average fruit weight. g	Diameter in two planes. mm
Akademik	10.08 ± 5	55.3	40.5	42/37
Amur	10.08 ± 2	39.1	26.2	39/38
Gritikaz	20.08 ± 5	41.0	25.8	39/40
Kazmar	25.08 ± 3	36.0	26.0	40/41
Ranniy Marusicha	25.07 ± 3	35.2	38.1	37/35
Titan	05.08 ± 4	40.5	33.2	41/35
Khabarovskiy	24.08 ± 2	45.2	31.2	40/34

According to the chemical composition, apricot fruits of the FE ARI varieties contain 6–8% sugars, titrated acids - 1.7 - 3%, dry substances 10–16%, ascorbic acid 7–9 mg, % (Table 2).

Table 2. Chemical composition of fruits of apricot varieties selected by FE ARI.

Variety	Dry matter. %	The sum of sugars. %	Titratable acids. %	Ascorbic acid. mg/100g
Akademik	16.6	11.8	2.3	9.8
Amur	15.4	12.3	1.7	9.7
Gritikaz	16.6	8.5	1.8	7.7
Kazmar	11.3	8.7	1.7	7.6
Ranniy Marusicha	11.3	6.5	1.6	7.3
Titan	10.1	9.1	1.8	8.0
Khabarovskiy	16.1	12.3	1.8	7.9

Their composition is close to the primary standard apricot variety of the southern regions of Russia, Krasnoshcheky. Fruits of this variety contain from 7.5 to 8.5% of sugars and 1.7% of titrated acid. The best of the varieties of FE ARI – Akademik and Khabarovskiy in some years contained sugars and titrated acid even slightly more than Krasnoshcheky [11].

The accumulation of sugars in fruits in local conditions depends on the heat resources and precipitation. In cold and rainy growing seasons, sugar accumulation in fruits is reduced by 1–1.5%.

According to the taste of the fruit, local apricot varieties, although inferior to southern ones, are pretty suitable for fresh consumption. They are pretty harmonious in taste, have a pleasant apricot flavor, a well-developed kernel, and most varieties have a sweet seed kernel.

Late-maturing varieties can be stored in freezers for quite a long time (from 2 to 3 months) and are available for transportation.

3.2 Plum

Plum is one of the essential fruit plants in the Russian Far East. Excellent quality of fresh fruits is well combined with high productivity and rapid fruitfulness, the ability to adapt to a complex of adverse environmental factors. With the appropriate selection of varieties with different fruit ripening periods, the consumption of fresh fruits can be extended for two months.

The most widespread in the Far East are Ussuri plum (Prunus ussuriensis Kov. Et Kost.) and Chinese plum (Prunus Salicina Linde). Many taxonomists believe that the Ussuri plum is a subspecies of the Chinese plum; their only difference is the higher winter hardiness of the Ussuri plum trees. The Ussuri plum is the most adapted species to the harsh conditions of the Far East. Almost all varieties were obtained with the participation of this species. In recent years, adaptive varieties with increased resistance to bio- and abiostressors, rapid fruit production, yield, weak growth, and rare crown, as well as high commercial qualities of fruits, have been obtained in the FE ARI (Table 3).

Table 3. Technical indicators of fruits of plum varieties selected by FE ARI.

Variety	Ripening period	Average weight. g	Maximum fruit weight. g	Fruit cover color
Amurskaya roza	01.08 ± 5	45.1	48.5	Pink
Egorovna	05.08 ± 5	39.2	42.5	Pink and yellow
Lara	25.08 ± 3	41.1	39.5	Dark red
Miss Khabarovsk	15.08 ± 5	52.5	63.4	Dark red
Nelya	10.08 ± 7	41.0	43.6	Dark red
Smuglyanka	28.08 ± 7	34.2	36.0	Maroon
Khabarovskaya desertnaya	10.08 ± 5	42.4	46.5	Dark red

According to the average long-term data, plum ripening begins in early July; the varieties with an early ripening period include Amurskaya roza and Egorovna, the varieties with an average ripening period include Miss Khabarovsk, Nelya, and Khabarovskaya desertnaya, the late ones - Lara and Smuglyanka.

According to the size of the fruits and the varieties studied can be attributed to relatively large-fruited, the largest-fruited variety is Miss Khabarovsk (63.4 g). A mandatory requirement for a modern industrial variety is the dark color of the fruit. In varieties with these parameters, the marketable appearance of the fruit is better preserved; most of the varieties presented correspond to this parameter.

The high taste and dietary advantages of plum fruits are determined by the successful combination of sugars and acids. In addition, they contain vitamins and biologically active substances, increasing their importance in human nutrition. Fresh plum fruits of the FE ARI varieties contain, on average, from 13 to 18% of dry substances, sugars from 9 to 13%, titrated acids from 0.7 to 1% (Table 4).

Table 4. Chemical composition of fruits of plum varieties selected by FE AR.

Variety	Dry matter. %	The sum of sugars. %	Titratable acids. %	Ascorbic acid. mg/100g
Amurskaya roza	14.6	10.2	1.2	4.5
Egorovna	13.4	12.8	1.1	6.7
Lara	16.2	11.3	0.7	4.2
Miss Khabarovsk	15.5	9.4	0.9	4.3
Nelya	17.0	11.9	0.7	3.9
Smuglyanka	15.7	9.4	1.1	5.7

(continued)

Table 4. (*continued*)

Variety	Dry matter. %	The sum of sugars. %	Titratable acids. %	Ascorbic acid. mg/100g
Khabarovskaya desertnaya	15.2	12.4	0.8	6.4

Chemical analysis of the fruits showed significant differences in the varieties. So, according to the content of dry substances, the varieties Lara (16.2%) and Smuglyanka (15.7%) are distinguished. According to the sugar content, the varieties Egorovna (12.8%), Nelya (11.9), and Khabarovskaya desertnaya (12.4%) were distinguished. Fruits of all varieties are not rich in Vitamin C from 4 to 6.5 mg/100 g. The chemical composition of plum fruits changes significantly under the influence of environmental conditions. In dry years, increased content of dry substances, sugars, and acids accumulates, while ascorbic acid content decreases.

3.3 Apple Tree

The apple tree is a common seed crop that grows in the territory from the south of Russia to the Far East and has several advantages: early maturity, yield, fruits are different in taste and maturity. The existing assortment of numerous varieties of transition apple (*Malus transitoria*) of the Far East meets the requirements of this region: high winter hardiness, adaptability to harsh climate conditions. All apple varieties of the selection of the FE ARI are divided into three groups according to the ripening period: summer, autumn, and winter.

The total consumption period of apple fruit is about 55 days, from August 10 to October 10. Considering the local climatic conditions, the size of the apple fruit can be attributed as average (Table 5).

Table 5. Technical indicators of fruits of apple varieties selected by FE ARI.

Variety	Ripening period	Average weight, g	Maximum fruit weight, g	Fruit cover color
Aborigen	10.08 ± 5	55.1	129.8	Streaked blush
Amurskoe krasnoe	20.09 ± 7	60.2	112.3	Burgundy
Dalnevostochnoe zolotistoe	15.08 ± 7	82.5	125.7	Golden
Lyubitelskoe	25.09 ± 5	45.3	61.2	Dark red
Nadezhnoe	20.08 ± 5	42.9	65.8	Streaked blush
Ranneye Bolonyaeva	5.08 ± 5	54.7	98.8	Streaked blush

The varieties with early maturation dates include Aborigen (10.08–17.08) and Rannee Bolonyaeva (05.08–10.08), autumn – Dalnevostochnoe zolotistoe (15.08–25.08), Nadezhnoe (15.08–20.08), combs with late maturation dates include Amurskoe krasnoe (15.09–25.09) and Lyubitelskoe (22.09–30.09). The fruiting duration of varieties is 7–15 days; the beginning and duration of fruiting can vary depending on the climatic conditions during the fruiting period.

The chemical composition of apple fruit depends on the variety, climatic conditions, time of collection, terms, and storage methods (Table 6).

Table 6. Chemical composition of the fruits of apple varieties selected by FE ARI.

Variety	Dry matter. %	The sum of sugars. %	Titratable acids. %	Ascorbic acid. mg/100 g
Aborigen	14.7	11.9	0.7	11.1
Amurskoe krasnoe	15.5	10.5	1.1	17.3
Dalnevostochnoe zolotistoe	15.2	11.0	0.6	11.4
Lyubitelskoe	15.3	10.2	1.0	15.1
Nadezhnoe	15.1	10.8	0.8	17.0
Rannee Bolonyaeva	14.9	13.2	0.6	11.0

The varietal difference in the chemical composition of apple fruits is quite significant. The dependence of soluble solids content on the ripening period of fruits was noted. In varieties with an early maturation period, this value is average 14.8% over many years, in autumn – 15.1%, and in winter – 15.4%. The total amount of sugars in our conditions ranges from 10.2 to 13.2%, while in varieties with an early ripening period, this indicator is higher. Titratable acids in apples contain from 0.6 to 1.1%. In varieties with a late ripening period, this component is higher. The content of ascorbic acid (vitamin C) in apple fruits ranges from 11 to 17.3 mg % or more per 100 g of raw weight. In different years, the content of ascorbic acid varies depending on climatic conditions. The fruits of the apple tree of the winter ripening period have a harder skin and thick peel and are characterized by higher consumer qualities (storage, transportation, processing). In terms of the size of the fruits of the studied varieties of apple trees of the Far Eastern selection, they are significantly inferior to varieties growing in the Central and Southern zones of horticulture of Russia, 2-2.5 times. But in terms of the content sum of sugars, titratable acids, ascorbic acid, they surpass the southern varieties [12].

4 Conclusion

Studies of the gene pool of individual varieties of fruit crops with different maturation periods, bred in the conditions of the Far Eastern region, showed the specifics of varietal characteristics of crops. Conducted technical assessment of fruits of 6 varieties of apricot,

6 varieties plums and 6 apple varieties obtained in the Far East, allows us to conclude that all the fruits of the studied varieties can be attributed to the first commercial variety. The obtained data on the biochemical composition of fruits allow selecting fruits for food and processing.

References

1. Kulikov, I.M., Minakov, I.A.: Problems and prospects of development of horticulture in Russia. Horticul. Viticul. **6**, 40–46 (2018). https://doi.org/10.31676/0235-2591-2018-6-40-46
2. Nenko, N.I., Kiseleva, G.K., Ulyanovskaya, E.V.: Comparative physiological biochemical characterization of the resistance of apple varieties in different ecological and geographical origin to abiotic stresses. Hortic. Viticul. **1**, 29–33 (2016)
3. Paliy, I.N., Gubanova, T.B., Paliy, A.E.: Physiological and biochemical parameters of apricot varieties characterized by various drought resistance. Hortic. Vitic. **1**, 23–28 (2020)
4. Fan, X., Zhao, H., Wang, X., Cao, J., Jiangc, W.: Sugar and organic acid composition of apricot and their contribution to sensory quality and consumer satisfaction. Sci. Horticul. **225**, 553–560 (2019)
5. Kochubey, A.A., Zaremuk, R.Sh.: Complex evaluation of varieties of plum (Prunus Domectica L.) for the quality of fruits in the conditions of south gardening. Agrarian Sci. **3**, 62–65 (2019). https://doi.org/10.32634/0869-8155-2019-323-3-62-65
6. Eremin, G.V.: New varieties of plums to produce prunes. Hortic. Vitic. **2**, 8–15 (2016). https://doi.org/10.18454/VSTISP.2016.2.1089
7. Musacchi, S., Serra, S.: Apple fruit quality: overview on pre-harvest factors. Sci. Hortic. **234**, 409–430 (2018)
8. Ershova, I.V.: Bioactive compounds in fruits of Altai apple varieties. Hortic. Vitic. **6**, 48–53 (2017). https://doi.org/10.18454/VSTISP.2017.6.8430
9. Motyleva, S.M., Borisova, A.A.: Comparative biochemical composition of apple fruits of domestic and foreign varieties. Hortic. Vitic. **6**, 12–18 (2018). https://doi.org/10.31676/0235-2591-2018-6-12-18
10. Nikitin, A.L., Makarkina, M.A.: Dependence of the content of ascorbic acid in apples at the beginning and at the end of storage on the hydrothermal conditions of the growing season. Hortic. Vitic. **5**, 24–31 (2020). https://doi.org/10.31676/0235-2591-2020-5-24-31
11. Zaurov, D.E., Molnar, T., Eisenman, S., Ford, T., Mavlyanova, R.F., Capik, J., Goffreda J.: Genetic resources of apricots (*Prunus armeniaca L.*) in central Asia. HortScience **48**(6), 681–691 (2013). https://doi.org/10.21273/HORTSCI.48.6.681
12. Nenko, N.I., Kiseleva, G.K., Ulyanovskaya, H.V.: Physiologico-biochemical evaluation of combined stability in different ecologo-geographical origin cultivars of apple to abiotic stress factors of summer period in southern region of Russia. Hortic. Vitic. **1**, 27–32 (2015). https://doi.org/10.31676/0235-2591-2015-1-27-32

Comparative Assessment of Promising Soybean Varieties at the Final Selection Stage

Ekaterina Butovets(✉), Ludmila Lukyanchuk, and Evgenia Vasina

FSC of Agriculture Biotechnology of the Far East named after A.K. Chaiki, Timiryazevskiy Settlement, 692539 Ussuriysk, Russia

Abstract. The purpose of the research was to identify promising soybean varieties at the final stage of breeding using a multi-criteria assessment, studying resistance to local populations of phytopathogens, assessing the resistance of samples to stressful growing conditions, and establishing correlations between economically valuable traits. Strong positive corrrelationships between yield and mass of seeds per plant and number of beans and seeds per plant (r = 0.79–0.88) were established. Promising soybean varieties Primorskaya 1625 and Primorskaya 1595 have the best integrated assessment for a set of traits (−1.4 and 2.2, respectively). Based on the testing results, small-seeded vegetable soybean variety Primorskaya 1628 was submitted for testing to the State Commission for the Protection of Breeding Achievements. According to the results of the study of soybean disease resistant varieties against the natural background of the development of phytopathogens (Septoria glycines Hemmi, Cercospora sojina Hara, Peronospora manshurica Naum.), resistant (11–25% of damage) and medium resistant (26–50%) samples were identified. After forced infection, their immune status changed to moderately resistant and susceptible (51–75%), respectively. When assessing resistance to stressful growing conditions, soybean varieties with high adaptive potential were identified: Primorskaya 1628 (−1.5), Primorskaya 1551 (−7.7) and Primorskaya 1623 (−10.7). The efficiency and objectivity of the application of biological statistics methods in the comparative assessment of promising soybean breeding material is noted.

Keywords: Soybean · Cultivar · Integral assessment · Breeding process · Correlation · Adaptability · Resistance · Fungal diseases

1 Introduction

Glycine max (L.) Merr. is a strategic culture of versatile use, and its production is growing steadily throughout the world and in the Russian Federation [1, 2]. Most of our country's soybean production is located in the Far East. Positive dynamics of the growth of soybean acreage in some regions of Russia was achieved due to the emergence of new early-maturing high-yield varieties [3, 4]. The main task of breeders is to create soybean varieties that best meet the requirements of intensive farming [5–8].

It became necessary to apply a phenomenological approach, mathematical statistics, decision theory and other branches of applied mathematics for the development of

A. Muratov and S. Ignateva (Eds.): AFE 2021, LNNS 353, pp. 219–229, 2022.
https://doi.org/10.1007/978-3-030-91402-8_26

breeding methods for multivariate analysis of experimental data. For this reason, our research goals are to conduct a comparative assessment of promising soybean varieties at the final stage of selection, using biological statistics methods.

2 Materials and Methods

The work was carried out in 2018–2020 in the fields of the laboratory of soybean selection of the FSBSI "FSC of Agricultural Biotechnology of the Far East named after A.K. Chaika", located near the city of Ussuriysk, 120 km from the regional center of Vladivostok on the Prikhanka plain. The region is characterized as the warmest in the Far East, humid with severe winters. The sum of active temperatures (above 10 °C) fluctuates within 2400–2600 °C, hydrothermal coefficient (HTC) is 1.6–2.0. During the experiments the meteorological conditions were contrasting, but basically they corresponded to the biological needs of soybean, which are formed at the HTC 1.3–2.0. The soil of the experimental site is meadow-brown bleached with heavy texture.

2.1 Field Experiment

Soybean varieties of the final stage of the breeding process (competitive testing) were taken as objects of the study. Mid-season variety Primorskaya 4, zoned and approved for use in the Far Eastern region, ws taken as standard.

The experiment was carried out according to the methods B.A. Dospehov (2012). Plot area - 22.0 m^2, replication - three times, sowing and harvesting - mechanized. Evaluation of productivity and accounting for the main economically valuable traits was carried out in accordance with the guidelines for selection and seed production of soybeans (N.I. Korsakov et all, 1975). Protein and oil content in soybean seeds was determined on an Inframatic 9200.

The study of competitive testing soybean varieties resistance to local populations of pathogens was carried out against the background of natural and artificial development of diseases. For artificial infection of soybean plants, suspensions of harmful soybean pathogens were used - Septoria glycines Hemmi, Cercospora sojina Hara and Peronospora manshurica (Naum.) Syd. Syn. The prepared suspensions with spore concentration, giving an optimal infectious load, were used to infect soybean plants (D.V. Podkina, I.A. Kotlyarova, 2000). The treatment of soybean samples with suspensions was carried out on the 14th day of plant development.

Visual assessments and accounting of the lesion of the soybean leaf plate with fungal diseases (septoria, cercospora, peronosporosis) were carried out at the stage of plant development V5 and R6 on the basis of methodological guidelines (N.I. Korsakov et al, 1979).

2.2 Preparation of a Suspension of Pathogen Spores

Cultures of the fungi Septoria glycines Hemmi and Cercospora sojina Hara were identified and isolated in the soybean breeding laboratory of the FSC of Agricultural Biotechnology of the Far East named after A.K. Chaika. For transplantation, a culture medium was used — potato-dextrose agar (PDA); a pure culture of the pathogen was grown in Petri dishes at a temperature of +24–26 °C in a thermostat (Binder, Germany) for 6–7 days until the appearance of conidia. Since Peronospora manshurica (Naum.) Syd. Syn. is an obligate parasite, spores were washed off infected green soybean leaves. The suspensions were diluted to obtain the required concentration of 10 septoria conidia, 3 cercospora conidia, and 30 peronosporosis conidia in the field of view of the microscope. The density of the suspensions was determined by viewing at least 10 drops under a Motik Microscores 1.3 MP microscope at 80x magnification, counting the number of conidia in the field of view (N.I. Korsakov et al., 1979).

2.3 Statistical Analysis

Statistical analysis of the results was carried out by methods of variance and correlation analysis using the Microsoft Excel software package.

The integral assessment of competitive soybean varieties was calculated according to the method of S.P. Martynov (1987). In the program of multicriteria assessment, when setting the ideal, the following indicators were taken into account: according to the characteristics of yield and quality, the highest values were desirable; according to the signs of resistance, there was a restriction from the bottom (at least 70%), the height of the plants should be no more than 80 cm, and the duration of the growing season should not be longer than 120 days; according to the yield characteristics, the confidence intervals of the least significant difference were calculated (HCP 0,05); weight coefficients and standard deviation were calculated. The grades are ranked relative to a given ideal (the integral assessment is a deviation from the ideal). The variety with the lowest grade is considered as the best.

The assessment of the samples for flexibility, compensatory ability, and stress resistance of yield was carried out according to the method of V.A. Zykin et al. (1984). The index of environmental conditions was assessed according to the method proposed by S.A. Eberhart and W.A. Russell as presented by V.Z. Pakudin (1979).

3 Results

3.1 Correlations Between Economically Valuable Traits of the Studied Varieties

Determination of correlations between the traits of soybean varieties (Table 1) [9] is of no small importance for increasing the efficiency of material analysis. According to the results of the correlation analysis, strong positive relationships were revealed between yield and mass of seeds per plant (productivity), the number of beans and seeds per plant ($r = 0.79$–0.88). With an increase in the duration of the growing season, the height of plants increases slightly ($r > 0.33$ - weakk positive relationship).

Inverse relationship was observed between the protein content in seeds with the duration of the growing season ($r = -0.73$) and yield ($r = -0.13$), which confirms the difficulty of combining high rates of these traits in one soybean genotype. Medium positive relationships were established between plant height and elements of the structure of the soybean yield (number of branches, beans and seeds) ($r = 0.42$–0.57).

Table 1. Correlation coefficients (r) between economically valuable traits of soybean varieties of the competitive testing, 2018–2019.

Indicator	Yield. centner /ha	Vegetation period. days	Plant height. cm
Mass of seeds per plant. g	0.88**	0.12	0.50
Plant height. cм	0.04	0.58*	–
Lower bean attachment height, cm	0.28	0.08	0.18
Number of branches. pcs/plant	0.39	0.27	0.42
Number of beans. pcs/plant	0.82**	0.25	0.57
Number of seeds. pcs/plant	0.79**	0.29	0.55
Protein content in seeds. %	−0.13	−0.73**	0.22
Oil content in seeds. %	0.34	0.39	−0.16

* – the correlation coefficient is reliable at the level of significance $p \leq 0.05$.
** – the correlation coefficient is reliable at two levels of significance ($p \leq 0.05$ and $p \leq 0.001$).

3.2 Complex Assessment of Promising Soybean Varieties

In terms of yield, five soybean varieties exceeded the standard Primorskaya 4 (22.7 centner/ha) by more than 15.0%, in terms of protein content in seeds - nine samples exceeded by 1.5–4.6% (Table 2). The most promising varieties in terms of yield can be considered Primorskaya 1551 (28.2 centner/ha) and Primorskaya 1625 (29.7 centner/ha). 1000 seed weight index varied from 120 to 180 g. The greatest value of the trait (178–180 g) was noted in the soybean varieties Primorskaya 1606, Primorskaya 1625 and Primorskaya 1627. In terms of vegetation period, 61.1% of the samples belong to the mid-ripening group (114–119 day), 33.3% to the mid-ripening (120–121 days) and 5.6% - to mid-season (105 days).

As can be seen from the data given in Table 2, Primorskaya 1625 (−1.4) and Primorskaya 1595 (2.2) have the best integral estimates. Primorskaya 1623 was in 16th place (although exceeding the standard by 1.8 centner/ha in terms of yield), since its parameters are far from ideal (1000 seed weight and resistance to diseases indicators are low). As a result of the multicriteria analysis, Primorskaya 1623 and Primorskaya 1602 will be excluded from the breeding process, because their integral estimate is below the level of Primorskaya 4.

The small-seeded soybean variety Primorskaya 1628 received the minimum rating in the test due to its low yield and 1000 seed weight, which directly depend on the seed size. Small-seededness (seed size up to 5.0 mm) is a valuable indicator of soybeans with the likelihood of using it in the vegetable direction (growing soybean seedlings).

Table 2. Integral estimation of soybean varieties of competitive testing, 2018–2020.

Variety	Yield. centner /ha	1000 yield weight.g	Plant height. cm	Vegetation period. days	Content in seeds. %		Resistance to diseases. %	Integral estimation
					oil	protein		
Primorskaya 4. standard	22.7	155	85	116	22.8	37.5	74.4	10.9
Primorskaya 1625	29.7	180	63	121	22.5	39.1	86.7	−1.4
Primorskaya 1595	26.3	175	62	115	24.2	36.8	75.7	2.2
Primorskaya 1551	28.2	157	72	119	22.2	38.8	76.9	3.0
Primorskaya 1627	24.9	180	92	114	21.7	41.6	89.7	3.7
Primorskaya 1596	26.3	162	69	118	24.1	37.0	75.4	5.2
Primorskaya 1555	26.3	170	82	121	21.7	39.4	74.9	5.8
Primorskaya 1419	25.7	170	82	120	22.9	38.6	75.3	5.9
Primorskaya 1606	23.0	178	67	120	23.8	36.9	77.4	6.2
Primorskaya 1607	25.4	153	59	117	24.4	36.0	78.6	6.6
Primorskaya 1626	23.8	170	71	121	22.2	39.5	88.3	6.6
Primorskaya 1576	25.1	162	74	116	23.0	38.2	72.8	6.7
Primorskaya 1575	25.3	157	84	116	23.1	39.0	71.9	7.7
Primorskaya 1622	24.5	150	78	115	22.1	41.8	86.7	8.0
Primorskaya 1621	24.1	155	92	117	21.6	42.1	86.0	9.1
Primorskaya 1623	24.5	145	85	119	22.8	39.2	69.0	11.3

(*continued*)

Table 2. (*continued*)

Variety	Yield. centner /ha	1000 yield weight.g	Plant height. cm	Vegetation period. days	Content in seeds. %		Resistance to diseases. %	Integral estimation
					oil	protein		
Primorskaya 1602	20.8	160	64	120	22.3	38.1	78.0	11.3
Primorskaya 1628	20.2	120	55	105	21.4	41.4	80.0	14.5
$LSD_{.05}$	2.4							
Standard deviation	2.3	14.8	11.3	3.8	0.9	1.8	5.9	–
Weight coefficient	2.275	2.275	−0.700	−0.700	0.350	0.350	0.350	–
(Model) ideal variety	30	190	80	120	20	38	70	–

3.3 Resistance of Soybean Varieties Against the Background of Natural and Artificial Development of Diseases

The search for optimal genotypes that combine high productivity and resistance to fungal pathogens is one of the important directions in crop production and selection [9].

Practically all studied soybean varieties proved to be moderately resistant to *Septoria* glycines Hemmi - their damage varied from 27 to 42% (Fig. 1). Against an artificial background of development, the varieties showed susceptibility to the pathogen; after infection, the degree of leaf damage increased by 18–43%. It should be noted that soybean samples have a higher resistance to disease in comparison to standard. Primorskaya 1626 showed a high responsiveness to forced infection with septoria (68%), although, under natural conditions of pathogen development, the percentage of damage among the studied samples was very low (25%). This fact indicates the inability of the variety to withstand infectious pressure.

According to the scale for determining disease resistance (N.I. Korsakov et all, 1979), all varieties showed resistance against the natural background of the development of the pathogen *Cercospora* sojina Hara (the degree of damage did not exceed 25%) (Fig. 2). Against an artificial infectious background, according to immunological assessment, 50% of the samples were assigned to the groups of moderately resistant and susceptible.

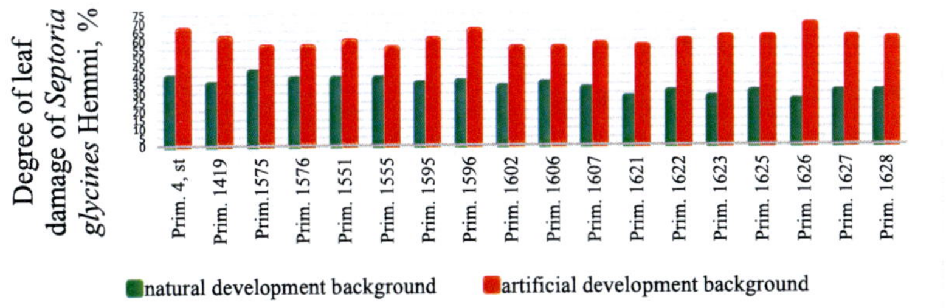

Fig. 1. Degree of leaf damage of soybean varieties by the pathogen *Septoria glycines* Hemmi on natural and artificial development backgrounds, 2018–2020.

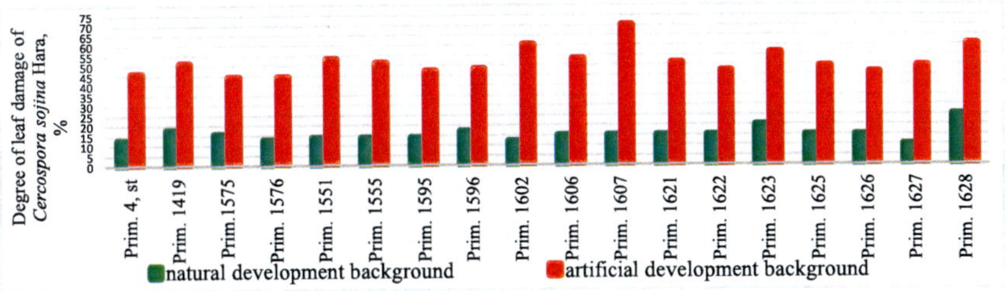

Fig. 2. Degree of leaf damage of soybean varieties by the pathogen *Cercospora sojina* Hara on natural and artificial development backgrounds, 2018–2020.

Analyzing the data of a competitive test of varieties resistance to *Peronospora manshurica* (Naum.) Syd. Syn. under natural conditions of the development of fungal infection, it was noted that the damage did not exceed 20%, while in artificial conditions it reached 67% (Fig. 3). At a high infectious load, the immune system of Primorskaya 1622 could not fully resist additional pressure by the pathogen, and the degree of damage increased from 15% (natural background) to 67% (artificial background). In terms of resistance to peronosporosis forced infection, 33% of the samples were characterized as moderately resistant, 67% - as susceptible.

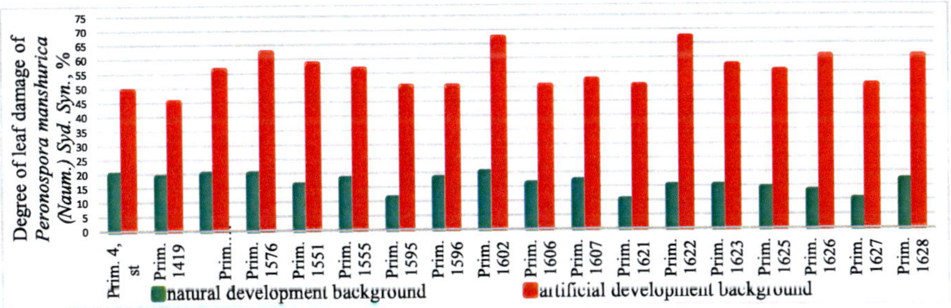

Fig. 3. Degree of leaf damage of soybean varieties by the pathogen *Peronospora manshurica (Naum.) Syd. Syn.*on natural and artificial development backgrounds, 2018–2020.

3.4 Assessment of the Soybean Varieties Potential for Resistance to Stress in Primorsky Krai

Identification of samples resistant to changing weather conditions and calculation of statistical parameters characterizing the adaptive potential in terms of yield are of current importance (Table 3).

Table 3. Evaluation of soybean samples by genotypic variability, stress resistance, and flexibility, centner/ha (average values in 2018–2020).

Sample variety	Stress resistance ($X_{min} - X_{max}$)	Genetic flexibility ($X_{min} + X_{max}$)/2	Genotypic variability ($X_{min} - X_{max}$)
Primorskaya 4. standard*	−11.9	20.8	14.9–26.8
Primorskaya 1419	−16.2	24.4	16.3–32.5
Primorskaya 1575	−14.0	23.6	16.6–30.6
Primorskaya 1576	−12.9	23.3	16.9–29.8
Primorskaya 1551	−7.7	25.6	21.8–29.5
Primorskaya 1555	−13.0	24.6	18.1–31.1
Primorskaya 1595	−14.2	24.5	17.4–31.6
Primorskaya 1596	−13.6	22.7	15.9–29.5
Primorskaya 1602	−18.1	20.6	11.6–29.7
Primorskaya 1606	−14.9	20.6	13.2–28.1
Primorskaya 1607	−15.7	23.5	15.7–31.4
Primorskaya 1621	−15.3	22.0	14.4–29.7
Primorskaya 1622	−12.5	22.4	16.2–28.7
Primorskaya 1623	−10.7	22.4	17.1–27.8
Primorskaya 1625	−11.1	27.6	22.1–33.2
Primorskaya 1626	−13.0	21.8	15.3–28.3
Primorskaya 1627	−16.6	22.9	14.6–31.2
Primorskaya 1628	−1.5	20.2	19.5–21.0

Note: X_{min} – yield formed under limited environmental conditions;
X_{max} – yield formed in favorable environmental conditions. Environmental condition index (Ij): in 2018 –0.3; in 2019 –6.3; in 2020 +6.6

According to the results of the studies, it was found that Primorskaya 1628 (−1.5), Primorskaya 1551 (−7.7), and Primorskaya 1623 (−10.7) have the highest indicator of resistance to stress and a wide range of use. Samples Primorskaya 1602 (−18.1), Primorskaya 1627 (−16.6), and Primorskaya 1419 (−16.2) are able to form high yield in years more favorable for the growth and development of soybeans.

Primorskaya 1625, Primorskaya 1551, Primorskaya 1555, Primorskaya 1595, and Primorskaya 1419 were characterized by greater genotypic flexibility (24.4–27.6), i.e. had a higher degree of genotype-environment dependence. Primorskaya 1628, Primorskaya 1602, Primorskaya 1606, and Primorskaya 4 had the smallest stress resistance indicator (20.2–20.8).

The index of environmental conditions (Ij) during the study period was very contrasting and varied from +6.6 to −6.3. In 2020, favorable conditions developed for realizing the soybean varieties yield potential, where the environment index was the highest positive value (6.6), with the highest average yield of 29.4 t/ha.

4 Discussion

Soybean is one of the most important crops in the world, along with corn and wheat [10]. The selection and creation of highly productive, adaptive to local conditions varieties resistant to diseases and pests is of great importance for the expansion of the soybean cultivation zone, which served as the basis for this breeding work [11].

Strong positive relationships were noted between yield and mass of seeds per plant, number of beans and seeds per plant (r = 0.79–0.88). These observations are similar to the results obtained in earlier publications, in which there was a similar relationship between productivity and number of beans and seeds per plant (r = 0.72–0.88) [9]. According to the results of our research, it was noted that with an increase in the duration of the growing season, the number of beans, seeds, and their weight per plant, as well as the height of plants, increase, which is confirmed by a number of authors [12].

Soybean varieties of the middle and middle late ripeness group were characterized by the highest yield and productivity. This fact was also noted in studies carried out in the conditions of the Middle Amur Region [13]. The effectiveness and objectivity of the application of integral estimation of soybean varieties are emphasized and also noted by breeders in work with various crops [14].

Septoria, peronospora, and cercosporiasis are the main common and harmful fungal diseases of soybeans in the Far East and the forest-steppe zone of Western Siberia [15]. In our studies, all studied soybean varieties showed themselves to be resistant and moderately resistant against the natural background of the development of pathogens. Responsiveness of the varieties to artificial infection with pathogens was observed.

Resistance to stress under the conditions of Primorsky Krai was shown by the varieties of competitive testing with both low and high yields. Studies carried out in the Tyumen Region reflect high stress resistance only of low yield soybean samples. Strong variability of this trait was also observed in the Rostov Region.

5 Conclusion

Strong positive relationships were revealed between yield and mass of seeds per plant, and number of beans and seeds per plant. Soybean varieties of the middle and middle late ripeness group were distinguished by the highest yield and productivity. According to the ranking, promising soybean varieties Primorskaya 1625 and Primorskaya 1595 have the best integral assessment for a complex of traits. Based on the results of testing soybean samples, it was decided to submit a small-seeded vegetable soybean variety Primorskaya 1628 for testing in the State Commission for the Protection of Breeding Achievements. Against the natural background of the development of phytopathogens (Septoria glycines Hemmi, Cercospora sojina Hara, Peronospora manshurica Naum.), stable and medium-resistant samples were recorded. After forced infection, their immune status changed to medium-resistant and susceptible, respectively. When assessing resistance to stressful growing conditions, soybean varieties with a high adaptive potential were identified: Primorskaya 1628, Primorskaya 1551, and Primorskaya 1623. The effectiveness and objectivity of the application of biological statistics methods in the comparative assessment of promising breeding material of soybeans was noted.

References

1. Lu, F., Hongyan, W., Xiaowei, M., Hongbo, P., Jianrong, S.: Modeling the current land suitability and future dynamics of global soybean cultivation under climate change scenarios. Field Crops Res. **263**, 108069 (2021). https://doi.org/10.1016/j.fcr.2021.108069
2. Watson, C.A., Reckling, M., Preissel, S., et al.: Grain legume production and use in European agricultural systems. Adv. Agr. **144**, 235–303 (2017). https://doi.org/10.1016/bs.agron.2017.03.003
3. Koester, R.P., et al.: Has photosynthetic capacity increased with 80 years of soybean breeding, An examination of historical soybean cultivars. Plant Cell Environ. **39,** 1058–1067 (2015)
4. Bandillo, N.B., et al.: Dissecting the genetic basis of local adaptation in soybean. Sci. Rep. **7**(1), 7195 (2017)
5. Jephter, B.F., Peter, M., Bokosi, J.M.: Assessment of the agronomic productivity and protein content in 16 soybean genotypes. Afr. J. Food Agr. Nut. Dev. **17**(4), 12600–12613 (2017). https://doi.org/10.18697/ajfand.80.15995
6. Finoto, E.L., Soares, M.B.B., de Anchieta Alves de Albuquerque, J., et al.: Selection of soybean genotypes for yield, size, and oil and protein contents. Austr. J. Crop. Sci. **15**(1), 48–50 (2021). https://doi.org/10.21475/ajcs.21.15.01.2390
7. Zhou, Z., Lakhssassi, N., Cullen, M.A., et al.: Assessment of phenotypic variations and correlation among seed composition traits in mutagenized soybean populations. Genes **10**(12), 975 (2019). https://doi.org/10.3390/genes10120975
8. Li, S., Xu, H., Yang, J., et al.: Dissecting the genetic architecture of seed protein and oil content in soybean from the yangtze and huaihe river valleys using multi-locus genome- wide association studies. Int. J. Molec. Sci. **20**(12), 3041 (2019). https://doi.org/10.3390/ijms20123041
9. Butovets, E.S., Vasina, E.A., Lukyanchuk, L.M.: Screening of soybean germplasm in Primorsky Krai. Ach. Sci. Technol. AIC **34**(8), 23–27 (2020). https://doi.org/10.24411/0235-2451-2020-10803
10. Matiu, M., Ankerst, D.P., Menzel, A.: Interactions between temperature and drought in global and regional crop yield variability during 1961–2014. PLoS ONE **12**(5) (2017). https://doi.org/10.1371/journal.pone.0178339
11. Zhang, S.-R., et al.: Photoperiodism dynamics during the domestication and improvement of soybean. Sci. China Life Sci. **60**(12), 1416–1427 (2017). https://doi.org/10.1007/s11427-016-9154-x
12. Seferova, I.V., Boyko, A.P., Perchuk, I.N., Shelenga, T.V., Sholukhova, T.A.: Results of studying soybean samples at the Adler experimental station of VIR in 2013–2015. Proc. Appl. Botany Genet. Select **179**(3), 143–151 (2018). https://doi.org/10.30901/2227-8834-2018-3-143-151

13. Shepel, O.L., Aseeva, T.A., Zvolimbovskaya, M.P.: Dependence of economic and biological characteristics of soybeans on the hydrothermal conditions of the Middle Amur region. Ach. Sci. Technol. AIC **34**(8), 16–22 (2020). https://doi.org/10.24411/0235-2451-2020-10802
14. Omelyanyuk, L.V., Asanov, A.M., Yusova, O.A., Churakov, A.A.: New early ripening soybean variety. Zaryanitsa Siber Condit. Oil Crops **3**, 161–168 (2020). https://doi.org/10.25230/2412-608X-2020-3-183-161-168
15. Zaostrovnykh, V.I., Kadurov, A.A., Dubovitskaya, L.K., Ryazanova, O.A.: Monitoring of the species composition of soybean diseases in different zones of soybeanb planting. Far East Agrar Bull. **4**, 51–67 (2018). https://doi.org/10.24411/1999-6837-2018-14081

Checking the Validity of Soy Threshing and Separation Models Using Experimental Data

Irina Prisyazhnaya(✉), Serafima Prisyazhnaya, Valentina Sinyegovskaya, and Mikhail Sinegovskiy

All-Russian Scientific Research Institute of Soybean, Amur Region, 19 Ignatievskoe sh, Blagoveshchensk, Russia675027
info@vniisoi.ru

Abstract. Harvesters on soybean harvesting, damage grain (up to 15%) and do not separate seeds by different qualities. These ripened seeds have a larger mass, increased growth energy, high germination and productivity. When harvesting soya, they fall into the total mass of grain, since the graff goes through the entire threshing stage with the first and second drums, cleaning on a grating mill and the isolated grain is collected in one bin. When harvesting soybeans, more than 60% of the seeds do not receive the resource for increasing the harvest laid down by nature, since different-quality seeds cannot be isolated during post-harvest mining. Threshing and further separation of soybean grains depends on the length of the deck. However, there is no reliable assessment of the effect of the deck length of the threshing-separating device of the two-phase threshing combine on its threshing and separating ability at present, since it can only be determined experimentally. To this end, mathematical models have been developed that allow us to generally describe the process of threshing and separation of soybean grains in a two-phase threshing combine. Evaluation of mathematical models for adequacy, using experimental data, based on the method of successive approximations.

Keywords: Soybeans · Harvesting · Two-phase threshing · Damage · Separation · Harvesting · Quality seeds

1 Introduction

In world grain production, soybeans occupy the fourth place in terms of the size of the sown area. Its production and consumption in the world is steadily growing. Over the past 15 years, the area of soybean crops has grown 1.5 times, the volume of harvested crops 1.6 times, and the volume of international trade 2 times [1, 2]. Currently, Russia's global soybean production is moving in the general global trend, gradually increasing the volume of production, processing and sales of this crop. In 2018, 50.6% of crops were planted in the Far East (1468.9 thousand hectares), of which 33.2% (960 thousand hectares) were in the Amur region [3–5].

A. Muratov and S. Ignateva (Eds.): AFE 2021, LNNS 353, pp. 230–237, 2022.
https://doi.org/10.1007/978-3-030-91402-8_27

Soybean production is the main system-forming element of agro-industrial production for the region, the share of which in the total volume of agricultural production annually amounts to more than 40%. The high demand for soybean cultivation among agricultural producers is due to its profitability and the presence of a processing base in the region [6].

Experience shows that soybeans play an important role in increasing the soybean yield, but their quality in the Amur region is not very high, since annually up to 15–20% of beans are sown substandard. At the same time, one of the main reasons for the decline in the quality of soybeans material is their significant crushing during harvesting and processing. Given these features, the issue of obtaining high-quality seeds remains relevant. It is possible to obtain high-quality soybean seeds directly during harvesting in case of improving the two-phase threshing harvester [7–11].

The threshing and separation of soybeans depends on many factors, including the length of the deck. A reliable assessment of the influence of the length of the deck of the threshing and separating unit of a two-phase threshing harvester on its threshing and separating ability is determined only experimentally [12].

Earlier there was developed a mathematical model of the technological process that describes the threshing and separation of soybeans along the length of the threshing and separating unto of a two-phase threshing harvester [13]. The aim of the research is to check the validity of the developed model of soybeans threshing and separation process in the two-phase threshing unit of the harvester for obtaining high-quality soybean seeds based on experimental data and to determine the error of the experiment.

2 Materials and Methods of Research

The verification of the equations [13] was carried out according to the zonal separation data obtained during the production test of a two-phase threshing harvester with beater drums, with an angle of 12800 and an intermediate four-bladed beater. The research was carried out in the Federal State Unitary Enterprise «SADOVOE» on the VNIIS–1 variety, with the soybeans moisture content of 10.3%, stems of 19.4%, leaves of 15.7%, weight of 1000 seeds equaled to 147 g. When threshing soybeans in the field, the rotation frequency of the first threshing drum was 460, of the second one it was 610 min^{-1}, the threshing gaps were 24/12 and 18/9 mm. Cleaning was removed from the experimental harvester and the entire threshing and separating part of the harvester was divided into four zones and samplers were fixed under them. The first zone received the beans from the first half of the first threshing drum, the second zone received the beans from the second half of the first threshing drum. In the third zone, there were the beans from the intermediate beater and in the fourth zone there were the beans from the second threshing drum. The chaff coming off the straw rack was collected in a separate sampler. The separation of soybeans in each zone, as well as the under threshing and the content of loose soybeans coming from the threshing drums were determined according to the generally accepted method.

3 Results

As a result of the conducted studies, it was found that most intensively soybean seeds are threshed and separated at the beginning of the threshing and separating device of the two-phase threshing harvester. For example, when sending 4.9 kg/s to the thresher, in the zone of the first half of the first threshing drum 37.5% of the beans is separated, and at the end of the first threshing drum 70.8% of the beans is separated on the whole, with the under threshing of 22.1% (Table 1).

Table 1. Results of experiments on the threshing and separation (the amount of soybeans sent to the threshing unit is 4.9 kg/s).

Thresher zone	Soybeans separation, %		The contents of soybeans coming from, %	
	Experimental	Theoretic	the first drum	the second drum
The first	37.5	37.6	$y_1 = 22.1$	$X_1 = 7.1$
The second	33.3	33.4		
The third	12.7	11.5	$y_3 = 2.37$	$X_3 = 0.43$
The fourth	13.7	14.5		

Note: y is loose soybeans; x is unthreaded soybeans

4 Discussion

Using the results of experiments, there were tested previously developed and published [13] models of the technological process of threshing and separation of soybeans. Determining the seven variables in the model equations by the method of successive approximations based on the general least squares method is very difficult, since a system of seven equations for these variables will contain them implicitly. To simplify calculations, soybeans separation coefficient μ1 = const was determined by sequentially refining the threshing coefficients -A1 = const, and the intensity of the threshing process μ01 = const along the length of the deck of the first drum by the method proposed by Professor S. A. Alferov and V. S. Braginets [12].

There is taken the value of the under threshing on the first bar of the deck of the first drum X11 = 0.2; 0.3; 0.4; 0.5 and 0.6.

From the equation X11 = e^{-A_1} the following is obtained:

$$\lg X_1^1 = -A_1 \lg \mathrm{e},$$

Where

$$\mathrm{A1} = \frac{\lg X_1^1}{\lg \mathrm{e}} \tag{1}$$

Knowing the preliminary approximate values of the threshing coefficient for the first deck bar of the first drum A1 with the value of the under threshing X11 = 0.2, 0.3, 0.4, 0.5, and 0.6, and using the available experimental data on under-threshing (X1 = 7.1%,) from the expression (3) [13], (calculating a logarithm from it and solving it taking μ01), there was determined the value of the intensity coefficient of the threshing process μ01, corresponding to the accepted values of the threshing coefficient for the first deck bar of the first drum A1:

$$\mu_{01} = -\frac{\ell g X_1 + A_1 \ell g e}{L_1 \ell g e} = -\frac{1}{L_1}\left(\frac{\ell g X_1}{\ell g e} + A_1\right) \quad (2)$$

The amount of soybeans that went through the deck of the first drum can be recorded as

$$Z_1 + e^{-\mu_1 L_1} - 1 = \frac{\mu_1 e^{-A_1}(e^{-\mu_{01} L_1} - e^{-\mu_1 L_1})}{\mu_{01} - \mu_1} \quad (3)$$

$$z_1 + e^{-\mu_1 L_1} - 1 = \sigma_1 \quad (4)$$

After finding the left and right sides of Eq. (3), there is obtained a system of Eqs. (4) and (5), from which the intensity of the process is found.

$$\frac{\mu_1 e^{-\mu_{01} L_1} - e^{-\mu_1 L_1}}{\mu_{01} - \mu_1} = \xi_1 \quad (5)$$

The value of the intensity of the separation process μ_1 will be determined by the intersection point of the curves $\sigma_1(\mu_1) u \xi(\mu_1)$ and $\sigma_1(\mu_1) u \xi(\mu_1)$ (Fig. 1a).

Given that the experimental values of curve of the soybeans going through the deck Z_1 will always contain an error δ_1, the Eq. (4) [13] can be written as:

$$1 - e^{-\mu_1 L_1} - Z_1 + \frac{\mu_1 e^{-A_1}(e^{-\mu_{01} L_1} - e^{-\mu_1 L_1})}{\mu_{01} - \mu_1} = \delta_1 \quad (6)$$

Where δ_1 is an error in sequential approximation.

e^{-A_1} and e^{-A_3} of separations μ1, corresponding to each of the intermediate values of the threshing coefficients A1 and the intensity of the threshing process μ01.

Using the expression (6), there is found the sum of the squared errors of the value of the soybeans going through the deck of the first drum Z1, for the accepted values e^{-A_1} and μ01 and μ1 corresponding to them.

Calculating the dependence $\sum \delta_1^2$ and e^{-A_1} (Fig. 2a), there is found the true value of the under threshing on the first bar of the deck of the first drum e^{-A_1}, which is determined by the minimum of the curve $\sum \delta_1^2$ and (e^{-A_1}).

By expression (1) there is determined the true refined value of the threshing coefficient A1, by expression (2) there is also determined the refined value of the intensity coefficient of the threshing process μ01, and from the system of Eqs. (4) and (5) there is determined the value of the intensity coefficient of the separation process μ1.

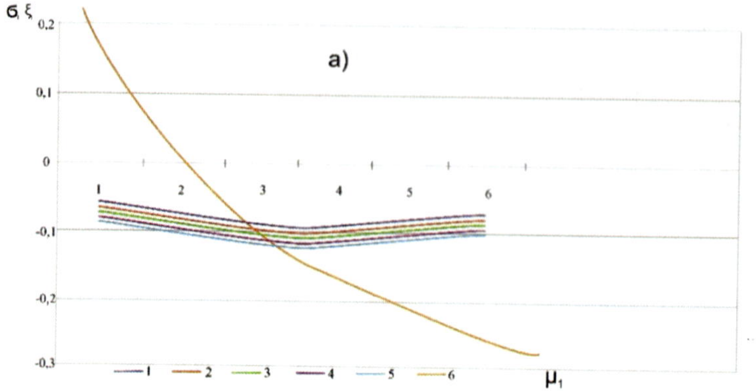

a) – 1, 2, 3, 4, 5 - ξ_1 (μ_1) at e^{-A_1} that equals to 0.2; 0.3; 0.4; 0.5; 0.6, respectively;

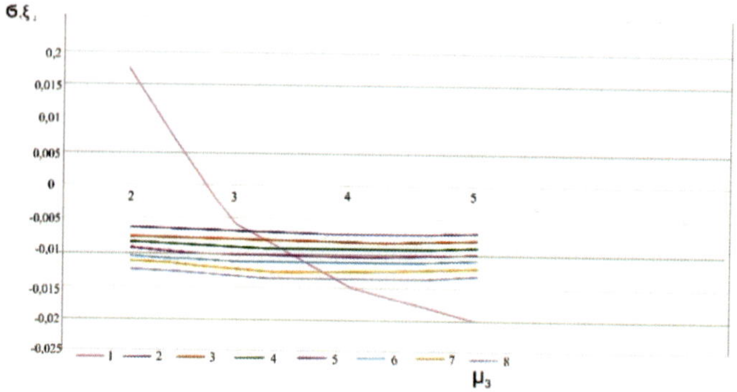

b) – 1, 2, 3, 4, 5, 6, 7 - ξ_2 (μ_3) at e^{-A_3}that equals to 0.02; 0.025; 0.03; 0.035; 0.04; 0.045; 0.05; curves 6 and 9 forσ_1 and σ_2;

Fig. 1. a, b – Dependence of the values of the separation coefficients μ_1 and μ_3 on the accepted values.

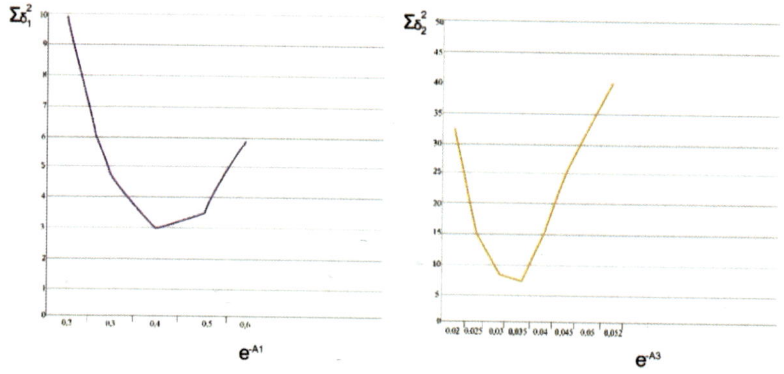

a) b)

Fig. 2. a, b – Dependence $\sum \delta_1^2$ and $\sum \delta_2^2$ on the accepted values e^{-A_1} and μ_1, e^{-A_3} and μ_3.

For section L_1, the decks of the first drum, using experimental data ($Z_1 = 0.708$; $e^{-A1} = 0.4$; $A_1 = 0.916$; $\mu_{01} = 2.7$; $\mu_1 = 2.81$), obtained the value of the intensity coefficients of the threshing process and soybeans separation along the length of the deck of the first drum ($\mu_{01} =$ const); and using $\mu_1 =$ const, a coefficient that takes into account the threshing at the first bar of the deck A_1, and experimental data on soybeans separation at the section L2 ($Z_2 = 0.835$) from expressions (8) [13],there is determined the value of the intensity coefficient of the soybeans separation process along the length of the separating grid of the intermediate beater $\mu 2 = 3.72$. The intensity coefficient of separation along the length of the deck of the second threshing drum μ_3 is determined by sequentially refining the threshing coefficients A_3 and the intensity of the threshing process μ_{03}.

In the first stage of threshing, the amount of under threshing in the second drum will be:

$$X_3^1 = X_1 e^{-A_3}$$

or

$$X_3^1 = e^{-A_1 - \mu_{01} L_1 - A_3} \quad (7)$$

The value of the under threshing on the first bar of the deck of the second drum is X31 = 0.02; 0.025; 0.03; 0.035; 0.04; 0.045; 0.05. For each of the accepted values, according to Eq. (7), there is determined the value of the threshing coefficient at the first bar of the deck of the second drum A3. Using the experimental data on the under threshing in the second drum (X3 = 0.43% Table 1) from the expression (11) [13] (calculating a logarithm from it and solving it taking $\mu 03$), there is determined the value of the intensity coefficient of the threshing process, with the second threshing drum $\mu 03$, corresponding to the accepted values of the threshing coefficient A3:

$$\mu_{03} = \frac{1}{L_3}\left(\frac{lgX_3}{lge} + A_1 + \mu_{01} L_1 + A_3\right) \quad (8)$$

The expression (17) [13] can be written as

$$\begin{aligned} & Z + e^{-\mu_1 L_1 - \mu_2 L_2 - \mu_3 L_3} - e^{-A_1 - \mu_{01} L_1 - \mu_3 L_3} \\ & + \frac{e^{-A_1 - \mu_2 L_2 - \mu_3 L_3}(\mu_1 e^{-\mu_1 L_1 -} \mu_{01} e^{-\mu_{01} L_1})}{\mu_{01} - \mu_1} - 1 \\ & = \frac{\mu_3 e^{-A_1 - \mu_{01} L_1 - A_3}(e^{-\mu_{03} L_3 - e - \mu_3 L_3})}{\mu_{03} - \mu_3} = \delta_2 \end{aligned} \quad (9)$$

Denoting the left expression of Eq. (9) by σ_2, and the right one by ξ_2, there is obtained a system of equations from which the values of the separation coefficients $\mu 3$ corresponding to each of the intermediate values of the values A3 and $\mu 03$ are determined. The values of $\mu 3$ will be determined by the intersection point of the curves $\sigma_2(\mu_3)$ and $\xi_2(\mu_3)$ (Fig. 1b). From the figure at the intersection $\sigma_2(\mu_3)$ and $\xi_2(\mu_3)$ depending on e^{-A_3} there is graphically found the intensity of the separation process through the deck of the second drum $\mu 3$.

Given that the experimental values of the amount of soybeans going through the deck of the second drum Z3 will always contain an error, equation [13] can be written as:

$$1 - Z - e^{(-\mu_1 L_1 - \mu_2 L_2 - \mu_3 L_3)} - e^{(-A_1 - \mu_{01} L_1 - \mu_3 L_3)}$$
$$- \frac{e^{-A_1 - \mu_2 L_2 - \mu_3 L_3}(\mu_1 e^{-\mu_1 L_1 -} \mu_{01} e^{-\mu_{01} L_1})}{\mu_{01} - \mu_1} \tag{10}$$
$$- - \frac{\mu_3 e^{-A_1 - \mu_1 L_1 - A_3}(e^{-\mu_{03} L_3 - e - \mu_3 L_3})}{\mu_{03} - \mu_3} = \delta_2$$

Where δ_2 is an error in sequential approximation.

Using expression (10), there is determined the sum of the squared errors of the amount of soybeans going through the deck of the second drum Z3 for the accepted values of the amount of the under threshing on the first bar of the deck of the second drum and the corresponding coefficients of the threshing intensity and the intensity of the separation process along the length of the deck of the second drum μ03 and μ3. Having calculated the dependencies on $\sum \delta_2^2$ from e^{-A_3} (Fig. 2b), there is found the true value e^{-A_3}, which is determined by the minimum of the curve $\sum \delta_2^2$ from e^{-A_3}. By expression (7) there is determined the true refined value of A3, by expression (8) there is also determined the refined value of μ03, and from the system of Eqs. (9) μ3 is determined.

For the data presented in Table 1 for the section of the deck of the second drum $Z_3 = 0.972$) and e^{-A_3}=0.033; $A_3 = 0.767$; $\mu_{03} = 3.186$; $\mu_3 = 3.33$ the following values of the coefficients of threshing and separation are obtained:

$$A_1 = 0.916;\ \mu_{01} = 2.7;\ \mu_1 = 2.81;\ \mu_2 = 3.72;\ A_3 = 0.767;\ \mu_{03} = 3.186;\ \mu_3 = 3.33$$

Using the obtained coefficients, there are determined the values of the dependencies of threshing X and separation Z of soybeans along the entire length of the two-phase threshing device of the combine harvester.

Studies of the process of threshing and separation of soybean along the length of the combine thresher with a constant supply to the thresher of 4.9 kg/s showed that at the ratio of the circumferential speed of the first and second drums beaters V1/V2 = 13.24/17.56 m/s, 70.8% of the soybeans are separated in the zone of the first deck. In the zone of the intermediate beater and the deck of the second drum, 12.7 and 13.7% are separated [13, 14]. Reducing the circumferential speed of the first drum beaters and keeping the speed of the second drum beaters at the same level leads to a decrease in the intensity of threshing at the first bar of the first drum deck and soybeans separation in the zone of the first deck of the first drum and an increase in soybeans allocation in the zone of the intermediate beater and the second drum deck.

5 Conclusion

1. The obtained theoretical expressions (1–10) and graphical dependencies (Figs. 1 and 2) give grounds to show an accurate (0.49%) real physical picture of the threshing

and separation process along the length of the thresher and objectively assess the effect of the length of the thresher on the quality performance of the threshing unit of the two-phase threshing combine harvester.

2. Quantitative indicators of threshing and separation of soybeans along the length of the threshing and separating machine helped to justify a rational scheme of the device for separating and collecting high-quality seeds in a separate combine hopper (Patent No. 2679508).

References

1. Qin, Y., Xiao, X., Li, X.: Mapping paddy rice planting area in cold temperate climate region through analysis of time series Landsat 8 (OLI), Landsat 7 (ETM+) and MODIS imagery. ISPRS J. Photogram Remote Sens. (2015)
2. Herrold-Menzies, M.: Fundamentals of Weed Science, 5th edn. (2018)
3. Hanna, H.M., Quick, G.R.: Grain Harvesting Machinery. Handbook of Farm, Dairy and Food Machinery Engineering 8, 3rd edn. (2019)
4. Bern Graeme, C.J., Herum, Q.F.L.: Harvesting and Postharvest Management Corn. Third Edition 5 (2018)
5. Chaab, R., Karparvarfard, S., Mohammadi, M.: Predicting header wheat loss in a combine harvester, a new approach. J. Saudi Soc. Agric. Sci. **19**, 179–184 (2018)
6. Hanna, H., Quick, G.: Grain Harvesting Machinery Handbook of Farm, Dairy and Food Machinery Engineering. Second Edition 6 (2013)
7. Reddy, P., Patil, J.: Breeding Methods for Winter Sorghum Improvement Genetic Enhancement of Rabi Sorghum 7 (2015)
8. Chaab, R., Karparvarfard, S., Mohammadi, M.: Predicting header wheat loss in a combine harvester, a new approach.J. Saudi Soc. Agric. Sci. **10** (2018)
9. Bern, C., Quick, G., Herum, F.: Harvesting and Postharvest Management Corn. Third Edition 5 (2018)
10. Chaab, R., Karparvarfard, S., Mohammadi, M.: Predicting header wheat loss in a combine harvester, a new approach.J. Saudi Soc. Agric. Sci. **12** (2018)
11. Prisyazhnaya IM, Prisyazhnaya SP, Prisyazhniy MM, Protsenko PP (2018) Improvement of process of the thresh, separation and transportation for improvement of quality of seeds at harvesting cleaning of energetically effective soy. ISBN 978–5–93493–319–8
12. Prisyazhnaya IM, Sinegovskaya VT, Prisyazhnaya SP, Sinegovskii MO (2020) Harvester and transporting device development for high-quality soybean seeds obtaining. AGRITECH-III-2020: IOP Conf Ser: Earth and Envir Sci 548:062078. DOI: https://doi.org/10.1088/1755-1315/548/6/062078
13. Prisyazhnaya IM, Sinegovskaya VT, Prisyazhnaya SP (2020) Gathering of polova at kombaynovy cleaning of soy, for creation of the food supply of livestock production. J of Agric and Envir 4 (16). DOI: https:/doi.org/https://doi.org/10.23649/jae.2020.4.16.6
14. Prisyazhnaya IM, Prisyazhnaya SP, Lipkan AV, Sakharov VA, Usanov VS (2021) Study of sowing qualities of soybean seeds in seed farms of Amur region. J of Agric and Env 3:(19). DOI: https:/doi.org/https://doi.org/10.23649/jae.2021.3.19.4
15. Gievsky AM, Chernyshov AV, Maslov DL, Milgunov VYu (2019) Substantiation of the operation mode of the threshing-separating device of the combine during soybean harvesting. Vestnik 1(60). DOI: https://doi.org/10.17238/issn2071-2243.2019.1.50
16. Dospekhov BA (2012) Methodology of field experiment. ISBN 978–5–458–23540–2

Adaptive Potential Assessment of the Potato Varieties in the Conditions of the Amur Region

Sergey Rafalsky(✉), Olga Rafalskaya, and Tatiana Melnikova

All-Russian Scientific Research Institute of Soybean, Ignatyevskoe shosse, 19Amur region, Blagoveshchensk 675027, Russia
rsv@vniisoi.ru

Abstract. The article presents the results of the study carried out in 2016–2020 and focused on the plasticity of adaptability, plasticity and stability of potato varieties in the natural and climatic conditions of the Amur region. The research was carried out in the selection crop rotation of the potato breeding laboratory of the Federal State Budgetary Scientific Institution Federal Scientific Center "All-Russian Scientific Research Institute of Soybean" in the village of Sadovoe in the Tambov region with 36 potato varieties of various ripeness groups. The aim of the study was to conduct a comprehensive agroecological assessment of potato varieties for cultivation in the region. The study was carried out in accordance with generally accepted methods for potato culture. The average yield of tubers over the years of research was 25.5 t/ha. According to the results of the collection assortment assessment of early-maturing, mid-early and mid-late maturing groups of potatoes, adaptive plastic and stable varieties were identified: Skoroplodnyi (bi–1.10, V–30.5%, Ca–1.12, S_i^2–100.4), Lyubava (bi–1.01, V–27.9%, Ca–1.12, S_i^2–88.37), Ocharovanie (bi–1.20, V–33.3%, Ca–0.98, S_i^2–119.49) и Vdokhnovenie (bi–1.04, V–26.7%, Ca–1.00, S_i^2–89.75).

Keywords: Potato · Variety · Ripeness group · Yield · Coefficient of variation · Adaptability · Plasticity · Stability

1 Introduction

Variety is one of the main factors determining the productivity and stability of potato production. However, not every variety is capable of providing high and stable yields. The role of a variety in the formation of yield is 50%. Its importance is assumed to remain in the future as high as it is now, and in some cases even increase [1]. Currently, there are more than 4 thousand varieties of potatoes in the world. As of January 1, 2020, 438 potato varieties are included in the State Register of Breeding Achievements Admitted for Use, 78 of which are proposed for cultivation in the Far East region [2]. For each region, a group of complementary varieties is needed that maximize the use of specific ecological and agrotechnical conditions, capable of withstanding unfavorable factors of the cultivation environment [3]. New generation potato varieties should have a wide degree of response to changing growing conditions, high plasticity and stability

A. Muratov and S. Ignateva (Eds.): AFE 2021, LNNS 353, pp. 238–245, 2022.
https://doi.org/10.1007/978-3-030-91402-8_28

of quantitative and qualitative traits, and stress resistance [4]. Plasticity as applied to a crop is defined as the ability of a variety to give stable yields under changing external conditions [5].

The potential possibilities of a variety can be realized only if the growing conditions meet its biological requirements [6].

It has been proven that the adaptability of plants to environmental conditions is expressed by adaptive potential. The adaptability degree of a variety depends not only on its adaptation, but also on the specifics of the ecological conditions created in the agrobiocenosis [7]. At the same time, to ensure high productivity, varieties must have both specific and general adaptive capacity [8–10].

As is known, the progress in breeding research is impossible without a deep study of the interaction in the "genotype – environment" system, the assessment of which is considered one of the actual tasks of breeding [11–13].

The challenging natural and climatic conditions of the Amur Region do not allow consistently high crop yields to be obtained annually. Therefore, when cultivating potatoes in the region, the ability of the variety to form high yields in various growing conditions comes out on top [14].

The purpose of the study was to conduct a comprehensive agroecological assessment of 36 potato varieties of various ripeness groups for cultivation in the Amur Region.

Research objectives were to determine the productive and adaptive potential of potato varieties in terms of "yield", to analyze the variability of the marked trait, to calculate the parameters of ecological plasticity and stability.

2 Conditions, Materials and Methods

Research was carried out in the potato breeding laboratory in 2016–2020 on the experimental field of the All-Russian Scientific Research Institute of Soybean in the village of Sadovoe, Tambov District, Amur Region. The object of research is 36 potato varieties of various ripeness groups.

The soil of the experimental plot is meadow chernozem-like, heavy in granulometric composition. The humus content was 4.5…4.7%, N-NH4 – 19…28 mg/kg of soil, N-NO3 – 30…56 mg/kg, P2O5 и K2O – respectively 46…49 and 130…190 мг/кг. pHsol. – 5.2. Volume weight – 1.04…1.1 g/cm^3. The porosity was 43… 46%.

The meteorological conditions of the growing seasons, according to the years of research, changed in terms of the main indicators (temperature regime and moisture supply), as evidenced by the value of the hydrothermal coefficient (HTC). So, in July and August 2016 and 2017, the HTC was 0.8 and 0.7, respectively (with the norm from 1.5 to 1.7). In 2018, due to early spring, the snow cover rapidly disappeared. Precipitation in June (HTC = 3.5) compensated for the lack of moisture in the soil in spring. The amount of precipitation for the period May - September in 2019 was above the multiyear level. The amount of precipitation in May was 66.6 mm, exceeding the long-term indicator by 2 times, in June it was 94.4 mm with a norm of 85 mm. Abundant precipitation in the third decade of July, the intensity of which was 74% of the monthly norm (HTC = 3.7), led to a short-term waterlogging of the soil. In 2020, the hydrothermal conditions of existence differed from the average long-term norm and were mainly characterized by waterlogging. Particularly abundant precipitation, which caused soil waterlogging, was noted in May, June, August and September with an excess of the norm, respectively, by month 33.5 mm, 55.0 mm, 72.2 mm and 44.3 mm. The HTC varied by months from 0.6 to 3.1.

Thus, weather conditions, due to frequent waterlogging of the soil caused by abundant rainfall during the periods of tuber formation, crop accumulation and ripening, negatively affected the potato yield and especially the quality of the tuber crop, reducing their marketability and increasing disease damage.

Cultivation agrotechnics in the experiment is generally accepted for the research zone [15]. Placement, carrying out the experiment, the study of the collection material of potatoes was performed according to the methods generally accepted for culture [16,17]. Mathematical processing of the data was carried out according to the method of B.A. Dospekhov [18]. The assessment of the variety productive and adaptive potential was carried out according to the method of L.A. Zhivotkov, Z.A. Morozova, L.I. Sekutaeva [19]. When analyzing the adaptive potential of varieties, we used the "average yield of the year" - this is the yield level in a particular year and region. Due to this it was possible to get an additional information about the value of specific varieties, as well as objectively assess the role of environmental factors in the formation of the yield of both the crop as a whole and individual varieties.

The adaptability coefficient (or coefficient of adaptability - Ca) was calculated by the formula:

$$Ca = \left(X_{ij} \cdot 100 :\right) : 100 \quad (1)$$

where Xij is the yield of the i-variety in the j-th year of testing; X is the average yield of the year.

The parameters of ecological plasticity (regression coefficient) and stability (standard deviation from the regression line) were calculated using the method of E.A. Eberhart. and Russell W.A., set out by V.A. Zykin. and others [20]. The method is based on the calculation of linear regression (bi), which characterizes the ecological plasticity of the variety, and the standard deviation from the regression lines (S_i^2), which determines the stability of the variety under environmental conditions.

The theoretical yield was determined by the dependence:

$$X_{ij} = X_i + b_i I_j \quad (2)$$

where Xi is the average yield of the i-variety sample over the years of testing; bi is the regression coefficient; Ij is the index of environmental conditions.

The deviation of theoretical yields from actual yields was calculated using the formula:

$$d_{ij} = X_{ij} - {}_{ij} \quad (3)$$

The variance characterizing the stability of the yield was calculated by the formula:

$$S_i^2 = \sum_j d_{ij}/n-2 \quad (4)$$

3 Results of the Study

It was found that the most favorable meteorological conditions for the growth and development of potatoes as well as for the formation of tuber productivity of the crop were observed in 2016 and 2018, when their average yield was 33.5 and 32.4 t/ha, respectively, and the environment index (Ii) reached 8.05 and 9.38 (Table 1).

Table 1. Characteristics of potato varieties in terms of adaptability (2016–2020).

Variety	Yield, t/ha					X_i2	bi	V, %	Ca	S_i2
	2016	2017	2018	2019	2020					
Early maturing group										
Udacha (st.)	29.7	19.2	29.7	25.1	17.6	24.2	0.68	21.0	0.95	34.40
Zhukovsky early	30.4	20.1	30.4	24.2	14.5	23.9	0.81	25.6	1.02	54.44
Kholmogorskii	31.9	19.4	30.5	18.0	17.4	23.4	0.94	27.2	0.91	73.32
Karatop	34.2	18.3	31.4	25.7	23.2	26.5	0.76	21.4	1.04	47.93
Latona	32.5	19.2	32.1	25.2	25.4	26.9	0.67	18.4	1.06	37.25
Red Ledy	33.9	20.7	32.4	29.5	18.4	26.9	0.78	23.2	1.06	50.49
Vitessa	32.8	21.4	32.8	16.9	19.2	24.6	0.99	27.7	0.95	81.33
Lyubava	34.5	22.3	33.2	22.9	28.6	28.3	0.68	27.9	1.12	38.37
Nikita	31.2	19.8	31.2	17.2	15.4	23.0	1.00	29.9	0.88	82.98
Odissey	31.8	17.8	33.8	18.5	29.1	26.2	0.79	25.7	1.03	51.79
Leonie	32.4	20.4	32.4	17.5	24.6	25.5	0.83	23.9	1.00	57.17
Ognivo	33.9	18.9	31.4	19.5	21.4	25.0	0.91	25.3	0.97	68.72
Rodrigo	32.4	21.4	32.3	22.3	17.9	25.3	0.87	23.6	0.98	62.81
Primadonna	33.1	20.8	33.1	17.6	22.5	25.4	0.92	25.4	0.99	70.24
Gala	33.7	21.2	33.7	23.2	22.8	26.9	0.83	20.7	1.05	57.17
Impala	31.2	21.4	31.2	17.0	19.8	24.1	0.86	24.7	0.94	61.37
Skoroplodny	37.2	24.6	35.2	22.6	15.1	27.0	1.10	30.5	1.12	100.41
Riviera	35.3	19.0	34.1	24.4	18.4	26.2	1.03	27.5	1.01	88.04
Middle-early group										
Nevsky (st.)	31.2	19.8	30.2	17.2	25.4	24.8	0.71	22.4	0.98	41.83
Archidea	34.8	21.3	33.4	21.2	20.6	26.3	0.95	24.5	1.02	74.89
Vulkan	31.4	21.0	32.0	18.4	16.2	23.8	0.96	24.4	0.92	76.48
Briz	30.5	20.4	31.4	23.7	19.8	25.2	0.72	19.5	0.99	43.02
Mustang	34.3	20.3	33.1	17.5	21.3	25.3	1.00	27.6	0.98	82.98
Natasha	31.2	19.8	30.6	24.0	15.4	24.2	0.83	25.3	0.94	57.17
Talisman	34.7	19.4	34.0	26.4	17.8	26.5	0.98	26.7	1.02	79.70
Ocharovanie	36.0	19.4	35.8	21.6	15.6	25.7	1.2	33.3	0.98	119.49
Zekura	32.9	21.0	31.8	16.4	18.4	24.1	0.99	28.6	0.93	81.33
Skarb	32.6	20.8	33.6	22.2	19.8	25.8	0.9	23.3	1.00	67.22
Rocco	32.3	20.6	31.9	13.7	18.6	23.4	1.04	31.7	0.90	89.75

(continued)

Table 1. (*continued*)

Variety	Yield, t/ha					X_i2	bi	V, %	Ca	S_i2
	2016	2017	2018	2019	2020					
Ryabinushka	31.4	20.7	32.4	25.1	20.6	26.0	0.74	19.4	1.02	45.44
Ketskiy	38.0	21.2	36.9	19.9	21.5	27.5	1.19	29.6	1.06	117.51
Mid-late group										
Lugovskoi (st.)	32.8	19.7	30.6	19.5	18.8	24.3	0.90	25.1	0.94	67.22
Zhuravinka	31.4	20.3	31.3	21.3	19.6	24.8	0.81	21.7	0.95	54.44
Vdokhnovenie	36.9	21.7	34.5	24.3	18.4	27.2	1.04	26.7	1.05	89.75
Zolsky	31.2	21.4	32	20.8	17.7	24.6	0.86	23.7	0.96	61.37
Chajka	34.3	20.8	33.2	17.8	17.8	24.8	1.08	29.9	0.95	96.79
Average varietal yield	33.5	20.4	32.4	21.6	19.7					
Environment index (Ij)	8.05	5.52	9.38	–5.16	–6.25					

The most extreme conditions for potato cultivation were noted in 2017, 2019 and 2020, when the environment index (Ii) was negative and amounted to: (-5.52), (-5.16) and (-6.25), respectively. The average yield per variety in 2017 was 20.4 t/ha, in 2019 - 21.6 t/ha, and in 2020 - 23.5 t/ha. When using the indicator "average yield of the year", it was possible to objectively assess the role of environmental factors in the formation of this trait for the culture as a whole and for individual varieties. Comparison of the yield of potato varieties with the average varietal yield in favorable years allowed us to judge their potential yield [21].

From the entire collection assortment of early-maturing, mid-early and medium-late maturing potato varieties, 36 samples were isolated, providing for five years (2016–2020) the formation of a tuber yield, on average, above 20 t/ha. 18 samples were noted with a tuber productivity of 25… 26 t/ha, and above (27 t/ha) - four varieties of potatoes: Skoroplodnyi (27 t/ha), Vdokhnovenie (27.2 t/ha), Ketsky (27.5 t/ha) and Lyubava (28.3 t/ha). The average yield of standard varieties for this period was: early-maturing Udacha - 24.2 t/ha, mid-early Nevsky - 24.8 t/ha, mid-late Lugovskoy - 24.3 t/ha. The coefficients of variation (V) of the yield of each potato variety were determined, indicating the phenotypic variability of the trait caused by environmental factors [22]. A low indicator was observed in potato varieties Latona (18.4%), Ryabinushka (19.4%) and Briz (19.5%). The rest of the varieties had a high coefficient of variation from 20.7% to 33.3%. These potato varieties, possessing a significant potential for yield in favorable years, in the absence of appropriate conditions, sharply reduce its level [23].

As a parameter characterizing the plasticity of a potato variety, we used the regression coefficient (bi) of the yield of varieties, which shows their response to a change in growing conditions. According to the results of the assessment of varieties for this parameter, the corresponding groups were identified. The most valuable are the varieties of the intensive type, characterized by a fairly high yield and responding well to improved growing conditions (bi >1). This group includes potato varieties Ocharovanie (bi–1.20), Ketsky (bi–1.19), Skoroplodny (bi–1.10), Rocco (bi–1.04) and Riviera (bi–1.03). The most plastic were 17 numbers, for which the regression coefficient bi is close to one: Nikita,

Mustang, Vitessa, Talisman, Zekura and others. Varieties with a regression coefficient significantly lower than one belong to the neutral type, are characterized by low plasticity and respond poorly to changes in environmental factors. However, under unfavorable conditions, their yield decreases to a lesser extent in comparison with varieties of the intensive type [24]. This group includes potato varieties with a regression coefficient (bi <1) Karatop (bi – 0.76), Nevsky (bi – 0.71), Latona (bi – 0.67), Zolsky (bi – 0.56) and others, which under unfavorable or extreme conditions of cultivation do not reduce productivity and under conditions of intensive farming do not provide an increase in productivity [25].

To obtain stable crop yields, the indicator of the variety's adaptability to local natural and climatic conditions is of great importance. According to the obtained average coefficient of adaptability (Ca), we can judge the productive capabilities of the studied varieties [26]. In our studies, this indicator varied from 0.88 to 1.12. On average, over five years, of the studied varieties, the adaptability coefficient was higher than one among 18 varieties of potatoes of the early maturing group - 9 (50%), from 13 medium early - 4 (31%), from 5 studied varieties of the middle late group - 2 (40%). The most adaptive to growing conditions from the early ones turned out to be the potato varieties Lyubava, Skoroplodny, Latona, Red Lady, Karatop. From medium early varieties - Ketsky, Ryabinushka, Talisman, Ocharovanie and Archidea. The potato variety Vdokhnovenie with Ka = 1.05 stood out from the middle late ripeness group.

Table 2. Theoretical yield of potato varieties (2016–2020).

Variety	Theoretical yield					Deviation of the actual yield				
	2016	2017	2018	2019	2020	2016	2017	2018	2019	2020
Early maturing group										
Udacha	19.95	20.45	30.58	20.69	19.95	−4.25	−3.75	6.38	-3.51	-4.25
Zhukovsky early	30.42	19.43	31.50	19.72	18.84	6.52	−4.47	7.60	−4.18	−5.06
Kholmogorskii	30.97	18.21	32.22	18.55	17.53	7.57	−5.19	8.82	−4.85	−5.88
Karatop	32.62	22.30	33.63	22.58	21.75	6.12	−4.20	7.13	−3.92	−4.75
Latona	32.29	23.20	33.18	23.44	22.71	5.39	−3.70	6.28	−3.46	−4.19
Red Ledy	33.18	22.59	34.22	22.88	22.03	6.28	−4.31	7.32	−4.02	−4.88
Vitessa	32.57	19.14	33.89	19.49	18.41	7.97	−5.46	9.29	−5.11	−6.19
Lyubava	33.77	24.55	34.68	24.79	24.05	5.47	−3.75	6.38	−3.51	−4.25
Nikita	31.05	17.48	32.38	17.84	16.75	8.05	5.52	9.38	−5.16	−6.25
Odissey	32.56	21.84	33.61	22.12	21.26	6.36	−4.36	7.41	−4.08	−4.94
Leonie	32.18	20.92	33.29	21.22	20.31	6.68	−4.58	7.79	−4.28	−5.19
Ognivo	32.33	19.98	33.54	20.30	19.31	7.33	−5.02	8.54	−4.70	−5.69
Rodrigo	32.30	20.50	33.46	20.81	19.86	7.00	−4.80	8.16	−4.49	−5.44
Primadonna	32.81	20.32	34.03	20.65	19.65	7.41	−5.08	8.63	−4.75	−5.75

(continued)

Table 2. (*continued*)

Variety	Theoretical yield					Deviation of the actual yield				
	2016	2017	2018	2019	2020	2016	2017	2018	2019	2020
Gala	33.58	22.32	34.69	22.62	21.71	6.68	−4.58	7.79	−4.28	−5.19
Impala	31.02	19.35	32.17	19.66	18.73	6.92	−4.75	8.07	−4.44	−5.38
Skoroplodny	35.86	20.93	37.32	21.32	20.13	8.86	−6.07	10.32	−5.68	−6.88
Riviera	34.49	20.51	35.86	20.89	19.76	8.29	−5.69	9.66	−5.31	−6.44
Middle-early group										
Nevsky	30.52	20.88	31.46	21.14	20.36	5.72	−3.92	6.66	−3.66	−4.44
Archidea	33.95	21.06	35.21	21.40	20.36	7.65	−5.24	8.91	−4.90	−5.94
Vulkan	31.53	18.50	32.80	18.85	17.80	7.73	−5.30	9.00	−4.95	−6.00
Briz	31.00	21.23	31.95	21.48	20.70	5.80	−3.97	6.75	−3.72	−4.50
Mustang	33.35	19.78	34.68	20.14	19.05	8.05	5.52	9.38	−5.16	−6.25
Natasha	30.88	19.62	31.99	19.92	19.01	6.68	−4.58	7.79	−4.28	−5.19
Talisman	34.39	21.09	35.69	21.44	20.38	7.89	−5.41	9.19	−5.06	−6.13
Ocharovanie	35.36	19.08	36.96	19.51	18.20	9.66	−6.62	11.26	−6.19	−7.50
Zekura	32.07	18.64	33.39	18.99	17.91	7.97	−5.46	9.29	−5.11	−6.19
Skarb	33.05	20.83	34.24	21.16	20.18	7.25	−4.97	8.44	−4.64	−5.63
Rocco	31.77	17.66	33.16	18.03	16.90	8.37	−5.74	9.76	−5.37	−6.50
Ryabinushka	31.96	21.92	32.94	22.18	21.38	5.96	−4.08	6.94	−3.82	−4.63
Ketskiy	37.08	20.93	38.66	21.36	20.06	9.58	−6.57	11.16	−6.14	−7.44
Mid-late group										
Lugovskoi	31.55	19.33	32.74	19.66	18.68	7.25	−4.97	8.44	−4.64	−5.63
Zhuravinka	31.32	20.33	32.40	20.62	19.74	6.52	−4.47	7.60	−4.18	−5.06
Vdokhnovenie	35.57	21.46	36.96	21.83	20.70	8.37	−5.74	9.76	−5.37	−6.50
Zolsky	31.52	19.85	32.67	20.16	19.23	6.92	−4.75	8.07	−4.44	−5.38
Chajka	33.49	18.84	34.93	19.23	18.05	8.69	−5.96	10.13	−5.57	−6.75

In the breeding process at the present time, the stability indicator (Si2) of the studied trait, that is, the ability of the variety to maintain the level of the trait in various soil and climatic conditions, acquires special importance [26]. Varieties with a higher index are presented as more stable, that is, adapted to these conditions. The highest stability coefficient was observed in the potato varieties Ocharovanie (119.49), Ketsky (117.51) and Skoroplodny (100.41). Low stability was characterized by 19.4% of the studied set of varieties (Si2 = 34.40… –41.83). The greatest deviation of the theoretical yield (TY) from the actual yield (AY) was observed in the potato varieties Vulkan (+9.0… –6.0 t/ha), Chaika (+10.13… –6.75 t/ha), Natasha (+9.79… −5.19 t/ha), Mustang (+9.38… −6.25 t/ha) and Nikita (+9.38… −6.25 t/ha) (Table 2).

The significant difference in the value of theoretical and actual indicators shows the instability of these genotypes. The less the standard deviation of the actual indicators from the theoretical expected ones, the more stable the variety is.

4 Conclusions

Assessment of 36 potato varieties of different ripeness groups allowed us to identify varieties with high adaptability, plasticity and stability, which include early maturing varieties: Skoroplodny (bi – 1.10, V – 30.5%, Ca – 1.12, Si2–100.4), Lyubava (bi – 1.01, V – 27.9%, Ca – 1.12, Si2 –88.37), which exceeded the yield of the standard variety Udacha by 2.8 and 4.1 t/ha, respectively, and the average yield by 1.5 and 3.1 t/ha. Medium early potato variety Ocharovanie (bi – 1.20, V – 33.3%, Ka – 0.98, Si2–119.49), which formed the yield of tubers higher than the standard variety Nevsky by 1.9 t/ha and the average yield by 0,2 t/ha. The potato variety Vdokhnovenie from the middle late group with indicators (bi – 1.04, V – 26.7%, Ca – 1.00, Si2–89.75) exceeded the productivity of the standard Lugovskoy potato variety by 2.9 t/ha and surpassed average yield by 1.7 t/ha.

References

1. Vlasenko, G.P.: Adaptive potential of grades of potatoes in the conditions of Kamchatka Krai. Vestnik DVORAN **3**, 17–20 (2019). https://doi.org/10.25808/08697698.2019.205.3.002
2. Zhevora, S.V.: Environmental adaptability of russian promising potato varieties and economic evaluation of their cultivation. Zemledelie **5**, 30–35 (2019). https://doi.org/10.24411/0044-3913-2019-10508
3. Devaux, A., Kromann, P., Ortiz, O.: Potatoes for sustainable global food security. Potato Res. **57**(3–4), 185–199 (2014). https://doi.org/10.1007/s11540-014-9265-1
4. Streltsova, T.A., Opleukhin, A.A., Okasheva, N.A.: Researching ecological variability of agronomic traits in potato varieties when they are introduced into severe environments of the Altai mountains. Works Appl. Bot, Gen Breed **176**(1), 110–123 (2015). https://doi.org/10.30901/2227-8834-2015-1-110-123
5. Kordabovsky VY (2016) Promising hybrid populations in potato breeding in the north of the far east Russia. Int. Res. J. **9**(51), 119–120. https://doi.org/10.18454/IRJ.2016.51.131
6. Galichenko, A.P., Fokina, E.M.: The source material estimation of early-maturing group soybeans by adaptability parameters. E3S Web of Conf 254:01028 (2021). https://doi.org/10.1051/e3sconf/202125401028
7. Goryanina, T.A.: Comparative evaluation of winter triticale varieties by adaptability and stability. Achiev Sci. Tech. Agr-Ind Comp. **1**, 37–41 (2020). https://doi.org/10.24411/0235-2451-2020-10107
8. Yusova, O.A., Nicolaev, P.N., Aniskov, N.I., Safonova, I.V.: Adaptability of barley varieties by the weight of 1000 grains under forest-steppe conditions of the Omsk region. Achiev Sci. Technol. Agr-Ind Comput. **2**, 24–28 (2020). https://doi.org/10.24411/0235-2451-2020-10205
9. Rafalsky, S.V., Rafalskaya, O.M., Melnikova, T.V.: Results of comparative evaluation of potato varieties according to productivity and adaptability parameters in the conditions of Priamurye. Vestnik DVORAN **4**, 81–85 (2020). https://doi.org/10.37102/08697698.2020.212.4.013
10. Potapova, G.N., Galimov, K.A., Zobnina, N.L.: Productivity and adaptability of winter rye varieties in the Middle Urals. Achiev Sci. Technol. Agr-Ind Comput. **10**, 28–33 (2020). https://doi.org/10.24411/0235-2451-2020-11004

Changes in the Yield and Quality Grain of Soybean When Using Liquid Fertilizers

Sergey Fokin(✉) and Olga Piletskaya

Far Eastern State Agrarian University, 86, Politeknicheskaya Str., Blagoveschensk 675005, Russia

Abstract. On meadow chernozem-like medium-thick soils of the southern agricultural zone of the Amur region, the influence of the methods of using liquid fertilizers from the Berluga company (Hungary): Nertus Start, Fotosintez and Bor on the productivity and quality grain of soybeans was studied. The research was conducted in 2018–2020 on the experimental field of Far Eastern State Agrarian University in the village of Gribskoie in Blagoveshchensk district. The object of research was a soybean variety of an early ripe group Umka breeding by the FRC All-Russian Scientific Research Institute of Soybean. In the experiment, liquid fertilizers were used in two ways: Nertus Start when pre-sowing seed treatment and Nertus Fotosintez and Bor when treatment vegetative soybean plants in the budding phase. The positive effect of liquid fertilizer Nertus Bor in the treatment of vegetative plants of soybeans and Nertus Start in the pre-sowing treatment of seeds on the yield of soybean has been established. In these variants, the maximum productivity was obtained over the years of research. Results of field studies in the conditions of 2018–2020 on meadow chernozem-like soil of the southern agricultural zone of the Amur region show that feeding with liquid fertilizer Nertus Bor in the budding phase of soybeans contributes to an increase in the protein content in seeds for all years of the experiment. The greatest effect was shown by spraying vegetative soybean plants in the budding phase with liquid fertilizer Nertus Bor.

Keywords: Soybean · Liquid fertilizers · Variety · Yield · Mass of 1000 seeds · Protein content

1 Introduction

In recent years, the oil and fat industry has become one of the leading in the world. Production volumes are increasing in Russia. According to Rosstat in 2019, about 3.04 million hectares were occupied under soy. In the same 2019, a record crop of this crop was harvested - 4.3 million tons. The average soybean yield at the same time is 15 c/ha. The Ministry of Agriculture of the Russian Federation was counting on an increase of at least 10 per cent under that valuable culture. Scientists from all over the world predict that soe play an important role in providing the population with inexpensive and full-fledged nutrition [1].

Soybeans have a unique biochemical composition. Soybeans have a unique combination of nutrients, which makes it possible to consider this culture a leader among

A. Muratov and S. Ignateva (Eds.): AFE 2021, LNNS 353, pp. 246–254, 2022.
https://doi.org/10.1007/978-3-030-91402-8_29

the world's common agricultural crops in terms of protein content (35–45%), balanced and easily soluble, oil (20–25%) having a favorable fat-acid composition, carbohydrates (20–25%), mineral salts (5–6%) and vitamins of groups A, B, C, D, E, K [2].

The use of various elements of the agricultural system (mechanical processing, plant protection, croporots) can partially increase the productivity of arable land, but does not solve the problem of unbalanced mineral nutrition [3].

Современное
сельское хозяйство предусматривает активное использование различных агрохимикатов. Наиболее известными являются минеральные удобрения, применение которых позволяет увеличить продуктивность пашни в очень короткие сроки. Оптимизация минерального питания под определенные сельскохозяйственные культуры наиболее полно раскрывает потенциал почвенного плодородия. Особенно это актуально для земель с низким содержанием питательных веществ и гумуса.

Soybean, in addition to the main nutrients, which are nitrogen, phosphorus, potassium, actively assimilates microelements that play an important role in the physiological and biochemical processes of plants (boron, cobalt, molybdenum, copper). Boron improves metabolism, plays an important role in the processes of fertilization, fruiting and seed ripening. Cobalt enhances the photosynthetic activity of plants, nitrogen fixation, and protective functions against disease. Molybdenum enhances nitrogen fixation, nitrogen metabolism, absorption of mineral nutrients by plants; increases plant resistance to climatic stress. Copper significantly affects the symbiotic fixation of molecular nitrogen by legumes, the formation of nodule bacteria and biochemical processes in nodules, and nitrogen, phosphorus, and carbohydrate metabolic processes in plants. When growing soybeans under different soil-climatic conditions, as indicated by studies of scientists in this field, the highest grain yields of this crop are achieved on highly fertile, organic-rich soils, with a slightly acidic and neutral reaction of the soil solution, which have good aeration and water permeability [4–6].

Studies of the effect of liquid fertilizers on soybean were carried out throughout Russia and abroad.

Research on the use in the technology of soybean cultivation for two years (2018–2019) of micro-fertilizers of the Metallocene series for crop formation and mineral nutrition of two soy varieties was carried out on the basis of FSBOU VO Kazan GAU. Treatment of crops was carried out in the butonization-initiation phase with preparations containing copper, zinc and manganese in the recommended norms. The reaction of soybean varieties to non-root feeding with various fertilizers was established. The maximum protein harvest with soybean harvest was obtained using manganese fertilizer. Application of all micro-fertilizers on both grades of non-indigenous application led to an increase in their content in grain and an increase in economic removal with a crop of macroelements [7].

Field experiments conducted in 2012–2014 in the fields of LLC "Urta Saba" of the Sabinsky municipal district of the Republic of Tatarstan with the drug ZhUSS-2, which contains microfertilizers in a chelated form, activated the work of nodule bacteria, increased the fixation of atmospheric nitrogen, which in turn was accompanied by an increase in the yield of soybean grain by 0,25 t/ha compared to control [8].

As a result of studies carried out in 2019 at the Research and educational production center "Integratsiya" of the Orlov State Agrarian University, it was found that the use of foliar feeding with liquid fertilizers Bazaplant, Nutrimiks and water-soluble fertilizer NPK 20–20-20 had a positive effect on the yield and quality of soybeans of OAK Prudens. The higher yield of soybean grain was due to the large number of beans and seeds per plant [9].

When testing liquid fertilizers in 2018, KAS-32 KAS-32 with a sulfur content and a feedstock solution containing sulfur (RPS) in field and production experiments of the Agroakademie MIP (Samara region, Kinelsky district, Ugorye village) it has been found that the use of sulfur containing fertilizers on soils with a low sulfur content has led to a significant improvement in the development of soybean plants and the formation of a higher yield and quality compared with the introduction of ammonium nitrate. The maximum yield effect was obtained with the introduction of fertilizer KAS-32 + RPC, the yield, compared with the control, increased by 6.9 c/ha or by 31%. The balance of the nutrition of soybean with nitrogen and sulfur was pain-neck important than the unilateral use of ammonium nitrate [10].

In the 2019 experiments, on the basis of the NOPC "Integration," FSBOU VO Oryol GAU, in conditions of different mine nutrition, the biological efficiency of fertilizer systems, yield and quality of soybean grain were determined. As a result of the studies, the use of mineral fertilizers together with LCD contributed to increase biomass growth, leaf area and increase in raw protein grain compared to control [11].

In studies on the use of microfertilizers conducted by V.N. Bosak (Republic of Belarus), when cultivating soybeans on sod-podzolic sandy loam soil, foliar treatment with microfertilizers in the budding phase increased grain yield by 1,8–3,7 c/ha, protein content – by 2,8–5,1% with a total grain yield 25,8–27,7 c/ha, protein content 30,7–33,0%, fat – 17,5–18,2% [12].

In studies conducted on the gray soils of the Almaty region near Taldykorgan (Kazakhstan) in 2015–2016 on the use of liquid environmentally biofertilizers "BioEcoGum" and "Rizovit-Aks" for soybean culture, a positive effect established of liquid preparations on growth, development, productivity and plant yield. The average yield of soybean obtained in fertilized variants is 12,7–31,8 c/ha higher than the control without treatment [13].

The aim of the research is to study the effect of the use of liquid fertilizers Nertus Start, Fotosintez and Bor on the yield and quality of soybean grain.

2 Objects and Methods of Research

The research was conducted in 2018–2020 on meadow chernozem-like medium-thick soils in the southern agricultural zone of Amur region on the experimental field of Far Eastern State Agrarian University in the village of Gribskoie in Blagoveshchensk district.

The object of research was a soybean variety of an early ripe group Umka included in 2015 in the State Register of Breeding Achievements of the Russian Federation for use in the Far East region. The originator of the variety is the All-Russian Scientific Research Institute of Soybean. The advantages of the variety are high productivity, early ripening, drought resistance, optimal protein and fat content in seeds 38,8… 41,1 and

22,6… 22,8%, respectively. When the plants are over-mature, the beans do not crack. Medium resistant to bacterial and fungal diseases. Average seed yield 3,0 t/ha, potential 3,84 t/ha.

The studied fertilizers, in addition to amino acids of natural origin, contain microelements, and each of them performs their own functions.

Nertus Start is a complex liquid fertilizer for treating agricultural seeds. The fertilizer compensates for the deficiency of nutrients in the initial phases of plant growth, ensures the development of a powerful root system and is recommended to be used for treatment agricultural seeds, together with insecticidal and fungicidal treatment agents.

Nertus Fotosintez is a universal liquid fertilizer with a set of micro- and macroelements in an easily available chelated form for foliar feeding of crops. Provides plants with essential nutrients for optimal performance. The structure of the fertilizer is closest to natural compounds based on polyphosphates.

Nertus Bor is a liquid fertilizer with a high boron content in organic form, intended for foliar feeding. Provides the needs of plants in boron, promotes more active formation of pollen grains, the movement of plastic substances from vegetative organs to generative ones, thereby determining an increase in the yield of the quality of the products obtained. Ethanolamine used in the preparation is a natural antistressant for plants.

Agrotechnics in the experiments is generally applicable in the agricultural conditions of Amur region. The sowing was performed by SN-1,6 seeder with the sowing rate of 500 000 viable seeds per 1 hector (ha). Treatment of soybean seeds before sowing was carried out with liquid fertilizer Nertus Start at the rate of 0,4 l/t of seeds and spraying of vegetative plants in the budding phase with liquid fertilizers Nertus Fotosintez and Bor at the rate of 2 l/ha.

The experience was designed according to the following scheme (Table 1):

Table 1. Scheme of the field experiment.

Experiment variant
control without fertilizers
N_7P_{30} (background)
background + Nertus Start (seed treatment)
background + Nertus Start (seed treatment) + Fotosintez (spraying in the budding phase)
background + Nertus Start (seed treatment) + Fotosintez + Bor (spraying in the budding phase)
background + Nertus Fotosintez (spraying in the budding phase)
background + Nertus Fotosintez + Bor (spraying in the budding phase)
background + Nertus Bor (spraying in the budding phase)

The repetition in the experiments is 4-fold, the accounting field area is 24 m^2.

Statistical data processing was carried out according to the method of B.A. Dospekhov. Analysis of dispersive of digital material was performed and SSD was determined (smallest significant difference). SSD is a value that indicates the limit of possible random fluctuations in an experiment, which is considered significant at the 5% significance level (SSD_{05}).

Statistical analysis allows to systematize extensive digital material and prove the reliability of experimental data.

3 Research Results and Their Discussion

As a result of the research carried out in 2018–2020 to study the effect of liquid fertilizers in the cultivation of soybean on meadow chernozem-like soil, it was envisaged to determine their effect on the macroelement background.

Improving the nutrition of soybean with the use of liquid fertilizers, methods of seed treatment before sowing and spraying vegetative plants culture in the budding phase provided an increase in the yield of soybean on average for three years from 0,28 to 0,49 t/ha (17,8–31,2% in relation to the control without the use of fertilizers).

When comparing the yield of soybean by years of research, some differences should be noted. In 2019 saw the highest yield compared to 2018 and 2020. The maximum increase in relation to the control this year was determined in the variant with the use of liquid fertilizer Nertus Bor on vegetative soybean plants in the budding phase – 0,43 t/ha.

In 2018, the grain yield of the studied crop was noted in the control without the use of fertilizers – 1,53 t/ha. The largest value of this indicator was also determined in the variant with the use of liquid fertilizer Nertus Bor on vegetative soybean plants in the budding phase – 2,08 t/ha, that relatively above control variant on 0,55 t/ha and background on 0,48 t/ha.

The smallest soybean yield, in almost all variants of the experiment, was obtained in 2020. Agrometeorological conditions of this year were unfavorable for the cultivation of soybean and significantly differed from the average long-term norm, mainly characterized by waterlogging (Table 2).

An uneven temperature background was observed throughout the growing season. The precipitation was uneven – in some months periods of excessive moisture prevailed, which led to the death of plants. The maximum yield value this year was noted in the variant with the use of liquid fertilizers Nertus Start when processing seeds before sowing and Nertus Fotosintez during the growing season, as well as in the variant with the application of Nertus Fotosintez during vegetation – 1,92 t/ha.

Thus, the positive effect of liquid fertilizer Nertus Bor in the treatment of vegetative plants of soybean and Nertus Start in the pre-sowing treatment of seeds on the yield of soybean has been revealed.

Along with the yield of soybean, the quality of the resulting grain is of great importance from an economic point of view.

The mass of 1000 seeds is one of the main economic indicators. Many environmental factors affect mass of 1000 seeds. In order to increase the weight of the grain, it is necessary to provide the plants with sufficient moisture and nutrients.

When using liquid fertilizers, the assessment of soybean seeds showed that the best average for 2018–2020 by the mass of 1000 grains (189,3g), there was a variant with the use of Nertus Fotosintez during vegetation with an indicator in the control of 173,3 g.

Table 2. Influence of the use of liquid fertilizers on the yield of soybean, t/ha.

Experiment variant	Year			Average for 3 years
	2018	2019	2020	
control without fertilizers	1.53	1.86	1.33	1.57
N_7P_{30} (background)	1.60	1.97	1.79	1.79
background + Nertus Start (seed treatment)	1.78	2.19	1.83	1.93
background + Nertus Start (seed treatment) + Fotosintez (spraying in the budding phase)	1.94	2.15	1.92	2.00
background + Nertus Start (seed treatment) + Fotosintez + Bor (spraying in the budding phase)	2.06	2.19	1.84	2.03
background + Nertus Fotosintez (spraying in the budding phase)	1.96	1.98	1.92	1.95
background + Nertus Fotosintez + Bor (spraying in the budding phase)	1.69	2.09	1.76	1.85
background + Nertus Bor (spraying in the budding phase)	2.08	2.29	1.81	2.06
SSD_{05}	0.30	0.24	0.24	0.19

From the use of Nertus Start in seed treatment + Fotosintez during vegetation and Nertus Bor, almost the same weight of 1000 grains was noted – 182,8 and 182,4 g, that above control without the use of fertilizers by 9,5 and 9,1 g respectively (Table 3).

The mass of 1000 seeds varied over the years of the study, and the largest indicator were determined in 2019. In the control variant, the mass of 1000 seeds was 211,8 g. The maximum value of this indicator was noted in the variants with the use of Nertus Start, Fotosintez and Bor together and Nertus Bor during vegetation – 219,2 g, that above control without the use of fertilizers by 7.4 g.

The smallest values were obtained in 2020 due to the prevailing unfavorable agrometeorological conditions for the growth and development of soybeans in this year. The largest mass of 1000 seeds was determined in the variant with the use of Nertus Fotosintez in the form of spraying vegetative soybean plants in the budding phase – 181,0 g, that exceeded the control variant by 42,3 g.

In 2018 the value of the mass of 1000 soybean seeds in the control was 169,4 g. The maximum indicator was noted in the variant background + Nertus Start (seed treatment) + Fotosintez (spraying in the budding phase) – 182,1 g, that is higher than the control variant by 12,7 g.

Thus, according to the data of determining the mass of 1000 seeds, a positive effect of liquid fertilizers was noted, as when applied together and separately, increasing this indicator of the quality of soybean grain relative to the control variant without the use of fertilizers.

Table 3. Influence of the use of liquid fertilizers on the mass of 1000 soybean seeds, g.

Experiment variant	Year			Average for 3 years
	2018	2019	2020	
control without fertilizers	169.4	211.8	138.7	173.3
N_7P_{30} (background)	169.5	211.2	158.0	179.6
background + Nertus Start (seed treatment)	179.7	212.4	134.4	175.5
background + Nertus Start (seed treatment) + Fotosintez (spraying in the budding phase)	182.1	215.1	151.3	182.8
background + Nertus Start (seed treatment) + Fotosintez + Bor (spraying in the budding phase)	164.1	219.2	134.1	172.5
background + Nertus Fotosintez (spraying in the budding phase)	176.9	209.9	181.0	189.3
background + Nertus Fotosintez + Bor (spraying in the budding phase)	170.1	210.7	129.8	170.2
background + Nertus Bor (spraying in the budding phase)	173.1	219.2	155.0	182.4
SSD_{05}	9.2	5.8	2.8	6.0

Due to its rich and varied chemical composition, soybean are widely used as food, fodder and industrial crops. According to FAO, soy protein is 3–4 times cheaper than wheat, corn or barley, pea protein – 7 times, alfalfa – 8 times, fodder yeast and fish meal protein – 29 times.

The use of liquid fertilizers in the treatment of soybean seeds before sowing (Nertus Start) and during vegetation in the budding phase (Nertus Fotosintez, Bor) contributed to an increase in the protein content in seeds in all years of research.

On average, over the years of the field experiment, the protein content in soybean seeds increased relative to the control variant from the use of liquid fertilizers by 0,1–1,4%. The maximum value of this indicator was noted in the variant with the use of Nertus Bor on vegetative plants – 41.0% (Table 4).

In 2018–2020 the effect of the studied liquid fertilizers on the level of protein accumulation in seeds was approximately equal, but in 2019 they contributed to a significant increase in the protein content in seeds by 0,2–2,2% in relation to the control. This year, the largest protein harvest was obtained in the variant with the use of Nertus Bor during vegetation plants – 42,9%. In 2018 and 2020, the maximum protein content of soybean seeds was also determined in this variant – 40,2 and 40,0%, respectively.

Results of field studies in the conditions of 2018–2020 on meadow chernozem-like soil of the southern agricultural zone of the Amur region show that feeding with liquid fertilizer Nertus Bor in the budding phase of soybeans contributes to an increase in the protein content in seeds for all years of the experiment.

Table 4. Influence of the use of liquid fertilizers on the protein content in soybean grain, %.

Experiment variant	Year			Average for 3 years
	2018	2019	2020	
control without fertilizers	39.6	40.7	38.5	39.6
N_7P_{30} (background)	39.8	40.9	38.5	39.7
background + Nertus Start (seed treatment)	39.7	41.1	38.2	39.7
background + Nertus Start (seed treatment) + Fotosintez (spraying in the budding phase)	39.3	41.9	39.3	40.2
background + Nertus Start (seed treatment) + Fotosintez + Bor (spraying in the budding phase)	39.6	41.7	38.4	39.9
background + Nertus Fotosintez (spraying in the budding phase)	38.4	41.5	38.6	39.5
background + Nertus Fotosintez + Bor (spraying in the budding phase)	38.5	42.3	38.9	39.9
background + Nertus Bor (spraying in the budding phase)	40.2	42.9	40.0	41.0
SSD_{05}	0.6	0.6	1.2	0.8

4 Conclusion

Thus, it should be noted that the use of each type of liquid fertilizer separately and their combined use had a positive effect on the yield and individual indicators of the quality of soybean grain, increasing the studied indicators relative to the control variant without the use of fertilizers.

Results of field studies in the conditions of 2018–2020 on meadow chernozem-like soil of the southern agricultural zone of the Amur region show that most effective was feeding with liquid fertilizer Nertus Bor during vegetation soybean plants in the budding phase, that contributes to an increase in yield, mass of 1000 seeds and protein content in grain for almost all years of the experiment.

References

1. Eliseeva, L.V., Glinsky, I., Filippova, S.V.: Humic preparations impact on soybean seeds productivity and quality. Bull. KrasGAU **7**, 3 (2021). https://doi.org/10.36718/1819-4036-2021-7-3-10
2. Lozhkin, A.G., Yeliseyev, L.V., Filippov, S.V.: Influence of ways of crops and microfertilizers on efficiency of soy. Bull. Ulyanovsk Agric. Acad. **1**, 38–44 (2020). https://doi.org/10.18286/1816-4501-2020-1-38-44
3. Sherstobitov, S.: The results of the differential mineral fertilization in the automatic mode according to the task map. IOP Conf. Ser: Mater. Sci. Eng. **537**, 062011 (2019)

4. Demina, O.N., Eremina, D.V.: Influence of the level of mineral nutrition on the elements of the structure of the yield of spring wheat in the forest-steppe zone of the trans-urals. Bull. KrasGAU **3**, 34–40 (2021). https://doi.org/10.36718/1819-4036-2021-3-34-40
5. Khatkov, K., Mamsirov, N.I.: The effect of mineral fertilizers and methods of basic soil tillage on the productivity of new promising soybean varieties. New Tech. **5**, 87–94 (2020). https://doi.org/10.47370/2072-0920-2020-16-5-87-94
6. Shabaldas, O.G., Pimonov, K.I., Esaulko, A.N., Borodychev, V.V., Vaytsekhovskaya, S.S.: Agrochemical and economic assessment of use of mineral fertilizers and Rizotorfin on grades of soy of various groups of ripeness in the conditions of irrigation. NV AUK News **2**, 209–222 (2021). https://doi.org/10.32786/2071-9485-2021-02-22
7. Sharipova, G.F., Kolesar, V.A., Safin, R.I.: Effectiveness of application of fertilizers with trace elements on various soy variety. Fertility **3**, 9–11 (2020). https://doi.org/10.25680/S19948603.2020.114.02
8. Prokina, L.N.: Dependence of yield and quality of soybean grain on macro and micro fertilizers against the background of liming aftereffect. Agric. Sci. Euro-NorthEast **4**, 417–424 (2020). https://doi.org/10.30766/2072-9081.2020.21.4.417-424
9. Bobkova, Y.: Reaction of OAK Prudence soybeans to foliar fertilization with macro-and micro-fertilizers in the conditions of the Orel region. Bull. Agrarian Sci. **5**, 11–18 (2020). https://doi.org/10.17238/issn2587-666X.2020.5.11
10. Milyutkin, V.A., Petrov, A.M.: Technical and technological application of liquid nitrogen and nitrogen-sulfur fertilizers based on KAS-32 in crops of grain and leguminous crops. Niva Volga Region **4**, 79–84 (2019). https://doi.org/10.36461/NP.2019.52.3.012
11. Shiryaeva, N.A., Beregovaya, Y., Petrova, S.N.: Efficiency of use of complex mineral fertilizers in a soy agrocnosis. Mess Agric. Sci. **5**, 66–72 (2020). https://doi.org/10.17238/issn2587-666X.2020.5.66
12. Bosak, V.N.: The use of microfertilizers in the cultivation of leguminous plants. Agrochem. Bull. **2**, 24–25 (2012)
13. Seytmenbetova, A.T., Kan, V.M., Asimzhanov, N.: Effect of liquid biologics "Bioecogum" and "Rizovit-AKS" on productivity of soya of "Decabit" variety in conditions of serous soils of Almaty region. Soil Sci. Agrochem. **1**, 42–49 (2020)

Cost Management of Soybean Production in the Amur Region, Russia

Anastasiia Boiarskaia[1], Boris Boiarskii[2(✉)], Mikhail Sinegovskii[2], and Hideo Hasegawa[1]

[1] Niigata University, 8050 Ikarashi 2-no-cho, Nishi-ku, Niigata 950-2181, Japan
[2] All Russian Scientific Research Institute of Soybean, Amur Region, Ignatievskoe Shosse 19, Blagoveshensk 67-5000, Russia

Abstract. Soybean production in Russia occupies one of the main niches in agriculture. An increase in the gross harvest and an increase in the efficiency of cultivation of agricultural crops within the framework of the state strategy of import substitution and strengthening food security are priority directions for developing the Russian agro-industrial complex in the context of the new agri-food policy. In Russia, the price of soybeans increases annually due to the ban on GM soybean production, the unstable situation in the world soybean market, the increase in Russian exports, and the interest of the world importer of soybean - China. Over 65 years, there have been significant changes in the structure of costs for soybean production in such items as the share of costs for seeds, agrochemicals, and fuel. Nowadays, among the considered enterprises producing soybeans in the Amur Region, on average, total costs were $ 406 per hectare in 2019. For the enterprise's profitability, it is important to properly invest in costs to obtain high yields, which will bring profit to producers. This article analyzed the pricing of 4 farms in the Amur Region and overviewed the situation of the soybean production. The authors made a prospective picture of soybean production in Amur Region and its price structure based on the survey data and statistical data.

Keywords: Agriculture · Far East · Price · Russia · Soybean

1 Introduction

For the last period, there is a clear upward trend in the expansion of soybean production due to multipurpose use as human food, livestock feed, industrial purposes, and as a source of bioenergy [1]. The soybean seeds contain a high protein content with rich biochemical content with great potential to meet the demand of protein deficiency challenge and considerably cheaper than other sources of high protein agriculture commodities [2]. Globally, the total cultivated area accounted for 120 million ha and total production of 333 million tons in 2019, with an average yield in the world of 2.7 tons per hectare, according to FAO statistics.

In Russia, the soybean-producing industry has trended upwards, and it seems it will continue in the same direction. That is because of soybean production's profitability, and

A. Muratov and S. Ignateva (Eds.): AFE 2021, LNNS 353, pp. 255–263, 2022.
https://doi.org/10.1007/978-3-030-91402-8_30

constant demand from the China and processing sector of the soybean crop and livestock sectors demand in feed. In this paper, the Amur region soybean production industry is observed since its national leader lasts a long time. The costs of producing a product are directly related to the price of the product. Price, in turn, plays an important role in the formation of competitive advantage in the market over competing firms. It is making the product more attractive to consumers accordingly. Soybean production in Amur Region involves different farming inputs [3].

Moreover, Amur Region's strategic location is also a feature that makes it a suitable opportunity for Russia to expand soybean export commodity and processed soybean products in addition to farmers' preference of the profitable crop harvesting. However, realizing the above potentials and strategic plan is to expand soybean production vertically and horizontally, unfortunately, without understanding its production economics and resource use efficiency [4].

In this paper, the authors divided expenses into the cost of seeds, agrochemicals, fuel costs, machinery, labor, and other costs. Accordingly, this research focused on identifying and estimating production costs and their components under four typical soybean producer company's conditions in Russia's central soybean production area, Amur Region, for the 2017–2019 period.

2 Materials and Methods

2.1 Study Site

The Amur Region is a federal subject located in the southeast of the Asian part of the Russian Federation and is part of the Far Eastern Federal District (Fig. 1). The region's total area is 361.9 thousand square kilometers (11.7% of the area of the Far Eastern Federal District). The population in the region is 793.2 thousand people, including an urban population of 535.6 thousand people and a rural population of 257.6 thousand people, with a population density of 2.2 people per 1 square kilometer, according to Amur statistical service. Its administrative center is Blagoveshchensk city. Amur Region borders with Yakutia in the north, with the Khabarovsk Territory in the east, with the Jewish Autonomous Region in the southeast. With the Trans-Baikal Territory in the west. The southwestern border of the Amur Region is the Russian Federation's state border and borders the People's Republic of China. The closest Chinese city, Heihe divided from Blagoveshchensk city by the Amur River. The closest distance is 750 m [5].

The plant growing and animal husbandry represent the agro-industrial complex of the Amur Region. The area of agricultural land is 2369.9 thousand hectares (6.7% of the entire territory of the region), including 1532.8 thousand hectares of arable land, 280.7 thousand hectares of hayfields, 354.0 thousand hectares of grazing, 195.4 thousand hectares of long-term uncultivated arable land. The cultivated area in the region has increased by 76% under all crops in farms of all categories from 2008 to 2018. For the most part, this growth was due to an increase in acreage for soybeans due to the plowing of fallow lands and reduced acreage for other crops. So, from 2014 to 2018, the sown area under grain crops increased by 5.0%, under fodder crops decreased by 4.5%, vegetables - by 6.7%, and under potatoes - by 9.3%. At the same time, the size of soybean acreage increased by 28.2%.

Fig. 1. The location of the soybean producers on the Far East of Russia, 1 – Amur Region, 2 – Primorskii krai, 3 – Khabarovskii krai.

2.2 Data Analysis and Data Survey

In this study, the economic-statistical, economic-mathematical, monographic, computational, and constructive research methods were used. The data for this study were taken from statistical government reports of the Amur Region. The analysis is based on a study of relevant documents: government regulations, government programs related to the agricultural sector, and an official international statistics database.

For the survey conducted in the Amur region, data were collected from primary sources through field survey, particularly at four different agricultural companies which are Ltd "Amurskaya zernovaya company", Joint-stock company "Luch", Kolkoz "Volkovo", Ltd "Priamurie". The surveyed variables included: net gross, hectares planted, average yield, and cost structure spend per ha (i.e., seeds cost, machinery, agrochemicals, labor, fuel) for 2017–2019.

3 Results and Discussion

3.1 Dynamics of Price Changes and Cost Production for Soybeans

For agricultural enterprises, the components of the economic efficiency of soybean production are price and production cost. For processing enterprises, soybean acts as raw material. Therefore, its price is formed by these enterprises, taking into account market prices for soybean processed products, processing costs, market conditions, and products of its processing (determined mainly by imports). The existing price level reflects the action of many interdependent factors and the competitive struggle.

Over the past 20 years, the price of soybean has fluctuated in the world. According to the Chicago Mercantile Exchange, soybean prices in 2020 compared to 2000 have grown 2.5 times. The highest levels of soybean prices reached in August 2012 $ 622 per ton. In recent years, there has been a decrease in world prices for soybeans, largely due to the high volumes of the world's gross harvest. According to the World Bank, in 2019, nominal prices for soybeans amounted to USD 368.3/t, for oil - USD 764.8/t, for meals - USD 347.0/t. High prices for soybeans in the world stimulate producers to increase their acreage and intensify production to obtain a larger yield.

In Russia, since 2011, there has been a clear upward trend in average annual prices for soybeans. This is due to the development of the livestock industry, which has grown rapidly over the past decade and created the need for an equally large amount of animal feed. Moreover, the annual increase in prices is associated with expanding the soybean processing plant in Russia.

Figure 2 shows the average annual prices for soybean in the Far East and the Amur Region since 2015. Over 15 years, the price had grown 3.5 times in Russia, from 91 US dollars per ton in 2005 compared to 2020 when the average price in Russia was USD 372 per ton. There was no large price discrepancy. However, there is a rapid increase in soybean production volume in the Far East and the Amur Region due to the growing demand for soybean from China. Therefore, from 2015 to 2020, there was an increase in soybean prices by an average of 35 US dollars per year.

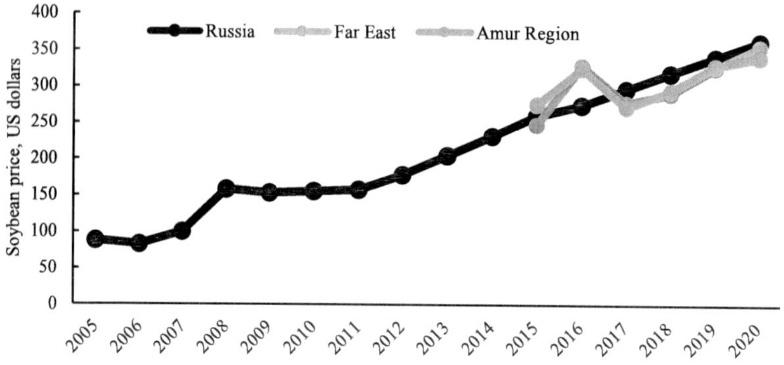

Fig. 2. The average price of soybean in Russia from 2005 to 2020.

According to FAS, three main factors caused the rise in prices on the Russian domestic market of soybeans and soybean meal. The first is the changes in the regulatory framework. In 2016, the Rosselkhoznadzor banned the import of hybrid GMO soybean meal into Russia. It turned out to be challenging to replace the falling volume of soybean meal supplies since GMO soybeans account for about 90% of the world harvest, and imports of its processed product provided up to 20% of its consumption in Russia. In this regard, the volume of available soybeans decreased, which was the first reason for the increase in prices for domestic products.

The second is the unfavorable situation in the global market. From time to time, Argentina, being one of the largest soybean producers, is losing a significant part of the harvest due to weather conditions, which has led to an increase in prices for soybeans and meals on commodity exchanges since March 2018. In addition, after the reorientation of Chinese processors to new markets due to the trade war with the United States, the cost of Latin American soybeans for Russian buyers increased. Additional reasons include the devaluation of the ruble.

Furthermore, the third is the export factors. Non-GMO soybeans grown in Russia have been increasing demand in China for several years. For example, at the end of 2017, 89% of all Amur soybeans were supplied to China, which was an important reason for the rise in prices. Following the global upward trend in prices, Russian soybean gradually became more expensive and reached an average value of 435 US dollars per ton in the third quarter of 2018, 33% higher than the average monthly cost of soybeans first quarter of 2018. As expected, the increase in prices for raw materials was accompanied by an increase in prices for processed products [6].

The FAS sees several ways to stabilize the market for further development: first of all, it is proposed to develop exchange trading to form fair market prices. Secondly, it is necessary to adopt guidelines for molecular genetic studies of GMO organisms promptly. In addition, it is necessary to create conditions for the development of breeding aimed at soybean seed varieties with high protein and oil content, providing a high quality of the protein component in soybean raw materials and meals [7, 8].

3.2 Comparison of Cost Structure Components of the USSR Period and Nowadays Typical Farm

With the 55 years, the cost items structure of soybean production has changed dramatically. The structure showed in percentage for the appropriate analysis since the prices are not comparable in monetary terms. The share for machinery expenses has a significant share and did not change much, 25.8% in 1963 to 24.6% in 2018. The great change has occurred in seeds, agrochemicals, fuel cost items shares for soybean production in a typical company. In the USSR, the lion share was for the seed item cost after half a century it declined to 11% in the 2018 year (Figs. 3 and4). The reason for that can be the significant work of the researchers and breeders of soybean varieties to make seed costs more affordable prices and productive cultivation.

Comparing analysis of soybean production costs now and then show agrochemicals cost in nowadays company to be about two times above that of the USSR period. Recent studies show that fertilization has a positive effect on increasing yields and, as a result of profit, the share of agrochemicals has increased several times. As for the share of usage of fuel cost, it has increased through time from 3.9% of the total cost to 16.4% accordingly. The share of labor and total cost has not changed significantly over time, staying at almost the same level (Figs. 3 and4.).

3.3 Soybean Cost Structure in Case of Amur Region Farmers Soybean Crop Producers in 2017–2019

As shown in Fig. 5 in the Amur Region, the average soybean production costs per hectare are increasing with the average price for soybean crop year by year. For three years, the Kolkhoz "Volkovo" has the highest average cost per one-hectare cost of production. Ltd "Priamurie", oppositely during the observation period, had the lowest average costs amount in monetary terms. The mean index for four companies accounted for 345 US dollars per ha in 2017, 339 US dollars per ha in 2018, and 403 US dollars per ha in 2019. According to the data, the average cost is increasing year by year, it can be associated with the increase of fuel price and applied agrochemicals.

The above figures of soybean production cost structure in four typical producer companies in the Amur region were represented for three years from 2017 to 2019. From Fig. 6, it can be seen that the higher yield is in Ltd "Amurskaya zernovaya company"

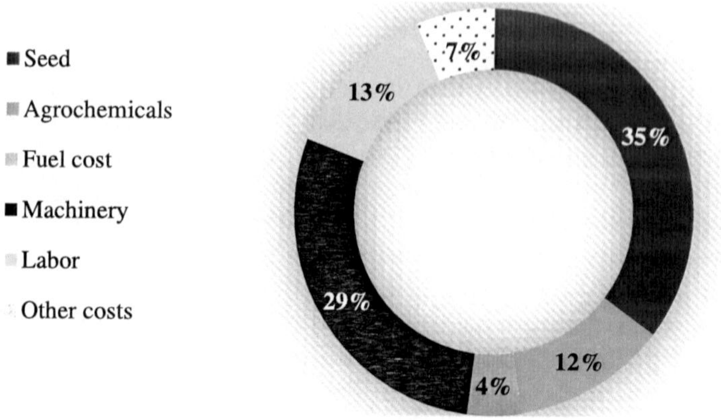

Fig. 3. The cost structure of the soybean production in a typical farm in USSR 1983 year.

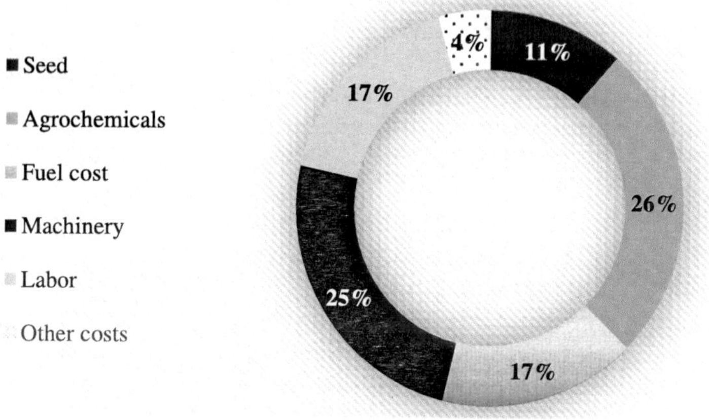

Fig. 4. The cost structure of the soybean production in a typical farm in the 2018 year.

during the observed period. Agrochemicals cost item was the main input in soybean production, following the cost associated with machinery for Ltd "Amurskaya zernovaya company". The lowest yield index was in Ltd "Priamurie". For the period under review, in all four companies, the largest share of expenses falls on agrochemicals (fertilizers and pesticides). Unlike other companies in Ltd "Amurskaya zernovaya company". The authors also noticed that machinery cost accounted for a significant part of total expenses.

The expenses for labor were minor in all companies except for kolkhoz "Volkovo". In the structure of the prime cost of soybeans, the largest share belongs to the costs associated with agrochemicals. Also, the price for fertilizers and plant protection is growing every year. However, some farms still apply a large amount of fertilizer. With an increase in the cost of agrochemicals, it shows an increase in soybean yield in practice. The next item of costs that takes a significant share in the cost structure are expenses assisted with fuel. These costs are directly related to the provision of a tractor fleet. Differences in the cost structure of enterprises do not differ significantly. However,

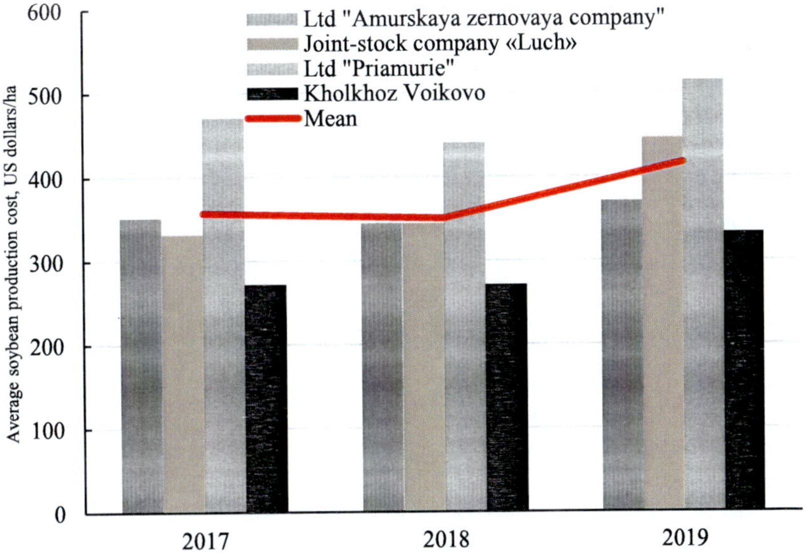

Fig. 5. Average soybean production cost in 4 typical companies in Amur Region, Russia for the 2017–2019 years.

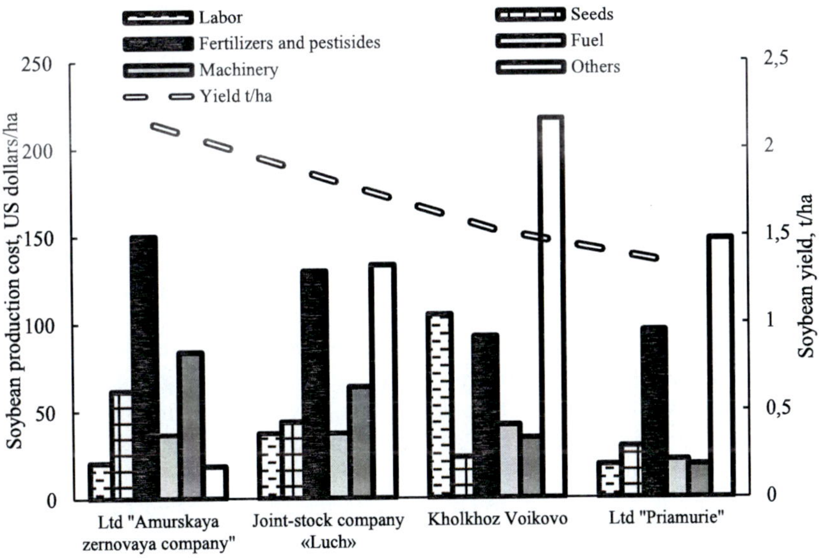

Fig. 6. Soybean cost structure in case of Amur Region farmlands in 2019.

there are significant differences associated with the technology used in the production of soybeans. For example, weed control can be mechanical or chemical, which entails increasing or decreasing certain costs, respectively. In this regard, enterprises need to pay more attention to decision-making related to the choice of soybean cultivation technology based on the initial material and technical equipment for enhancing soybean production and obtaining a greater financial result.

4 Conclusions

This study revealed the world prices for main soybean trend, the situation in the Russian market, and Amur Region, comparing analysis of cost structure of the USSR period soybean producers and nowadays, moreover the average costs of soybean production in the Amur Region in the 2017–2019 year. The soybean industry in Amur Region has covered more than 35% of the total cultivated land in the country. In this paper, the authors surveyed that the planted area of observed farmlands ranges from 4 thousand hectares to 24 thousand hectares.

The price of soybean plays a great role in soybean production. High prices for soybean in the world stimulate the production of this crop in many countries as well as in Russia. In Russia, the average annual price for soybeans is growing from year to year. In 2020 has reached 372 dollars per ton. The reasons for the growth may be the following factors: the ban on the cultivation of GMO soybeans in Russia since 2016, the unfavorable situation in the world market, also an increase in soybean exports in Russia and, the great interest from China in Russian soybeans, which is the absolute world leader in soybean imports.

A comparative analysis of the structure of costs of the production of soybeans in modern soybean-producing enterprises and of the USSR time has revealed considerable changes in the following cost items as seeds, agrochemicals, fuel costs. At the same time, the cost structure of the soybean production associated with machinery, labor, and other costs has not changed dramatically.

An analysis of the cost structure of four soybean farmlands in the Amur Region showed that the average annual cost of producing one hectare of soybeans varied from $ 345 per hectare in 2017 to $ 403 per hectare in the 2019 year. The Kolkhoz Volkovo had the lowest total cost of producing one hectare of soybeans among the observed farmlands, while the Ltd "Priamurie" company, on the contrary, had the highest.

In connection with the above, it is worth emphasizing the economic issues. The efficiency of soybean production should be given the greatest attention. Agricultural enterprises should take a detailed approach to the formation of production costs and primary to the choice of soybean production technology. The level of productivity allows covering the costs incurred and earning funds for further expansion, and the selected reproduction and technology depend on production costs.

In the future, it is important to conduct a factor analysis of cost items and their impact on profit-making to fully understand the situation and that will help local farmers and enterprises to choose the direction of solutions that increase the profitability of the industry and the general welfare of agriculture in the region.

References

1. Boiarskaia, A.I., Hasegawa, H., Boiarskii, B.S., Lyude, A.V.: History of development of Soybean Production in the Amur Region and Far East District in the USSR. IOP Conf. Ser. Earth Environ. Sci. **548**(2), 022079 (2020). Doi: https://doi.org/10.1088/1755-1315/548/2/022079
2. Boiarskii, B., Hasegawa, H., Lyude, A., Kolesnikova, E., Sinegovskaia, V.: Current situation and perspectives for soybean production in amur region, Russian federation. AMA, Agric. Mechanization Asia, Africa Latin Am. **51**(2), 33–38 (2020)
3. Karlova, N., Serova, E.: Prospects of the Chinese market for Russian agri-food exports. Russian J. Econ. **6**(1), 71–90 (2020). https://doi.org/10.32609/j.ruje.6.50824
4. Potenko, T.A., Zhupley, I.V., Schmidt, Y.I., Kvashko, L.P.: Agricultural policy and import substitution opportunities in the Russian Far East. IOP Conf. Ser. Earth Env. Sci. **548**(2) (2020). https://doi.org/10.1088/1755-1315/548/2/022011
5. Sinegovskii, M.O., Malashonok, A.A., Sinegovskaya, V.T.: Assessment of the export potential of Russian soybeans. IOP Conf. Ser. Earth Env. Sci. **677**(2) (2021). https://doi.org/10.1088/1755-1315/677/2/022025
6. Sinegovskii, M., Yuan, S., Sinegovskaya, V., Han, T.: Current status of the soybean industry and research in the Russian Federation. Soybean Sci. **37**(01), 1–7 (2018)
7. Sinegovskiy, M.O.: Perspectives of soybean production in the far East Federal District. Vestnik Rus. Agric. Sci. **1**, 13–16 (2020). https://doi.org/10.30850/vrsn/2020/1/13-16
8. Yao, H., Zuo, X., Zuo, D., Lin, H., Huang, X., Zang, C.: Study on soybean potential productivity and food security assessment in China under the influence of the COVID-19 outbreak. Geogr. Sustain. **1**(2) (2020). https://doi.org/10.1016/j.geosus.2020.06.002

Check for updates

Peopleware of the Region's AIC Industry

Lyudmila Chausova[1](✉), Nikolai Chausov[2], and Natalia Ogorodnikova[3]

[1] Kaluga Branch of the Russian State Agrarian University of the Timiryazev Moscow Agricultural Academy, 27, Vishnevsky Str., Kaluga 248007, Russia

[2] Kaluga State University named after K.E. Tsiolkovsky, 26, Stepan Razin Str., Kaluga 248023, Russia

[3] Ministry of Agriculture of the Kaluga Region, 5, Vilonova Str., Kaluga 248000, Russia

Abstract. This study examines the staffing of the agro-industrial complex based on regional statistics. The results show that a set of staffing measures for agro-industrial complex enterprises is being implemented in the Kaluga region. It includes: improving training and retraining of personnel; Continuous professional development; public funding for the main tasks of staffing of the agro-industrial complex; Improving social living conditions in the countryside; Staffing information support; enshrining qualified personnel in the countryside. Studies have shown that the measures applied are not sufficient to achieve the goals of staffing in the region. The level of staffing in 2020 was 90.7%. To rectify the situation, an additional set of personnel measures in the region has been substantiated. The scientific novelty of the study is to address the problem of human resources in the new conditions, which are manifested by the presence of large investors in the agricultural market. Practical significance is in the possibility of using the recommendations to ensure the development of agriculture in other regions.

Keywords: Personnel support of the agro-industrial complex · Training and retraining of personnel · Professional development · Personnel consolidation · Personnel security · A set of personnel support measures

1 Introduction

The level of development of the rural economy is largely determined by the availability and efficient use of modern jobs, which in turn leads to the formation of a rational employment structure for the rural population and an improvement in its standard of living. The key success factor here is the staffing of the branch.

The term "peopleware" became widely used in modern scientific literature after the publication of its work in 1987 by Tom De Marco and Timothy Lister. They have considered the problem of personnel influence on software product development [1]. In this context, human resources are seen as a combination of factors in production activities, along with information, technology, programmes and others. The another approach to define the essence of the "staffing" category is related with consideration of solving staffing problems mainly for individual business entities through the purposeful management impact on the processes of working with staff. At the same time, the purpose

A. Muratov and S. Ignateva (Eds.): AFE 2021, LNNS 353, pp. 264–274, 2022.
https://doi.org/10.1007/978-3-030-91402-8_31

of management impact is to ensure that the organization has human resources fully in line with its needs, both quantitatively and qualitatively. Also, to such works can be attributed, for example, the publications of both domestic and foreign researchers, whose work focuses on socio-psychological, motivational factors, as well as improving the working conditions of employees [2–8].

In the context of this study, we will use The Definition of Anisimova N.Yu. [9]. According to which, staffing is an inter-industry process of the formation and efficient functioning of the labour force, aimed at the quantitative and qualitative balance of the labor market and its subjects. The formation of the efficient staffing system and ensuring sustainable agricultural development on its basis in the agricultural and industrial complex in general and in the agriculture, as its leading industry, is one of the priority national tasks. However, despite the recent measures taken to solve staff problems of agriculture, the intended goals are not fully achieved. The researches show us that it is mainly due to demographic factors of the staff potential formation and the staffing policy, which is not enough efficient both at the state level and the level of organizations [10–13].

It is assumed that the reform of traditional management methods and the continuous improvement of the efficiency of public administration of agricultural production in the field of staffing are an adequate response to this challenge. Many scientists pay attention to the problems above.

Based on the study of foreign experience they conclude that the staff training system should be carried out on the basis of forecasting situation in the future and respond to changing needs of the labor market. Continuous learning can range from short-term refresher courses to second vocational education. In addition, as N.Yu. Anisimova notes, the practical implementation of the competencies obtained in the preparation process needs the knowledge of the specific features of the agricultural industry and the region.

Thus, despite the many and varied publications on the subject of staffing in the agricultural-industrial complex, many questions, taking into account the regional specifics and the formation of clear, accessible and effective tools, remain open. The above defines the relevance and significance of the topic of the study.

2 Materials and Methods

The purpose of the research: to investigate the features of the staffing of the Kaluga region and to formulate on this basis the recommendations for its improvement.

The study looked at the following issues:

1. What does the complex of peopleware activities include in the agro-industrial complex?
2. What is the current level of peopleware in the region's agro-industrial complex and are there sufficient measures to achieve its acceptable values?
3. What measures can improve the personnel situation in the region?

The theoretical and methodological basis of the study was the dialectical method of cognition and systemic approach, the works of domestic and foreign scientists on peopleware. The research was conducted on the basis of abstract-logical, calculating-constructive, comparative methods using analysis of official statistical information. The information and empirical basis was the data of the annual statistical reports of the Federal State Statistics Service for the Kaluga region, the monitoring of the social and labor sphere of the regional level, materials of departmental surveys of personnel composition of agricultural enterprises. The object of the study is the agricultural-industrial complex of the Kaluga region, the subject is its staffing.

3 Results

All modern strategic personnel programs of the AIC industry were based on the Russian Government's Resolution No. 117 of February 10, 2000 "About improving the peopleware of the agro-industrial complex." The document speaks about the need to develop the system of training, retraining, training and use of workers of the agro-industrial complex by the Ministry of Agriculture, together with the executive authorities of the subjects of the Russian Federation.

In order to improve the peopleware of agriculture by managers and specialists, as well as by personnel of mass professions in the Kaluga region, the Kaluga Region Agricultural Personnel Program for 2001–2005 was developed with the participation of authors. The main provisions of this program are still in force. These include: improved personnel training and retraining; continuous professional development; public funding of the main tasks of peopleware of the agro-industrial complex; Improving social living conditions in the countryside; Staffing information support; mechanism for securing qualified personnel in the countryside.

The work on training highly qualified personnel of the agro-industrial complex of the region in accordance with the Program was based on agreements concluded between the Ministry of Agriculture and Food of the Russian Federation, the administration of the Kaluga region and the Kaluga branch of the Russian State Agricultural University - the Moscow Agricultural Academy named after K.A. Timiryazev about the provision of agro-industrial complex by qualified professionals with higher education; between the Ministry of Agriculture and Food of the Russian Federation and the Kaluga Region Administration to provide the agro-industrial complex with qualified professionals with secondary vocational education.

In accordance with the agreements above, part of the expenses of agricultural educational institutions was covered by the regional budget. A total of 20 million and 299.7 thousand rubles were paid from the region's budget for this purpose between 2000 and 2004.

During 5 years, 316 managers and 1,893 agricultural specialists have passed through the training system in the region, of which more than 1,000 have improved their skills without leaving their area. The field study, carried out under the 72-h training program on the basis of the Kaluga Training Center of the AIC, has allowed to cover the training of the vast majority of working specialists and managers of the agro-industrial complex of the region.

In accordance with the regional government's decree of 20 March 2002, No. 50 from the 1 January of 2002, state support for the human resources potential of agricultural production was established, which, taking into account the annual amendments made to the Use of Budget funds for these purposes, was carried out in the form of supplementary salaries of newly elected (appointed) managers of agricultural organizations over the course of two years, depending on the financial and economic results of the work of their farms, as well as co-payments to the official salaries of young professionals for 3 years in the amount of 0.5 to 2.5 thousand rubles, depending on what education they received in the educational institution - primary, secondary or higher professional. In addition, monthly co-payments of 500 rubles were made to young specialists studying part-time studies at universities and graduates of primary vocational educational institutions who were employed in agricultural organizations.

On November 21, 2003, the Regional Government established a one-time state allowance of 10 thousand rubles, paid to young professionals after the expiration of one year of their work in the agricultural organization.

These measures have significantly improved the employment and consolidation of young people in agricultural production, and in 2003 they were employed in farms more than twice as much as in 2002. In 2004, the total number of young professionals in agricultural organizations in the region increased to 30% and reached 179 people.

The amount spent on government support for human resources has increased annually since 2002. In total, the regional budget spent 11,758.5 thousand rubles on state support of human resources capacity.

During this period, the Kaluga region began to develop and implement the Kaluga Region Social and Economic Development Strategy. The main provisions of the strategy were laid down in the Kaluga Region Act of 20.10.2004 No. 362-OL "About the program to improve the quality of life of the population of the region for 2004–2010." This allowed the region to move to a qualitatively new level of "human is a centre of investment" when the economic development of the region was supposed to be carried out not only through investment, but primarily due to the activation of the human factor of economic development, increasing the level of social and political unity in the region, increasing the civil political participation of the population, the formation and development of civil society institutions. The tourism, investment and innovation industries have been adopted as the main potentials for the development of the Kaluga region as a "new development region." In addition, the fundamental provisions of the Kaluga Region's Social and Economic Development Strategy «human is a center of investment» until 2030 were the implementation of cluster policies in the Kaluga region and the formation of networked economic communities. The main goal of the strategic development of the Kaluga region is to achieve the competitiveness of the region and the sustainability of the region's economic development to ensure the growth of the quality of life of its population.

Although the transformations began in industrial production, they immediately led to changes in the agro-industrial complex. The outflow of labour force from agricultural enterprises to the enterprises of the "new economy" occurred because of the higher level of wages in the latter. This, in turn, necessitated increased productivity, which was enabled by improving the skills of staff and purchasing more modern and productive machinery and equipment. Additional investments in agricultural production were required. The consequence was a reform of production and a change in property relations, a gradual transition of enterprises to large agricultural investors who own property in many regions of the country, changes in the requirements for personnel and the content of the peopleware in industry.

Currently, the agro-industrial complex of the Kaluga region includes 256 organizations engaged in agricultural activities, 45 large and medium-sized enterprises of food and processing industry, 750 peasant (farm) farms, 110.0 thousand personal auxiliary farms.

The availability of personnel in the Kaluga region is evidenced by the data of the departmental statistical observation of the Ministry of Agriculture of the Kaluga region (Table 1).

Table 1. Information about the composition and number of agricultural workers in the Kaluga region, people.

Positions	Actually working human			
	2017	2018	2019	2020
Total employees, holding executives and specialists positions	2764	2818	3048	3054
Leaders of agricultural organizations	212	210	197	190
Chief specialists	489	479	514	482
Including chief agronomists	62	58	51	54
Chief zootechnicians	61	55	50	37
Chief veterinarians	49	51	54	46
Chief engineers	96	103	160	150
Chief power engineers and electricians	28	29	31	30
Chief economists	32	30	27	20
Chief Accountants (Chief Accountants, Financial Directors)	138	136	121	115
Other positions of chief professional	23	17	20	30
Specialists (except majors)	1234	1240	1406	1391
Total agricultural workers	1240	12395	12692	12623
Including agricultural production	7428	8193	8791	8744
Fish farming industry	80	73	74	58
Servicing agriculture	171	214	132	112
Food and processing industry	4354	3905	3695	3709

Agricultural organizations in the region are currently experiencing a shortage of permanent employees. According to the monitoring of the Ministry of Agriculture of the Kaluga region, the provision of agro-industrial production organizations by managers and specialists in 2020 (as of January 1, 2021) is 91.6%, and workers - 87.8%. The shortage of personnel in the categories the employees is 281 and 1317, respectively.

From among the workers for the spring field and harvesting period of 2020, the shortage of mechanics was, respectively, 137 and 178 people, or 11.2% and 23.67% of their need. The seasonal shortage of personnel was covered by the transfer to other jobs of personnel of other professions of agricultural organizations, the involvement of employees of other organizations of agro-industrial complex and industrial enterprises, students of the Kaluga branch of the Russian State Agrarian University of the Timiryazev Moscow Agricultural Academy named after K.A. Timiryazev and even employees of other regions. In total, 494 employees, holding positions of managers and specialists and 5,150 workers, resigned in 2020. The main reasons for the dismissal were: the seasonality of agricultural production; related to job cuts or liquidation of the enterprise: harmful and dangerous working conditions.

There is also a high level of turnover of agricultural managers. This, among other things, is evidenced by the performance of their work. Thus, more than a quarter of managers of their total number have work experience up to three years.

The data from the departmental statistical observation of the Ministry of Agriculture of the Kaluga region, shows that out of 3,054 people who replace the positions of managers and specialists in 2020, only 1,849 people (60.5%) have higher education and 1,010 people (33.1%) have secondary education. More than a quarter of agricultural chiefs do not have a specialized agricultural education, which certainly reduces their effectiveness (Table 2).

Table 2. Indicators of education and erudition of agricultural chiefs.

Indicators	Years							
	2017		2018		2019		2020	
	Humans	%	Humans	%	Humans	%	Humans	%%
Agricultural organizations chiefs. total	212	100	210	100	197	100	190	100
Have basic economic or managerial education	43	20.3	44	21	37	18.8	80	42.1
Have non-core (non-agricultural) education	44	20.7	43	20.5	41	20.8	51	26.8
Have a degree	4	1.9	4	1.9	7	3.6	3	1.6
Have management training (qualification "Management of the organization")	3	1.4	6	2.9	6	3.0	0	0.9
Passed advanced training(at least 72 h)	2	0.9	4	1.9	19	9.6	1	0.5
The number of reserve managers	14	–	5	–	4	–	7	–
From the reserve appointed (elected) managers	0	–	0	–	0	–	0	–

Training of personnel with higher education for the village is carried out, first of all, by the Kaluga Branch of the Russian State Agrarian University of the Timiryazev Moscow Agricultural Academy. The annual production of the branch is about 200 people. Additional social support measures in the form of quarterly cash payments and annual cash allowance have been established to enshrine young professionals in the Kaluga Region's Law «About Additional Measures to Support Young Agricultural Professionals in the Kaluga Region». These measures include the appointment of quarterly cash payments of 9,387 to 15,645 rubles to young professionals, depending on the level of education they receive, as well as an annual cash allowance of 31,290, 52150 and 78,225 rubles after the first, second and third year of work in the agricultural organization, respectively.

In order to increase the prestige of agricultural specialties on the initiative of the Ministry, 7 scholarships named after the outstanding leader, former chairman of the collective farm "Russia" Georgiy Ivanovich Sonin, which are presented annually in a solemn atmosphere to the best students. Agricultural educational organizations successfully operate services to assist in the employment of graduates. The services are run by a bank of personal data of specialists, planned for release and have already graduated from an educational institution, on the basis of which the operational selection of qualified personnel for employers is carried out. Information about the availability of specialists allows to keep "on control" every job seeker and to search for vacancies and prompt information to non-employed graduates. However, these activities do not provide the expected level of staff consolidation in the countryside.

In order to determine the factors influencing the consolidation of young people in agricultural organizations, according to the Kaluga region data for 2016–2019, multiple correlation and regression analysis was conducted. As a dependent variable, the share part of workers aged 20–29 in the total number of agricultural staff has been accepted. Five factor indicators are considered: migration and natural population growth, average wages of workers, the amount of capital investment per 100 hectares of agricultural land; the amount of subsidies paid to agricultural organizations (territories) from the region's budget. They determine 72% of the variance of the result attribute.

The highest level of wage (regression ratio 0.61) has the greatest impact on the fluctuation of the share part of young people in agricultural organizations in the region. The other most important attributes are the amount of investments per 100 hectares of agricultural land (K 0.52) and migration growth (K0.40). The rest of the indicators do not have a significant impact.

In 2021, 381.3 million rubles of budget funds are planned to be allocated to the implementation of the Kaluga region's state program "Complex Rural Development in the Kaluga Region", approved by the Kaluga Region Government's resolution of 31.01.2020 No. 63.

However, these measures are clearly not sufficient to remedy the situation. Thus, if the existing state of the region's human resources management mechanism is maintained, it is necessary to predict the stagnation of the personnel situation in the agro-industrial production of the Kaluga region.

4 Discussion

The crisis of human resources in agro-industrial production in the Kaluga region is characterized by the following signs: agro-industrial enterprises of the region are currently experiencing a shortage of permanent workers. The level of personnel security in 2020 was 90.7%. Managers and specialists have a lack of education, 6.3 per cent of them do not have vocational education. More than a quarter of the heads of agricultural organizations do not have a specialized agricultural education, which, of course, reduces the effectiveness of their activities, at the level of organizations there is an irrational arrangement of personnel – among the number of workers in 2020, 1,254 people had an excessive higher education, and 3,679 people - secondary vocational education. The crisis is accompanied by demographic problems in the countryside, manifested in low birth rates, high mortality, natural loss of population, its aging, migration of young people to cities, as well as a significant proportion of jobs with harmful and/or dangerous working conditions. Replacing indigenous people with migrants from neighbouring countries does not solve the problem of human resources.

The studies have shown that the arrival of large investors in agriculture, to which can be attributed, first of all, the agro-industrial holding Miratorg and EcoNiva-AIC Holding, which operate in many regions, significantly changed the situation in the labor market. Previously, a significant number of agricultural executives owned most of their business assets and were able to make decisions on their behalf. At the same time, as a rule, they were native residents of Kaluga region and were connected by long-term business and interpersonal relations with both their employees and other parties interested in the development of human resources: state and municipal bodies, educational institutions, representatives of civil society, consumers.

At the present, business leaders are appointed with limited functions, especially in terms of strategic decision-making. The heads of EcoNiva-AIC Holding, for example, have as their head not only the owner and president of EcoNiva-AIC Holding, but also the regional director for the Kaluga region. Manufacturing managers, as hired managers, require, first of all, executive and accuracy in financial transactions to gain investor confidence.

In this regard, the lack of agricultural education is not an obstacle for candidate to gain a post of the head of an agricultural organization. In turn, the appointed managers on such criteria choose the management apparatus of the enterprise. However, the lack of agricultural education sometimes leads to a decrease in the efficiency of the production activities of organizations, which in turn leads to personnel reshuffles.

Judging by the results of the study, the mechanism of personnel management of the AIC industry in the region still requires improvement and consideration of the current situation. This is also evident in traditional methods of working with vocational education institutions and with a staffing pool. For example, Kaluga agricultural enterprises are smaller than in other regions. Managers of organizations prefer to have as a chief specialist an employee with experience. In some cases, an employee of a certain profession (economist, zoo technician, agronomist, engineer) due to the small size of the enterprise, preference is given to an experienced specialist who came from another region, rather than a graduate of the Kaluga branch of the agricultural university. Sometimes managers even leave posts vacant for some time. In these conditions, it becomes justified to solve the problem of providing managers and specialists by attracting to these positions persons with higher and secondary vocational education, the annual formation of a regional target order for training for enterprises, organizations of the AIC, the personal distribution of graduates of agricultural educational institutions, trained at the expense of the federal budget for targeted contract training. In addition, investors working in many regions of the country have established relationships with agricultural universities in other regions. Consequently, the regional Ministry of Agriculture should ensure the development of an inter-regional public-private partnership involving investors, coordination of the activities of all stakeholders in the process of personnel support of the region. This will also solve the problem of efficiency of work with the regional personnel reserve. At present, the formation of a human resources reserve is not adequately linked to the interests and objectives of investors and therefore there are no appointments to senior positions. Thus, funds from the regional budget aimed at creating a personnel reserve are spent in idle.

The studies have shown the relationship between the consolidation of personnel in the countryside on the one hand, with the level of wages, investments in the social and economic development of the village and the possibility of moving to the countryside on the other. In this regard, the tasks of state and municipal authorities are: the improving of the demographic situation in the countryside on the basis of integrated rural development; enshrining rural youth, including the way of promoting agricultural work and creating agricultural classes in schools, vocational training facilities in which workers can be acquired, providing them with modern high-performance jobs.

Agricultural educational institutions should continuously monitor technological and socio-economic changes in the region, assess the existing and prospective need for workers and employees, constantly interact with agricultural enterprises, state and municipal authorities, and coordinate activities with them. With this in mind, it is necessary to adjust the quantitative training plans for specialists and workers. The public authorities should help to create conditions for a significant improvement in the quality of personnel and their competitiveness in the labour market by strengthening the logistics of the agricultural education institutions, establishing compliance with the mechanism of forecasting and regulating the training of the AIC staff.

5 Conclusion

The research of peopleware in the region's agro-industrial complex shows that a set of measures is being implemented in the Kaluga region, which includes: improving training and retraining; continuous professional development; public funding for the main tasks of staffing in the agro-industrial complex; improving social living conditions in the countryside; staffing information support; enshrining qualified personnel in the countryside.

Also, the studies have shown that the measures applied are not enough to achieve the goals of staffing in the region. Agro-industrial enterprises in the region currently lack permanent employees. The level of personnel security in 2020 was 90.7%. There is a lack of education for managers and specialists, 6.3per cent of whom do not have vocational education. More than a quarter of agricultural organizations' heads do not have a specialized agricultural education, which certainly reduces the effectiveness of their activities. At the same time, among the number of workers in 2020, 1,254 people had an excess higher education, and 3,679 people - secondary vocational education.

To correct the current situation, an additional set of measures was reasoned. It includes: solving the problem of providing managers and specialists by attracting to these positions persons with higher and secondary vocational education, the annual formation of a regional target order for training for enterprises, organizations of the AIC. With this in mind, to adjust the quantitative plans of training specialists in the branch of the agricultural university, technical schools, vocational schools; Improving rural demographics through integrated rural development; enshrining rural youth in the countryside and providing them with modern, high-performing jobs; Personal distribution of graduates of agricultural educational institutions who have been trained by the federal budget for targeted contract training; Creating conditions for a significant improvement in the quality of staff and their competitiveness in the labour market by strengthening the logistics of agricultural education institutions; Establishing a compliance mechanism for forecasting and regulating the training of the agricultural sector with the working conditions of the labour market; ensuring a guaranteed level of wages for managers of economically weak farms working on a contract basis with the Regional Administration; Development of inter-regional public-private partnerships involving investors; coordinating the activities of all stakeholders in the region's staffing process.

References

1. DeMarco, T., Lister, T.: Peopleware: Productive Projects and Teams. Addison-Wesley Professional, New York (2013)
2. Gerasimov, A.N., Gromov, E.I., Lelikova, E.I.: Trends in provectione agriculturalis laboris resources in Stavropol regions. AIC Econ. Manag. **11**, 14–25 (2020). https://doi.org/10.33305/2011-14
3. Rybkina, M.V., Kangro, M.V., Pirogova, E.V.: Razvitiye rynka truda v usloviyakh stanovleniya tsifrovoy ekonomik. Vestnik universiteta **11**, 36–41 (2019). https://doi.org/10.26425/1816-4277-2019-11-36-41
4. Suvalov, O.S., Suvalova, T.V.: Motivatsiya personala kak instrument effektivnoy kadrovoy politiki organizatsii. Vestnik universiteta **7**, 22–27 (2020). https://doi.org/10.26425/1816-4277-2020-7-22-27
5. Ecirli, A., Dobre, E.-M., Dobrescu, E., Danetiu, M.I.: Human resources in European market in the past decade–a sociological overview. Proc. Soc. Behav. Sci. **150**, 320–329 (2014)
6. Salehi-Isfahani, D., Mostafavi-Dehzooeic, M.H.: Cash transfers and labor supply: evidence from a large-scale program in Iran. J. Dev. Econ. **13**, 349–367 (2018). https://doi.org/10.1016/j.jdeveco.2018.08.005
7. Stathakopoulos, V., Kottikas, K.G., Theodorakis, I.G., Kottika, E.: Market-driving strategy and personnel attributes: top management versus middle management. J. Bus. Res. **104**, 529–540 (2018). https://doi.org/10.1016/j.jbusres.2018.09.020
8. Teruyama, H., Goto, Y., Lechevalier, S.: Firm-level labor demand for and macroeconomic increases in non-regular workers in Japan. Japan World Econ. **48**, 90–105 (2018)
9. Anisimova, N.Yu.: Organizational and economic mechanism of quality management of staffing of agro-industrial complex of the region. Eur. Sci. J. **6**(10) (2018). https://doi.org/10.15862/64ECVN618

10. Blinova, T.V.: Demographic ageing of rural areas of Russia. AIC Econ. Manag. **2**, 76–80 (2021). https://doi.org/10.33305/212-76
11. Semakin, A.G.: AIC Econ. Manag. **5**, 56–57 (2020). https://doi.org/10.33305/2012-56
12. Potenko, T.A., Zhupley, I.V., Schmidt, Y.I., Kvashko, L.P.: Agricultural policy and import substitution opportunities in the Russian Far East. IOP Conf. Ser. Earth Environ. Sci. **548**, 022011 (2020). https://doi.org/10.1088/1755-1315/548/2/022011
13. Burtseva, T.A., Gagarina, S.N., Chausov, N.Yu.: Quality of life assessment of the older population at the substantiation of active longevity strategies in the context of structural demographic changes. Vestnik universiteta **2**, 5–12 (2019). https://doi.org/10.26425/1816-4277-2019-2-5-12

Assessment of the Adaptive Ability of Corn Hybrids in Primorsky Krai

Irina Danilenko, Natalia Kraskovskaya, and Ekaterina Butovets(✉)

Federal Scientific Center of Agricultural Biotechnology in the Far East named after A. K. Chaika, Timiryazevskiy settlement, 692539 Ussuriysk, Russia

Abstract. Stable corn grain production is possible if there are hybrids with high productivity potential and environmental conditions for realizing this potential. An important property of the studied hybrids is ecological stability - the ability to slightly reduce yield under poor growing conditions. The aim of this work was to study and assess the parameters of adaptability and stability of maize hybrids in Primorsky Krai. The article presents an analysis of three-year data from corn hybrid tests of Russian and foreign breeding in the conditions of Primorsky Krai. The plasticity and stability of hybrids in terms of yield of an ecological harvesting environment have been determined. Various climatic conditions during the years of research made it possible to obtain a comprehensive assessment of the studied hybrids. The index of environmental conditions during the period of the study was characterized by a sharp contrast, as the indicator changed from +17.2 to −9.1. The highest resistance to stress and the range of use were shown by early and medium-early hybrids - Baikal, Ladozhsky 181 MV, NK Gitago, R 7054, R 7709, R 8816. hybrids Varieties Si Rotango, Delitop, Si Novatop, R 8523, and R 9074 were noted as capable of producing high grain yields in the years more favorable for corn growth and development.

Keywords: Hybrid · Corn · Yield · Grain moisture · Adaptability · Stress resistance

1 Introduction

Corn is characterized by versatile use and high yield and takes the leading place in terms of yield and gross harvest in global grain production. About 20% of corn grain is used for food purposes, about 15% - for industrial purposes, and about two-thirds is used as fodder. Corn is the leader among fodder grain crops due to its high calorie content and high content of fodder units in the grain [1, 2].

In recent years, the area under corn and grain yield has increased. Growth in corn grain production will occur not only due to an increase in its share in crops, but also mainly due to an increase in productivity based on modern technology, the introduction of hybrids of different ripeness groups corresponding to very diverse growing conditions.

The Primorsky Krai agricultural producers harvested 297 thousand tons of corn grain in 2020, which is 2.4% more than a year earlier. In this regard, it is possible to

A. Muratov and S. Ignateva (Eds.): AFE 2021, LNNS 353, pp. 275–284, 2022.
https://doi.org/10.1007/978-3-030-91402-8_32

note the economic feasibility of growing corn for grain in Primorsky Krai. To achieve the set objectives, maize hybrids are needed, which, along with high grain yield and low harvesting moisture, are characterized by resistance to environmental stress factors that limit the formation of potential productivity [3–5]. For this reason, our research goal was to study and evaluate the parameters of adaptability and stability of maize hybrids when grown in Primorsky Krai.

2 Materials and Methods

The work was carried out in 2017–2019 in the fields of the laboratory of selection and primary corn seed production of the FSBSI "FSC Agricultural biotechnology of the Far East named after A.K. Chaika", near the city of Ussuriysk, Primorsky Krai, located at latitude 43.8°, longitude 132.02°.

The area is characterized by a warm, humid climate with severe winters. The sum of active temperatures (above 10 °C) fluctuates within 2400–2600 °C, the hydrothermal coefficient (HTC) - 1.6–2.0 (1.0 - characterizes extremely insufficient moisture conditions; 2.0 - excessive). The hydrothermal coefficient, which characterizes the average state of thermal moisture supply of the territory during the years of the experiments, corresponded to weather conditions favorable for growth and development of corn (Fig. 1). The HTC of 1.3–2.0 is required for normal growth and development of corn. In some decades, the HTC indicators were below 1, but the plants did not experience lack of moisture during these periods due to the abundant amount of precipitation that fell in the previous decades of the month.

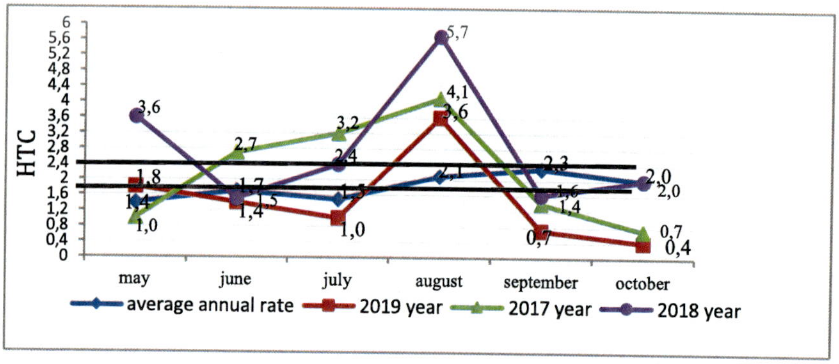

Fig. 1. HTC during the study according to the agrometeorological Timiryazevskiy station.

The soil of the experimental site is bleached meadow-brown with a heavy loam texture. The thickness of the root layer is 20–25 cm, the arable horizon is underlain by heavy water-impermeable loams [6].

2.1 Field Experiment

Corn hybrids for grain use, different in origin and group of ripeness, were taken as the objects of study. The mid-season hybrid population Slavyanka, approved for use in the Far East region, was taken as standard.

The used cultivation technique is generally accepted in Primorsky Krai, the row spacing is 70 cm, the seeding rate is 70 thousand plants per ha. The ecological test was repeated three times in the nursery, the location of the plots was randomized, the plot area was 28 m^2. The test was carried out according to the method of the field experiment by B.A. Dospekhov (2012). Phenological observations, field assessment of plants, and counts for the main economically valuable traits were carried out according to the methodological recommendations G.E. Shmaraev et al. (1985) and B.P. Sokolov et al. (1982). The harvest moisture content of corn samples was determined on a Pharmpoint moisture analyzer (Denmark).

2.2 Statistical Analysis

Statistical processing of the experimental data was carried out by the analysis-of-variance method, calculation of the coefficient of variation using Microsoft Excel.

The assessment of the ecological plasticity and stability of yield indicators and moisture content of corn grain was carried out according to the methodology V.Z. Pakudin (1979). A measure of the response degree of the maize genotype to changes in environmental conditions is the regression coefficient (bi), which gives an estimate of plasticity and stability (S2di). The first shows the response of the genotype to an improvement in growing conditions, and the second characterizes the variance of the stability of the sample under various environmental conditions. Hybrids with a regression coefficient significantly lower than 1 are of the neutral type (with low ecological plasticity). 0 or close value of the regression coefficient shows that the sample does not respond to changes in the environment (V.A. Zykin et al. 2008; G.P. Vlasenko 2017).

The assessment of the samples for flexibility, compensatory ability and stress resistance of yield was carried out according to the method of V.A. Zykina et al. (1984). The index of environmental conditions was assessed according to the method proposed by S.A. Eberhart, W.A. Russell as presented by V.Z. Pakudin (1979).

3 Results

3.1 Ecological Plasticity of Corn Hybrids by Yield

With the introduction of new corn hybrids with a high productivity potential into production, the question of stabilizing their yield arose: only due to the poor adaptability of hybrids to specific environmental conditions and the level of agricultural technology, yield losses reach 30% or more. Therefore, it is very important to evaluate hybrids not only by the average yield over the years of research, but also by their ecological plasticity and stability (Table 1). The analysis of corn hybrids for three years of research made it possible to distinguish four groups according to the degree of reaction of the genotype to environmental conditions.

Table 1. Parameters of ecological plasticity of corn hybrids by yield, 2017–2019.

Hybrid population variety*	Yield centner/ha	Decreasing in yield in unfavorable conditions.%	Regression coefficient (b_i)	Stability coefficient (S^2d_i)	Variation coefficient (V). %
Slavyanka*(standard)	44.9	42.9	0.80	85.7	16.8
Birsu*	36.0	58.3	0.63	276.9	19.4
NUR	29.6	54.2	1.08	6.4	19.3
R 7709	77.4	20.1	0.62	17.5	11.7
R 7054	68.9	22.6	0.78	1.7	13.9
Baikal	65.5	15.1	0.44	2.6	8.3
Ladozhsky 175	71.1	28.8	1.00	0.10	18.4
R 8451	85.7	27.8	1.16	24.3	21.2
Ladozhsky 181 MV	50.8	25.5	0.52	14.6	10.0
NK Falcon	89.5	23.6	0.88	46.2	16.9
Si Rotango	75.9	33.8	1.39	3.3	24.8
NK Gitago	80.4	18.7	0.66	12.9	12.2
Delitop	90.4	30.0	1.26	13.5	22.7
Si Nivatop	90.8	32.4	1.28	55.8	23.9
Ladozhsky 250 MV	90.2	20.9	0.76	21.5	14.3
R 8400	83.1	30.5	1.07	51.3	20.2
R 8523	80.7	33.7	1.14	79.7	22.0
R 8688	79.6	30.3	1.10	17.5	19.9
R 8816	83.9	19.7	0.78	0.03	13.8
R 9074	92.6	34.5	1.59	7.1	28.3
R 9175	86.6	27.9	1.12	40.0	20.7
R 9241	90.0	43.5	1.91	73.3	35.0
Average yield in the experiment	74.6	-	-	-	-
$LSD_{0.95}$	14.0	-	-	-	-

The first group is represented by one corn hybrid Ladozhsky 175 with a regression coefficient equal to 1, which indicates a high responsiveness of the sample to changes in external factors. The second and third groups include samples with low ecological plasticity (bi < 1), which showed good results in yield under unfavorable cultivation conditions, but only one of them (R 8816 hybrid bred by the Pioneer Company) is stable (S2di close to 0). Eleven hybrids were assigned to the fourth group (regression coefficient greater than 1) and characterized as plastic with unstable behavior (S2di above 0), i.e. specimens are able to progressively increase the yield under the influence of improved cultivation conditions, but show dependence on the conditions of the year. This combination of reactions of genotypes causes difficulties in predicting their behavior.

The variation coefficient (V) shows the variegation or evenness of a trait or property. The performed calculation based on the yield data showed that significant variability (V over 20%) was in the hybrids of the fourth group. The average variability of the trait (10–20%) was observed in 54.5% of the samples from the total number. Baikal, the hybrid selection of the FSBSI All-Russian Research Institute of Corn, was characterized by a slight variation in yield (below 10%) over the years.

The decrease in the yield of genotypes R 7709, Baikal, NK Gitago, Ladozhsky 250 MV, and R 8816 in extreme conditions was below 21.0% (15.1–20.9%), which is 1.7–43.2% less than in other studied hybrids and standard.

3.2 Assessment of Corn Samples for Genotypic Variability, Stress Tolerance, and Genetic Flexibility

Taking into account the specific features of the climate of Primorsky Krai and the use of seeds of hybrids of foreign selection in corn cultivation, it is very important to study the resistance of the samples to changing weather conditions. Table 2 shows the main statistical parameters characterizing the adaptive property of hybrids in terms of yield.

Early and mid-early hybrids are Baikal, Lazhsky 181 MV, NK Gitago, R 7054, R 7709, R 8816 with values: −9.8; −14.9; −17.0; −18.3; −17.4; −18.9 have the highest stress resistance and widest range of use. Syngenta Company genotypes - Si Rotango, Delitop, Si Novatop; Pioneer Company genotypes - R 8523, R 9074, R 9241 are capable of producing high seed yields in more favorable years for the growth and development of corn.

Ladozhsky 250 MV, NK Falcon, Delitop, Si Novatop, R 9074, and R 9241 have high values of the flexibility index (more than 90), which confirms their significant degree of genotype-environment dependence. The minimum indicators (45.6; 33.4; 33.7) for this feature were noted in the Slavyanka, Birsu, and NUR samples, proving their adaptive ability to change meteorological and edaphic conditions.

The weather and climatic conditions during the test years varied, which made it possible to more objectively evaluate the corn samples, taking into account the existing external environmental conditions. The environmental index (Ij) during the study period was characterized by a sharp contrast, the indicator varied from +17.2 to −9.1. In 2017, there were favorable conditions for realizing the potential of the yield of hybrids and population varieties of corn, where the index of the environment was the highest positive value (17.2), with the highest value of the average yield of samples 89.9 center/ha.

Table 2. Adaptive indicators of corn hybrids, 2017–2019.

Variety name	Originator	Stress resistance ($X_{min} - X_{max}$)	Genetic flexibility ($X_{min} + X_{max}$)/2	Genotypic variability ($X_{min} - X_{max}$). centner/ha
Slavyanka (standard)	FSC of Agricultural Biotechnology in the Far East named after A. K. Chaika"	−24.9	45.6	33.1–58.0
Birsu	FSBSI "Far Eastern research Institute of Agriculture"	−27.5	33.4	19.7–47.2
NUR	FSBSI All-Russian Research Institute of Corn	−25.1	33.7	21.2–46.3
Baikal	FSBSI All-Russian Research Institute of Corn	−9.8	66.0	61.1–70.9
Ladozhsky 175	NGO "Seeds of Kuban"	−25.1	74.3	61.8–86.9
Ladozhsky 181 MV	NGO "Seeds of Kuban"	−14.9	50.9	43.5–58.4
Ladozhsky 250 MV	NGO "Seeds of Kuban"	−21.1	90.8	80.2–101.4
NK Falcon	Syngenta Company	−24.5	91.3	79.1–103.6
Si Rotango	Syngenta Company	−32.9	80.7	64.3–97.2
NK Gitago	Syngenta Company	−17.0	82.2	73.7–90.7
Delitop	Syngenta Company	−32.8	92.8	76.4–109.2
Si Novatop	Syngenta Company	−35.5	91.8	74.1–109.6
R 7054	Pioneer Company	−8.3	71.7	62.6–80.9
R 7709	Pioneer Company	−17.4	77.7	69.0–86.4
R 8400	Pioneer Company	−30.1	83.6	68.6–98.7
R 8451	Pioneer Company	−28.9	89.4	75.0–103.9

(continued)

Table 2. (*continued*)

Variety name	Originator	Stress resistance ($X_{min} - X_{max}$)	Genetic flexibility ($X_{min} + X_{max}$)/2	Genotypic variability ($X_{min} - X_{max}$). centner/ha
R 8523	Pioneer Company	−32.8	80.8	64.4–97.2
R 8688	Pioneer Company	−29.1	81.3	66.8–95.9
R 8816	Pioneer Company	−18.9	86.3	76.9–95.8
R 9074	Pioneer Company	−40.2	96.4	76.3–116.5
R 9175	Pioneer Company	−29.1	89.6	75.1–104.2
R 9241	Pioneer Company	−51.5	92.5	66.8–118.3

Note. Xmin is the yield formed under limited environmental conditions; Xmax - yield. formed in favorable environmental conditions
Environmental index (Ij): 2017 +17.2; 2018 −8.2; 2020 −9.1

3.3 Ecological Plasticity of Corn Hybrids by Harvesting Grain Moisture

One of the main factors limiting grain production in corn-growing zones is the harvesting moisture content of grain, which in turn dictates the technological parameters of harvesting and the very possibility of grain use of corn.

When evaluating the hybrids according to the harvesting moisture content of grain, some differences were noted in response to growing conditions in the tested years. According to the regression coefficient, three groups of samples were identified (Table 3).

The group of high-plasticity samples included samples with a regression coefficient significantly higher than one (1.37–1.69): R 8451, NK Falcon, Delitop, Si Novatop, Ladozhsky 250 MV, R 8523, R 8688 showed the best results in the experiment show in favorable environmental conditions. The second group of ecologically plastic samples includes nine genotypes of different maturity groups with a regression coefficient close to one (the range of values was 0.81–1.14), i.e. they respond well to changes in meteorological and edaphic conditions.

The third group is represented by corn samples R 7709, Ladozhsky 175, Ladozhsky 181 MV, Si Rotango, R 8400, R 8816, which belong to the neutral type (low ecological plasticity). They do not respond well to changes in environmental factors, in conditions of intensive farming, they are not able to achieve high results, but under unfavorable conditions their indicators decrease less in comparison with hybrids of the intensive type. Eight corn hybrids proved to be highly stable (S2di close to zero, range from 0.05 to 0.53) according to the variance value; the rest of the samples are characterized by low stability in terms of grain moisture.

The maximum valuc of increasing the harvesting moisture content of grain in extreme conditions was observed in the R 8688 genotype (43.8%), minimum - Ladozhsky 181 MV (3.0%). A slight variation (below 10%) of harvesting grain moisture over the years was recorded in the hybrids Ladozhsky 175, Ladozhsky 181 MV, Si Rotango, and R

Table 3. Parameters of ecological plasticity of corn hybrids by harvesting grain moisture, 2017–2019.

Hybrid. population variety	Average value of grain moisture during harvesting. %	Increasing harvesting grain moisture in unfavorable conditions. %	Regression coefficient (b_i)	Stability coefficient (S^2d_i)	Variation coefficient (V). %
Slavyanka (standard)	28.3	29.5	1.02	26.0	19.5
Birsu	25.6	37.1	1.14	22.8	20.5
NUR	22.5	30.2	0.97	0.16	14.3
Baikal	24.6	26.4	0.21	26.6	13.3
Ladozhsky 175	29.0	27.6	1.02	15.4	16.5
Ladozhsky 181 MV	25.6	37.6	1.09	0.11	21.1
Ladozhsky 250 MV	30.0	12.6	0.49	0.07	7.3
NK Falcon	32.8	35.8	1.57	19.1	25.5
Si Rotango	32.7	3.0	0.08	0.3	1.8
NK Gitago	29.3	32.6	1.43	0.4	20.9
Delitop	29.8	10.2	0.32	1.7	5.7
Si Novatop	29.3	22.1	0.94	0.9	14.0
R 7054	30.1	32.4	1.47	1.0	21.7
R 7709	28.7	35.4	1.43	10.1	22.4
R 8400	30.5	36.1	1.46	31.2	25.5
R 8451	27.6	17.9	0.65	0.5	9.6
R 8523	24.0	37.6	1.37	0.1	20.0
R 8688	24.4	43.8	1.69	0.9	24.8
R 8816	25.3	24.6	0.76	9.3	13.5
R 9074	27.9	30.5	1.04	28.1	20.2
R 9175	31.0	19.3	0.81	0.0	11.9
R 9241	31.8	28.4	1.05	17.4	18.6
Average grain moisture in the experiment	28.2	-	-	-	-
$LSD_{0.95}$	4.7	-	-	-	-

8400. Average variability of the trait (10–20%) was noted in 36.4% of the total samples. Significant variability (V over 20%) was observed in the hybrids of the first and second groups.

4 Discussion

Corn (Zea mays L.) is the third most important cereal among cereals and occupies at least 30% in the world grain balance [7]. Corn grain production reaches 14 million tons in Russia and 1,171 million tons worldwide. There has been an increase in acreage and gross harvests of corn for grain in recent years in Russia. The expansion of corn sowing and an increase in productivity are the result of breeding progress, due to which the productivity of hybrids has significantly increased and their adaptability to a lack of heat has improved [8]. Ecological stability - the ability to slightly reduce the yield when growing conditions deteriorate - is an important property for a hybrid [9].

Revealing the features of the adaptive response is the most important condition for the effectiveness of hybrids in the system of environmental tests for adaptability to stress factors of cultivation. We have proved the reliable influence of environmental conditions and the genotype-environment interaction on yield and harvesting grain moisture of the studied corn hybrids, which is confirmed by the results of other researchers [10, 11]. For a more complete characterization of the adaptive properties of the studied genotypes to changes in cultivation conditions, a number of statistical indicators were calculated, which made it possible to select the most valuable hybrids. Similar studies are presented in the works of other authors.

5 Conclusion

It was revealed that the intensity of modern corn hybrids should be considered as a combination of high potential yield and low grain moisture, stably retaining these characteristics, regardless of growing conditions, having a high adaptive ability to change the meteorological and edaphic conditions of Primorsky Krai. Various climatic conditions during the research made it possible to give an objective assessment of the studied hybrids. The environmental index during the study period was characterized by a sharp contrast, the indicator of which varied from +17.2 to −9.1. High resistance to stress and a wide range of use were shown by early and medium-early hybrids Baikal, Ladozhsky 181 MV, NK Gitago, R 7054, R 7709, R 8816. Hybrids capable of producing high grain yield in more favorable years for the growth and development of corn have been identified - Si Rotango, Delitop, Si Novatop, R 8523, R 9074. The Slavyanka, Birsu, and NUR samples have a high adaptive ability to changes in the meteorological and edaphic conditions of Primorsky Krai. The maximum value for increasing the harvesting moisture content of grain in extreme conditions was observed in the R 8688 genotype (43.8%), the minimum - in Ladozhsky 181 MV (3.0%).

References

1. Fernandes, J., et al.: Influence of hybrid, moisture, and length of storage on the fermentation profile and starch digestibility of corn grain silages. Anim. Feed Sci. Tech. **271**, 114707 (2021). https://doi.org/10.1016/j.anifeedsci.2020.114707
2. Baldin, M., et al.: Characterization of linoleic acid (C18:2) concentration in commercial corn silage and grain hybrids citation data. J. Dairy Sci. **101**(1), 222–232 (2018). https://doi.org/10.3168/jds.2017-12972
3. Ward, J.K., Henry, W.B., Nock, M.W.: Transaction of the ASABE. Am. Soc. Agric. Boil Eng. **59**(5), 1111–1115 (2016)
4. Guan, W.Zh., et al.: A generic risk assessment framework to evaluate historical and future climate-induced risk for rainfed corn and soybean yield in the U.S. Midwest. Weather Clim. Extrem **33**, 100369 (2021). https://doi.org/10.1016/j.wace.2021.100369
5. Kuang, W., Xianjiang, Y., Xiuqing, C., Yafeng, X.: Experimental study on water production function for waterlogging stress on corn. Proc. Eng. **28**, 598–603 (2012). https://doi.org/10.1016/j.proeng.2012.01.775
6. Kostenkov, N.M., Oznobikhin, V.I.: Soils and soil resources in the southern Far East and their assessment. Eur. J. Soil Sci. **39**(5), 461–469 (2006). https://doi.org/10.1134/S1064229306050012
7. Dermail, A., Suriharn, B., Chankaew, S., Sanitchon, J., Lertrat, K.: Hybrid prediction based on SSR-genetic distance, heterosis and combining ability on agronomic traits and yields in sweet and waxy corn. Sci. Hortic. **259**, 108817 (2020). https://doi.org/10.1016/j.scienta.2019.108817
8. Glennie, E., Anyamba, A.: Midwest agriculture and ENSO: a comparison of AVHRR NDVI3g data and crop yields in the United States corn belt from 1982 to 2014. Int. J. Appl. Earth Obs. Geoinf. **68**, 180–188 (2018). https://doi.org/10.1016/j.jag.2017.12.011
9. Salvo, J.I.D., Lee, C., Salmerón, M.: Regional multi-environment analysis of corn productivity and yield stability as impacted by hybrid maturity. Field Crops Res. **262**, 108025 (2021). https://doi.org/10.1016/j.fcr.2020.108025
10. Lychuk, T.E., et al.: Evaluation of climate change impacts and effectiveness of adaptation options on crop yield in the Southeastern United States. Field Crops Res. **214**, 228–238 (2017). https://doi.org/10.1016/j.fcr.2017.09.020
11. Ma, Y., Zhang, Z., Yang, H.L., Yang, Z.: An adaptive adversarial domain adaptation approach for corn yield prediction. Comput. Electr. Agricu. **187**, 106314 (2021). https://doi.org/10.1016/j.compag.2021.106314

Problems of State Regulation of Intra-industry Integration of Business Structures of the Primorsky Region

Alexander Latkin[1], Olga Fedoreeva[2(✉)], Dina Mukhina[2], and Ekaterina Saveleva[2]

[1] Vladivostok State University of Economics and Service, 41 Gogol Street, Vladivostok 690014, Russia

[2] Primorskaya State Academy of Agriculture, 44 Blucher Avenue, Ussuriisk 692510, Russia

Abstract. The article provides the results of perennial studies of the authors of the intra-industry integration of entrepreneurial structures of the agro-industrial complex (AIC). The basis for the generalization of the theoretical ideas of foreign and Russian scientists on the economic essence of this process, as well as a multifactor assessment of the trends in the cost of food, especially in the Far Eastern Federal District was based.

Keywords: Intra-industry integration · Business structures · Regions · Agricultural products · Agricultural enterprises · Integration processes · Fish flour · State regulation · Primorsky Territory · Fishing companies

1 Introduction

The world community, like Russia, is carrying out large-scale work in 2020–21 to overcome the consequences of the coronavirus pandemic, the main of which is to significantly reduce the pace of economic growth while increasing food prices [1].

The priority direction of solving this problem, many scientists and specialists consider improving the efficiency of agricultural production by reducing the cost of obtaining the final product [2–4]. Special relevance to the implementation of this area acquires for the regions of the Russian Far East, where the cost of most basic foods exceeds the average Russian level, reducing the already poor quality of the population, increases its outflow in the Western regions of the country.

Among the many components of the formation of a pricing system for agricultural products in the Far East, a high role belongs to the intra-industry integration of agricultural enterprises, including agricultural and fish industrial structures, which are carried out by the Ministry of Agriculture within the framework of relevant strategies and programs of prospective development. It should be added that in the socialist period of business, it is such integration that ensured the high efficiency of using raw water resources of the secondary order in reducing the cost and improving the quality of agricultural products, primarily in animal and poultry farming and pig breeding.

A. Muratov and S. Ignateva (Eds.): AFE 2021, LNNS 353, pp. 285–292, 2022.
https://doi.org/10.1007/978-3-030-91402-8_33

According to scientists, the integration of business structures of the agro-industrial complex helps to obtain economic benefits [5, 6].

In a number of scientific publications, it is noted that intra-industry interaction is activated by innovative and investment activity, and has a positive impact on the development of a competitive environment [7–10].

At the same time, O. Yu. Frolova, L. V. Fomina and Zh. N. Shmeleva believe that the market mechanism is not able to resolve the confrontation of production forces and relations and focus on vertically integrated interaction between business entities and state management bodies [11].

It is important to note that world and domestic experience shows the need to use various models and tools for state regulation of agricultural production processes in the market economy [12, 13].

In particular, this is confirmed by the practice of the intra-industry interaction of APK enterprises in a recent period of planned socialist economic management.

Taking into account the current requirements for the development of national agricultural production, as well as the theory and practice of its organization in Russia and other countries, the President of the Russian Federation in March 2021 instructed the government and the leadership of the Far Eastern region to analyze the level and dynamics of prices for the main types of agricultural products and food products and take measures aimed at reducing them. The meaning of this decision is extremely important for preserving and increasing the potential of the Russian Far East.

2 Materials and Methods

The purpose of the study is to substantiate the need for state regulation of intra-industry integration between agricultural and fishing enterprises of the agro-industrial complex in the regions of the Russian Far East, as a basis for reducing prices for agricultural products.

The methodological basis of the research is a systematic and comprehensive analysis of the problem under consideration using general economic and economic-statistical methods, as well as publications of Russian and foreign scientists.

Information base of research - materials of the Federal State Statistics Service (Rossatat) of the Russian Federation, regulatory documents regulating the functioning of agricultural enterprises, analytical reports of agricultural and fish industrial enterprises APK, the results of the expert survey of managers and specialists.

3 Results

In recent years, the Far East has been a challenge of globalization much sharper than the central regions, due to the tasks of its accelerated development [14]. Such a reaction is due to the National Program of Socio-Economic Development of the Far East approved in 2020 for the period up to 2024 and for the future until 2035 of this strategic important region for Russia.

Realizing the urgent need for its successful implementation, the President of the Russian Federation instructed the government and leaders of the Far Eastern regions, to analyze the level and dynamics of prices for the main types of agricultural products and food products in the Far Eastern Federal District and take measures aimed at reducing them to the average Russian Federation.

To aware of the scale and complexity of this problem, the authors made a comparative analysis of the value of the main agricultural products in all Far Eastern regions compared to the general Russian level. The data of this analysis in the Primorsky Territory and in general, the Far East of the Russian Federation, according to the results of 2020, are shown in Table 1.

Table 1. The cost of basic foods in the Russian Federation, Far Eastern Federal District and Primorye Territory as of December 31, 2020, Rub.

	Beef (except for boneless meat), kg	Pork (except for boneless meat), kg	Chickens, kg	Milk, l	Butter creamy, kg	Sunflower oil, l	Bread and bakery products, kg	Vegetables, kg	Eggs, 10 pieces	Fish and fish products, kg
Russian Federation	516.6	269.1	153.9	59.0	645.2	125.0	85.1	112.5	77.1	209.4
Far Eastern Federal District	576.5	333.5	208.4	82.4	881.7	143.1	134.9	149.5	90.6	138.9
Primorye Territory	629.3	330.6	182.2	87.6	843.2	137.4	107,3	148.4	89.0	137.4
Food value index in DFOs to the level in the Russian Federation, %	111.6	123.9	135.4	139.7	136.7	114.5	158.5	132.8	116.7	66.3
Food value index in the Primorsky Territory to the level in the Russian Federation, %	121.8	122.9	118.5	148.5	130.1	109.9	126.1	131.9	115.6	65.6

These data allow us to conclude that there is a significant excess of food prices in the regions of the Far East in comparison with the all-Russian ones. So in the Primorsky Territory, the prices of bread and bakery products increased by 58.5%.

Unfortunately, the analysis carried out by the authors revealed a steady trend of increasing the gap between these prices (Far Eastern and average Russian) in the last 5 years.

The indicated trend according to the study was due, first of all, an annual increase in prices for fuel and energy resources, agricultural equipment, fertilizers, as well as related to the fall of the national currency, an increase in the value of imported food from the PRC, Republic of Korea, Vietnam and other countries.

It is especially important to note the imperfection of the state support of agricultural enterprises of the Far Eastern regions, including the processes of their integration with fishing enterprises, which are now the lion's share of waste processing products export abroad and sell to Siberian regions of the country.

State regulation worldwide is one of the key factors for the development of agriculture. In the Russian Federation, it is carried out on the basis of FZ No. 264 of December 29, 2006. "On the development of agriculture in the Russian Federation" and the state program for the development of agriculture and regulating the markets for agricultural products, raw materials and food for 2020–2027. The main goal is to form a competitive and stable agro-industrial production and based on this provision of improving the standard of living of the population.

Table 2. List of subprogrammes of the State Program of the Primorsky Territory "Development of agriculture and regulation of markets for agricultural products, raw materials and food" at 2020–2027 and their financing, in %.

Name of the event	2020	
	plan	fact
Subprogramme 1. Technical and technological upgrades of the agro-industrial complex	13.73	9.58
Subrogram number 2. Increased financial stability, of which:	50.99	57.58
– stimulation of investment activity in the agro-industrial complex (subsidies for compensation of part of the cost of interest and for reimbursement of a part of interest rates on investment loans (loans) in the APC)	50.69	57.30
– Reducing risks in agricultural production (payment of the insurance premium accrued under the Treaty of SK Insurance)	0.20	0.28
Subprograms 3. Development under the branch of crop production, processing and sale of crop production	9.16	5.38
Subprogramme 4. Development of extension of animal husbandry, processing and sale of livestock production	18.61	17.93
Subprogramme 5. Development of the food and processing industry	3.77	2.81
Subprogramme 6. Development of agricultural cooperation and small forms of management	1.24	1.17
Subroutine 7. Development of agricultural land reclamation	0.76	0.53
Subprogramme 8. ensuring the functions of managing the implementation of the state program	0.17	0.16

(continued)

Table 2. (*continued*)

Name of the event	2020	
	plan	fact
Subroutine 9. Export of products of the agro-industrial complex	1.35	4.61
Subprogramme 10. Creating a farmers support system and the development of rural cooperation	0.15	0.13
Subprogramme 11. Complex Development of Rural Territories	0.04	0.18
State program, total	100	100

The state program of the Primorsky Territory "Development of agriculture and regulation of markets for agricultural products, raw materials and food" was developed and operated in Primorsky Territory, 2020-20-20, which includes subprogrammes aimed at food provision of residents of the edge of high-quality and environmentally friendly products, and creating conditions for the effective development of agricultural production. Financing of activities of this program is carried out at the expense of the federal budget and budgets of the subjects of the Russian Federation (Table 2).

From the data of the table, it follows that more than 50% of budgetary funds are aimed at increasing financial stability, most of which have to pay interest on investment loans and loans, on the development of crop 9.16% and animal husbandry 18.16%, the amount of state support for technical and technological upgrades amounted to 13.73%. Agricultural enterprises in the absence of the necessary state support do not receive profits for their development, and this is one of the reasons for the intra-industry integration of the business structures of the agro-industrial complex, which increases the efficiency of agricultural production by reducing the cost of the final product.

During the planned economy, intra-industry integration processes had their own specifics, which were carried out vertically between fishing and agricultural enterprises and were regulated by higher administrative bodies, so horizontal ties with other enterprises were weak. Enterprises of the fishing industry of the Primorsky Territory supplied small fish and fish meal for fattening poultry and livestock to livestock farms and animal farms, the entire process was regulated by the regional authorities. During the transition to market relations, the state planning system was destroyed, the previously existing economic ties were lost and at the moment have not been restored.

Currently, a large "Russian Fishing company" engaged in fish production in the Far East produces 16.7 thousand tons of fish meal, which is supplied to the foreign market in China, Vietnam (90%), to the domestic market in Novosibirsk, Krasnoyarsk and local markets (10%).

Based on statistical data on the number of livestock's grown livestock in the Primorsky Territory and their average daily needs, the authors are estimated the potential need for the fish flour of agricultural producers of the region, which is about 10 thousand tons per year.

Due to the expert survey of the leaders of the agrofirms of the Primorsky Territory, enterprises were identified that use fish flour in the production of livestock production:

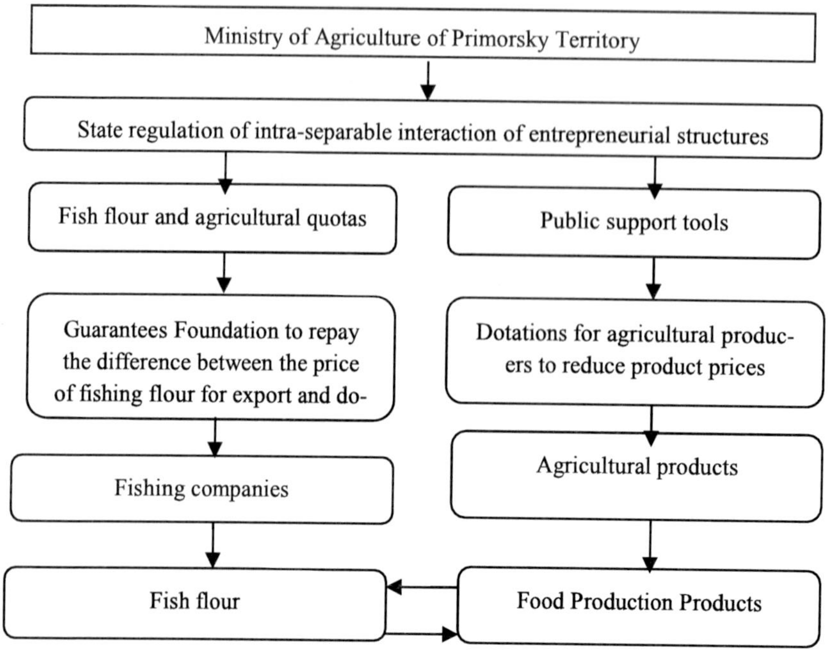

Fig. 1. Diagram of the intra-industry integration of entrepreneurial structures in the apk of Primorsky Territory.

LLC "Poultry Farm" Ussuriyskaya will feed the chickens before slaughter, IP (KFH) Ryachenkov R.Yu.- Cords of the calves, AgroFond P LLC acquired On the trial for fattening pigs.

These data allow us to conclude that it is necessary to develop intra-industry cooperation between agricultural enterprises, which is not yet included in the existing state programs and subprograms for the development of agriculture in the Primorsky Territory. Schematically, the process of intra-industry integration of entrepreneurial structures of the agro-industrial complex of the Primorsky Territory is shown Fig. 1.

4 Discussion

This paper examines the problems of state regulation of intra-industry integration of agricultural enterprises, as one of the important components of reducing prices for agricultural products in the Far Eastern region. This experience of the socialist period of economic activity was used as the main prerequisites for solving this problem, when fishing enterprises supplied producers of agricultural products, this process was regulated by state bodies, as well as the established practice of supplying these products from the Far Eastern basin to the regions of Siberia and abroad.

The expert survey of managers and specialists of agricultural enterprises of the Primorsky Territory conducted by the authors confirmed the need for the entire agricultural sector of the Far East economy to develop intra-industry integration, starting with the use of fish meal and other types of unused products of fishing companies as feed.

In our opinion, the organization of this process is possible if the existing system of state regulation of intra-industry integration of agricultural enterprises is changed in a single Ministry of Agriculture of Russia and the Primorsky Territory.

Of course, this is not limited to a fairly wide range of possible forms and types of intra-industry integration, which, in our opinion, should be the subject of further research.

The effect of intra-industry cooperation will be expressed in obtaining a significant economic effect due to the combination and cooperation of fishing and agricultural enterprises of the Primorsky Territory, rational use of resources, and solving the problem of rising food prices in the region.

5 Conclusion

Summing up the results of the study, it is important to note the revealed negative trend of annual increase in the cost of agricultural products in the regions of the Russian Far East, at a rate significantly higher than the national level of inflation, while reducing the average income of the population. This trend leads to an increase in the gap between the Far Eastern and average Russian food prices, which lead to a reduction in the demographic potential of the region that is strategically important for Russia.

Thus, the commissioning of the President of the Russian Federation should be considered extremely relevant to the end of 2021 years of food prices in the Far Eastern Federal District until the average Russian level.

References

1. Polbin, A.V., Sinelnikov-Murylev, S.G., Trunin, P.V.: The economic crisis of 2020: reasons, policies to deal with and further development of the Russian economy. Voprosy Ekonomiki **6**, 5–21 (2020). https://doi.org/10.32609/0042-8736-2020-6-5-21
2. Borisov, A.I., Danilova, S.S.: Current Russian agricultural development trends. IOP Conf. Ser. Earth Environ. Sci. **548**, 022031 (2020). https://doi.org/10.1088/1755-1315/548/2/022031
3. Zhupley, I.V., Potenko, T.A., Gubarkov, S.V., Tretyak, N.A., Grafov, R.A.: Structural shifts and reform of the agrarian sector of the Russian economy under the conditions of the import substitution. Policy Space Cult. India **6**(4), 25–35 (2018). https://doi.org/10.20896/saci.v6i4.385
4. Yakovlev, V.A., Stepanova, A.N.: Analysis and prospects for the development of agribusiness: regional aspect. IOP Conf. Ser. Earth Environ. Sci. **548**, 022034 (2020). https://doi.org/10.1088/1755-1315/548/2/022034
5. Krasnova, T.G., Plotnikova, T.N., Dulesov, A.S., Karandeev, D.J.: Agro-industrial integration in modern conditions. IOP Conf. Ser. Earth Environ. Sci. **548**, 022014 (2020). https://doi.org/10.1088/1755-1315/548/2/02201
6. Huseynov, E.N., Semenovich, V.S., Zhantemirov, S.A.: Some challenges development of agro-based integration of the cluster approach. RUDN J. Agro Anim. Ind. **4**, 12–20 (2013). https://doi.org/10.22363/2312-797X-2013-4-12-20
7. Simachev, Y., Kuzyk, M.G., Fedyunina, A.A., Zaytsev, A.A., Yurevich, M.A.: Labor productivity in the non-resource sectors of the Russian economy: what determines firm-level growth? Vopr. Ekon. **3**, 31–67 (2021). https://doi.org/10.32609/0042-8736-2021-3-31-67

8. Fojtikova, L., Stanickova, M., Vahalik, B.: Economic integration impact on the EU trade structure. Act Probl. Econ. **182**(8), 41–50 (2016)
9. Tekeeva, F.: State regulation in the APK. Trends Dev. Sci. Educ. **13–2**, 38–39 (2016). https://doi.org/10.18411/LJ2016-4-38
10. Chebokchinova, H.M., Kapsargina, C.A.: Some aspects of national experience of the cluster approach in agro-industrial complex. IOP Conf. Ser. Earth Environ. Sci. **548**, 022048 (2020)
11. Frolova, O.Y., Fomina, L.V., Shmeleva, Z.: The importance of the agrarian sector in the socio-economic systems development: methodological aspect. IOP Conf. Ser. Earth Environ. Sci. **548**, 022023 (2020). https://doi.org/10.1088/1755-1315/548/2/022023
12. Aksyutina, S.V., Mironenko, N.V.: Current issues of the agricultural sector state regulation. Ekonomicheskie otnosheniya **10**(4), 1469–1490 (2020). https://doi.org/10.18334/eo.10.4.111047
13. Potenko, T.A., Zhupley, I.V., Schmidt, Y.I., Kvashko, L.P.: Agricultural policy and import substitution opportunities for the Russian Far East. IOP Conf. Ser. Earth Environ. Sci. **548**, 022011 (2020). https://doi.org/10.1088/17551315/548/2/022011
14. Latkin, A.P., Kharchenkova, E.V.: Russian Far East: prospects for innovation breakthrough, The Territory of New Opportunities. Herald Vladivostok State Univ. Econ. Serv. **11**(1), 6–16 (2019). https://doi.org/10.24866/VVSU/2073-3984/2019-1/006-016

Phytosanitary Monitoring of Grain Crops and Soybeans in the Khabarovsk Territory

Anna Tishkova(✉) and Olga Imtosimi

Far Eastern Agricultural Research Institute, Khabarovsk Federal Center, Far Eastern branch, Russian Academy of Sciences, Klubnaya street, 13, v. Vostochnoe, Khabarovsk District, Khabarovsk 680521, Russia

Abstract. The main reason for the global decline in crop productivity is the development of plant diseases. Fungal and bacterial diseases develop on the roots, leaves, stems, ears and panicles of grain and soybean crops in the Khabarovsk Territory. Inspection of crops facilitates the determination of the type of disease and the level of its development and spread. A decision is made, based on the monitoring data, regarding the methods of protecting the culture. The air temperature and amount of precipitation greatly influence the development of plant pathogens.

Keywords: Khabarovsk territory · Grain crops · Soybeans · Monitoring of plant diseases · Intensity of development and distribution · Temperature regime · Amount of precipitation · Growing season

1 Introduction

Plant diseases are the main reason for the decline in global agricultural production. Approximately 20–50% of the annual global grain losses are due to the development of diseases [1–7]. The symptoms of the diseases adversely affect the growth and development of crops, reduce yields and render agricultural products unusable. Environmental conditions are key factors in the spread and incidence of leaf-stem infections in plants. Warm (24.0–29.0 °C) and rainy weather (relative air humidity ≥80%) contributes to the development of most phytopathogens [8–10].

The primary task of modern crop production is to increase the volume. To solve this issue, it is necessary to implement protective measures at all levels of agricultural production management. Integrated plant protection techniques are planned and conducted based on the results of the annual phytosanitary monitoring of agricultural crops, which provides information on the dynamics of the development, distribution and danger of pests.

The hydrothermal conditions of the southern territories of the Russian Far East (Khabarovsk Territory) satisfy the biological needs of grain crops (oats, wheat and barley) and soybeans, and are also favourable for the development and spread of pathogens of various aetiologies.

The main pathogens of cereal crops are netted spot of barley (*Drechslera teres* [Sacc.] Shoem.), dark brown spot (*Bipolaris sorokiniana* (Sacc.) Shoem.), striped spot

A. Muratov and S. Ignateva (Eds.): AFE 2021, LNNS 353, pp. 293–302, 2022.
https://doi.org/10.1007/978-3-030-91402-8_34

of barley (*Drechslera graminea* [Rabenh.]), red-brown oat speck (*Drechslera avenae* [Eidam.] Scharif.), brown rust (*Puccinia triticina* Erikss.), powdery mildew (*Blumeria graminis* [DC] Speer f. sp. *tritici* March.), barley smut (*Ustilago nuda* [Jens.] Kell & Swing), wheat dust smut (*Ustilago tritici* [Pers.] Rostr.), oat dust smut (*Ustilago avenae* [Pers.] Rostr.), ear and panicle fusarium (*Fusarium* spp.), black ear (*Alternaria* spp., *Cladosporium herbarum* Lk) [12].

According to the literature [11], significant damage to soybean crops is caused by root rot. The causative agents of the root rot can be the fungi *Fusarium spp.*, *Cylindrocarpon destructans* (Zins.) Scholter, *Helminthosporium sativum*, *Gliocladium roseum* (Lk.) Thom., *Pythium ultimum* Tr., *Corynespora cassiicola* (Berk. Et Curt.) Wei, *Thielaviopsis basicola* (Berk. Et Br.) Ferr., *Rhizoctonia solani* Kuehn., *Ascochyta sojiecola* Abr. and others., Septoria (*Septoria glycines*Hemmi), Peronospora (*Peronospora manshurica* [Naum.] Syd.), Cercospora (*Cercospora sojina* Hara.)Yardn.), Ascochytosis (*Ascochyta sojaecola* Abr.), Fusarium (*Fusarium* spp.). Moreover, the most common bacteria that cause root rot are bacterial angular spotting (*Pseudomonas glycineum* Coerper.) and pustular spotting (*Xantomonas phaseoli* var. *sojense* Hedges). A high natural infectious background hinders the productive potential of grown crops. Therefore, the average grain yield ranges from 1.5 to 2.5 ton and the average soybean yield ranges from 0.9 to 2.5 ton.

The purpose of the research was to conduct annual phytosanitary monitoring of grain crops and soybeans in the Khabarovsk Territory to identify the level of development of the most common ground-air infections and analyse the pathological process of the dominant infections under various hydrothermal conditions.

2 Materials and Methods

The screening of phytopathogens was conducted annually from 2000 to 2020 on agrocenoses of grain crops (oats, barley, wheat) and soybeans located in the southern regions of the Khabarovsk Territory (the Khabarovsk, Bikinsky, Vyazemsky districts and the Lazo district). The areas surveyed for each crop are presented in Table 1.

The climatic resources of the southern agricultural regions of the Khabarovsk Territory are monsoon in nature due to the proximity of the Pacific Ocean. The winter monsoon ensures a cold winter with little snow. The duration of the frost-free period (in the air) is 114–164 days. In summer, the highest air temperatures are observed between July and August, and the average monthly air temperatures are +21.1 and +20.0 °C, respectively). The monthly precipitation is unevenly distributed; the winter months account for 19% of the annual precipitation, 27% of the precipitation occurs between April and June and 54% of the precipitation occurs between July and September. The amount of precipitation during the warm period ranges from 360 to 1048 mm [12].

Route surveys of oat, barley and wheat crops were conducted three times during the period of active plant growth: when the full shoots appeared, during the heading and flowering phases, and before harvesting. During the period of tillering-milk ripeness of the grains, diseases on the leaves and stems were taken into account. These included rust diseases, powdery mildew and helminthosporium spots of the grain crops.

Diseases of the leaves and stems of the grain crops were counted along a stepped diagonal by inspecting 10 plants at 10 points.

Table 1. Total surveyed and affected area, thousand hectares.

Year	Oats	Barley	Wheat	Soybean	Total area
2000	7.1	1.9	0.8	7.8	17.6
2001	5.7	2.7	2.2	7.5	18.1
2002	6.1	2.6	2.6	6.6	17.9
2003	4.9	2.2	2.2	9.45	18.75
2006	5.1	1.5	1.45	8.0	16.05
2007	6.7	1.4	2.3	7.7	18.1
2008	7.77	1.8	2.3	7.2	19.87
2011	4.7	1.93	3.4	8.3	18.6
2012	3.3	0	1.1	6.0	10.4
2013	5.35	1.42	1.0	2.11	9.88
2014	3.0	1.5	0.5	8.7	13.7
2015	2.2	0.56	0.24	8.4	13.4
2016	1.7	0.7	0.2	8.6	11.2
2017	1.24	1.0	1.0	6.54	9.78
2018	1.4	0.5	0.5	6.5	8.9
2019	2.125	0.425	0.7	6.4	9.65
2020	1.73	0.44	0.63	9.71	12.51

The degree of damage to the plants by rust diseases (*P. triticina*) was determined using the Peterson scale (Fig. 1) to compare diseased leaves or internodes.

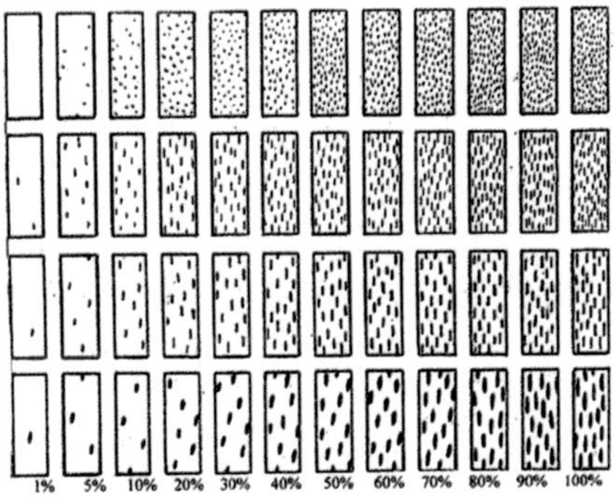

Fig. 1. A scale for assessing the degree of damage to grain crops by types of rust (Peterson et al.).

The intensity of the lesion of the mealy race (B. graminis) was determined using a scale. 0 - no signs of damage (no disease); 1 - very weak, up to 10% of the leaf surface is affected (a light bloom or single pads of the fungus on the leaves of the lower layer;

2 - weak, 11–25% of the leaf surface is affected (a moderate number of pads on the leaves of the lower layer); 3 - medium, 25–50% of the leaf surface is affected (abundant development of the fungus on the lower leaves, on the upper, local, scattered pads); 4 - strong, greater than 50% of the leaf surface is affected (all leaves are strongly affected, pads are well expressed, merging, with abundant hyphae).

Determination of the defeat of grain crops by helminthosporium spots (D. teres, B. sorokiniana, D. graminea, D. avenae) was established for each disease separately based on an eye assessment using a points scale: 0 - there are no signs of damage; 1 - up to 10% of the leaf surface is affected; 2 - 11–25% of the leaf surface is affected; 3 - 26–50% of the leaf surface is affected; 4 - greater than 50% of the leaf surface is affected.

Spike and panicle diseases caused by Fusarium spp., Alternaria spp. and C. herbarum Lk. were counted along the diagonal of the field, where 50 ears were examined at 20 points. To determine the degree of damage, a points scale was used: 0 - there are no signs of damage; 1 - the manifestation of the disease covers up to 10% of the surface of the ear; 2 - the manifestation of the disease covers up to 25% of the ear surface; 3 - the manifestation of the disease covers up to 50% of the surface of the ear; 4 - the manifestation of the disease covers over 50% of the surface of the ear.

The infestation of cereal crops by smut (U. nuda, U. tritici, U. avenae) was conducted separately by species. A total of 50–100 plants were examined at 10 points from fields of up to 100 ha in size. When examining an area greater than 100 ha, an additional two sheaves were taken. The prevalence of smut, expressed as a percentage, determined the lesion group: 0.01–0.2% - single (point 1); 0.21–1.0% - weak (point 2); 1.01–3.0% - average (point 3); 3.01 and greater- strong (point 4).

The phytosanitary monitoring of soybean crops was conducted during the full germination phase, when three to four trifoliate leaves appeared, at the beginning and end of flowering and during the development of the fruits and seeds.

The assessment of root rot infestation was conducted on a brown area and was achieved by digging out 10 plants at 10 equidistant points, followed by washing of the roots in the laboratory. The degree of damage was determined using a points scale: 0 - healthy plants; 1 - weak lesion (not more than 25% of the root area is affected); 2 - medium damage (damage up to 50%); 3 - severe damage (up to 75% of the root is affected); 4 - very severe damage (greater than 75% of the root area is affected) and the presence of dead plants.

Route surveys of soybean crops were conducted to establish the intensity of the development and spread of diseases and were performed on a diagonal at equal distances. A total of 10 plants were examined at 10 points on an area of up to 10 ha, at 20 points on an area of 11–25 ha, at 30 points on an area of 26–50 ha and at 50 points on an area of 51–100 ha. The level of development of leaf spots was determined using a points scale: 0 - there are no signs of damage; 1 - very weak lesion, spots occupy up to 10% of the leaf surface; 2 - weak lesion, spots occupy up to 25% of the leaf surface; 3 - medium lesion, spots occupy up to 50% of the leaf surface; 4 - severe damage, spots occupy up to 75% of the leaf surface; 5 - very severe damage, spots occupy greater than 75% of the leaf area.

The prevalence (number of diseased plants, expressed as a percentage) was calculated using the following formula:

$$P = \frac{n}{N} \times 100\%, \quad (1)$$

where P – is the prevalence of the disease (%); N – is the total number of plants in the samples; n – is the number of diseased plants.

The degree of development of the diseases was calculated using formula 2:

$$R = \frac{\sum(a \times b)}{\sum n \times k} \times 100\% \quad (2)$$

where R – is the degree of development of the disease (%); Σ (a × b) – is the sum of the products of the number of affected plants by the corresponding damage score; Σn – is the total number of plants counted (healthy and diseased) in the variant; k – is the highest score of defeat.

3 Results and Discussion

The mass development of phytopathogens is facilitated by high air temperatures combined with high air humidity and heavy rainfall. The average monthly air temperature during the ontogenesis of the agricultural crops was higher than the average long-term values by 0.2–1.8 °C. The warmest months during the growing season were July and August, and the temperatures did not differ significantly from the norm (Table 2).

Table 2. Temperature regime during the growing season in the study.

Year	Month						Average over 6 months
	April	May	June	July	August	September	
Average for 2000–2020	5.3	13.7	18.1	21.5	19.7	14.0	15.4
Average long-term value	4.5	11.9	17.5	21.2	19.7	13.2	14.7

The amount of precipitation was unevenly distributed throughout the months, particularly in July and August (Table 3).

During the analysis of the phytosanitary monitoring data, the main diseases of the spring grain crops were identified as powdery mildew (B. graminis [DC] Speer), Pyrenophora tritici-repentis (Died.) Drechsler, barley netted spot (D. teres [Sacc.]), dark brown spotting of cereals (B. sorokiniana [Sacc.] Shoem.), striped spotting of barley (D.graminea [Rabenh.] Shoemaker.), reddish-brown spotting of oats (D. avenae [Eidam.] Scharif.), brown spotting of oats (D. avenae [Eidam.] Scharif.), P. triticina Erikss., crown rust of oats (Puccinia coronifera Kleb.), dust smut of barley (U. nuda [Jens.] Kell &

Table 3. The amount of precipitation during the growing season in the study years.

Year	Month						The amount of precipitation during the growing season
	April	May	June	July	August	September	
Average for 2000–2020	38.6	72.3	88.1	147.3	143.0	82.2	571.5
Average long-term value	41.7	60.7	85.0	133.0	147.0	86.5	553.9

Swing), dust smut of wheat (U. tritici [Pers.] Rosttilago), dust smut of oats (U. avenae [Pers.] Rostr.), Fusarium spike and panicle (Fusarium spp.), ear moles (Alternaria spp., C.herbarum Lk) and grain ergot (Claviceps purpurea [Fr.] Tul.).

Powdery mildew in the Khabarovsk Territory is a relatively rare phenomenon. In 2010, barley plants were susceptible to the pathogen. In the phase of milky ripeness, the disease exhibited the characteristics of an epiphytotic in terms of the degree of spread and intensity of the plant damage. In our studies, the prevalence of powdery mildew on cereal crops was observed from 2010 to 2013. The damage to the plants varied from 33.9–100% and the disease development ranged from 6.6–60% (Fig. 2).

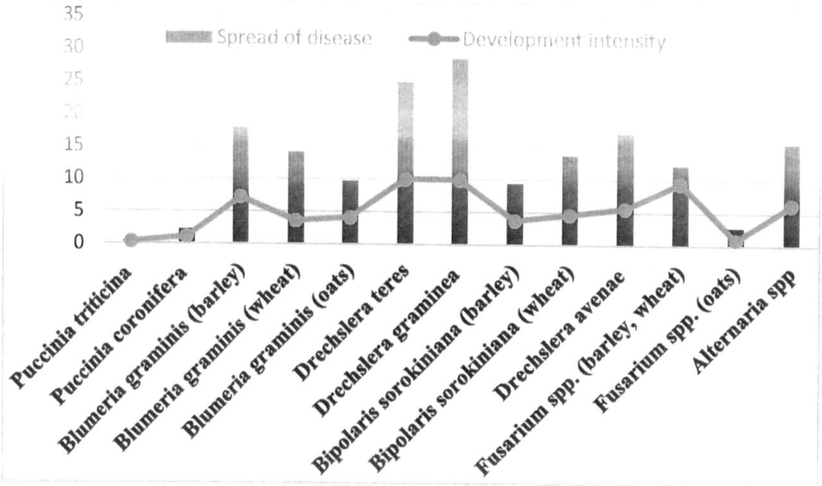

Fig. 2. Weighted average percentage of the disease prevalence on grain crops.

Helminthosporium spots of cereals are diagnosed annually. Barley was the crop that was most affected by helminthosporiosis in the Khabarovsk Territory. Netted spot showed the strongest distribution among the barley crops and, on average, affected 24.9% of plants with a development of 10%. This was followed by striped spot of barley which, on average, damaged 28.6% of the plants and showed a development of 9.9%. Moreover, in certain years, dark brown spotting of barley was noted, and the weighted average percentage of prevalence for the studied time interval was 9.4% with a development of 3.7%. In 2000, of the surveyed crops of striped barley, 100% of the plants were affected, and a development of 65% was observed. Additionally, during the first half of July, the disease had developed to an epiphytotic level. In 2009, in the crops of barley, a complex infection of plant lesions with netted and striped spots was noted; the infection level ranged from 39–86% and a maximum of 100% of the plants were affected. The average development of the disease was 42.5% and the maximum ranged from 80–85%. A high level of development of leaf-stem diseases (helminthosporiosis) was associated with the accumulation of infection in the soil and plant debris. Despite the almost annual manifestation of dark brown spotting of wheat and reddish-brown spotting of oats, the dynamics of the development of the disease was approximately the same. The weighted average prevalence rate for the studied time interval was 13.7% for wheat and 17.1% for oats.

Brown rust of wheat was recorded in 2014 on an area of 0.1 thousand hectares of a total of 0.7 thousand hectares of cereal crops surveyed. A total of 6.5% of the plants were affected, with an intensity of damage of 5%.

Due to the weather conditions of 2010, 2012 and 2014, weak development (from 1–12.7%) and spread (8.5–21.2%) of oat crown rust were observed. In 2010, in certain fields, focal development of the disease was observed with damage to 30–80% of the plants. The degree of damage was 20–25%. In subsequent years, oat crown rust was not recorded.

The development of smut diseases over 18 years of observation was focal in nature. Head smut was observed in crops of barley, wheat and oats and ranged from 0.01%–0.7%, but in most cases, it consisted of a single lesion. No dust smut was observed during the past 5 years of surveys of oat crops. Hard smut was not observed in the spring grain crops throughout the study period.

Fusarium blight of cereals has spread globally. It exhibits the characteristics of a pandemic, and under favourable weather conditions (high temperature and humidity during the flowering period), it develops continuously. An analysis of the nature of the spread of fusarium revealed that the disease showed an epiphytotic development in 2006, and61% of the ears of wheat and 29% of the ears of barley were affected. In 2011, 70–75% of the ears were affected in all wheat crops surveyed. In other years, the disease showed a weak degree of damage.

The defeat of the generative organs of grain crops by moles occurs annually, with an average prevalence of 15.7% and an intensity of 6.3%. In 2012, the mob of a spike and panicle developed on plants that had been exposed to waterlogging and stagnated due to the untimely harvesting of the fields. In 2018, the development of the mob was epiphytotic, and 32.6% of the ears were affected, with 21.7% intensity.

In 2014, for the first time, ergot was noted on wheat of the Altayskaya bezostaya 70 variety in the Khabarovsk region on an area of 100 ha; 0.1% of the ears were affected.

The results of the phytosanitary monitoring of the soybean crops conducted between 2000 and 2020 revealed that the main plant diseases in the Khabarovsk Territory consisted of root rot, the main pathogens of which were *Fusarium* spp., *R. solani*, *A. sojiecola*, etc., Septoria, Peronosporosis, Cercosporiasis, Bacteriosis (*P. glycineum*) and Ascochitis. The degree of development and spread of the diseases was dependent on the hydrothermal conditions during the growing season and varied greatly from year to year.

The development of root rot was recorded annually. The average rate of the intensity of the development over 18 years was 5.4 and7.8% of plants were affected (Fig. 3). Heavy rainfall and higher air temperatures in May 2003, 2006, 2016 and 2019 contributed to the increased development of rot on the roots of soybeans. On average, 22.3% of the root surface was affected in 14.6% of the plants.

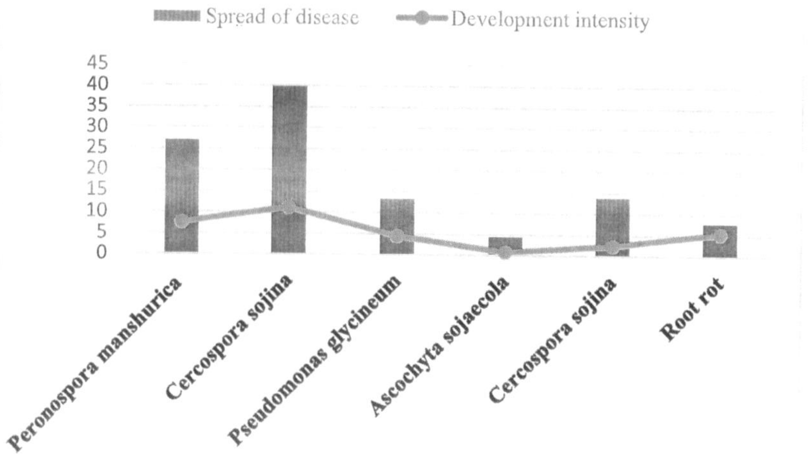

Fig. 3. Weighted average percentage of the development and spread of soybean diseases.

During the study, the development of leaf-stem infections of soybeans generally began in late June to early July, when the temperature and humidity of the air reached optimal values for the start of the pathological process.

The optimal environmental conditions for the development of Septoria on soybean leaves are an air temperature of 26–28 °C and relative humidity of 80–90%. July 2011 was hot and humid and provoked the epiphytotic development of this disease. The entire surveyed area of soybean crops was affected, with an average intensity of 45%. The annual weighted average rust spot rate was 11%, with traits present on 39.8% of the plants.

The average development of downy mildew on soybean leaves in the Khabarovsk Territory during dry years (2001, 2003, 2007) ranged from 0.7 to 3.8%. During the period of excessive waterlogging (2011, 2016, 2019–2020), there was a significant increase in the prevalence of Septoria, and its development on the leaves ranged from 10.0 to 29.8%.

Analysis of the data showed that the average level of bacteriosis development in the region was 4.4%. The angular spotting reached maximum development during the formation of the beans and seeds in years with abundant precipitation (2013, 2019, 2020), which caused intense waterlogging of the upper soil layer. The average prevalence of bacteriosis was 36.4% and the degree of development was 10.5%.

Until 2011, the development of Cercosporosis was depressed (distribution 0.5%, development 0.2%). However, in 2011–2013 and 2015, the weather conditions provoked increased development of the disease throughout the survey area. Signs of the disease were recorded on 50% of the plants, which, on average, occupied11.6% of the leaf area. In subsequent years, the infection of soybean plants with Cercospora was at low levels and was, on average, 0.8%, with a spread of 5%.

During the study period (18 years), the development of Ascochitosis was depressed and, on average, was recorded at 0.72%, and4.4% of the plants were affected. The fungus showed the greatest development in 2011 and 2020 (5% and 7.4%, respectively), with an average spread of 28% over the two years.

The phytosanitary monitoring of soybean plants, conducted over 18 years, demonstrated that, in the monsoon climate of the region, the main pathogens of soybean diseases were root rot of complex aetiology (*Fusarium* spp., *P. ultimum* Tr., *A. sojiecola* Abr., etc.), Septoria on the leaves (*S. glycines* Hemmi), Peronosporosis on the leaves (*P.manshurica*[Naum.] Syd.), Cercospora on the leaves (*C.sojina* Hara.), Ascochitis (*A.sojaecola* Abr.) andbacteriosis (*P. glycine*).

The level of development and spread of the pathogens in soybean agrocenoses was dependent on the hydrothermal conditions of the ontogeny period and varied markedly over the years. Thus, 2011, which was hot and humid, became the most favourable year for the development of fungal diseases, and the presence of moisture was the main condition for the development of bacterial diseases.

The pathocomplex of fungal soybean diseases should be monitored annually in the Khabarovsk Territory to limit the damage and develop effective methods of protecting the soybean crops.

4 Conclusions

The phytosanitary monitoring of the spring crops and soybeans, conducted over 18 years, demonstrated that, in the monsoon climate of the region, the pathogens present were *B. graminis, D. teres, B. sorokiniana, D. graminea, D. avenae, P. triticina, P. coronifera, U. nuda, U. tritici, U. avenae, Fusarium* spp., *Alternaria* spp., *C. herbarum, C. purpurea, P. ultimum, A. sojiecola, S. glycines, P. manshurica, C. sojina* and *P. glycineum.*

The dynamics of the spread of epiphytotic infectious diseases were dependent on the hydrothermal regime during the period of active plant growth and varied noticeably from year to year. Thus, 2011, which was hot and humid, was the most favourable year for the development of fungal diseases on the soybean plants, and the presence of moisture was the main condition for the development of bacterial diseases. The weather conditions in 2009 contributed to the intense damage (via spots) to the leaves of the cereal crops.

Based on the results obtained from these surveys on the agricultural crops in the Khabarovsk Territory, it is necessary to implement protective measures in the region that will optimise the phytosanitary situation of agrocenoses to prevent the development of epiphytoties and increase grain yield.

References

1. Reznikov, S., et al.: Evaluation of the efficacy and application timing of different fungicides for management of soybean foliar diseases in northwestern Argentina. Crop Prot. **124**, 104844 (2019). https://doi.org/10.1016/j.cropro.2019.104844
2. Kandel, Y.R., Mueller, D.S., et al.: Wise Impact of fluopyram fungicide and preemergence herbicides on soybean injury, population, sudden death syndrome, and yield. Crop Protect **106**, 103–109 (2018). https://doi.org/10.1016/j.cropro.2017.12.009
3. Pérez Pizá, M.C., Prevosto, L., Zilli, C., Cejas, E., Kelly, H., Balestrasse, K.: Effects of non–thermal plasmas on seed-borne Diaporthe/Phomopsis complex and germination parameters of soybean seeds. Innov. Food Sci. Emerg. Technol. **49**, 82–91 (2018). https://doi.org/10.1016/j.ifset.2018.07.009
4. Lidwell-Durnin, J., Lapthorn, A.: The threat to global food security from wheat rust: ethical and historical issues in fighting crop diseases and preserving genetic diversity. Glob. Food Sec. **26**, 100446 (2020). https://doi.org/10.1016/j.gfs.2020.100446
5. Sharma-Poudyal, D., Sharma, R.C., Duveiller, E.: Control of Helminthosporium leaf blight of spring wheat using seed treatments and single foliar spray in Indo-Gangetic Plains of Nepal. Crop Protect **88**, 161–166 (2016). https://doi.org/10.1016/j.cropro.2016.06.017
6. Kaur, J., Bala, R., Kaur, H., Pannu, P.P.S., Kumar, A., Bhardwaj, S.C.: Current status of wheat diseases in Punjab. Agric. Res. J. **55**(1), 113–116 (2018). https://doi.org/10.5958/2395-146X.2018.00018.2
7. Gangwar, O.P., Bhardwaj, S.C., Singh, G.: Barley diseases and their management- an Indian perspective. Wheat Barley Res. **10**(3), 138–150 (2018). https://doi.org/10.25174/2249-4065/2018/83844
8. Stumbriene, K., Gudiukaite, R., Semaskiene, R., Svegzda, P., Jonaviciene, A., Suproniene, S.: Screening of new bacterial isolates with antifungal activity and application of selected Bacillus sp. cultures for biocontrol of Fusarium Graminearum under field conditions. Crop Protect **113**, 22–28 (2018). https://doi.org/10.1016/j.cropro.2018.07.006
9. Silva, O.S., Santos, H.A.A., Dalla, P.M., May-De, L.L.: Mio Potassium phosphite for control of downy mildew of soybean. Crop Protect **30**, 598–604 (2011). https://doi.org/10.1016/j.cropro.2011.02.015
10. Figueroa, M., Hammond-Kosack, K.E., Solomon, P.S.: A review of wheat diseases - a field perspective. Mol. Plant Pathol. **19**(6), 1523–1536 (2018). https://doi.org/10.1111/mpp.12618
11. Zaostrovnykh, V.I., Kadurov, A.A., Dubovitskaya, L.K., Ryazanova, O.A.: Monitoring of the species composition of soybean diseases in different zones of soybean. Far East Agric. Herald **4**(48), 51–67 (2018). https://doi.org/10.24411/1999-6837-2018-14082
12. Aseeva, T.A., Trifuntova, I.B., Zenkina, K.V., Imtosimi, O., Tishkova, A.G., Savchenko, N.E.: Features of the manifestation of fungal diseases on grain crops in the monsoon climate of the Far East. Acrh. Sci. Technol. Agribus. **34**(12), 12–18 (2020). https://doi.org/10.24411/0235

Influence of Alfalfa Changeable (Medicago Varia Mar.) on the Fertility of Meadow-Brown Bleached Soil of the Primorsky Krai

Elena Ivanova(✉) and Lyubov Yayuk

Sakhalin Research Institute of Agriculture, 22, Gorky Alley, 693022 Yuzhno-Sakhalinsk, Russia
kirena2010@yandexl.ru

Abstract. A very important step for preserving and reproducing the fertility of the meadow-brown soil of the Primorsky Krai is an increase of agriculture's biologization and ecologization and growth of the input level of fresh organic matter into the soil through the introduction of perennial legumes, in particular alfalfa changeable (Medicago Varia Mar.), into field crop rotations. In the first year of alfalfa, the soil was characterized as moderately supplied with P_2O_5, by the third year - an increased content of P_2O_5, by 4–5 years - high and very high. The content of K_2O under alfalfa 2–5 years of life showed a clear positive trend, similar to P_2O_5. According to the availability of K_2O, the soils were of average and increased availability in the first years of the alfalfa herbage life and very high availability - by the fifth year of life. The accumulation of humus in the arable layer with five years of use of alfalfa was 1.1–1.2%. The content of readily hydrolyzable nitrogen under alfalfa in the long-term cycle increased - before the experiment was laid, the content of readily hydrolyzable nitrogen corresponded to a low supply, from the fall of the first year and in subsequent years - to an average supply. The calcium content increased from 10.9 mg equiv./100 g before setting up the experiment to 16.7 mg equiv./100 g by the fifth year of alfalfa life in spring and autumn, respectively, from 11.6 to 17.2 mg equiv./100 g of soil.

Keywords: Alfalfa changeable · Fertility · Agrochemical indices of soil

1 Introduction

Soil fertility is the most important condition for the existence of life on Earth in general and the level of agricultural productivity in particular. There is nothing that can replace the soil, the material basis of human existence. The arable soils of the Far East with a low level of natural fertility are in dire need of cultivation and an increase in fertility. The results of monitoring the fertility of arable land in the Far Eastern Federal District show a high proportion of acidic soils, with a low content of mobile phosphorus (more than 50%), and a low content of organic matter (20–60%).

The soils of the south of the Russian Far East are pronc to waterlogging, destructive processes of a natural and anthropogenic nature: slope runoff, floods (the intensity of

A. Muratov and S. Ignateva (Eds.): AFE 2021, LNNS 353, pp. 303–310, 2022.
https://doi.org/10.1007/978-3-030-91402-8_35

soil washout is 5–40 t/ha per year), a decrease in fertility (the loss of humus is 0.25–0.45 t annually), overconsolidation, pollution, etc. [1].

The preservation and reproduction of soil fertility in the region (especially in conditions of low availability of arable land with agrochemical resources) determine the relevance and necessity of introducing perennial leguminous grasses into field crop rotations. This will allow solving a set of major priority tasks: increasing soil fertility and crop yields, improving the environmental situation, reducing energy consumption, increasing the economic efficiency of the agro-industrial complex, ensuring food security in the Far Eastern region, etc.

The monograph "Scientific Foundations of Agriculture in the Far East of Russia" summarizes the long-term studies of scientists from the Far East. It emphasizes that one of the conditions for the expanded reproduction of the fertility of seasonal permafrost soils in the zone of monsoon agriculture is the development of crop rotations with perennial and leguminous grasses. This helps to increase the yield of all crops. Perennial grasses of two years use contributed to the reproduction of humus. The best results for the reproduction of soil fertility in Primorye were obtained in crop rotation in the presence of 20% of perennial grasses [2]. According to S.N. Inshakova and A.N. Emelyanov, the content of the main nutrients in the soil under perennial crops increases with the duration of their life. This trend is especially clear pronounced in legumes - alfalfa and clover. The organic matter content increases under alfalfa by 0.20%. It is under alfalfa and clover that the most favourable conditions are formed for increasing soil fertility, increasing the content of humus, improving the water-physical and agrochemical properties of soils, increasing microbiological activity [3]. Unfortunately, there is a significant reduction in the sown area of forage crops in the Far Eastern Federal District - in 2018 there were 203.3 thousand hectares, while in 1990–1196 thousand hectares (forage crops remained only 17% of the 1990 indicators) [4].

Leguminous herbs have a beneficial effect on the physicochemical properties of the soil, contribute to an increase in its fertility, and have a positive effect on the structure-forming process and soil fertility. The cultivation of leguminous grasses is cheaper, and they make the most of the bioclimatic resources [5]. A powerful root system of perennial grasses has a beneficial effect on the physical and agrochemical indicators of soil fertility. E.P. Denisov and A.M. Kosachev (2011) have established that alfalfa, goat's rue and sainfoin increased soil fertility the most intensively of all leguminous crops in the Saratov region, and their cultivation is energetically and economically efficient [6]. It is necessary to improve the structure of crops and increase soil fertility due to biological factors, including the expansion of crops of perennial legumes, especially high-protein alfalfa [7].

In scientifically grounded farming systems perennial grasses are a source of carbon and nitrogen to replenish humus reserves, and a factor in protecting soil from erosion. The source of reproduction of organic matter is root and crop residues [3]. Alfalfa changeable accumulates more root and post-cut residues than annual grasses; due to their death and decomposition the reserves of mobile nutrients in the soil increase. In the Primorsky Krai, an increase in the mass of alfalfa roots in the first to fifth years of life was established from 4.17 t/ha to 24.62 t/ha, the increase was 3.78–8.34 t/ha [8]. The composition of root and crop residues contributes to better humification of organic matter, which creates conditions for a deficit-free or positive humus balance. Alfalfa prevents the leaching of nutrients outside the root layer, which is important for excessively humid periods of the Far Eastern zone.

In field crop rotations, with a small number of livestock, soil replenishment with organic matter is possible only with perennial grasses [9]. In Central Yakutia one of the methods of increasing the fertility of permafrost soils in the permafrost zone is the cultivation of perennial legumes, which increases the content of soil organic matter by 0.6% [10]. In the Central Non-Black Earth Region, alfalfa from the first to fourth years of life provides an annual increase in humus in the arable soil layer of 0.4–2.6 t/ha, at a high rate in the first two years.

The enrichment of the soil with biological nitrogen by alfalfa-growing is 255.7 kg/ha; this gives an additional increase in the yield of spring wheat 1.6–3.0 t/ha [11]. The increase in nitrogen nutrition after leguminous grasses is conditioned by the decomposition of stubble-root residues, which increases the mineralization of organic matter in the soil and enhances the effect of the action of the layer of legumes [13]. Also, during the decomposition of its roots, the soil is enriched with phosphorus, calcium and potassium. By the quantity and quality of plant residues, by their influence on the structure of the soil and its fertility, alfalfa is not inferior to organic fertilizers. The creation of a strong lumpy structure increases the yields of all crops sown on the layer of alfalfa and its mixtures with cereal grasses. The phytosanitary and phytomeliorative role of alfalfa is great.

After perennial grasses, the number of structural particles increases significantly, a more complete mineralization of organic matter occurs, i.e. improves the nutritional background for crops [14]. Alfalfa is a natural soil structure improver. Experiments carried out in the Primorsky Krai found that for five years of alfalfa vegetation, the soil structure was characterized by the predominance of agronomically valuable particles [15]. Under the conditions of Udmurtia, the cultivation of alfalfa changeable increased the coefficient of soil structure to 2.05 compared to 0.86 in the variant with grain crops [16].

The invaluable effect of alfalfa on increasing soil fertility is also confirmed by foreign researchers. Alfalfa is the leader in the accumulation of organic matter in the soil due to nitrogen fixation [17–23].

The purpose of the research is to establish the influence of the alfalfa changeable (Medicago Varia M.) on the indicators of soil fertility of meadow-brown bleached soil.

2 Materials and Methods

The objects of study were the arable layer of meadow-brown bleached soil, heavy in granulometric composition, and alfalfa changeable varieties Vega for 87 different years of life.

The soil was sampled in experimental plots under alfalfa changeable 1–5 years of life (in spring and autumn 2010) on the territory of the collective farm of the feed production department of the Primorsky Research Institute of Agriculture.

Soil samples were analyzed: pH_{salt} - according to the CINAO method (GOST 26483-85), mobile phosphorus and potassium compounds - according to the Kirsanov method in the CINAO modification (GOST R 54650-2011), humus - according to the Tyurin method in the CINAO modification (GOST 26213-91), easily hydrolysable nitrogen - by the Tyurin and Kononova method, exchangeable calcium - by the CINAO method (GOST 26487-85). The assessment of agrochemical indicators was carried out according to the regional scale.

3 Results and Discussion

We have established an increase in the supply of nutrients with prolonged exposure to alfalfa (Table 1).

Table 1. The content of nutrients in the soil under alfalfa changeable 1–5 years of life.

Variants	pH	P_2O_5	K_2O	Humus, %	N	Ca
		Spring				
Alfalfa 1 year	4.9	37	125	3.0	5.6	10.9
Alfalfa 2 year	5.0	41	150	3.6	6.1	12.1
Alfalfa 3 year	5.4	45	150	3.2	5.9	13.4
Alfalfa 4 year	5.5	85	185	3.9	6.4	13.4
Alfalfa 5 year	6.0	108	215	4.1	6.9	16.7
		Autumn				
Alfalfa 1 year	4.6	30	90	3.2	6.1	11.6
Alfalfa 2 year	4.9	31	95	3.2	6.4	12.0
Alfalfa 3 year	5.1	40	100	3.5	6.8	11.6
Alfalfa 4 year	5.2	73	160	4.2	6.8	15.1
Alfalfa 5 year	6.0	95	195	4.4	6.9	17.2

The agrochemical properties of soils are associated with the dynamics of soil processes. Their intensity, direction, cyclicity depend on physicochemical interactions, the vital activity of microorganisms, and the development of plants. The seasonal dynamics of the content of mobile nutrients in the soil is well known.

Acidity was less susceptible to seasonal fluctuations than other agrochemical indicators of meadow-brown bleached soil. Analyzing the data in Table 1, we note that the acidity of soils under alfalfa in the long-term cycle varies from moderately acidic to slightly acidic and close to neutral.

The availability of phosphorus mobile forms is one of the main indicators of soil fertility. We noted a decrease in the content of mobile phosphorus during the growing season of alfalfa, which is explained by biological (alfalfa is characterized by an increased need for the content of nutrients in the soil and, first of all, phosphorus) and physicochemical absorption.

The cultivation of alfalfa contributes to an increase in the content of mobile phosphorus in meadow-brown soil, and in the first years the growth is gradual, and in the fourth or fifth years, a significant jump is observed. Alfalfa plants consume different amounts of phosphorus at different periods of their life. In the first year of alfalfa, the soil was characterized as moderately supplied with P_2O_5, increased content of P_2O_5 - by the third year of life, high and very high - by 4–5 years of age.

Insufficient potassium content in the soil can lead not only to a decrease in the productivity and quality of cultivated crops but also to the loss of the ecological and economic functions of the soil. Soil scientists and agrochemists of the Primorsky Krai note the high importance of the degree of anthropogenic impact on the formation of the potassium state of agriculturally transformed soils.

Heavy soils can fix significantly more potassium than light soils.

Lack of potassium leads to a weakening of the nitrogen-fixing activity of alfalfa nodule bacteria. The nature of the vegetative dynamics of the content of mobile potassium is similar to the seasonal dynamics of the content of mobile phosphorus since these elements have similarities in their physicochemical nature. Changes in hydrothermal conditions (desiccation-moistening-waterlogging) promote the fixation of potassium in a non-exchangeable form. Intensive absorption by plants during the period of intensive growth increases the deficiency of exchangeable potassium in the soil by the end of the growing season (by autumn). The content of mobile potassium under alfalfa 2–5 years of age showed a clear positive trend - from 125 to 215 mg/kg in the spring and from 90 to 125 mg/kg in the autumn. According to the availability of K_2O, the soils in the experiment were of average and increased availability in the first years of the life of the alfalfa herbage and very high availability - by the fifth year.

Humus is the most important organic compound, which impacts the physicochemical properties of the soil and its food and water regimes. The seasonal dynamics of the humus content under alfalfa of different years is not unambiguous: by the end of the first and the third-fifth years of life an increase is observed, of the second - a decrease. The decrease in the humus content is most likely due to an increase in the rate of humus oxidation, the intensity of redox processes, which depend on aeration and soil moisture. In autumn, the recovery processes begin to prevail because of the low aeration at the increased moisture, and accordingly, the humus content increases. The important role of plant life in increasing the humus content should be emphasized. With a five-year period of alfalfa use, the accumulation of humus in the arable layer was 1.1–1.2%.

The yield and quality of crops largely depend on the content of available nitrogen-containing substances in the soil. By the amount of hydrolyzable nitrogen in the soil, it is judged about the supply of plants with nitrogen, if its content in the soil is more than 60 mg, such soils are referred to as little need for nitrogen fertilizers. At the same time, in addition to the content of easily hydrolyzed nitrogen, it is necessary to take into account the cultivation of the soil, the characteristics of the cultivated crop, the content of nitrates in the arable layer of the soil.

As for the seasonal dynamics of the content of easily hydrolyzed nitrogen, by autumn its content under alfalfa grows in comparison with the spring values. The content of easily hydrolyzed nitrogen under alfalfa in the long-term cycle increased - from 5.6 to 6.9 mg/100 g of soil in the spring and from 6.1 to 6.9 mg/100 g in the autumn. Before setting up the experiment, the content of easily hydrolyzed nitrogen corresponded to a low supply. From the fall of the first year and in subsequent years - average security.

Calcium helps to optimize soil structure. The cultivation of alfalfa provides a consistent increase in the calcium content in the bleached meadow brown soil. Its content in our experiment varied from 10.9 mg equiv./100 g before setting the experiment to 16.7 mg equiv./100 g by the fifth year of alfalfa life in spring and autumn, respectively, from 11.6 to 17.2 mg.eq./100 g of soil.

The reserves of nutrients in the soil under alfalfa during long-term cultivation are shown in Table 2.

Table 2. Total reserves of nutrients in soil under alfalfa changeable in a long-term cycle, kg/ha.

Variants	P_2O_5	K_2O	Humus	N
	Spring			
Alfalfa 1 year	117.48	396.88	95.25	177.8
Alfalfa 2 year	134.28	491.25	117.90	199.8
Alfalfa 3 year	148.50	495.00	105.60	194.7
Alfalfa 4 year	286.88	624.38	131.63	216.0
Alfalfa 5 year	369.90	736.38	140.43	236.3
	Autumn			
Alfalfa 1 year	95.25	285.75	101.60	193.7
Alfalfa 2 year	101.53	311.13	104.80	209.6
Alfalfa 3 year	132.00	330.00	115.50	224.4
Alfalfa 4 year	246.38	540.00	141.75	229.5
Alfalfa 5 year	325.38	667.88	150.70	236.3

Nowadays it is believed that the stock of mobile phosphates is one of the main characteristics of the overall cultivation of the soil. Phosphorus reserves in the arable horizon largely depend on its capacity. By the fifth year of life of alfalfa, the reserves of mobile phosphates increase 3.2–3.4 times, while the reserves of potassium - 1.8–2.3 times. Humus reserves in the soil under alfalfa crops, similar to P2O5 and K2O, increased more intensively for 3–5 years of its life. The reserves of easily hydrolyzable nitrogen on the meadow-brown soil under the alfalfa changeable in the second year of life amounted to 205 kg/ha, by the fifth year of life - 236 kg/ha (which is consistent with the data of E.P. Sinelnikov, Yu.I. Slabko, 2005).

4 Conclusion

Thus, for restoring, preserving and reproducing the fertility of the meadow-brown soil of the Primorsky Krai, an increase of agriculture's biologization and ecologization and growth of the input level of fresh organic matter into the soil due to the introduction of perennial legumes, in particular alfalfa, into field crop rotations is important. Long-term cultivation of alfalfa changeable contributes to the enrichment of the soil with organic matter and nutrients. The acidity of soils under alfalfa in the long-term cycle varies from moderately acidic to slightly acidic and close to neutral. In the first year of alfalfa's life the soil was characterized as moderately supplied with P2O5, by the third year - an increased content of P2O5, by 4–5 years - high and very high. The content of K2O under alfalfa 2–5 years of life showed a clear positive trend similar to P2O5. According to the availability of K2O, the soils were of average and increased availability in the first years of the life of the alfalfa herbage and very high availability - by the fifth year. The accumulation of humus in the arable layer during five years of use of alfalfa was 1.1–1.2%. The content of readily hydrolyzable nitrogen under alfalfa in the long-term cycle increased - before the experiment was laid, the content of readily hydrolyzable nitrogen corresponded to a low supply, from the fall of the first year and in subsequent years - to an average supply. The calcium content increased from 10.9 mg equiv./100 g before setting up the experiment to 16.7 mg equiv./100 g by the fifth year of alfalfa life in spring and autumn, respectively, from 11.6 to 17.2 mg equiv./100 g of soil.

Cultivation of alfalfa changeable is a significant factor in regulating the balance of organic matter, the cheapest and easiest way to replenish soil organic matter. The enrichment of the soil with the main nutrients and calcium contributes to an increase in soil fertility, the yield of cultivated crops, and an increase in the efficiency of agriculture.

References

1. Surzhik, M.M., Mukhina, N.V., Avramenko, A.A., Naumova, T.V., Falko, V.V.: Ecological sustainable of agricultural soils of Russian Southern Far East. Trends Dev. Sci. Educ. **63**(2), 119–122 (2020). https://doi.org/10.18411/lj-07-2020-50
2. Kosolapov, V.M., Kostenko, S.I., Tyurin, Y., Shamsutdinova, E.Z., Piskovskii, Y.: Perennial forage grasses - the basis for greening agricultural production. IOP Conf. Ser. Earth Env. Sci. **663**, 012022 (2021). https://doi.org/10.1088/1755-1315/663/1/012022
3. Li, D.C., Velde, B., Li, F.M., Zhang, G.L., Zhao, M.S., Huang, L.M.: Impact of long-term alfalfa cropping on soil potassium content and clay minerals in a semi-arid loess soil in China. Pedosphere **21**(4), 522–531 (2011). https://doi.org/10.1016/S1002-0160(11)60154-9
4. Bekuzarova, S.A., Khatsaeva, F.M., Tebieva, D.I., Bekmurzov, A.D., Kebalova, L.A., Gobeev, M.A.: Soil toxicity reduction by phytoindicators. IOP Conf. Ser. Earth Environ. Sci. **677**, 042100 (2021). https://doi.org/10.1088/1755-1315/677/4/042100
5. Ivanova, E.P.: Change of meadow-brown bleached-out soil aggregate compound under long-term use of lucerne. Var. Pac. Sci. Rev. **14**(3), 313–314 (2012)
6. Zhu, R.F., Tang, F.L., Liu, J.L., Liu, F.Q.: Co-inoculation of arbusculr mycorrhizae and nitrogen fixing bacteria enhance alfalfa yield under saline conditions. Pak. J. Bot. **48**(2), 763–769 (2016)
7. Ramírez-Bahena, M.-H., Vargas, M., Martín, M., Tejedor, C., Velázquez, E., Peix, Á.: Alfalfa microsymbionts from different ITS and nodC lineages of Ensifer meliloti and Ensifer medicae symbiovar meliloti establish efficient symbiosis with alfalfa in Spanish acid soils. Appl. Microbiol. Biotechnol. **99**(11), 4855–4865 (2015). https://doi.org/10.1007/s00253-014-6347-6
8. Gu, Y.J., Han, C.L., Kong, M., Shi, X.Y., Zdruli, P., Li, F.M.: Plastic film mulch promotes high alfalfa production with phosphorus-saving and low risk of soil nitrogen loss. Field Crops Res. **229**, 44–54 (2018). https://doi.org/10.1016/j.fcr.2018.09.011

9. Jáuregui, J.M., Mills, A., Black, D.B.S., Wigley, K., Ridgway, H.J., Moot, D.J.: Yield components of lucerne were affected by sowing dates and inoculation treatments. Eur. J. Agron. **103**, 1–12 (2019). https://doi.org/10.1016/j.eja.2018.10.005
10. Lu, J., Liu, M., Yang, M., Xie, J., Yang, H., Li, L.: Leaf resorption and stoichiometry of N and P of 1, 2 and 3 year-old lucerne under one-time P fertilization. Soil Til. Res. **197**, 104481 (2020). https://doi.org/10.1016/j.still.2019.104481
11. Sun, T., Li, Z.: Alfalfa-corn rotation and row placement affects yield, water use, and economic returns in Northeast China. Field Crops Res. **241**, 107558 (2019). https://doi.org/10.1016/j.fcr.2019.107558

Fixed Capital Investments in the Russian Federation: Territorial Differences

Natalia Ivanova(✉) and Lubov Pashina

Far Eastern State Agrarian University, 86, Politeknicheskaya Street, Blagoveschensk 675005, Russia

Abstract. This article presents a comparative analysis of the dynamics of indicators characterizing investments in fixed assets in the context of the federal districts of the Russian Federation. The dynamics of the volume of investments in fixed assets based on the volume of the gross regional product, capita and land area in the context of the federal districts of the Russian Federation is considered. Due to significant territorial and sectoral differentiation of federal districts, as well as ongoing economic processes in the Russian Federation, for the period 2000–2019. noted an increase in investment in fixed assets, changes in the regional and sectoral structure of investments, an increase in investment in agriculture, forestry, hunting, fishing and fish farming, as well as changes in the structure of ownership of investments in fixed assets.

Keywords: Investments in fixed assets · Federal district · Dynamics · Structure · Industries · Land area · Population · Gross regional product · Form of ownership

1 Introduction

Fixed capital investments are an important factor for successful and stable economic development, stimulating production growth and increasing its economic efficiency. A feature of modern economic growth in certain regions and countries is the increasing influence of intraregional differentiation of the socio-economic development of small territories on it [1].

The problems of differentiating regions in terms of investment activity, investment climate and investment attractiveness have been studied by many researchers [2, 3]. According to some authors, the investment activity of individual regions is influenced by the conditions of the region's existence at the macro level and its natural dependence on the external environment: the conditions of the unsteady economy of the Russian Federation and the unstable foreign policy environment increase the risks for foreign investors [4].

The growth of investments in fixed assets contributes to an increase in the gross regional product (GRP). Previously, researchers have identified the dependence of the influence on the GRP of the number of innovative enterprises in the regions, the development and concentration of production and technologies, the volume of trade,

A. Muratov and S. Ignateva (Eds.): AFE 2021, LNNS 353, pp. 311–320, 2022.
https://doi.org/10.1007/978-3-030-91402-8_36

the availability of resources and infrastructure [2, 5]. Therefore, the choice of an investment-friendly location is of paramount importance for Russian firms [6].

According to the authors, poor management quality reduces the return on investment and is an important factor in determining economic activity at the regional level [7].

The investment market in Russia is highly concentrated, as evidenced by the Herfindahl-Hirschman coefficient (2.205 with a threshold of 1.800) [8]. The greatest concentration of investments is achieved in the Central Federal and Volga Districts, against the background of a lack of investment in the Northwest, followed by the Caucasian, Ural and Far Eastern Federal Districts [8].

The diversity and complexity of the natural and climatic conditions of the regions affect the level of agricultural production, while large agribusiness is more profitable and attracts investors from outside [9].

Currently, the problem of regional differentiation is still relevant. However, the problems of regional differentiation of investment and production, excessive concentration of production, uneven distribution of investments, including in the agricultural sector, remain unresolved [8].

The following abbreviations are used in the text: RF - Russian Federation; Central Federal District - CFD; Northwestern Federal District - NFD; Southern Federal District - SFD; North Caucasian Federal District - NCFD; Volga Federal District - VFD; Ural Federal District - UFD; Siberian Federal District - SFD; Far Eastern Federal District - FEFD; GRP - gross regional product; SLHOR - branches of agriculture, forestry, hunting, fishing and fish farming.

2 Materials and Research Methods

For the research, open data from the Federal State Statistics Service were used. Research period 2000–2019. Purpose of the research: by analyzing statistical data, to trace the change in the volume of investments in fixed assets by federal districts. The study of the materials was carried out by economic and statistical methods. To analyze the dynamics of differentiation of investment volumes in the context of the Federal Districts of the Russian Federation, the following conditions were studied: the structure of investments in fixed assets by Federal Districts; the amount of investments as a whole, per 1 ha of the total area of the district, per capita; the share of investments in fixed assets in the gross regional product; structure of investments by forms of ownership and by industries and spheres of activity; the volume of investments in the sectors of agriculture, forestry, hunting, fishing and fish farming and the share of the latter in the total volume of investments in the federal district.

3 Research Results

As studies of statistical data had shown, from 2000 to 2019, the structure of investments in fixed assets by regions of the Russian Federation has undergone changes. Thus, the share of investments in a number of federal districts, such as the Volga Federal District, the Ural Federal District, the Southern Federal District, decreased, while in other regions it increased (Fig. 1).

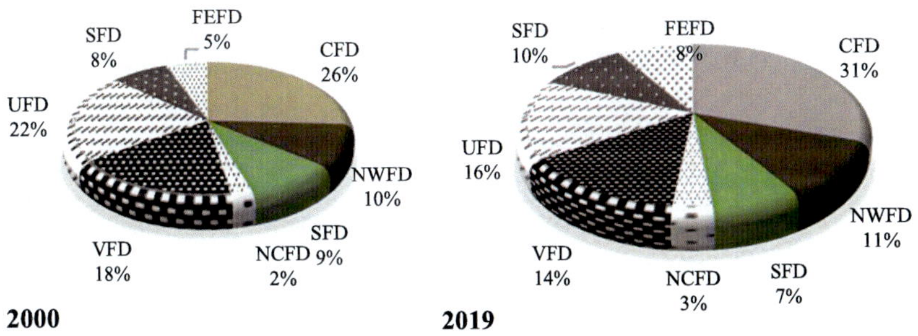

Fig. 1. Structure of investments in fixed assets in the Russian Federation by federal districts.

In 2019, one third of the total volume of investments in fixed assets fell on the Central Federal District. Less than 10% accounted for the Far Eastern Federal District, the Southern Federal District and the North Caucasian Federal District: 8.7 and 3%, respectively (Fig. 1).

For the period 2000–2019. there is an increase in the volume of investments in fixed assets in all regions of the Russian Federation. As the analysis of the dynamics of investment investments in 2019 compared to the indicators of 2000 showed, the number of investments in the Siberian Federal District, the North Caucasian Federal District and the Far Eastern Federal District increased by 20, 23 and 24 times, respectively (Fig. 2).

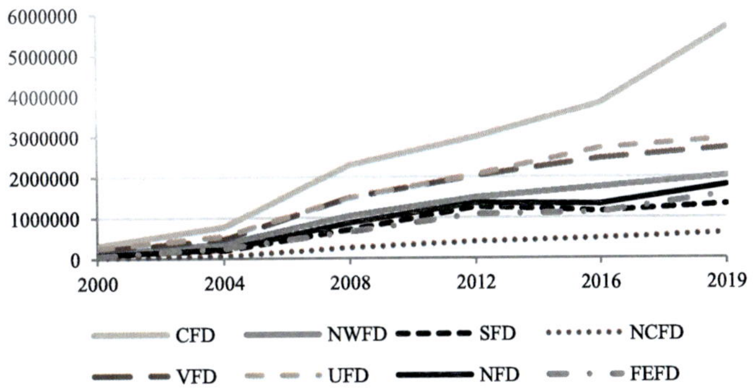

Fig. 2. Dynamics of investments in fixed assets by federal districts of the Russian Federation for 2000–2019, million rubles (in actual prices).

The trend of active investment in the CFD continues as the most promising in terms of capital return. On average in the Russian Federation, the growth in investment in fixed assets over the specified period increased by more than 16 times, which amounted to more than 19.32 trillion in 2019. rubles (Fig. 2).

The districts of the Russian Federation, due to their location, are significantly different in land area and during 2015–2019 remained practically unchanged. Thus, the largest Far Eastern Federal District and Siberian Federal District account for 66.1% of the total land area of the Russian Federation. For the period 2015–2019. the positive dynamics of the ratio of the volume of investments in fixed assets per unit area in all federal districts of the Russian Federation was noted (Table 1). The largest increase was noted in the Siberian Federal District - 66.10%, in the Central Federal District - 58.26%, in the Far Eastern Federal District - 37.42%.

Table 1. Dynamics of the volume of investments in fixed assets per 1 ha of the total land area of the Federal District of the Russian Federation, thousand rubles/ha (at current prices).

Region	2015	2016	2017	2018	2019	2019 to 2015
CFD	55.03	58.37	65.23	76.87	87.09	158.26
NWFD	8.52	10.33	11.16	13.69	11.93	139.98
SFD	28.94	25.74	32.23	32.51	29.45	101.76
NCFD	27.92	28.51	29.11	31.95	35.84	128.37
VFD	23.76	23.51	23.42	24.02	25.94	109.21
UFD	12.97	14.78	15.58	16.31	16.01	123.46
SFD	2.47	2.58	2.75	3.61	4.10	166.10
FEFD	1.65	1.82	2.09	2.08	2.27	137.42
RF (average)	8.12	8.61	9.36	10.38	11.28	139.01

However, in the most populated, industrially developed regions that produce the largest volume of GRP (Central Federal District, Ural Federal District, Volga Federal District), the volume of investments in fixed assets is several times higher than this indicator in remote Siberian and Far Eastern Federal Districts. So, in 2019 in the Central Federal District, per 1 ha of land area, the volume of investments amounted to 87.09 thousand rubles, then in the Far Eastern Federal District - 2.27 thousand rubles. This is a consequence of the presence of a significant number of undeveloped, hard-to-reach territories both in the Far Eastern Federal District and the Siberian Federal District (Table 1).

On the territory of the federal districts of the Russian Federation, the uneven distribution of the population is pronounced: for example, in 2019. 46.9% of the population of the Russian Federation lives in two federal districts - Central and Volga, the smallest number of residents is in the Far Eastern Federal District - 5.6%. During the study period, population growth was observed in five of the eight federal districts, of which in the Southern Federal District and the North Caucasus Federal District the increase in population was 19.0 and 9.9%, respectively, in the Central Federal District - 3.5%, in the North-West Federal District and the Ural Federal District - about 1%. A decrease in the number of residents was observed in the Siberian Federal District, the Volga Federal District and the Far Eastern Federal District by 1.6, 3.8 and 4.5%, respectively.

Analysis of the dynamics of investments in fixed assets per capita showed their increase in the regions of the Russian Federation for the period 200–2019. 10–24 times (Fig. 3). The largest increase in relative terms was observed in the Far Eastern Federal District - 24.7%, the Siberian Federal District - 21.6% and the North Caucasus Federal

District - 20.6%, the smallest - in the Southern Federal District - by 10.4%. Thus, due to the absence of significant population growth and growth in investment volumes, the trend towards an increase in the volume of investment investments per capita persisted throughout the entire period of research.

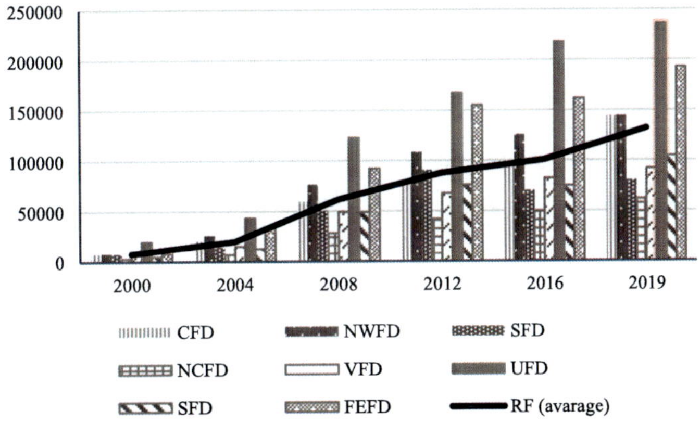

Fig. 3. Dynamics of investments in fixed assets per capita by federal districts of the Russian Federation for 2000–2019, rubles (in actual prices).

Above the national average values of this indicator in 2019 were noted in four regions: Ural Federal District - 235.6 thousand rubles per capita, Far Eastern Federal District - 192.7 thousand rubles per capita; Central Federal District and Northwestern Federal District - about 144 thousand rubles (Fig. 3).

The growth of investment in fixed assets of regions must inevitably be reflected in the size of the gross regional product (GRP), since investments in fixed assets are included in the GRP and stimulate the production of goods and services for end use. Analysis of the share of investments in fixed assets of GRP showed that on average in the Russian Federation in 2019. This indicator is 22.7%, in dynamics during the research period it increased by 2.5%, which corresponds to the "Forecast of the socio-economic development of the Russian Federation until 2024". Despite the increase in GRP for the period 2000–2019 in all regions by 12–18.5%, an increase in investment volumes during the period under review did not in all regions led to an increase in the share of investments in fixed assets in GRP: in the Ural Federal District and the Southern Federal District, this ratio decreased by 6.1 and 10.5% respectively (Table 2). In the Far Eastern Federal District, in 2019, the highest share of investments in GRP is noted - 30.3%, which is associated with the introduction of innovative projects and zones of advanced economic development.

Analysis of the dynamics of investments in fixed assets by type of property showed that for the period 2005–2019 in the RF there was an increase in the share of Russian property by 6.6 pp. and a decrease in the share of foreign ownership of 2.4% points, which amounted to 87.2 and 5.8%, respectively (Table 3). In the structure of Russian property for the specified period in all federal districts, a decrease in investments of state and mixed forms of ownership in favor of the private form was noted, which indicates an

Table 2. Dynamics of the share of investments in fixed assets in the gross regional product in the context of federal districts of the Russian Federation for the period 2000–2019, % (at current prices).

Region	2000	2004	2008	2012	2016	2019	Deviation (±)	
							2019 to 2000	2019 to 2016
CFD	16.5	16.7	18.0	17.0	15.7	19.3	2.7	3.5
NWFD	20.2	24.4	30.7	28.3	22.6	22.3	2.2	−0.2
SFD	33.0	26.4	35.2	39.4	23.1	22.6	−10.5	−0.5
NCFD	24.7	22.4	35.7	33.3	27.3	31.5	6.8	4.1
VFD	19.9	20.3	27.9	25.6	23.6	21.6	1.6	−2.0
УрФО	28.9	23.9	30.8	28.7	28.4	22.8	−6.1	−5.6
SFD	13.9	15.4	27.5	28.2	20.0	21.5	7.6	1.4
FEFD	17.7	30.0	36.4	34.9	26.8	30.3	12.6	3.5
RF (average)	20.3	20.5	25.9	25.2	21.3	22.7	2.5	1.4

increase in the infusion of private Russian capital into the economy of the regions. Thus, among the federal districts of the Russian Federation in the Central Federal District, the highest share of Russian state investments (19.3%) and the smallest - of private Russian property (52.6%), while the share of foreign investments is also quite high - 8%.

In the North Caucasus Federal District, the ratio of Russian state and private property is 17.3 and 72.2%; the number of foreign and joint (Russian-foreign) investments is extremely low, just over 1%. The largest volume of foreign investment during the research period was observed in the Far Eastern Federal District: in 2005, more than a third of investments in fixed assets accounted for foreign property, by 2019 this figure had decreased to 11.2%, but is still the highest among the federal districts of the Russian Federation (Table 3).

Analysis of the structure of investments in fixed assets by industry and spheres of activity for the period 2000–2019. showed the following. In 2000 the largest share of investments fell on the industries, transport activities and housing - 38.5, 21.1 and 11.6%, respectively. In 2010, in connection with the informatization and digitalization of economic processes, the volume of investments in activities in the field of communications increased sharply: if in 2000. the share of investments in this industry was 2.6% of the total investment in fixed assets of the Russian Federation, then in 2010 it increased to 16.54%. In 2010, fixed assets in the field of mining and manufacturing accounted for 9.39 and 8.80% of investments. Investment in activities related to public administration, military security and social insurance was 8.23%. In 2019, compared to 2010, there is a twofold increase in the share of investments in the areas of mining and manufacturing activities: in absolute terms, the share increased by 11.55 and 7.52%. During this period, there has been an increase in the share of investments in fixed assets for activities related to real estate operations from 0.79 to 6.13%. In 2019, in connection with the development of science, technology and technology in the structure of investment in fixed assets, areas related to professional, scientific and technical activities were identified; the share of investments in these sectors was 4.53%. Reduced investments related to public administration, military security and social security: their share fell to 2%, in numerical terms by 918 billion rubles, which amounted to 24.32% of the 2010 investment volume.

Table 3. Structure of investments in fixed assets by type of ownership in the context of federal districts of the Russian Federation, %.

Region	Russian	Including				Foreign	Cooperative (Russian and foreign)[1]
		State	Municipal	Private	Mixed		
2005							
RF	80.6	18.8	3.8	44.9	12.9	8.2	–
CFD	74.4	15.8	2.3	38.7	17.3	11.2	–
NWFD	79.3	23.2	2	44.2	9.9	8.4	–
SFD	91.9	16.5	4.6	62.5	8.1	1.3	–
NCFD	93.7	35.3	3.1	48	7.1	0.6	–
VFD	88.1	21.3	4.3	43.2	19.2	3.2	–
UFD	83.9	11.6	6.8	57.8	7.6	2.3	–
SFD	87.3	21.3	5.4	52.5	7.9	2.4	–
FEFD	57.4	24.2	2.2	18.5	12.4	37.7	–
2019							
RF	87.2	14.6	2.6	63.3	5.6	5.8	7
CFD	85.4	19.3	2.1	52.6	9.4	8	6.6
NWFD	85.6	18.3	1.9	59.6	3.7	6.5	7.8
SFD	93.7	23.3	3.9	62.4	3.8	3.9	2.4
NCFD	97.6	17.3	3.1	72.2	4.5	1.1	1.3
VFD	89.1	12	3.3	66.7	6.5	3.9	7
UFD	82.5	5.1	2.7	71.7	2.8	4.9	12.6
SFD	87.4	13.2	3.1	65.8	4.9	3.1	9.5
FEFD	85.6	15.9	3.3	62.4	3.9	11.2	3.2

Note: 1 - no data for 2005.

For the period 2000–2010, in the structure of investments in fixed assets as a whole across the Russian Federation, the share of economic activity associated with agriculture, forestry and fisheries remained practically unchanged - 2.1–2.7% (Table 4). In 2019, there is an increase in this indicator to 3.66%, which in monetary terms compared to 2010 amounted to 229.8 billion rubles with an increase of 73.9%. It is noteworthy that the total volume of investments in the Russian Federation in this period increased to a lesser extent - by 61.2%.

In the context of federal districts in 2010, the largest volume of investments in agriculture, forestry, hunting, fishing and fish farming fell on the Central Federal District and the Volga Federal District: 64.7 and 50.1 billion rubles, and 4.40 and 5.41% of the total the volume of investment in the district, respectively. In the North Caucasus Federal District, the share of investments in fixed assets of the SLHOR industries is the largest among the Federal District and amounted to 6.09% of the total investment capital. In the Far Eastern Federal District, less than 1% of investment capital was directed to the SLHOR industries (Table 4).

Table 4. Investments in fixed assets in the sectors of agriculture, forestry, hunting, fishing and fish farming (SLHOR) in fixed assets in the context of federal districts of the Russian Federation.

	CFD	NWFD	SFD	NCFD	VFD	UFD	SFD	FEFD	RF
2010									
Investments. billion rubles. total	1472	895	620	140	927	1197	699	675	9151
Incl. in SLHOR	64.7	21.6	28.8	8.5	50.1	13.5	21.1	6.7	306.1
Structure of investments in SLHOR. %	38.54	8.85	9.56	6.15	17.39	4.29	7.31	7.92	100.0
Share of investments in SLHOR in the total volume of investments. %	4.40	2.41	4.65	6.09	5.41	1.13	3.01	0.99	3.40
2019									
Investments. billion rubles. total	4601	1664	935	268	1813	2611	1372	1488	14753
Incl. in SLHOR	208.3	47.8	51.7	33.2	94.0	23.2	39.5	42.8	540.6
Structure of investments in SLHOR. %	21.08	6.73	9.37	2.76	16.38	4.38	6.88	1.45	100.0
Share of investments in SLHOR in the total volume of investments. %	4.53	2.87	5.53	12.40	5.18	0.89	2.88	2.88	3.66

By 2019, in all federal districts, an increase in investments in the SLHOR industry was noted in the range from 1.7 (UFO) to 9.6 (Far Eastern Federal District) times. This year, more than a third of all investments in the Russian Federation in the SCLOR industry were in the Central Federal District and over 17% in the Volga Federal District. A significant increase in the share of investments in the SCLOR industry in the total volume of investments in the Okrug is observed in the North Caucasus Federal District from 6.05 to 12.4% (more than 2 times), and in the Far Eastern Federal District: from 0.66 to 2.88% (more than 4 times), which indicates a redistribution of the directions of capital investment between the sectors of the national economy.

4 Discussion

The period from 2000 to 2019 for the Russian Federation is associated with the development of the country's economy, an increase in investment in fixed assets. The territory of Russia is vast, which causes a significant difference between the federal districts in terms of natural, climatic, social and economic conditions. Depending on the available resources of the region, certain industries are developing, where most of the investments in fixed assets are directed in order to obtain the greatest benefits. Therefore, the more economically developed a region is, the greater the share of investments it accounts for. There is a decrease in the share of state and private Russian investments. There is a significant differentiation in the federal districts of investment volumes per capita and per unit of land area. The development of market relations and competition predetermined the allocation of investments in industries related to the development of science, technology, technology, communications and infrastructure. The sectoral specialization of the regions determines the unevenness of investment volumes in the context of federal districts, both in the sectors of the economy as a whole and in the SLHOR industry.

5 Conclusion

Territorial differences in the distribution of investments in fixed assets in the Russian Federation are due to the development of both external and internal economic relations. We believe that in the future there will be further differentiation of the federal districts of the Russian Federation in terms of investments in fixed assets in connection with the development of new promising industries and activities, the development of science and technology, and the further growth of the industry specialization of the regions. Therefore, the state, with the help of various management mechanisms, should regulate the distribution of investments in fixed assets to maintain a balance of economic development between technologically related industries and spheres of activity, which is an interesting material for further research.

References

1. Glinskiy, V., Serga, L., Novikov, A., Bulkina, A., Litvintseva, G.: Investigation of correlation between the regions sustainability and territorial differentiation. Procedia Manuf. **84**, 323–329 (2017). https://doi.org/10.1016/j.promfg.2017.02.041
2. Kravchenko, N.A.: Assessment of diversification in the regional economy (case study of subjects of the Siberian Federal District). Reg. Res. Russ. **6**(3), 65–89 (2016). https://doi.org/10.1134/S2079970516030035
3. Nagoev, A.B., Rokotyanskaya, V.V., et al.: Mobilization of innovation and investment resources for environmental management in Russian federation regions. Environ. Qual. Manage. **27**(4), 47–54 (2018). https://doi.org/10.1002/tqem.21564
4. Kvon, G.M., et al.: Strategic priorities of regional investment activity. Eurasian J. Anal. Chem. **12**(7b), 1099–1106 (2017). https://doi.org/10.12973/ejac.2017.00235a
5. Mayorova, A.N., Panasenko, S.V., Nikishin, A.F., et al.: Analyzing regional differences in the condition and development of trade in Russia. Entrepreneurship Sustain. Issues **6**(2), 927–938 (2018). https://doi.org/10.9770/jesi.2018.6.2(30)

6. Dikova, D., Panibratov, A., Veselova, A.: IBR **28**(4), 625–663 (2019). https://doi.org/10.1016/j.ibusrev.2018.12.007
7. Kuzmina, O., Volchkova, N., Zueva, T.: Foreign direct investment and governance quality in Russia. J. Comp. Econ. **42**(4), 874–891 (2014). https://doi.org/10.1016/j.jce.2014.08.001
8. Vorozheykina, T., Demichev, V., Maslakova, V., Leshcheva, M.: Algorithmization for processes of regional differentiation and concentration of investments in Russian agriculture. Int. Trans. J. Eng. Manage. Appl. Sci. Technol. **11**(6), 1–13 (2020). https://doi.org/10.14456/itjemast.2020.114
9. Uzuna, V., Shagaidaa, N., Lermanb, Z.: Russian agriculture: growth and institutional challenges. Land Use Policy **83**, 475–487 (2019). https://doi.org/10.1016/j.landusepol.2019.02.018

Findings of the Study of Potato Varieties of Different Geographical Origin in the Kamchatka Territory

Anna Ivashchenko(✉), Tamara Sherstyukova, and Galina Vlasenko

Kamchatka Research Institute of Agriculture, Sosnovka Settlement, Yelizovsky District, 684033 Kamchatka, Kamchatka Territory, Russia
ivashchenkoanna@bk.ru

Abstract. The purpose of the study is to identify the parent material with high adaptation and yielding capacity potential and a complex of agronomic characters for inclusion in a hybridization program based on the study of a collection of potato varieties. In the collection nursery, 86 varieties of various geographical origins were studied. The average yield of the studied varieties for the period 2018–2020 was 38.4 t/ha and ranged from 15.4 t/ha (2018) to 55.0 t/ha (2020). The potentially high-yield varieties include Provento, Impala, Geyser, Amalia, Rannespelyi, Elizaveta and Varmas having an adaptability coefficient (Ca = 1.28–1.01). The flexible varieties (bi = 0.84–1.14) include Zhukovsky rannii, Geyser, Amalia, Fresco, Provento, Alyi mestnyi, Impala, Anosta, Alvara, Elizaveta, Sibiryachka, Solnyshko and Sante. The intensive-type varieties (bi = 1,21–1,26) include Signal, Madeleine and Red Scatlett. The following varieties had a marketability of 83–97%: – Alvara, Anosta, Scarlet mestnyi, Vulkan, Varmas, Elizaveta, Zhukovsky rannii, Madeleine, Monoliza, Masai, Rannespelyi Sante, Solnyshko, Scarlet, Sibiryachka, Signal, Fiesta, Charoit, Evolution. Varieties Du-nun, Kamchatka, Otrada, Pamyati Rogacheva, Ryabinushka, Lazar were distinguished by a high tuber starch content (13.3–16.6%), while the reference value was 12.4%. Excellent tuber taste qualities (8.5–9.0 points) were obtained by the varieties Anosta, Aksamit, Bridzh, Bellarosa, Vulkan, Varmas, Zekura, Zenimaru, Kamchatka and Yubilei Zhukova. Sources that combine high yielding capacity and resistance to viral diseases and early blight were selected: Alvara, Vulkan, Elizaveta, Radonezhsky, Rannespelyi, Sappho, Solnyshko, Sibiryachka and Penzenskaya skorospelka. The selected varieties represent a valuable parent material for inclusion in the hybridization program.

Keywords: Selection · Cultivar · Potatoes · Collection · Adaptability · Plasticity · Initial material · Economically valuable traits

1 Introduction

The complexity of potato breeding at the current stage of development is constantly increasing following the increase in the number of target characteristics and specific

A. Muratov and S. Ignateva (Eds.): AFE 2021, LNNS 353, pp. 321–328, 2022.
https://doi.org/10.1007/978-3-030-91402-8_37

requirements for new-generation varieties [1]. Potato varieties shall ensure the yield stability and product quality, resistance to all the most common and harmful diseases and pests, which complicates breeding in Kamchatka, where temperatures provide only a minimum of the crop's biological requirements [2]. Potatoes are planted in the south-eastern zone of Kamchatka on June 7–15 when the soil is warmed at a depth of 10 cm to +3... 4 °C. Tuber formation begins at the end of July and continues until the end of the growing season, which is interrupted by frost in the first and second decade of September. The natural death of the tops does not occur even in the most early-maturing potato varieties. In terms of moisture availability, the zone is considered to be sufficiently moist and precipitation is distributed unevenly over the months [3]. The most adapted varieties meet the requirements of consumers in the conditions of the Kamchatka Territory in full. The limited heat during the growing season allows cultivating early and middle-early varieties with a short growing season [4].

Potato breeding results depend directly on the objective evaluation of the studied parent material based on valuable characteristics and a wide variety of harmful pathogens [5].

The most efficient way to control diseases and pests in all agricultural plants is cultivating resistant varieties. Breeding work to increase resistance to diseases and pests requires a comprehensive study of the existing varieties and the search for new resistant forms. The most harmful in the conditions of Kamchatka are various viral diseases: potato rugose and streak mosaic, leaf-rolling mosaic and potato leaf-rolling. Fungal diseases include early blight, late blight and Rhizoctonia blight The selection and evaluation of samples from genetic collections, the selection of the best sources of agronomic characters and their preparation for crossbreeding is given in the work by B. Flis [6]. One of the most dangerous potato diseases is potato nematode disease resulting in immense damage to potato growing in many countries of the world, caused by two types of cyst nematodes: golden – Globodera rostochiensis (Woll., 1923) and pale – Globodera paallida (Stone, 1973). Golden potato nematode has five pathotypes (R_{01}–R_{05}), R_{01} pathotype is found in Russia [7]. Aggressive pathotypes of potato nematodes are common in most countries of Western Europe, for example, in the Netherlands – about 30% of potato nematode populations contain aggressive pathotypes, and in Germany – more than 15%. [8]. The problem of breeding nematode-resistant potato varieties in Russia is acute and is one of the priorities in domestic breeding [9]. An effective way to protect potatoes from pathogens is to develop and produce resistant varieties of local breeding, obtained on the basis of highly resistant forms selected from the collection, and tested in different weather conditions of the region [10].

The purpose of the research is to identify the parent material with high adaptation and yielding capacity potential and a complex of agronomic characters for inclusion in a hybridization program based on the study of nursery potato collection varieties.

2 Materials and Methods

In total, 86 varieties of domestic and foreign breeding were studied in the period from 2018 to 2020 (Austria, Belarus, China, Germany, the Netherlands, Poland, Ukraine). The research was performed in the breeding three-course rotation of the FSBSI Kamchatka

Research Institute of Agriculture. The preceding crop is the green-manured fallow. The soil is ochreous volcanic. The potato cultivation technology is the technology generally accepted for the study area [11]. As a reference, an early variety of Fresco potatoes, recognized in the Kamchatka Territory, is accepted. Records and observations were performed according to the potato research method of the Lorkh Federal Potato Research Centre [12]. Among the potato fungal diseases, late blight and early blight were observed, registered by a 9-point scale (9 – no lesion, 0 – all plants are affected). Viral diseases were identified visually in the field. The assessment of potato tubers for quality characteristics was performed according to generally accepted methods: N. I. Vavilov All-Russian Institute of Plant Genetic Resources and the Lorkh Federal Potato Research Centre. The flexibility and stability of the studied varieties were assessed using the method of S. A. Eberhart and V. A. Russel as expounded by V. A. Zykin [13]. The analysis of the variety's adaptation potential in terms of yield was performed according to the method of L. A. Zhivotkov, Z. A. Morozova, L. I. Sekutaeva [14]. When analyzing the yielding capacity and adaptation potential of the varieties, the indicator of the variety-average yield of the year was used—this is the yield level in a particular year and in a particular region. The adaptability coefficient (Ca) was calculated for each year and variety according to the formula:

$$\mathrm{Ca} = (\mathrm{Xij} * 100 : \mathrm{X}) : 100, \tag{1}$$

where Xij is the yield of the i-variety in the j-th year of the study; X is the variety-average yield of the year. In this method, the variety-average yield is accepted as 100% [15].

Meteorological conditions of the growing season by year (2018–2020) differed in terms of heat and moisture availability. Less favorable for the potato plant formation and development were in 2018 and 2019. The heat distribution over the decades was uneven, with the sum of active temperatures above 10 °C was 1002 °C, 1009 °C with a normal value of 1092 °C Atmospheric precipitation during the growing season in 2018 amounted to 449.5 mm, in 2019–271.5 mm, with a normal value of 269 mm. The vegetation period of 2020 was the most favorable in terms of meteorological conditions. The sum of active temperatures above 10 °C was – 1121 °C with an average long-term temperature of 1092 °C.

3 Results and Discussion

Following the study of the potato variety collection in the Kamchatka Territory, as a parent material, varieties were identified with a yield higher than the average in the experiment: Amalia – 41.6 t/ha, Geyser – 41.5 t/ha, Impala – 45.1 t/ha and Provento – 46.9 t/ha. The average yield of the studied varieties for the period 2018–2020 was 38.4 t/ha (Table 1). The yielding capacity of the studied varieties was evaluated by the average adaptability coefficient (Ca). In our studies, it ranged from 0.69 to 1.24. If the adaptability coefficient of a variety exceeds 1, then such a variety is consequently potentially adaptive or potentially high-yielding. During the three years of the study, only 7 of the 19 potato varieties presented had an adaptability coefficient higher than 1: Provento (1.28), Impala (1.20), Geyser (1.17), Amalia (1.09), Rannespelyi (1.04), Elizaveta (1.02), Varmas (1.01). The lowest adaptability coefficient was in the varieties Sibiryachka (0.88),

Fresco (0.87), Red Scarlett (0.86). Varieties with a regression coefficient significantly higher than 1 belong to the intensive type, they respond well to improving growing conditions, and in unfavorable years and with a low agricultural background, yielding capacity decreases sharply. In our studies, Signal (bi = 1.26), Madeleine (bi = 1.24), and Red Scarlett varieties, which can be attributed to the intensive type, were the most responsive to the conditions of the year. In our studies, the flexible varieties are Zhukovsky rannii, Geyser, Amalia, Fresco, Provento, Alyi mestnyi, Impala, Anosta, Alvara, Elizaveta, Sibiryachka, Solnyshko, Sante (bi = 0.84–1.14).

Table 1. Potato varieties distinguished by yield, ecological flexibility, and adaptability (average for 2018–2020).

Variety	Yield, t/ha				Adaptability coefficient (Ca)	Flexibility (regression coefficient bi)
	2018	2019	2020	Average for three years		
Fresco (st)	13.7	35.8	51.2	33.6	0.87	0.91
Amalia	15.0	71.4	38.5	41.6	1.09	0.88
Alvara	15.0	43.0	57.1	38.7	0.99	1.04
Anosta	17.5	28.0	66.0	37.2	0.99	1.04
Alyi mestnyi	17.0	28.0	64.4	37.0	0.97	1.02
Vulcan	17.0	41.0	43.0	33.7	0.93	0.69
Varmas	17.5	48.0	44.4	36.6	1.01	0.76
Geyser	22.6	43.0	59.0	41.5	1.17	0.87
Elizaveta	16.0	40.0	63.3	39.7	1.02	1.11
Zhukovsky rannii	15.0	41.6	47.7	34.8	0.92	0.84
Impala	19.3	60.0	56.0	45.1	1.20	1.03
Madeleine	11.1	43.0	62.0	38.7	0.94	1.24
Provento	22.1	63.0	55.5	46.9	1.28	0.96
Rannespelyi	17.5	55.0	42.0	38.2	1.04	0.76
Sante (st)	13.7	40.0	61.4	38.4	0.97	1.14
Solnyshko	12.0	47.7	55.0	38.2	0.95	1.12
Red Scarlett	11.2	30.0	65.0	35.4	0.86	1.21
Sibiryachka	8.1	57.0	46.0	37.0	0.88	1.11
Signal	11.9	31.0	68.3	37.1	0.90	1.26
Variety-average yield	15.4	44.6	55.0	38.4	-	-

The regression coefficient is significantly lower than one (bi = 0.69–0.76) in Vulkan, Varmas, and Rannespeplyi varieties, these varieties respond poorly to changes in weather conditions and are of the neutral type (with low environmental flexibility). Of all the studied collection samples of potatoes, 18 (20.9%) varieties with tubers marketability of 83–97% were identified: Alvara, Anasta, Alyi mestnyi, Vulkan, Varmas, Elizaveta, Zhukovsky rannii, Madeleine, Monoliza, Masai Rannespelyi, Sante, Solnyshko, Scarlet, Sibiryachka, Signal, Fiesta, Charoit, Evolution, Geyser, Amalia, Impala, Provento, Penzenskaya skorospelka. The average weight of the marketable tuber of the studied varieties was in the range of 54–110 g, with 91 g in Fresco (reference). The large-tuber varieties (100–110 g) include Sappho, Zhukovsky rannii, Sibiryachka, Irbitsky, Kosmos, Fermer, and Elizaveta. According to the results of biochemical analyses, the tuber starch content varied in the range of 8.5–16.6% (Table 2). Varieties Du-nun, Kamchatka, Otrada, Pamyati Rogacheva, Ryabinushka, Lazar were characterized by high starch content (13.3–16.6%) at 12.4% in the reference and dry matter content of 19.1–21.6%. Varieties such as Arizona, Elizaveta, Zhukovsky rannii, Zhavoronok, Kosmos, Lileya, Madeleine, Masai, Peterburgsky, Tyoma, Chaika, Evolution are classified as low starch (less than 10%).

An increased content of vitamin C in potato tubers (11.6–16.3 mg%) at 8.0 mg% in the Fresco reference variety was observed in Vulkan, Zhukovsky rannii, Crocus, Monoliza, and Raya varieties. Varieties Astilla, Delfine, Elizaveta, Kamchatkskyi rannii, Cosmos, Roko, Sante, and Evolution were characterized by low content of vitamin C (1.6–2.6 mg%). The varieties Anosta, Aksamit, Bridzh, Bellarosa, Vulkan, Varmas, Zekura, Zenimaru, Kamchatka, and Yubilei Zhukova had excellent tuber taste qualities (8.5–9.0 points). 50 (58.1%) varieties had a good taste of boiled potatoes (7.0–8.5 points), and 26 (30.2%) varieties had a satisfactory taste (5.0–7.0 points).

The wide variety and the presence of numerous virus strains challenge breeding for comprehensive resistance to viral diseases, so it is of great importance to select forms that are highly sensitive to certain viruses, as well as varieties with field resistance. Of the viral diseases of potatoes in Kamchatka, the greatest harm is caused by the following: mosaic leaf-rolling (PVM), ordinary mosaic (PVX), leaf-rolling (PLRV), crinkle, etc. The registration of plants affected by viral diseases was determined by external signs in the field during 2018–2020. The most widespread was mosaic leaf-rolling (PVM), this disease affected 27 (31.4%), ordinary mosaic (PVX)—31 (36%) varieties, including the standard reference Fresco variety. Plant damage by leaf-rolling (PLRV) was observed in Zenimaru, Kosmos, and Masai varieties Varieties free from viral diseases were identified: Apis, Alvara, Astilla, Alyi mestnyi, Agata, Bellarosa, Vulkan, Varmas, Vasilek, Dalvas, Du-nun, Elizaveta, Zhavoronok, Irbitsky, Lileya, Otrada, Provento, Penzenskaya skorospelka, Rannespelyi, Radonezhsky, Solnyshko, Sibiryachka, Siren, Sappho, Fermer, Yugra. One of the most common and dangerous fungal diseases in the Kamchatka Territory is early blight. The analysis of field resistance to early blight showed that a strong lesion (1–3 points) was observed in the varieties Aksenovsky, Vasilek, Darik, Zenimaru, Karina, Lileya, Proventa, Ryabinushka, Scarlet, Snegir, Siren, Signal, the average lesion (4–6 points) was registered in 15 (17.4%) varieties – Apis, Anosta, Alyi mestnyi, Avgustin, Vital, Varmas, Du-nun, Elizaveta, Zhavoronok, Zekura, Impala, Kamchatka, Kamchatskyi rannii, Kamensky, Paul Wagner, a weak lesion (7–8 points) was

Table 2. Potato varieties of the collection nursery distinguished by the tuber quality characteristics (average for 2018–2020).

Variety	Country of origin	Content in tubers			Taste, points
		Starch, %	Dry matter, %	Vitamin C, mg%	
1	2	3	4	5	6
Fresco*(St)	the Netherlands	12.4	17.4	8.0	8.3
Amalia	Germany	11.6	16.6	7.9	7.0
Alvara*	Germany	13.7	18.8	6.3	8.0
Anosta*	the Netherlands	12.2	17.3	4.7	8.6
Alyi mestnyi	Russia	12.4	17.4	8.6	8.0
Arizona*	the Netherlands	9.4	13.8	4.7	5.0
Aroza	Germany	11.5	16.9	13.8	7.0
Bellarosa*	Germany	12.1	17.1	5.2	8.6
Vulcan*	Russia	12.6	17.6	11.6	8.8
Varmas	Russia	10.8	15.7	3.4	8.6
Vasilek	Russia	13.3	19.1	2.7	8.0
Geyser	Russia	11.6	16.6	5.1	8.2
Dalvas	Russia	12.8	17.1	3.9	7.5
Du-nun	China	12.9	18.0	3.4	6.0
Zhukovsky rannii*	Russia	8.5	13.6	14.0	6.5
Zhavoronok	Russia	9.0	13.7	4.2	7.5
Zekura*	Russia	11.8	16.9	8.6	8.5
Kamchatka	Russia	14.2	19.3	4.2	9.0
Ketsky*	Russia	11.4	16.4	8.4	7.6
Crocus*	Poland	11.5	16.9	14.1	6.5
Lazar	Russia	16.6	21.6	4.2	8.3
Madeleine*	the Netherlands	9.5	14.5	6.0	5.5
Masai*	the Netherlands	8.2	13.2	3.4	8.0
Otrada*	Russia	14.2	19.2	5.4	7.6
Pamyati Rogacheva	Russia	14.2	19.2	6.7	7.0
Peterburgsky	Russia	9.3	14.5	4.1	4.6

(continued)

Table 2. *(continued)*

Variety	Country of origin	Content in tubers			Taste, points
		Starch, %	Dry matter, %	Vitamin C, mg%	
1	2	3	4	5	6
Raya*	the Netherlands	12.1	17.1	12.6	6.5
Ryabinushka*	Russia	13.6	18.9	3.0	8.0
Rasinka*	Russia	12.9	18.0	4.1	7.5
Sappho*	Russia	9.9	14.9	8.1	5.5
Solnyshko	Russia	11.4	16.5	5.5	7.6
Tyoma	Russia	9.7	14.7	8.8	7.0

* - varieties resistant to the golden cyst potato nematode.

observed in 43 (50.0%) varieties. Only 16 (17.4%) potato varieties were not affected by early blight – Alvara, Arizona, Bridzh, Vulkan, Peterburgsky, Pushkinets, Penzenskaya skorospelka, Radonezhsky, Rodrigo, Roko, Rucheek, Rannespelyi, Sappho, Solnyshko, Fermer, Charodey.

Russian breeding does not have enough varieties that are resistant to the golden potato nematode. Our working collection includes nematode-resistant potato varieties of various geographical origin, created by Russian breeders and breeders from Germany, the Netherlands, Austria, Belarus, namely Avurora, Alvara, Anosta, Artemis, Arosa, Arizona, Bellarosa, Vital, Gala, Geyser, Delfin, Zhukovsky rannii, Zekura, Irbit, Ket, Cosmos, Latona, Lilea, Ladoga, Mayak, Madeline, Masai, Mustang, Nikita, Otrada, Pushkinets, Proventa, Rasinka, Rodrigo, Radonezh, Rucheek, Ryabinushka, Raya, Sappho, Sante, Scarlet, Teterev, Evolution, Yubilei Zhukova, Yubiliar, Yugra, which serve as a good parent material for breeding for nematode resistance.

4 Conclusion

Following the study of the collection material in the Kamchatka Territory, genetic sources of various geographical origin were identified, combining potentially high tuber yielding capacity (36.6–46.9 t/ha), flexibility – Amalia, Warmas, Geyser, Elizabeth, Impala, Rannespelyi and high marketability (87–98%). The material for breeding with a high starch content - Du-nun, Kamchatka, Otrada, Pamyati Rogacheva, Ryabinushka, Lazar and with high tuber taste qualities - Anosta, Aksamit, Bridzh, Bellarosa, Vulkan, Varmas, Zekura, Zenimaru, Kamchatka, Yubilei Zhukova was selected. Varieties that combine high yielding capacity and resistance to viral diseases and early blight were selected: Alvara, Vulkan, Elizaveta, Radonezhsky, Rannespelyi, Sappho, Solnyshko, Sibiryachka, and Penzenskaya skorospelka. Varieties distinguished by a complex of agronomic characters, generally, have the ability to bloom and represent a valuable parent material for including them into hybridization program.

References

1. Bonneie, E.: Potato breeding: a challenge as always! Potato Res. **51**, 327–332 (2008). https://doi.org/10.1007/s11540-008-9116-zV.A
2. Sherstyukova, T.P., Ivashchenko, A.D.: Otsenka evaluation of source material for use in potato breeding in Kamchatsk region. DV agrar vestnik **4**(56), 48–53 (2020). https://doi.org/10.24411/1999-6837-2020-14048
3. Vlasenko, G.P.: Plasticity and stability of potato varieties in the Kamchatka territory. Arch. Sci. Tech. Agro-Ind. Comp. **32**(4), 44–46 (2018). https://doi.org/10.24411/0235-2451-2018-10410
4. Sherstyukova, T.P., Ivashchenko, A.D.: The results of a comprehensive assessment of collection potato varieties in the Kamchatka Territory. DV Agrar Vestnik **3**(51), 64–68 (2019). https://doi.org/10.24411/1999-6837-2019-13037
5. Rafalsky, S.V., Rafalskaya, O.M., Melnikova, T.V.: The results of a comparative assessment of potato varieties in terms of yield and adaptability parameters in the Amur region. Bull. Far East Branch Russ. Acad. Sci. **4**, 81–85 (2020). https://doi.org/10.37102/08697698.2020.212.4.013
6. Chekushkina, T.N., Fisenko, P.V., Kim, I.V., Raifegerst, T.R.: Identification of genetic markers of resistance to Potato virus Y and Golden potato cyct nematode in promising potato varieties by means of PCR. IOP Conf. Ser: Earth Environ. Sci. **548**(4), 042028 (2020). https://doi.org/10.1088/1755-1315/548/4/042028
7. Flis, B.: Selection of potato parental lines with complex resistances to potato pathogens and pests. Plant Breed Seed Sci. **76**, 57–63 (2017). https://doi.org/10.1515/plass-2017-0022
8. Klimenko, N.S., Antonova, O.Y., Kostina, L.I., Mamadbokirova, F.T., Gavrilenko, T.A.: Marker - direct breeding of domestic potato varieties with markers of genes for resistance to potato golden nematode (pathotype RO1). Proc. Appl. Bot. Genet. Breed. **178**(4), 66–75 (2017). https://doi.org/10.30901/2227-8834-2017-4-66-75
9. Simakov, E.A., Mityushkin, A.V., Zhuravlev, A.A.: Selection of potato varieties resistant to nematodes. Zashhita i karantin **10**, 20–22 (2016)
10. Shesteperov, A.A., Morozov, M.A.: Demonstration field experiment on testing nematode-resistant potato varieties in the Vladimir Region. Russ. J. Parasit. **15**(2),113–122 (2021). https://doi.org/10.31016/1998-8435-2021-15-2-113-122
11. Rafalsky, S.V., Rafalskaya, O.M., Melnikova, T.V.: Sources of increased productivity of tubers for potato breeding in the Far East. Veget. Crops Russ. **3**, 94–97 (2020). https://doi.org/10.18619/2072-9146-2020-3-94-97
12. Pradel, W., Gatto, M., Haro, G., Pandey, S.K., Bharduay, V.: Adoption of potato varieties and their role in adapting to climate change in India. Clim. Risk Manag. **23**, 114–123 (2019). https://doi.org/10.1016/j.crm.2019.01.001
13. Ryakhovskaya, N.I., Gainatulina, V.V., Vlasenko, G.P.: Sistema zemledeliya Kamchatskogo kraya. Monograph. National Grain Center named after P. P. Lukyanenko. Petropavlovsk-Kamchatsk (2015)
14. Zykin, V.A., Belan, I.A., Yusov, V.S.: Methods for calculating the ecological plasticity of agricultural plants in the discipline "Ecological genetics": method. recommendations. Omsk: Omsk GAU (2008)
15. Molyavko, A.A., Maruhlenko, A.V., Borisova, N.P.: Koefficient adaptivnosti sortov kartofelya opredelyaet ego produktivnost. Kartofel' i ovoshchi **3**, 10–11 (2012)

Application of Bio-Mix Organic Fertilizer in Potato Cultivation in the Kamchatka Territory

Vera Gainatulina, Olga Khasbiullina(✉), and Roman Khasbiullin

Kamchatsky Scientific Research Institute of Agriculture, Elizovo District, Kamchatka Territory, 604033 Sosnovka, Russia
khasbiullina@kamniish.ru

Abstract. The effect of Bio-Mix organic fertilizer from aquatic organisms when applied to the soil jointly with mineral fertilizer in doses of $(NPK)_{90}$ and $(NPK)_{60}$ was studied. The positive effect of the preparation on the agrochemical parameters of the soil, potato yield, its quality and Rhizoctonia solani J. G. Kuhn incidence in plants and tubers were identified. When introducing Bio-Mix at a dose of 5.0 l/ha at $(NPK)_{90}$ by the end of the growing season, the amount of ammonia and nitrate nitrogen in the soil increased by 45.0 and 5.7%, at $(NPK)_{60}$ – by 23.2 and 37.4% in relation to background values. At $(NPK)_{90}$ and $(NPK)_{60}$ by the end of the growing season, Bio-Mix preparation contributed to an increase in the plant-available phosphorus in the soil by 29.3 and 35.0 mg/kg (26.8 and 33.3%), and of the exchangeable potassium by 25.4 and 16.3 mg/kg (21.4 and 11.9%). The yield of potatoes increased by an average of 10.6 and 14.4% compared to the background values following the introduction of Bio-Mix. The highest amount of ascorbic acid in tubers was obtained at $(NPK)_{90}$ and $(NPK)_{60}$ when introducing Bio-Mix at a dose of 5.0 l/ha -16.59 and 16.06 mg%. The introduction of the Bio-Mix preparation provided for the decrease in the degree of development and incidence of rhizoctoniosis before harvesting against background by an average of 8.8% and 32.8%, however, compared to the budding period, these indicators were higher by an average of 3.9 and 13.1%.

Keywords: Potato · Biostimulator · Fertilizer · Yield · Quality · Soil agrochemical indicators

1 Introduction

The maintenance and improvement of soil fertility remain the key issue in agriculture. The current level of agricultural production requires the improvement of cultivation technologies with the application of advanced fertilizer systems that meet the needs of plants in macro-and micro-fertilizers, to reduce the anthropogenic burden. Today, alternative forms of fertilizers emerge on the market, which are successfully included in the crop cultivation process cycle [1, 2]. A significant increase in the number of publications devoted to their use indicates the prospects for the application of organic preparations

A. Muratov and S. Ignateva (Eds.): AFE 2021, LNNS 353, pp. 329–338, 2022.
https://doi.org/10.1007/978-3-030-91402-8_38

in crop production [3–5]. The ochreous volcanic soils of the Kamchatka Territory are characterized by low natural fertility, so the use of mineral fertilizers on such lands is a decisive factor in improving the yield. To date, biological methods of enhancing plant productivity are actively applied in agriculture. Organic fertilizers are of great importance for improving soil fertility, increasing crop yields and the nutritional value of crops [6]. As the amount of fertilizers is insufficient, the question is raised about their application in reasonable doses, and the involvement of local resources as plant nutrition sources. Currently, more and more organic preparations of various compositions and properties are emerging on the fertilizer market. The demand for such fertilizers increases every year. The most essential organic fertilizers are: fish emulsion, seaweed, soil processed by earthworms, bone meal, seaweed flour, and natural phosphates. These natural fertilizers promote plant growth and improve the soil. The key advantage of such fertilizers is their ability to activate plant defences in extreme conditions, improve product quality, reduce maturation time, and increase stress resistance. Currently, the Kamchatka Territory has established the production of organic fertilizers based on fish emulsion and seaweed. The preparations consist only of natural substances, they are based on long-term processing of Pacific northern breeds of pelagic fish (sea urchin, crab, starfish, plaice, halibut, cod), red-brown algae. The preparations contain over 20 natural amino acids and more than 50 chelated microelements which are 99% absorbed by plants [7]. Organic fertilizer Bio-Mare of grades Bio-Fish, Bio-Algo and Bio-Mix includes substances resulting from the processing of Pacific northern breeds of pelagic fish (sea urchin, crab, starfish, plaice, halibut, cod), red-brown algae. Bio-Mix is a complex preparation of organic origin, containing a balanced composition of micro-, meso- and macronutrients of the Far Eastern aquatic flora and fauna. A considerable amount of trace elements and amino acids determines the relevance of studying the effectiveness of organic fertilizers in crop cultivation technology in the extreme conditions of Kamchatka.

2 Materials and Methods

Research on the effect of Bio-Mix organic fertilizer was performed in 2018–2020 in the fields of the Federal State Budgetary Scientific Institution Kamchatka Research Institute of Agriculture, located in the edaphic-climatic zone of the Elizovo district, Kamchatka Territory. The study object was organic fertilizer Bio-Mare by Bio-Mix from aquatic organisms [8]. The effect of the preparation was studied on Sante, a middle-early potato variety, seed category – primary, at two mineral nutrition backgrounds $(NPK)_{90}$ and $(NPK)_{60}$. Two doses of Bio-Mix organic fertilizer of 2.5 and 5.0 l/ha were studied when introduced into the soil by spraying. As a reference, the non-fertilized soil was accepted, and the recommended reference value was represented by the mineral nutrition background $(NPK)_{120}$ (recommended mineral fertilizer dose for potato cultivation in the Kamchatka Territory). The experiment was of two factors, with systematic plot allocation and quadruplicated replication. The term of planting potatoes – the second ten-day period of June, harvesting – the second ten-day period of September. The preceding crop – a layer of perennial grasses. The potato cultivation technology in the experiment is the one generally accepted for the Kamchatka Territory [9]. Potatoes were harvested with a KTH-2 potato raiser, and the tubers were picked manually. The yield from each plot was registered.

The soil type is ochreous volcanic, light in granulometric composition, the initial agrochemical data at the time of trial establishment were characterized by the following indicators: humus 6.6% (according to Tyurin), average mobile phosphorus content (mg/kg of dry soil) – 60–81, exchangeable potassium content – 110–123 (according to Kirsanov method in Pryanishnikov All-Russian Agrochemical Institute modification), nitrate nitrogen content – 19.5–28.8 (GOST 26488–85), and ammonium nitrogen content – 7.0–9.0 (GOST 26716–85). Hydrolytic soil acidity was 5.82 (according to Kappen).

The effectiveness of Bio-Mix preparation in potato cultivation was evaluated according to Lorkh Federal Potato Research Centre methodology and potato protection from diseases, pests, weeds and immunity [10, 11]. The following indicators were analyzed: the accumulation dynamics of plant uptake forms of nitrogen, phosphorus and potassium in the soil; tuber sprouting; biometric indicators; phytopathological observations; crop yield weight and marketability; starch, ascorbic acid and nitrate content in the tubers. The biological effectiveness of Bio-Mix preparation was calculated [12]. The results of the studies were statistically processed by the method of variance analysis according to B. A. Dospekhov [13].

The meteorological conditions in the summer-autumn period of 2018–2020 differed in terms of temperature and moisture availability. According to the totality of hydrothermal indicators in 2018, 2019 and 2020, the warmest was July, the average monthly temperature was 14.2 °C, 13.1 °C and 14.0 °C with an average annual temperature of 12.5 °C; precipitation in June 2018 and 2019 was 52.2% and 82%, in July – 14.1% and 7.7%, in the first ten days of September – 71.3% and 7.6% over the norm. In 2020, precipitation in June amounted to 33.4 mm, by 47%, in July – by 52.4%, in September – by 76.1% below normal, and in August, precipitation was higher than the long-term average by 109.4%.

The sum of active temperatures >10 °C from May to September was 1002 °C in 2018, 1009.5 °C in 2019, and 1121 °C in 2020, with an average annual temperature of 1092 °C. 2020 was a more favorable year for the growth and development of potatoes in terms of temperature, and 2018 and 2019 were characterized by a low sum of active temperatures and high humidity, which contributed to the occurrence of rhizoctoniosis.

3 Results and Discussions

The balanced composition of macro- and microelements of algae cultures and aquatic organisms of Bio-Mix preparation promotes the development of micro-organisms' organic activity, creating favorable conditions for nutrients assimilation. The application of Bio-Mix organic fertilizer from aquatic organisms in potato cultivation has impacted the content of the key nutrients in the soil in plant uptake form. The content of ammonium nitrogen in the soil at the beginning of potato growing season in all variants was low and amounted to 8.74–11.88 mg/kg of soil, during potato growth until the end of the growing season, the amount of NH4 decreased to 4.23–7.93 mg/kg (Table 1). The trend shows an increase in ammonia nitrogen when applying Bio-Mix at a dose of 5.0 l/ha at $(NPK)_{90}$ during the period of mass sprouting by 20.9%, by the end of the growing season by 45.0% against the background.

At $(NPK)_{60}$ in the mass sprouting stage, the content of ammonia nitrogen in the soil was at the reference level, but by the end of the growing season, an average increase of 23.2% was observed. As per the potato development stages, the nitrate nitrogen content ranged from average values of 23.15–27.15 mg/kg to close to elevated values of 25.4–62.7 mg/kg during budding and low values of 6.05–19.9 mg/kg by the end of the growing season, due to nitrogen consumption by potato plants. Their introduction of Bio-Mix resulted in an average increase in nitrate nitrogen content at (NPK)90 and (NPK)60 by 5.7% and 37.4%, respectively, by the end of the potato growing season, and an increase of 8.6% against the recommended reference value.

When introducing Bio-Mix preparation into the soil, the content of plant-available phosphorus in the soil during mass sprouting and budding was average (95.5–140), by the end of the growing season – increased (117.0–145.5) mg/kg, the content of exchangeable varied from high in the period of mass sprouting (172.4–192.4) mg/kg to increased by the end of the growing season (137.8–153.5) mg/kg (Table 2), since the plant-available potassium is utilized a lot for tuber formation.

By the end of the growing season, when applying organic fertilizer Bio-Mix at (NPC)90 and (NPC)60, an increase in plant-available phosphorus is observed on average by 29.3 and 35.0 mg/kg (26.8 and 33.3%), and the exchangeable potassium – by 25.4 and 16.3 mg/kg (21.4 and 11.9%) compared to the mineral-only background value, respectively. Against the recommended reference value, the plant-available phosphorus was registered higher by an average of 51.2%, and the exchangeable potassium was 11.3% higher due to the introduction of Bio-Mix.

Table 1. Dynamics of ammonia and nitrate nitrogen content depending on the application of Bio-Mix organic fertilizer, mg/kg (average for 2018–2020).

Experiment variants	Potato development stage					
	Mass sprouting		Budding		End of growing season	
	N-NH4	$N\text{-}NO_3$	$N\text{-}NH_4$	$N\text{-}NO_3$	$N\text{-}NH_4$	$N\text{-}NO_3$
Reference (no fertilizers)	8.67	22.95	6.50	28.65	6.17	10.15
(NPK)120 (recommended reference value)	11.43	26.55	4.56	32.45	4.22	13.65
Bio-Mix – 2.5 l/ha	9.48	27.15	6.63	25.40	6.67	6.05
Bio-Mix – 5.0 l/ha	11.01	23.15	6.31	26.80	6.36	8.05
Background 1 – $(NPK)_{90}$	9.07	23.50	6.82	45.65	5.47	13.10
Background 1 + Bio-Mix – 2.5 l/ha	8.74	26.85	7.70	54.85	4.23	13.80
Background 1 + Bio-Mix – 5.0 l/ha	10.97	24.80	7.91	62.70	7.93	13.90
Background 2 – $(NPK)_{60}$	11.43	27.05	8.49	42.00	6.17	11.50
Background 2 + Bio-Mix – 2.5 l/ha	10.94	25.40	9.85	55.90	7.58	16.65
Background 2 + Bio-Mix – 5.0 l/ha	11.88	26.05	9.38	46.70	7.62	19.90

Table 2. Dynamics of exchangeable phosphorus and potassium in the soil depending on Bio-Mix organic fertilizer application, mg/kg (average for 2018–2020).

Experiment variants	Potato development stages					
	Mass sprouting		Budding		End of growing season	
	P_2O_5	K_2O	P_2O_5	K_2O	P_2O_5	K_2O
Reference (no fertilizers)	124.0	189.0	110.0	167.0	86.0	144.5
(NPK)120 – recommended reference value	84.5	190.0	83.5	146.0	92.0	133.6
Bio-Mix – 2.5 l/ha	113.0	187.5	93.5	187.0	117.0	151.0
Bio-Mix – 5.0 l/ha	99.0	189.2	110.5	168.0	125.0	153.5
Background 1 – $(NPK)_{90}$	112.0	182.5	112.0	158.0	109.0	118.6
Background 1 + Bio-Mix – 2.5 l/ha	113.0	192.4	93.0	167.2	145.5	137.8
Background 1 + Bio-Mix – 5.0 l/ha	119.5	193.5	115.5	178.0	131.0	150.1
Background 2 – $(NPK)_{60}$	91.0	175.0	83.5	158.0	105.0	137.2
Background 2 + Bio-Mix – 2.5 l/ha	95.5	172.4	88.5	175.1	139.0	153.5
Background 2 + Bio-Mix – 5.0 l/ha	103.5	183.1	86.0	171.7	141.0	153.4

Bio-Mix organic fertilizer based on a balanced composition of micro- and macronutrients, the necessary set of vitamins and amino acids, is a natural stimulant for plant growth and development. Combined application of mineral fertilizer at reduced rates and Bio-Mix organic fertilizer resulted in mass potato sprouting four days earlier, budding seven days earlier, and the growing season of potatoes was 74 days, which is four days longer than the mineral-only background value. Field emergence increased by 2.0–3.5% compared to the reference, an increase in field emergence by 0.5–1.3% was registered provided by organic fertilization. Bio-Mix preparation, activating the immune system and increasing the plant's stress resistance, stimulated their growth and improved their resistance to diseases. It was determined that the use of Bio-Mix organic fertilizer in combination with mineral fertilizer contributed to an increase in the linear height of plants at (NPK)90 by 4.4–7.7 cm, at (NPK)60 – by 2.2–7.6 cm. At (NPK)90 and (NPK)60 when applying Bio-Mix at a dose of 5.0 l/ha, the maximum increase in plant linear height was obtained – by 7.7 and 7.6 cm against the background value and by 3.8 and 1.8 cm against the recommended reference value.

The introduction of Bio-Mix organic fertilizer into the soil before planting reduced the intensity of the rhizoctoniosis development. The introduction of the Bio-Mix solution into the soil by spraying the potato bed contributed to a decrease in the development of rhizoctoniosis during budding by an average of 3.6% compared to the background and by 11.4% against the recommended reference value (Table 3).

The incidence of the disease in these variants decreased by 11.7% and 41.9%, respectively. The lesion of sprouts during the sprouting period was reduced by an average of

1.3% by applying Bio-Mix preparation. Prior to potato harvesting at (NPK)90, the development of rhizoctoniosis and the incidence of the disease were lower on average by 9.0% and 32.9%, the lesion of sprouts – by 0.9%, of tubers – by 1.8%; at (NPK)60 – by 8.5 and 32.7%, 0.9%, 2.9, respectively, compared to the background values. Against the recommended reference value, these indicators were lower by an average of 7.3% and 31.3%, 1.3% and 3.1%.

Table 3. Development and incidence of rhizoctoniosis depending on Bio-Mix organic fertilizer application to the soil, % (average for 2018–2020).

Experiment variants	Rhizoctoniosis development, %		Rhizoctoniosis incidence, %		Rhizoctoniosis lesions, %	
	Budding stage	Prior to harvesting	Budding stage	Prior to harvesting	Sprouts	Tubers
Reference (no fertilizers)	18.8	28.2	71.9	80.6	2.3	10.6
$(NPK)_{120}$ – recommended reference value	16.3	16.1	61.6	64.1	2.2	8.4
Bio-Mix – 2.5 l/ha	11.3	16.8	42.2	61.7	0.9	8.9
Bio-Mix – 5.0 l/ha	10.8	15.1	40.8	57.0	0.8	7.6
Background 1 – $(NPK)_{90}$	8.1	13.5	28.5	48.1	1.7	7.5
Background 1 + Bio-Mix – 2.5 l/ha	4.3	7.5	16.7	25.5	0.8	5.7
Background 1 + Bio-Mix – 5.0 l/ha	4.8	10.5	20.8	40.4	0.8	5.7
Background 2 – $(NPK)_{60}$	8.8	14.1	34.3	55.4	1.3	7.8
Background 2 + Bio-Mix – 2.5 l/ha	5.6	9.2	22.8	34.5	0.4	5.1
Background 2 + Bio-Mix – 5.0 l/ha	4.6	7.9	18.6	30.9	0	4.7

The average biological effectiveness of Bio-Mix preparation in the budding stage was 68.9% compared to the recommended reference value, at $(NPK)_{90}$ – 38.6%, at $(NPK)_{60}$ – 40.9%, prior to harvesting – 45.1%, 32.4%, 40.4%, respectively.

It has been noted that the pattern of reducing the incidence of rhizoctoniosis between variants remains, but before harvesting, the degree of development and incidence of rhizoctoniosis increases compared to the data obtained during budding from an average of 4.9 to 8.8% and from 19.7 to 32.8% for variants with Bio-Mix introduced with the background of mineral fertilizer, and the biological effectiveness of the preparation - from 68.9% to 45.1%. Since the effect of the Bio-Mix preparation decreases by the end of the growing season, the incidence of plant rhizoctoniosis increases, and for more effective plant protection, another treatment for vegetating plants is required as established by the research conducted at the Federal State Budgetary Scientific Institution Kamchatka National Scientific and Research Institute of Agriculture [8].

The research has shown that when applying a Bio-Mix to the soil at $(NPK)_{90}$ and $(NPK)_{60}$, the yield of potatoes increases by an average of 10.6 and 14.4%, while more marketable and seed tubers were formed by 1.3 and 5.6% and 1.9 and 2.4% compared to the background values, respectively (Table 4). At $(NPK)_{90}$, when applying Bio-Mix organic fertilizer at a dose of 5.0 l/ha, the potato yield was 31.9 t/ha and was at the level of the recommended reference value $(NPK)_{120}$; with the introduction of this method, the dose of mineral fertilizer can be reduced by 30%.

Table 4. The effect of Bio-Mix organic fertilizer on the yield, marketability, and biochemical parameters of potato tubers (2018–2020).

Experiment variants	Yield, t/ha	Marketable tubers, %	Seed tubers, %	Content in tubers		
				Starch, %	Ascorbic acid, mg%	Nitrates, mg/kg
Reference (no fertilizers)	12.5	63.9	36.5	10.7	8.80	64.7
$(NPK)_{120}$ – recommended reference value	30.4	81.8	39.4	10.5	13.36	55.5
Bio-Mix – 2.5 l/ha	13.1	68.0	42.5	10.7	9.52	46.6
Bio-Mix – 5.0 l/ha	14.0	66.9	43.0	10.6	8.63	46.2
Background 1 – $(NPK)_{90}$	28.2	78.0	37.2	10.6	10.27	52.7
Background 1 + Bio-Mix – 2.5 l/ha	30.5	78.6	42.2	10.6	15.02	51.7
Background 1 + Bio-Mix – 5.0 l/ha	31.9	80.1	43.3	10.6	16.59	50.4

(*continued*)

Table 4. (*continued*)

Experiment variants	Yield, t/ha	Marketable tubers, %	Seed tubers, %	Content in tubers		
				Starch, %	Ascorbic acid, mg%	Nitrates, mg/kg
Background 2 – $(NPK)_{60}$	26.3	80.3	39.6	10.8	8.89	48.2
Background 2 + Bio-Mix – 2.5 l/ha	29.5	81.1	40.9	11.1	11.63	46.6
Background 2 + Bio-Mix – 5.0 l/ha	30.6	83.3	43.0	11.3	16.06	45.6
HCP05 total Factor A reference Factor B $_{variants}$	1.2 0.5 0.8					

Bio-Mix organic fertilizer from aquatic organisms contributed to the improvement of potato tuber quality indicators. The amount of nitrates in tubers in all variants differed within the range of 45.6–64.7 mg/kg, when using Bio-Mix organic fertilizer, it did not exceed 51.7 mg/kg, which is 4.6 times lower than the maximum allowable concentration (MAC = 240). When introducing Bio-Mix at a dose of 5.0 l/ha with mineral fertilizer, the maximum amount of ascorbic acid in tubers was obtained – 16.59 and 16.06 mg%, which is higher than the background variants by 6.32 and 7.17 mg%, and the recommended reference value by an average of 2.96 mg%. A trend of increase in tuber starch content from 10.5% in the recommended reference value to 11.1–11.3% when applying Bio-Mix at $(NPK)_{60}$ is noted.

According to the results of the phytopathological analysis held in spring, the application of Bio-Mix organic fertilizer has affected tuber survivability. The minimum incidence of diseases in tubers during storage was observed when using Bio-Mix organic fertilizer at a dose of 5.0 l/ha, which is lower than the mineral-only background values by 7.8 and 3.7%, the tuber survivability was the highest and amounted to 89.5 and 89.1%. The greatest tuber lesion during storage was by skin spot and rhizoctoniosis, but due to the use of organic fertilizer, the disease incidence in tubers decreased in comparison with the mineral-only background values by an average of 2.0 and 3.1%, and against the recommended reference value – by 4.3 and 1.7%. The lesion of tubers with foma-fusarium and soft rot when using Bio-Mix did not exceed 1.0% and 0.5%. The survivability of tubers when using Bio-Mix against the background of mineral fertilizer was higher than the reference by an average of 8.7%, the recommended reference value – by 6.2% (Table 5).

Table 5. The effect of Bio-Mix organic fertilizer on the potato tuber survivability (spring 2019–2020).

Experiment variants	Tubers with lesions, %					Tuber survivability, %
	Rhizoctoniosis	Foma-fusarium rots	Soft rot	Skin spot	Total	
Reference (no fertilizers)	12.9	0.9	1.0	20.1	18.3	79.9
$(NPK)_{120}$ – recommended reference value	7.8	1.1	0	8.7	17.6	82.4
Bio-Mix – 2.5 l/ha	7.3	0.5	0.6	4.1	12.5	87.5
Bio-Mix – 5.0 l/ha	6.6	1.0	0	5.4	13.0	87.0
Background 1 – $(NPK)_{90}$	9.4	1.6	0.1	7.9	19.0	81.0
Background 1 + Bio-Mix – 2.5 l/ha	6.1	0.7	0.5	4.6	11.9	88.1
Background 1 + Bio-Mix – 5.0 l/ha	5.5	0.7	0.1	4.2	10.5	89.5
Background 2 - $(NPK)_{60}$	9.0	0.6	1.0	4.8	15.4	84.6
Background 2 + Bio-Mix – 2.5 l/ha	6.6	0.7	0.2	5.0	12.5	87.5
Background 2 + Bio-Mix – 5.0 l/ha	6.1	0.9	0.1	3.8	10.9	89.1

4 Conclusions

In the Kamchatka Territory, the combined use of Bio-Mare organic fertilizer by Bio-Mix and mineral fertilizers in doses reduced by 30% – at (NPK)90 and 50% – at (NPK)60 from the recommended reference value – (NPK)120, had a positive effect on the content of basic nutrients in the soil. The increase in exchangeable forms of phosphorus and potassium in the soil is conditioned by the chemical composition of Bio-Mix preparation, which, in comparison with soils, has a lower content of plant-available phosphorus and potassium, but the necessary composition of trace elements, which stimulates the balance of the plant's nutritional regime and improves soil properties. Thus, the use of Bio-Mix at (NPK)90 and (NPK)60 resulted in an increase in the plant-available phosphorus in the soil by 29.3 and 35.0 mg/kg (26.8 and 33.3%), and in the exchangeable potassium by 25.4 and 16.3 mg/kg (21.4 and 11.9%) compared to the mineral-only background values, respectively. In relation to the recommended reference value – (NPK)120, the introduction of Bio-Mix resulted in registering higher values of the plant-available phosphorus by an average of 51.2%, and the exchangeable potassium – by 11.3%.

The introduction of Bio-Mix preparation into the soil contributed to an increase in potato yield up to 14.4%, improved product quality in the form of an increase in tuber ascorbic acid content by 6.7 mg%, a decrease in nitrate content by 4.6 times, and the reduction of rhizoctoniosis development and incidence by an average of 3.6 and 11.7% during budding, and 8.8 and 32.8% prior to harvesting, as compared to the mineral-only background values. The tuber survivability when using Bio-Mix increases by an average of 8.7% against the reference, and by 6.2% against the recommended reference value.

References

1. Salwinder, S.D., Sharma, V., Mandal, A., Rama, K.N., Verma, G.: Improving soil micronutrient availability under organic farming. Adv. Org. Farm. **7**, 93–114 (2021). https://doi.org/10.1016/B978-0-12-822358-1.00002-X
2. Vasilenko, T.T., Astafyeva, et al.: O katastroficheskom sostoyanii plodorodiya pochv selskokhozyajstvennykh ugodii v Kamchatskoi oblasti. Petropavlovsk-Kamchatsky (2005)
3. Craigie, J.S.: Seaweed extract stimuli in plant science and agriculture. J. Appl. Phycol. **23**, 371–393 (2011). https://doi.org/10.1007/s10811-010-9560-4
4. Jramas, E.O.A.G.G., Gonzalezetal, R.L.: Respose of potatos cultivation (Solanum tuberosum L.) to the combination of the ecological fertilizer Herba Green with chmical fertilizer. Centra Agricola **44**(1), 80–89 (2017)
5. Yakhin, O.I., Lubyanov, A.A., Yakhin, I.A., Brown, P.H.: Biostimulants in plant science: a global perspective. Front. Plant Sci. **7**(2049), 1–32 (2017). https://doi.org/10.3389/fpls.2016.02049
6. Gainatulina, V.V., Khasbiullina, O.I.: Effektivnost' primeneniya biopreparatov i fungitsidov v bor'be s rizoktoniozom kartofelya. J. Vestn. Dalnevostochnogo otdeleniya RAN **4**, 93–99 (2020). https://doi.org/10.37102/08697698.2020.212.4.015
7. Sorokin, M.A., Popov, I.A.: Seaweed-based fertilizer and its production method. Invention Patent RUS 2161599, 6 Dec 1998
8. List of pesticides and agrochemicals approved for application on the territory of the Russian Federation 2021. Reference edition. Moscow (2021)
9. Ryakhovskaya, N.I., Gainatulina, V.V., Vlasenko, G.P., et al: Sistema zemledeliya Kamchatskogo kraya. Kamchatka National Scientific and Research Institute of Agriculture. Petropavlovsk-Kamchatsky (2015)
10. Volovik, A.S., Trofimets, L.N., Dolyagin, A.B., Glez, V.M.: Metodika issledovanij po zashhite kartofelya ot boleznej, vreditelej, sornyakov i immunitetu. Lorkh Federal Potato Research Centre, Rosselkhozakademia, Moscow (1995)
11. Metodology for conducting agrotechnical experiments, accounting, observations and analyzes on potatoes. Lorkh National Potato Research Centre, Moscow (2019)
12. Anisimov, B.A., Uskokov, A.I., Simakov, E.A., Varitsev, Y., Zeyruk, V.N., Alyabeva, A.V.: Metodika provedeniya polevykh issledovanij i posleuborochnogo kontrolya kachestva semennogo kartofelya. Ikar Publishing House, Moscow (1989)
13. Dospekhov, B.A.: Metodika polevogo opyta: (s osnovami statisticheskoj obrabotki rezultatov issledovanij). Kolos, Moscow (1985)

Study of Breeding Lines of Soft Spring Wheat in a Competitive Variety Trial in Irkutsk Region

Natalia Klimenko(✉), Irina Abramova, and Elena Kuznetsova

Irkutsk State Agrarian University Named After A.A. Ezhevsky, Molodezhny, 1/1, Irkutsk District, 664038 Irkutsk Region, Russia

Abstract. The yield potential of the plant is the most essential property of the variety and is usually defined as the key factor in the breeding objectives. The work presents an assessment of soft spring wheat breeding lines in a competitive variety trial. The competitive variety trial included 13 lines selected in hybrid swarms and propagated in breeding and reference nurseries. 2 varieties of soft spring wheat were taken as the reference: Iren in the early group, and Tulunskaya 11 in the middle-early group. The evaluation of the studied lines was performed according to the key indicators that are of utmost importance in breeding practice. The growing season of wheat is a very important indicator, since the climate of the Irkutsk region is severe, the duration of the frost-free period in the region is 70–110 days only. The observations enabled identification of the AA-67-14 line with a shorter growing season, it can be attributed to the group of early wheat. AA-41-15 line was at the reference level in the middle-early group. This line was marked as the highest yielding. In terms of protein and gluten content, it was at the reference level.

Keywords: Soft spring wheat · Competitive variety trial · Line · Variety · Earliness · Grain · Productivity

1 Introduction

Spring wheat is one of the most essential food crops. It is one of the primary grain crops cultivated in the Irkutsk region. The area occupied by spring wheat in the region in 2020 amounted to about 400 thousand hectares [1]. Breeding research and methodology layout are determined by the soil and climatic conditions of the region. The Irkutsk region belongs to the zone of risky farming, which requires a particularly thorough approach to the cultivation technology and the optimal selection of varieties.

The climate of the Irkutsk region is markedly continental with cold winters and relatively warm summers, as well as a wide range of fluctuations in the daytime and nighttime temperatures. The frost-free period in the region is 70–110 days. Average annual precipitation: 250–350 mm [2]. The water availability of the main agricultural crops in most areas is 70–100% of their biological demand, in steppe and forest-steppe areas it decreases to 55–65%. The main limiting factor is the lack of heat. The sum

A. Muratov and S. Ignateva (Eds.): AFE 2021, LNNS 353, pp. 339–346, 2022.
https://doi.org/10.1007/978-3-030-91402-8_39

of the active temperatures varies from 1,300 °C to 1,700 °C. According to agroclimatic calculations, midseason spring wheat varieties have good heat supply only in areas where the sum of above-zero temperatures is 1,500 °C-1,700 °C, of which there are no more than 10% in the region. [1].

In the Irkutsk region, grey forest soils with a low humus content and sod-carbonate soils with slightly higher fertility are most common. In general, it shall be noted that the entire territory of the Irkutsk region is divided into 6 soil and climatic zones distinct in terms of temperature and moisture regime. The peculiarity of natural and climatic conditions requires a differentiated approach to the spring wheat breeding. The practice has shown that in all areas of the region up to 40% of this crop sown area shall be allocated to early wheat varieties, and in the sub-taiga and taiga areas, this value amounts to up to 100%. Therefore, the key breeding activity in the Irkutsk State Agrarian University is the development of early climate-resistant varieties characterized by good processing and baking properties, as well as resistant to spring and summer droughts and adapted to growing conditions, capable of generating a sound and stable yield in any weather conditions of the year [1, 2]. With the right cultivation technology, the share of the variety in the total yield increase in the region is 35% to 50% [3].

A strong variety and a well-developed cultivation technology are among the most important factors in increasing the yield of agricultural crops, including wheat. Therefore, there is a need in developing new high-yielding varieties. In the Irkutsk region, 15 varieties of soft spring wheat (Iren, Tulunskaya 12, Tulunskaya 11, etc.) are recognized [1, 4], as having a high potential yield, which, with utilizing the appropriate varietal technology, can generate high yields. The climate of the Irkutsk region is characterized by spring and summer droughts. Thus, the lack of resistance to spring-summer drought and extreme growing conditions is a significant disadvantage of all recognized varieties [1, 2].

Research purpose: assessment of promising spring soft wheat breeding samples in competitive variety trial.

Research objectives conducting a comprehensive assessment of the lines in the competitive variety trial, and selecting the best samples.

2 Materials and Methods of Research

The research was performed at Molodezhny research and trial site of the Irkutsk State Agrarian University. The agroclimatological zoning of the site territory is part of the Irkutsk-Cheremkhovo plain of the Predsayansky foredeep. Terrain is flat, slope steepness is 2–5° [5]. Trial site soil: grey forest, subtype: light grey, weakly-podzolized. According to the granulometric composition, they are characterized as being at the margin of heavy and medium loam with 2.03% humus content in the horizon of 0–20 cm. As per the mobile phosphorus content, the soil belongs to the group of high-availability, the exchangeable potassium – low-availability, the pH value of the salt extract is 5.0 [6].

Sowing in the nursery of the competitive variety trial was performed in the second ten days of May. In the competitive variety trial, 13 lines were studied, selected in hybrid swarms, and propagated in breeding and reference nurseries. The assessment of the studied lines was performed by the following indicators:

- growing season duration;
- resistance to lodging, disease, and grain shatter;
- yield, including the yield formula elements;
- grain protein and gluten content.

Two varieties of spring wheat were taken as the reference: Iren in the early group and Tulunskaya 11 in the middle-early group.

The area of plots in the competitive variety trial was 75 m^2. Seeding rate: 7.0 million of germinating grains per hectare. The tillage was performed according to the farming system generally accepted in the region. Preceding crop – fallow.

The assessment of the breeding material was performed by the economic and biological characteristics of highest significance as per the procedure adopted at the state breeding stations [7, 8]. Sowing was performed in the first ten days of May, with grain of Grade 1, 100% purity and 99.5% germination rate. During the growing season of soft spring wheat plants, phenological observations were conducted [8].

Wheat harvesting in the nursery of the competitive variety trial was performed in the first ten days of September. For the yield formula analysis, plants from 1 m^2 were selected. In the laboratory, the elements of the yield formula were determined by agronomic characters. Mathematical processing of the results obtained was performed according to the generally accepted method by B. A. Dospekhov [9].

3 Results

The growing season duration and its structure determine the suitability of the variety for the growing conditions in a particular area and are crucial in future yield generation. In the Irkutsk region, the duration of the frost-free period according to the average long-term statistical values is 90–98 days. This value is determined following the phenological observations, i.e., consideration of each phase onset timing. The line in which the development rhythm corresponds to the local climatic conditions, and the critical period does not coincide with the time of adverse factor occurrence, will be more adapted to unfavorable environment [8, 10].

The key interphase periods in spring wheat are vegetative (seedlings – heading) determining the vegetative mass growth pattern of plants, plant habitus, and generative (heading – waxy maturity) determining the nature of the propagation organs development [8, 11].

Table 1 shows the data on the growing season duration in soft spring wheat lines.

The shortest growing season was identified in AA-67-14 line, and it can be attributed to the early group. AA-41-15 line was at the reference level in the middle-early group. The remaining breeding varieties by this attribute were between the references in early and midseason groups.

Table 1. Growing seasons of spring wheat varieties and lines (days).

Seq. No	Line, variety	Seedlings – heading	Heading – waxy maturity	Growing season
1	AA-708	41	54	95
2	AA-708-15	40	54	94
3	AA-3-15	41	55	96
4	AA-31-15	40	56	96
5	AA-38-15	41	53	94
6	AA-105-14	40	54	94
7	AA-109-15	41	55	96
8	AA-87-14	42	54	96
9	AA-41-15	43	56	99
10	AA-67-14	40	52	92
11	AA-59-15	42	54	96
12	AA-120-14	42	53	95
13	AA-92-15	41	54	96
14	Tulunskaya-11	43	56	99
15	Iren	40	52	92

In addition to studying the growing season duration, the breeding samples were assessed for resistance to lodging. The analysis of this indicator revealed that all the breeding lines studied were resistant to lodging.

The wheat yield formula depends on the number of spikelets and grains in the head, their weight, and, accordingly, the head length and the weight of 1,000 grains. These indicators are closely related to each other. One of the key indicators determining the yield of spring soft wheat is the number of grains on the main head. Their number varies and quite strongly depends on the environmental and growing conditions. The number of grains in the head can vary from a hollow spikelet to several hundred grains or more in the main ear, which is mostly associated not with the genotype attribute manifestation, but with abiotic environmental factors, which are expressed both by climatic and soil conditions [5, 12–14].

Table 2 provides data on the yield formula elements. The largest number of grains in the main head was identified in lines AA-41-15 and AA-38-15. High main head productivity was marked in line AA-41-15, and this breeding sample also had an increased grain size. Two lines were distinguished by the weight of 1,000 grains: AA-708 and AA-120-14, they were at the reference levels, and the other varieties were inferior to the references by this attribute.

Yield determination of the studied lines is of great interest. The highest and most reliable yield increase (+2.07 t/ha) is identified in the sample AA-41-15.

In addition, two samples (AA-67-14 and AA-105-14) were almost at the level of the reference from the mid-season group (Tulunskaya 11) in terms of yield.

Table 2. Analysis of yield formula in spring wheat lines.

Seq. No	Line, variety	Head length (cm)	Number of spikelets per head (pcs.)	Number of grains per head (pcs.)	Weight of grain in the head (g)	Weight of 1,000 grains (g)	Average yield, t/ha
1	AA-708	6.6	13.1	24.2	0.97	46.3	3.58
2	AA-708-15	6.9	12.7	23.8	0.91	38.3	3.95
3	AA-3-15	6.4	12.7	23.8	0.91	38.3	3.36
4	AA-31-15	6.1	12.1	23.2	0.82	35.6	3.03
5	AA-38-15	6.1	12.9	30.6	0.96	31.3	3.55
6	AA-105-14	6.1	12.3	27.7	1.09	39.5	4.03
7	AA-109-15	6.1	13.2	25.3	0.90	35.7	3.33
8	AA-87-14	5.5	11.1	21.6	0.85	39.5	3.15
9	AA-41-15	7.6	13.7	33.6	1.72	51.3	6.36
10	AA-67-14	7.7	13.5	27.2	1.11	41.0	4.10
11	AA-59-15	6.9	12.0	27.8	1.04	37.4	3.84
12	AA-120-14	5.8	11.8	24.8	0.99	46.3	3.66
13	AA-92-15	5.8	13.6	28.5	1.00	35.1	3.70
14	Tulunskaya 11 (reference)	6.0	12.2	27.8	1.16	41.7	4.29
15	Iren	7.1	12.6	23.2	1.10	47.4	4.07
	HCP_{05}						0.33

The most early-maturing line, AA-67-14, was somewhat less productive than Tulunskaya 11, but exceeded Iren in this respect.

Thus, following the variety trial of the best soft spring wheat lines, three breeding varieties have been identified subject to further work.

At the current stage, when developing new breeding varieties, the breeding material shall be assessed not only in terms of the yield amount but also in terms of its quality. [15]. Data on this attribute are provided in Table 3.

Table 3. Grain quality of spring soft wheat lines in CVT, %.

Line, variety	Protein	Hardness	Gluten	Starch	Phosphorus	Crude fat	Crude fiber
AA-41-15	15.58	32.80	32.61	55.83	0.31	2.29	4.22
AA-109-15	16.31	33.69	35.17	46.44	0.33	2.24	4.01
AA-38-15	16.44	31.31	36.35	47.96	0.33	1.94	3.87
AA-31-15	16.53	33.38	35.86	45.67	0.33	2.22	4.10
AA-3-15	15.86	32.26	33.77	47.05	0.33	2.18	4.07
AA-708-15	15.11	32.51	30.99	49.11	0.32	2.18	4.01
AA-708	15.12	32.95	31.15	49.87	0.31	2.23	4.05
AA-92-15	16.59	31.40	36.47	46.85	0.33	1.95	4.03
AA-120-14	16.19	35.39	33.57	45.31	0.33	2.28	4.00
AA-59-15	15.94	32.84	33.77	49.83	0.32	2.19	4.43
AA-67-14	15.51	29.08	32.91	49.81	0.31	2.03	4.29
AA-87-14	16.02	36.49	33.05	46.05	0.33	2.38	4.08
AA-105-14	16.49	31.48	36.10	50.82	0.32	1.90	3.81
Iren	15.83	33.31	33.69	46.96	0.34	2.23	3.91
Tulunskaya 11	16.09	32.77	34.58	52.35	0.32	2.24	4.18

It shall be noted that six lines exceeded the reference (Tulunskaya 11) in terms of protein content, and 5 – in terms of gluten content. The line AA-41-15 has shown high yield and, at the same time, had values within the reference level in terms of protein and gluten content. AA-105-14 line was slightly inferior to the reference variety in terms of yield, however, in terms of the quantitative protein and gluten content it exceeded the reference. In terms of starch, phosphorus, crude fat, crude fiber content, and grain hardness, the samples studied were at the reference levels or slightly differed from them to one end or the other.

4 Conclusions

The study of a competitive variety trial allowed for obtaining the following results:

As previously noted, the growing season duration largely depends on the prevailing weather conditions. Thus, the studies showed that line AA-67-14 was characterized by a shorter growing season and it can be attributed to the group of early-maturing varieties. AA-41-15 line was at the reference level in the middle-early group.

Line AA-41-15 was identified as having the highest productivity of the main head. This breeding sample also had an increased grain size. The lowest results were obtained for lines AA-708-15, AA-708, AA-67-14.

Grain size is a qualitative indicator that depends on the cultivation conditions and is related to the most essential agronomic character. In breeding research, it is given major attention. According to the study, by this indicator, two lines were identified: AA-708 and AA-120-14, and the other breeding varieties were at the reference level.

The qualitative indicators that were determined in the grain of the studied lines provided for determining that the best results in terms of protein content were shown by lines AA-109-15, AA-38-15, AA-31-15, AA-92-15, AA-105-14. They exceeded the reference variety by 0.22–0.5%.

In the competitive variety trial, two lines were distinguished – the earliest-maturing AA-67-14 and the highest-productive AA-41-15, due to their increased values of the yield formula elements Sample AA-41-15 significantly and reliably outperformed Tulunskaya 11 reference variety in terms of yield and in terms of grain quality, it can be classified as "strong" wheat.

Thus, following the soft spring wheat lines variety testing, we have identified three breeding varieties subject to further work.

References

1. Akman, H., Karaduman, Y.: Evaluating technological quality of cultivated Triticum species, interspecific, and intergeneric hybrids for wheat-based products and breeding programs. J. Cereal Sci. **99**, 103188 (2021). https://doi.org/10.1016/j.jcs.2021.103188
2. Yushkevich, L.V., Pakhotina, I.V., Shchitov, A.G.: Agrotechnological methods of soft spring wheat cultivation use efficiency to increase productivity and grain quality in the Omsk region. Vestn. KrasGAU **7**, 26–34 (2021). https://doi.org/10.36718/1819-4036-2021-7-26-34
3. Abramova, I.N., Klimenko, N.N.: Evaluation of breeding lines of soft spring wheat in the control nursery. Sci. Pract. J. Vestn, IrGSHA **3**(104), 6–15 (2021). https://doi.org/10.51215/1999-3765-2021-104-6-15
4. Fedosenko, D.F., Nikitina, V.I.: The effect of vegetation and genotype conditions on the variability of the main quantitative features of spring soft wheat in Krasnoyarsk forest steppe. Vestn. KrasGAU **1**, 43–48 (2021). https://doi.org/10.36718/1819-4036-2021-1-43-48
5. Abramov, A.G., Abramova, I.N., Klimenko, N.N.: Estimation of breeding material jf spring soft wheat in the forest-steppe zone of the Pre-Baikal region. East Siber. J. Biosci. **102**, 8–17 (2021). https://doi.org/10.51215/1999-3765-2021-102-8-17
6. Solodun, V.I., Amakova, T.V.: Efficiency of direct sowing technology for grain crops depending on the predecessors and backgrounds of the main tillage. East Siber. J. Biosci. **101**, 76–81 (2020). https://doi.org/10.51215/1999-3765-2020-101-76-81
7. Yashchenko, S., Loginov, Y., Kazak, A.: Influence of preceding crop on yield and quality of seeds of wheat varieties in northern forest-steppe of Tyumen region. Vestn. Buryat STSTE Acad. Agric. V. Philippov **1**(62), 47–57 (2021). https://doi.org/10.34655/bgsha.2021.62.1.007
8. Arguello, M.N., Mason, R.E., et al.: Performance of soft red winter wheat subjected to field soil waterlogging: grain yield and yield components. Field Crops Res. **194**, 57–64 (2016). https://doi.org/10.1016/j.fcr.2016.04.040
9. Taranova, T.Y., Kincharov, A.I., Demina, E.A., Mullayanova, O.S., Chekmasova, K.Y.: Selection evaluation of soft spring wheat input material by productivity and its elements. Vestn. KrasGAU **5**, 81–88 (2021). https://doi.org/10.36718/1819-4036-2021-5-81-88
10. Ghulam, H., Ahmed, M.D., et al.: Early selection of bread wheat genotypes using morphological and photosynthetic attributes conferring drought tolerance. J. Integr. Agric. **18**, 2483–2491 (2019). https://doi.org/10.1016/S2095-3119(18)62083-0
11. Muhammad, S., et al.: Genome-wide association analysis for stripe rust resistance in spring wheat (Triticum aestivum L.) germplasm. J. Integr. Agric. **19**, 2035–2043 (2020). https://doi.org/10.1016/S2095-3119(19)62841-8

12. Guzmán, C., Autrique, E., et al.: Genetic improvement of grain quality traits for CIMMYT semi-dwarf spring bread wheat varieties developed during 1965–2015: 50 years of breeding. Field Crops Res. **210**, 192–196 (2017). https://doi.org/10.1016/j.fcr.2017.06.002
13. Zhu, Z., Hao, Y., et al.: Breeding wheat for resistance to Fusarium head blight in the Global North: China, USA, and Canada. Crop J. **7**, 730–738 (2019). https://doi.org/10.1016/j.cj.2019.06.003
14. Kefale Sewenet, H., Menzir Anley, A., Alemayehu Getie, M.: Performance evaluation and participatory varietal selection of improved bread wheat (Triticum aestivum L.) varieties, the case of Debre Elias District, Northwestern Ethiopia. Ecol. Genet. Genomics **19**, 100086 (2021). https://doi.org/10.1016/j.egg.2021.100086
15. Haoa, B., Ma, J., et al.: Wheat breeding in China over the past 80 years has increased grain zinc but decreased grain iron concentration. Field Crops Res.**271**, 108253 (2021). https://doi.org/10.1016/j.fcr.2021.108253

Aspects of the Productive Longevity of Cows in the Dairy Farms of the Middle Priamurye

Natalia Klyuchnikova(✉), Michael Klyuchnikov, and Elena Klyuchnikova

Khabarovsk Federal Research Center of the Far Eastern Branch of the Russian Academy of Sciences, Separate Division of the Far Eastern Branch of the Agricultural Research Institute, 107, Karl Marx str., 680031 Khabarovsk, Russia

Abstract. Analysis of zootechnical and breeding account shows that the age of first calving influences on the duration of productive period and the size of life milk yield. The calvings before 24 months age in the region conditions reduce a duration of productive period and life duration. 34,8% of first-calf heifers from 2218 had calvings in 26,1–28,0 months age. The tendention to decrease the productive cows longevity as it increases the age of first calving in 28,1–30,0 months age is observed clearly. For the first time in the region nditions we identified the negative influence of spring-summer calvings on the duration of productive period and the size of life milk yield. In the region conditions summer calvings come to increase by 7,8% of birth and post-natal complications, that leads to culling of animals, death of newborn calves and influences negatively on the dairy productivity and the duration of productive period. The season of first calving or birth influences significantly on the number of cows, who has lived more than five calvings. Sire bulls have the significant influence on the duration of productive longevity and the economic use of cows. The assessment of economically useful traits of animals with different types of haptoglobin and ceruloplasmin. It was noted, that by all economically useful indicators the animals of the phenotype Hp 2-2 and Cp BB were superior than animals the same age of other phenotypes of haptoglobin and ceruloplasmin.

Keywords: Productive longevity · Life milk yield · Age of first calving · Season of birth and first calving · Reproductive ability · Sire bulls · Haptoglobin · Ceruloplasmin

1 Introduction

The biological features of cattle determine the priorities of dairy cattle breeding in the specific natural and economic conditions of the southern parts of the Far East. In the last decades, the misallocation of resources, use of industrial animal husbandry methods, and violation of optimal environmental parameters have led to a developing ecological crisis expressed as an increase in morbidity, a decrease in the duration of animals' productivity, and a decrease in the rate of reproduction. The total number of cattle decreased by 16.6 thousand head from 2010 to 2016, decreasing annually by 1.59%. The total number of

A. Muratov and S. Ignateva (Eds.): AFE 2021, LNNS 353, pp. 347–356, 2022.
https://doi.org/10.1007/978-3-030-91402-8_40

livestock animals decreased by 1.7% in 2016, ending at 8,263,732 head. Agricultural operations lose 10–15% of calves annually, or 200 thousand heifers per year, due to low reproductive health. Direct losses in cases of infertility include the death of calves; a decrease in the milk production of cows; the cost of treatment, both in drugs and payment for specialists; the loss of culled milk during the treatment period; and the cost of keeping non-productive animals. Indirect losses include an increase in the number of inseminations; an increase in the number of diseases; premature culling of cows; and a decrease in reproductive ability. Despite these costs, infertility problems often remain outside the daily concerns of farmers, perhaps because temporarily infertile individuals tend to produce milk within two years, and because the assessment of damage is based on relative losses of milk and issues which are difficult to perceive in real time [1].

The biological limits of the productive longevity of cows, at 10–12 lactations or more, offer high enough potential. In practice, this potential is realized extremely unsatisfactorily. In many farms of the Russian Federation, cows are productive for 2–2.5 lactations. Cows are removed during their most productive period or even before this point, a practice commonly applied in dairy cattle breeding in the Far East. Dairy cattle bred in the Far East show indicators of low economically useful traits, including high barrenness and a short period of economic utility — on average, fewer than three lactations [2]. An increase in cow culling and reduction of the length of the productive period is a global tendency in dairy farming. The average duration of use of cows in the United States is 2.8 lactations [3]. The average survival rate of cows to the second calving is 73%; to the third, 50%; to the fourth, 32%; and from the fifth to the eighth, 19%, 10%, 5%, and 2%, respectively [4]. According to the data presented by J.S. Brickell and co-authors, in English dairy farms, 11% of Holstein-Friesian replacement heifers, selected at 1 month of age, do not live to their first calving. Nineteen per cent of calves are rejected during the first lactation and 24% during the second lactation. Only 55% of the replacement heifers complete their first lactation successfully [5].

The main reasons cows are retired are infertility; different udder, extremity, and metabolic diseases; abortions; stillbirths; difficulty calving; and other causes. Of retired cows, 2.7–15% are rejected for old age; 7.2–27% for diseases of the udder; 7.7–24.6% for diseases of the extremities; 9.0–20.3% for low productivity; and 23–31.8% for barrenness [6–11]. Barrenness is thus one of the main causes of the early culling of animals. Only in isolated farms of the Russian Federation the release of calves per 100 cows per yearis recorded, at 86%; pregnancy from primary inseminations averages 39% with a range of 26–56% at an insemination index of 1.9–5.6; barrenness of cows ranges from 59–67%; the release of calves per 100 cows per year from 57–67%; the cow interbody period is 414–470 days; and the service period is 100–174 days [2]. Barrenness and the infertility of cows cause enormous damage to agricultural operations that lose milk and live weight of cattle because of the low weaning rate of younger animals and the reduction in heifers entering the main herd. The tendency to increase the culling of cows and shorten their productive periods because of low fertility is observed in other regions and countries with well-developed cattle breeding due to the high milk yield of cows, exceeding 9–10 thousand liters of milk per lactation. Because of reproductive dysfunction, 52% of dairy cows in Germany, 16% of Holstein cows in the US, 27% in Israel, and 30% in Switzerland drop out. The productivity of an average annual dry cow is about 50% that of a cow inseminated on time [12–15], proven both by science and practice. This connection is manifested in lower indicators of productivity under the region's conditions. Both genetic and paratypic factors influence the duration of productive longevity. Preservation and high productivity, reproductive ability, productive longevity, and viability are pressing problems in dairy cattle breeding.

2 Materials and Methods

The research was performed in dairy farms of the Middle Priamurye. The data were collected through tribal animal cards and the personal investigations and observations of the authors. Milk productivity was assessed by the amount of milk produced over 305 days of the first lactation and lifetime milk yield. The longevity of cows was assessed by the number of calvings, lactations, life duration, and cows culled after the first calving. Reproductive ability was assessed by the duration of the service period and the percentage of barren individuals.

3 Results and Discussion

The data showing the main indicators of the productivity of first-calf heifers depending on their age at first calving are presented in Table 1. Of 2,218 first-calf heifers, 34.8% calved at 26.1–28.0 months of age. Their productivity did not raise the average indicators of the herd. The proportion of cows surviving more than five lactations was 9.85%, but in older groups, this number was doubled. At the same time, the productivity of the first-calving heifers increased by 700–800 kg in comparison with the animals mated early.

The data resulting from analysis of the primary zootechnical registration of 1,108 Australian Holstein cows and cows of the same age from the 'Zarya' farm in the Khabarovsk district are of undoubted interest (Tables 2 and 3). The results of the analysis mark some features of the Holstein cows' productivity as due to the age of first calving. Regardless of the breeding conditions of heifers up to two years of age, only 45% of animals, including 3.5% of the local cows and 3.5% of the Australian Holstein cows, calved, and 40% of calving occurred when the heifers were older than 30 months. In the local cattle herd, the proportion of such heifers exceeded 50%, and that among the imported animals was less than 30%. The productivity of first-calf heifers up to two years of age was the lowest among all age groups, with the imported animals showing clear superiority. On average, over the whole study group, this difference was 24% (842 kg). However, the Australian animals were slower to become impregnated and more difficult to fertilize. In total, the period from calving to the next pregnancy for imported cows was on average 208.4 days, and in local cows, 161.5 days. The proportion of barren individuals was, respectively, 81.4% and 71.8%. The data showing lifelong productivity of early-calving first-calf heifers are of special interest. Of the Australian animals, 3.34% belonged to this group with a herd average of 2.48%. Of the local animals, 1.86% calved in this group with a herd average of 2.49%, and the number of lactations was 2.89%, 2.02%, 1.65%, and 1.90%, respectively. The lifetime milk production of imported cows calving first at two years of age was 17,315.2 kg of milk, while that of the equivalent local cows was 6,095.2 kg, or 65% less. Regardless of the age of first calving, each imported cow produced an average of 11,048.2 kg of milk over its lifetime, an amount 20% higher than that of local cows of the same age.

Table 1. Influence of the age of first calving on productivity use of first-calf heifers.

Indicators	Age of first calving. months					
	<24	24.1–26.0	26.1–28.0	28.1–30.0	30.1–32.0	≥32.0
n	83	432	772	595	244	92
%	3.8 ± 0.4	19.5 ± 0.8	34.8 ± 1.0	26.8 ± 0.9	11.0 ± 0.7	4.1 ± 0.4
Milk yield for 305 daysof first lactation. kg	4028.0	4294.0	4683.7	4763.8	4834.6	4895.0
Life duration. Days	1991.7	2279.5	2220.0	2595.4	2076.4	2184.0
Productiva period. Days	1191.1	1305.4	1511.2	1797.8	1392.5	1240.0
Number of calvings	3.57	3.50	3.55	4.77	3.31	2.60
Namber of lactations	3.00	3.50	3.46	4.23	2.85	2.32
Jield of milk for life.kg	11734.4	14380.0	19212.6	24081.6	18071.1	15914.4
Cows leaved after first calving. %	14.3	0.0	0.0	5.9	17.6	34.40

Table 2. Age of the first calving and the indicators of productive use of local Holstein cows.

Indicators	Age of first calving. Days					
	up to 720	721–811	812–902	903–993	994>	Average
Cows	21	105	171	212	93	602
Average age of calving. days	661.0	783.1	859.0	945.8	1130.4	935.3
Milk yield for 305 days of first lactation. kg	3119.2	4030.6	3415.8	3516.4	3332.1	3530.0
Service-period first-calf heifers. days	171.6	167.7	164.5	152.7	166.7	161.5
Barrenness. %	85.7	67.6	73.7	65.6	83.9	71.8
Milk yield for life. kg	6099.2	11380.1	8699.5	8370.0	9333.1	9216.1
Number of calvings	1.86	2.99	2.49	2.44	2.27	2.49
Number of lactations	1.65	2.57	1.94	1.88	1.71	1.96

Table 3. Age of the first calving and the indicators of productive use of Australian Holstein cows.

Indicators	Age of first calving. Days					
	up to 720	721–811	812–902	903–993	994>	Average
Cows	27	139	192	103	45	506
Average age of calving. days	680.3	777.4	857.6	943.3	1106.8	851.9
Milk yield for 305 days of first lactation. kg	3900.4	4443.2	4495.8	4261.0	4132.6	4371.7
Service-period first-calf heifers. days	181.2	184.2	202.0	210.4	304.5	208.4
Barrenness. %	81.0	71.6	86.1	82.1	92.5	81.4
Milk yield for life. kg	17315.2	10530.0	11121.2	10927.4	9053.0	11048.2
Number of calvings	3.34	2.54	2.43	2.42	2.18	2.48
Number of lactations	2.89	1.97	2.02	2.02	1.60	2.00

The seasonal timing of calving is a serious problem for dairy cattle breeding in the Far East region, especially in the south of the region. More than 60% of calving occurs in the winter-to-spring period. This feature is also seen in the range of heifers' calving dates (Table 4). The duration of the economic utility and the lifetime milk yield of first-calf heifers depending on the calving season are also shown in Table 4.

Table 4. Duration of the economical use and milk yield for the life of first-calf heifers in dependence on calving season.

Indicators	Season of calving				Year
	Winter	Spring	Summer	Autumn	
Counted calvings	433	584	213	307	1537
Age of first calving. Days	900.3	917.9	966.5	896.2	914.6
Leaved in first month after calving. %	14.5 ± 4.4	17.5 ± 4.8	19.4 ± 7.0	11.8 ± 4.5	15.4 ± 2.5
Milk yield for 305 days of first lactation. kg	4701.1 ± 150	4487.4 ± 187	4264.6 ± 210	4968.5 ± 201	4614.0 ± 200.0
Service-period. days	134.1	122.4	126.0	142.3	132.4 ± 5.0

(*continued*)

Table 4. (*continued*)

Indicators	Season of calving				Year
	Winter	Spring	Summer	Autumn	
Barrenness. %	81.1 ± 5.0	71.1 ± 5.7	88.0 ± 5.8	93.3 ± 3.6	82.3 ± 2.7
Life duration. days	2357.3	2086.2	2046.7	2305.8	2215.6 ± 315.0
Milk yield for life. kg	18260.8	15372.5	14968.7	18722.2	16896.7

Of 1,537 recorded calvings, 66.2% were in the winter-to-spring period, and the more productive autumn calvings accounted for 20.0%. The annual milk yield of first-calf heifers varied from 4,968.5 kg from autumn-calving heifers to 4,264.6 kg from summer-calving heifers. The average interseasonal difference was 703.9 kg or 16.5%. For the first time in the region, we exposed the negative influence of spring-to-summer calving on the duration of the productive period and the lifetime milk yield. In the region, summer calvings increase birth and post-natal complications by 7.8%, leading to the culling of animals and the death of newborn calves. The proportion of individuals that have survived five or more calvings may serve as one of the indicators of cows' viability (Table 5).

Table 5. Cows are met at age of five and more calvings.

Season of birth	Season of first calving				Year
	Winter	Spring	Summer	Autumn	
Winter	139 (17)	154 (8)	164 (27)	95 (10)	552 (62)
Spring	212 (18)	205 (16)	250 (29)	92 (12)	759 (75)
Summer	104 (10)	95 (13)	42 (0)	73 (16)	314 (39)
Autumn	116 (24)	157 (21)	61 (12)	45 (14)	379 (71)
Year	571 (69)	611 (58)	532 (86)	290 (34)	2004(247)

Note: In brackets – number of long-lived cows

The results of the analysis of primary registration showed that the season of first calving or birth significantly influences the number of cows surviving more than five calvings. With an average index of 12.3%, this proportion varied from 9.5% among spring-calving individuals to 16.3% among autumn-calving individuals. The birth season of cows shows an analogous influence. However, the combined influence of these factors noticeably increases the variability of outcomes. While there were no animals of five or more calvings in the group of 42 cows born and calving in summer (denoted 'summer-summer'), the proportion of 'old residents' among 45 autumn-autumn individuals was 31.1%. The other seasons, 'winter-winter' and 'spring-spring', occupied intermediate positions of 12.2% and 7.8%, respectively. The received data offers great practical value in the organization of herd reproduction with the stall-camp system of maintaining cattle. Further study of these results in all farms of the south of the Far East is needed to account for the annual reduction of the duration of cows' economic utility in those farms.

The sire bulls significantly influence the productive longevity and economic utility of cows. We researched the influence of cows' sires on their productivity and longevity (Table 6). Bulls siring no fewer than 30 daughters were included in the analysis and processing of the data.

Table 6. Productivity and longevity of cows-daughters of different sire bulls.

Indicators	Sire-bulls			
	Printz 4440	Lebed 109	Dzhut 4449	Mif 4445
Number of daughters	35	33	34	83
Age of first calving. Days	916.6 ± 21	895.0 ± 19	873.0 ± 17	913.7 ± 8
Leaved after first calving. %	6.1	6.7	91.2	13.2
Milk yield for 305 days of first lactation. Kg	4329.2 ± 87	3980.8 ± 91	1924.5 ± 67	3880.6 ± 15
Life duration. days	2288.2 ± 138	2723.4 ± 147	1173.6 ± 99	2114.4 ± 135
Milk yield for life. kg	16080.4 ± 1880	19540.5 ± 1600	2653.0 ± 580	15170.4 ± 1554

By the main indicators, lifetime milk yield and longevity, Lebed 109's daughters took first place. The intergroup difference in lifetime milk yield was 4370.1 kg, and in longevity was 609 days or 28.8%. The intergroup difference in the number of calvings was 29.8%.

Choosing animals by statistical methods alone is insufficiently effective due to the low heritability and relatively late display of traits or their display only under certain conditions. Modern scientific development allows selection not only through the assessment of the pedigree value by statistical methods but also by genetic markers [16–19]. We measured the polymorphism of the blood proteins haptoglobin and ceruloplasmin as the genetic markers of assessment and prognosis of the length of cattle's productive life. The economically useful traits of animals with different haptoglobin phenotypes, assessed over a long period, allowed us to define their lifetime milk yield and the duration of their productive period (Table 7).

Table 7. Economically useful traits of cows with different phenotypes of haptoglobin.

Indicators		Phenotype of haptoglobin		
		Hp 1–1	Hp 2–1	Hp 2–2
Counted cows		165	439	219
Age of first calving. days		846.5 ± 7.0	866.4 ± 4.1	862.8 ± 12.0
Leaved during first lactation	n	22	53	14
	%	13.3	12.1	6.40
Milk yield for 305 days of first lactation. kg		4404.4 ± 170	4250.0 ± 31	4502.0 ± 120
Milk yield for life. kg		18017.2 ± 501	17343.6 ± 297	23485.9 ± 1344
Duration of productive period. Days		2123.2 ± 85	2183.3 ± 28	2575.3 ± 138
Amount of milk for life day. kg		8.45 ± 0.35	7.94 ± 0.55	9.15 ± 0.37

Overall, the animals with phenotype Hp 2–2 excelled over cows of the same age in other groups by all indicators. The animals with phenotype Hp 2–2 had higher milk yields in the first lactation and surpassed individuals of other phenotypes in lifetime milk yield by 30.4–35.4% or 5.5–6.1 t of milk. This difference is due not only to lengthier productive periods (averaging 392–45204 days) but also to higher productivity. These animals produced an average of 9.15 kg of milk per day, compared with averages of 8.45 and 7.94 kg of milk per day from animals with other phenotypes of haptoglobin. Significant intergroup differences in economically useful traits allow the selection of animals for breeding at an early age, avoiding the expensive assessment of their own productivity and further expense to assess their lifetime milk yield. Selecting heifers with phenotype Hp 2–2 at 8–10 months of age to restock the herd allows an increase in the duration of cows' economic utility by 1.2 years, a 29–35% increase in lifetime milk productivity, a 1.3-fold increase in calf production, and a reduction in cow morbidity by ovarian cyst.

Ceruloplasmin phenotypes in the blood serum of 842 cows were determined simultaneously with haptoglobin. Three phenotypes were identified: Cp-AA-198, Cp-AB-483, and Cp-BB-161. The economically useful traits of animals with different ceruloplasmin phenotypes are presented in Table 8.

The creation of lifetime production indices exposes differences in the average outcomes of many economically useful traits (Table 8). The animals with phenotype Cp BB had the highest milk yield in the first lactation and surpassed individuals of other phenotypes in lifetime milk yield by 32.2–44% or 5.9–7.4 t of milk. Significant intergroup differences are evident in the number of barren first-calf heifers. Barrenness among animals with phenotype Cp BB occurred 7.8–9.9% less often compared with animals of the same age displaying other phenotypes. The individuals heterozygous for ceruloplasmin type had the highest rate of reproductive issues including stillbirths, loss of cattle, and forced slaughter of calves before 1 month of age.

Table 8. Productive qualities of cows of different phenotypes of ceruloplasmin.

Indicators		Phenotype of ceruloplasmin		
		Cp AA	Cp AB	Cp BB
Counted cows		198	483	161
Age of first calving? days		852.5 ± 12.0	836.9 ± 9.6	867.8 ± 12.0
Leaved during first lactation	n	49	54	0
	%	24.7 ± 3.0	8.5 ± 1.0	0
Milk yield for 305 days of first lactation. kg		4157.4 ± 92.0	4337.3 ± 100	4605.2 ± 126
Milk yield for life. kg		16809.0 ± 2725.0	18305.6 ± 1326.0	24198.8 ± 1101.0
Duration of productive period. Days		1928.7 ± 156	2872.2 ± 92.0	2732.6 ± 78.0
Amount of milk for life day. kg		8.72 ± 0.37	8.03 ± 0.55	8.86 ± 0.38

4 Conclusion

We have identified correlations between the length of the productive period and the longevity of cattle, as well as high productivity and adaptability to the monsoon-prone climate conditions of the Far East, and paratypic and genotypic factors that may be used in practice to predict the lifetime value of dairy cattle in the future. The season of the first calving significantly affects the number of cows that survive more than five calvings. With an average of 12.3%, the proportion of cows surviving more than five calvings varied from 9.5% among spring-calving individuals to 16.3% among autumn-calving individuals. The season of the cows' birth has a similar effect. However, the combined effect of these factors markedly increases variability in outcomes. These results make it possible to select animals by haptoglobin and ceruloplasmin type for breeding at an early age, avoiding an expensive assessment of each animal's productivity and even more expensive assessment of their lifetime milk yield.

References

1. Shatalina, O.S.: Associations between blood groups and reproductive rates in cattle. Agric. Biol. **53**(2), 309–317 (2018). https://doi.org/10.15389/agrobiology.2018.2.309rus
2. Klyuchnikova, N.F., Klyuchnikov, M.T., Klyuchnikova, E.M.: Artificially acquired infertility of a cows on dairy farms in Primorye and Priamurye. Bull. Rus Agric. Sci. **4**, 63–66 (2021). https://doi.org/10.30850/vrsn/2021/4/63-66
3. Wright, J.R., Van Radeon, P.M.: Genetic evaluation of dairy cow livability. J. Anim. Sci. **94**(5), 178 (2016). https://doi.org/10.2527/jam2016-0368

4. Hare, E., Norman, H.D., Wright, J.R.: Survival rates and productive herd life of dairy cattle in the United States. J. Dairy Sci. **89**(9), 3713–3720 (2006). https://doi.org/10.3168/jds.S0022-0302(06)72412-2
5. Brickell, J.S., Watches, D.C.: A descriptive study of the survival of Holstein-Friesian heifers through to third calving on English dairy farms. J. Dairy Sci. **94**(4), 1831–1838 (2011). https://doi.org/10.3168/jds.2010-3710
6. Lopez de Maturana, E., Ugarte, E., Gonzales-Recio, O.: Impact of calving ease on functional longevity and herd amortization costs in Basque Holsteins us9ing survival analysis. J. Dairy Sci. **90**(9), 4451–4457 (2007). https://doi.org/10.3168/jds.2006-734
7. Le Cozler, Y., et al.: Rearing strategy and optimizing first-calving targets in dairy heifers: a review. Animal **2**(9), 1393–1404 (2008). https://doi.org/10.1017/S1751731108002498
8. Sewalen, A., et al.: Relationship between reproduction traits and functional longevity in Canadian dairy cattle. J. Dairy Sci. **91**(4), 1660–1668 (2008). https://doi.org/10.3168/jds.2007-0178
9. Bell, M.J., Wall, E., Russel, G., Roberts, D.J., Simm, G.: Risk factors for culling in Holstein-Friesian dairy cows. Veter. Rec. **167**, 238–240 (2010). https://doi.org/10.1136/vr.c4267
10. Sawa, F., Bogucki, M.: longevite of cows depending on their first lactation yield and herd production level. Ann. Anim. Sci. **17**(4), 1171–1183 (2017). https://doi.org/10.1515/aoas-2016-0096
11. Sewalem, A., Miglior, F., Kistemake, G.J.: Analysis of the relationship between workability traits and functional longevity in Canadian dairybreeds. J. Dairy Sci. **93**(9), 4359–4365 (2010). https://doi.org/10.3168/jds.2009-2969
12. Schuermann, Y., St-Yves, A., Dicks, N., et al.: The transition cow; may the odds be ever in her favor. J. Dairy Sci. **94**(5), 234–235 (2016). https://doi.org/10.2527/jam2016-0488
13. Koketsu, Y.: Longevity and efficiency associated with age structures of female pigs and herd management in commercial breeding herds. J. Anim. Sci. **85**(4), 1086–1091 (2006). https://doi.org/10.2527/jas.2006-493
14. Kaupe, B., et al.: Joint analysis of the influence of CYPIIBI and DGATI genetic variation on milk production, somatic cell score, conformation, reproduction, and productive lifespan in German Holstein cattle. J. Anim. Sci. **85**(1), 11–21 (2007). https://doi.org/10.2527/jas.2005-753
15. Engle, B.N., et al.: Genome-wide association study for stayability measures in Nellore-Angus crossbred cows. J. Anim. Sci. **94**(4), 142 (2016). https://doi.org/10.2527/jas2016.94supplement4142x
16. Kaupe, B., et al.: Joint analysis of the influence of CYPI IBI and DGATI genetic variation on milk production, somatic cell score, conformation, reproduction, and productive lifespan in German Holstein cattle. J. Anim. Sci. **85**(1), 11–21 (2007). https://doi.org/10.2527/jas.2005-753
17. Nayeri, S., Sargolzaei, M., Abo-lsmail, M.K., et al.: Genome-wide association study for lactation persistency, female fertility, longevity, and lifetime profit index traits in Holstein dairy cattle. J. Dairy Sci. **100**(2), 1246–1258 (2016). https://doi.org/10.3168/jds.2016-11770
18. Zhang, Q., et al.: Genome-wide association study for longevity with whole-genome sequencing in 3 cattle breeds. J. Dairy Sci. **99**(9), 7289–7298 (2016). https://doi.org/10.3168/jds.2015-10697
19. Engle, B.N., Herring, A.D., et al.: Genome-wide association study for stayability measures in Nellore-Angus crossbred cows. J. Anim. Sci. **94**(4), 142 (2016). https://doi.org/10.2527/jas2016.94supplement4142x

Prospects of Using Whole Grain Flour from Recognized Selection Wheat Varieties of the Far Eastern State Agrarian University in Food Technologies

Elena Gartovannaya(✉) and Anna Ermolaeva

Far Eastern State Agrarian University, Polytechnicheskaya St., 86, Blagoveshchensk, Russia

Abstract. Breeding of cereals, in particular wheat, is a rather difficult task, because newly developed varieties can only become widespread if they provide stable high yields and exceed the standard in quality. The quality physical indicator is grain volume weight [4, 11]. Another of the most important quality characteristics of grain is the falling number (FN), which characterizes the α-amylase activity. If we compare the activity of α-amylase over the three years of study, the highest values were obtained in 2018, although many researchers believe that excessively high values deteriorate the bread quality. The use of whole grain flour in formulations does not require changes in the technological process or equipment and allows to enrich baked goods with fiber, vitamins, and minerals. As a result of this study, we confirmed the possibility of using whole grain flour from three Experiment data has shown that the replacement in the bread formula of up to 30% of the top-grade wheat flour with whole-grain flour from the studied varieties decisively improves the products without detriment to their organoleptic and physicochemical characteristics. However, the experiments also established that each variety has a different effect on bread quality. In the authors' opinion, a bread sample using the DalGAU-4 variety was among the best.

Keywords: Experiment · DalGAU-4 · Falling number · Wheat

1 Introduction

The Far Eastern State Agrarian University is rightfully considered one of the oldest field-specific institutions of higher education in the Russian Far East. Agricultural technologies have been its primary focus since the very beginning. The university laboratories have been engaged in wheat breeding for many years because this crop is one of the main cereals in the world. The climatic conditions of the Amur Region are extremely tough for the cultivation of crops due to sharp daily fluctuations in temperature and precipitation in different years. The new varieties of grain should have both high yields and excellent technological properties consistently manifesting in a variety regardless of the prevailing weather conditions. This challenging task requires rational selection of parental pairs and thorough laboratory analysis of grain variety samples. The use

A. Muratov and S. Ignateva (Eds.): AFE 2021, LNNS 353, pp. 357–365, 2022.
https://doi.org/10.1007/978-3-030-91402-8_41

of the grain of selection wheat varieties in baking depends on various characteristics, such as vitreousness, grain volume weight, as well as gluten quantity and quality. Each characteristic has a great impact on the general understanding of baking advantages of a variety. A significant indicator of baking properties of grain and flour is autolytic activity, reflecting the activity of amylases or amylolytic enzymes contained in grain or flour determined by the falling number (FN). If the amylase activity is increased, the activity of proteolytic enzymes that destroy gluten will also be high. If the α-amylase activity is reduced, then proteases will have a weak effect on gluten proteins. This indicator is used to assess the flour quality that can vary widely. Per state standard for bakery wheat flour, the lower FN limit for weak wheat flour is 120 s, for other grades of bakery wheat flour — 220–200 s. The standard establishes only the lower limit for the falling number; the upper limit of this indicator is not regulated. That said, the excessively high FN negatively affects the bread quality [6, 7].

This study aims to explore the possibility of use in formulations of bakery products the whole grain flour from the Amur selection varieties of soft spring wheat — DalGAU-3, DalGAU-4, and KSI-27, yields of the last 3 years (2018, 2019, and 2020).

2 Materials and Methods of Study

The study has been conducted at the Research Laboratory of Grain Crops Breeding (RLGCB) of the Federal State Budget Education Institution of Higher Professional Education "Far Eastern State Agrarian University". The studied material was three samples of the grain of the Amur selection varieties of soft spring wheat — DalGAU-3, DalGAU-4, and KSI-27, yields of 2018–2020.

Field experiments were carried out in the crop rotation of the RLGCB. For wheat cultivation in field experiments, we implemented the standard agrotechnology recommended by the climatic cropping pattern of the Amur Region. The quality characteristics (gluten quantity and quality per GOST R 54478-2011, mash acidity per GOST 10844-74, starch quantity per GOST 10845-98, and humidity per GOST 13586.5-2015) were assessed in a laboratory setting. The gluten quantity was determined using the Perten Glutomatic system, the gluten quality — using the IDK-5m device. For the determination of vitreousness, a diaphanoscope was used. The alpha-amylase was analyzed using the PChP-3 instrument.

The organoleptic quality assessment of the studied product was based on GOST 5667-65 "Bread and bakery products". The moisture content was determined per GOST 21094-75 "Bread and bakery products. Moisture determination method". Per this method, the moisture is extracted by heating with subsequent indirect determination. The acidity was determined per GOST 5670-96 "Bakery products. Acidity determination method". This method involves the extraction of acids from the product and their subsequent titration with an alkali solution. The porosity is determined per GOST 5669-96 "Bakery products Porosity determination method". The method is based on the fact that the porous mass of bread has relatively constant density. Knowing the volume and mass of a piece of bread crumb and the density of the non-porous mass, it is possible to calculate the volume occupied by the pores.

3 Results

The nutritional and technological value of grain and seeds of various crops is directly dependent on the variety, agricultural technology, climatic factors, conditions, methods and time of harvesting, transportation and storage. The Far Eastern State Agrarian University has been breeding varieties of spring soft wheat that are currently used only as forage.

Grain production is the main source of raw materials from which flour, bread, cereals, pasta and confectionery are made, so the quality of this raw material is sufficiently important [2, 3]. In this age of high-paced life, environmental issues and decrease in quality of nutrition and raw materials, the introduction of whole-grain flour in formulations of bakery products is of great current interest.

The study aimed at proving the possibility of using whole grain flour from three Amur selection wheat varieties in the baking industry.

Having analyzed three samples of grain, the crop breeders got the quality indicators represented in Table 1. The baking or technological properties of each grade of wheat flour determine its purpose.

Table 1. Grain quality of spring wheat varieties bred by the RLGCB.

Sample	Year	Grain volume weight, g/l	Vitreousness, %	Weight of 1000 grains, g	Gluten amount/Quality group	Falling number, seconds
DalGAU-3	2020	740	59	27.4	25/2	205
	2019	790	69	29.4	24/1	383
	2018	765	92	26.8	31/1	412
DalGAU-4	2020	730	47	25.4	29/2	238
	2019	775	59	28.3	28/2	423
	2018	780	97	25.0	38/1	465
KSI-27	2020	690	14	26.6	25/1	237
	2019	740	25	25.3	21/2	398
	2018	725	72	22.9	32/1	438

Vitreousness is a specific quality indicator of wheat grain that characterizes the endosperm consistency, the degree of protein binding with endosperm starch. This is indicative of the technological (milling) properties of wheat [10]. The higher the vitreousness of a grain batch, the higher the overall flour yield and the yield of flour of the highest grades.

According to the data obtained, in terms of vitreousness, the studied soft wheat samples can be classified as follows: DalGAU-3 and DalGAU-4 are highly vitreous varieties (over 60%), KSI-27 is medium in vitreousness.

The raw gluten content in the wheat grain of Top class, Classes 1 and 2 must be, respectively, at least 36%, 32%, and 28%; the gluten quality must be at least Group 1. The quantitative and qualitative requirements for the wheat grain of Classes 3 and 4 are as follows: 23% and 18% gluten content respectively; gluten quality — at least Group 2. The wheat with raw gluten content above 28% is considered highly glutenous. Quality raw gluten is sufficiently cohesive (not breaking down into separate fragments), elastic, and moderately stretchable so that dough and finished products have good rheological and consumer properties. All the above points characterize the technological properties of wheat flour and determine the relevance of this study.

In this study, all samples of wheat belong to Class 2 in terms of gluten content and to Group 2 in terms of gluten quality. The gluten of these wheat varieties has elastic properties.

To obtain flour from the presented samples, close grinding was used. The lower the flour grade, the higher the amount of vitamins and mineral elements. The content of albumin and globulins containing essential amino acids also increases [1]. However, products from low-grade flour have a darker color and inferior bakery properties. For this reason, the bakery industry often uses a combination of different flour batches [5].

When using composite flour for bread-making, it is beneficial to replace part of wheat flour with whole-grain flour. In pilot batches, we made different bakery products using various percentages. The results of test baking are presented in Fig. 1.

The best sample had the following mixing rate: 30% of whole grain flour and 70% of top-grade flour (Fig. 1).

The experiment also established that flour from different wheat varieties provides for different indicators.

Fig. 1. Pilot batches.

Table 2 contains a five-point scale assessment of the quality of various bread samples obtained from a composite flour mixture: Sample 1 (wheat flour and whole grain flour from the DalGAU-4 variety), Sample 2 (wheat flour and whole grain flour from the DalGAU-3 variety), Sample 3 (wheat flour and whole grain flour from the KSI-27 variety).

Table 2. Score assessment of the quality of bread from composite flour.

Samples	Score	Quality characteristics of bread
Shape accuracy		
Sample 1	4	Bread with a noticeably convex top crust
Sample 2	5	Bread with a domed top crust
Sample 3	3	Bread with a slightly convex crust; breaks in the crust
Crust color		
Sample 1	5	Dark golden to brown
Sample 2	5	
Sample 3	5	
Crust surface condition		
Sample 1	5	Flawlessly smooth, free from bubbles, cracks, seams, or burst marks, exceptionally glossy
Sample 2	3	Slightly bubbly, rough, noticeable small cracks and bursts, barely noticeable seams
Sample 3	1	Burst crust with an emerging crumb
Crumb color		
Sample 1	4	Grayish
Sample 2	4	
Sample 3	4	
Porosity		
Sample 1	5	Small, thin-walled pores evenly distributed over the crumb cut surface
Sample 2	3	Pores of different sizes and average thickness, unevenly distributed
Sample 3	2	Large, slack baked, thick-walled pores; a small quantity of dense non-porous areas, minor voids, noticeable crumb separation from the crust, significant voids
Structural and mechanical properties of the crumb		
Sample 1	5	Very soft, tender, elastic crumb
Sample 2	2	Visibly compacted, but elastic, crumbling crumb with signs of under-kneading
Sample 3	3	Satisfactory soft (slightly compacted) but elastic crumb with signs of under-kneading
Aroma (smell)		
Sample 1	5	Very strong, characteristic of bread
Sample 2	3	Mild, characteristic of bread
Sample 3	4	Strong, characteristic of bread

(*continued*)

Table 2. (*continued*)

Samples	Score	Quality characteristics of bread
Taste		
Sample 1	4	Strong, characteristic of bread
Sample 2	3	Mild, characteristic of bread
Sample 3	4	Strong, characteristic of bread
Chewability		
Sample 1	5	Very tender, juicy, chews well
Sample 2	4	Sufficiently tender, dryish, chews well
Sample 3	3	Slightly rough, dryish, slightly lumpy

The taste is evaluated as follows: a 6–8 mm thick slice is cut from the product. The taster bites of samples of 1–2 g off the top, bottom and side crusts and the crumb, chews them, and keeps them in the mouth for 3–5 s. Then the taste of the sample is compared with the description specified in the standard. When chewing, the taster pays attention to the presence or absence of crunching on the teeth. The smell is assessed after two deep inhalations of air from the surface of the whole and cut product. The appearance of the bakery product (color, shape, sectional appearance, surface of the product, crumb condition) is evaluated by visual inspection [9]. The tester assesses the shape and color of the crust, the presence or absence of cracks on the surface. The doneness is assessed by touching the crumb surface in the middle of the product with fingertips. The kneading quality and porosity are determined based on a visual inspection and a comparison to the description in the standard. The elasticity of the crumb is determined by pressing on it: the tester observes how quickly and to what extent the crumb assumes its original state [8].

The analysis of the obtained results shows that bread sample 1 from composite flour has had the best organoleptic characteristics. After the general sensory evaluation of bread, this sample had the highest score among others.

Physicochemical indicators of the obtained samples are presented in the diagrams (Figs. 2, 3 and 4).

4 Discussion

The obtained experimental data confirms the possibility of use of using whole grain flour from selection wheat varieties. Among the all studied varieties, the flour sample from DalGAU-4 variety had the best score. This flour also differs from the rest of the samples in chemical composition (Table 3).

The obtained chemical composition data and the results of test baking confirm that the Amur selection variety DalGAU-4 can be used in baking. The pilot batches of bread and bakery products with lower moisture content (doughnut products) and amended formulations introducing parts of whole-grain flour from this and other varieties of soft

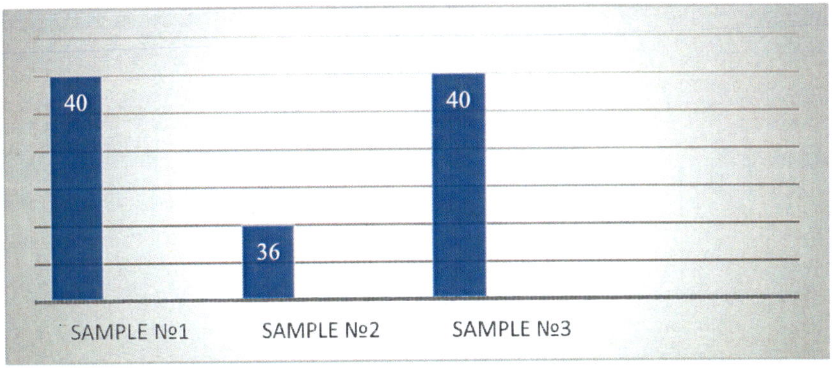

Fig. 2. Effect of the composite flour on the moisture content of wheat bread, %.

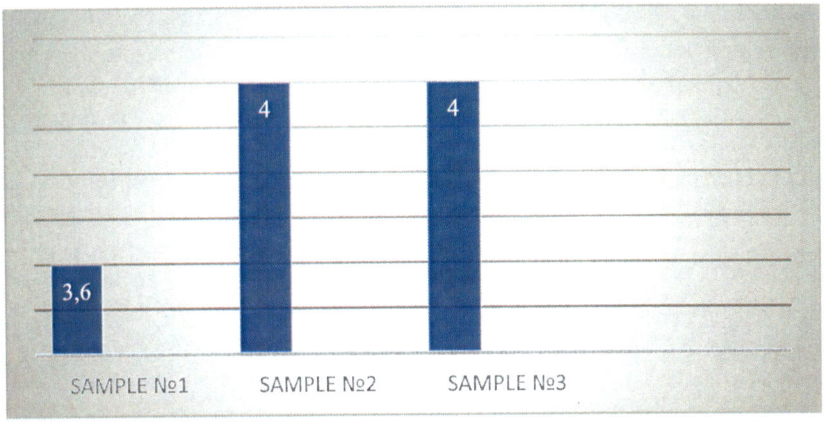

Fig. 3. Effect of the composite flour on the acidity of wheat bread, degree.

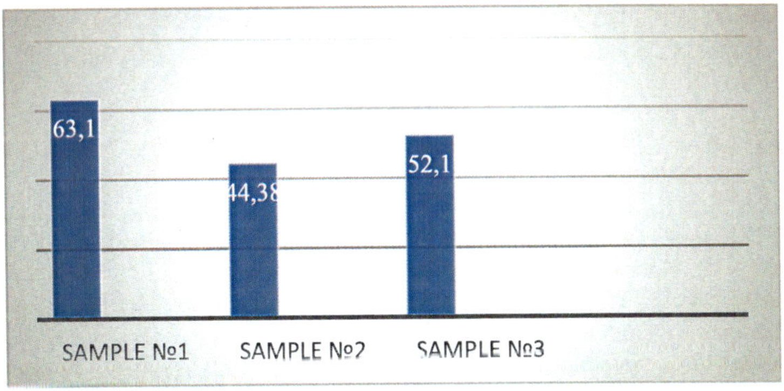

Fig. 4. Effect of the composite flour on the porosity of wheat bread, %

Table 3. Chemical composition of flour.

Graded flour sample	Protein, %	Fat, %	Dry matter, %	Ash, %	Fiber, %
DalGAU-4	15.17	1.25	93.09	2.89	2.7
DalGAU-3	12.49	1.31	87.75	2.11	2.5
KSI-27	14.39	1.37	92.95	2.90	3.11

spring wheat were produced in the food laboratory of the Far Eastern State Agricultural University. The physicochemical characteristics of these products conformed to the requirements of reference documentation: humidity — 18.4% (standard value — up to 19%), acidity — 1.2° (standard value — up to 3), swelling ability — 2.5.

5 Conclusions

Summing up the experimental data obtained, we can conclude that per the reference documentation currently in force in Russia (in particular, GOST 34702-2020 "Baking wheat. Specifications"), varieties of soft spring wheat bred by the Far Eastern State Agrarian University (DalGAU-4, DalGAU-3, and KSI-27) are filler (quality) wheat in terms of gluten content and the falling number. However, the growing conditions and the specific climate characteristics of the region can negatively affect the quality indicators of grain. As a result, the baking properties of flour may also change leading to the deterioration of products. Therefore, further research on this topic is needed.

References

1. Bukhovets, V., Yefimova, D., Davydova, L.: New production technology for nutritionally enhanced bakery products. Equip. Tech. Food Prod. **49**(2), 193–200 (2019). https://doi.org/10.21603/2074-9414-2019-2-193-200
2. Kiryuhina, A., Grigor'eva, R., Kozhevnikova, A.: Bread production and bakery products in Russia: current state and prospects. Equip. Tech. Food Prod. **49**(2), 330–337 (2019). https://doi.org/10.21603/2074-9414-2019-2-330-337
3. Kirdyashkin, V., Kandrokov, R., Andreeva, A., Shchebelev, V.: Obtaining highly dispersed buckwheat flour for baby food using infrared processing. Storage Process. Farm Prod. **4**, 43–54 (2020). https://doi.org/10.36107/spfp.2020.357
4. Rifna, J., Sushil, S., Madhuresh, D.: Process analytical technology for bakery industry: a review. J. Food Process Eng. **42** (2019). https://doi.org/10.1111/jfpe.13143
5. Hui, Y.H., Corke, H., De Leyn, I., Nip, W.K., Cross, N.: Bakery Prod. Sci. Technol. (2007). https://doi.org/10.1002/9780470277553
6. Batura, N.G., Tipsina, N.N., Chaplygina, I.A.: The development of bread technology using crushed grain of wheat and rye. Krasnoyarsk State Agrarian Univ. **11**(152), 133–137 (2019). https://doi.org/10.36718/1819-4036-2019-11-133-137
7. Ilyukhina, N.V., Kolokolova, A.Y., Trishkaneva, M.V., Kryukova, E.V., Goryacheva, E.D., Berketova, L.V.: Study of the dynamics of inhibition of the microflora of plant materials as a result of treatment with ultraviolet radiation. Storage Process. Farm Prod. **1** (2021). https://doi.org/10.36107/spfp.2021.194

8. Novikova, Z., Lavrinovich, D.S., Sergeeva, S.M.: Modelling of bread and flour product recipes with the use of chia seeds. Int. Res. J. **8–1**(110), 93–97 (2021). https://doi.org/10.23670/IRJ.2021.110.8.013
9. Romanchikov, S.A., Yukhnik, I.P.: Expanding the use case grain processing products in bakery products. Mod. Sci. Innov. **1**(33), 90–94 (2021). https://doi.org/10.37493/2307-910X.2021.1.13
10. Shaimerdenova, D.A.: Influence factors on the formation of technological capabilities of grain soft wheat. Process VSUET **79**(1), 205–208 (2017). https://doi.org/10.20914/2310-1202-2017-1-205-208
11. OECD: Challenges for global agricultural research. OECD Publishing (2010). https://doi.org/10.1787/9789264090101-en

Results of Spring Barley Hybridization in the Amur Region

Alexandra Kuznetsova(✉)

Far Eastern State Agrarian University, 86, Politeknicheskaya Str, Blagoveschensk 675005, Russia

Abstract. One of the most cost-effective ways to increase crop yields is to use new varieties. It is the selection that makes it possible to resolve the issue with the range of varieties that will maximally meet the requirements of farmers. For each region, its own varieties, adapted to weather and climatic conditions, should be proposed accordingly. For the Amur Region, spring barley is one of the main grain fodder crops. At the moment, the State Register of Breeding Achievements contains 6 varieties admitted to cultivation in the Amur fields, and only two of them are the varieties of own selection. Breeding work needs to be continued to increase the number of local varieties. The article presents data on the number of crossbreeding combinations carried out in the research laboratory of the Far Eastern State Agrarian University during the period from 2010 to 2020. During the study period, a total of 112 varieties were used as parents, of which 62.5% were domestic breeding; 25.9% of foreign selection and 11.6% of the Near Abroad countries. Among the Near Abroad countries, Ukrainian varieties are of the greatest interest. About 500 combinations of crosses have been carried out. A variety from competitive variety trials, which is currently the most promising, is under development. The hybrid grain of this sample was obtained by crossing the Acha and Sh-2178 varieties in 2010.

Keywords: Spring barley · Grade · Grade sample · Hybridization · Individual selection · Productivity

1 Introduction

In the agro-industrial complex of the country the Amur region occupies a leading place among other regions of the Far East. Since 2003, there has been stable growth in agricultural production. In crop production, soy has been the undisputed leader in terms of the cultivated area for many years [1, 2]; but it should be noted that there is a tendency to increase cultivated areas under grain crops, including those occupied by spring barley. The gross harvests of spring barley also increased. Our region occupies the 46th place in the all-Russian rating for this crop production. The average yield in the region in 2020 was 22 centners/ha.

The share of animal husbandry in the structure of the region's agriculture is just over 30%. To develop this direction actively, it is necessary to increase the fodder base. The

A. Muratov and S. Ignateva (Eds.): AFE 2021, LNNS 353, pp. 366–375, 2022.
https://doi.org/10.1007/978-3-030-91402-8_42

Amur Region is a zone of risky farming, the climate is sharply continental with signs of a monsoon. In recent years, June has been characterized by low air temperatures; and July, on the contrary, by high temperatures, often accompanied by heavy rains and hail. In some years crops fall under the hurricane wind.

It is not possible to exclude negative weather factors, but with the help of selection it is possible to expand the range of varieties adapted to the complex agro-climatic conditions of the Amur region, with high productivity, resistance to lodging and diseases, that form grain with good technological properties.

Classical breeding has been the main method of using new crop varieties in many research institutions for many years. An integral step is hybridization, which includes the castration of mother plants and their subsequent pollination. The results of many years of breeding work in other scientific institutions show that the most effective breeding methods are hybridization and subsequent selection [3–5]. In addition, with proper equipment, it is possible to combine hybridization with the methods of biotechnology, genetics and mutagenesis [6–8]. The Scientific Research Laboratory of the Grain Crop Selection (SRL GCS) of the Far Eastern State Agrarian University" (Blagoveshchensk, Amur region) has experience in selection of mutant forms of spring barley.

Hybridization allows solving two main problems: obtaining new traits by combining crosses; obtaining transgressive forms by systematic crossing. Moreover, the prospects for obtaining transgressions are higher if the crossed varieties differ as much as possible genetically, and the less their relationship. In this regard, varieties of different ecological and geographical origin are involved in hybridization in our laboratory.

When working with spring barley, employees face a number of difficulties: the ear of the mother plant, suitable for castration, is still on the flag leaf. When releasing it, you must be very careful and accurate not to damage the fragile ear. In some years, weather conditions develop in such a way that the difference in flowering time between varieties does not allow pollination according to the established crossing pattern.

The purpose of the research. To show the volume of breeding work in the spring barley hybridization nursery over the past 10 years, to identify countries and regions, if there are any, the varieties of which give the largest number of useful hybrids for breeding work.

2 Materials and Methods

The breeding material of hybridization nurseries for the period of 2010–2020 was analyzed. The material for the research was the varieties previously studied in the collection nursery, participating in crosses as maternal or paternal forms, and the hybrid grain F0. Sowing was carried out in the optimal agrotechnical terms for our region (2–3 ten-day periods of April), harvesting as it ripens in the first ten-day period of August. For castration, a normally developed ear is selected, whose spines just appeared from the flag leaf. It is in such ears that the stamens have not yet matured and the process of self-pollination has not begun. 1 2 less developed lower ears and a part of the upper ears are removed with tweezers in such a way that only 10 flowers on one ear are castrated. When working with both two-row and multi-row varieties, the lateral flowers are also removed with tweezers. The tops of the flower scales are cut off with scissors and the green anthers are carefully removed with tweezers. After castration, the ear is wrapped in a flag leaf for additional protection and nutrition, and then placed under a parchment insulator. Pollination was carried out with paternal forms with well-developed (yellow) anthers after 2–3 days. The flower scales of paternal ears above the top of the anther were trimmed with scissors. When the anthers appeared from the ear, we started to pollinate. We pollinate in two ways: limited free and twirl-method.

3 Results

In the period from 2010 to 2020 in SRL GCS 484 combinations of crosses were carried out, 205 of them were productive and a hybrid grain was obtained. Every year a new crossing scheme is drawn up. Since the entire work takes place in natural field conditions, and different weather conditions develop every year, we will consider the results of hybridization for years in more detail.

In 2010, the sowing of varieties for crosses was carried out in two terms. The varieties zoned in the Amur region (Acha, Pervenets), as well as varieties from other regions, including those from China, were taken as maternal forms. The pollinators were mainly varieties from other regions and countries. In total, 27 combinations of crosses were carried out, 18 of them were productive. 145 hybrid grains were received (Table 1). The ovary rate was 10.7%. The air temperature was 5 °C above normal during castration and pollination. Probably, such an excess had a negative impact on the pollination process.

The largest number of grains was obtained by crossing the following varieties: Acha × Timerlan (34 pcs.), Acha × Arat (28 pcs.), Acha × Timirkhan (16 pcs.), Acha × Novosibirskiy 80 (11 pcs.).

Table 1. Combinations of crosses, in which hybrid grain was obtained in 2010. (pcs.).

Maternal form	Paternal form	Hybrid seeds	Maternal form	Paternal form	Hybrid seeds
Acha	Sh-2178	6	Acha	Karat	9
	Sh-2140	3		[Erofey × Ida]	1
	Sh-2113	10		Arat	28
	Sh-2257	4		Timerlan	34
	Sh-2226	10		Timirkhan	16
	Novosibirskiy 80	11	[Erofey × Ida]	Acha	7
	Moscovskiy 2	1		Novosibirskiy 80	1
	Ladik	4	([Keyston × Luch] × Vityaz)	Bonanza	1
	Alexis	2		Suzdalets	7

Sowing of varieties for intraspecific paired crosses in 2011 was carried out in two terms. As maternal forms, both zoned varieties of the region (Pervenets, Acha) and varieties from the nursery of preliminary variety trials of own selection, as well as varieties of Primorsk selection were taken. The pollinators were mainly varieties from other regions and foreign varieties. 48 combinations of crosses were carried out, and 41 of them were productive (Table 2). A total of 1088 hybrid grains were obtained. The ovary rate was 32%.

The largest number of grains was obtained by crossing the following varieties: Pervenets × Simon (63 pcs.), Odesskiy 100 × Primorskiy 103 (58 pcs.), Amur × Primorskiy 103 (56 pcs.), Odesskiy 100 × Simon (53 pcs.). As well the varieties from Far Eastern Agricultural Research Institute were good as pollinators.

In 2010, the hybridization nursery was founded in two terms. 58 combinations of crosses were performed, 14 of them were productive.

A total of 45 hybrid grains were obtained (Table 3). The ovary rate was 1.6%. This is probably due to the lock of the mother plant system and insufficient heat.

The largest number of hybrid grains was obtained in the following combinations: Amur × Sh-2154 (10 pcs.), Amur × Sh-2093 (6 pcs.), Odesskiy 100 × Murom (6 pcs.).

2013 was an extremely difficult year - prolonged waterlogging of the soil, flooding of fields; in some places flooding negatively affected the growth and development of plants. There was a massive spread of diseases and decay of the root system. The hybridization nursery was founded in two terms. 22 varieties were sown in the first and second terms. 41 combinations of crosses were carried out, 22 of them were productive (Table 4). A total of 124 hybrid grains were obtained. The ovary rate was 5%.

The largest number of hybrid grains was obtained in the following combinations: Adamovskiy 1 × Zevz (18 pcs.), Auxiniai 3 × Ukrainskiy 11 (18 pcs.), Simon × Yuzhnyy (13 pcs.), Russell × Deth (12 pcs.).

The hybridization nursery was founded in 2014 in two terms (April 17 and April 23). A total of 28 varieties were sown. The following varieties were directly used as maternal forms: Kimberly, Vikont, Ecolog, Ladik, Yak-401, Primorskiy 153, Betina, Donetskiy 6, Clark, Chernigovskiy 96, Murom, Adamovskiy 1, Sh-2134, and Sh-2120. The paternal forms were: Bonanza, Vikont, Ecolog, Ladik, Yak-401, Primorskiy 153, Betina, Donetskiy 6, Clark, Odesskiy 100, Murom, Adamovskiy 1, Vorsinskiy, Novosibirskiy 80, Sh-2120, and Talovskiy 9.

As a result of crossing, a hybrid grain was obtained in the following combinations: Sh-2120 × Murom (3 pcs.); Novosibirskiy 80 × Clark (4 pcs.); Primorskiy 153 × Ladik (4 pcs.). The ovary rate was extremely low (0.6%) due to the increased temperature regime (3 °C above the climatic norm) during castration and pollination.

In 2015, the last soil frosts were recorded on May 25 (6 days later than the long-term average date), in the air - on May 16 (11 days later than the long-term average date). The hybridization nursery was founded in two terms (April 14 and April 24), 28 varieties were sown. Castration and pollination were carried out at the end of the 2nd ten-day period and at the beginning of the 3rd ten-day period of June. This period was characterized by a high temperature regime and a lack of precipitation. 60 combinations of crosses were carried out, 24 of them were productive. A total of 203 hybrid grains were obtained (Table 5). The ovary rate was 6.7%.

Table 2. Combinations of crosses, in which hybrid grain was obtained in 2011 (pcs.).

Maternal form	Paternal form	Hybrid seeds	Maternal form	Paternal form	Hybrid seeds
[Keyston six-raw × Luch]	Sh-2139	24	Primorskiy 103	Bonanza	7
	Sh-2239	41	Odesskiy 100	Simon	53
	Sh-2233	34		Kimberly	53
	Sh-2145	10		Bonanza	40
	Sh-2092	32		Primorskiy 103	58
	Roland	53		Sh-2093	13
	Naran	25		Sh-2154	10
	Vorsinskiy	47		Ukrainskiy 3	25
[Erofey × Ida]	Primorskiy 1103	56	Pervenets	Primorskiy 1103	13
	Sh-2107	35		Sh-2093	18
	Sh-2120	34		Sh-2120	22
	Sh-2154	37		Sh-2154	3
	Murom	43		Murom	11
Acha	Sh-2093	3		Simon	63
	Sh-2120	29	Naran	[(Keyston six-raw × Luch) × Vityaz]	5
	Ukrainskiy3	6	Ukrainskiy 3 3	[Pervenets × Erofey]	52
	Murom	18	Sh-2154	[Nosovskiy 9 × Vityaz]	4
	[Erofey × Ida]	8	Sh-2093	[Erofey × Pervenets]	30
[Pervenets × Erofey]	Sh-2107	13	SH-2120	Kimberly	12
	Sh-2154	20		[Pervenets × Erofey]	28

The largest number of hybrid seeds was obtained in the following combinations: KNR 1 × Andrey (32 pcs.), Gate × Visit (23 pcs.), Nikita × Kozak (16 pcs.), Kozak × Erofey (15 pcs.)

In the spring of 2016, there was an increased thermal regime, dry winds in May, thunderstorm activity with hail. The hybridization nursery was founded in one term: sowing was on April 7 and the first sprouts were recorded on April 20. Castration began

Table 3. Combinations of crosses, in which hybrid grain was obtained in 2012 (pcs.).

Maternal form	Paternal form	Hybrid seeds	Maternal form	Paternal form	Hybrid seeds
Odesskiy100	Murom	6	Amur	Odesskiy 100	1
	Roland	1		Sh-2093	6
	Naran	2		Sh-2154	10
Bonanza	Sh-2145	4		KNR 2	2
Pervenets	Vorsinskiy	1		Kimberly	2
Sh-2145	Vorsinskiy	2		Roland	1
Sh-2107	Ukrainskiy 3	4			
	KNR 2	3			

Table 4. Combinations of crosses, in which hybrid grain was obtained in 2013 (pcs.).

Maternal form	Paternal form	Hybrid seeds	Maternal form	Paternal form	Hybrid seeds
Nosovskiy 9	Roland	1	Alexis	Talovskiy 9	1
	Alexis	1		Simon	8
Chernomorets	Karat	5	Talovskiy 9	Ratnik	3
	Berezinskiy	4	Karat	Alexis	1
Adamovskiy 1	Zevs	18	Halikko	Klages	6
	Ukrainskiy 11	3	3	Ukrainskiy 11	18
Russell	Auxiniai 3	3	Volgar 1	Russell	3
	Deth	12	Simon	Yuzhnyy	13
Paragon	Nosovskiy 9	2	Clark	Klages	4
	Kazer	7	Berezinskiy	Adamovskiy 1	2
Naran	Alexis	1	Primorskiy 89	Paragon	8

on June 6. 38 combinations of crosses were carried out, 21 of them were productive. 174 hybrid grains were obtained. The percentage of ovary was low and amounted to 9.6%, which is associated with a large number of days with precipitation during the period of pollination and grain ovary.

The greatest number of productive combinations was in variants where the Acha, Vikont and Sh-2127 varieties acted as maternal forms. The best paternal forms are Acha, Vorsinskiy, Russell and Patricia.

In the spring of 2017, there was a sharp change in cold and warm periods, early snow cover melting, late frosts in the air and soil. These unfavorable conditions held back the growth and development of barley. The hybridization nursery was founded in one term, sowing was on April 27. A total of 28 varieties were sown. Hybridization was carried out from June 19 to June 25. 53 combinations of crosses were performed, 17

Table 5. Combinations of crosses, in which hybrid grain was obtained in 2015 (pcs.).

Maternal form	Paternal form	Hybrid seeds	Maternal form	Paternal form	Hybrid seeds
Amur	KNR 2	5	Nikita	Kozak	16
	Odesskiy22	1		Vizit	1
	Naran	14	Kozak	Erofey	15
	Alexis	3		Vizit	7
	Vizit	6	Alexis	Sencis	12
KNR 1	Petr	5		Petr	6
	Pivdennyy	4	Naran	Conchita	2
	Primorskiy89	5		CI-9825	2
	Andrey	32	Sh-2103	Amur	2
Gate	Kozak	7		KNR 2	4
	Amur	14	KNR 2	Etienne	14
	Vizit	23	Conchita	Albright	3

of them were productive (Table 6). 88 hybrid grains were obtained. The ovary rate was 3.3%. Probably the low percentage of ovary is due to the fact that periods of cold snaps were abruptly replaced by intense warming. The largest number of hybrid grains was obtained in following combinations: KSI-13 × Zeppelin (17 pcs.), Amur × Vadim (12 pcs.), Cruzer × KSI-5 (10 pcs.).

Table 6. Combinations of crosses, in which hybrid grain was obtained in 2017 (pcs).

Maternal form	Paternal form	Hybrid seeds	Maternal form	Paternal form	Hybrid seeds
Vadim	Amur	9	Propino	Amur	3
	KSI-5	3		KSI-5	2
Izumrud	KSI-15	2	Kruzer	KSI-5	10
Askold	KSI-5	2	Amur	Vadim	12
Despino	Amur	2		Miar	6
KSI-13	Zeppelin	17		Stepan	5
	Acha	3		Prizer	3
KSI-15	Expioer	2			
	Propino	3			
	Stepan	4			

In 2018 the hybridization nursery was established in one term, sowing was carried out on April 16. A total of 36 varieties were sown. Hybridization was carried out from 18 to 25 June.

Table 7. Combinations of crosses, in which hybrid grain was obtained in 2018 (pcs.).

Maternal form	Paternal form	Hybrid seeds	Maternal form	Paternal form	Hybrid seeds
Askold	Narymskiy 6 × Sh-2180	1	Prizer	Askold	8
Acha	Propino	8		Greys	12
Greys	Amur-270-15.18	2		[Narymskiy 6 × Sh-2180]	2
Despino	Amur-270-15.18	4	Expirior	Amur-270-15.18	2
Krasnoyarskiy 6	Amur-270-15.18	1		Amur-12-15.13	3
Miar	Narymskiy 6 × Sh-2180	1	[Narymskiy 6 × Sh-2180]	Askold	7
	Amur-12-15.13	3		Vadim	9
Amur 270-15.18	Izumrud	1		Greys	12
	Prizer	1		Despino	9
Amur-300-Л-12.13	Vadim	2		Izumrud	16
Pionier	Narymskiy 6 × Sh-2180	2		Krasnoyarskiy 6	4
	Acha × KNR 1	3		Prizer	3
				Stepan	8

In the second ten-day period of June 123 mm of precipitation fell, which led to waterlogging of the soil and a decrease in temperature during the period of castration and pollination to 16.1 °C. 53 combinations of crosses were carried out, 25 of them were productive. As a result, 124 hybrid grains were obtained. The ovary rate was 29.8%. Table 7 shows the productive cross combinations.

The maximum number of hybrid grains was obtained in the following combinations: [Narymsky 6 × Sh-2180] × Izumrud (16 pcs.), [Narymsky 6 × Sh-2180] × Grace (12 pcs.), Prizer × Grace (12 pcs.).

In 2019 hybridization was not carried out due to a lack of experienced specialists.

The hybridization nursery in 2020 was founded in one term, sowing was on April 16. A total of 21 varieties were sown. Hybridization was carried out from 23 to 29 June.

45 combinations of crosses were carried out, 20 of them were productive, 188 hybrid grains were obtained. The data are presented in Table 8. The percentage of ovary was at an average level and amounted to 34%, which is associated with a large number of days with precipitation during pollination and grain ovary.

Table 8. Combinations of crosses, in which hybrid grain was obtained in 2020 (pcs.).

Maternal form	Paternal form	Hybrid seeds	Maternal form	Paternal form	Hybrid seeds
Amur	Primorets	4	KSI-29-19	Primorets	10
	Alei	4		KSI-18-19	19
	KSI-22-19	5		Alei	15
Kudesnik	KSI-29-19	14		Amurets	23
	Amurets	4	Margaret	Kudesnik	8
	KSI-18-19	11	Kalcum	KSI-9-19	5
	Amur	14		KSI-29-19	4
	KSI-22-19	15	Alei	Amur	6
Primorets	KSI-29-19	10		KSI-22-19	1
	KSI-9-19	4		KSI-18-19	12

The largest number of productive combinations was in variants where the Kudesnik and KSI-29-19 varieties were used as parent forms. The best paternal forms are KSI-22-19, KSI-29-19 (KSI-29-19 × Amurets, 23 hybrid; KSI-29-19 × KSI-18-19, 19 hybrid grains; KSI-29-19 × Alei, 15 hybrid grains; Kudesnik × KSI-22-19, 15 hybrid grains).

In 2021 the Amurets spring barley variety was included in the State Register of Breeding Achievements (patent No. 11068 dated 05/21/2020), bred by individual selection from the Amur variety. The new variety is mid-season and medium-sized (80–85 cm), and resistant to lodging. Its growing season is 80–85 days. The positive qualities of new variety: field resistance to loose smut and fusarium head blight, as well as resistance to shattering and germination of grain in the ear.

In 2020, the average yield of the Amurts variety on the experimental fields was 27.5 centner/ha, at the Svobodnensky SVP - 23.5 centner/ha, at the Tambov SVP - 40.2 centner/ha, with a potential yield of 60 centner/ha.

In order to start breeding the Amurts variety, it is necessary to increase the production capacity of the primary seed production department, provide a sufficient amount of equipment, and also have qualified personnel.

Currently, a promising cultivar KSI-24-20 [Acha × Sh-2178] is on preliminary propagation. A hybrid seed of this number was obtained in crosses in 2010. The height of KSI-24-20 plants is 90 cm, resistance to lodging is 7–8 points, resistance to grain diseases at the level of the Amur standard (fusarium and helminthosporiosis resistance - 7 and 6 points respectively). The yield is slightly lower than the Amur variety and amounted to 27 centner/ha in 2020, the weight of 1000 grains is 37.7 g.

4 Conclusion

1. The sowing of the varieties in a hybridization nursery in two terms is advisable.
2. To increase the efficiency of castration and pollination, and to increase the quantity and quality of hybrid seeds, it is necessary to have greenhouses and climatic chambers with a wide range of regulation.
3. In most cases, the largest number of hybrid grains was obtained in combinations where varieties from the regions and countries neighboring the Amur Region acted as the parent form, and the paternal form was as geographically remote as possible in origin.

References

1. Kim, L.V., Vdovenko, A.V., Nazarova, A.A.: Problems and prospects of crop production in the Far East. Far Eastern Agrarian Herald **3**(51) (2019). https://doi.org/10.24411/1999-6837-2019-13031
2. Sinegovskaya, V.T., Lyovina, A.N.: Influence of daylight duration on soybean growth, development and productivity. Far Eastern Agrarian Herald **2**(54) (2020). https://doi.org/10.24411/1999-6837-2020-12021
3. Raggi, L., et al.: Evolutionary breeding for sustainable agriculture: selection and multi-environmental evaluation of barley populations and lines. Field Crops Res. **204** (2017).https://doi.org/10.1016/j.fcr.2017.01.011
4. Howard, T.P., Fahy, B., Leigh, F., Howell, P.: Use of advanced recombinant lines to study the impact and potential of mutations affecting starch synthesis in barley. J. Cereal Sci. **59**(2) (2014). https://doi.org/10.1016/j.jcs.2013.12.012
5. Daszkowska-Golec, A., Skubacz, A., Sitko, K.: Mutation in barley (*Enhanced Response to ABA1*) gene confers better photosynthesis efficiency in response to drought as revealed by transcriptomic and physiological analysis. Environ. Exp. Bot. **148** (2018). https://doi.org/10.1016/j.envexpbot.2018.01.003
6. Summanwar, A., Basu, U., Rahman, H.: Non-coding RNAs as emerging targets for crop improvement. Plant Sci. **297** (2020).https://doi.org/10.1016/j.plantsci.2020.110521
7. Bouchetat, F., Aissat, A.: Evaluation of the genetic determinism of an F1 generation of barley resulting from a complete diallel cross between autochthones and introduced cultivars. Heliyon **5**(11) (2019). https://doi.org/10.1016/j.heliyon.2019.e02744
8. Pourkheirandish, M., et al.: Evolution of the grain dispersal system in barley. Cell **162**(3) (2015). https://doi.org/10.1016/j.cell.2015.07.002

Actual Problems of State Support of Entrepreneurship in the Agro-Industrial Complex of the Far East

Alexander Latkin[1], Dina Mukhina[2(✉)], Olga Fedoreeva[2], and Ekaterina Saveleva[2]

[1] Vladivostok State University of Economics and Service, 41, st. Gogol, Vladivostok 690014, Russia
[2] Primorskaya State Academy of Agriculture, 44, Blucher Avenue, Ussuriysk 692510, Russia

Abstract. The article deals with certain issues of state support for entrepreneurship in the agro-industrial complex, provides an overview of the directions and tools of state regulation of the agro-industrial complex of the Russian Federation and the Far East, and indicates the problems that exist in this area. On the basis of statistical data and scientific research, a conclusion is formulated about the inefficiency of state regulation of the agro-industrial complex. In conclusion, the authors propose some measures to improve state support for entrepreneurship in the agro-industrial complex of the Far East.

Keywords: Government support · State regulation · Agro-industrial complex · The Far Eastern Federal District · State programs · The tools of state regulation

1 Introduction

To support entrepreneurship in the agro-industrial complex in 2021, a number of state programs are working, many of them have been in operation for several years, and this is certainly a positive moment. However, the presence of state support does not yet indicate an effective economic policy in the field of agriculture. The question arises: "how effective is the state support of entrepreneurship in the agro-industrial complex, and whether it meets the modern conditions of management". An effective state policy in the field of agriculture should contribute to a significant increase in gross output, improve the financial condition of enterprises, develop social conditions in rural areas, ensure food security, economic accessibility of food products, etc.

The problems of state regulation of the agro-industrial complex are presented in the studies of domestic and foreign theorists and practitioners. Problematic aspects in the field of state regulation in the context of food import substitution, integration and clusterization of the agro-industrial complex are revealed in the works of P. V. Burkovsky, N. V. Vlasova, R. A. Godlevsky [1], A. P. Latkin [2, 3], O. V. Shimuk [4], E. Alimkulov, D. Aitmukhanbetov [5], E. Lovchikov [6], Yu. Schmidt [7] and others.

The authors [1] reveal the problematic aspects in the field of state regulation on the basis of the program-target management method. To improve the process of regulating

A. Muratov and S. Ignateva (Eds.): AFE 2021, LNNS 353, pp. 376–385, 2022.
https://doi.org/10.1007/978-3-030-91402-8_43

the development of the agro-industrial complex in the region and to increase the effectiveness of the implementation of the program-target management method, the authors propose tools to stimulate the processes of inter-farm cooperation on the basis of tax incentives and accelerated depreciation methods. The methodological scheme of the municipal agricultural cluster of the full production cycle is proposed.

Aksyutina S. V., Mironenko N. V. [8] consider some issues of state regulation of the Russian agricultural sector. They analyze the financial support of the industry and conclude that it is insufficient and significantly lags behind the leading foreign countries.

The experience of state support for the income of agricultural producers in the EU, USA and China is presented in the works of Kelemetov E. M., Yakubovich Ya. N. [9], H. Genkins [10], S. Gorlov, G. Panaedova, A. Borodin [11], M. Holodova [12], M. Hejazi, M. Marchant [13] and others [14].

At the same time, the theoretical and practical problems of state regulation of the agri-food complex in the context of the pandemic and taking into account the regional conditions of the Far East seem to be insufficiently developed and studied due to the lack of adequate theoretical, methodological and methodological tools for their formation and development.

2 Materials and Methods

The purpose of the study is to develop theoretical and methodological provisions and practical recommendations for the formation and development of state support for entrepreneurship in the agro-industrial complex of the Far East.

The object of the study is the system of state support for entrepreneurship in the agro-industrial complex.

The subject of the study is theoretical, methodological and practical problems related to the development of the system of state support for entrepreneurship in the agro-industrial complex of the Far East.

The theoretical and methodological basis of the study was the works of domestic and foreign scientists devoted to the problems of state support for the development of the agro-industrial complex.

The research is based on the analysis of statistical, legislative, regulatory and other materials from open sources.

In the course of the study, various methods were used: abstract-logical, dialectical, analytical, economic-statistical, monographic.

3 Results

In 2006, the implementation of the (priority) national project "Development of the agro-industrial complex" began. Since then, the development of the industry has been given great importance at the state level. This project became the launch pad of the State Program for the Development of Agriculture and Regulation of Markets for Agricultural Products, Raw Materials and Food for 2008–2012. In accordance with the State Program, regional programs have been adopted in the Far Eastern regions of the Russian Federation. In the future, this program was extended.

In 2021, a number of programs aimed at state support of entrepreneurship in the agro-industrial complex are being implemented – The State Program for the Development of Agriculture and Regulation of Markets for Agricultural Products, Raw Materials and Food", the State Program "Integrated Development of Rural Areas", the State Program for the Effective Involvement in the Turnover of Agricultural Land and the Development of the Reclamation Complex of the Russian Federation, the State Program for the Development of Agriculture, the Development of the Fisheries Complex, the Federal Scientific and Technical Program for the Development of Agriculture and others. The state provides entrepreneurs various types of assistance:

Financial-support programs imply the allocation of "compensating" and "stimulating" subsidies, compensation for part of the costs.

Property-entrepreneurs get the opportunity to use state property free of charge or on preferential terms (rent of premises, land plots).

Informational – by creating federal and regional information systems, official websites to provide business entities with up-to-date information.

Consulting – support for small businesses involves providing assistance in the form of professional advice.

Educational – development of training programs for specialists, professional development of employees.

State regulation of the agro-industrial complex within the framework of the implemented programs is carried out in various areas, including:

— development of agro-industrial sectors;
— promotion of investment activity in the agro-industrial complex;
— technical and technological modernization, innovative development;
— development of agricultural land reclamation;
— integrated development of rural areas;
— creation of a system of support for farmers and development of rural cooperation;
— protection of the interests of domestic producers in the implementation of foreign economic activities.

In addition, the tools of state regulation of the agro-industrial complex are also diverse:

— prices (target, guaranteed, collateral);
— special tax regime for agricultural producers;
— loans (concessional, commodity, interest support on loans from commercial banks);
— budget (compensating and stimulating subsidies, compensations, subsidies, leasing);
— investments in the production of agricultural products, raw materials and food;
— insurance;
— customs duties for the export and import of agricultural products, raw materials and products, raw materials and food;
— grants.

State support is provided in the form of subsidies from the federal and regional budgets. Within the framework of the implemented programs, 157636.35 million rubles were allocated for the development of the agro-industrial complex in 2020, of which 5.2% was allocated for the development of the agro – industrial complex in the Far Eastern Federal District-8247.19 million rubles. Currently, the absolute leader in the level of financing of agriculture in the Far Eastern Federal District is the Republic of Sakha (Yakutia). In 2020, the federal budget allocated more than 2 million rubles for the development of the agro-industrial complex (Table 1).

Table 1. Subsidies and other inter-budget transfers to budgets for the development of the agro-industrial complex in 2020, million rubles.

The name of the subject	Total	Federal budget	Budget subject of the Russian Federation
Russian Federation	157636.35	132094.01	25542.34
Far Eastern Federal region	8247.19	7421.22	825.97
Republic of Buryatia	739.01	696.59	42,43
Yakutia	2 315.19	2 145.65	169.53
Trans-Baikal Territory	860.19	813.24	46.95
Kamchatska Territory	157.99	146.51	11.48
Primorskye Territory	1 681.74	1 443.26	238.48
Khabarovsk Territory	510.76	433.64	77.12
Amur Region	1 535.88	1 405.85	130.03
Magadan Region	82.16	75.28	6.88
Sakhalin Region	162.85	72.29	90.56
Jewish Autonomous Region	127.43	120.46	6.97
Chukotka Autonomous Area	73.99	68.45	5.55

On December 12, 2013, Russian President Vladimir Putin identified the development of the Far East as a national priority for the entire XXI century. Since 2014, the state program of the Russian Federation "Socio-economic Development of the Far Eastern Federal District" has been implemented. The goal of the new state policy is to create globally competitive conditions for investing and doing business in the Far East, thanks to which private investments necessary for its advanced development will come to the region, followed by new jobs and a new quality of the social sphere, forming attractive conditions for people's lives. To achieve this goal, we have developed and launched such innovative mechanisms as advanced development territories, the free port of Vladivostok, targeted infrastructure support for investors, the Far Eastern Hectare program, and others. For the Far East, a unified system of special development institutions has been formed, designed to help private investors work more easily and efficiently in the difficult conditions of the macroregion.

As a result of the implementation of this program, the territories of advanced development were created (TAD) – these are separate production sites in which the state at its own expense creates the necessary infrastructure for investors, provides them with tax benefits and the necessary public services in a simplified manner. As of May 2021, 22 priority development territories have been created in the Far East in 8 regions of the Far East (with the exception of the Magadan Region). Some of them have an agro-industrial orientation: TAD "Yuzhnaya" (agriculture, food production), Sakhalin region; TAD "Belogorsk" (agro-industrial orientation), Amur region; TAD "Nadezhdinsky" (light and food industry, transport and logistics orientation), Primorskye Territory; TAD "Mikhailovsky" (animal husbandry, crop production, food production), Primorskye Territory; TAD "Kuriles" (fishing industry), Sakhalin region.

State regulation of the agro-industrial complex is applied in almost all countries of the world, especially in the United States, Western Europe, Canada, Japan, and China. In no other sector of the economy of these countries is there such a significant, deep and multilateral regulation as in agriculture. The size of state subsidies reaches 70–80% of the cost of agricultural products (Norway, Japan, Switzerland), 49% (EU), 45% (Canada), 30 – 35% (USA) [11, 12, 14]. The agricultural policy of the United States, China, and Western European countries is based on strong state regulation and financing of agriculture. State support for domestic agriculture is much lower than in the EU, USA, and China. Financial support is especially necessary for the regions of risky agriculture, which include the Far Eastern Federal District [15, 16].

4 Discussion

To assess the effectiveness of state support for the agro-industrial complex of the Far East, consider a number of indicators (Table 2).

Table 2. Agricultural products (in farms of all categories; in actual prices; millions of rubles).

The name of the subject	2015	2016	2017	2018	2019	2019/2015%	Position held in the Russian Federation at the time of 2019
Russian Federation	4794615	5112356	5109475	5348803	5801410	21.0	
Far Eastern Federal region	180454	187831	199642	198823	192651	6.8	8
Republic of Buryatia	16034	15599	15013	16214	16493	2.9	65
Yakutia	20723	21930	24972	25781	26121	26.0	60
Trans-Baikal Territory	19519	21464	22512	22904	22510	15.3	62

(*continued*)

Table 2. (*continued*)

The name of the subject	2015	2016	2017	2018	2019	2019/2015%	Position held in the Russian Federation at the time of 2019
Kamchatska Territory	6721	7378	7916	8249	10129	50.7	73
Primorskye Territory	37482	39390	42393	39963	38140	1.8	49
Khabarovsk Territory	17439	16347	16702	17099	14586	−16.4	66
Amur Region	43567	47012	50420	47637	44736	2.7	41
Magadan Region	1981	2259	2521	2723	2787	40.7	81
Sakhalin Region	10840	10388	10526	11147	12157	12.1	68
Jewish Autonomous Region	5751	5486	5216	5772	3344	−41.9	78
Chukotka Autonomous Area	397	578	1453	1334	1649	315.4	83

The share of the Far Eastern Federal District in the gross agricultural output of Russia over the past 5 years has not changed much and varies from 3.3 to 3.9% over the years. There is a slow growth in the value of gross output. In the Khabarovsk Territory and the Jewish Autonomous Region, there was a decrease in the value of gross agricultural output.

Products of animal husbandry in most subjects of DFOs have become cost-effective, crop production remains unprofitable. It should be noted that the agricultural producers of the Far East, while at a significant distance from the centers of the production of equipment and fertilizers, the means of protection are in an unequal position with Western colleagues. In terms of objective reasons, the cost of production is higher, the realization costs for the delivery of products to the central part of the country are extremely high, while the establishment of a preferential transverse tariff does not solve the problem. The introduction of customs duties for soybeans for the Far East can have critical consequences, both economically and social and even geopolitical. In addition, the established tariff limits affected the execution of previously concluded export contracts. Failure to fulfill contractual obligations leads to additional expenditures of exporters, a decrease in business profitability. This decision negatively affects well-established trade relations with foreign companies, reduces the level of confidence in doing business with Russian suppliers, the investment attractiveness of crop production (Table 3).

Table 3. The main financial indicators of organizations operating in crop production and animal husbandry.

The name of the subject	Salded financial result (profit minus loss) organizations. million rubles							
	Crop production				Animal husbandry			
	2015	2017	2018	2019	2015	2017	2018	2019
Russian Federation	136159	68532	101749	104258	123055	102224	102932	13012
Far Eastern Federal region	1929	1130	521	−42	810	−128	−187	1791
Republic of Buryatia	17	18	15	40	357	236	402	499
Yakutia	−14	−14	−10	−32	58	−3	−217	61
Trans-Baikal Territory	38	82	−18	−123	−13	16	6	−4
Kamchatska Territory	7	–	–	–	273	110	86	60
Primorskye Territory	−985	−575	−806	−1544	−856	−647	−343	611
Khabarovsk Territory	7	−20	−90	−40	79	−249	−404	82
Amur Region	2336	750	1022	1327	821	−2	268	655
Magadan Region	–	–	–	–	−1	−0,4	−1	−0,2
Sakhalin Region	437	875	371	217	−41	170	−110	−309
Jewish Autonomous Region	86	14	39	115	−1	−22	−17	–
Chukotka Autonomous Area	−0.2	−0,5	−2	−2	134	263	143	136

There are low rates of involvement of unused agricultural land in turnover. At the same time, in a number of subjects, DFO declined sown areas: Republic of Buryatia, Trans-Baikal Territory, Kamchatka Territory.

In the Far East of Russia, there is a constant outflow of the population compared to other subjects of the Russian Federation. The share of rural population falls every year. Rural territories are still lagging behind social furnishings, domestic comfort, infrastructure development, there are no necessary conditions for the development of entrepreneurship. Rural territories are described, as a rule, the low level of availability and quality of educational, medical, housing and communal and socio-cultural services. This tendency leads to an acute deficit of qualified personnel [6] (Table 4).

Table 4. Sowing areas of agricultural crops (in farms of all categories; a thousand hectares).

The name of the subject	2015	2016	2017	2018	2019	2019/ 2015, %
Russian Federation	78634.8	79311.9	80048.7	79633.7	79880.2	101.6
Far Eastern Federal region	2216.2	2262.6	2385.3	2444.8	2299.3	103.7
Republic of Buryatia	150.5	142.0	141.8	130.9	118.3	78.6
Yakutia	45.1	44.4	46.8	47.2	47.1	104.4
Trans-Baikal Territory	208.6	200.3	202.8	206.3	198.4	95.1
Kamchatska Territory	20.0	19.7	20.7	20.2	19.5	97.5
Primorskye Territory	402.6	422.6	462.3	478.7	484.7	120.4
Khabarovsk Territory	69.6	68.6	75.5	80.7	72.0	103.4
Amur Region	1165.4	1213.8	1252.7	1282.0	1180.2	101.3
Magadan Region	5.9	6.6	7.0	6.8	6.9	116.9
Sakhalin Region	24.2	26.0	28.3	28.8	29.0	119.8
Jewish Autonomous Region	124.3	118.6	147.3	163.2	143.1	115.1
Chukotka Autonomous Area	0.0	0.0	0.0	0.0	0.0	0.0

In addition, the essential obstacle to the development of business and attracting investments in the Far East are high rates for energy resources and services of infrastructure companies. In the Far East, the highest prices for petroleum products that are used both in the energy sector and transport are almost everywhere in the macroregion. Thus, in April 2021, the cost of automotive gasoline in the Far East exceeded the average Russian value by 6%, and diesel fuel – by 13.5%. At the same time, in the northern regions of the district, 15–35% reaches excess. As a result, the overestimated cost of petroleum products affects the cost of products of all other sectors of the economy.

Tariffs for rail transportation of goods are also inhibited. So, in the cost of production, the cost of railway transport can reach 50%. In general, the share of transport costs in the GRP of the Far East is 2 times higher than the similar average Russian indicator.

In the implementation of state programs, a number of problems are maintained: a decrease in investment in agriculture, reducing the financial support of state support for agricultural development, low yield and high degree of importance of commodity producers of the agro-industrial complex, significant dependence on imports of seed, planting and tribal material, lack of meliorated land, Low rates of unused agricultural land in turnover, insufficient provision of agricultural enterprises agricultural machinery and equipment, high level of depreciation of fixed assets, insufficient level of infrastructure of the agrobridge market, minor agricultural insurance, carried out with state support, insufficient use of innovative technologies in agricultural production, unstable

Development of rural areas, demographic problem of rural settlements, lack Personnels [6, 7].

5 Conclusion

It is obvious that it is necessary to increase the efficiency, effectiveness and basic state programs for the development of the agro-industrial complex. It is necessary to significantly increase direct state support for the agro-industrial complex. So far, this support is small. To do this, at the legislative level, fix the minimum spending limit in the federal budget to support the agro-industrial complex.

The main share of state support is received by large agricultural producers, which leads to the monopolization of the market, reducing the efficiency of the functioning of small businesses. To improve the mechanism for providing state support, it is necessary to make it easier and more transparent, to integrate it with public services, so that subsidies, grants, soft loans, preferential leasing – all these support measures can be issued in electronic form. It is necessary to work out measures to reduce the cost of electricity for farmers and agricultural cooperatives.

To increase food security in the Far Eastern Federal District and increase the availability of food, it is necessary to initiate and support integration processes in the agro-industrial complex, which will reduce prices by eliminating numerous intermediaries.

For the Far East, for which soybean production is a traditional direction for the agricultural economy, the production of which employs a significant part of the rural population, the adoption of tariff regulation measures in terms of establishing and increasing export duties on soybeans is a threat to the economy of the region as a whole, and will have serious economic and political risks associated with the severance of trade relations with the main trading partner-the PRC.

Such restrictions can be established only if the possibility of its own processing is ensured (it will take 2–3 years to implement the relevant investment projects), and a fair price for soybeans is established throughout the Russian Federation (taking into account transport costs).

Currently, we propose to introduce a quota for duty-free soybean exports at the rate of one ton per hectare of acreage, i.e., at the end of the reporting year, a quota is provided to the exporter who purchases soybeans from producers with confirmed volumes linked to land plots.

Thus, it is possible to interest agricultural producers to increase the area of sowing, and hence the volume of production, in addition, there will be traceability of production and sales of soybeans for export operations. This will allow us to get away from the gray schemes of selling soybeans abroad, as well as use up to 40 percent of the soybean crop to provide the domestic market of the Far East in seeds, raw materials for factories and feed.

It is necessary to continue subsidizing producers of domestic agricultural machinery. This is necessary both for the agro-industrial complex, and for the Russian industry as a whole, for the domestic machine-building industry, and for other related industries.

References

1. Burkovskiy, P., Vlasova, N., Godlevskiy, R.: The effectiveness of the program-target method of state regulation of the agro-industrial complex of the region. Nat. human Stud. **29**(3), 92–100 (2020). https://doi.org/10.24411/2309-4788-2020-10236
2. Latkin, A.: Comparative cost of living as a key factor in the development of the Russian Far East. Territoriya novyh vozmozhnostej **3**(38) (2017). https://doi.org/10.24866/VVSU/2073-3984/2017-3/47-56
3. Latkin, A., Korotkih, E.: Programmatic approach to state support for agricultural economy. AIC Econ. Manage. **07**, 4–12 (2018)
4. Shimuk, O.: Rural development management: evolution of methodological approaches and improvement of program tools. Ars Administrandi **12**(1), 44–60 (2020). https://doi.org/10.17072/2218-9173-2020-1-44-60
5. Alimkulova, E., Aitmukhanbetova, D.: State regulation of agricultural sector. Prob. AgriMarket **4**, 47–53 (2020). https://doi.org/10.46666/2020-4-2708-9991.05
6. Lovchikova, E.: Impact of state regulation on agricultural staffing. Bull. Acad. Knowl. **5**(40), 244–249 (2020). https://doi.org/10.24412/2304-6139-2020-10623
7. Chistyakova, M., Schmidt, Y.: Improving state support for small businesses in the agriculture of the Orel region. Bull. Agrar. Sci. **1**(88), 135–142 (2021). https://doi.org/10.17238/issn2587-666X.2021.1.135
8. Aksyutina, S., Mironenko, N.: Current issues of the agricultural sector state regulation. Ekonomicheskie otnosheniya **10**(4), 1469–1490 (2020). https://doi.org/10.18334/eo.10.4.111047
9. Kelemetov, E., Yakubovich, E,: Transformation of state income support practices for agricultural producers in the EU and the USA. Moskovskij ekonomicheskij zhurnal **5**(2), 22–37 (2018). https://doi.org/10.24411/2413-046X-2018-15054
10. Jenkins, H.: A 'business opportunity' model of corporate social responsibility for small- and medium-sized enterprises. Bus. Ethics A Eur. Rev. **18**(1), 21–36 (2009)
11. Gorlov, S., Panaedova, G., Borodin, A.: Features of foreign experience state support for agriculture and its application in Russia. Vestnik Severo-Kavkazskogo federal'nogo universiteta **6**(75), 52–60 (2019). https://doi.org/10.37493/2307-907X-2019-75-6-52-60
12. Holodova, M.: About features of state support transformation of agriculture in Russia. Ekonomika i ekologiya territorial'nyh obrazovanij **4**(1), 14–24 (2020). https://doi.org/10.23947/2413-1474-2020-4-1-14-24
13. Hejazi, M., Marchant, M.A.: China's evolving agricultural support policies. CHOICES **32**(2), 1–7 (2017)
14. Gale, F.: Growth and Evolution in China's Agricultural Support Policies ERR-153 Economic Research Service/USDA 54 (2013). https://www.researchgate.net/publication/259338481_Growth_and_Evolution_in_China's_Agricultural_Support_Policies
15. Zhupley, I., Schmidt, J., Tretiak, N.: Analysis of structural changes in food accessibility assurance in Russia and its macroregions through food securit. IOP Conf. Ser. Earth Environ. Sci. **666**, 052028 (2021). https://doi.org/10.1088/1755-1315/666/5/052028
16. Potenko, T., Zhupley, I., Schmidt, Y., Kvashko, L.: Agricultural policy and import substitution opportunities in Russian Far East. IOP Conf. Ser. Earth Environ. Sci. **548** (2020).https://doi.org/10.1088/1755-1315/548/2/022011

Ecological Plasticity of Varieties *Triticum Aestivum L.*, *Triticum Durum Desf.* And *Hordeum Vulgare L.* Of Various Origins in the Conditions of Primorsky Krai

Galina Murugova(✉), Alexey Klykov, Polina Bogdan, Inna Konovalova, and Nadezhda Kryuchkova

Federal Scientific Center of Agricultural Biotechnology of the Far East named after A.K. Chaika, 30 Volozhenina Street, Timiryazevsky Stl., 692539 Ussuriysk, Primorsky Krai, Russia

Abstract. This article presents the results of ecological trials of varieties of various origins *Triticum* and *Hordeum* in the steppe zone of Primorsky Krai. The most important tasks of breeding are the creation and introduction into production of new high-yielding varieties that are resistant to stressful environmental conditions and high quality grain varieties adapted to the monsoon climate and the requirements of modern agricultural production. We studied 38 varieties of *Triticum aestivum L.* and 31 varieties of *Hordeum vulgare L.* The studied varieties of barley and wheat were characterized by a wide range of variability (b_i) from 0.5 to 1.9 and the yield of 2.5–4.9 t/ha. Based on a comprehensive assessment of yield and adaptability parameters, the best varieties of *Triticum aestivum L.* are Nikolskaya, *Triticum durum Desf.* - Aleiskaya; *Hordeum vulgare L.* - Primorets. Of great importance are stable samples that respond well to improving growing conditions ($b_i = 1{,}0$ and $S_2d_i = 0{,}0$): *Triticum aestivum L.* - Primorskaya 39, Feng Qlang 11; *Triticum durum Desf.* - Aleiskaya. The varieties selected for yield and ecological plasticity are recommended for use in breeding for adaptability in the monsoon climate.

Keywords: Triticum durum desf · Triticum aestivum L. · Hordeum vulgare L. · Ecological plasticity · Stability · Yield

1 Introduction

The Far East of Russia is characterized by a monsoon climate with high air humidity, frequent fogs, contributing to the increased development of diseases, a decrease in grain quality, and resistance to lodging [1]. To create competitive varieties, it is necessary to have a genetically diverse and comprehensively studied source material. When breeding work, it is important not only to create a variety, but also to identify the most favorable zone for its cultivation [2–6].

A. Muratov and S. Ignateva (Eds.): AFE 2021, LNNS 353, pp. 386–396, 2022.
https://doi.org/10.1007/978-3-030-91402-8_44

Over the past 50 years, the productivity of wheat, barley and other agricultural crops became at least two times bigger. Breeding undoubtedly played a decisive role: in many countries, the contribution of the variety to ensuring crop yields began to amount to over 50%. At the same time, resistance to biotic and abiotic environmental factors decreases. According to scientists, the latter circumstance is due to the fact that agriculture in general, and breeding in particular, aimed mainly at achieving maximum productivity, significantly weaken the protective systems of plants and, as a consequence, reduce their resistance to adverse conditions [7].

Therefore, in recent years, the most relevant task in breeding is the creation of adaptive varieties of grain crops, which will reduce the impact of negative anthropogenic factors on the environment and provide the population with environmentally friendly products. The need to develop adaptive varieties for agricultural production is determined both by the contrast of the natural and climatic conditions of Russia and by global climate changes [8, 9].

There are more than 14 billion hectares of agricultural land in the world, of which only 10% are characterized by good conditions for cultivated crops. On the rest of the lands, high yields are limited by various limiting factors [10].

An increase in the level of productivity and stability is possible with the cultivation of highly adapted varieties that are capable of providing high and stable productivity in various environmental conditions [11].

Following the above, a comprehensive assessment of the varieties Triticum and Hordeum in terms of yield and adaptability parameters is relevant.

The purpose of this work is to assess wheat and barley varieties in terms of yield and adaptability parameters based on the results of their testing in different years, and to select the best varieties for the conditions of Primorsky Krai.

2 Materials and Methods

The work was carried out in the field and laboratory conditions of the laboratory of selection of grain and cereal crops of the FSBSI "FSC of Agricultural Biotechnology of the Far East named after A.K. Chaika" in 2015–2020. 34 wheat varieties were the object of the study, of which 18 are soft wheat (Triticum aestivum L.), 16 are durum wheat (Triticum durum Desf.), and 31 varieties of spring barley (Hordeum vulgare L.), of which 18 are two-row (Hordeum vulgare L. subsp distichum) and 13 are multi-row (Hordeum vulgare L. subsp.vulgare).

3 Results

The main method of creating a new source material for grain crops at the Federal State Budgetary Scientific Institution FSBSI "FSC of Agricultural Biotechnology of the Far East named after A.K. Chaika" is intraspecific hybridization with subsequent individual selection. Parental forms are selected according to adaptive properties and ecological-geographical principle by means of mathematical analysis of quantitative traits that provide directions in breeding.

In the ecological test, 18 varieties of common wheat (Triticum aestivum L.) were studied, including 4 - "two-handed", of which 72.2% are of Russian selection, 16.7% are from China (Asian countries), and 11.1% are from Sweden (Western Europe), represented by the species lutescens and erythrospermum. 93.8% of the studied durum wheat varieties (Triticum durum Desf.) are from different regions of Russia (Omsk Region, Krasnodar Territory, Altai Territory, Rostov Region) were varieties of hordeiforme and 6.2% - leucurum.

The barley species represented by the genus Hordeum L. are widespread and grow in a variety of climatic conditions. The most widespread and practical use is the cultivated barley Hordeum vulgare L. [12]. The ecological test is represented by two subspecies of cultivated barley: two-row (Hordeum vulgare L. subsp distichum) and multi-row (Hordeum vulgare L. subsp. Vulgare). The main part of the samples belongs to the nutans variety (two-row): Russia - 12 pcs., Western European countries - 3 pcs. (Table 1). Multi-row varieties of barley: pallidum - 10 samples (Russia - 3 pcs., Western European countries - 1 pc., Asian countries - 6 pcs.); Ricotense - 3 samples from Russia.

Table 1. Origin of Triticum aestivum L., Triticum durum Desf., and Hordeum vulgare L.

Specimen	Origin		
	Russia and CIS countries	Asia	Western Europe
Triticum aestivum L			
Lutescens	8	–	
Erythrospermum	5	3	2
Total	13	3	2
Triticum durum Desf			
Leucurum	1	–	–
Hordeiforme	15	–	–
Total	16	–	–
Hordeum vulgare L. *subsp. vulgare*			
Pallidum	3	6	1
Ricotense	3	–	
Total	6	6	1
Hordeum vulgare L. *subsp distichum*			
Nutans	12	–	3
Medicum	1	–	–
Submedicum	1	–	–
Nudum	1	–	–
Total	15	–	3

When studying varieties in years of different meteorological conditions, information on plasticity can be obtained, which shows the peculiarities of the reaction of the genotype to changes in environmental conditions. In this regard, they can be divided into intensive and extensive varieties. Breeding to unfavorable environmental factors presupposes the presence of an ecologically plastic starting material; therefore, its comprehensive assessment is necessary in order to obtain more complete information on the reaction of varieties to environmental conditions.

On the basis of statistical methods for calculating the parameters characterizing the degree of reaction of the genotype to changes in environmental conditions, many different methods have been developed to assess the genotype-environment interaction [13].

Evaluation of ecological plasticity according to the method of S.A. Eberhart and W.A. Russell makes it possible to single out the linear regression indicator (bi), which characterizes the variety's responsiveness to changing conditions and the non-linear regression indicator (stability variance (S_2d_i), which shows the degree of variation in the productivity of the variety over the years of testing.

The coefficient of ecological plasticity (bi), showing the responsiveness of varieties to changes in growing conditions, takes on a value greater than, less than or equal to one. If bi $\geq$ 1, the variety is more responsive, bi $\leq$ 1 - reacts weaker to changes in environmental conditions, with bi = 1 there is a correspondence between changes in yield and changes in growing conditions. Stability (Sd2) - resistance to limiting environmental factors, the ability to give not very high, but stable yield in any growing conditions - is an integral property of adaptability. The smaller the deviation, the more stable the variety.

The prevailing meteorological conditions during the years of research differed in the amount of heat and precipitation, which made it possible to fully assess their influence on the formation of the yield and its quality in the studied varieties of spring wheat and barley.

When assessing the parameters of the adaptability of the yield of Triticum and Hordeum, it was noted that the regression coefficient (bi) had a wide range of variability from 0.5 to 1.9, which indicates a different response to changes in growing conditions. According to the results of studies of the ecological test variety according to the indicators of plasticity and stability, the studied varieties were divided into 5 groups, the characteristics of which are presented in Table 2.

Table 2. Characteristics of the obtained samples of *Triticum aestivum L., Triticum durum Desf.,* and *Hordeum vulgare L.* in terms of ecological plasticity.

Parameters		Characteristic	Varieties
b_i	$S^2{}_d$		
1	0	Responds well to improving conditions, stable	*Triticum aestivum* L.: Primorskaya 39, Feng Qlang 11; *Triticum durum* Desf.: Alciskaya, *Hordeum vulgare* L.: Primorets

(*continued*)

Table 2. (*continued*)

Parameters		Characteristic	Varieties
b_i	$S^2{}_d$		
<1	0	Has the best results in adverse conditions, stable	*Triticum aestivum* L.: Sella, Feng Qlang 7; *Triticum durum* Desf.: Omsky zircon, Omsky izumrud, Pamyati Yanchenko, Voronezhskaya 7; *Hordeum vulgare* L.: Kazminsky
<1	>0	Has better results in adverse conditions, unstable	*Triticum aestivum* L.: Nikolskaya; *Hordeum vulgare* L.: Vostochny
>1	0	Has the best results in favorable conditions, stable	*Hordeum vulgare* L.: Primorsky 98
>1	>0	Has the best results in favorable conditions, unstable	*Hordeum vulgare* L.: Erofey, Ecolog, Pamyati Rodinoy, Leningradsky, Master

As follows from the Eberhart - Russell calculation model, the most valuable varieties are those in which bi > 1, and stability (S2d) tends to zero. Based on the calculations obtained, it was found that this model corresponds to the variety Hordeum vulgare L. - Primorsky 98. Stable samples that respond well to improving growing conditions (bi = 1.0 and S2di = 0.0) are also important: Triticum aestivum L. - Primorskaya 39, Feng Qlang 11, Triticum durum Desf. - Aleiskaya, Hordeum vulgare L. - Primorets.

However, for Primorsky Krai, whose climate is unstable, interest is acquired by varieties that form stable yield of high quality grain and in unfavorable growing conditions (bi > 1, S2di = 0): Triticum aestivum L. - Sella, Feng Qlang 7, Triticum durum Desf. - Omsky zircon, Omsky izumrud, Pamyati Yanchenko, Voronezh 7, and Hordeum vulgare L. - Kazminsky.

The starting point in breeding for increasing the yield and adaptability is to analyze the structure of the crop for its elements, which allows the use of scientifically based selection of parental pairs when crossing. Over the years of research, we have identified varieties of various ecological and geographical origin, possessing a complex of economically valuable traits that are used in the breeding program (Table 3). Tillering capacity, grain number per ear, 1000 grain weight, grain weight per ear and per plant are interconnected and often compensate each other under unfavorable conditions.

One of the important indicators of adaptability is stress resistance (Ymin-Ymax). The smaller the gap between the maximum and minimum yields, the higher the stress resistance of the variety and the wider the range of its adaptive capabilities. Resistance to stress was shown by varieties Triticum aestivum L. - Nikolskaya (−1.2), Sella (−1.2); Triticum durum Desf. - Pamyati Yanchenko (−1.0); Hordeum vulgare L. - Vostochny (−0.7).

Table 3. Characteristics of *Hordeum vulgare* L., *Triticum aestivum* L., *Triticum durum* Desf. in terms of ecologically valuable traits.

Variety	Plant height. cm	Tillering capacity	Ear length. cm	Grain number pereaR. pcs	Grain weight. g		1000 seed weight. g	Lodging resistance. point
					Per ear	Per plant		
Triticum aestivum L								
Primorskaya 39 *var. Lutescens*	105.7	1.5	7.6	25.8	0.9	1.2	28.2	7
Nikolskaya *var.erythrospermum*	105.6	1.7	7.8	25.0	0.8	1.3	37.3	9
Sella *var.erythrospermum*	68.3	1.9	7.3	21.1	0.6	1.1	35.3	9
Feng Qlang 7 *var.erythrospermum*	85.5	1.3	7.2	21.3	0.9	1.1	43.1	9
Feng Qlang 11 *var.erythrospermum*	96.3	1.5	9.1	23.4	0.9	1.2	39.1	9
Triticum durum Desf								
Omsky zircon *var.hordeiforme*	95.9	1.8	6.4	32.5	1.4	1.8	44.8	9
Omsky izumrud *var. hordeiforme*	108.2	1.4	6.6	27.0	0.1	1.4	42.4	9
Aleiskaya *var. hordeiforme*	112.9	1.8	6.8	29.5	1.0	1.5	35.8	9
Pamyati Yanchenko *var. hordeiforme*	109.8	2.0	5.8	22.3	0.9	1.5	36.9	9
Voronezhskaya 7 *var. hordeiforme*	107.3	1.8	6.4	24.9	0.9	1.2	37.8	9
Hordeum vulgare L. *subsp distichum*								
Vostochny *var. nutans* (st)	93.4	2.7	6.6	20.2	0.9	2.2	43.8	9
Primorsky 98 *var. submedicum*	82.2	2.8	7.5	18.2	0.7	1.7	39.9	9–7
Primorets *var. nutans*	82.6	2.6	5.9	19.3	0.7	2.0	38.2	9
Erofey *var. nutans*	76.7	1.9	7.3	20.5	1.0	1.7	46.1	7–8

(continued)

Table 3. (*continued*)

Variety	Plant height. cm	Tillering capacity	Ear length. cm	Grain number pereaR. pcs	Grain weight. g		1000 seed weight. g	Lodging resistance. point
					Per ear	Per plant		
Ecolog *var. nudum*	89.2	2.5	6.6	36.3	1.4	2.6	35.9	7–8
Pamyati Rodinoy *var. nutans*	71.5	2.7	6.8	20.2	0.9	2.6	43.3	9
Hordeum vulgare L. *subsp. vulgare*								
Leningradsky *var. pallidum*	87.1	1.7	6.0	29.7	1.3	1.9	31.7	9
Kazminsky *var. rikotense*	66.1	1.8	5.7	32.9	1.1	2.0	34.1	9
LSD_{05}	7.1	0.1	0.3	2.1	0.1	0.1	3.0	

The average yield of varieties under contrasting (stress and non-stress) conditions (Y1 + Y2/2) characterizes their genetic flexibility. The maximum ratio between genotype and environmental factors was noted in Triticum aestivum L. - Nikolskaya (3.9), Triticum durum Desf. - Omsky izumrud (3.6) and Voronezh 7 (3.5); Hordeum vulgare L. - Pamyati Rodinoy (4.9) and Primorets (4.5).

Homeostasis (Hom) is one of the important indicators reflecting the ability of the genotype to minimize the effects of exposure to adverse external conditions. The breeding value indicator (Sc) is the most suitable for practical breeding assessment of varieties. The greatest homeostaticity and breeding value was noted in the varieties Triticum aestivum L. - Nikolskaya (Hom = 37.7, Sc = 2.7), FengQlang 7 (Hom = 27.8, Sc = 2.5); Triticum durum Desf. - Omsky izumrud (Hom = 29.3, Sc = 2.2), Pamyati Yanchenko (Hom = 36.3, Sc = 2.2); Hordeum vulgare L. - Ecolog (Hom = 37.0, Sc = 3.4), Kazminsky (Hom = 40.0, Sc = 3.6), Vostochny (Hom = 27.7, Sc = 3.8). It was noted that varieties with a low coefficient of variation and high stability of the trait have high indicators of homeostaticity and breeding value. (Table 4).

Table 4. Adaptability parameters of highly productive varieties *Hordeum vulgare* L., *Triticum aestivum* L., *Triticum durum* Desf.

Variety	b_i	S^2_d	Yield. t/ha		Ecological stability. Y_{min}-Y_{max}	Genetic Flexibility($Y_2 + Y_1$)/2	Coefficient of variation(V). %	Homeostaticity (Hom)	Breeding value indicator(S_C)
			Min (Y_2) –Max (Y_1)	Average					
Triticum aestivum L									
Primorskaya 39 (st) *var. lutescens*	1.0	0.0	2.2–4.0	3.0	−1.8	3.1	29.9	20.1	1.6
Nikolskaya *var.erythrospermum*	0.6	0.2	3.3–4.5	3.7	−1.2	3.9	17.8	37.7	2.7
Sella *var.erythrospermum*	0.7	0.0	2.1–3.3	2.7	−1.2	2.7	22.2	24.3	1.7
Feng Qlang 7 *var.erythrospermum*	0.9	0.0	2.5–4.1	3.5	−1.6	3.3	24.0	27.8	2.5
Feng Qlang 11 *var.erythrospermum*	1.0	0.0	2.1–3.9	2.9	−1.8	3.1	31.6	18.0	1.5
Triticum durum Desf									
Omsky zircon *var.hordeiforme*	0.9	0.0	2.1–3.8	3.0	−1.7	2.9	28.7	20.7	1.6
Omsky izumrud *var.hordeiforme*	0.9	0.0	2.8–4.4	3.5	−1.6	3.6	23.4	29.3	2.2
Aleiskaya *var.hordeiforme*	1.0	0.0	1.6–3.4	2.5	−1.8	2.5	36.6	13.5	1.1
Pamyati Yanchenko *var.hordeiforme*	0.5	0.0	2.5–3.5	3.0	−1.0	3.0	16.6	36.3	2.2

(*continued*)

Table 4. *(continued)*

Variety	b_i	$S^2{}_d$	Yield. t/ha		Ecological stability. Y_{min}-Y_{max}	Genetic Flexibility($Y_2 + Y_1$)/2	Coefficient of variation(V). %	Homeostaticity (Hom)	Breeding value indicator(S_C)
			Min (Y_2) –Max (Y_1)	Average					
Voronezhskaya 7 *var.hordeiforme*	0.8	0.0	2.8–4.2	3.5	−1.4	3.5	30.5	16.1	1.4
Hordeum vulgare L. *subsp distichum*									
Vostochny *var. nutans* (st)	0.6	0.6	4.0–4.7	4.5	−0.7	4.3	14.2	27.7	3.8
Primorsky 98 var. *submedicum*	1.7	0.0	3.5–4.6	4.1	−1.3	3.9	17.1	18.4	2.9
Primorets *var. nutans*	1.0	0.0	3.6–5.5	4.6	−1.9	4.5	20.7	11.7	3.0
Erofei *var. nutans*	1.9	0.2	3.4–5.2	4.4	−1.8	4.3	26.7	7.0	2.6
Ecolog *var. nudum*	1.1	0.1	3.5–5.2	4.5	−1.7	4.4	10.7	37.0	3.4
Pamyati Rodinoy *var. nutans*	1.9	0.1	4.1–5.7	4.9	−1.6	4.9	13.4	24.0	3.6
Hordeum vulgare L. *subsp. vulgare*									
Leningradsky *var. pallidum*	1.7	0.1	3.3–5.0	4.2	−1.7	4.2	14.4	20.0	2.9
Kazminsky *var. rikotense*	0.2	0.0	2.8–5.0	3.9	−2.2	3.9	11.6	40.0	3.6

4 Discussion

The problem of resistance to environmental factors limiting the formation of potential productivity is especially relevant in areas with a sharp manifestation of unfavorable for plants climatic conditions [11].

In a monsoon climate, it is most advisable to cultivate several varieties with different responses to growing conditions - the most adapted to various conditions of the environment, characterized as plastic and responsive to changes in growing conditions. When evaluating breeding material for adaptability and selection of parental pairs for hybridization, it is recommended to take into account the parameters of ecological plasticity and stability in the created varieties.

5 Conclusion

As a result of the study of varieties of various ecological and geographical origin, in order to create new varieties with high ecological plasticity and stability samples advisable to use in breeding as a source material were selected.

1. *Hordeum vulgare L.* - Primorsky 98 has the most valuable $b_i > 1$, and its stability (S_2d_i) tends to zero. Such varieties are classified as an intensive type, they are highly responsive and demanding of a high level of agricultural technology.
2. *Triticum aestivum L.* - Primorskaya 39, Feng Qlang 11, *Triticum durum Desf.* - Aleiskaya, and *Hordeum vulgare L.* - Primorets are stable samples responding well to improving growing conditions ($b_i = 1{,}0$ and $S_2d_i = 0{,}0$)
3. Of great importance are semi-intensive specimens, which are best grown in an extensive background, where they will produce higher yields of *Triticum aestivum L* - Feng Qlang 11; *Hordeum vulgare L* - Erofey, Ecolog, Pamyati Rodinoy, Leningradsky, Master.
4. New adapted varieties *Triticum aestivum L.* - Nikolskaya, *Hordeum vulgare L.* - Primorets zoned in the Far Eastern zone since 2021 have been created.

References

1. Klykov, A., Murugova, G., Bogdan, P., Konovalova, I., Timoshinova, O., Kryuchkova, N.: Use of genetic resources in white straw and cereal crops breeding in the monsoon climate. IOP Conf. Ser. Earth Environ. Sci. **548**(6), 062014 (2020). https://doi.org/10.1088/1755-1315/548/6/062014
2. Dzybenko, N.: Vavilov's collection of worldwide crop genetic resources in the 21st century. Biopreserv. Biobank **16**(5), 377–383 (2018). https://doi.org/10.1089/bio.2018.0045
3. Milner, S.: Genebank genomics highlights the diversity of a global barley collection. Nat. Genet. **51**(2), 319–326 (2019). https://doi.org/10.1038/s41588-018-0266-x
4. Gudzenko, V.: Statistical and graphical (GGE biplot) evaluation of the adaptive ability and stability of winter barley breeding lines. Vavilov J. Genet. Breed. **23**(1), 110–118 (2019). https://doi.org/10.18699/VI19.469

5. Kincharov, A., Demina, E., Taranova, T., Chekmasova, K.: Study of collection samples of spring soft wheat by precocity. Int. J. Hum. Nat. Sci. **10**(1), 136–141 (2018). https://doi.org/10.24411/2500-1000-2018-10082
6. Des Marais, D., Hernandez, K., Juenger, T.: Genotype-by-environment interaction and plasticity: Exploring genomic responses of plants to the abiotic environment. Ann. Rev. Ecol. Evol. Syst. **44**, 5–29 (2013). https://doi.org/10.3389/fpls2016/01123
7. Rybas, I.: Improving adaptability in the selection of grain crops. Agric. Biol. **51**, 617–626 (2016). https://doi.org/10.15389/agrobiology.2016.5.617rus
8. Hassan, M., Mohamed, G., El-Said, R.: Stability analysis for grain yield and its components of some durum wheat genotypes (Triticum durum L.) under different environments. Asian J. Crop Sci. **5**, 179–189 (2013)
9. Dopierala, P., Kordas, L.: The effects of genotype - environment interaction on the yield and its structure in some winter cereals. Biul Lnst Hodowli Aklimat Rosl **253**, 165–173 (2009)
10. Fethi, B., Mohamed, E.: Epistasis and genotype-by-environment interaction of grain yield related traits in durum wheat. Genet. Molec. Biol. **33**(1), 125–130 (2010). https://doi.org/10.1590/S1415-47572010000100021
11. Sapega, V., Tursumbekova, G.: Productivity, ecological plasticity and stability of spring soft and durum wheat varieties in the southern forest-steppe of the Tyumen region. Agric. Sci. Euro-North-East **21**(2), 114–123 (2020). https://doi.org/10.30766/2072-9081.2020.21.2.114-123
12. Chernov, V., Pendin, G.: Comparative assessment of callus formation and regeneration in different types of barley. Agric. Biol. **1**, 44–53 (2011)
13. Cheshkova, A., Stepochkin, P., Aleynikov, A., Grebennikova, I., Ponomarenko, V.: A comparison of statistical metods for assessing winter wheat grain yield stability. Vavilov J. Genet. Breed **24**(3), 267–275 (2020). https://doi.org/10.18699/VJ20.619

Telescopic Seeder Coulter Suspension

Sergey Inshakov[1](✉), Anna Redkokashina[1], Aleksander Redkokashin[1], and Viktor Balabanov[2]

[1] Primorskaya State Agricultural Academy, 44, Blyukhera Street, Ussuriysk 692510, Russia
inshakov_serg@bk.ru

[2] Russian Timiryazev State Agrarian University, 49, Timiryazevskaya Street, Moscow 127550, Russia

Abstract. The suspension of the seeder working bodies affects the quality of the technological process of seeding and to a large extent determines the operational characteristics of the sowing unit. The purpose of this study was to develop and technically substantiate a new design of the seeder coulter suspension, which has technological and operational advantages over existing analogues, including simplicity, possibility of adjusting the pressing force against the soil in a wide range, as well as reducing the kinematic length of the machine and tractor unit. The classic coulter suspension represents a pendulum linkage with an elastic element that provides a vertical force for penetration. New promising designs of the coulter suspension were analyzed based on a search in the databases of Russian patents for inventions and utility models. All the analyzed structures of the suspension of the working bodies had significant dimensions that inevitably led to an increase in the overall dimensions of the seeder and deterioration in its performance. An obvious advantage of the telescopic coulter suspension developed by the authors of this article consists in the fact that moving of the seeding unit is accompanied by copying the microrelief of the field surface by moving the coulter with the press wheel in the vertical plane that is provided by the telescopic connection of the guide tube and the barrel.

Keywords: Sowing method · Seeder · Soshnik · Pendant mechanism · Telescopic mechanism

1 Introduction

Sowing seeds has always been the most important operation in the field, and the design of the sowing machine determines distribution uniformity and planting depth of seeds in the row. These indicators affect the yield [1–3]. A characteristic feature of modern seeders manufactured in Russia and other countries is the presence of sections of working bodies that ensure the formation of a groove and proper seeding [4–7]. The designs of these seeders, as a rule, represent a frame on wheels with a corresponding number of sowing sections on it, where each section corresponds to a seeding device, and devices for planting seeds into the soil are represented by coulter as well as covering and rubbing-down working bodies. A high sowing quality is provided by the correct operation of all

A. Muratov and S. Ignateva (Eds.): AFE 2021, LNNS 353, pp. 397–403, 2022.
https://doi.org/10.1007/978-3-030-91402-8_45

mechanisms of the seeder. One of the most important mechanisms is the suspension of the working bodies interacting with the soil to ensure covering of seeds.

Suspension of working bodies affects not only the quality of the technological process of seeding, but also, being an integral part of the design of the sowing machine, to a large extent determines the operational characteristics of both the machine and the seeding unit, making up a part of the kinematic length of the seeder.

The purpose of this study was to develop and technically substantiate a new design of the seeder coulter suspension, which has obvious technological and operational advantages over existing analogues, including simplicity, the possibility to adjust the pressing force against the soil in a wide range and to decrease the kinematic length of the machine and tractor unit.

2 Materials and Study Methods

The classic coulter suspension represents a pendulum linkage with an elastic element that provides a vertical force for penetration. Many seeders (SZ-3.6A, SZT-3.6A, SZS-6, etc.) are equipped with a radial coulter suspension, consisting of a lever and an elastic element. However, this design, for all its simplicity, does not provide a constant vertical approach angle of the coulter when changing the dimensions of the elastic element. A parallelogram-type coulter suspension is also widely used in the design of sowing machines (in the seeders SUPN-8, SST-12, SO-4.2, SUPO-6, etc.) It consists of two levers and an elastic element, which is somewhat more difficult to manufacture, but at the same time it ensures the constancy of the vertical approach angle of the coulter, and, accordingly, the high quality of the bottom of the furrow for placing the seeds. To ensure the best seeding quality and performance of sowing machines, their design is constantly improved [8–13]. New promising designs of the coulter suspension were analyzed based on search in the databases of Russian patents for inventions and utility models with a 15-year retrospective.

Suspension of the coulter for scattered sowing of grain seeds registered under the Russian patent No. 68227 contains a rack with a V-shaped sweep and a mechanism for adjusting the angle of entry of the sweep into the soil. The sweep is attached to the frame through a spring-loaded parallelogram mechanism and a press wheel. The design of this mechanism is complex; specific amount of metal per structure is high; the speed of the coulter in the vertical plane is slow; the kinematic length of the machine-tractor unit is great.

The coulter suspension registered under the Russian patent No. 2312484 is rigidly connected to the frame and a spring-loaded disc knife; the spring tension is adjusted by moving the mount on the rack of the V-shaped sweep. The drawback of the design is the inability to copy the relief with a V-shaped sweep, high specific amount of metal per structure of the structure, and great kinematic length of the machine-tractor unit.

The disc-anchor coulter with a combined suspension registered under the Russian patent No. 2464764 is equipped with a lever, on which the left-side and right-side single disc coulters are mounted. They are arranged in a block and attached to the lower part of the rear link of the parallelogram suspension by means of elastic elements. The considered mechanism is characterized by a sufficiently large specific amount of metal per structure due to the complexity of the design that makes the cost of the entire machine unreasonably high and increases the kinematic length of the machine-tractor unit.

The section of working bodies of the seeder registered under the Russian patent No. 168825 is designed in the form of a block mounted on the frame using a parallelogram spring-loaded suspension and a seeding coulter in the form of a chisel anchor with a spring-loaded press wheel that allows adjusting the depth of seeding. However, this design is rather complicated to manufacture, and hence it is characterized by high cost, high specific amount of metal per structure, low coulter stroke in the vertical plane, and great kinematic length of machine-tractor unit.

The coulter suspension registered under the Russian patent No. 130778 consists of body, coulter on a rack, damper mechanism in the form of a spring connecting the coulter rack with the body, as well as articulated arcuate rods connecting the coulter rack with the body and arched towards the coulter. Due to the complexity this design is characterized by high specific amount of metal per structure, small coulter stroke in the vertical plane, and great kinematic length of the machine and tractor unit.

The coulter registered under the Russian patent No. 125017 contains a furrow former connected to the beam of the traction frame by means of a bracket fixed through elastic elements made in the form of triangular prisms located between the planes of the supporting square bar and a square bracket bordering the vertices of the supporting square bar in the middle of the sides, while the middle part of the rack is arced opposite to the movement of the coulter. The drawbacks of the device are small stroke of the coulter in the vertical plane causing constant tension of the elastic elements in operation, withdrawal of the coulter from the vertical position due to the residual deformation of the elastic elements, and significant kinematic length of the machine-tractor unit.

The seeder registered under the Russian patent No. 123629 is technically interesting since it consists of a body with a pivotally fixed coulter rod with the press wheel beam, coulter, damper mechanism in the form of a spring mounted with the lower end on a support rigidly fixed to the rod coulter with a threaded rod inside the spring, one end of which is pivotally attached to the body and the other end is attached to the support of the upper end of the spring, while the beam is equipped with a mechanism for adjusting the height of the press wheel relative to the end of the coulter. The drawbacks of the known device are complexity of manufacturing and high specific amount of metal per structure, as well as great kinematic length of the machine-tractor unit.

3 Results

All of the above-mentioned designs of working bodies' suspensions have a common disadvantage – significant dimensions. This inevitably leads to an increase in the overall dimensions of the seeder. This drawback is especially pronounced when designing combined sowing machines that perform several technological operations in one pass of the unit. An increase in dimensions leads to an increase in the kinematic length of the machine and the entire unit, which determines the maneuverability of the unit and the width of the headland. The length of the sowing machines can be reduced by decreasing the dimensions of the coulter suspension, which occupies a significant part in the design of many machines, especially in seeders for row crops. Significant dimensions of the suspensions are due to horizontal articulated levers, both radial and parallelogram, and some promising combined designs contribute to an even greater increase in dimensions. To increase the maneuverability of sowing machines we need to replace the horizontal suspension elements of the working bodies with a vertical slider equipped with elastic elements.

A wide variety of design versions of the suspension of the working bodies of the sowing machines are classified according to the features presented in Fig. 1.

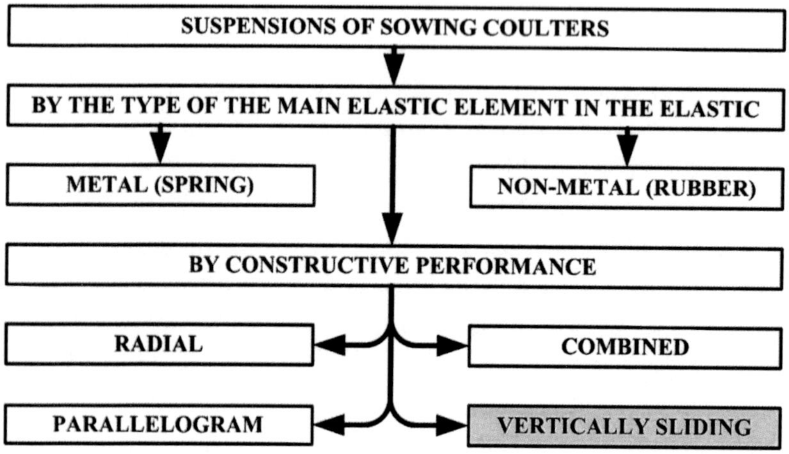

Fig. 1. Classification of suspensions of the working bodies of sowing machines.

In connection with the above, the telescopic coulter suspension developed by the authors of this article has obvious advantages. The device consists of housing, coulter rack with a seed tube, two telescopic and paired guide tubes and barrels connecting the coulter rack with the body, mount beam of press wheel, and press wheel itself. Spring-shaped damper mechanism is located inside a telescopically connected guide tube and barrel (Fig. 2) [14].

The body of the telescopic suspension is mounted on the transverse bar of the seeder shaped as a square tube in cross-section; its immobility is ensured by fasteners. When lowering the sowing machine frame with a cross bar, on which the suspension is installed, the coulter touches and then deepens into the soil layer. The final depth of its placement in the soil layer and, accordingly, the depth of the sowing furrow and the placement of seeds are determined by the position of the press wheel. Further lowering of the sowing machine frame triggers the telescopic mechanism of the coulter suspension, which moves the barrel over the surface of the guide tube.

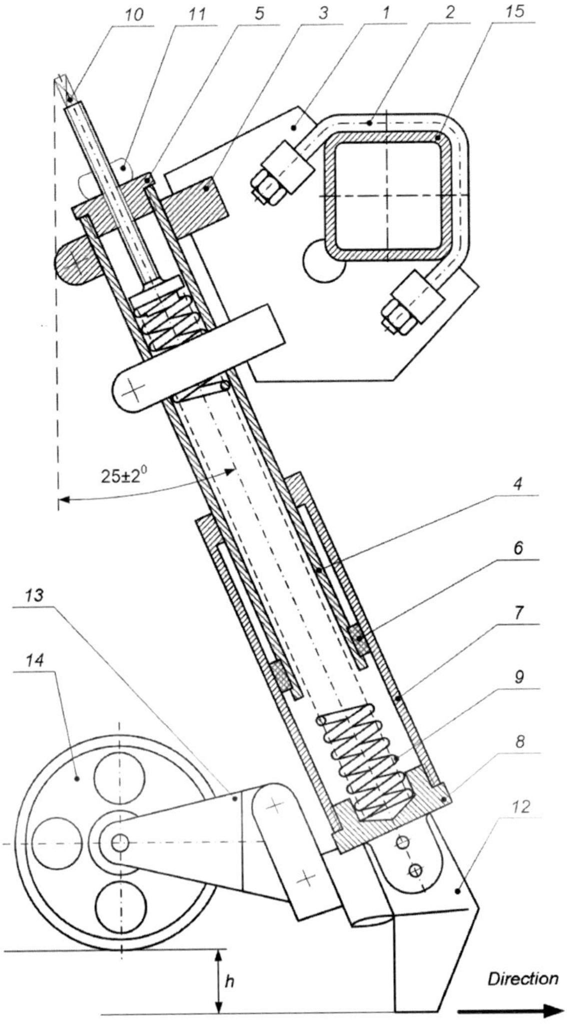

Fig. 2. Telescopic coulter suspension.

Where: 1 – body; 2 – fasteners; 3 – clamp; 4 – guide tube of the telescopic mechanism; 5 – guide tube nut; 6 – split sleeve; 7 – barrel of the telescopic mechanism; 8 – barrel nut 9 – cylindrical spring; 10 – threaded rod; 11 – lock nut; 12 – coulter rack with a seed tube; 13 – beam; 14 – press wheel.

4 Discussion

The coulter suspension is in a balanced state when the gravity of the sowing machine frame is compensated by the spring rate of the damper mechanism. The spring rate can be changed by changing the position of the threaded rod. When the mechanism is in the balanced state, the barrel can both move upward and downward along the guide tube, which, when copying the microrelief of the field, ensures the normal operation of the coulter in recessed places. When the sowing unit moves, the microrelief of the field surface is copied by moving the coulter with the press wheel in the vertical plane. This is possible owing to the telescopic connection between the guide tube and the barrel. Part of the coulter cuts into the soil and opens a narrow furrow, into which seeds fall from the seed tube located behind the coulter. The seeds are then covered by soil from the walls of the furrow and the following press wheel compacts the soil layer above the seeds, thereby improving their contact with the soil.

Obviously, in the best way the proposed telescopic suspension will interact with a disc coulter, which digs a sowing furrow with simultaneous rolling of the discs. The telescopic suspension is best combined with a disc coulter for two-row sowing [15, 16] consisting of symmetrically arranged paired discs that open the furrow for each seed row and include one disc located with its plane along the moving direction and the other disc located at an angle to the moving direction.

The two-row sowing method allows arranging seeds in rows, with an alternating distance, while the distance between closely spaced rows should be 0.866 of a pitch in a row; the distance between remote rows should be 3–5 times greater than the distance between close rows. In this case, the pitch should be 0.8–1.0 of the diameter of the crown of plants in the phase of full development.

The operability of the telescopic mechanism of the disc coulter suspension during the cultivation of soybeans in the Primorsky Territory was studied by the authors in the experimental field of the Primorsk State Agricultural Academy (45°52′13.7″N 131°55′59.1″E) in the field seasons of 2018–2020. All the mechanisms of the experimental seeder, including new structural elements, proved to be highly reliable and stable, while the suspension provided uniform penetration of the disc coulters and seeding to a depth of 5.0 ± 0.8 cm.

5 Conclusion

The use of a telescopic coulter suspension in the seeding unit makes it possible to simplify the design and to reduce its specific amount of metal per structure by replacing a complex lever or parallelogram mechanism with several hinged joints with a simple telescopic mechanism as well as to reduce the kinematic length of the machine-tractor unit, which helps to reduce deviations from the straightness of the sowing furrows, to reduce the size of the headlands in the field, to facilitate the transfer of the sowing machine to the transport position.

References

1. Blokhin, V.D., Moiseenko, A.A., Stupin, V.M.: Scientific foundations of agriculture in the Far East of Russia. Dalnauka, Vladivostok (2011)
2. Yechale, M., Alemayehu, A., Abaynew, J.J.: Sowing methods and seeding rates effects on yield and yield components of Tef (Eragrostis tef [Zucc.] Trotter) at Adet, North West Ethiopia. Heliyon **7**(3), e06519 (2021). https://doi.org/10.1016/j.heliyon.2021.e06519

3. Kirkegaard, N.S., Munkholm, L.J., Lamandé, M., Nørremark, M., Edwards, G.T.C., Green, O.: Seed drill depth control system for precision seeding. Comp. Electr. Agric. **144**, 174–180 (2018). https://doi.org/10.1016/j.compag.2017.12.008
4. Kirkegaard, N.S., Nørremark, M., Green, O.: Sensor and control for consistent seed drill coulter depth. Comp. Electr. Agric. **127**, 690–698 (2016). https://doi.org/10.1016/j.compag.2016.07.029
5. McLaughlin, N.B., Campbell, A.J., Owen, G.T.: Performance of hoe and triple disc furrow openers on no-till grain drills in a fine sandy loam soil. Soil Till Res. **195**, 104373 (2019). https://doi.org/10.1016/j.still.2019.104373
6. Malasli, M.Z., Celik, A.: Disc angle and tilt angle effects on forces acting on a single-disc type no-till seeder opener. Soil Till Res. **194**, 104304 (2019). https://doi.org/10.1016/j.still.2019.104304
7. Hasimu, A., Chen, Y.: Soil disturbance and draft force of selected seed openers. Soil Til Res. **140**, 48–54 (2014). https://doi.org/10.1016/j.still.2014.02.011
8. Mogilnay, O., Paschenko, V., Kharchenko, S., et al.: East-Eur. J. Enter Tech. **1**(94), 37–46 (2018). https://doi.org/10.15587/1729-4061.2018.139846
9. Kokoshin, S.N., Kirgintsev, B.O., Tashlanov, V.I.: EMMFT 2019 J. Phys. Conf. Ser. **1614**, 012051 (2020). https://doi.org/10.1088/1742-6596/1614/1/012051
10. Bai, J., Hao, F., Cheng, G., Li, C.: Machine vision-based supplemental seeding device for plug seedling of sweet corn. Comp. Elect. Agric. **188**, 106345 (2021). https://doi.org/10.1016/j.compag.2021.106345
11. Gebresenbet, G., Jönsson, H.: Performances of seed drill coulters in relation to speed, depth and rake angles. J. Agric. Eng. Res. **52**, 121–145 (1992). https://doi.org/10.1016/0021-8634(92)80055-W
12. Zhiwei, Z., Chen, Y.: The performance of a fluted coulter for vertical tillage as affected by working speed. Soil Til Res. **175**, 112–118 (2018). https://doi.org/10.1016/j.still.2017.08.006
13. Endrerud, H.: Dynamic performance of drill coulters in a soil bin. J. Agric. Eng. Res. **74**(4), 391–401 (1999)
14. Inshakov, S.V., Redkokashina, A.V., Redkokashin, A.A., Inshakov, R.S., Krotov, I.S.: Russian Patent No. 179016 MPK A01C 7/20. Telescopic suspension mechanism of the seed coulter suspension; applicant and patentee: Inshakov S.V. – No. 2017145211 – Applied 21 December 2017 (2018)
15. Redkokashina, A.V., Redkokashin, A.A., Inshakov, S.V., Inshakov, R.S.: Russian Patent No. 2604928 MPK A01C 7/00. Method of sowing seeds of leguminous crops. Applicant and patentee: Primorskaya State Agricultural Academy. – No. 2015140790/13; Applied 24 September 2015 (2016)
16. Inshakov, S.V., Redkokashina, A.V., Redkokashin, A.A., Borodin, I.I., Savel'eva, E.V.: The efficiency of innovative soybeans planting. Int. J. Pharm. Res. **12**(3):2367–2372 (2020). https://doi.org/10.31838/ijpr/2020.12.03.329

The Research of Genotoxic Potential of Water Samples Taken from Agricultural Areas Based on Test-System of the Soybean

Sergey Shishlov[1](✉), Natalia Bykovskaya[2], Marina Shishlova[2], and Natalia Chugaeva[1]

[1] Primorsky State Agricultural Academy, Blucher Av., 44Primorsky Krai, Ussuriysk, Russia
[2] Branch of the Far Eastern Federal University (School of Pedagogy), Nekrasova St., 35Primorsky Krai, Ussuriysk, Russia

Abstract. Analysis of water resources plays an important role in controlling the conditions of production of raw materials and finished agricultural products. Assessment of the genotoxic situation in water bodies exposed to anthropogenic impact is becoming an important task at the present time. Heavy metals such as zinc (Zn), copper (Cu), nickel (Ni) and lead (Pb) are genotoxicants. Genetic monitoring of river water in the suburban area of Ussuriysk (Primorsky Territory) was carried out, taking into account its contamination with heavy metals. To study the genotoxic potential of river water, a plant genetic test system «Somatic mosaicism of soybeans» was used, which allows registering a number of cytogenetic disorders at once. Biotesting showed that the studied river water induces cytogenetic disturbances, therefore, it contains genotoxicants.

Keywords: Monitoring · Heavy metals · Cytogenetic water activity · Genetic test-system «Somatic mosaicism of the soybean»

1 Introduction

Water resources is a necessary step in the chain of actions aimed to perform environment protection and its rational and reasonable using, it plays an important role in the process of controlling of the conditions in which raw materials and agricultural products are produced.

The research of pollutants' toxic influence on the biota of the water objects used in agriculture industry requires physicochemical analysis as well as the assessment of genotoxic processes in rivers and other water bodies exposed to agricultural impact [1]. In river's ecosystems heavy metals are strong genotoxicants.

The research is aimed at genetic monitoring of the river waters taking into account the fact of their pollution with heavy metals in the suburban aria of the town of Ussuriysk.

A. Muratov and S. Ignateva (Eds.): AFE 2021, LNNS 353, pp. 404–414, 2022.
https://doi.org/10.1007/978-3-030-91402-8_46

2 Material and Methods

The research object is the water of the Gluhovka river (the right tributary of the Komarovka), the water samples have been taken in Gluhovka village (station №1), the water of the Rakovka taken in two stations Rakovka village (station №2) and Timiryazevskiy village (station №3) (Fig. 1).

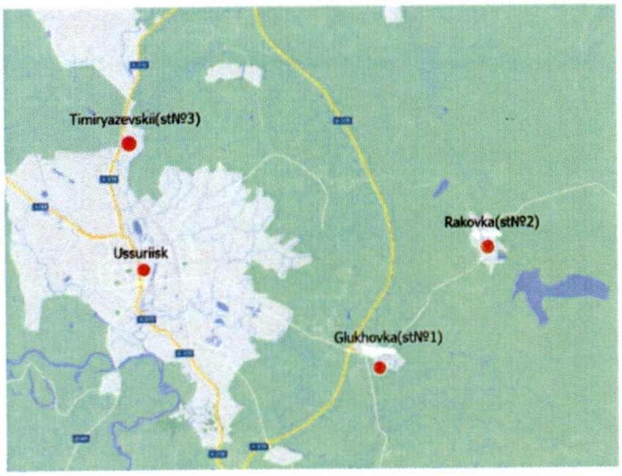

Fig. 1. Scheme of the area of work.

The current research was conducted by genetic test- system «Somatic mosaicism of the soybean», which allows to register a list of cytogenetic disorders such as somatic crossing-over, direct and reverse gene (point) mutations, chromosomal aberrations (only chromosome deletion), aneuploidy. The subject of the test-system is performed by specially selected line of soybean Glycine max (L.), which is heterozygous by gene y11 – yellow11 localized in nuclear DNA [2, 3].

The tested line chlorophyll synthesis depends on geneY11, recessive mutant alleley11 gets chlorophyll production blocked. Thus, homozygous dominant plants Y11Y11 are characterized with the dark-green colour of leaves, whereas heterozygous plants Y11y11 have light-green leaves, and homozygous recessive allele y11y11 plants are chlorophyll lethal mutants with yellow cotyledons and first. As the result of self-pollination, the progeny of heterozygous plants (Y11y11 x Y11y11) is presented by three types of leaf colour (Fig. 2), types correlation corresponds to Mendelian segregation with semidominance.

Fig. 2. Splitting by phenotype: dark green, light green, yellow plants.

The test is based on the record and analysis of combinations of various types of spots appearing on soybean leaves after the seeds' being treated by genotoxicants. The combination of spots on the top surface of three first leaves of soybeans with different colour, i. e. the character of somatic mosaicism, allows to recognize possible genetic mechanisms of these leaves appearance (Table 1). Possible mechanism of the spots appearance has been discovered by the following way: while using other genetic test-systems it was found that some mechanisms of cytogenic disorders were caused by some mutagens. The type of spots got stimulated by the same mutagens connected their actions with specific mechanisms. There are some generally accepted mechanisms of soybean spotting such as somatic crossing over, aneuploidy, chromosome aberrations, point mutation [4]. Chloroplast DNA – induction of spots with no precise boarders.

Table 1. Somatic mosaicism induction of soybean leaves effected by the water samples.

Possible mechanism	Effect of spots on plant leaves				
	light- green $Y_{11}\ y_{11}$			Dark- green $Y_{11}\ Y_{11}$	Yellow $y_{11}\ y_{11}$
	Dark- green spots	Yellow spots	Double- spots	Light- green spots	Light- green spots
Somatic crossing-over (SC)	+	+	+	–	–
Point mutations (PM)	+	+	–	+	+
Chromosome aberration (CA)	–	+	–	+	–
Aneuploidy (AP)	+	+	–	+	–
Changes in chloroplast DNA (chl. DNA)	+	+	–	+	+

Note: « +» - more than three-fold induction, «-» - no effect

To perform the research heterozygous soybean seeds have been put and kept in the samples of tested water at room temperature for 24 h, while the control seeds were put into the distilled water under the same conditions. After water steeping the seeds got cleaned with running water and then sown in the boxes filled with soil, the seeds were let grow till the appearance of the second compound ternate leaf. To make the same conditions for the control and test seeds, both were planted in the soil of the same sample. The plants grown from the test seeds are distinct with the spots on the leaves.

The spots from the top surface of two simple leaves and first compound leaf, whereas compound leaf is considered to be relevant to three simple ones [5].

Selection, filtration and conservation of the water samples were conducted according to GOST (Russian State Standard) 31861-2012. At the place of water samples taking there was hydrogen ion exponent (pH) measured by the portable pH analyzer pH-009(I) (Barry Century, China). Chemical analysis of the water samples has been conducted by the method of atomic absorption spectroscopy (AAS) on atomic absorption spectrometer MGA - 915 according to ecological security normative documents 14.1:2.214-06 [5]. The method has been widely used to measure the level of natural objects pollution with heavy metals. The method of calibration chart was used to determine the elemental composition.

All the experiments have been performed in three independent repetitions. Statistical analysis has been executed in Microsoft Excel, StatPlus and WINPIPI. In all cases we found and determined the average index and errors in mean. Reliability of differences was estimated by Student's test. The differences were considered reliable at confidence coefficient of 0,95 ($p < 0{,}05$).

3 Results

The research focuses on several certain heavy metals such as Zn, Cu, Ni, Cd, Pb, Sn, Fe. The choice of the elements for the research was based on the metals' biological significance (Cu, Fe), metallogenic features of the region (Sn, Pb, Ni, Zn, Cu), character of natural and industrial effluents (Zn, Cu, Ni, Cd, Pb, Cr, Fe), the surface waters in Ussuriysk suburban aria and character of air emissions of suspended substances containing metal elements (Zn, Cu, Ni, Cd, Pb, Fe) [6].

Background metals content in streams depends on natural factors, primarily the composition of local rocks and soils, which are different from each other and can be quite high [7]. The metals such as iron and zinc are contained in the soils of agricultural or industrial works in concentration of 5 mg/kg of soil and more background metal content in the streams characterized by soil washing of the mentioned metals can reach high points (Table 2). Cadmium is contained in soil only in impurity trace (0.06 mg/kg soil) and its concentration in water is caused only by technological activities.

Table 2. Average content of trace elements in soil [8].

Metal	Content, mg/kg	Metal	Content, mg/kg
Antimony	6.0	Zinc	7.0
Arsenic	6.0	Selenium	0.2
Barium	500	Silver	0.1
Beryllium	6.0	Tellurium	10.1
Cadmium	0.06	Tin	10.0
Lead	10.0	Vanadium	100

The source of aquatic environment pollution with heavy metals can be the washout of agricultural land being worked with mineral fertilizers or wastewater mud. Table №3 reveals the range of several metals' concentrations in the mud samples. Metals' concentration in the sediment of wastewater significantly exceeds their natural contain in soils. Zinc is concentrated in the mud in 300 times more, chromium in 8 times, cadmium more than in 1000 times.

Agricultural land gets heavy metals pollution as the result of using mineral fertilizers. Phosphate as well as orthophosphoric acid fertilizers have been discovered in the bigger amounts in the soils and also containing heavy metals. For instance, superphosphate contains cadmium, chrome, cobalt, copper, lead, vanadium and zinc. At fertilizing soil with superphosphate (90 kg/ha.) soil gets 11.3 gm copper, 55.8 gm lead, 1 gm cadmium [9, 10].

Table 3. The content of some metals in the dry matter of sewage sludge [11].

Metal	Content, mg/kg	Metal	Content, mg/kg
Silver	5–150	Iron	6000–62000
Boron	15–1000	Nickel	20–5300
Cadmium	60–1500	Lead	120–3000
Chromium	40–8800	Zinc	700–49000
Copper	200–8000	Cobalt	2–260

During the laboratory research and by using the soybean test-system we have studied genotoxic activity of heavy metals salines such as lead acetate, sulphates of copper and nickel, zinc chloride in concentrations meeting 1 TLV, 5 TLV standards. The character of somatic mosaicism (combinatorics of spots on leaves of different genotype soybeans) allowed to establish the most frequent mechanisms of spotting.

It was found out that the solution of copper sulphate in concentration 1 TLV (TLV_{Cu} = 0.1 mg/l) does not obtain mutagenic effect. Solution of copper sulphate in concentration of 5 TLV induces all types of spots, consequently the whole range of cytogenic disorders such as somatic crossing over, chromosome aberrations, point mutations (direct and reverse), aneuploidy, changes in chloroplast DNA (Table 4). Direct point mutations phenotypically result in light-green spots of dark-green plants (Table 4). Appearance of double spots on the leaves of heterozygous light-green soybeans and can be used as the index of somatic crossing-over, whereas light-green spots on yellow soybean plants prove the presence of direct mutations. Spots with no distinct borders connected with the change of chloroplast DNA have been found on the yellow plants.

Nickelous sulphate solution in concentration 1 TLV (TLV_{Ni} = 0.1 mg/l) contains no mutagenic effect (Table 4). Nickelous sulphate solution in concentration 5 TLV provokes somatic crossing-over, aneuploidy, chromosome aberrations. Single yellow and dark-green spots can appear as the result of somatic crossing-over at the loss of one of compound leaf's components and also because of chromosome non-disjunction with the marker gene (aneuploidy) as well as deletion of chromosome's part containing the marker gene (chromosome aberration). There was discovered slight induction of spots two – threefold in comparison with the control (Table 4).

Lead acetate solution in concentration 1 TLV (TLV_{Pb} = 0.1 mg/l) causes chromosome aberrations and aneuploidy, but the solution 5 TLV provokes all the possible cytogenic disorders to appear in soybean plants, they are direct and reverse mutations, somatic crossing-over, chromosome deletions, aneuploidy leading to somatic mosaicism in soybeans (Table 4). The increase of lead concentration in solution from 0.1 to 0.5 mg/l raise the number of direct and reverse mutations in 1.5-time, somatic crossing-over in 1.5 time. Besides causing somatic crossing-over in soybean plants lead acetate leads to mutagenic effect.

Earlier it has been found out that zinc chloride solution in concentration 1 TLV (TLVZn = 1.0 mg/l) induces somatic crossing-over. Zinc at % TLV level provokes somatic crossing-over, chromosome aberrations, aneuploidy. Thus, tested zinc chloride solution showed weak recombinogenic activity at 1 TLV level as well as weak recombinogenic and mutagenic activities at 5 TLV level [12]. On the grounds of the research actions, it was discovered that even at level 1 TLV lead, zinc show weak mutagenic (Pb) and recombinogenic (Zn) activities on test soybeans. By using plant test-systems other researchers have shown mutagenic effect caused by separate heavy metals and mixture of heavy metals in river water and wastewater samples [13].

Table 4. Induction of somatic mosaicism on soybean leaves under the action of solutions of salts of copper, nickel, lead.

Variant	Effect of spots on plant leaves					Possible mechanism
	Y11y11			$Y_{11}Y_{11}$	$y_{11}y_{11}$	
	dark-green spots	yellow spots	double-spots	light- green spots	light-green spots	
$CuSO_4$ 1 TLV	–	–	–	–	–	–
$CuSO_4$ 5 TLV	+	+	+	±	+	SC, CA, AP, PM, chl. DNA
$NiSO_4$ 1 TLV	–	–	–	–	–	–
$NiSO_4$ 5 TLV	+	±	±	–	–	SC, AP, CA
Pb $(CH_3COO)_2$ 1 TLV	–	±	–	–	–	CA, AP
Pb $(CH_3COO)_2$ 5 TLV	+	+	+	–	+	SC, CA, PM, AP, chl. DNA

Note: «I» - more than three-fold induction; «+» - induction is weak (two-three times); «-» - no effect; AP – aneuploidy; SC – somatic crossing-over; PM – point mutations; CA – chromosome aberration; chl. DNA – changes in chloroplast DNA

Chemical analysis of the river water taken from the Gluhovka and the Rakovka (pic. 1) in May and October 2020 revealed that none of the tested water samples contained threshold limit value (TLV) of Zn, Cu, Ni and Pb. hygiene standards (TLV) for water objects of household and drinking as well as cultural and general purposes has been used to get comparative evaluation [2]. pH measuring of the river water samples characterizes it as subacid. The results of chemical analysis of river water samples are shown in Table 5.

Test-system «Somatic mosaicism of the soybean» allowed us to study cytogenic (mutagenic and recombinogenic) activities of tested river water, the results are given in Table 6.

Table 5. Concentration of metals in river water of Rakovka.

TLV, mg/l № station	Zn	Cu	Ni	Pb	pH
	1.0	0.1	0.1	0.1	
May 2020					
1	$\frac{0.015}{0.001}$	$\frac{0.003}{0.0006}$	$\frac{0.001}{0.0002}$	$\frac{0.002}{0.0001}$	6.0
2	$\frac{0.018}{0.001}$	$\frac{0.003}{0.0006}$	$\frac{0.0016}{0.0002}$	$\frac{0.003}{0.0002}$	6.5
3	$\frac{0.032}{0.003}$	$\frac{0.004}{0.0008}$	$\frac{0.004}{0.0006}$	$\frac{0.0045}{0.00025}$	6.3
October 2020					
1	$\frac{0.012}{0.0025}$	$\frac{0.0028}{0.0006}$	$\frac{0.0010}{0.0001}$	$\frac{0.0015}{0.00015}$	5.8
2	$\frac{0.016}{0.001}$	$\frac{0.0025}{0.0010}$	$\frac{0.0015}{0.0002}$	$\frac{0.003}{0.0002}$	6.2
3	$\frac{0.023}{0.006}$	$\frac{0.0030}{0.0010}$	$\frac{0.0025}{0.0003}$	$\frac{0.004}{0.0002}$	6.5

Note: the numerator is the average value, and the denominator is the standard deviation

Table 6. Induction of somatic mosaicism on soybean leaves under the influence of river water.

№ station	Effect of spots on plant leaves					Possible mechanism
	Y11y11			$Y_{11}Y_{11}$	$y_{11}y_{11}$	
	dark- green spots	Yellow spots	Double- spots	Light- green spots	light- green spots	
May 2020						
1	–	±	–	–	–	CA, AP
2	–	+	–	–	–	CA, AP
3	–	+	–	±	–	PM, AP, CA

(*continued*)

Table 6. *(continued)*

№ station	Effect of spots on plant leaves					Possible mechanism
	Y11y11			$Y_{11}Y_{11}$	$y_{11}y_{11}$	
	dark- green spots	Yellow spots	Double-spots	Light- green spots	light- green spots	
October 2020						
1	–	–	–	–	–	–
2	–	+	–	±	–	PM, AP, CA
3	–	±	–	±	–	PM, AP, CA

Note: «+» - more than three-fold induction; «+» - induction is weak (two-three times); «-» - no effect; AP – aneuploidy; PM – point mutations; CA – chromosome aberration

4 Discussion

Combination of different spot types in soybeans of different genotypes allows to establish possible mechanisms of genetic disorders which cause those spots (Table 6). The water sample taken at the st. №1 (October, 2020) does not cause spots in the tested line soybeans, consequently it bears no mutagenic or recombinogenic activities. Station №1 is located on the Gluhovka near the village of the same name, the place which does not get intensive man impact especially in late October.

The water samples from stations №1, 2 and 3 (May, 2020) and №2, №3 (October, 2020) provoke two or threefold increase of yellow spots on light-green plants in comparison with the test soybeans. At the absence of any other types of spots it can be explained by chromosome deletion (fallout, loss) of the chromosome part containing dominant allele Y_{11} of the marker gene (chromosome deletion is one of the types of chromosome aberrations) as well as aneuploidy (monosomy by chromosome the mutant allele y_{11} in this situation).

The water samples taken from the st. №3 in May, 2020 and stations №2 and №3 in October, 2020 induce yellow spots on light-green plants and light-green spots on dark-green plants. In such a combination the appearance of yellow spots on light-green plants and light-green spots on dark-green soybean plants can be resulted in by direct mutations of the gene marker (point mutations) and monosomy (aneuploidy) by chromosome with the gene marker (monosomy by chromosome with gene y_{11} for the light-green plants and monosomy by chromosome with gene Y_{11} for dark-green plants) and also by the loss of chromosome's part containing allele Y_{11} (chromosome aberration).

Over a period of years, we were monitoring cytogenic activity and concentration of heavy metals at st. №3, the results are presented in Tables 7 and 8.

Table 7. Induction of somatic mosaicism on soybean leaves under the influence of water samples from st. №3.

Data	Effect of spots on plant leaves					Possible mechanism
	Y11y11			$Y_{11}Y_{11}$	$y_{11}y_{11}$	
	dark- green spots	yellow spots	double- spots	light- green spots	light- green spots	
May 2016	–	–	–	–	–	–
September 2016	±	–	–	±	–	AP, PM, CA
May 2017	–	±	–	–	±	CA, AP, PM, chl. DNA
September 2017	+	+	+	±	±	SC, CA, AP, PM, chl. DNA
May 2018	–	–	–	–	–	–
October 2018	±	±	–	±	–	PM, AP
May 2019	–	–	–	–	–	–
October 2019	±	–	–	±	+	PM, AP, CA, chl. DNA

Note: «+» - more than three-fold induction; «+» - induction is weak (two-three times); «-» - no effect; AP – aneuploidy; SC – somatic crossing-over; PM – point mutations; CA – chromosome aberration; chl. DNA – changes in chloroplast DNA

Table 8. Concentration of heavy metals in river water from st. №3 (mg/l).

TLV mg/l № station	Zn	Cu	Ni	Cd	Pb
	1.0	0.1	0.1	0.01	0.1
May 2016	$\frac{0.03}{0.003}$	$\frac{0.0032}{0.0006}$	$\frac{0.0045}{0.0008}$	$\frac{0.00070}{0.00005}$	$\frac{0.0038}{0.0010}$
October 2016	$\frac{0.0211}{0.0028}$	$\frac{0.0025}{0.0005}$	$\frac{0.0028}{0.0005}$	$\frac{0.00019}{0.00003}$	$\frac{0.00350}{0.0010}$
May 2017	$\frac{0.0530}{0.0046}$	$\frac{0.0024}{0.0005}$	$\frac{0.0040}{0.0005}$	$\frac{0.00060}{0.00055}$	$\frac{0.0036}{0.0010}$
October 2017	$\frac{0.035}{0.0038}$	$\frac{0.0018}{0.0004}$	$\frac{0.0019}{0.0003}$	$\frac{0.00024}{0.00003}$	$\frac{0.0026}{0.0009}$
May 2018	$\frac{0.016}{0.006}$	$\frac{0.0020}{0.0004}$	$\frac{0.0031}{0.0005}$	$\frac{0.00035}{0.00004}$	$\frac{0.003}{0.001}$

(*continued*)

Table 8. (*continued*)

TLV mg/l № station	Zn	Cu	Ni	Cd	Pb
	1.0	0.1	0.1	0.01	0.1
October 2018	0.0111 / 0.0039	0.001 / 0.0001	0.002 / 0.0003	0.00012 / 0.00003	0.0025 / 0.0008
May 2019	0.036 / 0.0032	0.0038 / 0.0006	0.0030 / 0.0005	0.00055 / 0.00005	0.0046 / 0.0010
October 2019	0.025 / 0.003	0.0022 / 0.0003	0.0017 / 0.0003	0.00025 / 0.00004	0.0036 / 0.0010

Note: the numerator is the average value, and the denominator is the standard deviation

The water samples taken in May, 2016, 2018 and 2019 did not cause any spots on the soybean plants of testing line, consequently there was no mutagenic and recombinogenic activities as long as station №3 does not get strong technogenic impact from the town, while the water samples taken from the stations of the rivers of Rakovka and Komarovka situated in the urban part of Ussuriysk are constantly characterized by mutagenic and recombinogenic activity. Despite the fact that concentration of the researched metals Zn, Cu, Ni, Cd, Pb does not exceed TLV, the water samples taken in September, 2016, May and September, 2017, October, 2018 and October, 2019 showed mutagenic effect. That is why we estimated the genotoxic effect of the polluted water in general taking synergistic effect of heavy metals into account [14].

In that way biotesting showed cytogenetic disorders induced by the tested river water, consequently it contains genotoxicants [15]. The suggested by this research method of soybean (Glycine max (L.)) genetic test-system, which is mutant by the gene of chlorophyll deficiency yellow11, localized in nuclear DNA, makes it possible to establish genotoxic effect of chemical factors polluting the environment and is used in genetic screening and monitoring.

References

1. Zhang, H., Zhang, F., Song, J., et al.: Pollutant source, ecological and human health risks assessment of heavy metals in soils from coal mining areas in Xinjiang. China Envrion. Res. **202,** 111702 (2021). https://doi.org/10.1016/j.envres.2021.111702
2. Bykovskaya, N.V., Shishlova, M.A., Shishlova, T.M.: Use of the plant genetic test system of soybeans for assessing the state of watercourses in urbanized territories. Adv. Modern Natural Sci. **4**, 68–72 (2020). https://doi.org/10.17513/use.37364
3. Hong, C.S., Yang, S., Zhuang, Z.: Application of peptide nucleic acid-based assays toward detection of somatic mosaicism. molecular therapy. Nucleic Acids **5**, e314 (2016). https://doi.org/10.1038/mtna.2016.22
4. Xu, K., Zhang, X., Fan, C., et al.: A callus transformation system for gene functional studies in soybean. J. Integra Agric. **16**, 1913–1922 (2017). https://doi.org/10.1016/S2095-3119(16)61621-0
5. Bykovskaya, N.V., Shishlova, M.A., Shishlova, T.M., et al: Chemical and genetic study of water in urbanized territories for the content of heavy metals. In: AIP Conference Proceedings: Proceedings of the XXVII Conference on High-Energy Processes in Condensed Matter,

dedicated to the 90th anniversary of the birth of RI Soloukhin. Novosibirsk: AIP Publishing LLC, p 030095 (2020). https://doi.org/10.1063/5.0028711

6. Zhang, J., Li, X., Guo, L., et al.: Assessment of heavy metal pollution and water quality characteristics of the reservoir control reaches in the middle Han River. China. Sci. Total Envrion. **799,** 149472 (2021). Doi: https://doi.org/10.1016/j.scitotenv.2021.149472
7. Ma, Y., Wang, Y., Chen, Q.: Assessment of heavy metal pollution and the effect on bacterial community in acidic and neutral soils. Ecolog. Indic. **117**, 106626 (2020). Doi: https://doi.org/10.1016/j.ecolind.2020.106626
8. Prabagar, S., Dharmadasa, R.M., Lintha, A., et al.: Accumulation of heavy metals in grape fruit, leaves, soil and water: a study of influential factors and evaluating ecological risks in Jaffna, Sri Lanka. Environ. Sust. Indic. **12**, 100147 (2021). https://doi.org/10.1016/j.indic.2021.100147
9. Li, C., Quan, Q., Gan, Y., et al.: Effects of heavy metals on microbial communities in sediments and establishment of bioindicators based on microbial taxa and function for environmental monitoring and management. Sci. Total Environ. **749**, 141555 (2020). https://doi.org/10.1016/j.scitotenv.2020.141555
10. Lovinskaya, A.V., Kolumbaeva, S., Suvorova, M.A., et al.: Comprehensive study of the potential toxicity and genotoxicity of water samples from natural sources in the suburban area of Almaty. Ecolog. Gen. **17**(2), 69–81 (2019). https://doi.org/10.17816/ecogen17269-81
11. Wolfram, J., Stehle, S., Bub, S., et al.: Water quality and ecological risks in European surface waters – Monitoring improves while water quality decreases. Environ. Int. **152,** 106479 (2021). https://doi.org/10.1016/j.envint.2021.106479
12. Karazhanova, A.B., Kerimbek, N.M., Lovinskaya, A.V., et al.: Genotoxicity of Esentai and Ulken Almaty rivers water on animal test-systems. Eur. J. Ecol. **1**(66), 18–28 (2021). https://doi.org/10.26577/EJE.2021.v66.i1.02
13. Duarte, L.F., Souza, C.A., Pereira, C.D., et al.: Metal toxicity assessment by sentinel species of mangroves: *In situ* case study integrating chemical and biomarkers analyses. Ecotoxicol Environ. Saf. **145**, 367–376 (2017). https://doi.org/10.1016/j.ecoenv.2017.07.051
14. Ignatenko, A.V.: Sample preparation and biotesting of the toxicity of sludge sewage sludge. Chem. Saf. **2**(2), 251–271 (2018). https://doi.org/10.25514/CHS.2018.2.14120
15. Sycheva, L.P., Zhurkov, V.S., Rakhmanin, Yu.A.: Actual problems of genetic toxicology. Genet **49**(3), 293 (2013). https://doi.org/10.7868/S0016675813030168

Improving the Efficiency of Combine Harvesters

Irina Lontseva(✉) and Vyacheslav Sennikov

Far Eastern State Agrarian University, Polytechnic House 86, Blagoveshchensk, Russia

Abstract. The aim of the work is to study ways to increase the productivity of combine harvesters. The research was carried out in production conditions in the fields of the Amur Region. As a result, it was found that the main factors that reduce productivity are agronomic, weather and operational. To increase productivity, it is necessary to make adjustments and adjustments of the harvester to the specified operating modes, timely carry out maintenance and repairs, use original and high-quality fuel and lubricants and consumables, and organize the well-coordinated work of the harvesting and transport complex as a whole. As a result of research, it was found that the optimal settings and adjustments of combine harvesters can save up to 10% of fuel. The use of an accelerator drum in the classical design of the threshing-separating device allows to increase the threshing area, thereby improving the quality of the grain in the hopper. Combine harvester is one of the most expensive and sophisticated machines in agricultural production and the competent design of the harvesting and transport complex allows you to make it as efficient as possible.

Keywords: Combine harvester · Threshing drum performance increase · Grain transportation · Grain loss · Fuel consumption

1 Introduction

Combine harvester is one of the most complex and expensive machines, and the operating time is about 2–3 months a year. Therefore, the main task of using these machines is to maximize efficiency during the entire harvesting period.

Research aimed at improving the efficiency of combine harvesters is constantly being carried out by all scientists in the world. It is known that the higher the speed, the larger the harvested area, but the loss of grain also increases. A balance must be struck between speed and losses [1, 2]. Also, the amount of loss depends on the settings and adjustments [3–5]. All modern agricultural machinery is equipped and operated using information and communication technologies. It has been established that thanks to these technologies it is possible to optimize the transport and harvesting complex and the method of movement of the unit [6–8].

Another urgent task and problem is fuel economy and control of exhaust gas emissions into the atmosphere [9]. Scientists Spokas & Steponavičius, 2009, Spokas & Steponavičius, 2010 [10, 11] carried out experimental studies on wheat harvesting at different operating speeds, as a result of which it was found that fuel consumption per unit of time increased with increasing flow mass and decreased with increasing altitude.

A. Muratov and S. Ignateva (Eds.): AFE 2021, LNNS 353, pp. 415–423, 2022.
https://doi.org/10.1007/978-3-030-91402-8_47

In this study, a theoretical and experimental substantiation of methods for increasing the efficiency of combine harvesters for harvesting grain crops and soybeans as applied to the Russian Far East has been carried out.

2 Material and Methods

Modernization of the design of combine harvesters is aimed at improving: the quality of threshing, the level of environmental safety, working hours, the comfort of the operator; reduction of: fuel consumption, level of emissions of exhaust gases into the atmosphere, the amount of grain losses, breakdowns.

During harvesting, machine operators and engineers at agricultural enterprises strive to improve the efficiency of combine harvesters by:

- reducing fuel consumption; selection of optimal design parameters of the threshing-separating device;
- reducing the costs associated with maintenance and repair;
- reduction of losses during harvesting;

In addition, as follows from all that has been said - increased productivity.

Experimental studies aimed at increasing the productivity of combine harvesters during harvesting were carried out in agricultural enterprises of the Amur region (Russia). At the same time, a comparative assessment of equipment manufactured in Russia and in other countries was carried out.

Theoretical studies are aimed at describing the course of the process and the interaction of parameters.

2.1 Research of Fuel Consumption

Cutting Plants. When the segment-finger cutter is operating, the movement of the segment blade captures the stems, brings them to the counter plate and cuts them off. Since in the process of feeding the stalks are bent, due to which the stubble height is greater than the distance from the cutter bar to the soil surface, in order to avoid ear loss, especially when cutting undersized and lodged grain, the cutter bar is adjusted below the recommended ones. Since the most intense precipitation in the form of rains and thunderstorms occurs in july-august, which coincides with the onset of the wax ripeness phase of plants, the lower part of the plant stem is covered with a fine fraction of the soil. This fine fraction acts as an abrasive for the cutting device, as a result of which the wear of the elements of the cutting device increases and the average value of the shear resistance force R_{cp}.

$$T = R_{\text{cp}} = P_j + F \tag{1}$$

where T – force on the knife;

R_{cp} – average shear strength;

F – friction force of the knife on the finger bar.

From expression (1) it can be seen that an increase in the shear resistance force leads to an increase in the forces acting on the knife, and to overcome these forces, additional power is required.

Power N required to overcome the forces of resistance of the knife to movement,

$$N = T \cdot \vartheta \tag{2}$$

where ϑ – combine speed.

Fuel Quality. The operation of internal combustion engines depends on the quality of the fuel used. It affects not only the composition of exhaust gases, the amount of fuel consumed, but also the service life of the engine [12, 13]. If the engine is working properly, then the performance is high. The parameters characterizing the quality of the fuel are as follows:

- cetane number (CN). The most important fuel index, the rate of combustion of diesel fuel (DF) in the engine depends on it. A high-quality diesel fuel must have a CC index of at least 45–55. With lower parameters, diesel fuel will ignite much more slowly, which will lead to rapid engine wear. If the figures are too high, the fuel consumption and exhaust will be significant.
- fractional composition. This indicator indicates the evaporation temperature. With an overestimated index, oil dilution and sludge formation occurs faster.
- viscosity. Diesel fuel should not be too viscous or too liquid. For summer time, the optimal indicator is 3–6 cSt, for winter time 1.8–5 cSt.
- sulfur content. The content of this element in the fuel provokes the oxidation of the oil in the engine, which entails a strong pollution of the atmosphere with exhaust gases. The lower its amount in the fuel, the longer the service life of the catalytic converter. But even without this component, the fuel is not considered to be of high quality, sulfur is designed to lubricate engine components, reducing their wear. The optimum content is no more than 1.5%.
- low temperature qualities. Diesel fuel is divided into three main brands: summer - used at temperatures above 0 degrees; winter - not lower than -20 degrees; arctic - up to -50°C.
- degree of purity. The poor quality of diesel fuel is the presence of water. Once in the engine, it can damage the fuel pump. Water can appear in the fuel during transportation and in case of improper storage. The possibility of mechanical impurities entering the diesel fuel is not excluded.
- flash point. This indicator indicates the safety of using diesel fuel in engines.

2.2 Research Methods for the Design of a Threshing and Separating Device

Modern combine harvesters have 3 design options for threshing and separating devices (MSU):

Drum type - with a threshing drum across the movement of the combine and a keyboard straw walker. Such models are the most common and demanded by manufacturers of crop products.

Rotary type - with a longitudinal rotor along the length of the inner part of the combine. Such models are the least common.

Hybrid - combining a threshing drum and a rotor (rotors).

One of the main parameters of the drum and rotary type MSU is the drum diameter, which varies for different models from 450 to 800 mm.

In the classical scheme of a combine harvester in the absence of an accelerator drum, the mass supplied to the threshing space from the feeder chamber moves at different speeds. So at the entrance the speed of the grain mass is 3… 6 m/s, and at the exit it reaches 20 m/s (Fig. 1).

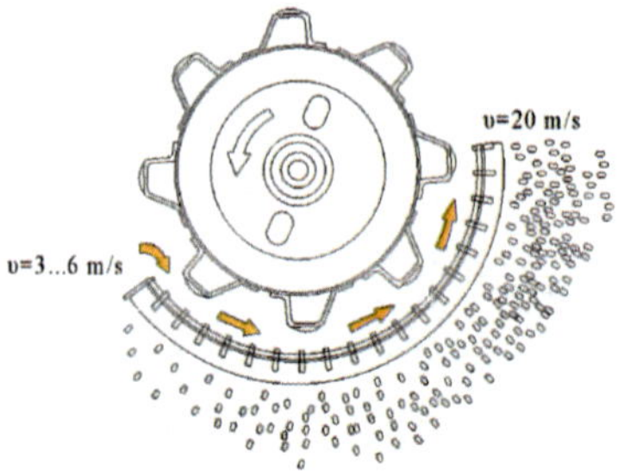

Fig. 1. The scheme of movement of the grain mass in the threshing drum.

The figure shows a diagram of grain threshing with a threshing drum. When the grain mass flow reaches a speed of 15 m/s, effective grain separation begins, that is, in the second half of the deck. For this purpose, an accelerator drum is installed in front of the threshing drum. It is designed to increase the speed υ of the grain mass moving towards the threshing drum and to distribute it evenly. In this case, the acceleration of the grain mass will be higher than in a combine harvester without an accelerator. Thus, without increasing the speed, you can thresh up to 90% of the grain only due to the threshing drum.

$$F = \frac{m \cdot v^2}{R} \quad (3)$$

where F – the force with which the threshing drum acts on the grain mass,

- m – mass supplied to the threshing device,
- υ – the speed of the grain mass in the area between the drum and the deck,
- R – the distance from the center of the threshing drum to the threshing mass, depending on the diameter.

2.3 Research Methods for the Design of the Grinder-Spreader

Modern combine harvesters are equipped with devices for chopping and evenly spreading straw over the surface of the field. The straw mass is a mulching layer and is an organic fertilizer for the future harvest. In the Amur Region, 85–90% of straw is mixed with soil for fertilization or used in animal husbandry (with subsequent introduction of straw-manure mass into the fields).

The design, shape, method of fastening, material of knives and the technology of their manufacture largely determine the quality of grinding and distribution of material, energy consumption for performing a technological operation, the possibility of reducing technological downtime during operation of the combine, and much more [14].

2.4 Maintenance

Manufacturing plants strive to create, and grain producers to purchase, such units that carry out the technological process as efficiently as possible.

The efficiency of use of combine harvesters R_e depends on the utilization factor K_{tu} and the trouble-free operation between services $P(t_g)$:

$$R_e = K_{tu} \cdot P(t_g) = \left[1 + \frac{t_b}{t_m} + \frac{t_r}{t_w}\right]^{-1} exp\left(\frac{t_m}{t_w}\right) \to max \quad (4)$$

where t_b - MTBF;

t_m - frequency of maintenance;
t_r - total downtime during troubleshooting;
t_w - working time of the combine.
In the same time

$$R_e = exp[-(1 + t_k \cdot m_k + t_o)] - 1 \quad (5)$$

where t_k - time spent on ensuring 1 h of trouble-free operation during maintenance;

t_o – time for 1 h of failure-free operation during maintenance;
m_k – number of TO.

2.5 Logistics of Grain Movement from the Field

Well-coordinated work of the harvesting and transport complex is possible if the condition

$$N_A \cdot W_A = N_K \cdot W_K \quad (6)$$

where N_A, N_K – number of cars and combine harvesters, pcs.

W_A, W_K – productivity of vehicles and grain harvesters, ha.

The maximum cleaning efficiency will be when N_A and $N_K \to min$, W_A and $W_K \to max$.

In fact, the condition of equality of the right and left sides of expression (6) is not met, because of this, the technique is idle. It is especially important to avoid downtime of combine harvesters while waiting for unloading. This is possible with a sufficient number of vehicles for transporting grain.

$$N_A = \frac{M \cdot T}{V \cdot t_K}, \quad (7)$$

where M – bunker capacity of a combine harvester, t

T – time for unloading the vehicle, h
V – carrying capacity of the car, t
t_K – time of filling the bunker of one combine, h.

$$t_K = L \cdot \vartheta = \frac{10^4 \cdot M \cdot \vartheta}{Q \cdot B} \tag{8}$$

where Q – yield, t/ha,

B – width of the header of the combine harvester, m
ϑ – speed of the combine, km/h

Carrying capacity of the vehicle accommodates grain from several combines. The number of combines unloading into one vehicle is determined from the expression:

$$N_1 = \frac{V}{M}, \tag{9}$$

Most often, KAMAZ trucks of various models are used to transport grain. The average carrying capacity of KAMAZ grain carriers with a trailer is 25 000 kg, and of grain harvesters – 7 000 kg. This means that 3 combine harvesters can be simultaneously unloaded into one vehicle. Vehicle performance must be considered

$$W_A = \frac{V \cdot \gamma_0}{t_r} \cdot K_g^A \tag{10}$$

where γ_0 – grain density, kg/m^3,

t_r – duration of one car trip, h
K_g^A – technical readiness ratio

Combine harvester performance

$$W_k = 0.1 \cdot B \cdot \vartheta \cdot Q \cdot \tau \cdot K_g^k \tag{11}$$

where τ – main work time, h
K_g^k – combine harvester technical readiness factor.

3 Results

3.1 Setting up the Reaping Units

Fuel consumption, the degree of wear of the working parts and productivity depend on the setting of the header. The height of grain crops in the Far East is less than 0.8 m, which means, according to agrotechnical requirements, the cutting height should be 10–12 cm.

In modern combine harvesters, the cutting height is set from the on-board computer, but it must be borne in mind that when using various adapters, it is necessary to calibrate the header [15].

Based on all of the above, we can conclude that the optimal setting of the reaping part will reduce the cost of fuels and lubricants and elements of the cutting device.

3.2 Use of Fuels and Lubricants

The use of fluids corresponding to the quality and technical characteristics is a necessary requirement for the operation of internal combustion engines.

Refueling in crop production organizations occurs once a day when working one shift and 2 times a day when working in two shifts. If the work is carried out at full engine load, the fuel consumption increases. In this mode, there is not enough fuel for the entire shift. To avoid downtime while waiting for refueling, the tractor drivers take with them a supply of fuel in cans. In this case, the trash impurities in the canisters enter the fuel tank.

Failure of an internal combustion engine is not a warranty case, which leads to additional costs for maintaining equipment in working order.

3.3 Selection of Optimal Design Parameters of the Threshing-Separating Device

The use of a threshing drum of a smaller diameter allows a greater acceleration of the threshing mass to be obtained. These technical solutions form the basis for the design of the Claas threshing devices.

With an increase in the flow of the mass flow, the value of the force with which it is necessary to influence the mass should remain the same, and not decrease; for this purpose, it is necessary to use a threshing drum of a larger diameter, such as in Rostselmash and Gomselmash combines.

3.4 Chopper Operation

Modern knives made of alloys of various metals are self-sharpening and designed for long-term operation. But they have a reserve. Taking into account the fact that the operating time of one combine per season can reach 500 h, this will lead to wear on the cutting edge of the knife (Fig. 2).

When the combine harvester has run 1 500 h, the fuel consumption increases by 6%, only due to the work of the chopping device. In addition, with 3 000 operating hours, fuel consumption increases up to 19%. When switching to corn harvesting, the cutting system must be reconfigured. For this, the number of knives is reduced and the rotor speed is reduced. If this is not done, when the cob is stuck, the knives will bend, break the fasteners and even damage the skin. This will cause major damage to the combine body.

In addition, during the daily maintenance, it is necessary to check the integrity of the chopping drum knives. Breaking the knife can lead to imbalance and damage to the mechanism.

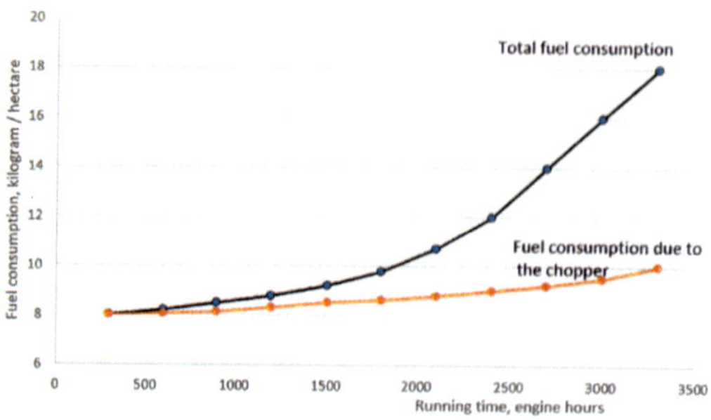

Fig. 2. Fuel consumption versus operating time.

3.5 Maintenance

The time spent on ensuring 1 h of trouble-free operation during maintenance of combine harvesters largely depends on the quality of the daily maintenance.

With the group method of harvesting, combine harvesters must be made up in teams of 4 - 6 harvesters, taking into account their technical characteristics. That is, one brigade should have all combines with similar technical characteristics. It should be borne in mind that if a large number of high-performance combine's work in one field, then during the working day they harvest a large area, therefore, the travel time of the vehicles changes. In addition, this leads either to downtime of equipment (harvesting or automobile). In this case, additional adjustments must be made, involving or releasing the vehicle.

4 Discussion

The article presents the factors of increasing productivity, which can be influenced by the engineering service and the machine operator, the basis of which is the technical equipment of the enterprise with combine harvesters and their competent operation.

5 Conclusion

The main ways to increase the productivity of combine harvesters are:

- timely and high-quality performance of maintenance and repair;
- setting and adjustment depending on the culture and harvesting conditions;
- recruiting of harvesting and transport teams;
- use of high-quality consumables and fuels and lubricants;
- control of crop losses for the working bodies of the combine;
- movement with optimal speed.

References

1. Mokhtor, S.A., Pebrian, D.E., Johari, N.A.A.: Actual field speed of rice combine harvester and its influence on grain loss in Malaysian paddy field. J. Saudi Soc. Agric. Sci. **19**, 422–425 (2020). https://doi.org/10.1016/j.jssas.2020.07.002
2. Sarkari, M.: Field evaluation of grain loss monitoring on combine JD 955. Adv. Environ. Biol. **4**(2), 162–167 (2010)
3. Karmulla, R., Seyyed, C., Karparvarfard, H., et al.: Predicting header wheat loss in a combine harvester, a new approach. J. Saudi Soc. Agric. Sci. **19**(2), 179–184 (2020). https://doi.org/10.1016/j.jssas.2018.09.002
4. Ni, Y., Jin, Ch. et al.: Computational model and adjustment system of header height of soybean harvesters based on soil-machine system. Comput. Electr. Agric. **183,** 105907 (2021). https://doi.org/10.1016/j.compag.2020.105907
5. Nian, C., Yanwei, H., Yubin, M.: Research on fuzzy control of harvester header height. Mechatronics **4**, 17–22 (2018)
6. Bochtisa, D., Claus, D., Sørensena, G.C., Busatob, P.: Advances in agricultural machinery management: a review. Biosyst. Eng. **126**, 69–81 (2014). https://doi.org/10.1016/j.biosystemseng.2014.07.012
7. Orfanoua, A., Busatob, P., Bochtisa, D.D., et al.: Scheduling for machinery fleets in biomass multiple-field operations. Comput. Electr. Agricul. **94**, 12–19 (2013). https://doi.org/10.1016/j.compag.2013.03.002
8. Nourbakhsh, S.M., Bai, Y., Maia, G.D.N., et al.: Grain supply chain network design and logistics planning for reducing post-harvest loss. Biosyst. Eng. **151**, 105–115 (2016)
9. Seki, E.: Evaluation of exhaust gas emissions for agricultural tractors based on actual operations. Nogyo kikai gakkaishi **68**(6), 130–137 (2006). https://doi.org/10.1016/j.biosystemseng.2016.08.011
10. Spokas, L., Steponavičius, D.: Optimization of fuel consumption during the harvest of wheat. TEKA Kom Mot Energ Roln OL PAN **9**, 298–303 (2009)
11. Spokas, L., Steponavičius, D.: Impact of wheat stubble height on combine technological parameters. J. Food Agric. Environ. **8**(2), 464–468 (2010)
12. Nelsona, P.F., Tibbettb, S.R., Day, S.J.: Effects of vehicle type and fuel quality on real world toxic emissions from diesel vehicles. Atmosph. Environ. **42**(21), 5291–5303 (2008). https://doi.org/10.1016/j.atmosenv.2008.02.049
13. Ulberth-Buchgraber, M., Charoud-Got, J., Held, A.: Certified reference materials for effective automotive diesel fuel testing. Fuel **286**(1), 119367 (2021). https://doi.org/10.1016/j.fuel.2020.119367
14. Rodimtsev, S.A., Kuznetsov, Y.A., Kolomeichenko, A.V., et al.: Theoretical grounding of rational design of working elements of straw disperser of grain combine harvester: Agricultural and mechanical engineering. In: International symposium, Bucharest (2017)
15. Srivastava, A.K., Goering, C.E., Rohrbach, R.P., Buckmaster, D.R.: Engineering principles of agricultural machines. American Society of Agricultural and Biological Engineers, USA (2006)

Management of Demographic Processes in the Countryside of the Far East of Russia

Yurij Shmidt[1], Irina Zhupley[2](✉), Yuliya Schmidt[3], and Lyubov Soldatova[4]

[1] Far Eastern Federal University, Russky Island, Ajax, Building G 22, Vladivostok 690902, Russia
[2] Primorskaya State Academy of Agriculture, 44 Blucher Avenue, Ussuriisk 692510, Russia
[3] Tver State Agricultural Academy, 7 Marshal Vasilevsky Street, Tver, Sakharovo 170904, Russia
[4] Kostroma State Agricultural Academy, 34 Educational Town, Karavaevo Village, Kostroma District, Kostroma 156530, Russia

Abstract. The study identified trends in the demographic development of the countryside of the Far East of Russia; The dynamics of the migration loss of the rural population is considered in the context of the subjects of the Federation as part of the Macro-region; trend models of fertility and mortality coefficients were constructed, the adequacy of models was proved; Forecast estimates of the main demographic indicators of the macro-region (fertility coefficients, mortality, natural growth, the economy of population reproduction) are found to the short term. The main demographic problems of the Far East of Russia are allocated. It is concluded that the low effectiveness of state regulation measures in the macro-region aimed at improving the demographic situation. Attention is drawn to the need to adjust the priorities of the demographic and socio-economic policies of the territory, taking into account the results of prospective planning and forecasting using the modeling method. It is indicated on the need for a significant change in the organizational structure and management of demographic and migration processes in the Far Eastern Federal District.

Keywords: Far Eastern Federal District · Rural areas · Demographic management · Demographic problems · Fertility coefficient · Mortality coefficient · Natural growth coefficient · Migration · Forecast

1 Introduction

In modern Russia, the decision of demographic problems has become one of the important vectors of the prospective development of the state. If until recently, long-term demographic trends rather favored the economic development of the Russian Federation (RF), now Russia comes into a long period unfavorable demographic changes for its economy.

Demographic problems, their negative impact on the socio-economic development of territories, as well as issues of effective management of demographic processes are relevant for many States.

A. Muratov and S. Ignateva (Eds.): AFE 2021, LNNS 353, pp. 424–433, 2022.
https://doi.org/10.1007/978-3-030-91402-8_48

The problems of socio-demographic development of rural areas are becoming particularly important in today's environment. This is due to the fact that the excess of the urban population over the rural population forms a production structure that negatively affects the reproduction processes in the agricultural sector, and, combined with the negative balance of demographic indicators, poses a real threat to food security [1, 2].

C. Simon, et al. [3] write that many industrialized countries face fertility rate below the level of reproduction in combination with a decrease in mortality, which is negatively affected by the trajectory of economic growth. The fertility models, mortality, their influence on the evolution of the population as a whole (on the example of France) are presented in the study of H. Boumezoued, et al. [4]. In the work of D. Croix et al. [5] the existence of a causal relationship between population density and fertility in developing countries has been proved. The influence of socio-economic, demographic and political factors at the birth rate in states with different income levels (in China and Bangladesh) considered Q. Wang and X. Sun [6]. On the need for systemic study of migration processes in analyzing reproduction and economic growth, many authors indicate in their works (S. Mishchuk [7, 8], T. Nefedova et al. [9], E. Parrado [10], S. Sardadvar and E. Vakulenko [11]).

But while there are many different publications on the demographic theme, issues of managing demographic processes in rural areas, taking into account regional specifics and the formation of accessible and effective tools, remain open.

Thus, on the one hand, the problems of demographic development are relevant both for Russia as a whole and for its individual regions. On the other hand, there is a shortage of practice-oriented techniques that can effectively manage the demographic development of the region and shape the content and priorities of demographic policy at the meso-level, as well as predict demographic indicators and determine the potential direction of demographic flows depending on certain management decisions.

Therefore, the research topic is relevant and has practical importance for regional governments.

2 Materials and Methods

The purpose of the study is to analyze the modern demographic situation and the development of demographic processes in the countryside of the Far Eastern Federal District (FEFD) of Russia and to carry out short-term forecasting of the main demographic indicators of the territory.

The object of the study: demographic processes in rural areas of the region. The subject of the study is the management of the demographic processes of the macro-region and management tools that allow for forward planning and forecasting of demographic indicators.

The study looked at the following issues:

- What is the modern level of demographic development of rural territories of the FEFD and the subjects of the Federation in its composition?
- What kind of models of demographic processes of the territory and what are the forecast estimates of FEFD demographic indicators for the short-term period?

– What measures can improve the demographic situation in the Far East of the Russian Federation?

The theoretical basis of this work is the study of foreign and Russian scientists on the analysis of demographic processes and the application of modeling methods in demographic studies [12–16].

The methodological basis of the study is a systematic and comprehensive analysis of the problem under consideration using general economic, economic and statistical methods and special demographic parameters (mortality coefficients, fertility, reproduction efficiency; turnover and population growth).

Information framework research – Statistical materials of the Federal State Statistics Service of the Russian Federation, regulatory documents regulating demographic policies at the federal and regional levels.

3 Results

We will fulfill the economic - mathematical modeling of socio-demographic indicators and highlight the main factors that have a negative impact on the demographic of the largest area and the very least populated Macro-region of Russia – the Far Eastern Federal District. FEFD occupies 40.6% of the country's area; the population of the Macro-region is 5.6% of the population of the Russian Federation; population density in the Far East is 1.17 people/km^2).

At first glance at the Far East of Russia, economic indicators are quite stable: an increase in industrial production at the beginning of 2019 in Macro-region amounted to 23% (against 8.3% in Russia as a whole); the average nominal wage is equal to 51.7 thousand rubles per month (against 43.7 thousand rubles on average by country); the average value-added per busy in the economy amounted to 1.172 million rubles (which is 22% higher than the all-Russian level).

In the Russian Far East (in addition to the All-Russian) adopted a whole range of program documents aimed at improving the socio-economic and demographic indicators of Macro Region: the Federal Target Program "Economic and Social Development of the Far East and the Baikal Region for the period up to 2025", Regional Programs "Far Eastern hectare, "Far Eastern Mortgage"; The concept of the demographic policy of the Far East for the period up to 2025, the concept of state migration policy of the Russian Federation for the period up to 2025, the concept of state policy regarding the young family, etc. However, despite all the measures taken, the FEFD is a macro-region in which the maximum is observed among all Russian macro-regions outflow of the population.

Indeed, the population of Macro-region for 2010–2019 decreased by 194 thousand or 2.32% (which exceeds the number of residents of Ussuriysk) and amounted to 8169 thousand people in 2019. This situation is explained by migration (including young people) in large cities of the European part of Russia. The population in the country as a whole for the period under review increased by 2.72% (Table 1).

Table 1. The population of the FEFD (thousands of people, in brackets -% of the total population).

	All population				Rural population			
	2010	2013	2016	2019	2010	2013	2016	2019
RF	142865	143667	146804	146749	37444 (26.2)	37118 (25.8)	37772 (25.7)	37186 (25.3)
FEFD	8363	8291	8246	8169	2363 (28.3)	2299 (27.7)	2253 (27.3)	2211 (27.1)
Republic of Buryatia	972	974	984	986	403.3 (41.5)	401.3 (41.2)	404.5 (41.1)	401.9 (40.8)
Republic of Sakha (Yakutia)	958	955	963	972	343.6 (35.9)	332.1 (34.8)	332.3 (34.5)	329.3 (33.9)
Trans-Baikal Territory	1106	1090	1079	1060	376.2 (34.0)	358.6 (32.9)	344.9 (32.0)	337.2 (31.8)
Kamchatka Territory	322	320	315	313	72.7 (22.6)	72.3 (22.6)	69.1 (21.9)	67.4 (21.5)
Primorye Territory	1953	1938	1923	1896	466.4 (23.9)	451.3 (23.3)	439.9 (22.9)	429.1 (22.6)
Khabarovsk Territory	1343	1340	1333	1316	244.2 (18.2)	245.8 (18.3)	239.1 (17.9)	235.7 (17.9)
Amur Region	829	811	802	790	274.7 (33.1)	266.9 (32.9)	262.3 (32.7)	254.9 (32.3)
Magadan Region	156	150	146	140	7.1 (4.6)	7.0 (4.7)	6.3 (4.3)	5.5 (3.9)
Sakhalin Region	497	491	487	488	100.3 (20.2)	92.5 (18.8)	88.6 (18.2)	86.2 (17.7)
Jewish Autonomous Region	176	171	164	158	56.8 (32.3)	54.7 (32.0)	51.5 (31.4)	49.6 (31.4)
Chukotka Autonomous Area	51	51	50	50	17.7 (34.7)	16.4 (32.2)	14.9 (29.8)	14.3 (28.6)

For the subjects of the Federation as part of the Russian Far East, the population reduction is noted for most territories (except Buryatia and Yakutia). The most decrease in the inhabitants in 2019 compared to 2010 was recorded in the Magadan region (by 10.26%) and in the Jewish Autonomous Region (by 10.23%).

The population of the rural population in 2010–2019 has decreased steadily and in Russia as a whole (by 0.69%), and at its Far East (by 6.43%), and in all subjects of the Federation as part of the FEFD. Five - "anti-leaders" in the loss of the rural population in the Far East as follows: Magadan region (the decrease was 22.54%),

Chukotka Autonomous Area (19.21%), Sakhalin (14.06%), Jewish Autonomous Region (12.68%), Trans-Baikal Territory (10.37%).

The share of the rural population in the Russian Federation and in the Far Eastern Federal District in 2019 amounted to 25.3% and 27.1%, respectively. The greatest share of the rural population in Buryatia (40.8%), Yakutia (33.9%), the Amur region (32.3%), the Trans-Baikal Territory (31.8%) and in the Jewish Autonomous Region (31.4%). At the same time, structural shifts in the amount of rural population at all administrative territories of FEFD are negative.

The solution to the problem of population loss at rural areas of the Far East, regional authorities see in attracting migrants and immigrants to the Far Eastern village [17, 18]. But, in reality, the migration flows of the rural population are characterized by a negative balance over the past ten years.

Figure 1 shows the histogram of the migration growth of the rural population in the context of the subjects of the Federation as part of the FEFD for 2010–2019. The figure shows that for all territories of the Russian Far East, the migration increase is located in a negative plane, which indicates a migration decline in the rural population. That is, the hopes for the replenishment of labor resources in the agricultural sector and the settlement of rural areas by international migrants (mostly from the former Soviet republics) are not entirely justified. Still, large cities and their suburbs remain as for international immigrants and interregional migrants.

To identify the nature of further demographic changes in the Far East of the Russian Federation, the authors built models of demographic indicators depending on the "time" factor and the corresponding forecast values were found.

The equation of the coefficient of fertility (K_F) in the Far East of Russia, depending on the time (t), has the form of the four-degree polynomial:

$$K_F = 0.0037 \cdot t^4 - 0.0779 \cdot t^3 + 0.4529 \cdot t^2 - 0.6488 \cdot t + 13.425. \quad (1)$$

The mortality coefficient trend equation (K_M) also takes the fourth degree polynomial:

$$K_M = 0.0011 \cdot t^4 - 0.0256 \cdot t^3 + 0.2235 \cdot t^2 - 1.021 \cdot t + 14.692. \quad (2)$$

Models (1) and (2) correctly describe the change in demographic parameters over time. This is confirmed by the determination coefficients (R^2) and the average error of approximation ($\overline{A}$): a) $R^2 = 0,97$ и $\overline{A} = 1,26\%$ – for model (1); б) $R^2 = 0,96$ и $\overline{A} = 0,89\%$ – for model (2). Therefore, these models can be used to predict the corresponding FEFD demographic indicators to the perspective.

Using the fertility and mortality rates of the population, the coefficient of natural growth and the coefficient of the reproduction of the population are obtained.

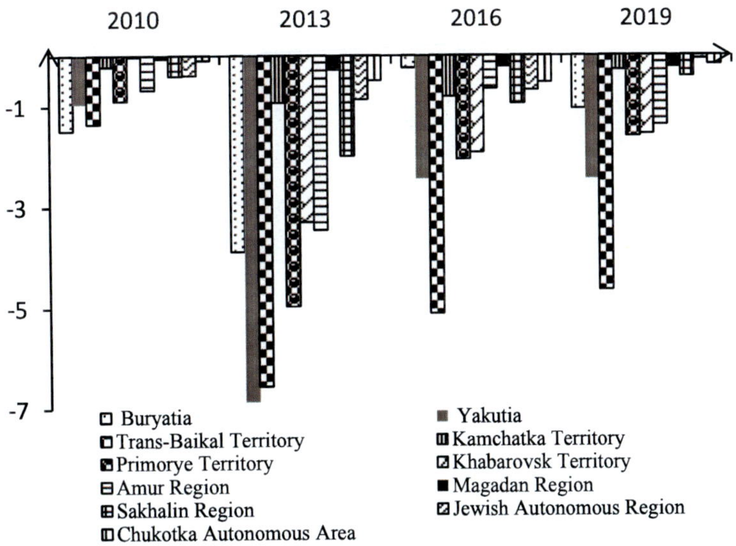

Fig. 1. Migration growth of the rural population in the FEFD.

The coefficient of natural population growth (K_{gr}^{nat}) located by formula:

$$K_{gr}^{nat} = K_F - K_M \tag{3}$$

The coefficient of the economy of population reproduction (K_{prof}^{rep}) determined by the formula:

$$K_{prof}^{rep} = \frac{K_F - K_M}{K_F + K_M} \tag{4}$$

Table 2 shows the real (for 2010–2019) and forecasts (for 2021–2022) the values of fertility coefficients, mortality, natural growth and economy of population reproduction for FEFD.

The negative population trend is based on a steady decline in fertility rate in Macro-region, starting from 2014 and to the present. The prediction with a high degree of reliability also shows a further fall in the fertility coefficient (until 11.576 in 2022).

The magnitude of the mortality coefficient for FEFD every year is slightly reduced: from 13.8 in 2010 to 12.2 in 2019. But, according to the forecast, the mortality rate will increase and will be 13.197 in 2022.

The coefficient of the natural growth of the population, which in 2010 was equal to minus 0.6, by 2012 increased to 0.9 and, after 2018, it turns back to the negative plane and will, according to the forecast, in 2022 minus 0.228.

The coefficient of reproduction efficiency reflecting the share of natural growth in the general turnover of the population since 2018 is in a negative field of values, the forecast indicators of this coefficient are also negative (Table 2).

Table 2. Dynamics of FEFD demographic indicators (in brackets - forecast error).

	Coefficient			
	Fertility	Mortality	Natural growth	Economy of population reproduction
2010	13.200 (0.34)	13.800 (0.51)	−0.600	−0.022
2011	13.200 (1.33)	13.500 (1.06)	−0.300	−0.011
2012	14.000 (1.78)	13.100 (0.47)	0.900	0.033
2013	13.900 (0.99)	12.600 (1.08)	1.300	0.049
2014	14.000 (0.56)	12.600 (0.49)	1.400	0.053
2015	13.900 (0.68)	12.600 (0.73)	1.300	0.049
2016	13.300 (0.45)	12.500 (1.15)	0.800	0.031
2017	12.100 (3.23)	12.100 (1.04)	0.000	0.000
2018	11.900 (1.20)	12.000 (1.34)	−0.100	−0.004
2019	11.100 (2.05)	12.200 (0.26)	−1.100	−0.047
2021[a]	11.576	12.536	−0.960	−0.040
2022[b]	12.969	13.197	−0.228	−0.009

a – forecast, found by model (1)
b – forecast, found by model (2)

Thus, while maintaining the existing demographic conditions and the current management system of demographic processes in Macro-region, it should be assumed to further deteriorate the demographic situation in the Far East of Russia as a whole, and, in rural areas, to a greater extent.

4 Discussion

Effective management of demographic processes on meso-level should provide conditions (legislative - legal, financial - economic, social, ecological, etc.) for the formation of rational demographic proportions in the population structure, achieve a positive population migration balance and sustainable population growth (especially in working age), and also transforming negative structural shifts in population into positive shifts.

When developing management decisions, regional and subregional differences should be taken into account due to the unequal level of development of territories. At the same time, it is impossible to ignore the presence of "timing" shifts characteristic of demographic statistics. You can take into account the effect of "timing" shifts using the method of economic and mathematical modeling and use in operational monitoring and planning of forecast estimates of demographic indicators.

The modern demographic situation in the Far East of Russia can be called a demographic crisis. The demographic crisis under FEFD is characterized by the following features: low birth rate; high mortality; Natural population decline; rapid aging of the population; shorter than in Russia as a whole, life expectancy; Sustained outflow of the population in Western regions of the country and abroad. What, in principle, naturally, since the spatial mobility of the population is one of the main ways to adapt to the conditions of territorial socio-economic contrasts. And, the Far East of the Russian Federation, unfortunately, becomes becoming less attractive territory for living. Analysis of actual and forecast demographic indicators shows that the measures taken by the authorities to improve the demographic "picture" of the Russian Far East today work inefficiently. In the formation of a complex of measures to optimize the population of the macro-region, the distribution of budget funds of state support for socio-economic programs should be processed from the ratio of "support/predicted result" ratio.

5 Conclusion

The study allows the following conclusions. First, it is necessary to adjust the priorities of the demographic and socio-economic policy of FEFD. Currently, the efforts of power structures are mainly focused on the development of integrated development projects that have to improve the quality of life of Russian citizens living in the Far East. Without a doubt, this is an important component of the demographic attractiveness of the region for the population. But no less important component is an innovative development of the economy, the creation of new industrial and agricultural enterprises. In fact, the real economy of the modern Far East is characterized by a reduction in the real sector of the economy on the one hand, and, on the other, - the expansion of financial and banking and trading and intermediary sectors. The region loses its able-bodied population due to the low supply in the labor market, especially for highly qualified personnel. The availability of education for the younger generation is reduced, due to a decrease in competition among educational institutions (including at the cost and quality of educational services) due to reducing the number of educational institutions of higher education and their branches. Therefore, you need a cardinal change of organizational structure and management of demographic and migration processes, it is necessary to restore the integrity of the management object, rebuild the structure and procedure for the interaction between its elements, deploy training in this area up to the municipal level.

Secondly, measures should be taken at the federal and regional levels to reduce mortality. The main reserve of mortality in Russia is the restoration of "lost capital of health" of the people, as well as the elimination of the consequences of negative shifts in the mortality structure and the reduction in the volume of "socially determined causes" mortality.

Thirdly, rural areas of the FEFD lose their population even more than cities and urban settlements. It requires well-thought-out effective state support for agriculture, the budget financial resources should be aimed at improving the living conditions of the rural population, improving infrastructure in rural areas, the development of the production sector, consideration of an active longevity as a factor in a decrease in demographic load. An effective mechanism for state support for rural producers should be developed and intensify measures to develop industries on a village that contribute to the expansion of rural employment.

References

1. Ma, C.C., Chen, H.S., Chang, H.P.: Crisis response and supervision system for food security: a comparative analysis between Mainland China and Taiwan. Sustainability **12**(7), 3045 (2020). https://doi.org/10.3390/su12073045
2. Zhupley, I.V., Potenko, T.A.: Challenges to agriculture in the Russian Far East: lessons from crises. IOP Conf. Ser. Earth Environ. Sci. **547**, 012048 (2020). https://doi.org/10.1088/1755-1315/547/1/012048
3. Simon, C., Belyakov, A.O., Feichtinger, G.: Minimizing the dependency ratio in a population with below-replacement fertility through immigration. Theoret. Popul. Biol. **82**, 158–169 (2012). https://doi.org/10.1016/j.tpb.2012.06.009
4. Boumezoued, H.L., Hardy, N., Arnold Karoui, S.: Cause-of-death mortality: what can be learned from population dynamics? Insur. Math. Econ. **78**, 301–315 (2018). https://doi.org/10.1016/j.insmatheco.2017.09.015
5. Croix, D., Gobbi, P.E.: Population density, fertility, and demographic convergence in developing countries. J. Dev. Econ. **127**, 13–24 (2017). https://doi.org/10.1016/j.jdeveco.2017.02.003
6. Wang, Q., Sun, X.: The role of socio-political and economic factors in fertility decline: a cross-country analysis. World Dev. **87**, 360–370 (2016). https://doi.org/10.1016/j.worlddev.2016.07.004
7. Mishchuk, S.N.: General characteristics and regional differences of migration processes in the Russian Far East in the Post-Soviet Period. Reg. Res. Russ. **10**, 86–96 (2020). https://doi.org/10.1134/S2079970520010074
8. Mishchuk, S.N.: Dynamics of internal and international labor migration in the Russian Far East. Nauka. Kultura. Obshchestvo **1**, 68–74 (2020). https://doi.org/10.38085/2308829X-2020-1-68-74
9. Nefedova, T.G., Starikova, A.V.: Migrations as a way of population adaptation to polarization of space at the center of Russia. Sotsiologicheskie issledovaniya **10**, 24–38 (2020). https://doi.org/10.31857/S013216250009567-6
10. Parrado, E.A.: Migration and Fertility (International Encyclopedia of the Social & Behavioral Sciences (Second Edition). Elsevier (2015). https://doi.org/10.1016/B978-0-08-097086-8.31128-X
11. Sardadvar, S., Vakulenko, E.: A model of interregional migration under the presence of natural resources: theory and evidence from Russia. Ann. Reg. Sci. **59**(2), 535–569 (2017). https://doi.org/10.1007/s00168-017-0844-3
12. Chen, Z., Xu, R., Yang, Z.: Numerical analysis of linear -methods with two-layer boundary conditions for age-structured population models. Math. Comput. Simul. **182**, 603–619 (2021). https://doi.org/10.1016/j.matcom.2020.11.016
13. Kalabikhina, E., Arkhangelsky, V.N., Nikolaeva, U.G., Kolotusha, A.V.: Short-term forecasting of demographic trends based on Google trends data. Prikladnaya informatika **6**(90), 91–118 (2020). https://doi.org/10.37791/2687-0649-2020-15-6-91-118
14. Omelyan, I.: Spatial population dynamics: beyond the Kirkwood superposition approximation by advancing to the Fisher-Kopeliovich ansatz. Physica A Statist. Mech. Appl. **544**, 123546 (2020). https://doi.org/10.1016/j.physa.2019.123546
15. Tang, Y., Ruan Ya, X., Yo, L.Y.: Feedback analysis of population, economy and pension: moderate scale of China's pension strategic reserve. Physica A Statist. Mech. Appl. **551**, 124043 (2020). https://doi.org/10.1016/j.physa.2019.124043
16. Shmidt, Yu., Ivashina, N., Lobodin, P., Kukhlevsky, A.: Forecasting of interregional migration flows. Econ. Reg. **13**(1), 126–136 (2017). https://doi.org/10.17059/2017-1-12

17. Yi, F., Gudaj, R., Arefieva, V., Mishchuk, S., et al.: How Chinese agricultural immigrants affect farmers in the Russian Far East. AJES **5**, 1387–1415 (2020). https://doi.org/10.1111/ajes.12361
18. Zhupley, I., Potenko, T., Gubarkov, S., Tretyak, N., Grafov, R.: Structural Shifts and Reform of the agrarian sector of the Russian Economy under the conditions of the import Substitution Policy. Space Cult. India **4**, 25–35 (2018). https://doi.org/10.20896/saci.v6i4.385

Check for updates

Phytoviruses in the Potato Field Tripartite Agroecosystem

Olga Sobko(✉), Nathalia Matsishina, Petr Fisenko, Irina Kim, and Natalia Boginskaya

Federal Scientific Center of Agrobiotechnologies in the Far East named after A. K. Chaika, 30, Vologenina st, 692539 Ussuriisk, Primorie Territory, Russia

Abstract. *Solánum tuberosum* L. does not grow in isolation in the field, it is surrounded on all sides by biotopes of segetal vegetation. Many weeds infect crops with viral infections and pests. We have identified a complex viral infection on potato, which increases the harmful effect on plants and leads in some cases to its death. The symptoms of viral infection are different and depend on both the viral load composition and the variety itself. The highest damage score of complex viral infection was observed in the varieties Smak - 2.4 points, Yubilyar - 1.63 points, Yantar - 1.47 points, and Laperla - 1.73 points. On these varieties, the external manifestation of complex viral infection was the most active. The lowest score was found in the Red Lady variety - 0.37 points and Belmonda variety - minor damage in certain specimens. The occurrence frequency of associative viral infection from the total number of registered plants is 30.51%. From 30 types of segetal vegetation, in potato plantings and along the field edge, 33.3% are reservoirs of viral infection. Among the insects, carriers of the viral infection are *Cicadella sp., Henosepilachna vigintioctomaculata, Dolycoris baccarum, Mythimna separata, Lygus pratensis u Rhopalosiphum padi.*

Keywords: Solanum tuberosum · Phytoviruses · Insects · Reservators · Agrobiocenosis · Phytophages · Weeds

1 Introduction

Potato is a valuable agricultural crop cultivated throughout Russia. The high incidence of viral diseases in potato seriously affects crop yields, leading to economic losses due to the culling of seed potato and increased storage losses. *Solanum tuberosum* L. does not grow in isolation on the field, it is surrounded on all sides by segetal vegetation biotopes, which are reservoirs of insect pests and diseases. In agroecosystems, as in artificially created ecosystems, there are peculiar relationships between cultivated, weedy plants and insects. Evolutionarily, plants produce more biomass to "feed" phytophages. In this sense, cultivated plants are no exception. However, in the course of long-term selection and careful human care, cultivated plants have decreased resistance to certain groups of insect pests and competitiveness to weeds [1]. Agroecosystems are artificially created and periodically recreated by humans, and should be considered as formations where

A. Muratov and S. Ignateva (Eds.): AFE 2021, LNNS 353, pp. 434–442, 2022.
https://doi.org/10.1007/978-3-030-91402-8_49

biocenotic regulation does not have any significant impact on maintaining the balance between their elements. However, to a certain extent, they are characterized by the ability to self-regulate, that is, to stabilize the dynamic balance of elements belonging to different trophic groups [2].

It is believed that it will be possible to achieve the phytosanitary stability of agroecosystems when 70–80% of the area will be sown with resistant varieties. But, first, absolutely immune plants that are adaptive to growth factors do not exist in nature, otherwise they would have long ago replaced all other species. Secondly, with the deterioration of the feed base quality, the adaptatiogenesis of phytophages and phytopathogens to resistant varieties accelerates. Third, the immune culture is only one dominant species in the system. In a complete agroecosystem of crop rotation, up to a dozen agricultural crops are cultivated, at least two to three dozen species of weeds grow. So agroecosystem can be rapidly modified in a short period of time in the direction of less stability due to the dynamics of the number and replacement processes of harmful organisms populations, against which varieties will not have immunity, and to protect plants from these objects will require more effort [3]. The main factors of crop production are the correct use of plant protection products against a complex of pests, diseases, and weeds, as well as the creation of a favorable biocenotic environment in agroecosystems, taking into account the role of beneficial insects in plant pest colonies [4].

Many weeds infect crops with phytopathogens and pests. Phytophages first feed on segetal plants, and then move from them to potato. Often weeds provide food for insects at a time when the potato have not yet risen, or vice versa, have already been removed from the field. Herbivorous insects on potato lead a secretive or free lifestyle, feeding on either one (monophages), or several (oligophages), or many plant (polyphages). The long growing season of potato allows most types of weeds to be seeded, which increases the risk of preserving phytoviruses in weed seeds [5].

Potato viruses are particularly dangerous due to the vegetative reproduction of the crop, in which there is a complete transmission of the infection to the offspring. In addition, the viruses are systemic pantropic immunosuppressants with a strong tendency to chronicity and recurrence of the process, and in some cases, a latent infection of plants [8]. Accumulating from year to year, they cause the degeneration disease, which reduces the potential yield of varieties by 30–80%. Viruses are easily transmitted from diseased potato plants to healthy ones by contact, by insects, fungi, and nematodes. This leads to a deterioration in the quality and productivity of the seed material with each subsequent field generation. The source of potato leafroll virus (PLRV) infection is usually seeds, as well as weeds and nearby crops. PLRV is persistently transmitted through aphids, there is no mechanical transmission, since PLRV is not able to stay outside the plant cell for a long time, without losing infectivity. Mosaic group viruses (PVY (potato virus Y) - various strains, PVA (potato virus A), PVM (potato virus M), PVX (potato virus X), and PVS (potato virus S) are non-persistently transmitted through aphids and other insects. They are also transmitted by contact from patients to healthy plants, during mechanical processing, through damage to the tops, and also spread with planting material. For PVM and PVA, mechanical transmission is not always possible, apparently due to the low concentration of the virus in the diseased plant tissues. To limit the spread of phytoviruses in the field, it is necessary to have reliable information about the reservoir plant species composition, the total vector activity, and the identification of possible ways of spreading infection in seed cultivation specific places. Such information is obtained in the course of phytomonitoring of migratory species of aphids and other phytophages, and the appearance of disease symptoms on potato. It is also important to identify factors which contribute or prevent the infection of plants with viruses, and the manifestation of diseases signs on various potato varieties [6].

The study of weed vegetation species composition and potato field entomofauna contributes to the identification of phytophages - carriers of viral and fungal diseases, as well as the development of more effective measures to control segetal vegetation and phytophage insects.

2 Materials and Methods

In a stationary field experiment, 13 potato varieties were used, planted on ridges of 90 × 30 cm at the rate of 37 thousand plants per 1 ha. The size of the plots was 25 cm^2. To determine the vegetation species, an Atlas determinant was used [7]. Insects collected in potato plantings and along the field edge were identified using a Determinant [8]. Symptoms and descriptions of viruses are given by S. Awasthi et al. 2016 [9].

To detect viruses using the PCR methods, potato and weed leaves were collected in separate biological bags made of filter paper, wrapped in plastic bags, and quickly frozen at –20 °C. Dirt from the leaves surface was removed with a non-woven material moistened with alcohol.

Total RNA was isolated from plants green parts and insects according using the commercial kit PhytoSorb (Syntol). The presence/absence of phytoviruses in the samples was performed by one-step RT-PCR with real-time fluorescence detection in a QuantStudio 5 amplifier (Applied Biosystems) using commercial kits of the "Potato Virus X, Y, M, L, S, A, PSTVd-PB" "Phytoscreen" series (Synthol) to detect RNA potato viruses.

3 Results and Discussions

During the, *Solanum tuberosum* L. growing season we observed phytoviruses clinical manifestations on leaves and inflorescences. The course of pathogenesis depended on the complexity of the infection, as well as the potato plants varietal timing. An important pathogenetic component of phytovirus disease for different varieties, which accompanies other symptoms, was chlorosis. Thus, on the Smak variety, chlorosis along the veins, mosaicity, leaf curl, not opened buds, yellowish, twisted flower petals were noted (Fig. 1), on the Yubilyar variety - chlorosis of the leaf plate and leaf edges curl.

A typical clinical picture of the phytovirus infection course in the Kazachok variety was the development of the leaf plate extensive chlorosis, convex, yellowish veins, curliness and deformity of the flowers, for Sante variety - tuberosity, in a pink border along the leaf edge, veins jaundice, leaf curl and high chlorotic level. For the Dachny variety curly, mosaicity, and leaf plate wavy edges are recorded, for Augustine variety - a slight curl and localized chlorosis. For the Yantar variety - mosaicity, extensive chloroses, pink border and undulation on the leaf edge. For Laperla variety dwarfism, necrosis, leaf plate convexity, curliness, chlorosis, leaf edge waviness, and mosaicity

Fig. 1. Healthy and diseased inflorescences *of Solánum tuberosum* on the Smak variety (photo by the authors).

were noted, for the Lilly variety - curliness, chlorosis, and leaf edge waviness. On the potato tops of the Queen Anna variety were observed chlorosis and light curliness, some plants were characterized by the leaves yellowing, on the Red Lady variety - dwarfism, curliness, mosaicity, and lack of turgor. For Labella variety - mosaicity, curliness, leaf plate chlorosis, the leaves were yellow as in autumn, a red border along the leaf edge (Fig. 2), for Belmonda variety - curliness in certain specimens, which were entwined with buckwheat bindweed. In general, chlorosis, curliness, and mosaicity were typical for the symptom complexes of viral infection, regardless of the variety, as described in the literature [9].

The study of the phytovirus infection pathogenesis symptoms showed that there can be various forms of interaction between the pathogen and the host, expressed in co-infection and superinfection. The pathogenesis of phytoviruses deep forms is ambiguous, and is determined by the properties of the virus and specific immune responses of the macroorganism that occur in response to the infection introduction. We analyzed about 2000 thousand potato plants by visual accounting, as well as plants with pronounced symptoms by PCR and ELISA methods to identify the infection origin. On potato varieties of Russian and foreign selection, we found viruses of four families: *Luteoviridae, Potyviridae, Flexiviridae* and *Pospiviroidae.* The *Luteoviridae* family is represented by the genus *Polerovirus* - PLRV. The *Potyviridae* family is represented by the genus *Potyvirus* - PVA and PVY. The *Flexiviridae* family is represented by two genera *Calavirus* and *Potexvirus.* The genus *Calavirus* includes PVM and PVS. The genus *Potexvirus* is represented by the PVX. To the *Pospiviroidae* family, the *Pospiviroid* genus belongs the potato spindle tuber viroid - PSTVd [9].

The highest coinfection score was observed in the Smak - 2.4 points, Yubilyar - 1.63 points, Yantar - 1.47 points, and Laperla - 1.73 points varieties. In the same varieties, the external manifestation of complex viral infection was the most active. The lowest - in the Red Lady - 0.37 points and Belmonda - minor damage in certain specimens, varieties (Table 1). The associative viral infection occurrence frequency from the total number of registered plants was 30.51%. In the Belmonda and Red Lady varieties, the PVY, PVS, PVM, PVA, and PSTVd viruses did not cause visible symptoms, which indicated a latent course of the disease. Viruses accumulated in the plant body, without appearing externally.

Fig. 2. Manifestation of complex viral infection (PVY, PVA, PVS, PVM) in Labella variety.

In this regard, strict control over the potato seed production is necessary. The higher the viral load in the seed material, the more serious the damage it can cause. To ensure the high quality of tubers, it is important to minimize the infection sources. It is necessary both to increase the requirements for the seed potato field inspection, and strict compliance by the subjects of seed production with such mandatory reception as health-improving phyto-cleaning [10].

Table 1. Damage to potato varieties by viral infection.

Potato variety	The score of the viral infection manifestation	Potato virus						
		PVY	PVX	PVA	PVS	PVM	PLRV	PSTDv
Smack	2.4	+	–	–	+	+	–	+
Yubilyar	1.63	+	–	+	+	+	–	+
Kazachok	1.2	+	–	+	+	+	+	+
Sante	1.6	+	–	+	+	+	+	–
Dachny	1.68	Undefined						
Augustine	0.82	+	+	+	+	+	–	–
Yantar	1.47	+		+	+	+	+	+
Laperla	1.73	+	+	+	+	–	+	–
Lilly	0.57	Undefined						
Queen Anne	0.81	+	–	+	+	+	+	+
Red Lady	0.37	+	+	–	+	+	+	+
Labella	0.87	+	+	–	+	–	–	–
Belmonda	–	+	–	+	+	+	–	+

Table 2. Plants - reservoirs of viral infection.

	PVY	PVX	PVA	PVS	PVM	PLRV	PSTDv
Segetal and ruderal plants							
Chenopodium album	+	–	–	+	+	+	–
Trifolium pratense	+	–	–	–	–	–	–
Plantago major	+	–	–	–	–	–	–
Matricaria perforata	–	–	–	+	–	–	–
Barbarea vulgaris	+	–	+	–	–	–	+
Ambrosia artemisiifolia	+	–	–	–	–	–	–
Sonchus arvensis	+	–	–	–	–	+	–
Galinsoga parviflora	+	–	–	+	+	–	+
Fam. Poaceae	+	+	–	–	+	+	+
Vegetable crops							
Solanum lycopersicum	+	–	–	+	+	+	+
Cucurbita pepo	+	–	–	+	–	–	–
Anethum graveolens	+	–	+	+	+	+	+
Armoracia rusticana	+	+	–	–	+	+	–

In our research, we studied different components of the potato field agroecosystem, including wild species - segetal and ruderal vegetation, phytophagous insects. All these components of the system are part of the phytoviruses circulation. The studied potato field biotopes included species from 13 botanical families, as previously shown in our studies [11]. From 30 types of weed vegetation in potato plantings and along the field edge, 10 species were viral infection reservoirs. These include *Chenopodium album, Trifolium pratense, Plantago major, Matricaria perforata, Barbarea vulgaris, Ambrosia artemisiifolia, Sonchus arvensis, Galinsoga parviflora, Poaceae* fam. (Table 2). Thus, we have determined the range of host plants of the *Solanum tuberosum* viral infection. Moreover, such plants as *Trifolium pratense, Plantago major, Ambrosia artemisiifolia* carried only PVY, *Matricaria perforata* - carried PVS. All other weeds had a complex viral infection (Table 2).

Vegetable crops, that are planted next to potato on private farmsteads, also are dangerous phytoviruses foci. We found potato mosaic viruses, as well as PLRV and PSTVd in *Solanum lycopersicum, Anethum graveolens, Armoracia rusticana*, and *Cucurbita pepo* (Table 2). Thus, for vegetable and allotment gardens a more frequent change of *Solanum tuberosum* seed material is relevant since the vegetable crops can also be sources of infection.

Among the insects living in the potato field agroecosystem, *Cicadella* sp., *Henosepilachna vigintioctomaculata, Dolycoris baccarum, Mythimna separata, Lygus pratensis,* and *Rhopalosiphum padi* were viral infection vectors. Moreover, the *Cicadella* sp. and *Rhopalosiphum padi* insects carried PVY, PVS, PVM, PVA, PLRV, and PSTVd (Table

Table 3. Insects-carriers of a viral infection.

	Potato virus						
	PVY	PVX	PVA	PVS	PVM	PLRV	PSTDv
Cicadella sp.	+	+	+	+	+	+	+
Henosepilachna vigintioctomaculata	+	–	–	+	+	+	+
Dolycoris baccarum	+	+	+	+	+	+	–
Mythimna separata	+	+	+	+	+	+	–
Lygus pratensis	+	–	+	+	+	+	+
Rhopalosiphum padi	+	+	+	+	+	+	+

3). In the majority of cases, the viral infection symptoms appeared on the potato plants damaged by insects. It should be noted that on the Belmonda variety, insect damage was isolated, and the infection symptoms manifestation was only near the food site. In the Smak, Yubilyar, Yantar varieties phytophagous damage was 4.5 points, in the variety Laperla 3.2 points, respectively, for the same varieties there was the viral infection highest score (Fig. 3).

According to our data, viral infection is common. We believe that the spread of a viral infection is facilitated by the presence of not one, but several vectors. Aphididae spp. in the monsoon climate conditions of the Primorsky Territory cannot compete with *Henosepilachna vigintioctomaculata* and *Lygus pratensis* in terms of harmfulness and activity of settlement on potato. In addition, the infection is spread mechanically: by insects that are not phytophages, but live in the potato field agroecosystem, for example, *Dolycoris baccarum*, agricultural machinery, people, as well as random animals. Also, we did not find a relationship between the potato maturation period and their viral infection level.

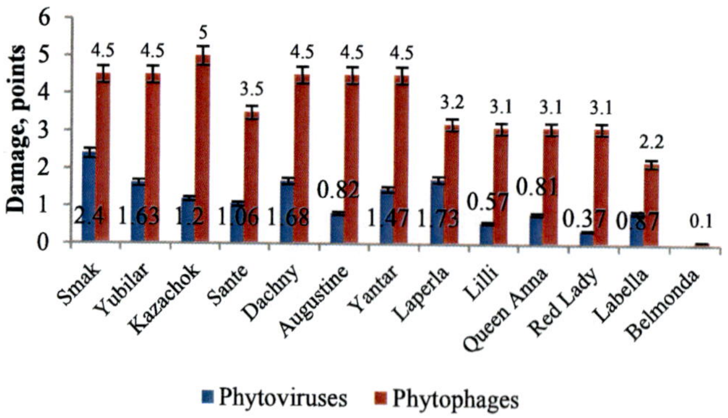

Fig. 3. The manifestation of viral infection symptoms and the score of insect damage on potato varieties.

4 Conclusion

According to the results of our research, a complex viral infection consisting of mosaic group viruses (PVX, PVY, PVM, PVA, PVS), PLRV and PSTVd is most often found in potato plantings. The incidence of a complex viral infection on *Solanum tuberosum* is 30.51%. The carriers of viral infection, in addition *to Solanum tuberosum*, are the following plant-reservoirs: *Chenopodium album*, *Trifolium pratense*, *Plantago major*, *Matricaria perforata*, *Barbarea vulgaris*, *Ambrosia artemisiifolia*, *Sonchus arvensis*, *Galinsoga parviflora*, fam. *Poaceae*, *Solanum lycopersicum*, *Anethum graveolens*, *Armoracia rusticana*, and *Cucurbita pepo*. And most of them are carriers of several potato viruses. Compared to the literature data, we identified a wider range of potato virus vectors, including *Cicadella sp.*, *Henosepilachna vigintioctomaculata*, *Dolycoris baccarum*, *Mythimna separata*, *Lygus pratensis*, and*Rhopalosiphum padi*. The formed potato field agroecosystem -*Solanum tuberosum*, ruderal and segetal vegetation, phytophage insects that feed on potato and weeds, has a favorable effect on the development and manifestation of viral diseases. Given the wide range of phytoviruses insect vectors in Primorye, the low effectiveness of measures to contain the spread of infection relevant for regions where aphids are the main vector should be noted. The vectors identified by us are broad polyphages and populate a variety of biotopes, including those recommended in the literature to contain the infection spread.

A set of preventive measures to protect potato from viruses should include plantings isolation; vector control; agrotechnical measures (planting and harvesting dates selection, weed control, etc.); selective measures (clone selection); biotechnological measures (improvement and reproduction in culture in vitro). The main way to combat viral infection remains selection for virus resistance [11, 12]. In the open ground, there is a rapid re-infection of the recovered material with viruses. In just 2–3 years, 50% to 80% of potato specimens can be re-infected with viruses [13]. To solve this problem, potato seed plantings should be placed where the main vectors of viruses have a low number. Within the boundaries of an isolated seed-growing territory, the possibility of planting potatoes in the vegetable gardens of the local population and suburban areas should be excluded [10]. Most viruses infect potatoes entomophilically, so it is necessary to develop effective methods of vector control relevant for the Primorsky Territory in the monsoon climate conditions.

Acknowledgments. The authors express their sincere appreciation to Romanciuc Riceard (Alecu Russo State University, Faculty of Philology, German and English interpreter) for assistance in the article translating.

References

1. Raccah, B., Fereres, A.: Plant virus transmission by insects. In: Encyclopedia of Life Sciences. Wiley, Chichester (2009). https://doi.org/10.1002/9780470015902.A0021525.a0000760.pub2
2. Santos, J.S., et al.: Landscape ecology in the Anthropocene: an overview for integrating agroecosystems and biodiversity conservation. Perspect. Ecol. Conserv. **19**(1), 21–32 (2021). https://doi.org/10.1016/j.pecon.2020.11.002

3. Boquel, S., Delayen, C., Couty, A., Giordanengo, P., Ameline, A.: Modulation of aphid vector activity by Potato virus Y on in vitro potato plants. Plant Dis. **96**, 82–86 (2012). https://doi.org/10.1094/PDIS-06-11-0499
4. Fingu-Mabola, J.C., Francis, F.: Aphid–plant–phytovirus pathosystems: influencing factors from vector behaviour to virus spread. Agriculture **11**, 502 (2021). https://doi.org/10.3390/agriculture11060502
5. Bertschinger, L., et al.: Incomplete infection of secondarily infected potato plants – an environment dependent underestimated mechanism in plant virology. Front Plant Sci. **8**, 74 (2017). https://doi.org/10.3389/fpls.2017.00074
6. Jones, R.A.C.: Global plant virus disease pandemics and epidemics. Plants **10**, 233 (2021). https://doi.org/10.3390/plants10020233
7. Zettler, J.A., et al.: To key or not to key: a new key to simplify and improve the accuracy of insect identification. Am. Biol. Teacher **78**(8), 626–633 (2016). https://doi.org/10.1525/abt.2016.78.8.626
8. Byng, J.W.: The Flowering Plants Handbook: A practical guide to families and genera of the world. Plant Gateway Ltd, Hertford (2014). www.plantgateway.com
9. Awasthi, S., Chauhan, R., Narayan, R.P.: Plant viruses: history and taxonomy. In: Gaur, R.K., Petrov, N.M., Patil, B.L., Stoyanova, M.I. (eds.) Plant Viruses: Evolution and Management, pp. 1–17. Springer, Singapore (2016). https://doi.org/10.1007/978-981-10-1406-2_1
10. Kumar, R., Jeevalatha, A.: Viral diseases and their management in potato production. In: Pandey, N.K., Singh, D.K., Kumar, R. (eds.) Summer School on Current Trends in Quality Potato Production, vol. 18, pp. 115–127. CPRI, Shimla (2014)
11. Sobko, O.A., et al.: Viruses in the agrobiocenosis of the potato fields. IOP Conf. Ser. Earth Environ. Sci. **677**(5), 052093 (2021). https://doi.org/10.1088/1755-1315/677/5/052093
12. Kreuze, J.F., Souza-Dias, J.A.C., Jeevalatha, A., Figueira, A.R., Valkonen, J.P.T., Jones, R.A.C.: Viral diseases in potato. In: Campos, H., Ortiz, O. (eds.) The Potato Crop, pp. 389–430. Springer, Cham (2020). https://doi.org/10.1007/978-3-030-28683-5_11
13. Cervantes, F.A., Alvarez, J.M.: Within plant distribution of potato virus Y in hairy nightshade (Solanum sarrachoides): an inoculum source affecting PVY aphid transmission. Virus Res. **159**, 194–200 (2011). https://doi.org/10.1016/j.virusres.2011.05.003

Comparative Evaluation of Different Potato Varieties for Their Suitability for Starch Processing

Dmitry Volkov(✉), Irina Kim, Aleksey Klykov, and Natalia Matsishina

Federal Scientific Center of Agricultural Biotechnology of the Far East named after A.K Chaika, 30, Volozhenin Street, Timityazevskiy Stl., 692539 Ussuriysk, Primorsky Krai, Russia
volkov_dima@inbox.ru

Abstract. The article presents the results of assessing collectible potato varieties of different maturity groups, for suitability for processing into starch. The cultivars grown at the experimental site of the FSBSI Federal Research Center for Agrobiotechnologies of the Far East in 2019–2020 were evaluated by productivity, marketability, starch and dry matter content, and morphology of starch granules. The study confirmed a higher starchiness in the mid-ripening (13.07%) and middle-late (13.53%) varieties with a longer growing season compared to the early-ripening (11.95%) and middle-early (12.44%) groups. The potato microstructure was studied by confocal laser scanning microscopy (CLSM). As a result of the starch granulometric analysis, significant varietal differences in the number and size of granules in the tuber cell were revealed. The largest number of granules in the cell was found in the Garant (9.6 pcs.), Adretta (9.5 pcs.) varieties, the smallest - in the Musinsky (6.2 pcs.), Svitanok Kievsky (6.9 pcs.) varieties. The varieties with the highest concentration of large- and medium-sized starch grains were identified: Svitanok Kievsky (75%), Musinsky (65.0%), Fregata (64.5%), Zarevo (63.3%), Fritella (59.3%), Dachny (56.6%), Solnyishko (50.3%). According to the complex of characteristics that meet the requirements of suitability for processing into starch, the varieties Musinsky, Svitanok Kievsky, Solnyishko, Fregata, Fritella are recommended.

Keywords: Potatoes · Starch · Dry matter · Tuber · Cell · Microstructure · Granulometric analysis · Size granule · CLSM

1 Introduction

Potato is one of the main human alimentary resources in the world, the importance of which is due to the presence of accumulated carbohydrates in the tubers, as well as a large number of essential amino acids. Compared to other starchy crops (rice, corn, and wheat), potato contains more vitamins (complex C and B), minerals (potassium, magnesium, calcium, phosphorus, and iron), and biologically active components (carotenoids, anthocyanins, and polyphenolic compounds) [1]. Potato tubers on average consist of

A. Muratov and S. Ignateva (Eds.): AFE 2021, LNNS 353, pp. 443–450, 2022.
https://doi.org/10.1007/978-3-030-91402-8_50

75% water and 25% dry matter. The main part of the tuber dry matter is carbohydrates, namely starch [2].

Starch is classified as one of the most widely available compounds and is widely used in both the food and non-food industries. Potato starch contains a reduced amount of protein and does not contain lipids, unlike grain starches, it consists of coarse grains, is characterized by higher quality, namely, the transparency of its paste, high gelling capacity, and the least amount of impurities [3].

The world leaders in the processing of potatoes are the United States, England, Germany, Holland, Denmark in these countries, up to 50% of food potatoes are processed. In Russia, this indicator at the end of 2019 amounted to 300 thousand tons, of which only 7% is processed for starch, flakes, granules, which only meets the market demand by 30%. One of the reasons for the low starch production in Russia is the lack of high-starch varieties, the required quality of tubers, and low yield [4].

Currently, one of the priority areas of the Federal State Budgetary Institution "Federal Scientific Center of Agricultural Biotechnology of the Far East named after A. K. Chaika" is the creation of potato varieties with a high starch content, suitable for processing. Technical potato varieties are subject to special requirements, namely: the starch mass fraction in the tubers is not less than 15%, and the high content (more than 50%) of medium- and large-sized starch grains with a size of 35 microns and higher [5, 6]. For potato processing it is important to combine both high yield and quality of tubers. This is important for achieving the maximum yield of end products and profit, both for potato producers and for industry [7].

The purpose of the research is to evaluate and identify potato varieties of different maturity periods that are suitable for processing for starch, as well as to select source varieties for potato breeding with high productivity and high starch content in tubers.

2 Materials and Methods

The research was carried out in the Department of Potato and Vegetable Growing of the Federal State Budgetary Institution "Far East Federal Scientific Center of Agrobiotechnology Named after A. K. Chaika" in 2019–2020. The research material was 179 potato varieties of various origins and maturity groups (53 early-ripening, 53 middle-early, 48 mid-ripening, 25 middle-late). The studied varieties were evaluated according to the characteristics that determine the suitability for processing for starch (productivity; starch content; size, shape and number of starch grains.). Productivity was determined by weighing [5], the mass fraction of starch and dry matter was determined by weighing the specific weight of tubers in air and water with a digital scale PW-2050.

2.1 Confocal Laser Scanning Microscopy (CLSM)

CLSM is a type of optical scanning technique that involves using laser beams to increase the contrast and resolution of an image by blocking unwanted planes of light. The use of CLSM in food research has been described previously [8, 9].

To visualize starch, cell walls, and protein in potato tubers, such dyes as Safranin O, acridine orange, and acid fuchsine are widely used [9].

In our studies, the morphological features of starch granules (size and shape) were determined by scanning the tuber pulp with a laser confocal microscope CLSM 800 (Zeiss, Germany) with a 10× lens. Excitation was carried out with an argon laser with a wavelength of 488, fluorescence was detected in two ranges, starch detection at 500–570 nm (green) and detection of cell walls and proteins at 590–700 nm (orange). To do this, an average tuber was selected for each potato variety and a 1 cm thick slice was cut from its middle. Through the center of this cut, a section was cut from one edge to the opposite one with a width of 1 cm. The section was cut into cubes of 1 cm. Then the cubes were colored in 0.1% acridine orange solution in 0.1 M phosphate buffer with pH 7. The cubes were washed three times for 1 min in a pure 0.1 M phosphate buffer with a pH of 7. At least 30 sites were scanned for each sample. The results were processed using the Zeiss Zen 2 software.

3 Results and Discussions

The assessment of productivity in the post-harvest period showed that, on average, for 2019–2020, the productivity among the studied samples varied from 508 to 1658 g/plant (Table 1). Variety with high productivity were: Kolymsky (1482 g/plant), Laperla (1406 g/plant), Pamyati Kulakova (1206 g/plant), Krepysh (1220 g/plant), Impala (1166 g/plant), Queen Anne (1101 g/plant), Matushka (1074 g/plant), Udacha (1057 g/plant), Zhukovskiy ranniy (1021 g/plant), Red Lady (1004 g/plant) – early-ripening group; Zoya (1424 g/plant), Sudaryinya (1249 g/plant), 7 For 7 (1231 g/plant), Arktika (1114 g/plant), Sante (1101 g/plant), Belmonda (1051 g/plant), Kamchatka (1010 g/plant) – middle-early group; Zlatka (1658 g/plant), Solntsesvet (1281 g/plant), st. Dachny (1196 g/plant), Alim (1183 g/plant), Slavyanka (1047 g/plant), Alaska (1020 g/plant) – mid-ripening group; Ragneda (1446 g/plant), st. Yantar (1178 g/plant), Kivi (1075 g/plant), Musinsky (1069 g/plant), Pobeda (1058 g/plant), Vetraz (1005 g/plant) – middle-late group.

Table 1. Productivity and marketability of potato tubers of different maturity groups.

Maturity group	Number of samples, pcs.	Productivity, g/plant			V,%	Marketability,%			
		min	max	$\bar{x}$		min	max	$\bar{x}$	V
Early-ripening group	53	548	1482	875	21.5	38.4	95.7	83.9	12.3
Middle-early group	53	508	1424	830	24.0	54.3	94.8	83.1	10.5
Mid-ripening group	48	518	1658	861	23.9	52.7	95.6	84.3	9.9
Middle-late group	25	530	1446	851	25.3	41.1	94.6	82.4	14.7

In general, in the collection nursery, the marketability of tubers of individual variety samples varied significantly and ranged from 38.4% to 95.7%, but the average marketability of tubers between the maturity groups varied from 82.4 to 83.9%. The lowest variability of this trait was observed in the varieties of the mid-ripening group (V = 9.9%).

The content of dry substances and the starch mass fraction in potato tubers characterizes their nutritional value, the degree of digestibility during cooking, as well as the profitability of using them as raw materials for the processing industry in the production of starch, alcohol, ethanol and other compounds [10, 11] (Table 2).

Table 2. The content of starch and dry matter in potato tubers of different maturity groups.

Maturity group	Number of samples, pcs.	Starch content, %				Dry matter content, %			
		min	max	$\overline{x}$	V	min	max	$\overline{x}$	V
Early-ripening group	53	6.56	15.66	11.95	15.78	12.84	22.18	18.65	11.39
Middle-early group	53	8.57	19.41	12.44	15.76	14.67	26.15	19.59	10.81
Mid-ripening group	48	9.61	17.28	13.07	14.97	15.76	24.92	19.84	11.62
Middle-late group	25	8.17	21.08	13.53	21.26	15.36	27.92	20.52	14.84

It is known that late-maturing varieties contain more dry matter and starch than early ones, which is also noted in our studies [12]. The average starch content over the years of study was in the samples of the early-ripening group (11.95%), middle-early (12.44%), mid-ripening (13.07%), and middle-late (13.53%). The greatest variability in starch content was observed in the middle-late group (V = 21.3%).

As a result of the analysis of the samples for the starch content in potato tubers, varieties with a content of more than 15% were identified, which is an indicator of their suitability for processing into starch. In particular, the highest starch content in the early-ripening group was observed in the Antonina (15.66%) and Cabo (15.01%); in the middle-early group in Svitanok Kievsky (19.41%), Solnyishko (15.04%); in the mid-ripening group in Nakra (17.28%), Garant (16.82%), Fritella (16.48%); in the middle-late group in the Zarevo (21.08%), Fregata (19.23%) and Musinsky (15.12%) varieties.

In the tuber, starch is found in the form of grains. In nature, the grain size of potato starch ranges from 1 to 110 microns with a predominance of 20–40 microns. The size of the starch grains affects the quality of processed products. In the food industry, starch with large grain size is highly valued. The size of the starch granules affects the hydrolysis of starch in industrial processes. The smaller size of the starch granules results in a higher percentage of hydrolysis. Large starch grains precipitate quickly and give a guaranteed starch mass, unlike small grains that can be carried away by washing water [13, 14, 15–17].

The analysis of starch granules in tuber cells using CLSM showed differences in the number of granules in the cell, their size, and shape (Table 3). Thus, the largest number of granules in the cell was observed in a mid-ripening variety - Garant (9.6 pcs. on average), the minimum was found in the Musinsky variety (6.2 pcs.). The size of the granules in the studied varieties varied from 4.09 microns to 108.91 microns.

Table 3. Morphological characteristics of starch granules in potato varieties with a high starch mass fraction.

Variety	Starch content, %	Average number of granules per cell, pcs.	Size of starch granules, microns			Share of starch granules larger than 35 microns, %	Productivity, g/plant	Starch yield, t/ha
			min	max	$\bar{x}$			
Early-ripening group								
Yubilyar st.	14.14±4.90	9.2±2.45	6.57±2.45	75.23±4.90	38.30±4.90	56.8±4.90	799±4.90	4.18±2.94
Antonina	15.66±4.90	8.5±2.45	5.87±2.94	72.06±4.90	30.36±4.90	31.8±4.90	978±4.90	5.67±2.94
Cabo	15.01±4.90	8.4±2.45	7.69±2.45	80.32±4.90	41.69±4.90	63.7±4.90	725±4.90	4.03±2.94
Middle-early group								
Adretta st.	14.03±4.90	9.5±2.45	4.09±2.94	108.91±4.90	33.61±4.90	33.9±4.90	765±4.90	3.97±2.94
Svitanok Kievsky	19.41±4.90	6.9±2.45	16.1±2.45	84.05±4.90	49.09±4.90	75.5±4.90	814±4.90	5.85±2.94
Solnyishko	15.04±4.90	9.1±2.45	9.04±2.45	84.01±4.90	36.10±4.90	50.3±4.90	865±4.90	4.81±2.94
Mid-ripening group								
Dachny st.	11.26±2.45	8.0±2.45	7.19±2.45	91.60±4.90	41.88±4.90	56.6±4.90	1196±4.90	4.87±2.94
Garant	16.82±4.90	9.6±2.45	6.89±2.94	96.18±4.90	34.30±4.90	37.2±4.90	803±4.90	5.00±2.94
Nakra	17.28±4.90	9.2±2.45	9.40±2.45	81.43±4.90	38.32±4.90	49.2±4.90	661±4.90	4.23±2.94
Fritella	16.48±4.90	7.5±2.45	15.79±4.90	82.93±4.90	41.27±4.90	59.3±4.90	913±4.90	5.57±2.94
Middle-late group								
Yantar st.	10.08±2.45	7.7±2.45	9.34±2.45	64.41±4.90	31.2±4.90	45.6±4.90	1177±4.90	4.39±2.94
Zarevo	21.08±4.90	7.0±2.45	19.42±4.90	80.57±4.90	43.74±4.90	63.3±4.90	604±4.90	4.71±2.94
Musinsky	15.12±4.90	6.2±2.45	12.72±2.45	84.62±4.90	45.99±4.90	65.0±4.90	1069±4.90	5.98±2.94
Fregata	19.23±4.90	8.6±2.45	10.29±2.45	98.32±4.90	46.62±4.90	64.5±4.90	920±4.90	6.55±2.45

Note: $\varepsilon = 0.95$

The largest size of starch granules in a fresh tuber cut was observed in the: middle-early group - Svitanok Kievsky (49.09 microns), middle-late group - Fregata (46.62 microns), Musinsky (45.99 microns) and Zarevo (43.74 microns) varieties.

In previous studies, the authors report that late varieties have a larger relative grain diameter compared to early-ripening varieties and the average size of starch granules has a positive correlation with the tubers specific weight, meaning that the dry matter content is higher when the variety tubers have large starch granules [14, 18]. In our studies, this pattern was not confirmed. According to the data obtained, the share of large and medium-sized grains (more than 50%) was observed in the following genotypes: early maturation - Yubilyar (56.8%), Kabo (63.7%); middle-early - Solnyishko (50.3%), Svitanok Kievsky (75.5%); mid-ripening - st. Dachny (56.6%), Fritella (59.3%); middle-late -Zarevo (63.3%), Fregata (64.5%), Musinsky (65.0%).

In our studies, the small granules were mostly rounded and oval in shape, while the medium and large granules were oval, elongated, and irregular in shape (Fig. 1).

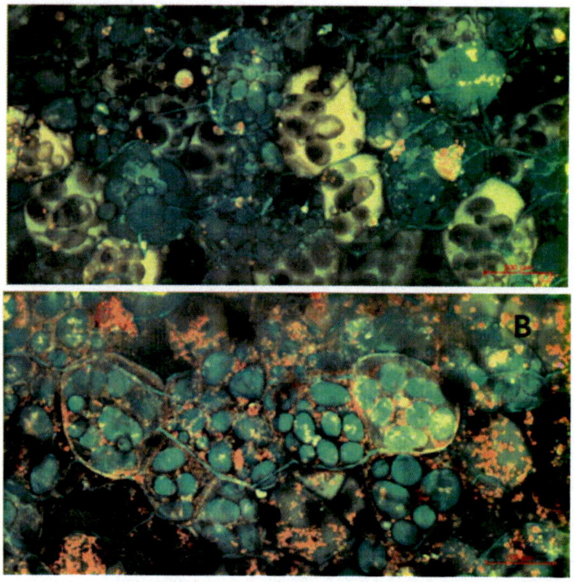

Fig. 1. Microscopic images of potato starch granules of Antonina (A) and Svitanok Kievsky (B) varieties.

An important indicator is the yield of starch per hectare, which is calculated taking into account the yield of the variety. According to our data, the average starch collection for 2 years of study in the tested cultivars varied from 3.97–6.55 t/ha. Of great practical interest are the varieties: Fregata (6.55 t/ha), Musinsky (5.98 t/ha), Svitanok Kievsky (5.85 t/ha).

4 Conclusion

As a result of the study of 179 potato cultivars of different maturity groups, source varieties were distinguished by economically valuable characteristics:

- with a productivity of more than 1000 g/plant: early-ripening group - Zhukovskiy ranniy, Impala, Kolymsky, Krepysh, Queen Anna, Laperla, Matushka, Pamyati Kulakova, Red Lady, Udacha; middle-early group – Arktika, Belmonda, Zoya, Kamchatka, Sante, Sudaryinya, 7 for 7; mid-ripening group - Alim, Alaska, st. Dachny, Zlatka, Slavyanka, Solntsesvet; middle-late group - Vetraz, Kivi, Musinsky, Pobeda, Ragneda, st. Yantar.
- with a starch content of more than 15%: early-ripening group-Antonina, Kabo; middle-early group - Svitanok Kievsky, Solnyishko; mid-ripening group - Garant, Nakra, Fritella; middle-late group - Zarevo, Musinsky, Fregata.
- with starch granules larger than 35 microns (above 50%): early-ripening group - Kabo, Yubilyar; middle-early group - Svitanok Kievsky (75%), Solnyishko (50.3%); mid-ripening group - st. Dachny (56.6%), Fritella (59.3%); middle-late group - Zarevo (63.3%), Fregata (64.5%), Musinsky (65.0%).

Thus, according to the characteristics complex (productivity, starchiness, high content of large- and medium-sized grains) that meet the requirements of suitability for processing into starch, the varieties Musinsky, Svitanok Kievsky, Solnyishko, Fregata, Fritella are recommended for production and selection.

References

1. Shin, E.-H., Baik, M.-Y., Kim, H.-S.: Comparison of physicochemical properties of starches and parenchyma cells isolated from potatoes cultivated in Korea. Food Sci. Biotechnol. **24**(3), 955–963 (2015). https://doi.org/10.1007/s10068-015-0123-y
2. Szpunar-Krok, E., Janczak-Pieniazek, M., Migut, D., Piechowiak, K.S.T., Pawlak, R., Balawejder, M.: Physiological and Biochemical properties of potato (Solanum tuberosum L.) in response to ozone-induced oxidative stress. Agronomy **10**, 1745 (2020). https://doi.org/10.3390/agronomy10111745
3. Abbaso, K.S., Qayyum, A., Mehmood, A., Mahmood, T., et al.: Analysis of selective potato varieties and their functional assessment. Food Sci. Technol. **39**(2), 308–314 (2019). https://doi.org/10.1590/fst.26217
4. Tessema, L., Mohammed, W., Abebe, T.: Evaluation of potato (Solanum tuberosum L.) varieties for yield and some agronomic traits. Open Agric. **5**(1), 63–74 (2019). https://doi.org/10.1515/opag-2020-0006
5. Zhevora, S.V., Anisimov, B.V., Simakov, E.A., Oves, E.V., Zebrin, S.N.: Potato: problems and prospects. Potato Veg. **7**, 2–7 (2019). https://doi.org/10.25630/PAV.2019.89.92.006
6. Gondwe, R.L., Kinoshita, R., Suminoe, T., Aiuchi, D., Palta, J.P., Tani, M.: Yield and quality characteristics of popular processing potato (Solanum tuberosum L.) cultivars in two contrasting soil types under grower management in Hokkaido, Japan. Potato Res. **63**(3), 385–402 (2020). https://doi.org/10.1007/s11540-019-09446-9
7. Jagadeesan, S., Govindaraju, I., Mazumder, N.: An insight into the ultrastructural and physiochemical characterization of potato starch: a review. Am. J. Potato Res. **97**(5), 464–476 (2020). https://doi.org/10.1007/s12230-020-09798-w

8. Das, S., et al.: Evaluation of quality parameters of seven processing type potato (Solanum tuberosum L.) cultivars in the eastern sub-himalayan plains. Foods **10**, 1138 (2021). https://doi.org/10.3390/foods10051138
9. Khlestkin, V., Eltsov, I.: Different reactivity of raw starch from diverse potato genotypes. Molecules **26**, 226 (2021). https://doi.org/10.3390/molecules26010226
10. Kim, I.V., Volkov, D.I., Klykov, A.G.: The study of bioresource collection of potato (Solanum Tuberosum L.) in the conditions of Primorsky Krai. IOP Conf. Ser. Earth Environ. Sci. **547**(1), 012013 (2020). https://doi.org/10.1088/1755-1315/547/1/012013
11. Lizarazo, H.S.P., Hurtado, R.G.G., Rodríguez, C.L.F.: Physicochemical and morphological characterization of potato starch (Solanum tuberosum L.) as raw material for the purpose of obtaining bioethanol. Agronomía Colombiana **33**(2), 244–252 (2015). https://doi.org/10.15446/agron.colomb.v33n2.47239
12. Niu, S., Li, X.-Q., Tang, R., et al.: Starch granule sizes and degradation in sweet potatoes during storage. Postharvest Biol. Technol. **150**(2), 137–147 (2019). https://doi.org/10.1016/j.postharvbio.2019.01.004
13. Yusuph, M., Tester, R.F., Ansell, R., Snape, C.E.: Composition and properties of starches extracted from tubers of different potato varieties grown under the same environmental conditions. Food Chem. **82**, 283–289 (2003). https://doi.org/10.1016/S0308-8146(02)00549-6
14. Zhang, J., et al.: Starch granule size variation and relationship with tuber dry matter content in heritage potato varieties. Sci. Hortic. **130**(3), 503–509 (2011). https://doi.org/10.1016/j.scienta.2011.08.006

The Effect of Tillage Minimization on the Weed Infestation of Soybean and Grain Crops in the Conditions of the Amur Region in Russia

Elena Zakharova(✉) and Aleksej Nemykin

Far Eastern State Agrarian University, 86, Politekhnicheskaya str, 675005 Blagoveshchensk, Russia

Abstract. Agriculture in the Amur Region has passed from moldboard tillage to non-moldboard tillage over the past two decades. The minimization is associated with the issue of weed infestation increase. Analysis of long-term monitoring data for the period from 1999 to 2020 revealed an increase in the number of perennial weeds in the structure of field plant communities of grain crops by 1.8 times, and soybean - by 3.2 times. In field experiments, it was found that replacing moldboard plowing with non-moldboard one leads to an increase in the number of perennial weeds by 2%, their weight increases by 1.5 times. Based on the results of field experiments, the use of non-moldboard loosening to a depth of 25–27 cm in the complex of basic tillage and moldboard loosening to a depth of 18–20 cm was substantiated as an effective method of minimizing pre-sowing soil cultivation to soybean. Taking into account the positive aspects of minimization as the basis of resource conservation and greening of agriculture, the identification of the most effective methods of soil cultivation in tackling weeds in the Amur Region will increase the productivity of agriculture in the region.

Keywords: Soybean · Grain crops · Weeds · Plowing · Tillage minimization · Direct sowing

1 Introduction

The minimum tillage technology is the main direction of conservation farming in the Amur region. It includes the replacement of moldboard tillage to non-moldboard, a decrease in the depth of tillage, a reduction in the number of technological operations, the use of combined and wide-cut agricultural machines and aggregates. The minimization of tillage in the Amur Region is of greatest importance, first of all, as an opportunity to reduce direct economic costs for the cultivation of agricultural crops (energy and labor resources).

According to numerous studies, the soil positive protective effect of tillage minimization helps to reduce the anthropogenic impact on the soil, provides protection against erosion, and preserves the health of the soil [1–3]. The replacement of moldboard tillage with non-moldboard, direct sowing technologies are acquiring particular importance in the context of overcoming the consequences of climate change [4–6].

A. Muratov and S. Ignateva (Eds.): AFE 2021, LNNS 353, pp. 451–459, 2022.
https://doi.org/10.1007/978-3-030-91402-8_51

At the same time, contradictions arise, leading to a decrease in the yield of agricultural crops. A decrease in the rate of mineralization of organic matter requires additional fertilization to provide crop plants with nutrients; soil compaction requires additional loosening to optimize the agrophysical properties of the soil; an increase in weediness of agricultural crops requires chemical weed control [7–9]. To overcome the contradictions, the choice of optimal technological operations for tillage is relevant [10]. The solution of weed control problems, depending on the tillage, is of regional nature [11–13].

Minimization in agriculture in the Amur Region is growing; direct sowing without tillage is spreading more widely. This is facilitated by a large share of soybean in the structure of sown areas (74%), which leads to repeated crops. With late harvesting and waterlogging, autumn tillage after soybean is impossible. In the spring, it is impractical to carry out moldboard processing under soybeans, since the soil thaws late, and dries up during intensive loosening. Therefore, the most suitable for spring processing are non-moldboard methods. Grain crops are good predecessors for soybeans in a crop rotation, after which it is possible to prepare the soil to soybean in the fall. But the cultivation of grain crops is not economically efficient for agricultural enterprises. Thus, the share of grain crops in rotational cropping of the Amur Region is small - 19%. One of the factors limiting the growth of soybean yields is weed infestation above the economic threshold of harm. Tackling weeds is carried out mainly by the chemical method, which poses a danger to the ecology of the region [14].

Therefore, the purpose of our study is to identify the methods of minimum tillage that are most effective for weed control of agricultural crops in the conditions of the Amur Region.

The following tasks were solved to achieve the purpose of our study:

1. To establish the change in the weed infestation of grain crops and soybean, depending on the minimization of tillage in the Amur region.
2. To evaluate the effect of non-moldboard tillage on the development of weeds during the cultivation of barley.
3. To substantiate the effectiveness of methods of minimization of tillage to soybean in tackling weeds.

2 Materials and Methods

The change in the weediness of spiked cereals (wheat, barley) and soybean, depending on the minimization of tillage in the Amur Region, was established by analyzing the data of long-term monitoring. In 1999, the fields of grain crops and soybean were examined on an area of 5778 hectares on the land use territory of JSC "Luch" in rural localities of Ivanovka in Ivanovsky district. In 2002 (4287.4 hectares) and 2006 (4539 hectares), a survey was carried out on the land use territory of the agrofirm "ARC" (until 2003 - the educational and experimental farm of the Far Eastern State Agrarian University) in rural localities of Gribskoye. The weed estimation was carried out by the quantitative-weight method. Samples were taken using a route method at each plot of crop rotation. For the analysis, the data were averaged from the entire surveyed area.

In 2005, 2007, 2008, a field experiment was carried out on the experimental field of the Far Eastern SAU in rural localities of Gribskoye in order to evaluate the effect of non-moldboard tillage on weed development during barley cultivation. The experiment scheme was the following: 1) plowing with the moldboard plow PLN-4-35 to a depth of 18–20 cm; 2) non-moldboard loosening with the cultivator KPE-3.8 to a depth of 16–18 cm. The placement of the plots was randomized. The repetition was fivefold. The plot area was 80 m^2. The soil was meadow-chernozemic medium-thick. The tillage was carried out in the year preceding the barley sowing in the third ten-day period of October after the soybean harvest. The barley was sown in the third ten-day period of April. The weed estimation was counted by the quantitative-weight method.

In 2015–2016 a field experiment was conducted in the production conditions of JSC "Luch" in order to substantiate the effectiveness of minimizing methods of the main soil cultivation to soybean in tackling weeds. The experiment scheme was the following: 1) direct sowing with Amazone Primera DMC-12000; 2) disk plowing with BDM-8× 4P to a depth of 10–12 cm; 3) non-moldboard loosening with the cultivator KUP-6 to a depth of 14–16 cm; 4) deep loosening with Salford 9715 CTS to a depth of 25–27 cm; 5) moldboard plowing with PLN-8-35 to a depth of 18–20 cm. The soil was meadow-chernozemic. The tillage after predecessor harvesting (grain crops) was in the first ten-day period of October. The soybean was sown in the third ten-day of May. The weed estimation was counted quantitatively.

In 2020, a field experiment was conducted in the production conditions of a peasant farm in the Belogorsk District of the Amur Region in order to substantiate the effectiveness of minimizing methods of the pre-sowing soil cultivation to soybean in tackling weeds. The experiment scheme was the following: 1) direct sowing with MAZUR 1100-6,0 D DUPLO; 2) non-moldboard loosening with the cultivator KÖCKERLING to a depth of 8–10 cm; 3) non-moldboard loosening with the subsoiler cultivator LEMKEN Karat 9 to a depth of 18–20 cm. The soil was alluvial soddy. The tillage before soybean sowing was in the first ten-day period of June. The weed estimation was counted by the quantitative-weight method.

3 Results

In 1999, the fields of spiked cereals were surveyed on an area of 2478 hectares. A high degree of weed infestation was identified. The total number of weeds was 168.4 pcs/m^2 (Fig. 1). Galeopsis bifida Boenn. and Echinochloa crus-galli L. were the predominant species, representing juvenile weeds, and made up 7.0% and 10.4% of the total number of plants respectively. The predominant species, representing perennial weeds were Elytrigia repens (L.) Nevski and Equisetum arvense L. (rootstock, 2.67 and 2.77% respectively). The cereal crops in the structure of the field plant community accounted for 72.6% of the total number of plants, juvenile weeds - 20.9%, perennials - 6.5%.

Where: A - The structure of the field plant community of spiked cereals, B - The structure of the field plant community of soybean.

In 1999, the fields of soybean were surveyed on an area of 3300 hectares. A high degree of weed infestation was revealed. The total number of weeds was 148.4 pcs/m^2 (Fig. 1). The predominant spccics was juvenile weed *Acalypha australis L.* The largest number among perennial weeds of plants made up *Equisetum arvense L.* The soybean in the structure of the field plant community accounted for 31.2% of the total number

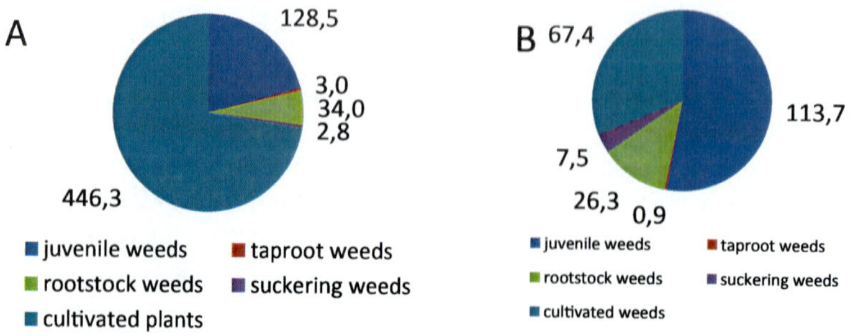

Fig. 1. The weed infestation of spiked cereals and soybean in the JCS "Luch" (1999), pcs/m^2.

of plants, juvenile weeds - 52.7%, perennials - 16.1%. The quantitative accounting was supplemented by weight accounting. It was found that in the grain crops sowings, juvenile weeds made up 2.7%, perennials - 2.5%, cultivated plants - 94.8% of the total mass. In soybean crops, juvenile weeds accounted for 11.7%, perennials - 9.9%, cultivated plants - 78.4% of the total mass.

In 2002, the fields of spiked cereals were surveyed on an area of 1786,7 hectares. A high degree of weed infestation was identified. The total number of weeds was 268.7 pcs/m^2 (Fig. 2). The predominant species was juvenile weed *Echinochloa crus-galli L.* The largest number among perennial weeds of plants made up *Equisetum arvense L.* and *Sonchus arvensis L.* The cereal crops in the structure of the field plant community accounted for 57.1% of the total number of plants, juvenile weeds - 36.6%, perennials - 6.4%.

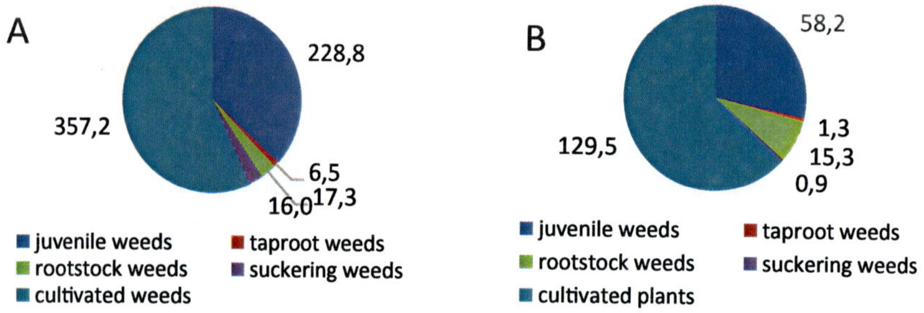

Fig. 2. The weed infestation of spiked cereals and soybean in the educational and experimental farm of the Far Eastern State Agrarian University (2002), pcs/m^2.

Where: A - The structure of the field plant community of spiked cereals, B - The structure of the field plant community of soybean.

In 2002, the fields of soybean were surveyed on an area of 2500,7 hectares. A high degree of weed infestation was revealed. The total number of weeds was 75.6 pcs/m^2 (Fig. 2). The predominant species were represented by juvenile weeds Echinochloa crus-galli L., which made up 15.5% of the total number of plants; and Chenopodium album L. The largest number among perennial weeds of plants made up Equisetum arvense

L. (rootstock weed, 5.6%). The soybean in the structure of the field plant community accounted for 63.1% of the total number of plants, juvenile weeds - 28.4%, perennials - 8.5%.

In 2006, the fields of spiked cereals were surveyed on an area of 2010 hectares. A high degree of weed infestation was identified. The total number of weeds was 209.8 pcs/m^2 (Fig. 3). The predominant species were represented by juvenile weeds Eriochloa villosa Kunth., which made up 17.8% of the total number of plants; and Setaria glauca (L.) Beauv. The largest number among perennial weeds of plants made up Sonchus arvensis L. The cereal crops in the structure of the field plant community accounted for 62.9% of the total number of plants, juvenile weeds - 31.8%, perennials - 5.3%.

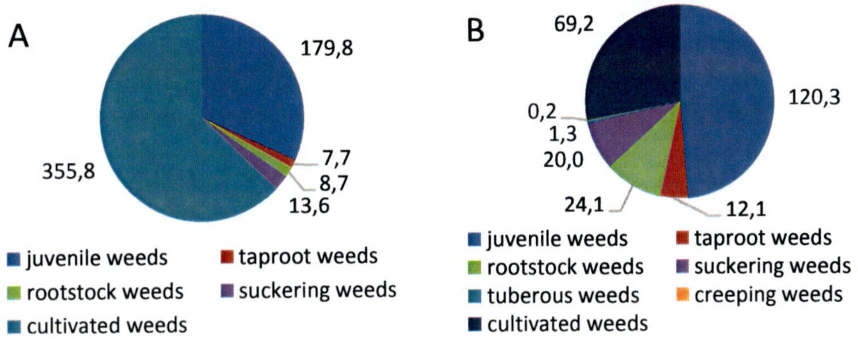

Fig. 3. The weed infestation of spiked cereals and soybean in the agrofirm "ARC" (2006), pcs/m^2.

Where: A - The structure of the field plant community of spiked cereals, B - The structure of the field plant community of soybean.

In 2006, the fields of soybean were surveyed on an area of 2529 hectares. A high degree of weed infestation was revealed. The total number of weeds was 177.9 pcs/m^2 (Fig. 3). The predominant species were represented by juvenile weeds Acalypha australis L., which made up 22.1% of the total number of plants; and Echinochloa crus-galli L. The largest number among perennial weeds of plants made up Equisetum arvense L. and Sonchus arvensis L. The soybean in the structure of the field plant community accounted for 28.0% of the total number of plants, juvenile weeds - 48.7%, perennials - 23.3%. The quantitative accounting was supplemented by weight accounting. It was found that in the cereal crops sowings, juvenile weeds made up 6.0%, perennials - 6.0%, cultivated plants - 88.1% of the total mass. In soybean crops, juvenile weeds accounted for 18.6%, perennials - 17.6%, cultivated plants - 63.8% of the total mass.

The current state of the field plant community was determined, based on the results of surveys of the FSBEI "RAC". Juvenile weeds were represented by the following early spring crops: Commelina communis L., Galeopsis bifida Boenn., Chenopodium album L.; and late spring crops: Acalypha australis L., Abutilon Theophrasti Medik., Setaria glauca (L.) Beauv., Echinochloa crus-galli L., Amaranthus retroflexus L. Annual weeds were represented by taproots: Artemisia absinthium L., Artemisia vulgaris L.; rootstocks: Elytrigia repens (L.) Nevski, Equisetum arvense L.; and suckering weeds: Sonchus arvensis L., Cirsium arvense (L.) Scop. It was found that 67.4% of grain crops and 69.6% of soybean had weed infestation above the economic threshold of harm. The total number of weeds in cereal crops sowings was 140.5 pcs/m^2, where juvenile weeds made up 50.1% and perennials - 49.9%. The total number of weeds in soybean sowings was 240 pcs/m^2, where juvenile weeds made up 58.3% and perennials - 41.7% (Fig. 4).

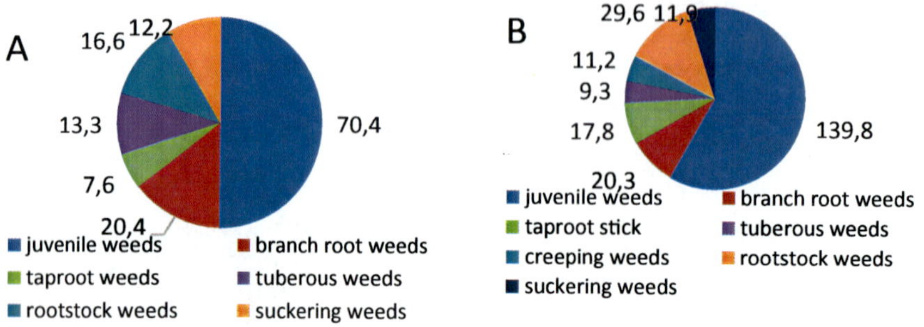

Fig. 4. The weed infestation of spiked cereals and soybean in the Amur region (2018–2020), pcs/m^2.

Where: A - The structure of the field plant community of spiked cereals, B - The structure of the field plant community of soybean.

In a field experiment, investigating the effect of non-moldboard tillage on the weed development during the barley cultivation, it was revealed that after plowing with a moldboard plow to a depth of 18–20 cm, a high degree of weediness is formed. The total number of weeds was 81.2 pcs/m^2. The predominant species, representing juvenile weeds Amaranthus retroflexus L. made up 5.2% of the total number of plants. The predominant species, representing perennial weeds was Equisetum arvense L. The barley in the structure of the field plant community accounted for 71.0% of the total number of plants, juvenile weeds - 19.6%, perennials - 9.4%. In the variant with non-moldboard loosening to a depth of 16–18 cm, a high degree of weed infestation is formed. The total number of weeds was 97.5 pcs/m^2. The predominant species, representing juvenile weeds Echinochloa crus-galli L. made up 5.95% of the total number of plants. The predominant species, representing perennial weeds was Equisetum arvense L. The barley in the structure of the field plant community accounted for 66.8% of the total number of plants, juvenile weeds - 20.3%, perennials - 13.0%. The weight of all weeds increased 1.4 times compared to moldboard plowing, especially the weight of perennials ones, which increased 1.5 times.

In a field experiment, investigating the effect of minimizing methods of basic tillage on the weed infestation, it was revealed that in all variants a high degree of weediness is formed. The predominant species of weeds were Commelina communis L.; Echinochloa crus-galli L., Acalypha australis L.; Equisetum arvense L. In the variant with direct sowing the total number of weeds was 92.7 pcs/m^2. This was 5.1–10.7 pcs/m^2 more than in variants with basic tillage with various agricultural machines. In the variant with direct sowing, there was an intensive development of perennial weeds (Equisetum arvense L., Artemisia absinthium L., Sonchus arvensis L., Cirsium arvense (L.) Scop.). These weeds accounted for 31.5% of the total number of plants. This was 4.5–15.2%

more than in other variants. The smallest number of perennials was in the variant with deep loosening with Salford 9715 CTS to a depth of 25–27 cm (16.3% of the total number of plants) and in the variant with plowing with PLN-8-35 to a depth of 18–20 cm (25.2%).

In a field experiment, investigating the effect of the minimization methods of the pre-sowing soil cultivation on the weed infestation of soybean, it was revealed that a high degree of weediness is formed during direct sowing. The total number of weeds was 131.2 pcs/m^2 (66.7% of the total number of plants). The weediness is less in the variant of pre-sowing loosening with the cultivator KÖCKERLING to a depth of 8–10 cm and the subsoiler cultivator LEMKEN Karat 9 to a depth of 18–20 cm than during direct sowing in half. The predominant species, representing juvenile weeds Amaranthus retroflexus L. made up 16.7% of the total number of plants in the variant of loosening to a depth of 18–20 cm. The predominant species, representing perennial weeds was Equisetum arvense L. The soybean in the structure of the field plant community accounted for 67.4% of the total number of plants, juvenile weeds - 18.1%, perennials - 14.5%. The predominant species, representing perennial weeds were Equisetum arvense L. and Sonchus arvensis L., in the variant of loosening to a depth of 8–10 cm. The soybean in the structure of the field plant community accounted for 70.9% of the total number of plants, juvenile weeds - 0.7%, perennials - 28.4%. The predominant species, representing juvenile weeds Acalypha australis L. made up 8.9% of the total number of plants in the variant of direct sowing. The predominant species, representing perennial weed were Sonchus arvensis L. and Equisetum arvense L. The mass of perennial weeds in the variant with direct sowing is twice the mass of cultivated plants. The soybean in the structure of the field plant community accounted for 33.3% of the total number of plants, juvenile weeds - 16.3%, perennials - 50.4%.

4 Discussion

Analysis of the data of long-term monitoring of weed infestation of spiked cereals and soybean for the period from 1999 to 2020 revealed an increase in the number of perennial weeds in the structure of field plant communities. Dangerous rootstock and suckering weeds develop especially intensively. In the production sowings of the Amur Region, the perennial weed infestation of grain crops increased by 1.8 times during this period, soybeans - by 3.2 times. Changes in the structure of field plant communities are due to the minimization of soil cultivation by replacing moldboard tillage with non-moldboard, decreasing in the depth of tillage, reducing the number of technological operations to direct sowing without tillage and abandoning inter-row soybean cultivation. The increase in weediness indicates insufficient effectiveness of herbicides against perennial weeds. The negative trend is intensifying against the background of an increase in the share of soybean in crop rotations, which leads to its cultivation. With late harvesting of soybean before the soil freezes, it is not possible to carry out a set of agrotechnical measures to combat weeds; soil preparation for sowing is postponed to the spring. In such circumstances, minimization is a must. This determines the relevance and significance of the research carried out. For grain crops, minimization under the conditions of the Amur Region is relevant because early sowing of grain crops and late harvesting of the predecessor limit the possibilities of carrying out a wide range of soil cultivation techniques aimed at tackling weeds. The agrotechnical role of grain crops in rotational cropping is increasing as a good predecessor for soybean. Replacing moldboard plowing with non-moldboard tillage is relevant from the point of view of resource conservation. At the same time, as established in our experience, the number of perennial weeds increases in the structure of the plant field community. Therefore, it is necessary to continue research in the direction of identifying the most effective minimizing tillage method to grain crops. Based on the results of the field experiment, the effectiveness of the control of vegetatively propagating perennial weeds of soybean by non-moldboard loosening to a depth of 25–27 cm was substantiated as a method of minimization of basic tillage. In the complex of methods of pre-sowing soil cultivation, the most effective is non-moldboard loosening to a depth of 18–20 cm. Over the past two decades, agriculture in the Amur Region has almost completely moved to non-moldboard tillage.

5 Conclusion

Analysis of the weediness monitoring data for the period from 1999 to 2020 revealed an increase in the number of perennial weeds in the structure of field plant communities of grain crops by 1.8 times, and of soybeans - by 3.2 times. Dangerous rootstock and suckering weeds develop especially intensively. In field experiments, it was found that replacing moldboard plowing with non-moldboard plowing leads to an increase in the number of perennial weeds by 2%, and their weight enlarges by 1.5 times. The use of non-moldboard loosening to a depth of 25–27 cm in the complex of basic tillage and non-moldboard loosening to a depth of 18–20 cm is substantiated as an effective method of minimizing pre-sowing soil cultivation for soybean. Taking into account the positive aspects of minimization as the basis of resource conservation and greening of agriculture, it is advisable to further search for the most effective methods of soil cultivation in tackling weeds in the Amur Region.

References

1. Nunes, M.R., et al.: Biological soil health indicators respond to tillage intensity: a US meta-analysis. Geoderma **369**, 114335 (2020). https://doi.org/10.1016/j.geoderma.2020.114335
2. Wu, F., et al.: Effects of deep vertical rotary tillage on the grain yield and resource use efficiency of winter wheat in the Huang-Huai-Hai Plain of China. J. Integr. Agric. **20**(2), 593–605 (2021). https://doi.org/10.1016/S2095-3119(20)63405-0
3. de Freitas, P.L., Landers, J.N.: The Transformation of agriculture in Brazil through development and adoption of zero tillage conservation agriculture. Int. Soil Water Conserv. Res. **2**(1), 35–46 (2014). https://doi.org/10.1016/S2095-6339(15)30012-5
4. Tran, D.Q., Kurkalova, L.A.: Persistence in tillage decisions: aggregate data analysis. Int. Soil Water Conserv. Res. **7**(2), 109–118 (2019). https://doi.org/10.1016/j.iswcr.2019.03.002
5. Chi, J., et al.: Assessing carbon and water dynamics of no-till and conventional tillage cropping systems in the inland Pacific Northwest US using the eddy covariance method. Agric. Forest Meteorol. **218**, 37–49 (2016). https://doi.org/10.1016/j.agrformet.2015.11.019
6. Kertész, A., Madarász, B.: Conservation agriculture in Europe. Int. Soil Water Conserv. Res. **2**(1), 91–96 (2014). https://doi.org/10.1016/S2095-6339(15)30016-2

7. Schwalbert, R., et al.: Soil tillage affects soybean growth and promotes heavy metal accumulation in seeds. Ecotoxicol. Environ. Safety **216**, 112191 (2021). https://doi.org/10.1016/j.ecoenv.2021.112191
8. Rusu, T.: Energy efficiency and soil conservation in conventional, minimum tillage and no-tillage. Int. Soil Water Conserv. Res. **2**(4), 42–49 (2014). https://doi.org/10.1016/S2095-6339(15)30057-5
9. Pittelkow, C.M., et al.: When does no-till yield more? A global meta-analysis. Field Crops Res. **183**, 156–168 (2015). https://doi.org/10.1016/j.fcr.2015.07.020
10. Chetan, C., Rusu, T., Chetan, F., Simon, A.: Influence of soil tillage systems and weed control treatments on root nodules, production and qualitative indicators of soybean. Procedia Technol. **22**, 457–464 (2016). https://doi.org/10.1016/j.protcy.2016.01.088
11. Feledyn-Szewczyk, B., Smagacz, J., Kwiatkowski, C.A., Harasim, E., Woźniak, A.: Weed flora and soil seed bank composition as affected by tillage system in three-year crop rotation. Agriculture **10**(5), 186 (2020). https://doi.org/10.3390/agriculture10050186
12. Gawęda, D., Haliniarz, M., Łukasz, J., Bronowicka-Mielniczuk, U.: Weed infestation and health of the soybean crop depending on cropping system and tillage system. Agriculture **10**(6), 1–20 (2020). https://doi.org/10.3390/agriculture10060208
13. Berbeć, A.K., et al.: Organic but also low-input conventional farming systems support high biodiversity of weed species in winter cereals. Agriculture **10**(9), 1–16 (2020). https://doi.org/10.3390/agriculture10090413
14. Shchegorets, O., Tikhonchuk, P., Bumbar, I., Yakimenko, A.: Innovation as a factor in increasing the efficiency of soybean production in the Amur region. In: E3S Web of Conferences, Ecological and Biological Well-Being of Flora and Fauna (EBWFF-2020), p. 05010 (2020). https://doi.org/10.1051/e3sconf/202020305010

Reproduction of Bee and Drone Brood in Bee Colonies of the Far Eastern Honey Bee (*Apis Mellifera* Linnaeus, 1758) in the Conditions of Primorsky Krai

Maxim Sharov(✉)

FSC of Agricultural Biotechnology of the Far East named after A.K. Chaika, Timiryazevsky Stl, 692539 Primorsky Krai, Ussuriysk, Russia

Abstract. This paper discusses the biological features of growing bee and drone broods of the Far Eastern honey bee. The development of bee individuals has two peak values, the first before the main honey harvest from linden - 230.1 squares and the second at the end of August during the period of building up the winter generation - 236.4 squares. The drone brood is reared simultaneously with the bee brood and reaches a maximum of 24.1 squares in mid-June and 20.5 squares in early August. During the July honey harvest, the minimum amount of brood is raised - 54.2 squares of bees and 1.3 squares of drones. This is due to the fact that the bees begin to limit the egg production of the queen bee by filling free cells with honey, thereby freeing the nurse bees from brood feeding and switching them to the preparation of nectar.

Keywords: Beekeeping · Bee colonies · Bee brood · Drone broods · Linden

1 Introduction

The biological ability of bee colonies to change their quantitative composition throughout the year allows to endure periods unfavorable for their lives at a lower cost. A colony of bees can increase its size by several times in a short period, which makes it possible to use honey harvest most efficiently and provide itself with food. The increase in the number of bees in a family is due to brood rearing. In the spring-summer period, when bees lead an active lifestyle, their average life span is 30–40 days. In the hive, old bees die every day, and young bees are born to replace them. If the yield of young individuals exceeds that of old ones - the colony grows, if the yield is equal to the yield of bees - no increase in its number is observed [1, 2].

The growth of bee colonies is conventionally divided into several periods, which differ both in the number of bees in the colony and in their physiological state. In early spring, the number of bees in a colony is minimal, and their calendar age is greatest. During this period, bees are engaged in the cultivation of the next generation. Further, the colonies grow intensively, accumulating bees by the beginning of the main honey harvest. During honey harvesting, the number of bees in the family decreases as a

A. Muratov and S. Ignateva (Eds.): AFE 2021, LNNS 353, pp. 460–468, 2022.
https://doi.org/10.1007/978-3-030-91402-8_52

result of intensive work on collecting and storing nectar. In the late summer period, bees intensively grow broods, but despite this, the number of individuals in the family continues to decrease. For the cultivation of brood, the quality of the honeycombs [3] and the temperature regime are of great importance [4].

Bee and drone brood facilitates the increase in the number of *Varroa destructor* mites. For this reason, brood cells should be carved out periodically to decrease the population of parasites [5–8].

The aim of the study is to study the growth and development of bee and drone broods in honey bee colonies (Apis mellifera Linnaeus, 1758) in the Far East under the conditions of Primorsky Krai [9, 10].

Fig. 1. Scientific and production apiaries.

2 Materials and Methods

The work was carried out in 2017–2019 at the FSBSI "FSC of Agricultural Biotechnology of the Far East named after A.K. Chaika". The experiment involved 10 bee families (Fig. 1). The brood amount was measured using a grid frame (Fig. 2) with cells 5 × 5 cm (1 5 × 5 cm square contains 100 bee cells and 67 drone cells): bee brood was measured once in 12 days, drone brood - once in 14 days, according to the standard methods of the Federal Beekeeping Research Center [11]. During the entire active period, the indicators of the nectar brought by the control family were taken into account daily, the timing of cold snaps, rainy periods, the main honey collection, flowering, and the nature of nectar release of the prevailing melliferous plants were recorded.

3 Results

The apiary is located in the foothill-forest-steppe zone of the Ussuriysk district of Primorsky Krai, the conditions of honey production are typical for this zone. Within the radius of productive bee flight, the territory abounds in such spring honey plants as varieties of willows, maples, honey plants, and garden pollen plants. For the secondary late summer honey harvest, the apiary is taken out to the forest-steppe area, where the

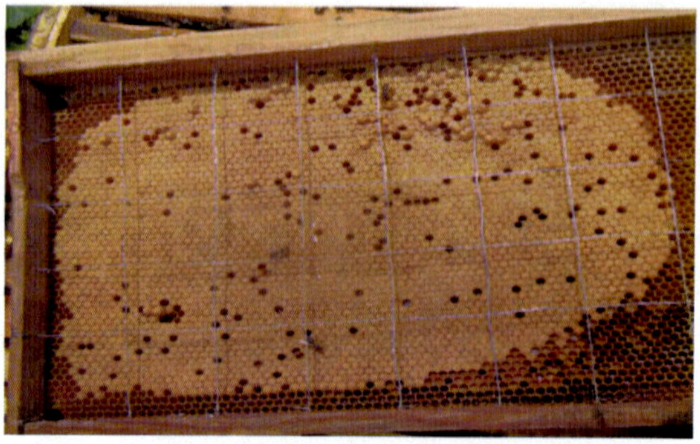

Fig. 2. Bee brood measurement.

main melliferous plants are field weeds (Canada thistle and field milk thistle), bicolor lespedeza, saw-wort, and other August melliferous plants. Table 1 presents the nature of nectar secretion and weather conditions during the research period.

The air during the day in March warmed up to +10 °C on average, and night temperatures dropped to −8 °C, it rained and snowed often. Due to low temperatures, spring honey plants produced insufficient amounts of nectar; the yield did not exceed 500 g per day in the control hive (Fig. 3), the balance was more often recorded. Willows (Salix siuzevii, Salix caprea, Salix rorida) have bloomed since mid-March. Maples followed the willow. Mono maple blooms first - in the first decade of May, followed by the bearded and Manchurian maples, the Amur maple blooms in the second half of June. The period of April 5–May 12 is the period of willow flowering in the conditions of the Ussuriysk district of Primorsky Krai.

On sunny days in April, bees bring pollen and nectar in large quantities. Amur cork tree (Phellodendron amurense Rupr.), Manchurian cherry (Padus Maackii Rupr.), and raspberries (Rubus idaeus L.) are also of great interest to bees.

The period of replacement of the bees kept during the winter by young bees in bee colonies ends by the beginning of May In the conditions of Primorsky Krai.

So, in our observations (Table 2), the sealed brood was found during the count on March 22 - 20.1 squares, and on May 1 this indicator increased 7.8 times - 156.1 squares. During May and the first half of June, the queens continued to lay eggs, and the size of sealed brood reached the highest value - 230.1 squares (Fig. 3). In the second half of June, due to low yields of nectar, the queen bee decreased egg production, this contributed to the manifestation of a swarming mood in the hives and, accordingly, led to a decrease in the brood by 23.7% (175.4 squares).

The fodder base of beekeeping in the south of the Far East is based mainly on wild-growing melliferous plants, but linden gives over 75% of honey in the region.

Vast lands of linden (Tilia), including three species - Taketa linden (T. Taqueti C. Schn.), flowers from June 26; the Amur linden (T. amurensis Kom.) - flowers from July 5 (Fig. 4a); Manchurian linden (T. mandschurica Rupr. et Max.), flowers from July 12 (Fig. 4b); the duration of flowering of each type of linden is 12–15 days, which makes the honey harvest not only strong, but also long-lasting - from June 26 to August 1.

Table 1. Weather honey collection and conditions in 2017–2019.

Posting date	Number of nectar brought days	Control hive indicators. kg	Weather conditions	
			Average air temperature. °C	Number of rainy days
21.03–31.03	0	−0.25	+1.1	2
1.04–10.04	4	−0.2	+3.4	3
11.04–20.04	2	−0.1	+7.0	0
21.04–30.04	5	0.8	+12.6	3
1.05–10.05	6	1.0	+10.7	3
11.05–20.05	3	0.8	+13.3	5
21.05–31.05	4	0.6	+13.8	4
1.06–10.06	5	0.2	+19.2	3
11.06–20.06	1	0.1	+15.9	7
21.06–30.06	4	5.6	+19.4	0
1.07–10.07	10	8.3	+21.3	0
11.07–20.07	7	6.8	+21.9	1
21.07–31.07	4	3.4	+23.3	4
1.08–10.08	4	2.5	+24.1	5
11.08–20.08	4	2.0	+19.9	4
21.08–31.08	4	1.4	+19.4	5
1.09–10.09	2	0.3	+21.1	7
11.09–20.09	0	−0.2	+12.4	5
21.09–30.09	0	−0.3	+14.7	4
1.10–10.10	0	−0.25	+12.7	3
11.10–20.10	0	−0.2	+6.6	4
21.10–31.10	0	−0.1	+7.6	3

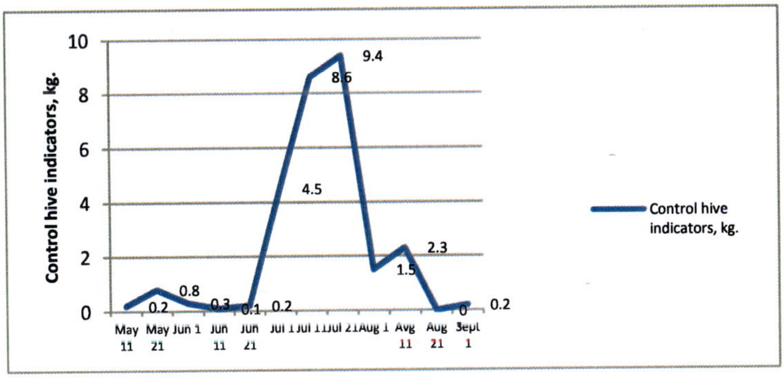

Fig. 3. Dynamics of the average daily honey yield over three years.

Table 2. Bee and drone brood growth, (n = 10).

Bee brood		Drone brood	
Posting date	Brood quantity. (M ± m) squares	Posting date	Brood quantity. (M ± m) squares
22.03	20.1 ± 0.35	–	–
7.04	63.6 ± 8.12	–	–
19.04	72.8 ± 8.86	20.04	1.5 ± 0.11
1.05	156.1 ± 10.41	4.05	6.5 ± 0.23
13.05	188.4 ± 11.23	18.05	15.3 ± 1.21
25.05	210.0 ± 13.14	2.06	22.5 ± 2.01
6.06	226.9 ± 14.54	16.06	24.1 ± 2.35
18.06	230.1 ± 15.36	30.06	22.6 ± 1.78
30.06	175.4 ± 11.75	14.07	10.4 ± 0.41
12.07	110.9 ± 9.01	28.07	1.3 ± 0.11
24.07	54.2 ± 6.12	12.08	15.4 ± 1.34
5.08	105.3 ± 7.10	26.08	20.5 ± 1.97
17.08	180.4 ± 11.63	10.09	2.0 ± 0.12
29.08	236.4 ± 13.42	–	–
11.09	161.2 ± 12.39	–	–
23.09	55.2 ± 4.12	–	–
5.10	12.3 ± 1.25	–	–

a b

Fig. 4. Amur linden *T. amurensis* Kom. (a) amd Manchu linden *T. mandschurica Rupr.* et Max. (b).

Instability is observed in the nectar release of the main honey plant linden. Thus, in the conditions of Primorsky Krai, there are low-yield years (honey yield of marketable honey per bee colony is less than 45 kg), moderate yield years (when 45–60 kg of honey are obtained from one family, and the yield of marketable products exceeds 60 kg.

The active flowering of linden and the abundant secretion of nectar of 4.5–9.4 kg (Fig. 3) contributed to the mobilization of workers for the preparation of forage. A feature of the Far Eastern bee is that the brought nectar fills all cells that are freed from brood, thereby limiting the egg-laying of the queen and the number of new generation is reduced.

Young bees performing the functions of nurses are released and go into the category of foragers, that is, they are actively switching to the preparation of carbohydrate feed. So, when measuring on July 24, the minimum amount of bee brood of 54.2 squares is observed.

During the period of late summer bee growth, the amount of brood in the hives increased and reached 236.4 squares by the end of August. The late summer active development of bee colonies allowed them to increase the required number of bees for the winter period and to prepare the required forage reserves.

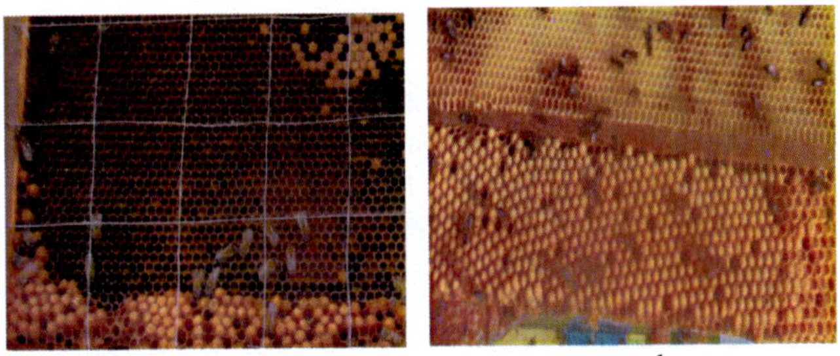

a b

Fig. 5. Drone brood in May (a) and August (b).

The appearance of drones in hives depends on the individual physiological state of a particular family. According to the data of long-term observations, in the central and northern regions of Russia, drone broods appear only in the third decade of May, and in the south - in the third decade of March.

The first drone brood was recorded during the measurement on April 20 (Table 2), on average, 1.5 squares (Fig. 5a). Bee colonies actively increased it during May and June, and by June 16, its number was 24.1 squares, which is 9.5 times less than that of bees in this period. The honey harvest from linden negatively affected the number of drone broods when measured on July 28, this figure was 1.3 squares.

After the end of the linden flowering, productive forage collection can be provided at the expense of melliferous plants of the Araliaceae Juss. family. In the Far East of Russia, the following species are of interest for beekeeping: Japanese angelica tree (Aralia elata (Miq.) Seem.), herbal aralia (A. cordata Thunb), continental aralia (A. continentalis Kitag.), castor aralia (Kalopanax septemlobus (Thunb.) Koidz.), Siberian ginseng (Eleutherococcus senticosus (Rupr. Et Maxim.) Maxim.), acanthopanax sessile-flowered

(Acanthopanax sessiliflorus (Rupr. Et Maxim.) Seem.), and devil's-club (Oplopanax elatus (Nakai) Nakai).

In August, favorable weather conditions favor the flowering of a large number of melliferous plants, which provide secondary honey harvest. Bicolor lespedeza (Lespedeza bicolor Toursz.), saw-wort (Serratula coronata L.) and many other flowering plants, including up to 150 species of nectar and pollen plants, are of predominant importance.

This activates the reproductive organs of the bee queen and the amount of bee brood reaches a peak value of 236.4 squares, at the same time drone brood is reared in large numbers with an indicator of 20.5 squares in the third decade of August (Fig. 5b).

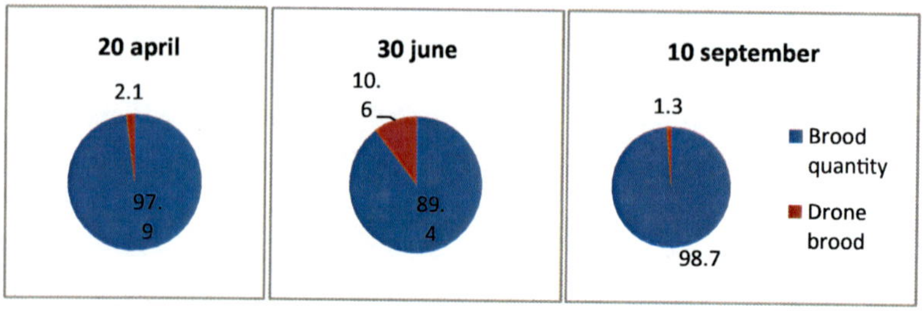

Fig. 6. The ratio between the number of bee and drone brood raised by colonies, %.

The ratio of bee and drone brood in the total number of reared larvae during the periods of minimum and maximum rearing is shown in Fig. 6. Thus, at the end of April and the beginning of September, the bee brood is bigger than drone brood at the end of April and the beginning of September and at the end of June by 97.9% and 89.4%, respectively.

Bees actively collect nectar until the end of August in Primorsky Krai. Brood rearing is gradually decreasing and ends in the first ten days of October. The last cells of the bee brood were observed on October 5 - 12.3 squares, and the last cells of the drone brood were noted on September 10 - 2.2 squares.

Fall in the Primorsky Krai is warm and dry, sunny and prolonged, frosts come in late September - early October, which has a positive effect on the flowering of melliferous vegetation and the growth of young bees in winter.

Fig. 7. Expelled drones.

In September, the number of the hive inhabitants gets reduced to 25–30 thousand individuals, and the drones are completely expelled.

In the conditions of Primorsky Krai, the mass expulsion of male bees (Fig. 7) is observed in the second half of September and they are completely absent in the hives at the end of October.

4 Discussion

It can be stated that the Far Eastern bees start growing bee broods at the beginning of the 3rd decade of March from 20.1 squares, reaching the first maximum value in early June - 230.1 squares. The second development peak was noted at the end of August, during the formation of the winter generation of bees - 236.4 squares. The first drone brood was recorded a month after the bees left their winter hive - at the end of April, reaching the maximum number of 24.1 squares at the end of June, and in the third decade of August - 20.1 squares. During the period of honey collection from linden, bees restrict the queen bee in oviposition and the bee brood is reduced to the minimum value of 54.2 squares, drone brood - to 1.3 squares. After the main honey harvest, the bee colonies leave with a minimum amount of brood and the bee colony begins active reproduction in order to accumulate the required number of individuals for long wintering.

References

1. Jensen, A.B., Evans, J., Jonas-Levi, A., Benjamin, O., Martinez, I., Dahle, B.: Standard methods for *Apis mellifera* brood as human food. J. Apic. Res. **58**(2), 1–28 (2019). https://doi.org/10.1080/00218839.2016.1226606
2. Ma, Z.-J., Yang, F., Dai, Y., Shen, Z.-J.M.: The migratory beekeeping routing problem: model and an exact algorithm. INFORMSJ Comput. **33**(1), 319–335 (2021). https://doi.org/10.1287/ijoc.2020.0958
3. AbdAl-Fattah, M.A.A.W., Yehia Ibrahim, Y., Ibrahim Haggag, M.: Some biological aspects of honey bee colonies in relation to the age of beeswax combs. J. Apic. Res. **60**(3), 405–413 (2021). https://doi.org/10.1080/00218839.2021.1899657
4. Szentgyörgyi, H., Czekońska, K., Tofilski, A.: Honey bees are larger and live longer after developing at low temperature. J. Therm. Biol. **78**, 219–226 (2018). https://doi.org/10.1016/j.jtherbio.2018.09.007
5. Lecocq, A., Foley, K., Jensen, A.B.: Drone brood production in Danish apiaries and its potential for human consumption. J. Apic. Res. **57**(3), 331–336 (2018). https://doi.org/10.1080/00218839.2018.1454376
6. Daher-Hjaij, N.Y., Alburaki, A.K.: Influence of brood type on infestation with varroa mite in colonies of local Syrian honeybees. Arab J. Plant Prot. **35**(2), 126–129 (2017). https://doi.org/10.22268/AJPP-035.2.126129
7. El Agrebi, N., et al.: Risk and protective indicators of beekeeping management practices. Sci. Total Environ. **79910**, 149381 (2021). https://doi.org/10.1016/j.scitotenv.2021.149381
8. van Dooremalen, C., van Langevelde, F.: Can colony size of honeybees (Apis mellifera) be used as predictor for colony losses due to varroa destructor during winter? Agriculture (Switzerland) **11**(6), 529 (2021). https://doi.org/10.3390/agriculture11060529
9. Sharov, M.A.: Honey bees (Apis mellifera L.) Far Eastern honey bee: Patent No. 9428 No. 8356497; application as of 12 Dec 2016 (2018)

10. Frunze, O., Brandorf, A., Kang, E.-J., Choi, Y.-S.: Beekeeping genetic resources and retrieval of honey bee Apis mellifera L. Stock in the Russian federation: a review. Insects **12**(8), 684 (2021). https://doi.org/10.3390/insects12080684
11. Delaplane, K.S., Van Der Steen, J., Guzman-Novoa, E.: Standard methods for estimating strength parameters of Apis mellifera colonies. J. Apic. Res. **52**(1) (2013). https://doi.org/10.3896/IBRA.1.52.1.0

The Dynamics of the Main Fertility Elements Content in a Long-Term Agrochemical Stationary Experiment

Roman Timoshinov(✉), Elena Kushaeva, Lyudmila Marchuk, Alexandr Dubkov, and Alexey Klykov

Federal Scientific Center of Agricultural Biotechnology of the Far East Named After A.K. Chaika, 30, Volozhenin Str.Timiryazevsky Stl, 692539 Ussuriysk, Primorsky Krai, Russia

Abstract. The article analyzes the changes in the main elements of fertility of meadow-brown bleached soils, which occurred during nine crop rotations in a long-term agrochemical stationary experiment with the use of various fertilization systems. Positive changes in the soil are observed under the influence of all studied fertilization systems: manure, manure + lime, manure + lime + mineral with a single dose of NPK, manure + lime + mineral with double dose of NPK, lime + mineral, and mineral (NPK). Studies have shown that a complex system that includes manure, lime, and a double dose of NPK is more effective. In this variant, a high content of humus, mobile phosphorus and exchangeable potassium, as well as the best pH_{KCl} indicator were noted during all rotations.

Keywords: Fertilizer systems · Manure · Lime · Phosphorus · Potassium · Humus · Crop rotation

1 Introduction

Soil fertility, its conservation, reproduction, and rational use are among the main conditions for sustainable agricultural production. Intensification of agricultural production has a great impact on the chemical composition, physical structure, content and qualitative composition of humus in almost all types of soils. This is due to significant changes in the functioning of the soil as a natural biological system. Microorganisms are indicators of changes in the soil. An important place is occupied by the development of effective fertilizer systems in the complex of measures to increase the productivity of agricultural crops, preserve and improve soil fertility. In many long-term experiments carried out in various natural conditions, the positive effect of organic and organomineral fertilizer systems has been noted. Soil biological activity is an indicator characterizing its fertility and the intensity of the processes in it, and any agrotechnical measures aimed at increasing crop yields and reproduction of soil fertility should have a soil microbiological substantiation [1, 2].

Soil fertility, its preservation, reproduction and rational use is one of the main conditions for conducting agricultural production. The intensification of agricultural production has a great impact on the chemical composition, physical structure, content and

A. Muratov and S. Ignateva (Eds.): AFE 2021, LNNS 353, pp. 469–478, 2022.
https://doi.org/10.1007/978-3-030-91402-8_53

qualitative composition of humus in all types of soils. This determines the main changes in the soil buds as a natural biological system. Microorganisms are indicators of changes in the soil. In a set of measures to increase the productivity of agricultural crops. In many long-term experiments carried out in various natural conditions, the positive effect of the action of organo-mineral fertilizer systems. An indicator characterizing soil fertility contributes to an increase in the quality of the processes taking place in them, is its biological activity, and any agrotechnical measures.

The introduction of organic fertilizers is one of the main methods of preserving and increasing soil fertility. They play an especially important role in maintaining a certain amount of humus in the soil composition. Experimental data of a number of authors [3, 4] showed that the humus content in the soil and its biological properties are so closely connected that a functional relationship between them is reasonable to speak about. In agrocenoses, there is a constant rejection of organic matter and nutrients by the biomass harvest. In order to prevent soil degradation, it is necessary to select measures to compensate for losses, maintaining a balance between the process of mineralization and synthesis of organic matter.

It is known that, in contrast to natural biogeocenoses with a relatively closed cycle of nutrients in agrocenoses, this cycle breaks due to the alienation of nutrients with the harvest, a decrease in their availability to plants, losses as a result of runoff, erosion, denitrification, infiltration, etc. the balance of nutrients in agriculture leads not only to a decrease in production and deterioration of its quality, but also to a decrease in the sustainability of agricultural landscapes. Systematic observation, study, and analysis of these processes and the adoption of the necessary measures form the basis for monitoring soil fertility.

Morphological differences between the soil profiles of crop rotation and permanent crops from the profile of perennial fallow lands are the result of long-term processing and In terms of the humus content, mobile phosphorus and exchangeable potassium, the profiles of differently fertilized soils of crop rotation, permanent crops and perennial fallow lands differ noticeably within the upper 40 cm layer. It is necessary to emphasize a specific feature in the intensive domestication action: the change in the agrochemical properties of the subsurface layer 20–40 cm is more significant than that of the arable layer [5, 6].

To obtain products at the lowest cost, it is necessary to optimize mineral fertilizers use against the background of the prolonged action of long-term applied fertilizer systems on meadow-brown bleached soil [7].

The purpose of this study is to investigate the effect of long-term use of various fertilizer systems on the change in the fertility of meadow-brown bleached soil.

2 Materials and Methods

The studies were carried out in an agrochemical station (field No. 7), established in 1941 on the basis of a nine-field crop rotation of the FSBSI "FSC of Agricultural Biotechnology of the Far East named after A.K. Chaika". The soil of the experimental site is meadow-brown, bleached, heavy loam, with a waterproof podzolic horizon. In August and the first half of September, typhoons passing over the Sea of Japan, accompanied by heavy rains, enter the South of the Far East. The abundant precipitation (more than 150–200 mm per day) during this period causes a strong waterlogging of clay soils, since a small arable layer does not allow accumulating the entire volume of precipitation and becomes waterlogged due to the fact that excess moisture in the soil cannot penetrate into underlying layers through the impenetrable podzolic horizon. Before the experiment was laid, the soil contained 12.5 mg/kg of soil of mobile phosphorus (P_2O_5), 60 mg/kg of soil of exchangeable potassium (K_2O), 4.84% humus, pH_{KCl} was 5.1.

The first crop rotation in the field no. 7 began in 1944, and in 2020 it is the sixth year of the ninth rotation. The rotation periods are as follows: 1. 1944–1952; 2. 1953–1960; 3. 1961–1969; 4. 170–1978; 5. 19–1987; 6. 188–1996; 7. 1997–2005; 8. 2006–2014; 9. 2015–2023. The fertilization systems taken for the study included the separate application of manure, lime and mineral fertilizers (single and double doses), as well as their various combinations. The following systems were studied: 1. Control (no fertilizers); 2. Manure; 3. Manure + lime; 4. Manure + lime + 1NPK; 5. Manure + lime + 2NPK; 6. Lime + NPK; 7. NPK. Manure and lime were applied in a seeded fallow (t/ha) at the beginning of each crop rotation, mineral fertilizers (kilograms of active agent) - annually. The total amount of fertilizers applied for the entire observation period is reflected in Table 1.

Table 1. The amount of applied organic and mineral fertilizers in 1944–2020.

No	Variant	Applied during the observation period				
		Manure, t	Lime, t	N, kg of active agent	P, kg of active agent	K, kg of active agent
1	Control (no fertilizers)	–	–	–	–	–
2	Manure	240	–	–	–	–
3	Manure + lime	240	17	–	–	–
4	Manure + lime + 1NPK	260	17	1120	1815	1245
5	Manure + lime + 2NPK	240	17	2988	3200	2700
6	Lime + NPK	–	23	1725	2058	1395
7	NPK	–	–	2854	3160	2385

Manure was applied from the first to the seventh rotation. In the eighth crop rotation, manure was excluded and an additional field of perennial legumes (red clover) was introduced. As a result, the input of organic matter began to be carried out due to the plowing of green mass of clover on green manure (one field) and clover crop residues (second field), and the following crop rotation was established: 1. 1st year clover; 2. Soybean; 3. Wheat; 4. Soybean; 5. Wheat with clover undersowing; 6. 1st year clover; 7. Wheat; 8. Soybean; 9. Wheat with clover oversowing (Table 2).

The total area of the plots is 250 m^2, the accounting area is 150 m^2, three time replication.

The following indicators were determined in soil samples: organic matter (humus) - according to Tyurin; pH_{KCl} - according to GOST 26483-85; mobile phosphorus (P_2O_5) and exchangeable potassium (K_2O) - according to the Kirsanov method (GOST R 54650-2011).

Table 2. Crop rotation in an agrochemical stationary.

Rotation				
I	II	III–IV	V–VII	VIII–IX
Black fallow	Green soybean fallow	Green clover fallow	Seeded fallow	Green clover fallow
Wheat	Wheat + perennial grass undersowing	Corn	Corn	Soybean
Oats + grasses	1st year perennial grasses	Wheat + perennial grass undersowing	Wheat (oats)	Wheat
1st year perennial grasses	2nd year perennial grasses	1st year perennial grasses	Soybean	Soybean
2nd year perennial grasses	Soybean	2nd year perennial grasses	Wheat + perennial grass undersowing	Wheat + perennial grass undersowing (clover)
Wheat	Wheat	Soybean	1st year perennial grasses	Clover fallow
Soybean	Soybean	Wheat	Soybean	Wheat
Wheat	Wheat	Soybean	Barley	Soybean
Oats	Oats + perennial grasses	Wheat	Soybean	Wheat + perennial grass undersowing (clover)

3 Results

Using mineral and organic fertilizers in crop rotation is one of the most important factors determining productivity and the state of agrochemical indicators of soil fertility. Phosphorus is one of the main elements of plant nutrition. A specific zonal feature of the main soil types in Primorsky Krai is that, with a relatively high content of gross phosphoric acid (0.08–0.20%), its reserves in a mobile form easily accessible to plants are negligible, therefore, there is a constant deficiency in phosphorus. The problem of phosphate nutrition of plants under the conditions of the region should be solved both by mobilizing natural reserves of phosphorus from tightly bound phosphate compounds, and by applying mineral phosphorus fertilizers. The solution to the problem was taken up by the employees of the Primorsky Regional Agricultural Experiment Station, who by the end of the 30s proposed a number of measures to improve the supply of local soils with phosphorus through liming and using mineral fertilizers.

In order to establish changes in the soil that can occur with prolonged use of various fertilizer systems in crop rotation and their effect on increasing soil fertility and crop yield, an agrochemical station was established in 1941 on the basis of a nine-field crop rotation. The effectiveness of various fertilizing systems on the 7th field has been studied over the past nine crop rotations. The most complete agrochemical survey was carried out from the fourth rotation. The results of analyzes of soil samples taken at the end of each rotation are presented. According to the given data of long-term experience, it was found that the concentration of mobile phosphorus in meadow-brown soil in the control variant was very low throughout the entire study period (5–26 mg/kg soil) (Fig. 1). Positive changes in the phosphate regime of the soil are observed under the influence of various fertilization systems. As a result of the research, it was revealed that the lime-mineral system is superior to the mineral and lime-manure, and they, in turn, are superior to the manure. Of the systems used in crop rotation, the complex system, which includes manure, lime and a double dose of NPK, has a higher efficiency. In this variant, the highest content of mobile phosphorus is noted (42–275 mg/kg of soil) and by the sixth rotation a very high level of soil supply with mobile phosphorus (165 mg/kg) was created, compared with a very low level in the control (21 mg/kg), and with the initial value before the experiment (12.5 mg/kg). The integrated system with a single dose of NPK in the seventh crop rotation increased the level of phosphorus supply to the soil to a very high level (163 mg/kg). The obtained data indicates the advantage of the combined application of manure, lime and mineral fertilizers. Probably due to the fact that lime enhances the mineralization of organic matter, and also promotes the mobilization of soil phosphates.

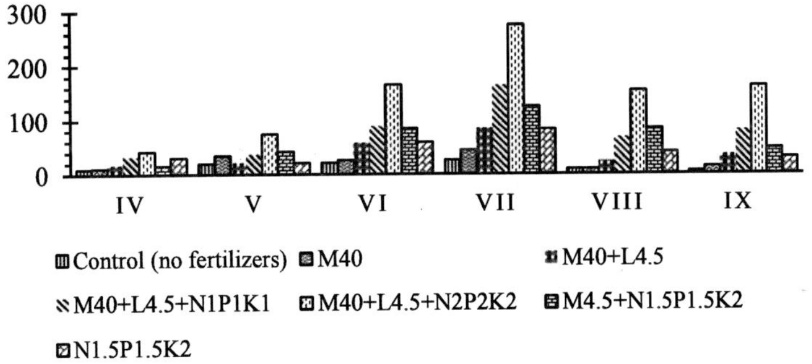

Fig. 1. Change in P_2O_5 content by crop rotation, mg/kg soil.

Potassium is essential for plant nutrition, as well as nitrogen and phosphorus. With its lack in the soil in plants, biological processes are disturbed, the synthesis of carbohydrates is weakened, yield and its quality decrease, as well as resistance to pathogens of fungal diseases during the period of plant growth and during product storage. In heavy soils, there is more potassium, since it is included in the minerals that present mainly in clay particles.

Prior to the trial establishment, the soil belonged to the group with a low supply of exchangeable potassium (60 mg/kg of soil). The use of different fertilization systems had a positive effect on the content of exchangeable potassium in the soil during all rotations of (Fig. 2). Due to this, the supply of exchangeable potassium increased to medium (95–120 mg/kg), increased (126–159 mg/kg), high (171–219 mg/kg) and very high (221–573 mg/kg). Complex fertilization systems, especially a system with a double dose of mineral fertilizers, had the greatest impact on the potassium soil regime. The highest concentration of exchangeable potassium (573 mg/kg soil) was noted in the fourth rotation of the fifth variant. Further, a decrease in its content in all fertilization systems was noted, reaching the lowest values for the entire period of the research (75–120 mg/kg of soil) by the ninth crop rotation.

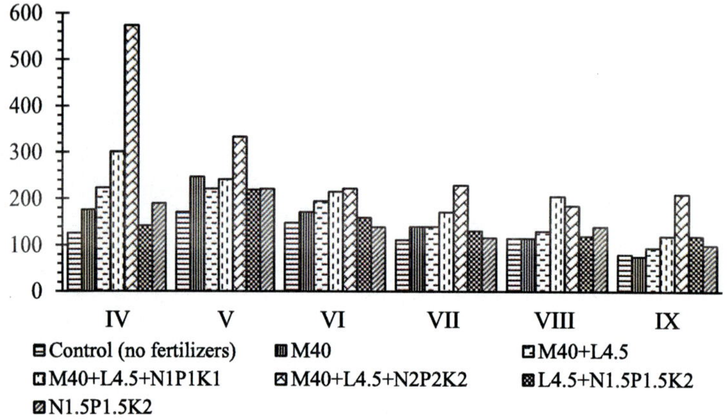

Fig. 2. Changes in the K_2O content in the soil according by crop rotation, mg/kg.

It is known that plant residues entering the soil undergo various processes, as a result of which a significant part of the organic material is destroyed (mineralized), and the other passes into a more stable form (humus). When cultivating agricultural crops without fertilizers, especially organic fertilizers, the destruction of soil organic matter significantly exceeds the accumulation (humification). As a result, its reserves decrease, which leads to a decrease in soil fertility and crop productivity. The systematic application of organic and mineral fertilizers to the soil against the background of liming promotes the activation of the humification process. Based on the studies carried out, it was found that the use of various fertilization systems created spatial diversity in the level of humus content in meadow-brown soil (Fig. 3). At the same time, the minimum values of the humus content were typical, mainly, for the control (2.96–3.69%), maximum - for the manure + lime + 2NPK (3.43–4.16%). In general, the analysis of different fertilization systems indicates an obvious decrease in the humus content during the study period, regardless of the use of different fertilization systems. The soil, which was characterized by an increased humus content during the establishment of the experiment, passed into the category of moderately-supplied.

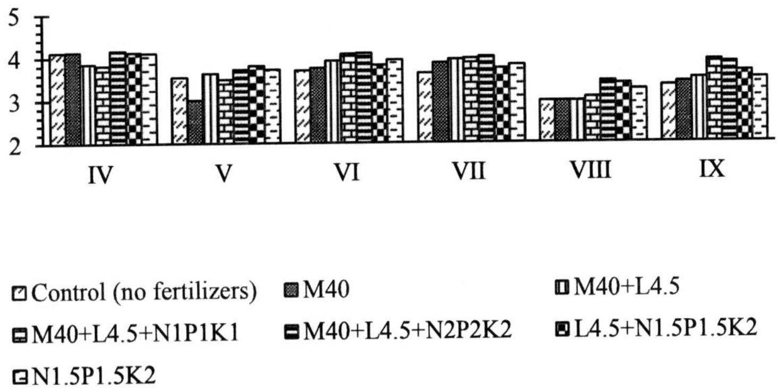

Fig. 3. Change in humus content in soil by crop rotation, %.

As a result of the systematic use of mineral fertilizers, the properties of soils can deteriorate due to the physiological acidity of some fertilizers, the absorption of fertilizer cations, and the displacement of hydrogen and aluminum from the absorbing complex, acidifying the reaction of the soil solution. However, the systematic use of mineral fertilizers is not always accompanied by a deterioration in soil properties, and the nature of these changes depends on soil and climatic conditions, forms and doses of fertilizers applied and the duration of their use. In our experiment, the reaction of the soil solution as a result of the application of mineral fertilizers for 77 years changed insignificantly, in comparison with the control, as evidenced by the 9 rotations data (Fig. 4). The pH_{KCl} value increased by 0.1 in the fourth and sixth rotations, in the fifth rotation - by 0.4 units, in the seventh it remained at the control level, and only in the eighth rotation it decreased by 0.1, and in the ninth - by 0.4 units compared to control. Periodic liming contributes to a significant weakening of the negative impact of mineral fertilizers on soil acidity. The systematic application of fertilizers against the background of lime is not accompanied by acidification of the soil solution, and, on the contrary, an increase in the pH value is observed. The application of manure and manure with lime also reduces the soil acidity. The best results are obtained by using complex fertilization systems. In these variants, the soil from the category of moderately acidic passes into the category of neutral (pH 6.2–6.5) and close to neutral (pH 5.7–6.0).

4 Discussion

Many scientists in different countries are studying the processes occurring in the soil in long-term stationary experiments. In the Hesbaye region (Belgium), studies begun

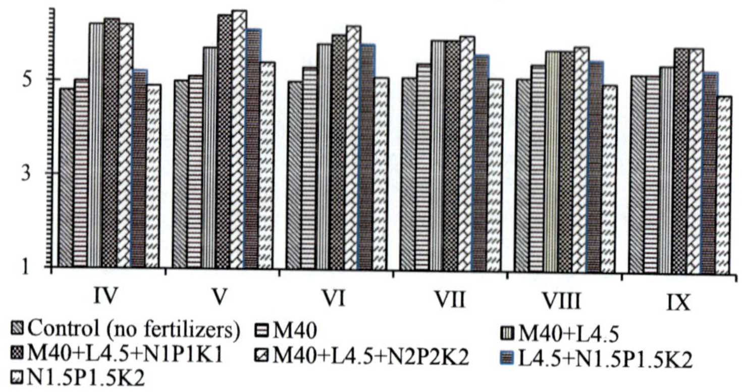

Fig. 4. Change in pH_{kcl} content in soil by crop rotation, %.

in 1959 proved the effect of crop production on the carbon (C) balance in the soil [8]. According to Hassani, A., Azapagic, A., Shokri, N. in conditions of climatic uncertainty, knowledge of spatio-temporal change and prediction of soil salinity is crucial for understanding land degradation and planning effective remediation strategies [9]. Long-term use of mineral fertilizers can affect not only soil pH, but also biodiversity and bacterial abundance [10]. At the same time, soil pH is not the main factor affecting nitrogen-fixing microorganisms - the addition of manure has a greater effect on their change, but it cannot eliminate the negative effect of chemical fertilizers on the abundance and diversity of soil bacteria [11]. Analysis of changes in soil acidity according to crop rotation of our experiment allows us to conclude that significant acidification due to prolonged application of mineral fertilizers was not observed, and systematic application of fertilizers against the background of lime is not only not accompanied by acidification of the soil solution, but on the contrary, there is an increase in the pH value. In Denmark, a field experiment was carried out to study the effect of using red clover as a plant fertilizer in comparison with bird manure on soil and nutrient availability, and the possibility of reducing the need for animal manure fertilization in organic farming was shown [12]. The use of manure is one of the important factors affecting the physical properties of the soil and the yield of agricultural crops [13, 14]. At the same time, the use of manure together with mineral fertilizers (NPK) gives the greatest increase in yield. In our studies, positive changes in the soil are observed under the influence of all studied fertilization systems, but a complex system that includes manure, lime and a double dose of NPK has a higher efficiency. In addition, depending on the type of land use, the quality and quantity of soil organic matter, as well as the microbiological parameters of the soil and the activity of enzymes, can change, which is a key indicator of soil health and fertility.

5 Results

The studies have shown that the systematic use of various fertilization systems for 77 years had a positive effect on the change in the fertility level of bleached meadow brown soils. The largest increase in the humus content (3.43–4.16%), mobile phosphorus

(42–275 mg/kg of soil) and exchangeable potassium (185–573 mg/kg of soil), as well as the most favorable acid regime (pH 5.8–6.5) during the entire research period provides an integrated system (variant 5), including manure, lime and a double dose of mineral fertilizers.

References

1. Jabborova, D., Ma, H., Bellingrath-Kimura, S.D., Wirth, S.: Impacts of biochar on basil (Ocimum basilicum) growth, root morphological traits, plant biochemical and physiological properties and soil enzymatic activities. Sci. Hortic. **03044238** (2021). https://doi.org/10.1016/j.scienta.2021.110518
2. Macůrková, A., et al.: Effect of round-shaped silver nanoparticles on the genetic and functional diversity of soil microbial community in soil and "soil-plant" systems. Appl. Soil Ecol. **09291393** (2021). https://doi.org/10.1016/j.apsoil.2021.104165
3. Řezáčová, V., et al.: Organic fertilization improves soil aggregation through increases in abundance of eubacteria and products of arbuscular mycorrhizal fungi. Sci. Rep. **20452322** (2021). https://doi.org/10.1038/s41598-021-91653-x
4. Wu, L., et al.: Fertilization effects on microbial community composition and aggregate formation in saline-alkaline soil. Plant Soil 0032079X (2021). https://doi.org/10.1007/s11104-021-04909-w
5. Behera, U.K., Sharma, A.R., Pandey, H.N.: Sustaining productivity of wheat–soybean cropping system through integrated nutrient management practices on the Vertisols of central India. Plant Soil **297**, 185–199 (2007). https://doi.org/10.1007/s11104-007-9332-3
6. Piazza, G., Pellegrino, E., Moscatelli, M.K., Ercoli, L.: Long-term conservation tillage and nitrogen fertilization effects on soil aggregate distribution, nutrient stocks and enzymatic activities in bulk soil and occluded microaggregates. Soil Tillage Res. **196**, 104482 (2020). https://doi.org/10.1016/j.still.2019.104482
7. Berezhnaya, V.V., Kushayeva, E.J., Timoshinov, R.V., Klykov, A.G.: Changes in the fertility of meadow-brown bleached soil with long-term use of different fertilizer systems. In: AGRITECH-IV-2020 IOP Conference Series: Earth and Environmental Science, vol. 677, p. 042049 (2021). https://doi.org/10.1088/1755-1315/677/4/042049
8. Hassani, A., Azapagic, A., Shokri, N.: Predicting long-term dynamics of soil salinity and sodicity on a global scale. Proc. Nation Acad. Sci. USA **117**(52), 33017–33027 (2021). https://doi.org/10.1073/PNAS.2013771117
9. Meena, A., Rao, K.S.: Assessment of soil microbial and enzyme activity in the rhizosphere zone under different land use/cover of a semiarid region, India. Ecol. Process. **10**(1), 1–12 (2021). https://doi.org/10.1186/s13717-021-00288-3
10. Zhou, J., Guan, D., Zhou, B., et al.: Influence of 34-years of fertilization on bacterial communities in an intensively cultivated black soil in northeast China. Soil Biol. Biochem. **90**, 42–51 (2015). https://doi.org/10.1016/j.soilbio.2015.07.005
11. Hu, X., et al.: Long-term manure addition reduces diversity and changes community structure of diazotrophs in a neutral black soil of northeast China. J. Soils Sediment. **18**(5), 2053–2062 (2018). https://doi.org/10.1007/s11368-018-1975-6
12. Hefner, M., Sorensen, J.N., Kristensen, H.L., Xie, Y., Petersen, K.K.: Replacing animal manure with plant-based fertiliser in organic blackcurrant (Ribes nigrum L.) production. Scientia Hortic. **285** (2021). https://doi.org/10.1016/j.scienta.2021.110168

13. Ding, X., Yuan, Y., Liang, Y., Li, L., Han, X.: Impact of long-term application of manure, crop residue, and mineral fertilizer on organic carbon pools and crop yields in a Mollisol. J. Soils Sediment. **14**(5), 854–859 (2014). https://doi.org/10.1007/s11368-013-0840-x
14. Buysse, P., Roisin, C., Aubinet, M.: Fifty years of contrasted residue management of an agricultural crop: impacts on the soil carbon budget and on soil heterotrophic respiration. Agric. Ecosyst. Environ. **167**, 52–59 (2013). https://doi.org/10.1016/j.agee.2013.01.006

Automating the Calculation of Damage Caused by Illegal Logging

Alexander Baranov(✉) and Victoria Sergeeva

Far Eastern State Agrarian University, Amur region, Gorky str., 92 Academic building No. 3, 675000 Blagoveshchensk, Russia

Abstract. The requirements of regulatory documents for determining the amount of damage caused to forest plantations due to illegal felling and other violations of forest legislation as a result of which the forest stand is destroyed are considered. Based on the analysis of the requirements of regulatory documents, an algorithm has been developed that makes it possible to structure the sequence of decisions made when determining the harm for the destruction of forest plantations. Work has been done to integrate the developed algorithm into the electronic environment and the commands for automatic calculation have been written based on the data entered by the system operator. Based on the data of different variations, test calculations of the amount of harm in an automated system were carried out. The obtained results of test calculations were compared with the calculations of experts. On the basis of a comparative analysis, conclusions were drawn about the correctness of the developed automated system for determining the amount of harm.

Keywords: Damage · Algorithm · Destruction of forest plantations · An automated system for determining the amount of harm

1 Introduction

Illegal logging is estimated to cost the international community US $ 30 billion annually and accounts for 90% of timber exports in some countries [1, 2]. According to some estimates, the losses in the Amazon alone amount to about 600 million US dollars [3]. The loss to the environment is largely unknown, although some have indicated that the problem of illegal logging directly affects the survival of certain species.

Illegal logging refers to logging activities in protected areas, logging of protected species, logging outside the boundaries of the concession, logging of crops in excess of the permitted yield, logging of oversized or stunted trees, and logging in areas where mining is prohibited, such as watersheds, steep slopes, and banks.

This comprehensive definition accurately focuses on the urgent problem of illegal logging to successfully combat vulnerable forest areas. This problem exists in many regions of the world. Specific examples include Ghana, Cameroon, the Democratic Republic of the Congo and Madasgakar from Africa to Bolivia, Peru [4] and the Brazilian Amazon [3] from South America to Bangladesh, the Philippines [5] and Indonesia [6],

A. Muratov and S. Ignateva (Eds.): AFE 2021, LNNS 353, pp. 479–486, 2022.
https://doi.org/10.1007/978-3-030-91402-8_54

in Asia, not spared illegal logging and Russia. Illegal use of forests poses a serious threat to the Russian timber industry. Among environmental crimes, the leading position is occupied by criminal acts in the forest sphere. Thus, the share of environmental acts in the number of criminal cases initiated accounts for 32–40%, while in Russia as a whole, this figure ranges from 32–44%. However, this indicator for the Far Eastern Federal District is very average and may differ significantly in different subjects, depending on the availability and reserve of resources. So, for example, in the Amur Region, this indicator ranges from 76–94%; in the Primorsky Territory – from 67 to 85%; in the Jewish Autonomous Region – from 60 to 72%; In the Khabarovsk Territory – from 25 to 30%; in the Kamchatka, Magadan, and Sakhalin regions-no more than 10%.

Research by many authors identifies illegal logging as a problem through broad and vague approaches that do not specifically address the immediate problem or high levels of technological complexity and knowledge that will take years and significant investment to develop [7].

However, illegal logging is high on the international forestry agenda, its definition is often ambiguous, and there is considerable uncertainty about its quantitative and qualitative characteristics. This makes it difficult to assess its social, economic and environmental consequences and does not allow making informed political conclusions [8].

Deforestation and illegal logging are closely linked. In many cases, illegal logging acts as a catalyst for the conversion of forest land to other uses. Illegal behavior is more likely if the benefits derived from breaking the law exceed the costs of non-compliance. In other words, when the expected net costs of legitimate activities are higher than the costs associated with clandestine operations, illegal logging will continue. The difference in value between legal and illegal production depends on the characteristics of the forest products markets. In the global timber export volume, 40% is used for basic energy needs, such as cooking and heating, and 60% is used as business roundwood.

In addition, some scholars suggest general legislative and international changes to standards that play an indirect role in the necessary changes in the fight against illegal logging, which may or may not be effective. All of these dynamics are important to take into account when addressing illegal logging, but by focusing on such components of the forest industry, opportunities are missed to address some of the more intimate features of the conditions in which illegal logging can flourish. Although most of the methods of countering illegal logging have already been described and implemented in the scientific literature and in practice. One of the most striking examples of the increase in illegal logging is the developed road network. Accessibility has been found to play an important role in the success of illegal logging of wood and other forest resources [9–13]. The increase in the volume of illegal logging in the Amur region, including also a large number of forest fires [14], especially those whose causes have not been identified (and may be intentional).

In the Russian Federation, for illegal logging, in addition to a fine and imprisonment, the violator must be compensated for the damage caused. The estimated amount of damage for illegal demolition, felling and/or destruction of trees is at least 50 times higher than the tax value of wood and depends on many indicators that need to be taken into account.

The amount of damage caused as a result of destruction, damage, demolition or illegal logging of forest stands is also important. So if the damage (damage) during illegal logging amounted to less than five thousand rubles, then this violation is qualified under Article 8.28 of the Code of Administrative Offenses of the Russian Federation (Administrative Code of the Russian Federation). Causing harm (damage) in the amount of more than five thousand rubles is punishable under Article 260 of the Criminal Code of the Russian Federation (Criminal Code of the Russian Federation).

In addition to law enforcement officers, the State forest inspectors are responsible for initiating and conducting administrative cases on illegal logging as part of an administrative violation. Consideration of administrative cases under Article 8.28 of the Administrative Code of the Russian Federation is entrusted to the executive authority of the subject of the Russian Federation, authorized in the field of forest relations. The initiation of criminal cases is not within the competence of the State forest inspectors. Thus, the division of powers for this type of violation is carried out depending on the amount of damage caused.

The development of an algorithm for calculating the damage from illegal logging and its integration into publicly available software, such as Excel, allows a specialist, without a specialized education in this field, to determine the amount of damage.

On the territory of the Russian Federation, thousands of cases of illegal logging are detected every year, the vast majority of which are committed by cutting down trees of basic species. Automation of the calculation of damage for illegal logging, demolition or damage to trees of basic species, the amount of taxes, which is presented in the Decree of the Government of the Russian Federation of 22.05.2007 No. 310 “On the rates of payment for a unit of forest resources and the rates of payment for a unit of area of a forest plot that is in federal ownership” will significantly reduce the time spent by employees of specialized organizations or completely do without their services when determining the amount of damage.

Thus, the purpose of our research is aimed at developing an automated system for determining the amount of damage (damage), taking into account the requirements of the current regulatory documents of the Russian Federation.

2 Materials and Methods

The main task of the user of the automated system is to enter data on the date of the violation, the species composition and the diameter of the felled trees. Based on the entered data, the system automatically calculates the amount of damage with an accuracy of one ruble.

The algorithm for determining damage fully complies with and takes into account the requirements of the methodology approved by the Decree of the Government of the Russian Federation No. 1730 of 29.12.2018 “On Approval of the Specifics of Compensation for damage caused to forests and natural objects located in them due to violations of forest Legislation”.

For the functioning algorithm in an electronic environment, it is necessary, in addition to writing it, to create databases on the cost of wood and sorting tables for each of the main species allowed for harvesting.

The calculation is carried out in the following sequence.

1. Enter data on the diameter, number of felled/destroyed trees and their species composition. Depending on the forest taxing area in which the felling of trees is carried out, the cost of wood varies significantly. At the same time, the tax value is always determined by the first category and the average category of size for business wood. When determining the diameter, it is necessary to measure it at a height of 1.3 m from the stump neck, and in the absence of a tree trunk, the actual diameter of the stump is taken as the taxational thickness indicator. If the actual diameter is less than 12 cm for coniferous and 16 cm for deciduous, its nominal value is taken as 12 cm and 16 cm, respectively.
2. The area of illegal logging or the area where trees are destroyed affects the method of determining the volume of harvested/destroyed wood. So with an area of up to one hectare, it is necessary to carry out a continuous enumeration. The exception in this case is the impossibility of carrying out a list due to the absence of stumps, in this situation, the volume is determined based on the data of the tax description. When felling/destroying trees on an area of more than one hectare, forest management materials are used or a tape list is used to determine the quantitative indicators of the stand.
3. Determining the volume. If the volume is not determined by the taxation indicators, then the sorting tables are used. The volume of felled / destroyed wood is determined by the first category of heights, and in the absence of the first category of heights in the sorting tables for any breed, the highest is accepted. Also, the volume of business wood is taken as the volume in the bark, that is, without taking into account wood-wood and waste.
4. When determining the tax value of wood, it is necessary to take into account the indexation coefficients, which are set annually by the Government of the Russian Federation, for 2021 is 2.72.
5. The very fact of cutting/destroying trees in accordance with the methodology indicates the need to increase the tax value by 50 times.
6. The period of felling/destruction of trees is not unimportant, since if the violation is committed in the period from November to January, the amount of damage is doubled.
7. If the destroyed trees were testes, plus trees, or felling was carried out on seed plantations, seed reserves or reserves, an additional factor of 2.0 is applied to the amount of damage already calculated at this stage.
8. Felling/destruction of trees in forests belonging to protective and/or specially protected areas the amount of damage increases by another 2 times. Special attention should be paid to the fact that this item does not include forests that are part of specially protected natural areas.
9. If the violation was carried out in the protective areas of protective forests belonging to the forest park zone of the green belt, then an additional coefficient of 3.0 is applied.
10. Felling/destroying trees within the boundaries of specially protected natural territories implies an increase in damage in a 5-fold amount.
11. The final action in determining the amount of damage is exclusively a mathematical decision on rounding the amount of damage to 1 ruble.

Table 1. The cost of business wood of the main breeds by the first category of taxes and the average category of size for the Amur forest taxation region and indexation coefficients for 2021–2023.

Breed	The cost per 1 cubic meter in accordance with the Decree of the Government of the Russian Federation No. 310 of 22.05.2007	Indexation coefficient (Decree of the Government of the Russian Federation No. 1318 of 12.10.2019)			Total tax cost per 1 cubic meter
		2021 year	2022 year	2023 year	
Pine tree	82.98	2.72	2.83	2.94	225.7056
Cedar	99.36	2.72	2.83	2.94	270.2592
Larch	66.42	2.72	2.83	2.94	180.6624
Fir	74.7	2.72	2.83	2.94	203.184
Fir tree	74.7	2.72	2.83	2.94	203.184
Oak	561.96	2.72	2.83	2.94	1528.5312
Maple	561.96	2.72	2.83	2.94	1528.5312
Ash tree	841.86	2.72	2.83	2.94	2289.8592
Birch tree	41.58	2.72	2.83	2.94	113.0976
Black alder	24.84	2.72	2.83	2.94	67.5648
Hornbeam	24.84	2.72	2.83	2.94	67.5648
Ilm	24.84	2.72	2.83	2.94	67.5648
Linden tree	24.84	2.72	2.83	2.94	67.5648
Aspen	8.28	2.72	2.83	2.94	22.5216
White alder	8.28	2.72	2.83	2.94	22.5216
Poplar	8.28	2.72	2.83	2.94	22.5216

Making decisions in accordance with the points described above without human participation is possible only after digitizing the reference data and numerical values of regulatory legal acts, followed by writing commands.

Digitized data on the cost of wood for the Amur forest taxation region is presented in Table 1, which also shows the indexation coefficients for 2021–2023.

Data on the volume in the crust for the highest category of heights are presented in Table 2.

The algorithm written in the electronic environment compares the data in Tables 1 and 2 and calculates the tax value taking into account the number of trees of a certain diameter and breed.

Table 2. The volume of one tree by the highest category of heights in the bark, cubic meters.

Breed	Thickness step (diameter), cm												
	12	16	20	24	28	32	36	40	44	48	52	56	60
Pine tree	0.1	0.19	0.31	0.48	0.68	0.92	1.2	1.5	1.86	2.2	2.58		
Larch	0.11	0.2	0.37	0.55	0.79	1.05	1.36	1.7	2.08	2.48	2.93	3.41	
Fir	0.09	0.18	0.32	0.5	0.73	1	1.32	1.68	2.1	2.54	3.04	3.6	
Fir Tree	0.1	0.19	0.31	0.46	0.64	0.84	1.07	1.39	1.62				
Oak	0.08	0.16	0.28	0.43	0.57	0.8	1.02	1.26	1.54	1.86	2.18	2.56	
White Birch	0.1	0.2	0.31	0.45	0.68	0.82	1.04	1.28					
Black Birch	0.07	0.14	0.23	0.34	0.47	0.63	0.8	0.98	1.21	1.46			
Aspen	0.1	0.22	0.32	0.51	0.71	0.95	1.23	1.52	1.86	2.23	2.64	3.04	3.47
Poplar	0.08	0.18	0.34	0.54	0.79	1.09	1.43	1.82	2.25	2.73			

Table 3. Test calculation of damage caused by illegal logging of 32 Scots pine trees.

Month of logging	Breed	Seedsdraw	Protective devices	Protective green belt	OOPT
March	Pine tree	No	No	No	No
Breed	Thickness step (diameter), cm	Number of stumps, trees, pcs	Volume, cubic meters	Cost, rub	Damage, rub
Pine tree	12	0	0	225.71	0
Pine tree	16	5	0.95	225.71	10721
Pine tree	20	1	0.31	225.71	3498
Pine tree	24	3	1.44	225.71	16251
Pine tree	28	0	0	225.71	0
Pine tree	32	1	0.92	225.71	10382
Pine tree	36	3	3.6	225.71	40627
Pine tree	40	6	9	225.71	101568
Pine tree	44	4	7.44	225.71	83962
Pine tree	48	8	17.6	225.71	198621
Pine tree	52	1	2.58	225.71	29116
Pine tree	56	0	0	225.71	0
Pine tree	60	0	0	225.71	0
Total:		32	44		494747

3 Results

We have tested the functioning of the algorithm prescribed in the electronic environment. In this case, data is entered in the cells indicated in green, and the data presented in the yellow cells are calculated automatically.

An example of calculating the damage for 32 Scots pine trees with different thicknesses is presented in Table 3.

In the corresponding cells, we provide data on the logging, including: the month of felling, the species, the number of trees of the appropriate diameter, and whether they belong to seed or protective forests of any purpose.

4 Discussion

Three different experts with experience in harm calculations identified various variations in the initial data for the calculation. In total, we conducted 50 test calculations. Comparing the values of automatic calculations with the data of calculations performed by experts, a discrepancy was found in two cases. The analysis of discrepancies showed inaccuracies made by experts, therefore, even with extensive experience in determining the amount of harm, the human factor has a place to be.

5 Conclusions

The algorithm developed by us allows us to record the sequence of actions when determining the amount of damage caused to forest stands as a result of illegal logging or destruction of trees of basic species.

The integration of the algorithm into the electronic environment made it possible to automate the calculation of the amount of damage and eliminate the human factor from almost all stages, except for data entry.

A test check of the functioning of the automated system and a comparison of the data with the calculations of experts allows us to speak about the full adequacy and possibility of using state forest inspectors, as well as other employees authorized to consider cases in the field of violations of forest legislation.

References

1. Miranda Montero, J.J., Wright, E.M., Khan, M.N.: Illegal Logging, Fishing, and Wildlife Trade: The Costs and How to Combat it World Bank (2019). https://doi.org/10.1596/32806
2. Goncalves, M.P., et al.: Justice for forests: improving criminal justice efforts to combat illegal logging. World Bank (2012). https://doi.org/10.1596/978-0-8213-8978-2
3. Gutierrez-Velez, V.H., MacDicken, K.: Quantifying the direct social and governmental costs of illegal logging in the Bolivian, Brazilian, and Peruvian Amazon. For. Policy Econ. **10**(4), 248–256 (2008). https://doi.org/10.1016/j.forpol.2007.10.007
4. Finer, M., et al.: Logging concessions enable illegal logging crisis in the Peruvian Amazon. Sci. Rep. **4**(1), 4719 (2014). https://doi.org/10.1038/srep04719
5. van der Ploeg, J., et al.: Illegal logging in the Northern Sierra Madre Natural Park, the Philippines. Conserv. Soc. **9**(3), 202 (2011). https://doi.org/10.4103/0972-4923.86991
6. Obidzinski, K., Chaudhury, M.: Transition to timber plantation based forestry in Indonesia: towards a feasible new policy. Int. For. Rev. **11**(1), 79–87 (2009). https://doi.org/10.1505/ifor.11.1.79
7. Dormontt, M., et al.: Forensic timber identification: It's time to integrate disciplines to combat illegal logging. Biol. Conserv. **191**, 790–798 (2015). https://doi.org/10.1016/j.biocon.2015.06.038

8. Hansen, C.P., Treue, T.: Assessing illegal logging in Ghana. Int. For. Rev. **10**(4), 573–590 (2008). https://doi.org/10.1505/ifor.10.4.573
9. Ferreguetti, A.C., et al.: One step ahead to predict potential poaching hotspots: modeling occupancy and detectability of poachers in a neotropical rainforest. Biol. Conserv. **227**, 133–140 (2018). https://doi.org/10.1016/j.biocon.2018.09.009
10. Radovani, N.I., et al.: Guanaco Lama guanicoe numbers plummet in an area subject to poaching from oil-exploration trails in Patagonia. Oryx **49**(1), 42–50 (2015). https://doi.org/10.1017/S0030605312001226
11. Risdianto, D., et al.: Examining the shifting patterns of poaching from a long-term law enforcement intervention in Sumatra. Biol. Conserv. **204**, 306–312 (2016). https://doi.org/10.1016/j.biocon.2016.10.029
12. Sharma, K., et al.: Tiger poaching and trafficking in India: estimating rates of occurrence and detection over four decades. Biol. Conserv. **179**, 33–39 (2014). https://doi.org/10.1016/j.biocon.2014.08.016
13. Watson, F., et al.: Spatial patterns of wire-snare poaching: implications for community conservation in buffer zones around National Parks. Biol. Conserv. **168**, 1–9 (2013). https://doi.org/10.1016/j.biocon.2013.09.003
14. Dyadchenko, O., Timchenko, N., Baranov, A.: Influence of permanent and variable factors on the number and area of forest fires in the Amur region. In: Web of Conferences, Ecological and Biological Well-Being of Flora and Fauna, EBWFF 2020, p. 03002 (2020). https://doi.org/10.1051/e3sconf/202020303002

Blueberry Ordinary as an Environmentally Friendly Product in the Amur Region

Irina Berkal, Natalia Yust(✉), and Elena Osipenko

Far Eastern State Agrarian University, ul. Politehnicheskaja, 86, 675005 Blagoveshchensk, Amur Oblast, Russia

Abstract. The Amur region is unique for its natural resources. Common blueberry (Vaccinium uliginosum L.) is a source of biologically active substances necessary for humans. In the Amur Region, it is the most valuable environmentally friendly natural product with healing properties. As for mineral elements, organic acids, toxic elements in the fruits of blueberries growing on the territory of the Amur Region, research has been carried out for the first time. The purpose of the experimental studies was to study the chemical composition of common blueberries growing in natural conditions. In the course of the experiments, it was found that when determining the mineral elements in the juicy fruits of blueberries collected in natural conditions, the greatest amount of potassium was found – 50,0; calcium – 14,0; magnesium – 10,0; sodium – 6,0 mg/100 g. Vitamin C – 21,4 mg/100 g. The content of organic acids is 0,58 g/dm^3 for Zeisky; 0,46 g/dm^3 Tyndinsky; 0,55 g/dm^3 Svobodnensky districts. The quantitative content of toxic elements in the fruits of Vaccinium uliginosum L is lead 0,3–0,5 mg/100 g; 0,01 – 0,05 cadmium; arsenic and mercury were not found, which makes this culture an environmentally friendly product, very promising in the conditions of the Amur region.

Keywords: Common blueberry · Research methods · Organic acids · Mineral and toxic elements

1 Introduction

On the territory of Russia, and in particular in the Amur region, one species of blueberry is found in nature – Common blueberry (Vaccinium uliginosum L.). The Amur region occupies the territory of the Amur Region, part of the Khabarovsk Territory and the Jewish Autonomous Region. The Amur Region is located in the southeast of the Russian Federation and is part of the Far Eastern Federal District. It is one of the largest constituent entities of the Russian Federation, occupying a long border position with the People's Republic of China. The border is almost 1250 km long (Fig. 1).

The temperate monsoon climate of the Amur region turns into a sharply continental one, winters are usually little snowy and cold, and summers are hot and humid in the second half.

A. Muratov and S. Ignateva (Eds.): AFE 2021, LNNS 353, pp. 487–494, 2022.
https://doi.org/10.1007/978-3-030-91402-8_55

Fig. 1. Administrative division of the Amur region.

The Amur Region is rich and unique in its natural resources. Along with the main products that forestry receives, an important place is also occupied by non-wood products, which include mushrooms, berries, nuts, medicinal herbs, birch bark and bark, honey from apiaries located in the forest zone, resin, as well as various plants, which include common blueberries [1, 2].

In the Amur Region, common blueberry grows mainly on natural lands in the central and northern regions, mainly in the Oktyabrsky, Romnensky, Belogorsky, Seryshevsky, Shimanovsky and Magdagachinsky districts and is an ecologically clean product.

The life form of the common blueberry (Vaccinium uliginosum L.) is a shrub or semi-shrub (in the northern regions and in swamps), with a standard height of 30–60 cm. Sometimes, there are populations where the plant height exceeds 80 cm (Fig. 2). Often, blueberries form underground shoots (rhizomes), with the help of which they spread vegetatively. It has white (sometimes pink) flowers, collected in cluster inflorescences [3, 4]. Fruits of blue color, round or oval, ripen in the Amur region in late July – early August, bears fruit abundantly, its yield is up to 2,7 t/ha.

Not afraid of changes in low and high temperatures. A number of authors note that the suitable and optimal conditions for the growth of blueberries in natural conditions are severe winters and short but hot summers, with a cool and temperate climate.

Blueberries are a source of biologically active substances necessary for humans. Due to the original chemical composition, the medicinal properties of blueberries have been confirmed for quite a long time [5].

Fig. 2. Common blueberry (Vaccinium uliginosum L.).

Berries can be used not only as a medicinal raw material, but also in the manufacture of various products. Decoctions are made from the leaves of the plant, in the production of confectionery products, jams are prepared from the fruits, raw blueberries are ground and covered with sugar, in marinades, like currant leaves. Numerous studies have shown that blueberries can be consumed both fresh and in the preparation of juices, compotes, and preserves.

Whereas, a prerequisite for picking blueberries is not to use watery and soft berries, as well as moldy fruits. Common blueberries should be blue, ripe, without reddish tints. In addition, blueberries should be sufficiently strong and firm in terms of quality characteristics.

Common blueberry juice is recognized as one of the most beneficial in terms of vitamins and antioxidant properties, much more useful for the body than pomegranate, apple or grape juice.

A juicy and ripe berry with blue fruits, common blueberry (Vaccinium uliginosum L.) in the Amur Region is the most valuable environmentally friendly natural product with healing properties.

Studies conducted to study the chemical composition of blueberries have shown that berries contain 88% water, about 8,0% sugar, 1,6% organic acids, 1,0% proteins, 1,2,0% fiber and 0,5% tannins, coloring and pectin substances (it is these substances that remove heavy radioactive elements) [6, 7].

Since studies of the chemical composition of common blueberry have not been carried out enough, the purpose of scientific research is relevant in studying the content of ascorbic acid, mineral elements, the quantitative content of metals and toxic elements in berries.

2 Materials and Methods

Scientific research on the chemical composition of common blueberries growing in natural conditions has been carried out in the laboratory of the Federal State Budgetary Institution Amur Reference Center of the Rosselkhoznadzor since 2018 in accordance with methodological guidelines in accordance with generally accepted methods. Blueberry fruit samples were taken in three districts (Zeisky, Tyndinsky, Svobodnensky) of the Amur region. The atomic absorption method was used to determine the mineral elements in the fruits of blueberries collected in natural conditions.

Ascorbic acid, other organic acids, and benzo (a) pyrene in blueberry fruits were determined by reversed-phase HPLC with a spectrophotometric detector in the ultraviolet region of the spectrum on a Shimadzu LC-20 liquid chromatograph. Determination of berry moisture content was carried out by gravimetric method.

The calcium content was studied by the titrimetric method with Trilon B.

3 Results

When determining the mineral elements in the fruits of blueberries collected in natural conditions, the greatest amount of potassium was found – 50,0; calcium – 14,0; magnesium – 10,0; sodium – 6,0 mg/100 g (Fig. 3).

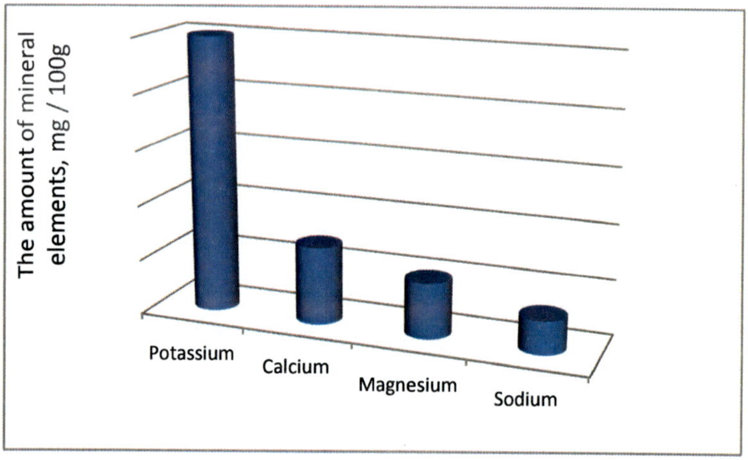

Fig. 3. Content of macronutrients in common blueberry, mg/100 g.

Ascorbic acid is a lactone of an acid similar in structure to glucose. It exists in two forms: reduced and oxidized. Both forms of ascorbic acid quickly and reversibly pass into each other (Fig. 4).

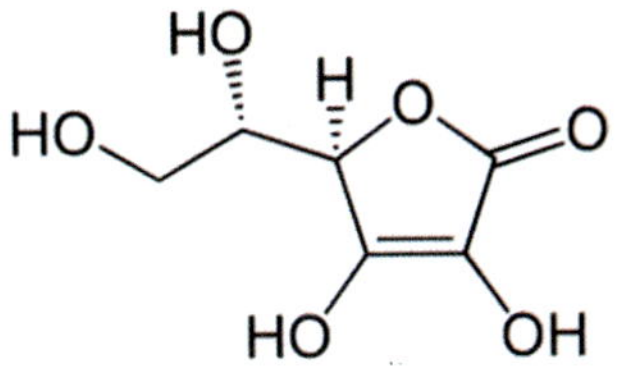

Fig. 4. Structural formula of ascorbic acid.

The results obtained in the process of the conducted scientific experiment indicate that the highest concentration of ascorbic acid was found in the samples of common blueberry from the Zeysky region and is 22,0 mg/100 g; in samples from the Svobodnensky region – 19,0 mg/100 g; in samples from Tynda region – 17,0 mg/100 g (Table 1).

Table 1. Concentration of ascorbic acid in common blueberry samples, mg/100 g.

Growing area	Zeisky district	Tyndinsky district	Svobodnensky district
Concentration of ascorbic acid	22.0	17.0	19.0

Studies to study the content of organic acids are presented in Fig. 5 and Table 3. Succinic, oxalic, malic and citric acids were found in all samples of common blueberry.

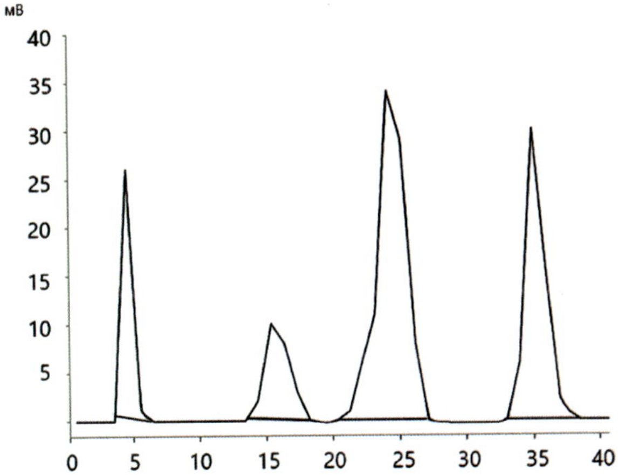

Fig. 5. Chromatogram of organic acids in the fruits of common blueberry.

The retention times of the desired organic acids obtained empirically, using the introduction of calibration solutions. The total content of four organic acids is 0,58 g/dm^3

for the Zeya region, 0,46 g/dm^3 for the Tynda region, 0,55 g/dm^3 for the Svobodnensky region.

Table 2. The content of organic acids in the fruits of common blueberries collected in different areas of the Amur region, g/dm^3.

Structural formula	Connection name	The concentration of acids in the sample common blueberry			HPLC retention time (min)
		Zeisky	Tyndinsky	Svobodnensky	
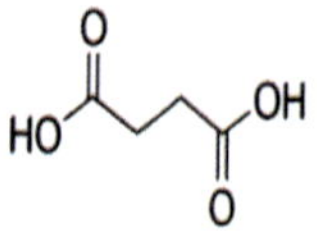	Succinic acid	0.25	0.18	0.22	33.8
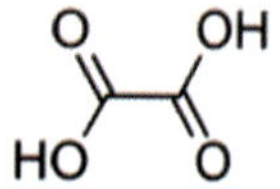	Oxalic acid	0.06	0.05	0.05	4.6
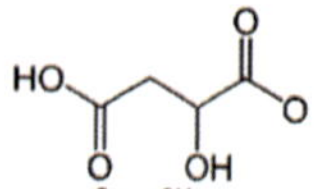	Apple acid	0.22	0.19	0.23	24.5
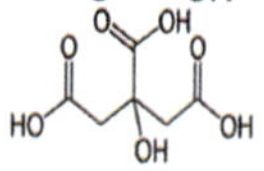	Lemon acid	0.05	0.04	0.05	15.5

When determining the moisture content in the fruits of common blueberries, collected in natural conditions, it was found that the berries growing in the Zeya region contain 69,1% moisture, in Tyndinsky – 69,5%, Svobodnensky – 71,4%.

As a result of studies on the content of metals and toxic elements in blueberry fruits, the following data were obtained, which are presented in Table 2.

It can be seen from the data in the table that the content of metals from blueberry samples from different regions was distributed as follows: cadmium varied from 0,01 in samples from Zeya region to 0,03 and 0,05 mg/100 g from Svobodnensky and Tyndinsky regions, respectively. Iron – 0,4 mg/100 g in berries from the Tyndinsky region and 0,5 and 1,0 mg/100 g from the Zeya and Svobodnensky regions, respectively. The sodium content ranged from 5,0 to 7,0 mg/100 g; The presence of magnesium from 5,0–9,0 mg/100 g. The presence of toxic elements arsenic and mercury was not detected in all three regions.

Benzo (a) pyrene – an aromatic compound, a member of the family of polycyclic hydrocarbons, has the strongest carcinogenic activity; extremely toxic. According to the degree of impact on the body, it belongs to the I (highest) hazard class (Fig. 6).

It is formed during the combustion of liquid hydrocarbon, solid and gaseous fuels (to a lesser extent, during the combustion of gaseous fuels).

The main sources of benzo (a) pyrene consumed by humans are: ambient air, tobacco smoke, heating (burning wood, coal or other biomass), road transport, asphalt, coal tar.

Table 3. Quantitative content of metals and toxic elements in blueberry fruits (element concentration mg/100 g).

The content of metals and toxic elements in blueberry fruits	Areas of common blueberry growth in the Amur region		
	Zeisky	Tyndinsky	Svobodnensky
Magnesium (Mg)	5.0	9.0	8.0
Sodium (Na)	5.0	7.0	7.0
Iron (Fe)	0.5	0.4	1.0
Cadmium (Cd)	0.01	0.05	0.03
Arsenic (As)	Below detection limit	Below detection limit	Below detection limit
Mercury (Hg)	Below detection limit	Below detection limit	Below detection limit

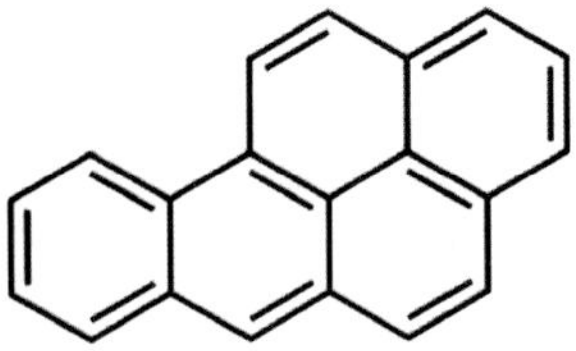

Fig. 6. Formula of benzo (a) pyrene.

In blueberries, this compound can accumulate from soil, air, the environment, especially in fire-hazardous areas. Benzo (a) pyrene analysis results are shown in Fig. 7.

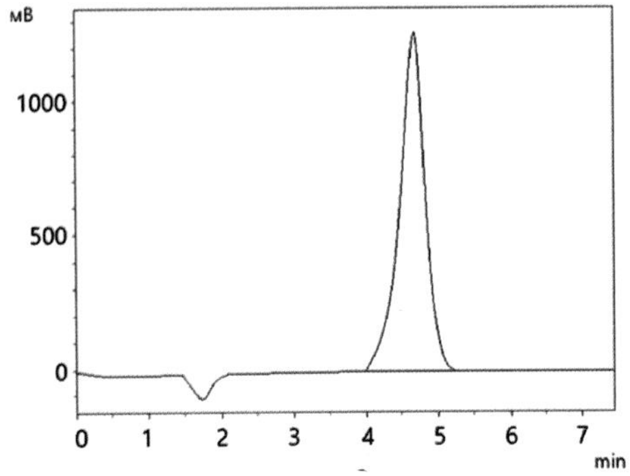

Fig. 7. Chromatogram of a standard sample of benzo (a) pyrene.

In the three required samples, there are no peaks at the retention time of benzo (a) pyrene of 4.4 min, therefore, in each of the samples, the content of benzo (a) pyrene is below the detection limit, less than 0,0001 mg/kg. Accordingly, the content of benzo (a) pyrene in the environment of blueberry growth is either minimal, or blueberry fruits do not accumulate this compound.

4 Discussion

Among the large number of non-woody natural resources growing in the Amur Region, blueberries are of great scientific and practical interest. Due to the original chemical composition, medicinal properties have been confirmed, the use of blueberries in the production of confectionery products, blueberries can be consumed both fresh and in the preparation of juices, compotes, jams and other products. Currently, it is customary to collect wild plants in conjunction with outdoor recreation. Therefore, research data on the chemical composition of blueberry fruits are important for the region, they contain mineral elements, vitamin C, have a set of organic acids, no toxic elements were found.

5 Conclusion

Research results show that weather conditions are suitable conditions for the growth, development and abundant fruiting of common blueberry (Vaccinium uliginosum L.) in natural conditions. Blueberries are rich in minerals and ascorbic acid. It is almost completely absorbed by the human body. When determining the mineral elements in the juicy fruits of blueberries collected in natural conditions, the greatest amount of potassium was found – 50,0; calcium – 14,0; magnesium – 10,0; sodium – 6,0 mg/100 g. Vitamin C – 21,4 mg/100 g. The content of organic acids is 0, 58 g/dm^3 for Zeisky; 0,46 g/dm^3 Tyndinsky; 0,55 g/dm^3 Svobodnensky districts. The quantitative content of toxic elements in the fruits of Vaccinium uliginosum L is 0,3–0,5 mg/100 g in lead; 0,01–0,05 for cadmium. Arsenic, mercury, and benzo (a) pyrene were not found, which makes this culture an ecologically clean product and is of great scientific and practical interest in the Amur region.

References

1. Lavrinenko, O.V., Lavrinenko, I.A.: Zonal vegetation of the plain East European tundras. Rastitel'Nost' Rossii. **32**, 35–108 (2018). https://doi.org/10.31111/vegrus/2018.32.35
2. Luczaj, L., Szymański, W.M.: Wild vascular plants gathered for consumption in the Polish countryside: a review. J. Ethnobiol. Ethnomed. **3** (2017). https://doi.org/10.1186/1746-4269-3-17
3. Liu, C., Bu, Z., Ma, J., et al.: Responses of forest ecosystems to increasing N deposition in China: a critical review. Chinese J. Ecol. **34**(10), 2711–2719 (2015)
4. Nestby, R., Hykkerud, A.L., Martinussen, I.: Review of botanical characterization, growth preferences, climatic adaptation and human health effects of Ericaceae and Empetraceae wild dwarf shrub berries in boreal, alpine and arctic areas. J. Berry Res. **9**(3), 515–547 (2019). https://doi.org/10.3233/JBR-190390
5. Ryzhakova, N.K., Babeshina, L.G., Kondratyeva, A.G., Sechnaya, D.Y.: Contents of macro-microelements and long-lived radionuclides in the medicinal plants belonging to the wetland community of Siberian region, Russia. Phytochem. Lett. **22**, 280–286 (2017). https://doi.org/10.1016/j.phytol.2017.10.001
6. Snakina, T.I., Kukushkina, T.A.: Variability of the chemical composition of the fruits of marsh blueberry (Vaccinium uliginosum L.) under conditions of introduction. Khimiya Rastitel'Nogo Syr'Ya **3**, 107–114 (2018). https://doi.org/10.14258/jcprm.2018033429
7. Stanoeva, J.P., Stefova, M., Andonovska, K.B., Vankova, A., Stafilov, T.: Phenolics and mineral content in bilberry and bog bilberry from Macedonia. Int. J. Food Prop. **20**, S863–S883 (2017). https://doi.org/10.1080/10942912.2017.1315592

The State of the Population of the Common Lynx (Lynx Lynx Stroganovi), Inhabiting the Territory of the Amur Region

Roman Chikachev(✉)

Far Eastern State Agrarian University, ul. Politehnicheskaja, 86, Blagoveshchensk, Amur Oblast 675005, Russia

Abstract. The article provides a general analysis of the state of the population of the common lynx of the Amur subspecies *(L.l. stroganovi)* inhabiting the territory of the Amur region. The data on the number of lynx for the last 11 years are summarized. The dynamics is determined both in the whole region and in the administrative districts. The main factors influencing the state of the population of the species in the Amur region – forage resources and food competitors-are identified. The limiting factor in the distribution of the lynx in the Amur region is the presence of its main food competitor – the wolf *(Canis lupus)*. Since the lynx is a predator of the highest ecological level, humans additionally took other factors that affect the number of species into account - the complex intraspecific structure of the population, long-term care of offspring, diseases, as well as economic use. For a clearer understanding of the places and types of habitat, the types of land suitable for habitation were bonitized with the calculation of the density of this species in these hunting grounds, the total area of land suitable for lynx habitation in the Amur region was calculated, and the elements of the habitat that are preferable for lynx were identified.

Keywords: Lynx · Population · Lynx numbers · Limiting factors · Feed base · Habitat typology

1 Introduction

The Amur subspecies of the common lynx lives on the territory of the Amur region (L.l. stroganovi). This subspecies of the lynx has a rather limited habitat (the species lives in the Amur region, Primorye, on Sakhalin Island and in North-Eastern China). At the moment, in the available scientific publications there are studies concerning the systematics of this view [1–3]. Currently, this subspecies is poorly studied and requires a deeper and more thorough description.

2 Materials and Methods

The material for the study was the data on the number of animals presented in the departmental documents of the Department for the Protection, Control and Regulation

A. Muratov and S. Ignateva (Eds.): AFE 2021, LNNS 353, pp. 495–504, 2022.
https://doi.org/10.1007/978-3-030-91402-8_56

of the Use of Objects of the Animal World and their Habitat of the Amur region, as well as the authors' own field and laboratory studies [4]. The bonitation of the grounds for the lynx, made according to the five-point scale of D. N. Danilov.

3 Results

The dynamics of the number of lynx in the Amur region by administrative districts over the past 11 years is presented in Table 1.

According to Table 1, it can be seen that in the period from 2010 to 2020, the minimum number of lynx is 427 individuals (2011), and the maximum is 1941 individuals (2016). Such a difference in fluctuations can indicate not only changes in the number under the influence of certain factors, but also the inaccuracy of the data. In the period from 2010 to 2012, the number of lynx was stable, but kept at a relatively low level. Further, since 2012, the number has gradually moved to the rise, with a tendency to increase, and reached its peak in 2016 (1941 individuals). At the moment, the number of lynx is stable and does not decrease below 1000 individuals.

The areas with the largest number of lynx are: Arkharinsky, Zeysky, Selemdzhinsky, Tyndinsky and Shimanovsky. During this time, the lynx was not registered in the Belogorsky, Ivanovo, Konstantinovsky and Tambov districts.

In the standard calculation of the density of the lynx, as well as of any other predator, rather small values were obtained per 1000 ha, analyzing which it would be difficult to estimate the density of the species [5]. Thus, for clarity of indicators, the density of this predator was calculated per 100 thousand hectares. (Table 2).

Thus, on average, the density of the lynx for 11 years varies from 0.6 to 14.5 oc./100 thousand hectares, depending on the habitat area. The highest density relative to the data is in the Arkharinsky district (14.5 ha/100 thousand ha). Over the past 5 years, in the Arkharinsky district, the lynx density index does not fall below 20.0 oc./100 thousand ha. The lowest density of lynx in the Zavitinsky district (0.6 oc./100 thousand ha), but the number in this area is very small. In such large areas as Zeysky and Tyndinsky, despite the high number, the density of the lynx is not very high. The average density of lynx in the Amur region as a whole is 3.6 os. per 100 thousand hectares.

According to Table 2 and Fig. 1, it can be seen that the density of the species in the study area is uneven and varies by year.

But for 11 years, the density of the lynx has remained stable, and in some areas it is increasing. Many factors are related to this indicator, which will be discussed below. The main factor affecting the number of lynx is the forage one. The presence of its main food in the land: white hare (Lepus timidus), siberian roe deer (Capreolus pygargus), as well as birds – capercaillie (Tetrao urogallus), black grouse (Tetrix lyrurus) and grouse (Tetrastes bonasia) it will determine its number. Animal species such as wild boar (Sus scrofa), izyubr (Cervus elaphus) and the reindeer (Rangifer tarandus) they also play a significant role in the nutrition of the lynx, but do not affect the number. As a rule, if the number of white hare and roe deer increases, the number of lynx increases accordingly [6, 7].

Certain basic food items of the lynx (Fig. 2, 3) do not clearly show the picture of the dependence of the species on the quantitative number of the lynx forage base. Although

Table 1. The number of common lynx in the Amur region for 11 years (2010–2020), individuals.

Name of the municipality	2010 g.	2011 g.	2012 g.	2013 g.	2014 g.	2015 g.	2016 g.	2017 g.	2018 g.	2019 g.	2020 g.	Average number over 11 years
Arharinskoe	56	37	54	22	161	416	326	304	312	333	262	207.5
Blagoveshchenskoye	9	12	14	25	19	17	48	37	44	26	43	26.7
Belogorskoe	–	–	–	–	–	–	–	–	–	–	–	–
Bureyskoe	6	6	6	3	41	26	–	11	10	7	–	10.5
Zavitinsk	2	3	3	4	–	–	3	2	–	2	3	2.0
Zeyskoe	46	69	58	68	258	71	138	174	180	167	213	131.0
Ivanovskoe	–	–	–	–	–	–	–	–	–	–	–	–
Konstantinovskoe	–	–	–	–	–	–	–	–	–	–	–	–
Magdagachinskoe	34	34	8	10	17	9	15	11	54	64	61	28.8
Mazanovskoe	12	20	12	43	46	59	77	37	–	42	17	33.2
Mikhailovskoe	9	10	19	–	–	–	–	–	40	–	–	7.0
October	9	7	1	–	11	6	–	4	–	–	3	3.7
Ramenskoye	12	12	7	3	16	63	4	11	7	4	23	14.7
Svobodnenskoe	41	34	10	37	43	65	64	24	39	15	25	36.0
Selemdzhinskoe	56	33	10	571	–	97	95	90	130	143	142	124.3
Seryshevskoe	18	18	18	–	–	1	–	–	5	1	5	6.0
Skovorodinskoe	11	39	118	11	9	11	247	115	26	41	33	60.0
Tambov	–	–	–	–	–	–	–	–	–	–	–	–
Tyndynskoe	86	50	13	7	–	56	783	269	197	214	140	165.0
Szymanowske	46	43	77	36	142	107	141	180	175	105	108	105.5
Total:	453	427	428	840	763	1004	1941	1269	1219	1164	1081	962.6

Table 2. Density of the common lynx in the Amur region for 11 years, individuals/100 thousand hectares.

№	Name of the municipality	2010	2011	2012	2013	2014	2015	2016	2017	2018	2019	2020	Average value
1	Arharinskoe	4.4	2.8	4.2	1.7	12.5	32.4	25.4	23.7	24.3	25.9	20.4	16.2
2	Belogorskoe	–	–	–	–	–	–	–	–	–	–	–	–
3	Blagoveshchenskoye	2.7	3.6	4.2	7.5	5.7	5.1	14.4	11.1	13.2	7.8	12.8	8.0
4	Bureyskoe	0.8	0.8	0.8	0.4	5.4	3.5	–	1.5	1.3	0.9	–	1.4
5	Zavitinsk	0.6	0.9	0.9	1.2	–	–	0.9	0.6	–	0.6	0.9	0.6
6	Zeyskoe	0.5	0.8	0.6	0.8	2.9	0.8	1.6	2.0	2.1	1.9	2.4	1.5
6	Magdagachinskoe	2.0	2.0	0.5	0.6	1.0	0.5	0.9	0.7	3.2	3.8	3.6	1.7
7	Ivanovskoe	–	–	–	–	–	–	–	–	–	–	–	–
8	Konstantinovskoe	–	–	–	–	–	–	–	–	–	–	–	–
9	Magdagachinskoe	–	–	–	–	–	–	–	–	–	–	–	-
10	Mazanovskoe	0.4	0.7	0.4	1.6	1.7	2.2	2.8	1.4	–	1.5	0.6	1.2
11	Mikhailovskoe	2.9	3.3	6.3	–	–	–	–	–	13.2	–	–	2.3
12	October	2.6	2.1	0.3	–	3.2	1.7	–	1.2	–	–	0.8	1.1
13	Romnenskoye	1.2	1.2	0.7	0.3	1.6	6.3	0.4	1.1	0.7	0.4	2.3	1.5
14	Svobodnenskoe	5.4	4.5	1.3	4.9	5.7	8.6	8.5	3.2	5.2	2.0	3.3	4.8
15	Selemdzhinskoe	1.3	0.7	0.2	12.8	–	2.2	2.1	2.0	2.9	3.2	3.2	2.8
16	Seryshevskoe	4.8	4.8	4.8	–	–	0.3	–	–	1.3	0.3	1.3	1.6
17	Skovorodinskoe	0.6	2.6	6.5	0.6	0.5	0.6	13.6	6.3	1.4	2.3	1.8	3.3
18	Tambovskoe	-	–	–	–	–	–	–	–	–	–	–	–
19	Tyndynskoe	1.0	0.6	0.2	0.1	–	0.7	9.3	15.1	2.3	2.5	1.6	3.0
20	Szymanowske	3.2	30	5.3	2.5	9.7	7.4	9.7	12.4	12.0	7.2	7.4	7.3

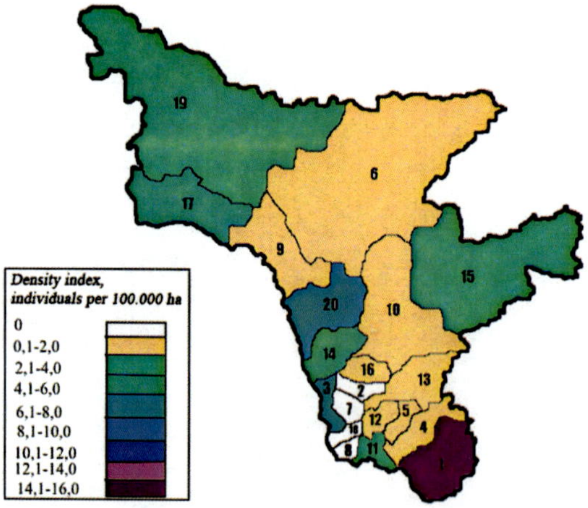

Fig. 1. Relative density of the lynx in the Amur region (2010–2020).

the year-long fluctuations in the number indicate this pattern. Thus, it can be concluded that the number of lynx food items (roe deer, white hare and black grouse) in the Amur region has high indicators and this has a positive effect on the number of lynx.

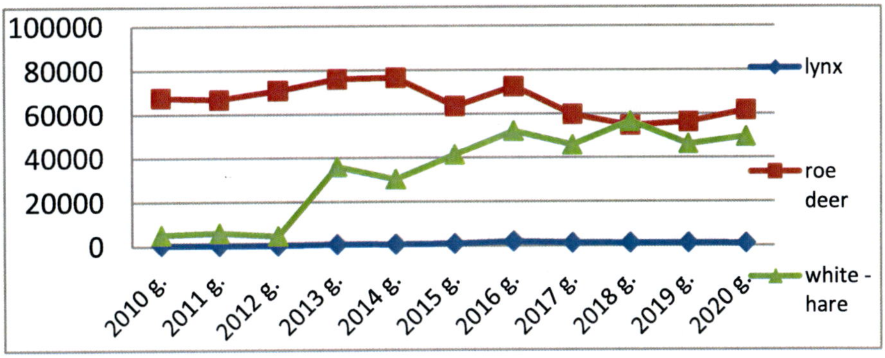

Fig. 2. The dependence of the number of lynx on the number of its main food items in the Amur region (roe deer, white hare).

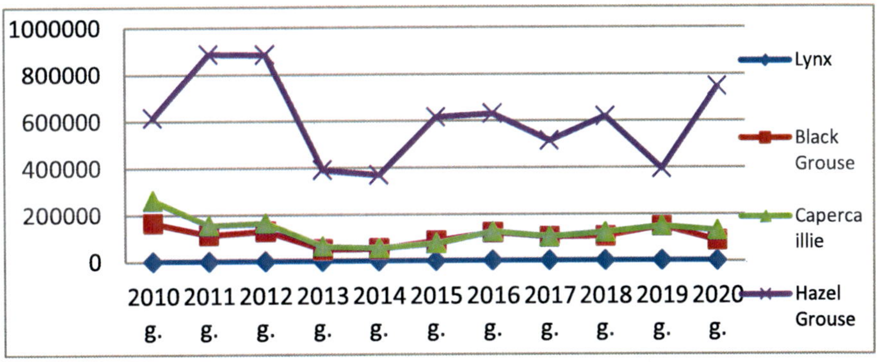

Fig. 3. The dependence of the number of lynx on the number of birds that are objects of its food in the Amur region.

The limiting factor in the distribution of the lynx in the Amur region is the presence of its main food competitor – the wolf (Canis lupus). For clarity, a comparison of their numbers was made.

In the period from 2010 to 2012, the number of lynx was kept at low levels, despite the fact that the number of wolves increased slightly (Fig. 4).

However, in 2013, the number of wolves decreased significantly, which led to a slight increase in the number of lynx. In the future, this dependence can be observed in 2015, 2017 and 2018. In 2016, the opposite happened – the numbers of both predators increased. This may be a consequence of good feeding and protective nesting conditions of the habitat. Since the wolf is an active predator with a high index of the heart index [10], which characterizes it as a predator actively hunting, leading intraspecific and interspecific struggle with competitors for areas with high protective-nesting, feeding conditions, where it is constantly present as a host and fights with competitors.

	2010 g.	2011 g.	2012 g.	2013 g.	2014 g.	2015 g.	2016 g.	2017 g.	2018 g.	2019 g.	2020 g.
Lynx	453	427	428	840	763	1004	1941	1269	1219	1164	1081
Wolf	2223	2504	2920	1380	2882	2450	2923	3208	4125	3208	3066

Fig. 4. Dependence of the number of lynx on the number of wolves in the Amur region.

The lynx is a specialized predator that reflects its ecological identity in morphology and is adapted to forage for food by stalking and stalking at short distances. Yielding in strength, he tries not to contact and leaves the territories with the constant presence of the wolf.

However, it is impossible to fully assert the direct dependence of the number of these predators. Since the lynx is a predator of the highest ecological level and the complex intraspecific structure of the population, long-term care of offspring, diseases and economic use of the species by humans do not allow to fully realize the potential of population growth.

Amur Lynx (L. l. stroganovi) it inhabits almost the entire forest area of the Amur region, with the exception of poorly forested agricultural areas. In this area, the species occurs only occasionally and is not taken into account when conducting accounting work. The calculation of the habitat area for the main types of land suitable for species is given in Table 3.

Thus, according to Table 3, the total area of land suitable for lynx habitat in the Amur region is 18911.51 thousand hectares, which is 53.0% of the total area of the elements of the habitat of hunting animals in the region. The lynx prefers mixed forests with a predominance of coniferous species (14.8%), coniferous deciduous forests (9.7%) and mixed forests with a predominance of small-leaved species (7.8%).

The least preferred areas for lynx in the Amur region are coniferous evergreen, broad-leaved, mixed forests with the presence of broad-leaved species, as well as young trees, floodplain complexes and damaged and transformed areas [8].

Table 3. Elements of the habitat suitable for the lynx in the Amur region.

Elements of the habitat of a hunting resource	Area in the Amur region, thousand hectares	Percentage of the total area of elements of the habitat of hunting animals in the Amur region, %
The woods		
Coniferous evergreen	67.81	0.2
Coniferous deciduous	3448.70	9.7
Small-leaved trees	1236.70	3.5
Broad-leaved trees	351.81	0.9
Mixed with a predominance of coniferous species	5290.64	14.8
Mixed with a predominance of small-leaved species	2786.41	7.8
Mixed with the presence of broad-leaved species	330.40	0.9
Young plants and shrubs		
Clearings and overgrown fields	3388.34	9.5
Deciduous shrubs	1853.47	5.2
Floodplain complexes		
With a predominance of forest	17.42	0.05
Mixed forest	41.59	0.1
Mixed shrub	96.68	0.3
Transformed and damaged areas		
Transformed and damaged areas (burning, peat processing, wind tunnels)	1.54	0.004
Total:	18911.51	53.0

Bonitirovka land for lynx, made according to the five-point scale and his formula in the Amur region:

$$\frac{(B_1 \cdot S_1) + (B_3 \cdot S_3) + (B_5 \cdot S_5)}{100}, \tag{1}$$

where
B_1, B_3, B_5, indicators of optimal population density of the species.
S_1, S_3, S_5, the specific weight of the area of this bonus in % of the total land area (Table 4).

Table 4. Elements of the lynx habitat in the Amur region, included in different classes of the bonite.

Elements of the lynx habitat	Bonus class
The woods	
Coniferous evergreen	1
Coniferous deciduous	3
Small-leaved trees	5
Broad-leaved trees	1
Mixed with a predominance of coniferous species	3
Mixed with a predominance of small-leaved species	3
Mixed with the presence of broad-leaved species	1
Young plants and shrubs	
Clearings and overgrown fields	5
Deciduous shrubs	5
Floodplain complexes	
With a predominance of forest	3
Mixed forest	3
Mixed shrub	5
Transformed and damaged areas	5

Table 5. Bonitirovka hunting grounds for lynx in the Amur region.

Bonitet		
1 class (Good grounds)	3 class (Middle lands)	5 class (Bad land)
750.02 thous. hectares	11584.76 thous. hectares	6576.73 thous. hectares

In general, according to Table 5, it can be seen that the lands of the Amur region for lynx are average (class 3 of the bonus).

4 Conclusions

Study of the features of the ecology of the common lynx (Lynx lynx) the study of the Amur region showed that the number of species in this territory for 2010–2020 has fluctuations from 427 to 1941 wasps. Density fluctuations from 0.6 to 14.5 oc./100 thousand ha. The highest density relative to the data falls on the Arkharinsky district (14.5 oc./100 thousand ha). The lowest density and number of lynx in the Zavitinsky district (0.6 os./100 thousand hectares). In such large areas as Zeysky and Tyndinsky, despite the high number, the density of the lynx is low. The average density of lynx in the Amur region as a whole is 3.6 os. per 100 thousand hectares.

The factors that affect the number and stability of the lynx population are determined, such as forage resources and the number of food competitors - the wolf. The main food items are identified: white hare, Siberian roe deer, capercaillie, black grouse and grouse. Animal species such as wild boar, wild boar, and reindeer also play a significant role in feeding the lynx, but do not affect the number. The lynx is a specialized predator, reflecting its ecological individuality in morphology, adapted to get food by sneaking up and chasing at short distances, inferior in strength to its food competitor, tries not to contact and leaves the territories with the constant presence of the wolf.

The main types of lynx habitat were identified: the total area of land suitable for the species in the Amur region is 18911.51 thousand hectares (53.0% of the total area of the elements of the habitat of hunting animals in the region). The lynx prefers mixed forests with a predominance of coniferous species (14.8%), coniferous deciduous forests (9.7%) and mixed forests with a predominance of small-leaved species (7.8%).The least preferred areas for lynx in the Amur region are coniferous evergreen, broad-leaved, mixed forests with the presence of broad-leaved species, as well as young trees, floodplain complexes and damaged and transformed areas.

The overall assessment of the species ' bonitation in the hunting grounds of the Amur region based on the results of the study is calculated as an average-class 3.

References

1. Squiresa, J.R., Holbrook, J.D., lson, L.E., Ivan, J.S., Ghormley, R.W., Lawrence, R.L.: A specialized forest carnivore navigates landscape-level disturbance: Canada lynx in spruce-beetle impacted forests. For. Ecol. Manag. **475**, 118400 (2020). https://doi.org/10.1016/j.foreco.2020.118400
2. Prentice, M.B., Bowman, J., Murray, D.L., Klütsch, C.F.C., et al.: Evaluating evolutionary history and adaptive differentiation to identify conservation units of Canada lynx (Lynx canadensis). Glob. Ecol. Conser. **20**, e00708 (2019). https://doi.org/10.1016/j.gecco.2019.e00708
3. Ovenden, T.S., Palmer, S.C.F., Travis, J.M.J., Healey, J.R.: Improving reintroduction success in large carnivores through individual-based modelling: How to reintroduce Eurasian lynx (Lynx lynx) to Scotland. Biol. Conserv. **234**, 140–153 (2019). https://doi.org/10.1016/j.biocon.2019.03.035
4. Chikachev, R.: The index of the hearts squad Carnivora, as an indicator of predator ecology. In: Ecological and Biological Well-Being of Flora and Fauna (EBWFF-2020), Blagoveshchensk, Russia, p. 01010 (2020). https://doi.org/10.1051/e3sconf/202020301010
5. Sarmento, P., Carrapato, C.: The use of spatially explicit capture-recapture models for estimating Iberian lynx abundance in a newly reintroduced population. Mamm. Biol. **98**(1), 11–16 (2019). https://doi.org/10.1016/j.mambio.2019.07.004
6. Soyumert, A., Ertürk, A., Tavşanoğlu, Ç.: The importance of lagomorphs for the Eurasian lynx in Western Asia: results from a large scale camera-trapping survey in Turkey. Mamm. Biol. **95**(1), 18–25 (2019). https://doi.org/10.1016/j.mambio.2019.01.003

7. Červenýa, J., Krojerová-Prokešová, J., Kuštaa, T., Koubek, P.: The change in the attitudes of Czech hunters towards Eurasian lynx: is poaching restricting lynx population growth? J. Nat. Conserv. **47**, 28–37 (2019). https://doi.org/10.1016/j.jnc.2018.11.002
8. Holbrook, J.D., Squiresa, J.R., Bollenbacher, B., et al.: Management of forests and forest carnivores: Relating landscape mosaics to habitat quality of Canada lynx at their range periphery. For. Ecol. Manag. **437**, 411–425 (2019). https://doi.org/10.1016/j.foreco.2019.01.011

Forestry and Economic Justification of Commercial Logging in the Amur Region

Olga Dyadchenko(✉), Natalya Ivanova, Alexander Baranov, and Natalya Timchenko

Far Eastern State Agrarian University, Blagoveshchensk, 92, Gorky Street, 675000 Amur Region, Blagoveshchensk, Russia

Abstract. The article analyzes and summarizes the domestic and foreign experience of commercial logging in mixed stands. The forest fund of the Amur region in need of commercial logging was identified, and its analysis was carried out by breed groups, completeness, forest areas, and the predominant economically valuable breeds. A comparative analysis of the tax value of stands during commercial logging and without them is carried out. The article presents the forestry and economic justification of commercial logging in the Amur region.

Keywords: Amur region · Forestry · Commercial logging · Thinning logging · Pass-through logging · Tax value of stands · Larch forests · Completeness · Intensity of logging

1 Introduction

Mixed forests form a significant part of both boreal and temperate forests. For example, 17% of woodlands in Canada are considered mixed in composition, with broadleaf species ranging from 25% to 74%. In Germany, 76% of mixed forests. In Sweden, the proportion of mixed species exceeds half. In Finland, the proportion of mixed forests is 46% [1]. In Russia, the exact number of mixed forests is not kept. On average, they account for up to half of the total area of the country's forest fund – this means forests where hardwoods predominate. If we talk about mixed stands by composition, that is, consisting of two or more species (it does not matter coniferous or deciduous), then these stands, as well as in European countries, will make up the bulk of forests, including in the Amur region.

The high proportion of mixed forests is in sharp contrast to the current forest management guidelines, which emphasize forest management in pure stands [2].

Forest management in mixed forests involves the implementation of maintenance felling, which allows you to increase the productivity of the stand by the age of the main felling. Care felling can be carried out without the removal of wood (clearing felling and clearing felling) and with the removal of wood (thinning felling and passing felling), the so-called commercial care felling.

Commercial logging operations allow forestry managers to perform wood production tasks. The productivity of mixed stands can be explained by the positive interaction

A. Muratov and S. Ignateva (Eds.): AFE 2021, LNNS 353, pp. 505–513, 2022.
https://doi.org/10.1007/978-3-030-91402-8_57

between the effects of thinning and the complementarity of niches. It is suggested that care cuttings that support the species diversity of mixed plantings can help align wood production goals with biodiversity concerns [3].

Research has also expanded the understanding of how biodiversity affects the functioning of ecosystems and the impact on the provision of ecosystem goods and services [4, 5]. Evidence suggests that certain combinations of species obtained, including from commercial logging, demonstrate additional resource use with positive consequences for productivity [6] accumulation of the largest stock [7], increasing the viability of ecosystems [4, 8], increasing the stability of stands [9], reducing economic risks as a result of intermediate harvesting of wood resources [10, 11] and to provide insurance against environmental changes [12].

Because of the intrinsic relationship between forest management and natural hazards, planted forest management strategies can be used as an effective risk management tool. For example, care felling, in particular commercial felling, can help reduce the risk of forest fires by reducing combustible materials and increasing the resilience of forests to restoration [13]. Thinning can also reduce the susceptibility of the stand to other natural disturbances associated with high stand density, such as pine beetles [14].

In the Amur region, a large number of fires occur, the largest number of them in all the analyzed time periods is registered in the Zeysky and Tyndinsky forest districts [15]. In this regard, the implementation of logging care in the forest areas of the region can contribute to their reduction.

Our research is aimed at meeting the obvious need to develop an economic understanding of the management of mixed stands, the feasibility of conducting commercial logging in them, increasing the forestry and economic characteristics of stands.

Currently, logging companies, including large tenants of forest plots, carry out logging only in mature and over-mature stands. Considering it unprofitable to perform any types of felling care, while carrying out some of them is economically feasible and in the future allows you to get a much higher yield of high-quality timber from each tree left on the root.

The purpose of the research is to substantiate the effectiveness of commercial types of logging (thinning and pass-through logging) on the lands of the state forest fund of the Amur region.

Deforestation is a key element for the transition to an intensive and sustainable forest management model. The correct use of logging allows you to increase the volume of harvested wood per hectare up to 1.5–2 times per farm cycle, and the yield of valuable varieties - from 1.5 to 3 times. As a result of the system of logging of forest care, the cost of harvested wood for the cycle of the farm increases several times. In addition, the intensive and sustainable forest management model takes into account the need to conserve biodiversity.

2 Materials and Methods

The object of our study was the stands in need of commercial logging of care and passed by logging of care. This work is based on the materials of the state forest inventory conducted by the Amur branch of FSUE "Roslesinforg" in 2018. The measurements of the state forest inventory sites were used to compare the tax value of stands (at the root) passed by thinning and passing felling and natural stands without anthropogenic interference in comparable periods of age. The data of the state forest inventory (forestry regulations and tax descriptions of forest areas, the forest plan of the Amur Region for 2019–2028) cover all areas of growth and types of stands and represent the entire territory of the Amur region. The disadvantage is that the history of plant management is little known and only periodic growth information is available. Nevertheless, the data from the state forest inventory can be used to analyze various types of forest management measures by selecting stands that are comparable in terms of forest growing conditions and other analyzed indicators. The tax value of stands of trees was calculated on the basis of the Decree of the Government of the Russian Federation No. 310 of 22.05.2007 and the Decree of the Government of the Russian Federation No. 1318 of 12.10.2019.

3 Results

One of the main criteria for the purpose of felling care is the completeness of stands. In this regard, the analysis of stands of the Amur region by groups of rocks was carried out. The following results were obtained: in the state forest fund of the Amur region, low-level stands predominate among coniferous and hard-leaved stands, medium-level stands predominate among soft-leaved stands. The high-field stands were divided into groups of rocks as follows:

- Coniferous-261.7 thousand ha-
- Hard-leaved-1.4 thousand ha-Soft
- leaved-256.1 thousand ha (Fig. 1)

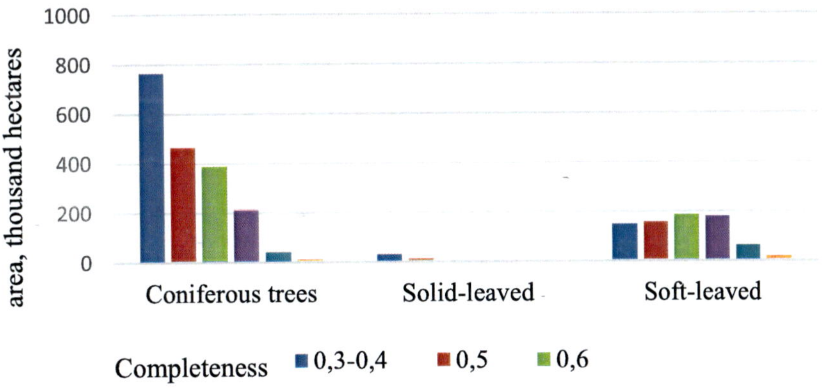

Fig. 1. Distribution of the forested area of the Amur region by completeness and groups of rocks.

13 forest districts are organized in the Amur Region. In each of them, it is necessary to carry out care felling. The area of stands in need of felling varies in each forest area: 50,765. 4 hectares of stands need thinning, 85,458. 3 hectares need passing felling (the smallest areas of these stands fall on the Zavitinsky forest area-27.4 ha and 8.6 ha, respectively, the largest on the Zeysky forest area-16,881. 7 ha and 41439. 2 ha, respectively) according to forestry regulations (Fig. 2).

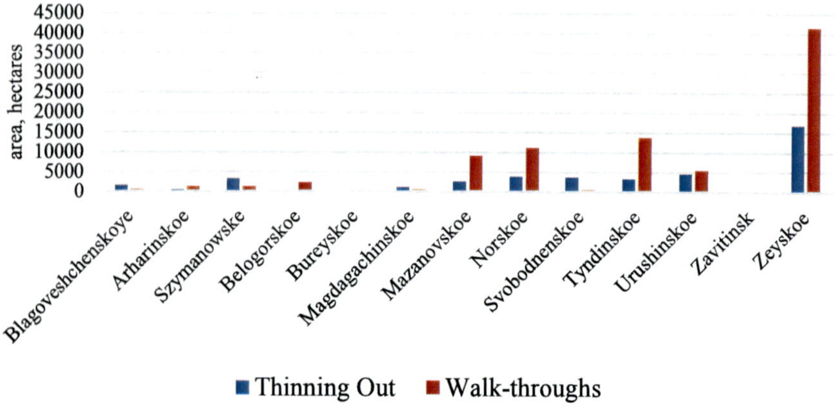

Fig. 2. Areas of stands in need of commercial logging.

The next stage of our work was to compare the annual allowable volume of wood withdrawal during commercial logging with the total area of stands in need of commercial logging. Table 1 shows that for the coniferous group of rocks, the annual allowable volume of withdrawal is 2797.2 hectares (which is 2.5% of the required area.). Even with a 10-year forest plan, only 25% of the commercial logging volume can be realized.

Table 1. Annual allowable volume of wood withdrawal during commercial logging.

Breed group	Area, ha	Reserve, thousand cubic meters	
		Liquid	Business
Coniferous trees	2797.2	38.631	21.078
Solid-leaved	–	–	–
Soft-leaved	127.1	3.14	1.35
Total:	2924.3	41.771	22.428

After analyzing the implementation of the previous forest plan, another problem was identified: none of the forest districts fulfilled the planned measures for cutting through and passing through logging over the past audit period (Fig. 3).In general, the planned volume for coniferous farming is 1531.6 hectares, the actual volume for coniferous farming is 175.8 hectares (which is only 11.5%).

The largest volumes of commercial logging were planned to be carried out in the Shimanovsky and Zeysky forest areas of the Amur region, 850,0 and 290.6 hectares, respectively. Shimanovskoye forestry completed 45.0 hectares (5.3%), Zeyskoye – 20.2 hectares (6.9%).

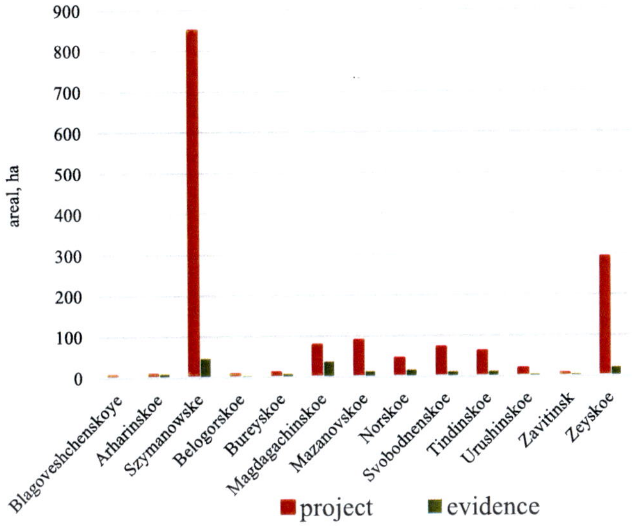

Fig. 3. Planned and actually completed volumes of commercial logging of coniferous care according to the previous forest plan (2009–2018).

Depending on the intended purpose of the forests, cuttings of care form economically feasible plantings in the conditions of the corresponding groups of forest types. Felling care should provide intermediate use in terms of quality and volume, without reducing the intended main use and the stability of plantings, as well as water protection, sanitary and hygienic and other useful properties of the forest. The result of commercial types of logging is as follows:

- When carrying out these types of logging, we can get liquid soft-leaved wood to meet part of the population's need for it.
- Later, the stands, having reached the age of ripeness, the feeding area of the main breed, as a result of which the increase in all taxational indicators, including the stock, will increase. In addition, this will lead not only to an improvement in quantitative characteristics, but also in quality, since many hardwoods can be carriers of fungal and viral diseases of conifers.

Larch forests that grow in large areas are very diverse in origin, combination of tree species, age, wood stock, its quality and other characteristics. The predominant (economically valuable) species in the forest fund of the Amur region is the Daurian larch. Most often it is accompanied by flat-leaved birch. Of all the forest formations in the Far East, larch forests occupy the largest territory. The share of larch in the composition of forest landscapes increases from south to north. Larch forests also predominate in most forest areas of the Amur Region, except for Blagoveshchensk, Belogorsky and Svobodnensky (southern regions of the Amur Region). Stands of this breed that require commercial logging of care amount to 101,454. 6 ha (Fig. 4). The second most important and economically valuable species of the Amur region is scots pine, but its area is much smaller compared to stands of Daurian larch. Stands of Scots pine in need of commercial felling care amount to 10196.2 hectares, which is ten times less than the same indicator in the larch stands of the region under consideration.

In connection with the above, we will analyze the economic effect of the behavior of commercial logging on the example of larch stands.

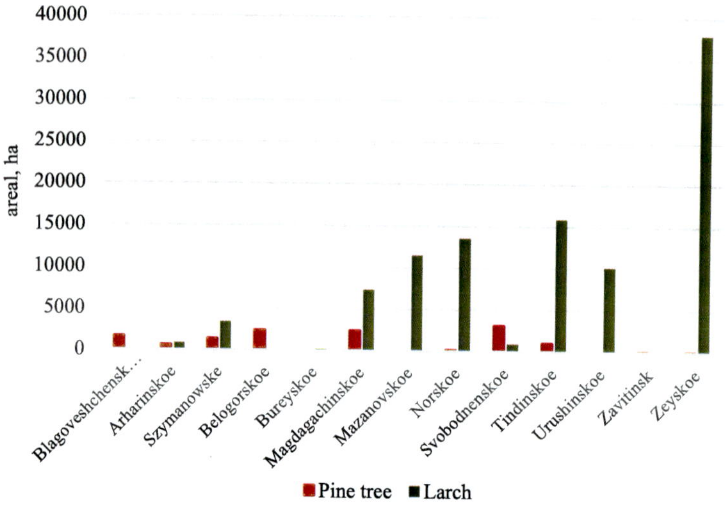

Fig. 4. Areas of economically valuable stands in need of commercial logging.

In order to justify the felling of care, an analysis was made of the average tax value of larch stands with flat-leaved birch, passed by commercial felling of care of various intensity and natural stands, without carrying out forestry measures in them.

Allotments with an intensity of 20%, 15%, and 10% of commercial logging were considered. As can be seen in the diagram, the tax value increases very sharply and reaches a maximum by the age of the main cutting, in addition, in comparison with natural stands in this period of time, it is 30–40% higher (Fig. 5).

The increase in the tax value of stands is due to:

- a decrease in the composition of the stand of soft-leaved species (since their tax value is significantly lower than coniferous)
- an increase in the average diameter of the main rock (as a result of an increase in the feeding area, and, consequently, an increase in the main taxational indicators).

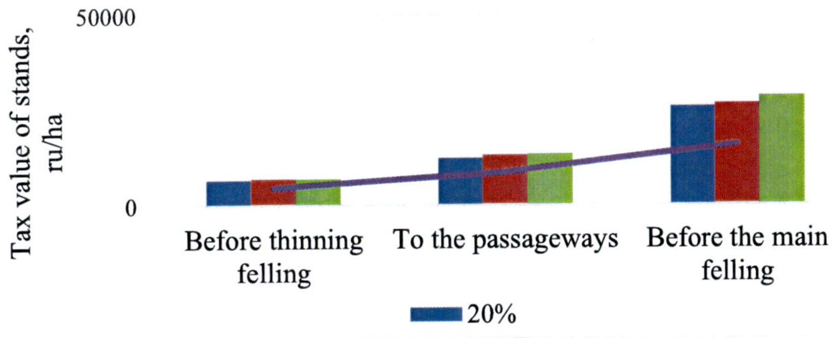

Fig. 5. Average tax value of larch stands.

The intensity of logging was regulated by their initial completeness, so in general it did not affect the final tax value of the stand.

In turn, stands that have not been cleared by logging have lower diameters and soft-leaved species take part in their composition, which significantly reduces their tax value.

4 Discussion

In the present study, we analyzed the areas of stands in need of commercial logging in the context of forest areas, the planned and actually performed volumes of thinning and pass-through logging on the territory of the Amur region over the past audit period.

According to the results of the data analysis, it was revealed that these types of felling care in the region are carried out in a small number.

In this regard, we have calculated the economic justification for their implementation on the example of the economically valuable larch forests prevailing in the region, in order to attract attention to their implementation not only by state institutions, but also to interest tenants who can get more economically valuable stands at the exit and at the same time meet the needs for small-scale soft-leaved wood harvested as a result of intermediate use.

5 Conclusions

1. When carrying out all types of care felling, the task of improving the sanitary condition of plantings is solved, in the process of felling, shrunken, damaged, weakened, affected by harmful insects and diseases trees are removed.
2. Depending on the intended purpose of forests, felling care forms economically feasible plantings in the conditions of the corresponding groups of forest types.
3. Maintenance cuttings provide intermediate use in terms of quality and volume, without reducing the intended main use and the stability of plantings, as well as water protection, sanitary and hygienic and other useful properties of the forest.

4. Logging is a key element for the transition to an intensive and sustainable forest management model. The correct use of logging allows you to increase the volume of harvested wood per hectare up to 1.5–2 times per farm cycle, and the yield of valuable varieties - from 1.5 to 3 times.
5. When carrying out these types of logging, we can get liquid soft-leaved wood to meet part of the population's need for it. Subsequently, in stands that have reached the age of ripeness, the feeding area of the main breed will increase, as a result of which the increase in all taxational indicators, including the stock, will increase. In addition, this will not only improve the quantitative characteristics, but also the quality, since many hardwoods can be carriers of fungal and viral diseases of conifers.
6. Carrying out these types of logging by loggers who have long-term lease of forest plots for timber harvesting will allow for the urgent and long-term rational use of the entire stock of raw-growing forest with economic benefits. By the age of cutting down mature and over-mature stands, the stand will reach the maximum supply of high-quality wood of economically valuable species.

References

1. Halbrittera, A., Deegena, P., Susaetab, A.: An economic analysis of thinnings and rotation lengths in the presence of natural risks in even-aged forest stands. For. Policy Econ. **118**, 102223 (2020). https://doi.org/10.1016/j.forpol.2020.102223
2. Kuuluvainen, T., Tahvonen, O., Aakala, T.: Even-aged and uneven-aged forest management in boreal fennoscandia: a review. Ambio **41**, 720–737 (2012). https://doi.org/10.1007/s13280-012-0289-y
3. Barrette, M., Tremblay, S., Auger, I.: Commercial thinning that maintained species diversity of a mixed black spruce–jack pine stand enhanced productivity. Scand. J. For. Res. **33**(8), 756–763 (2018). https://doi.org/10.1080/02827581.2018.1495254
4. Balvanera, P., et al.: Quantifying the evidence for biodiversity effects on ecosystem functioning and services. Ecol Lett **9**, 1146–1156 (2006). https://doi.org/10.1111/j.1461-0248.2006.00963.x
5. Gamfeldt, L., et al.: Higher levels of multiple ecosystem services are found in forests with more tree species. Nat. Commun. **4**, 1340 (2013). https://doi.org/10.1038/ncomms2328
6. Pretzsch, H., Schütze, G.: Transgressive overyielding in mixed compared with pure stands of Norway spruce and European beech in Central Europe: evidence on stand level and explanation on individual tree level. Eur. J. For. Res. **128**, 183–204 (2009). https://doi.org/10.1007/s10342-008-0215-9
7. He, Y., Qin, L., Li, Z., Liang, X., Shao, M., Tan, L.: Carbon storage capacity of monoculture and mixed-species plantations in subtropical China. For. Ecol. Manag. **295**, 193–198 (2013). https://doi.org/10.1016/j.foreco.2013.01.020
8. Jactel, H., Nicoll, B.C., Branco, M., et al.: The influences of forest stand management on biotic and abiotic risks of damage. Ann. For. Sci. **66**, 701 (2009). https://doi.org/10.1051/forest/2009054
9. Griess, V.C., Acevedo, R., Härtl, F., Staupendahl, K., Knoke, T.: Does mixing tree species enhance stand resistance against natural hazards? a case study for spruce. For. Ecol. Manag. **267**, 284–296 (2012). https://doi.org/10.1016/j.foreco.2011.11.035

10. Knoke, T., Ammer, C., Stimm, B., et al.: Admixing broadleaved to coniferous tree species: a review on yield, ecological stability and economics. Eur. J. For. Res. **127**, 89–101 (2008). https://doi.org/10.1007/s10342-007-0186-2
11. Griess, V.C., Knoke, T.: Bioeconomic modeling of mixed Norway spruce—European beech stands: economic consequences of considering ecological effects. Eur. J. For. Res. **132**, 511–522 (2013). https://doi.org/10.1007/s10342-013-0692-3
12. Schütz, J.-P., Götz, M., Schmid, W., Mandallaz, D.: Vulnerability of spruce (Picea abies) and beech (Fagus sylvatica) forest stands to storms and consequences for silviculture. Eur. J. Forest Res. **125**(3), 291–302 (2006). https://doi.org/10.1007/s10342-006-0111-0
13. Charnley, S., Spies, T., Barros, A., White, E., Olsen, K.: Diversity in forest management to reduce wildfire losses: implications for resilience. Ecol. Soc. **22**(1), 22 (2017). https://doi.org/10.5751/ES-08753-220122
14. Jenkins, M.J., Runyon, J.B., Fettig, J., Page, W.G., Bentz, B.: Interactions among the mountain pine beetle, fires, and fuels. For. Sci. **60**(3), 489–501 (2014). https://doi.org/10.5849/forsci.13-017
15. Dyadchenko, O., Timchenko, N., Baranov, A.: Influence of permanent and variable factors on the number and area of forest fires in the amur region. In: E3S Web of Conferences, Ecological and Biological Well-Being of Flora and Fauna, EBWFF 2020, p. 03002 (2020). https://doi.org/10.1051/e3sconf/202020303002

Restoration Prospects of Abies Holophylla Forests in the Southern Russian Far East

Alyona Kiselyova[1](✉), Ilona Rodnikova[1], Olga Ukhvatkina[2], Kirill Ganzei[1], and Nina Pshenichnikova[1]

[1] Pacific Geographical Institute FEB RAS, Radio Street, 7, Vladivostok, Russia
[2] Federal Scientific Center of the East Asia Terrestrial Biodiversity FEB RAS, Pr-t 100-let Vladivostoka, 159, Vladivostok, Russia

Abstract. The research focuses on the state of *Abies holophylla* plantations at the northern limit of its current distribution range and the comparative deendrochronological data revealing successful development of the planting trees. *Abies holophylla*-broadleaved forests are among the temperate zone richest vegetation types in terms of species diversity. The comparison between the plantation and the reference native forests shows the sustainable development of the plantations. As follows from our research, the climatic parameters of the south of the Russian Far East now provide the optimum growth conditions for *Abies holophylla* even considering the climate warming. We conclude that restoration of *Abies holophylla*-broadleaved forests in the south of the Russian Far East has great potential. Favorable environmental conditions coupled with the species adaptation ability may contribute to restoration success. In Russia *Abies holophylla* now comprises only a small fraction of the forested landscape in the southern Primorsky Krai. Using the assessment of the current state of *Abies holophylla* forest stands on Russky Island compared to that of reference native forests we estimate the prospects for *Abies holophylla* forest restoration under the current conditions. Our results indicate the patterns of woody species diversity in the study stands are different from those in the native forests. Therefore, sustainable development of the restored forests requires a simultaneous combination of restoration and protection techniques.

Keywords: *Abies holophylla* · Near threatened species · forest restoration · Radial growth · Podsolized burosems · Lichens

1 Introduction

Abies holophylla Maxim. (Pinaceae) arose in Miocene 13 million years ago. It is listed as Near Threatened species by Red List IUCN – the International Union for Conservation of Nature and Natural resources. It is one of the main forest-forming tree species in Northeast Asia, its northern limit is 44°N and the southern limit is near 33°N. The Abies holophylla range covers the south of the Russian Far East, the northeast of China and Korea [1]. Abies holophylla forms pure stands, mixed coniferous forests, and mixed coniferous deciduous forests with high species diversity. These forests are disturbed by fire and illegal cutting [2]. According to the recent research in the southern part

A. Muratov and S. Ignateva (Eds.): AFE 2021, LNNS 353, pp. 514–521, 2022.
https://doi.org/10.1007/978-3-030-91402-8_58

of the species distribution range, A. holophylla is expected to be vulnerable to global warming [3] and its seedling showed a high mortality rate under experimental warming because of increased air and soil temperatures and decreased soil moisture [4]. Although in the northern part of its distribution range the species demonstrated high survival rate in case of natural hazard, as well as it was introduced successfully northward of its current distribution range. Studies of Lee and Yao et al. [5, 6] indicate that physiological parameters of A. holophylla have different sensitivities and adaptabilities to changing environmental conditions.

Our previous research assessed environmental conditions of the Abies holophylla plantation on Russky Island and documented that in the past, Russky Island was partly covered with Abies holophylla forests [7]. The present study estimates the prospects for Manchurian fir forest restoration under the current conditions at its northern limit. Abies holophylla occurs sporadically on the islands Furugelm, Bolshoy Pelis, De-Livron, Rikord, Shkot, and it forms communities on the islands Stenin, Popov, Russky, Antipenko, Sibiryakov [8]. The goal of our research is to assess the current state of Abies holophylla plantations on Russky Island (biodiversity, soil cover conditions, and effect of the environment on planted trees during their growth compared to those of reference forests) and to determine the species restoration potential under the current environment conditions in the south of the Russian Far East.

2 Materials and Methods

Russky Island is incorporated into Vladivostok urban district. It is the largest island in Peter the Great Bay with the total land area of about 100 km2. The climate is characterized as monsoon with wet summers and dry winters. Annual precipitation rate is 724 mm, with maximum falls in August and September. The average annual temperature is 5,6 °C. The landscape of Russky Island is composed of low mountain landforms with gentle and mean steep slopes on granite and granitoid. Currently, forests cover over 75% of the island area. Most of the forested areas are high-closed polydominant forests on dark and typical burozems. A much smaller part is occupied by coniferous-broadleaved forests.

The field survey was carried out in 2016–2020. For the purpose of this study, we set 30 sample plots in 5 Abies holophylla plantations on Russky Island (Fig. 1). Six sample plots of 25 m × 25 m were set in each plantation. Landscape maps at a scale of 1:25000 were produced in ArcMap 10.5, based on satellite images provided by Google Earth Pro, combined with field work. Field surveys for landscape data were conducted on each sample plot. The modern landscapes were described in terms of landforms, development of geomorphologic processes, bedrocks, soil, and vegetation. Each plantation was assessed for slope degree, aspect, and total area. At each plot, all vascular plants, relative cover for each plant in each vegetation layer, height of trees, and total cover of each vegetation layer were recorded. The presence of adventitious plant species was recorded. We studied species diversity of epiphytic lichens at each plot. For each plot the ratio of anthropogenically modified to natural habitat species was measured, calculated on the basis of the number of taxa.

To reconstruct the disturbance history of the forest stands we used dendroecological evidence. To estimate radial growth and tree age, increment cores were taken using

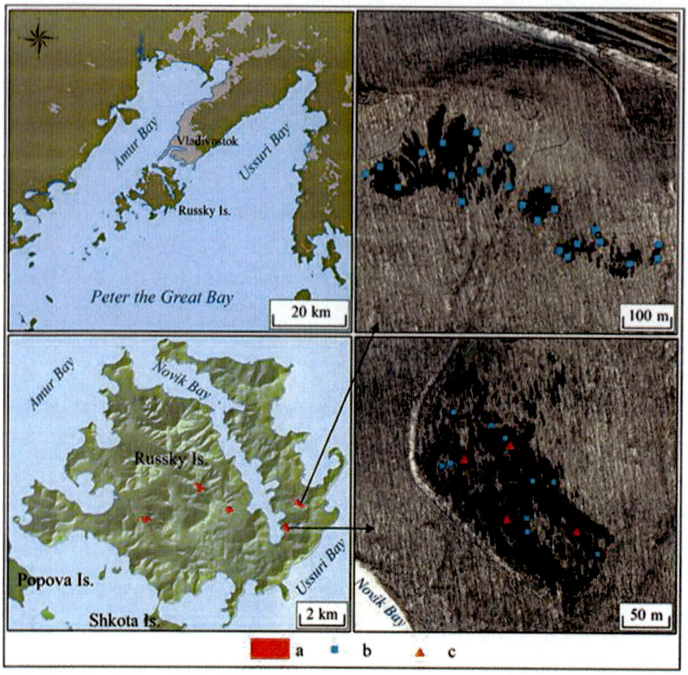

Fig. 1. The study stands of *Abies holophylla* on Russky Island: a – plantations of *Abies holophylla* on Russky Island, b – dendrochronological samples, c – soil samples.

Pressler increment borer at 30 cm above ground level at two plantations (Fig. 1). One core per tree was taken, 33 trees in total were cored. In the laboratory, all tree-ring samples were mounted, dried and progressively sanded to a fine polish until individual tracheid within annual rings were visible under the anatomical microscope according to the standard dendrochronological procedures. Preliminary calendar years were assigned to each growth ring, and possible errors in measurement due to false or locally absent rings were identified by the Skeleton-plot cross-dating method. The cores were measured using the semi-automatic Velmex measuring system (Velmex, Inc., Bloomfield, NY, USA) with a precision of 0.01mm. Then, the COFECHA program was used to check the accuracy of the cross-dated measurements. To mitigate potential trend distortion problem in traditionally detrended chronology [9, 10], we used a signal-free method [11] to detrend tree-ring series using RCSigFree program.

To date precisely and identify the missing rings, we used cores obtained from Abies holophylla forest on Muravyov-Amursky Peninsula in 2013. In total, 40 cores were taken from adult trees (two fir tree samples were 293 and 296 years old, respectively). To analyze radial growth-climatic relationships, age-related trends were removed from the raw tree-ring series using an age-dependent spline smoothing. The ratio method was used to calculate tree-ring indices, and the age-dependent spline was selected to stabilize the variance caused by core numbers. Statistical analyses of climate-growth relationships were carried out in R using the treeclim package [12] and STATISTICA software (StatSoft®).

Soil samples were collected at each plantation. The description of a soil profile to the depth of parent rock was used to characterize morphological structure and classify the soil type. Morphological characteristics of each horizon (thickness, color, humidity, texture, structure, inclusion, and neoformation) were identified and the horizon was classified according to Classification and diagnostics of soils of Russia. Soil samples were collected from each horizon. The collected samples were air-dried, ground and sieved through 1mm openings. Standard methods were used to examine the physical-chemical features of soil.

3 Results

The study plantations occur in low mountain landscapes of the south aspect at altitudes from 30 to 130 m a.s.l. The plantations are found on gentle slopes (7–10°) on granite and granitoid, locally on pebble conglomerate, sandstone, and siltstone. Our results show that, at present, the total island area reforested with Abies holophylla is 16.97 ha (0.16% of the total island area). Coniferous plantations on Russky Island were planted in rows at a step of about 2.5–3 m and a distance between rows of about 3–4 m. Analysis of Corona satellite images from 1962 to 1978 and the sampled cores of Tilia amurensis Rupr. (140 years old) indicate that Abies holophylla were planted in the understory of the secondary broadleaved forests (Table 1).

Table 1. Landscape, soil and floristic description of the *Abies holophylla* plantations on Russky Island.

Landscape	Low mountain denudation landscape subtype with gentle slopes on pebble conglomerate, sandstones, and siltstones
Tree layer cover (%)	80
Average height of tree layer (m)	12
Shrub layer cover (%)	40
Average height of shrub layer (m)	3
Herb layer cover (%)	40
Average height of herb layer (m)	0.8
Lianas cover (%)	6
Average length of lianas (m)	5
Total number of vascular plant species	49
Total number of lichens	41
Soil	Podsolized burozems
Thickness of humus horizon	8.9
Skeleton content of humus horizon (%)	16.8
Skeleton content of illuvial horizon (%)	84.8
Woody species (%)	30

The central parts of the plantations are pure stands of Abies holophylla without shrub and herb layers. Beyond these, the stands turn intoto fir-broadleaved forests with tree species such as Betula davurica Pall., Quercus mongolica Fisch. ex Ledeb., Acer pseudosiboldianum (Pax) Kom., Fraxinus rhynchophylla Hance, Kalopanax septemlobus (Thunb. ex Murray) Koidz., Tilia amurensis (Appendix). The average tree height is 12 m., the tree layer cover is 80%. The shrub layer is represented by Viburnum sargentii Koehne, Corylus mandshurica Maxim., Lespedeza bicolor Turcz., Philadelphus tenuifolius Rupr., Rhododendron mucronulatum Turcz., Ribes maximoviczianum Kom., with the average height of 3 m. The shrub layer covers 40% of plot size. The most abundant species in the herb layer are Dryopteris crassirhizoma Nakai, Maianthemum bifolium (L.) F.W.Schmidt, Arisaema amurense Maxim., Aquilegia oxysepala Trautv. et C. A. Mey., Aruncus dioicus (Walt.) Fern., Bistorta pacifica (V. Petrov ex Kom.) Kom., Carex lanceolata Boott, Cerastium holosteoides Fr., Hemerocallis middendorfii Trautv. et C.A. Mey., Onoclea sensibilis L., Smilacina hirta Maxim., Thalictrum amurense Maxim., with the average height of 0.8 m. This layer covers 40%. Woody vines such as Vitis amurensis Rupr., Celastrus orbiculata Thunb., Schisandra chinensis (Turcz.) Baill. are constantly present, with average height of 5 m. Adventitious species are found in all the plantations, but their occurrence is rather low (three species).

A total of 41 lichen species were recorded from the plantations. The epiphytic lichen flora comprises five Red-listed species: Anzia colpodes (Ach.) Stizenb., Coccocarpia palmicola (Spreng.) Arv. & D.J. Galloway, Menegazzia terebrata (Hoffm.) A. Massal., Parmotrema reticulatum (Taylor) M. Choisy, Pyxine sorediata (Ach.) Mont. The most common lichens are Anaptychia isidiza Kurok., Flavoparmelia caperata (L.) Hale, Gyalolechia flavorubescens (Huds.) Søchting, Frödén & Arup, Heterodermia speciosa (Wulfen) Trevis., Lecanora allophana (Ach.) Nyl., Lecidella euphorea (Flörke) Hertel, Myelochroa aurulenta (Tuck.) Elix & Hale, Parmelia saxatilis (L.) Ach. We found 7 species typical of anthropogenically modified habitats.

Abies holophylla trees are cut illegally every year at Christmas time. In December 2016, 32 trees were cut at the first study stand (5 trees – 25–35 years old, 22 – 35–45 y.o., and 5 – more than 50 y.o.). Their maximum height was 12 m. About 85% of the felled fir trees grew about 25–30 m. away from the road.

The radial growth of the adult trees sampled on Muravyov-Amursky Peninsula matches closely the radial growth of the planted fir trees and corresponds to natural developmental history. The cores, taken at the height of 30 cm, contain from 32 to 52 rings. It took from 4 to 15 years for the trees to reach the height of sampling, therefore the tree age could range from 36 to 67 years.

A visual inspection of the sampled cores shows the absence of fire events. Nevertheless, the average radial growth curve of the island trees reveals minimum growth rate years. To identify the cause of the low growth rate, we created the integrated chronology including all the examined cores and studied the climate-growth relationships. We used the records of Vladivostok Meteorological Station from 1941 to 2017: average monthly air temperature, minimum monthly air temperature, total monthly precipitation. The results show the absence of climate-growth relationships, except for weak correlation between the growth of A. holophylla and the minimum temperatures of September-November in the previous year ($R = 0.46$). The average radial growth curve of the island trees compared to minimum air temperature curve of September-November in the previous year reveals that the years with the minimum growth rate follow the years of minimum temperatures of September-November.

Soils under the Abies holophylla plantations are weakly-podsolized shallow highly skeletal burozems, that belong to Cambisol (Gleyic Dystric Cambisols) according to the World Reference Base for Soil Resources. The profile consists of the following horizons: AY-AYeBM-BMg-BMC. The texture of humus accumulative horizon AY is moderate clay and fine sand, with a bulk of herb and shrub roots as well as isolated tree roots present in the horizon. Skeletonless and loose consistency of the horizon contributes to its high hydraulic conductivity. Skeleton content increases greatly in the shallow weakly-podsolized AYeBM horizon and illuvial gleyed BMg horizon. Closely packed rock fragments result in poor drainage conditions (Table 2). High rainfall can contribute to overwetting and gleization of the illuvial part of the soil profile, as observed during soil sampling. Organic matter content decreases drastically down the soil profile, ranging from 7,85% in AY horizon to 2,74% in BMg horizon. Hydrolitic acidity level does likewise, with the values largely dependent on humus organic acid content. Calcium cations (9,5 mg-eq/100 g of soil) dominate the soil absorbing complex, with far less magnesium (3,5 mg-eq) and still less sodium (1,6 mg-eq) in the complex composition. Available phosphorus level holds stable throughout the soil profile and does not exceed 25 mg/kg. Available potassium level, high in the humus horizon (145 mg/kg), decreases sharply down the profile (50mg/kg) due to the element biological accumulation in the topsoil.

Table 2. Average physical and chemical properties of weakly podzolized burozem under the *Abies holophylla* plantatins on Russky Island.

Horizon	Depth, cm	pH KCl	Soil organic matter, %	Exchangeable soil acidity mmol/ 100g of soil	Mg-eq / 100 g of soil					Mg/ kg	
					Hydrolitic soil acidity	Exchangeable cations			Total absorbed bases	P_2O_5	K_2O
						Ca^{++}	Mg+	Na^+			
AY	5–15	4.6	7.85	0.36	7.92	9.5	3.5	1.6	14.8	25	145
BMg	20–60	4.5	0.71	0.18	2.74	5.0	3.0	1.7	8.8	25	50

4 Discussion

Pollen records indicate that Abies holophylla has been present in forest areas of Russky Island since the latter half of the Mid-Subboreal [7]. At present, Manchurian fir plantations on Russky Island are one of the most extensive populations of the species among the islands of Peter the Great Bay [8]. The floristic composition of the study fir stands resembles the composition of natural Manchurian fir forests of Muravyov-Amursky Peninsula – the closest natural reference forest stands. The species common to both the fir stands and the native fir forests are Fraxinus rhynchophylla, Quercus mongolica, Kalopanax septemlobus, Acer mono, A. pseudosieboldianum, Tilia amurensis, Philadelphus tenuifolius, Vitis amurensis. The fir-broadleaved forests of Muravyov-Amursky

Peninsula are diverse in terms of woody species – from 20 to 35. Nine species were found in the fir plantations under study. Native fir-broadleaved forests are characterized by a sparse shrub layer, and the shrub layer of the study plantations is not well-developed either (40%).

The tree age ranges from 36 to 67 years. The development of adult trees of Muravyov-Amursky Peninsula correlates well with the development of planted trees of Russky Island, indicating successful growth of the planted trees. Features of radial growth reveal the absence of disturbances over the developmental history of the fir stands. After 15 years, when the crown formation was completed, the annual growth rate became constant (2–3 mm/year) indicating favorable environmental conditions. This period of successful development started in 1978.

Such forests are not characterized by soil lichen cover because of competition with herbs or low light availability. By contrast, lichen communities thrive on the tree bark occupying from 15% to 100% of the substrate.

The study of the soil cover revealed that soil organic matter content in AY-horizon corresponds to that in soils with high humus level. Morphological, physical and chemical features of the soils are characteristics of favorable conditions for A. holophylla and they contribute to the successful plant establishment.

At present, the plantations of A. holophylla on Russky, Popov and Stenin islands, according to the scale of age development, refer to saplings with the transition of the oldest trees to the stage of intensive growth in height and diameter and active seed production. A. holophylla plantations are the "cores" of restoration of native island broad-leaved forests, which were formed at least from the Middle Holocene.

5 Conclusions

Island fir communities restored on Russky Island have the prospect of reconstructing coniferous Abies holophylla-Pinus koraensis forests. The other islands of Peter The Great Bay have low degree of recovery of fir forests. The practical basis supporting the Abies holophylla forests restoration is the forest role in maintaining high biodiversity. Basing on the study results we conclude that restoration of Abies holophylla forests in the study area has great potential. Morphological, physical and chemical parameters of weakly podzolized, highly skeletal burozems indicate a sufficiently high fertility, which provides favorable growing conditions for A. holophylla and successful forest restoration. The favorable environmental conditions considering climate warming coupled with the species adaptation ability may promote restoration success. But the success of the restoration of the Abies holophylla forests may be in part hampered by unrestricted recreational and economic activities. Therefore, sustainable development of the restored forests requires a combination of restoration and protection techniques.

References

1. Katsuki, T., Zhang, D., Rushforth, K.: Abies holophylla. The IUCN Red List of Threatened Species e.T42287A2969916 (2013). https://doi.org/10.2305/IUCN.UK.2013-1.RLTS.T42287A2969916.en

2. Krestov, P.V., Song, J.S., Nakamura, Y., Verkholat, V.P.: A phytosociological survey of the deciduous temperate forests of mainland Northeast Asia. Phytocoenologia **36**(1), 77–150 (2006). https://doi.org/10.1127/0340-269X/2006/0036-0077
3. Kim, Y.Y., Leem, H., Han, S., Ji, S.J., So, S.: Conservation measures and distribution of vulnerable species for climate change in Gayasan national park. Korean J. Plant Res. **30**(2), 167–175 (2017). https://doi.org/10.7732/kjpr.2017.30.2.167
4. Cho, M.S., Hwang, J., Yang, A.R., Han, S., Son, Y.: Seed germination and seedling survival rate of pinus densiflora and abies holophylla in open-field experim ental warming using the infrared lamp. J. Korean Soc. of Forest Sci. **103**(2), 203–210 (2014). https://doi.org/10.14578/jkfs.2014.103.2.203
5. Yao, Z., Xu, C., Chai, Y., Wei, H.: Effect of light intensities on the photosynthetic characteristics of Abies holophylla seedlings from different provenances. Ann. For. Res **57**(2), 181–196 (2014). https://doi.org/10.15287/afr.2014.266
6. Yao, Z., Chai, Y., Feng, B., Xu, C.: Differences in water use efficiency of Abies holophylla seedlings from seven provenances under water stress. J. Beijing For. Univ. **6**, 27–34 (2015). https://doi.org/10.13332/j.1000-1522.20140452
7. Ganzei, K.S., Kiselyova, A.G., Pshenichnikova, N.F., et al.: Vegetation fire control on protected areas. Geografiya i prirodnye resursy **2**, 59–68 (2019). https://doi.org/10.21782/GiPR0206-1619-2019-2(59-68)
8. Kiselyova, A.G.: Diversity of coniferous species and communities under natural and anthropogenic conditions on islands of Primorsky region. IOP Conf. Ser. Earth Env. Sci. **629**, 012011 (2021). https://doi.org/10.1088/1755-1315/629/1/012011
9. Anchukaitis, K.J., D'Arrigo, R.D., Andreu-Hayles, L., et al.: Tree-ring-reconstructed summer temperatures from Northwestern North America during the last nine centuries. J. Clim. **26**, 3001–3012 (2013). https://doi.org/10.1175/JCLI-D-11-00139.1
10. Melvin, T.M., Briffa, K.R.: Climatic effects on birch (Betula pubescens Ehrh.) growth in Fnjoskadalur valley, northern Iceland. Dendrochronologia **26**, 71–86 (2008). https://doi.org/10.1016/j.dendro.2007.12.001
11. Zang, C., Biondi, F.: Treeclim: An R package for the numerical calibration of proxy-climate relationships. Ecography **38**, 001–006 (2015). https://doi.org/10.1111/ecog.01335
12. Ukhvatkina, O.N., Omelko, A.M., Zhmerenetsky, A.A., Petrenko, T.Y.: Autumn–winter minimum temperature changes in the southern Sikhote-Alin mountain range of northeastern Asia since 1529 AD. Clim. Past **14**, 57–71 (2018). https://doi.org/10.5194/CP-14-57-2018

The Current State of the Northern Coniferous-Deciduous Forests (on the Example of the State Nature Reserve “BASTAK”)

Ekaterina Lonkina[1(✉)] and Tamara Rubtsova[1,2]

[1] The State Nature Reserve «Bastak» 679014, EAO, Sholom-Aleichem Street, 69a, Birobidzhan, Russia

[2] Institute for Complex Analysis of Regional Problems of the Far Eastern Branch of the Russian Academy of Sciences, EAO, Sholom Aleichem Street, 4, 679016 Birobidzhan, Russia

Abstract. Monitoring of the current state of vegetation cover is one of the priority tasks of the state nature reserve «Bastak». The purpose of this study is to identify the dynamic processes occurring in the phytocenoses of coniferous-broad-leaved forests of the nature reserve «Bastak». Revisions of four constant trial areas in the studied plant formations were carried out in 2007, 2013 and 2018 to fulfilling its purpose. The laying of trial areas and next revisions were carried out according to standard methods. This article presents the results of the re-description of these plant communities. It was revealed that the total number of trees has decreased and the process of natural thinning of the stand is observed in the plantings. Changes in the taxation indicators of the main tree species composing the plantings are analyzed. The analysis of the current state of the studied communities indicates that these plantings are phytocenoses of natural continuous development, which are at the stage of formation.

Keywords: Vegetation cover monitoring · Coniferous-broad-leaved forests · The constant trial area · The nature reserve «Bastak»

1 Introduction

Vegetation cover monitoring is a special long-term monitoring of the condition of flora and vegetation in the constant trial areas (CTA) and key areas. Its main purpose is to study the dynamics of vegetation under the influence of natural and anthropogenic factors. This is one of the most important areas of scientific research conducted in specially protected natural areas, including the nature reserve «Bastak», located in the southern part of the Russian Far East on the territory of the Jewish Autonomous Region.

The state nature reserve «Bastak» consists of two cluster plots: «Central» and «Zabelovsky». Coniferous-broad-leaved forests with a significant participation of Korean pine Pinus koraiensis Siebold et Zucc. grow in the northern and central parts of the plot «Central» on a total area of 11388.2 ha. This territory is characterized by an ultracontinental climate with manifestations of monsoon processes. The long-term average annual air temperature is + 1.7 °C. January is the coldest month of the year,

A. Muratov and S. Ignateva (Eds.): AFE 2021, LNNS 353, pp. 522–531, 2022.
https://doi.org/10.1007/978-3-030-91402-8_59

with an average monthly temperature of -21 °C. The warmest month of the year is July, its average monthly temperature was + 21 °C. The greatest amount of precipitation fell in 2009 (1670 mm), the least – in 2005 (539 mm). The humidification regime of the territory is characterized by a pronounced seasonality. The greatest amount of precipitation falls during the warm period of the year, when between 36% and 75% of all precipitation falls. The average multi-year date for establishing a stable snow cover is November 10. The number of days with a stable snow cover is on average 143 days [1]. Low temperatures in winter and low snow cover contribute to deep freezing of the soil, reaching 150–200 cm, which worsens the growing conditions of plant communities.

Most of the reserve's territory (71399.5 ha, 56%) is occupied by forests. A significant area (48430.8 ha) is occupied by swamps and other non-forest vegetation groups. Within the borders of the state nature reserve «Bastak», Korean pine grows at the northwestern limit of its distribution. Coniferous-deciduous forests occupy mainly gentle and sloping slopes of the southern, south-eastern, eastern and western exposures in a fairly wide altitude range from 150 to 600 m above sea level. These plant formations of the reserve are the most floristically rich. 201 species of vascular plants from 70 families and 138 genera have been recorded growing in these forests, which is 25.5% of the total number of vascular plants in the reserve. The composition and structure of broad-leaved Korean pine forests is described in sufficient detail in previous publications [2–4]. According to the geographical and genetic classification of B.P. Kolesnikov (1956), coniferous-broad-leaved forests belong to the facies of northern Korean pine forests (cold-wet hazel Korean pine forests with fir, linden and yellow birch, and according to the ecological and floral classification – to the order *Tilia amurensis Pinetalia koraiensis* of the class *Quarcetea mongolicae* [5].

Monitoring the situation of vegetation of coniferous-broad-leaved forests with the participation of Korean pine were started in 2002 by a team of geobotanists under the direction of corresponding member of the Russian Academy of Sciences P.V. Krestov. Four constant trial areas (CTA) of forest vegetation in the range of broad-leaved coniferous forests were laid. In the next period, the study of the age structure and dynamics of broad-leaved Korean pine forests of the reserve was carried out by Ph. D. in Biology A.S. Vozmishcheva, who analyzed the spatial and temporal patterns of renewal processes taking into account the age of the dendroflora.

The position of the studied plant communities at the distribution boundary explains their rather high vulnerability to external factors. The forests of the reserve before its creation were subjected to significant anthropogenic transformation, primarily-cutting of the main use and fires. Primary phytocenoses are preserved in small areas located in the mountainous part of a specially protected natural area.

The purpose of our researches is to identify the dynamic processes occurring in the coniferous-broad-leaved forests of the reserve «Bastak». To achieve this goal, revisionss of CTA were carried out in 2007, 2013 and 2018. The article presents the results of re-descriptions of the sample areas, as well as the characteristics of the main trends of changes in plant communities.

2 Materials and Methods

The studied plant communities are located in the northern and central parts of the cluster plot «Central» of the nature reserve «Bastak» in different types of forests with the participation of Korean pine. Constant trial areas are laid out in ecologically homogeneous conditions on areas from 0.64 to 1 ha, in accordance with the requirements for the selection of the site described in the guidelines [6]. At the corners, the CTA are marked with posts. The CTA has an accurate topographic reference, the coordinates are determined using a GPS navigator. Standage and young growth were measured by the method of constant measurement. Each tree in the CTA is numbered. All individuals of trees whose diameter at chest height was 6 cm or more are referred to the stand. Oppressed specimens of trees with a diameter of less than 6 cm are also assigned to the stand by visual signs. All diameter values are divided into steps of thickness of four centimeters. The height of trees is defined as the distance between the top of the tree and the base of the trunk using an altimeter. All height values are grouped into classes.

For the convenience of analyzing the obtained data, the undergrowth on each sample area is divided into three categories: small (individual height up to 0.5 m), medium (0.51–1.3 m) and large (1.31 m and more). The shrub and grass tiers are described by parcel, the quantitative characteristics are given for each parcel. Based on the results of field research, a database of permanent sample areas in the Microsoft Office Excel 2007 program was prepared. For the 2007, 2013, and 2018 revisions, the same methodology was used as for the permanent sample area tab. The materials were processed in Microsoft Office Excel 2007. The use of the same methodology at all stages of the work, the same measuring instruments allowed us to obtain well-comparable results. The frequency of revisions (once every five years) is met.

The full Latin names of taxa in the text of the article are given at the first use. The names of the species were adopted in accordance with the summary "Vascular plants of the Soviet Far East" (1985–1996). During the study, 11604 individuals were taken into account - representatives of 24 tree species and 29256 individuals of undergrowth belonging to 20 tree species.

3 Results and Discussion

The studied plant communities are represented by cold-wet Korean pine forests with fir, linden and yellow birch. For the entire period of observations (2002–2018), the species composition of stands of CTA did not change. The largest number species of trees (17 species) was recorded on CTA No. 4–2002, the smallest (14 species) - on CTA No. 1–2002 and 5–2002. In all CTA, the number of stems in the stand is dominated by Abies nephrolepis, Acer mono, A. ukurunduense, Betula costata, Picea ajanensis, Pinus koraiensis and Tilia amurensis.

The largest number of trees in the all CTA are healthy, without any signs of disease or depression. The largest number of individuals was recorded in 2002 (2563 individuals), the smallest - in 2018 (2144 individuals). The total number of healthy individuals decreased by 16.4%. The plantings are dominated by coniferous species, primarily Abies nephrolepis, Picea ajanensis, Pinus koraiensis, but their share of participation varies. The largest number of coniferous species was recorded in 2002, the smallest-in 2018. The number of soft-leaved species is stable during the entire observation period, as evidenced by the absence of significant changes in the participation of these species in the composition of stands. The number of hard-leaved species is increasing from 20% in 2002 to 24% in 2018. We analyzed the changes occurring in the stands of coniferous-deciduous forests, taking into account the dynamics of the number of tree trunks in the canopy (Table 1). The stand of CTA is multi-tiered. For ease of presentation of materials, all heights are combined into three canopies, the boundaries of the first (dominant) – 25 m or more, the co – dominant – 15–25, the subordinate-5–15 m. When laying the permanent test areas in 2002, the tree heights were not determined, and therefore, we provide data from the results of the 2007–2018 revisions.

The species composition of the canopy is stable throughout the entire observation period, but the number of trees has changed. The largest number of trees is noted in the subordinate canopy. During the period 2007–2018, the number of trees of the subordinate canopy decreased by 29.4% (from 1,840 individuals in 2007 to 1,299 in 2018). The largest reduction in the number was recorded in Abies nephrolepis, Picea ajanensis and Pinus koraiensis, while reducing the number of individuals of Pinus koraiensis and Picea ajanensis in the subordinate canopy, the number of these species in the co-dominant and dominant canopies increases. We believe that the reduction in the number of Abies in the subordinate tier is due to changes in both soil and climatic conditions and the competitive relationships between plants in the plant community. The number of deciduous species within the canopy varies, but quite slightly: a small growth is noted only in the yellow maple, the number of others either does not change or decreases. For deciduous species, as for coniferous species, changes in abundance are mainly observed only in the subordinate canopy, which can also be caused by both weather and climatic conditions and competitive relations in the plant community. The taxational characteristics of plantings also change (Table 2). The change in taxation indicators for the entire period of observations indicates that, despite the decrease in the number of coniferous species in the phytocenosis, the percentage of their participation in the wood reserve has not changed. This is primarily due to the growth of the remaining individuals. Such a tendency is typical for Pinus koraiensis, Picea ajanensis, Tilia amurensis, Betula costata. Betula platyphylla is characterized by an increase in the percentage of participation over the observation period and B. lanata (Regel) V. Vassil., Fraxinus mandshurica Rupr. – the situation is stable, for Abies nephrolepis – there is a significant reduction in participation. The average diameter of the plantings in 2002–2013 increased by 1 cm, and in 2018 a decrease of 3 cm was recorded (compared to 2013). Pinus koraiensis is characterized by the same trend. The average height of both the plant as a whole and Pinus koraiensis increases during the observation period. During 2002–2013, the wood reserve in the plantation is steadily growing, in 2018. A sharp decrease was recorded due to a 2.5-fold reduction in the reserve of Abies nephrolepis.

Table 1. Dynamics of the number of trees by canopy and observation periods.

Canopy	Species										
	An	Pa	Pk	Ld	Acm	Act	Acu	Bl	Fm	Ma	Pha
					Revision 2007						
I	28	6	35	9	7			5	14		2
II	201	61	119	5	33	7	8	30	26		8
III	599	279	281		57	50	135	68	47	11	14
Total	**828**	**346**	**435**	**14**	**97**	**57**	**143**	**103**	**87**	**11**	**24**
					Revision 2013						
I	38	17	56		9			12	26		4
II	178	67	134	9	35	5	5	39	20		6
III	537	230	253	6	45	50	144	49	40	10	6
Total	**753**	**314**	**443**	**15**	**89**	**55**	**149**	**100**	**86**	**10**	**16**
					Revision2018						
I	22	7	41	5	4			8	15		4
II	172	75	141	4	33	1	2	31	28		4
III	404	180	178	4	61	42	151	46	33	6	5
Total	**598**	**262**	**360**	**13**	**98**	**43**	**153**	**85**	**76**	**6**	**13**

						Revision 2007							
I	1		1			28	14	2			1		16
II	7	1	3	1		47	66	2			4	3	45
III	12		5	6	1	90	57		3	3			122
Total	**20**	**1**	**9**	**7**	**1**	**165**	**137**	**4**	**3**	**3**	**5**	**3**	**183**
						Revision 2013							
I	2		2			35	32						24
II	9		3	1		58	69	3	1	4	2	1	61
III	6	10	6	4	13	66	36		2				94
Total	**17**	**10**	**11**	**5**	**13**	**159**	**137**	**3**	**3**	**4**	**2**	**1**	**179**
						Revision 2018							
I	1		1			39	12	2			1		21
II	8		4	3		55	54	1			2		49
III	10	1	5		6	33	46		1	2		1	84
Total	**19**	**1**	**10**	**3**	**6**	**127**	**112**	**3**	**1**	**2**	**3**	**1**	**154**

Remark: An – Abies nephrolepis, Pa – Picea ajanensis, Pk – Pinus koraiensis, Ld – Larix cajanderi, Acm – Acer mono, Act – Acertegmentosum, Acu – Acer ukurunduense, Bl – Betulalanata (Regel), Qm – Quercusmongolica, Fm – Fraxinus mandshurica, Sa – Sorbus amurensis, Ul – Ulmus laciniata (Trautv.) Mayr, Alh – Alnus hirsuta, Are – Araliaelata (Miq.) Seem., Bc – Betulacostata, Bp – Betula platyphylla, Jm – Juglans mandshurica, La – Ligustrina amurensis Rupr., Pam – Padus maackii (Rupr.) Kom., Pt- Populus tremula, Sc – Salix caprea L., Ta – Tilia amurensis.

The second category in terms of the number of tree stems is dry-hardy individuals. The largest number of them was recorded when re-recording CTA in 2013 (431 individuals), the smallest number of dead trees – in 2002 (271 individuals). We believe that the significant shrinking of trees recorded in the re-description of 2013 is primarily due to an increase in the annual precipitation in the period 2008–2012, in this period it was 1374 mm, which is almost 2 times more than the climatic norm (600–800 m). In the stands of CTA, dry-hardy individuals of 17 tree species were noted. The most frequently recorded dry-hardy individuals are Abies nephrolepis, Picea ajanensis, Pinus koraiensis,

Table 2. Dynamics of the main taxation indicators of the tree stand for the observation period 2002–2018.

Composition of the tree stand, %	Average diameter, cm		Average height, m		Reserve, m3	
	All species	Including *Pinus koraiensis*	All species	Including *Pinus koraiensis*	All species	Including *Pinus koraiensis*
Revision 2002						
32Pk 25An 11Ta 7Bc 6Pa 6Bp 5Fm 3Bl 3 Acm	17	19	-	-	687.34	206.36
Revision 2007						
35Pk 17An 14Ta 9Bc 7Pa 6Bp 5Fm 4Acm 3Bl	17	20	11.4	11.8	776.8	259.5
Revision 2013						
34Pk 14An 13Ta 10Fm 9Bc 7Pa 6Bp 4Acm 3Bl	18	22	12.5	13.2	871.6	293.7
Revision 2018						
36Pk 15An 12Ta 11Bc 8Pa 5Bp 5Fm 5Acm 3Bl	15	19	12.7	14.3	482	165

Remark: An – Abies nephrolepis, Pa – Picea ajanensis,Pk – Pinus koraiensis, Acm – Acer mono, Bl – Betulalanata, Fm – Fraxinus mandshurica, Ul – Ulmus laciniata, Bc – Betulacostata, Bp – Betula platyphylla, Ta – Tilia amurensis.

which in different years account for 66–82% of the total number of dry-hardy individuals. The remaining tree species are represented by dry-hardy individuals in the number of 1 to 17 trees. The largest number of dead trees is Abies nephrolepis, Picea ajanensis recorded in the 2013 redescription, Pinus koraiensis – in 2018. We determined the ratio of the number of fallen and ingrown trees for the observation period (Table 3).

Table 3. Falling off the stand of trees and ingrowth a stand of trees on the constant trial areas.

Species	2007		2013		2018	
	Disappeared from the stand of trees	Grow into a stand of trees	Disappeared from the stand of trees	Grow into a stand of trees	Disappeared from the stand of trees	Grow into a stand of trees
Abies nephrolepis	214	106	243	59	180	46
Acer mono	6	6	10	6	5	15
Acertegmentosum	7	27	6	17	2	13
Acer ukurunduense	10	55	12	49	12	46
Araliaelata	0	1	0	13	1	3
Betulacostata	7	26	4	8	10	6
Fraxinus mandshurica	2	21	4	7	1	5
Maackia amurensis	1	3	1	2	0	3
Phellodendron amurense	0	4	5	0	0	0
Picea ajanensis	38	79	60	12	50	28
Pinus koraiensis	36	54	42	17	60	12
Quercusmongolica	0	1	2	0	0	0
Ligustrina amurensis	0	1	0	1	0	0
Tilia amurensis	2	17	13	16	7	7
Ulmus laciniata	0	1	0	2	0	3
Betulalanata	17	2	14	0	12	0
Betula platyphylla	7	3	13	0	26	0
Larix cajanderi	1	0	2	1	2	2
Total	**348**	**407**	**431**	**210**	**368**	**189**

During the observation period (2002–2018), the total loss of trees was 1147 trees, and 806 trees grew into the stand during the same period. In the description of 2007, it was recorded that fewer trees fell off in the plantings than they grew. The number of ingrown trees in subsequent years is twice less than the number of fallen trees. The main causes of this phenomenon, in our opinion, are changes in the climatic conditions of the study area (an increase in annual precipitation) and natural death. The growth of new individuals into the stand is gradual, but uneven in all CTA. The stand is replenished mainly with deciduous species, primarily Acer ukurunduense Trautv. et Mey. and A. tegmentosum Maxim. The drop-off and ingrowth rate is unbalanced (Table 3). Thus, the indicators of falling off and ingrowth of trunks indicate that plant communities are at the stage of formation. In the context of CTA, the highest stand shrinkage on CTA No. 1 was observed in 2007, on CTA No. 2-in 2002, on CTA No. 4 – in 2018, on CTA No. 5 – in 2013.

The general sanitary condition of the studied plantings can be assessed as weakened. This is evidenced by a significant number of faut trees with damaged trunks. The largest number of such individuals was recorded in 2002 and 2007 (222 and 230 individuals, respectively), the smallest number of damaged trees was recorded in 2018 (5 individuals) The main signs of damage plantings are the presence in the phytocenosis of individuals with the following characteristics necrotic-cancerous diseases, stem rot, windbreak, frost-breaking cracks, crown shrinkage [7]. The largest number of faut trees is represented by Abies nephrolepis (30–66% of the total number of faut trees), followed by Pinus koraiensis (11–19%). We analyzed the change in the number of faut stems of Abies nephrolepis and Pinus koraiensis over the years of observation. For example, the 147 fautal individuals of Abies nephrolepis recorded in 2002, when re-describing subsequent years, most (53%) are represented by dry individuals or these trees are not found. Only 22% of Abies nephrolepis individuals identified during the establishment of CTA were recorded during all revisions. The largest number faut trees of Pinus koraiensis were recorded in 2007 (44 individuals). In 2002, 33 individuals of this species showed signs of disease, depression, or mechanical damage. One-third of these trees turned out to be dry or not found in subsequent years, 6 individuals were noted in all re-descriptions as living individuals.

Natural renewal in the studied plant communities takes place unevenly. The species composition of the young growth corresponds to the stand. The number of young growth in 2007–2013 was characterized by high indicators (10691–13776 individuals), and in 2018 a sharp decrease in young trees was recorded to 4771 individuals. This trend is related both to the general decline in trees capable of producing seeds (from 831 individuals in 2002 to 618 in 2018), and to the trampling of young individuals by animals (formerly wild boars). Most of the undergrowth is represented by small individuals (from 39% to 60% of the total number of individuals recorded on CTA). The largest number of undergrowth is represented by hard-leaved species (primarily Fraxinus mandshurica and various Acer species). Among coniferous species, Abies nephrolepis is most often noted, but out of 553 individuals of this species recorded in 2013, only 46 individuals have grown into the stand of trees (the stand of trees grew only 46 individuals). Pinus koraiensis resumes fairly well in all CTA, but only one in ten individuals grows into a stand of trees.

The undergrowth of all CTA is quite dense (the total projective cover of shrubs is up to 50%), it is located in kurtins and is represented by *Corylus mandshurica*, *Lonicera maximowiczii* (Rupr.) Regel, *Sorbaria sorbifolia* (L.) A. Br., *Philadelphus tenuifolius* Rupr. et Maxim., among the extra-level vegetation *Actinidia kolomikta* (Maxim.) Maxim. dominates, which creates sometimes impassable thickets. The grass layer with a total projective cover of up to 70% is dominated by large ferns: *Dryopteris crassirhizoma* Nakai, *Pseudocystopteris spinulosa* (Maxim.) Ching, *Leptorumohra amurensis* (Christ) Tzvel., *Carex xyphium* Kom., *C. pallida* C. A. Mey, and *C. campylorhina* V. Krecz. In addition to these species, *Thalictrum filamentosum* Maxim., *Smilacina davurica* Fisch. et Mey are noted in the grass layer. et al.

4 Conclusions

Modern coniferous-broad-leaved forests with a significant participation of Pinus koraiensis, growing on the territory of the nature reserve «Bastak» are complex in composition and structure plant communities. They consist of a large number of tree species with different forest-biological characteristics. Prior to the creation of the reserve in the area of growth of these phytocenoses, logging was carried out for the main use. However, despite the anthropogenic intervention in plant communities, it did not lead to a change in the main species. Reforestation processes follow the scenarios typical for undisturbed forests. As a result of the study, the following was revealed:

1. In the plant communities of pine-broadleaf forests for the period 2002–2018, there was no change in the dominant tree stand. The species composition of the phytocenosis is quite stable. Traces of cutting were recorded on the CTA No. 4–2002. In our opinion, selective cutting was carried out on this territory, which did not lead to significant changes in the composition and structure of the phytocenosis. All plant communities can be attributed to phytocenoses of continuous development, which are at the stage of formation.
2. Despite the stability in the species composition of the stands, the total number of trees is decreasing. This allows us to talk about the processes of natural thinning of trees occurring in phytocenoses. The decrease in the number of trees in the all CTA is primarily due to their shrinkage in the subordinate canopy. The presence of the thinning process is also indicated by the analysis of the indicators of falling off and ingrowth, according to which trees grow into the plant community twice as much as they fall out.
3. The largest number of dry-hardy individuals is represented by Abies nephrolepis, Picea ajanensis and Pinus koraiensis. In the CTA, the highest stand shrinkage was observed at CTA No. 1 in 2007, at CTA No.2 in 2002, at CTA No. 4 in 2018, and at CTA No. 5 in 2013. The weakening and shrinking of stands occurs mainly as a result of the impact of adverse weather conditions, primarily increased precipitation and intraspecific competition.
4. The natural renewal of plant communities is extremely uneven. Among the young growth, small individuals predominate. The largest number of undergrowth is represented by hard-leaved species (primarily *Fraxinus mandshurica* and different species of *Acer*).

References

1. Lonkina, E.S., Lyamzina, L.V., Rubtsova, T.A.: Dynamics of changes in climatic indicators in the «Bastak» nature reserve in the period 2001–2019. Climatic changes and seasonal dynamics of landscapes: mater. vseros. nauch-prakt.konf. Yekaterinburg, 22–24 Apr. 2021. USPU Publishing House, Yekaterinburg (2021). https://doi.org/10.26170/KFG-2021-09
2. Kolobov, A.N., Lonkina, E.S., Frisman, E.Y.: Modeling and analysis of horizontal structure of a mixed tree stands (on example of sample plots in the «Bastak» nature reserve in the Middle Amur River area). Sib. For. J. **3**, 45–56 (2015). https://doi.org/10.15372/SJFS20150305

3. Lonkina, E.S.: The results of the monitoring of the north coniferous-broadleaf forests (on the example of the constant trial plot№ 5–2002) in the nature reserve «Bastak». Present Problems of Regional Development: Mater. of the VII All-Russian Scientific Conference on Birobidzhan, 09–11October 2018. Birobidzhan: ICARP FEB RAS (2018). https://doi.org/10.31433/978-5-904121-22-8-2018-179-182
4. Volkov, E.V.: Forests with Korean pine in the Bureya river basin. Russ. For. J. **5**(353), 34–43 (2016). https://doi.org/10.17238/issn0536-1036.2016.5.34
5. Krestov, P.V., Song, J.S., Nakamura, Y., Verkholat, V.P.: A phytosociological survey of the deciduous temperature forests of mainland Northeast Asia. Phytocoenologia **36**(1), 77–150 (2006). https://doi.org/10.1127/0340-269X/2006/0036-0077
6. Omelko, A., Ukhvatkina, O., Zhmerenetsky, A., Sibirina, L., Petrenko, T., Bobrovsky, M.: From young to adults trees: how spatial patterns of plants with different life strategies change during age development in an old-growth Korean pine-broadleaved forests. For. Ecol. Manage. **411**, 46–66 (2018). https://doi.org/10.1016/j.foreco.2018.01.023
7. Lonkina, E.S.: Nut productivity of Korean cedar stands in the «Bastak» nature reserve. Reg. Probl. **24**(2–3), 116–121 (2021). https://doi.org/10.31433/2618-9593-2021-24-2-3-116-121

Problems of Effective Forestry Fund Land Management in the Amur Region

Marina Makannikova(✉) and Lyudmila Lapshakova

Far Eastern State Agrarian University, ul. Politehnicheskaja, 86, 675005 Blagoveshchensk, Amur Oblast, Russia

Abstract. The study focuses on the current use of forestry fund land resources in the Amur Region. It has been established that the total area of the region's land is 36,190.8 thousand hectares, the largest part of which - 30,581.6 thousand hectares - is the forest land, accounting for 84% of the region's total land area. Current situation with the forest land use in the Amur Region has been analyzed. The total area of forest land is 25,477.2 thousand hectares, while 5,038.5 thousand hectares fall on non-forest land. Contractual relations are actively developing, under which the main types of permitted use include the geological study of subsurface and development of mineral deposits. However, sustainable forest management implies such maintenance and use of the forestry fund, which can preserve its productivity, recreational potential, biodiversity and thepotential to perform environmental, economic and social functions in the present and future.

Keywords: Forest fund lands · Forest and land resources · Lease · Efficiency · Amur region

1 Introduction

In modern socio-economic conditions of Russia, the use of natural resources, which include forests, depends to a large extent on rational, continuous and sustainable land use and forest management in the region. Consequently, an important component of efficient forestry land use is the correct value appraisal of lands and organization of its management on the basis of a systemic information complex (unified state register of real estate, monitoring of lands and forests, etc.). Taxes and rents, determined by the land value, supplement the region's budget and have a considerable impact on the economic activities of land and forest users. A more accurate assessment contributes to a more balanced distribution and turnover of lands, which increase the efficiency of land use.

Overall land and forest resources as well as human activities, based on these resources and aimed at employing the full spectrum of forest useful properties in accordance with the social, environmental and economic needs of society and individual citizens, should be economically viable and legally secured [1–3].

A. Muratov and S. Ignateva (Eds.): AFE 2021, LNNS 353, pp. 532–541, 2022.
https://doi.org/10.1007/978-3-030-91402-8_60

2 Materials and Methods

The research objective is to develop provisions for improving the management of the Amur Region forest resources with due account for the potential of the territory development.

The Amur Region is a constituent entity of the Russian Federation with total land area of 36,190.8 thousand hectares. As of January 1, 2021, the forestry fund lands with the area of 30,581.6 thousand hectares (84%) prevail in the structure of the land fund of the region [4]. There are 13 forestry districts on the lands of the forestry fund of the Amur Region, they, in turn, are divided into 92 local forestry districts; all forestries are managed by the Ministry of Forestry and Fire Safety of the Amur Region. The largest areas covered by forests are located in the northern and northwestern territories (Arkharinsky, Zeysky, Magdagachinsky, Tyndinsky, Shimanovsky, Skovorodinsky, Selemdzhinsky administrative districts). The total forest cover of the Amur Region as of 2019 was 65.1%, in 2020 it was 65.3%. There are coniferous, mixed and broad-leaved forests in the region, which replace each other gradually from north to south. Vertical zoning of vegetation cover is typical of mountainous areas. The total area of forests under forest management is estimated as 30,515.7 thousand hectares, including 25,477.2 thousand hectares of forest land and 5,038.5 thousand hectares of non-forest land.

Allocation of forests according to the designated purpose: shelter forests – 2,515.0 thousand hectares, including: forests, protecting natural and other objects - 583.7 thousand hectares; valuable forests – 1,931.3 thousand hectares; commercial forests – 26,956.4 thousand hectares; reserved forests – 1,044.3 thousand hectares [5]. Over the past few years, contractual relations have been actively developing in the Amur Region on the lands of the forestry fund. Types of forest use are provided for by article 25 of the Forest Code of the Russian Federation. As of January 1, 2021, 10% of forests have been handed over for use. More than 1070 agreements are currently in force pursuant to which the Ministry of Forestry and Fire Safety of the Amur Region makes forests available for use (lease, permanent (unlimited) use, free use).

102 forest plots on an area of 2979863.55 hectares are currently leased for the purposes of timber harvesting; 19 forest plots on an area of 48.13 hectares are used for recreational activities; 4 forest plots on an area of 25.30 hectares are contracted for processing wood and other forest resources; 4 forest plots on an area of 54.90 hectares are used for new forest plantations; 612 forest plots on an area of 34488.47 hectares are leased for geological exploration of subsurface and development of mineral deposits; 14 forest plots on an area of 1,453.48 hectares are exploited for creating hydraulic engineering structures; 309 plots on an area of 9348.38 hectares are used for construction, reconstruction and operation of linear facilities, etc. The number of contracts is decreasing year by year (from 1101, as of January 1, 2019, to 1070, as of January 1, 2021) because of the declining number of contracts for timber harvesting. In some cases the contracts are terminated, but the forest user takes a forest plot with an increased area or combines several contracts and takes one plot on lease.

The main types of use of forest lands in the Amur Region are the geological study of the subsurface; development of mineral deposits; construction, reconstruction and operation of linear facilities; timber harvesting and recreational activities. At the same time, according to the Ministry of Forestry and Fire Safety of the Amur Region, as of

January 1, 2021, 90 contractors, engaged in the main areas of activities, have rent arrears under current agreements (Table 1).

Table 1. Arrears under current lease agreements.

Type of use	Number of agreements, pcs	Arrears, rbs
Timber harvesting	11	7559145.34
Construction, reconstruction and operation of linear facilities	40	870804.09
Exploration of subsurface and development of mineral deposits	37	6227078.46
Recreational activities	2	5138.72
Total:	90	14662166.21

As of January 1, 2021, the amount in arrears under current lease agreements was estimated as 14,662,166.21 rubles, this amount was not paid or was not fully paid by the contractors within the period specified in the agreement and is subject to mandatory collection.

There are several reasons for the rent arrears: contractor's financial situation deteriorated dramatically; contractor went bankrupt; penalties were chargedfor overdue rent; force majeure commenced.

When forest leaseholders fail to pay the rent, the contract with them can be terminated in a judicial or extra-judicial procedure.In accordance with Article 98.1 of the Forest Code of the Russian Federation, the Federal Forestry Agency of the Russian Federation keeps a register of unscrupulous forest leaseholders with the purpose to improve the use of forestry fund lands while preserving their biodiversity, productivity, recreational capacity and potential for performing environmental, economic and social functions in the future. The register includes dataregarding the forest leaseholders whose contracts were terminated due to failure to execute rental payments. The records about unscrupulous leaseholders in the register prevent from concluding lease agreements with them in the future.

One of the most efficient and sustainable uses of the forestry fund land resources is their use for recreational purposes, which, therefore, implies thatprotective, sanitary and hygienic, health-improving, environment-forming, water protection and other useful functions of forests are preserved.

A package of measures should be taken to ensure that forests which have become recreational sitesare being preserved; that human recreational needs are met; and that recreational forest management is economically viable (for example, through the development of ecotourism, which is becoming increasingly popular and profitable for the economy and positively affects local employment).

3 Results

Only forest areas with considerable recreational potential should be used for recreational purposes. They also need to be resistant to recreational loads and be convenient to use.

Lands of the forestry fund can be leased on the following terms: lease, free use, short-term use. The transfer of forest land is carried out on the basis of a lease agreement (the maximum lease period for a forest plot used for recreational activities is 49 years) [6]. The lease agreement for a forest plot to be used for recreational activities can be concluded only on condition that this forest plot is registered in the national cadastre.

The recreational potential of a forest is an extent to which the forest can perform recreational functions resulting from its natural properties. Indicators of the forest recreational potential are important for estimating recreational loads on forest ecosystems, the calculation of which, in its turn, makes it necessary to determine the class of forest plantations resistance to such loads, as well as the stages of forest and vegetation degradation [7, 8]. The norms of permissible recreational loads can be set in accordance with the methodology, where their values are calculated depending on the types of forest and recreation, the species and age composition or the length of the road network.

Forests of the Amur Region are favorite places for public recreation, they perform two main functions characteristic of recreational forests, i.e. - environmental protection and recreation proper, which help people to regain physical power and spiritual strength and satisfy their emotional need to commune with nature.

Forestscan be used for recreational purposes throughout the regionon an area of 30515.7 thousand hectares. Recreational forests proper include green spaces covering an area of 52.8 thousand hectares and forest and parkland zones on an area of 2.4 thousand hectares. The only limitation for their use is set by Article 27 of the Forest Code of the Russian Federation.

The Amur Region has a high potential for the recreational use of forests. The region has special tourist sites, popular not only locally or nationally, but also worldwide. These include unique archaeological sites, former habitats of extinct dinosaurs, mineral water springs (Amurskaya, Gonzhinskaya, Konstantinovskaya), lotus lakes and cranes nesting areas, all ofwhich are particularly attractive for tourists.

One of the traditional types of recreational nature management in the Amur Region, as well as in the Far Eastof Russia generally, is educational recreationin nature.

Tourists get lasting impressions from communing with the nature of the region,its picturesque landscapes, placers of gemstoneson the riverbanks, meetings with representatives of the animal and plant kingdom of the Amur Region.

A significant part of local population tends to use their free time for cropping in taiga, hunting and fishing. However, available resources are underused due to economic factors (underdeveloped infrastructure, lack of financial resources), environmental pollution, floods, severe climate - all these hinder the development of tourism in the region.

To be actively used for recreational purposes forests should be located near large settlements, in places with picturesque landscapes,where the availability of forest stands and water bodies is combined with good transport infrastructure. Local people's weekly rest is a massive phenomenon, especially in summer. Most holidaymakers spend their days offon rivers and lakes, in summer cottages, on garden plots, in the forests [9].

Inthe Amur Region, the conditions mentioned above are best met by forest parks and green spacesin cities and towns, as well as forests located along rivers or near large lakes.

Forests of the Zeysky, Bureysky, Arkharinsky and Magdagachinsky forestry districts are most suitable for recreational use. Rivers the Zeya, the Bureya, the Arkhara flowing through these territories increase their recreational potential.

The main recreational objects in the Arkharinsky forestry district are water-conservation forests (restricted areas along the Arkhara river); the Khingano-Arkharinsky nature reserve; the Andreevsky nature reserve; the Ganukan state nature reserve; Rock paintings, a geological natural monument; Esaulovsky Mineral Spring, a specially protected hydrological area; Hot Springs, a hydrological natural monument.

The forestry has a good resource base for tourism: the rivers Arkhara, Solokachi, Uril, Khingan and Mutnaya are suitable for rafting.

Dinosaur fossils were discovered in Kundur forestry district during paleontological excavations.

There is a natural monument *Pisanitsy*(petroglyphs) on the bank of the Arkhara river, between the mouths of the Tatakan and the Dyda. *Pisanitsy*are rock carvings of the ancient people who lived on the territory of the present Arkharinsky administrative district. The petroglyphs date back as early as IV-II centuries BC.

Tourist destinations of the Bureysky forestry district are the Bureya and Dikan rivers; the *Malmalta* state botanical reserve; the *Andreevsky* state hunting reserve; a specially protected area - the *Bureysky* nature park; the *Kompaneisky* natural monument of regional significance. Special attention is being paid to the construction of recreational facilities aroundthe Bureysky and Nizhnebureyskywater reservoirs.

The Zeysky forestry districtalso has a good resource base for developing tourism. Popular tourist destinations include the rivers Urkan, Zeya, Umlekan, Tok, Gilyui, Dugda;the Zeysky water reservoir; the *Bekel-deul* hunting reserve; the *Verkhne-Depsky*fauna reserve;the Ust-Tygdinsky integrated nature reserve*Tokinsky* named after G.A. Fedoseev.The protective zone of the *Zeysky* nature reserve in the Zolotogorsky and Kirovskylocal forestries covers 34,000 hectares.Tourists can be attracted by the hydrological natural monument *Lake Ogoron*, the natural monument *Rock Paintings* on the riverOnoni, the natural monument *The Mouth of the Luchi River*, the natural monument *Gulikovsky Cliff*, the geological natural monument *Cliff on the Zeya River*.

The entire forest area of the Magdagachinsky forestry district can be used for recreational purposes, e.g., tourism, excursions, berry and mushroompicking, fishing, walking, etc.The *Magdagachinsky* state hunting reserve, the *Tolbuzinsky* state hunting reserve, the natural monument - the *Gonzhinsky Mineral Spring*, the protective zone of the *Ust- Tygdinsky* reserve are located here. Therecreationalareaisestimated as 1,308,852 hectares.

Forests located along highways can be used for development of recreation and tourism. Traditional recreational activities includewild harvesting, hunting, fishing, etc. [10].

Landscape conditions of the Amur Region are to a great extent determined by the structure of its surface - it is a mountainous-plainarea in the west of the Far Eastern Federal District.

Modern encyclopedias define a forest as an element of a geographical landscape, consisting of a set of trees, which have dominant position, shrubs, ground cover, animals and microorganisms,that are biologically interrelated in their development and affect each other and the external environment.

Picturesqueness of landscape or landscape aesthetics is at the core of the territory recreational potential. A special research area of landscape science is currently being developed, which studies the beauty and picturesqueness of natural and natural-anthropogenic landscapes and also the specific features of their aesthetic perception and evaluation.An important aspect of the aesthetic evaluation of a landscape is the subjectivity of its scenic beauty perception.For evaluation purposes researchers propose to use a list of features and point scales. Landscapes created by coniferous-broad-leaved forests and pine forests are considered to be the most beautiful. They are located in the south of the Amur-Zeya Plain and on the ridges of the Small Khingan.Quite picturesque are the landscapes of the mountain tundra, dwarf cedar belts, valleys and foothills covered with soft-leaved, oak and fir-spruce forests, meadows and areasalong the largest rivers - the Zeya, Bureya and Amur. All other types of landscapes - marshes, larch forests, soft-leaved and oak forests in valleys, agricultural lands- are classified as neutral.

Landscape characteristics of recreational forests of the Amur Regioninclude information about the types of landscapes, the degree of plant stability, the stages of recreational degradation of the territory, the passability and visibility of the plots, the recreational and aesthetic value of landscapes.

The term*forest-park landscape* should be understood as the type of forest plot, its structural form, the degree to which it is covered by trees, the nature of trees spatial distribution and the canopydensity. Closed landscapes are prevailing in the recreational forests of the region (60% of the forest area).

The stability of forest stands is their ability to cope with adverse conditions of growth and development, which lead to premature decay of tree stands and replacement of species. This indicator characterizes general condition of the plantation, the quality of growth and developmentand the level of natural regeneration.

Treestandsof the second stability class (60%) are prevailing in the recreational forests of the region, they include plantations with damped growth, where healthy trees account for 60–90%.

Recreational degradation of green zones plantation is characterized by severe stages of degradation and indicates the need for improvements in recreational nature management.

Evaluation of recreational degradation stages includes the description of recreational loads onthe territory in the past and present. Based on the nature of changes in the forest environment,caused by recreational use, recreational forests (forests of green zone) go through five stagesof recreational degradation - from absolutely undisturbed to disturbed and drying up.

Recreational forests of the Amur Regionare mainly onstages II and III (95%), i.e.,there arenotable signs of the disturbance of forest environment.

Green zone plantations are generally characterized by less attractive landscapes, mid-low recreational value and low stability.

Description of the ecological, sanitary and health potential of forests.

Recreational forests are natural resources used for recreation and tourism and for satisfaction of aesthetic and educational needs.

Forestries should monitor adherence to the regime of recreational forest use andoversee the development of recreational infrastructure.

When developing projects in such forests, it is necessary to implement a set of measures aimed atenhancing and preserving aesthetic, sanitary-and-hygienic and health-improving properties of plantations and maintaining the ecological balance of the forest environment. It is also important to ensure a rational distribution of visitors throughout the territorywith account for the permissible loads in its various parts.

The hardscaping and landscaping of the area is carried out in accordance with the project for the use of the leased forest area against a forest usage permit issued by the corresponding forestry. No other permissions are required.

The improvement of recreational forests' territory includes the construction and repair of roads, the development of recreationalfacilities, the creation of small architectural forms and forest sculptures, the construction of various forest park structures, etc. When identifying necessary improvements, it is importantto comply with the standards set out in the Catalogs and Design Standards of Small Architectural Forms and Improvement Elements for Use in Landscaping of Recreational Areas which determine the number of improvements per unit area of the forestry fund and the length of roads.

Knowledge of recreational loads is required to allocate areas of active recreation. A recreational load is the degree to which vacationers, their vehicles, construction of temporary housing, summer cottages and other buildingsaffectthe ecosystem or recreational facilities.This load is expressed in the number of people or person-days per unit area or per recreational facility over a certain period of time (usually a day or a year). The following types of recreational loads are distinguished: optimal, maximum (maximum permissible) and destructive (fatal).

The scale of recreational degradation and recreational evaluation of the forest plot developed by LLC*Lesproect* is used inforestry management for these purposes. The taxation also determines the type of landscape, aesthetic value, biological stability and passability of the forest plot.

4 Discussion

The development of tourism and recreational activities is one of the priorities of the socio-economic development of the Amur Region according to the Strategy for the Socio-Economic Development of the Amur Region for the Period up to 2025 and the National Program for the Socio-Economic Development of the Far East of Russia for the Period up to 2024 with a Perspective until 2035.

An essential issue in the development of the Amur Region territories is the use of the entire recreational resources for the selection of various types of recreation, which presents a range of activities aimed at restoring a person's working capacity. The Amur Region has a sufficient recreational potential to accomplish this task, mainly including natural, historical, and cultural sites.

One of the components of the region's recreational economy, which is actively developing, is suburban and urban gardening. Location of gardens and summer houses is affected by a number of factors, the most important among them is related to transportation. Specifically, the localization of the region's horticultural plots is linked with the main railways and highways. The density of gardens and summer houses decreases markedly with distance from large settlements.

Being one of the fundamental factors in the development of recreation and tourism, historical and cultural potential of the region is the basis for turning it into a major tourist center on a regional, national, and international scale.

Historical and cultural landmarks of the Amur Region are the most attractive sites in the complex of cultural and historical resources, they provide the basis for developing educational tourism routes. Educational tourism potential is closely related to archaeological heritage sites. One of the options for developing this area of tourism may be tourists' participation in archaeological excavations, under the guidance of professional archaeologists.

Settlements in the Amur Region represent, in architectural terms, a typical example of the former provincial towns, which, upon the approval of the regular city development plan, started turning from rural settlements into county centers. Architectural character of each of them reflects the specific features of their historical development and geographical location. It should be noted, that several buildings of religious architecture did not survive in their original form: some of them were periodically rebuilt, some others were irrevocably lost.

Unfortunately, the Amur Region suffers from shortage of infrastructure facilities designed for entertainment.

Cultural service provided in many settlements is of very low quality and does not meet the tourists' needs for entertainment, though the latter is an integral part of any tourist trip. This issue is currently pending and requires a detailed and careful consideration, since the availability of a network of entertainment enterprises is essential for the development and successful functioning of the tourist industry.

Recreational activities that are well developed in the Amur Region include hunting and fishing. The main areas of hunting and fishing tourism are Zeysky, Blagoveshchensky, Bureysky, Seryshevsky, Romnensky, Skovorodinsky and Magdagachinsky administrative districts. Hunting farms and hunting and fishing tourist bases with developed infrastructure and the highest quality of service are mainly located in these areas. The largest hunting and fishing farms in the region include the recreation centre Zeysky of the Directorate for Protection and Management of Wildlife and Special Nature Conservation Areas, located 10 km from the village of Natalino in Blagoveshchensky administrative district; cordon (fishing and hunting base) on the territory of Bureysky nature park of the Directorate for Protection and Management of Wildlife and Special Nature Conservation Areas, located 1.5 km from the Bureysky Hydro Power Plant in Bureysky administrative district; cordon Koster' on the territory of the state nature reserve of regional significance Voskresenovsky of the Directorate for Protection and Management of Wildlife and Special Nature Conservation Areas, located 15 km north-east of the meteorological station of Voskresenovka village in Seryshevsky administrative district; cordon Tashino on the territory of the Tashinsky State Nature Reserve of the Directorate for Protection and Management of Wildlife and Special Nature Conservation Areas, located 45 km southeast of Voskresenovka village in Romnensky administrative district; cordon on the territory of the state nature reserve Verkhne-Amursky, located 18 km east of the village Albazino in Skovorodinsky administrative district; cordon in Magdagachinsky State Nature Reserve of the Directorate for Protection and Management of Wildlife and Special Nature Conservation Areas, located 30 km from the village of Magdagachi in Magdagachinsky administrative district.

One of the most popular types of tourism in the Amur Region is sanatorium-and-health-resort tourism. As of the end of 2019, there were about 7 functioning sanatoriums, dispensaries, boarding houses, and recreation centers, which use the region's natural healing resources for treatment and prevention of various diseases. They are all fitted with up-to-date diagnostic equipment and provide medical services, including electrotherapy, hydrotherapy, thermotherapy, and mud therapy, based on modern research.

Ecological tourism is rapidly developing in the Amur Region. A special role in the development of ecological tourism in Russia is played by the unique specially protected nature conservation areas of federal significance: the Zeysky State Nature Reserve, located in Zeysky administrative district; the Khingansky State Reserve in Arkharinsky administrative district, and the Norsky State Nature Reserve on the territory of Selemdzhinsky administrative district.

5 Conclusions

1. Virtually all the territory of the region is suitable for the development of a recreational economy. Valuable recreational resources are concentrated in the south-west and north-east of the region, as well as in the valleys of large rivers, they form the basis for developing a network of children's health camps, vacation retreats, tourist centers and sanatoriums.
2. The combination of natural resources and specific natural and socio-economic factors make it possible to identify the most promising areas of the recreational development of the region:

   - medical-recreational and sanatorium-and-health-resort area is developing on the basis of climatic, balneological, water, plant and other recreational resources.
   - tourism – almost the entire territory of the region can be used for recreational activities.
   - suburban and urban gardening is actively developing in the areas adjacent to major settlements and transport routes.
   - hunting and fishing recreational activities havegood prospects for the developmentowing to the region's abundant natural resources and a greatnumber and diversity of game animals.

References

1. Mezenina, O.B.: Project, Economic and Information Support of the Territorial Management of Recreational Land Use in the Forestry Complex, p. 157. The State University for Land Use Planning Publ, Moscow (2013). https://doi.org/10.24411/2413-046X-2021-10138
2. Filipchuk, A.: Overview of the Forest Ownership in the Region. Presentation from Russia, Geneva (2014). https://doi.org/10.18551/rjoas.2017-10.28
3. Lerche, I.: Environmental Risk Assessment. Quantitative Measures, Anthropogenic Influences, p. 327. Human Impact Publ, Berlin (2010). https://doi.org/10.12737/21874
4. Annual Statistical Report on the Availability of Lands in the Amur Region and their Distribution by Type of Ownership, Categories, Lands and Users. Office of the Federal Service for State Registration of Cadastre and Cartography in the Amur Region, Blagoveshchensk (2020)
5. Makannikova, M.V.: Modern use of forestry fund lands. In: Proceedings of the International Scientific Conference "Biosphere and a Man", p 203–205. El ITPubl, Maikop (2019). https://doi.org/10.1088/1755-1315/547/1/012025
6. Antropov, D.V.: Some aspects of forest lands lease (the case of Kaluzhsky region). Ural Agrarian Bull. **5**, 75–79 (2014). https://doi.org/10.33920/sel-04-2105-03
7. Kvarchiya, O.V.: Resource potential of a territory as the basis of its competitive capacity. Fundam. Res. **9–2**, 345–352 (2016)
8. Kuklina, E.A.: Recreational capacity of a region as an essential element of its sustainable development. Manag. Consult. **4**(88), 116–124 (2016). https://doi.org/10.15826/vestnik.2016.15.3.021
9. Pisarevskii, I.M.: Tourist and recreational potential as an element of the territory competitive capacity. Bus. Inf. **12**(503), 148–154 (2019)
10. Blieva, M.V.: Methodology of project development in the area of tourism and recreation. science today: theory, practice, innovation. In: Proceedings of the XI International Scientific and Practical Conference, Moscow, Olymp, p 140–145 (2016). https://doi.org/10.1177/0307174x1404100505

Forest Burnability Analysis and Preliminary Fire Forecasting Dangers on the Territory of the Amur Region

Tatiana Molchanova[1](✉), Elena Grebenshchikova[1], Natalia Shelkovkina[1], Natalia Gorbacheva[1], and Evgeniy Khovanets[2]

[1] Far Eastern State Agrarian University, Polytechnicheskaya Street, 86, 675000 Blagovesh-chensk, Amur Region, Russia

[2] Main Directorate of the Ministry of Emergency Situations of Russia for the Amur Region, st. Pionerskaya, 47, 675000 Blagoveshchensk, Amur Region, Russia

Abstract. Forests of Russia occupy 69% of the country's territory. It is the main component of the natural environment, the regulator of the processes occurring in the biosphere, contributing to the survival of mankind. Forests contain 90% of the planet's organic matter. The purpose of the research: The forecast of the fire situation for the coming years and the adoption of measures to reduce the burning of forests. In this regard, protecting them from fires is the most important strategic task of the Amur region. During forest fires, the most valuable plantings, forest materials, communications, settlements, domestic and wild animals, and birds are lost or damaged. There are frequent cases of deaths when extinguishing forest fires and as a result of plane crashes caused by poor visibility due to smoke. The results of the analysis of forest burnability provide answers to the questions: where, when, why and in what quantities forest fires occurred.

Keywords: Natural fires · Forecast · Fire hazardous season · Long-term observations · Forest fire maximum

1 Introduction

The Amur Region is part of the Far Eastern Federal District. In the south and southeast, it borders the PRC. The length of the state border passing through the territory of the Amur region is 1250 km. Length of the territory: from west to east: 1150 km, from north to south –750 km. Center-Blagoveshchensk.

Date of formation of the region – October 20, 1932 (The date of formation of the intra-regional region is given, in 1948 the Amur region was separated from the Khabarovsk Territory into an independent region).

The territory is 361.9 thousand square kilometers. (72.1% – forests, 7.6% – agricultural land, surface water, including swamps-16.4%, other land-3.9%).

Distance to Moscow from Blagoveshchensk by rail-7985 km, by air-6480 km, State border: in the south – with the People's Republic of China. The Amur Region borders:

A. Muratov and S. Ignateva (Eds.): AFE 2021, LNNS 353, pp. 542–552, 2022.
https://doi.org/10.1007/978-3-030-91402-8_61

in the north – with the Republic of Sakha (Yakutia) in the west-with the Trans-Baikal Territory in the east-with the Khabarovsk Territory and the Jewish Autonomous Region.

A significant part of the territory of the region is occupied by forests. The total land area of the forest fund is 31247854.0 thousand hectares or 87% of its total land area. According to the decree of the Governor of the Amur Region No. 481ot on 22.12.08, the executive authority in the field of forest relations in the Amur region in the Far Eastern Federal District is the Forestry Department of the Amur Region. It manages forests located on the lands of 13 forest districts – 30525 thousand hectares or 97% of the total forest area of the region.

Forests that are not included in the forest fund (forests located on other categories of land) include forests on the lands of the Ministry of Defense of the Russian Federation – 144.6 thousand hectares (0.5% of all forests in the region, forests located on the lands of specially protected natural territories under the jurisdiction of the Federal Districts of Rosprirodnadzor-99.4 thousand hectares or 0.3% of the forests of the region (these are specially protected territories) and forests of settlements of 50.4 (0.2%). In general, the region's forest lands occupy 75% of its territory.

Of the total area covered by forest vegetation (as of 01.01.2018), coniferous stands account for 64.1%, soft–leaved stands for 24.2%, hard – leaved stands for 2.0%, and shrubs for 9.7% [1, 2].

The main forest-forming species: larch, spruce, pine, fir, oak, birch, in the south there are Amur velvet, Manchurian walnut, ash, linden, etc. Of the shrubs, the most common are bagulnik, rhododendron, Manchurian hazel, and Middendorf birch.

Despite the obvious reduction in the total area of burned territories in the world in the period from 1996 to 2015 [3, 4], the frequency and intensity of forest fires in both Russia and Canada increased significantly during the 20th and early 21st centuries [5–8]. These data are consistent with the output of the global and regional model, which predicts a further increase in the scale and severity of forest fires in the northern zone due to climate change [8–10].

In accordance with the scale of fire danger of plantings developed by Dalniilkh, the forest fund of the Amur region is differentiated by five classes of fire danger (Table 1). The average class of natural fire danger is 2.5, which indicates the average fire danger in the forests of the region. The first three classes of fire danger include 88.8% of the land area of the forest fund of the Amur region, that is, on 27.1 million hectares, low-level fires are possible during the entire fire season.

10.2% of the territory (3.2 million hectares) is classified as fire hazard Class I. Here, during the entire fire-dangerous period, riding fires are possible.

During periods of fire maxima, top fires are possible on another 21.6% of the territory (6.6 million hectares) covered with plantings classified as Class II fire hazard.

Only 11.2% of the forest fund of the region (3.4 million hectares) are areas of fire hazard classes IV and V, where fires are possible only during periods of fire maxima or after prolonged droughts.

Wildland fires are growing in the frequency of occurrence than ever before due to ever drier and hotter weather [14]. The fire season is a certain period of the year during which forest fires are possible in the territory under consideration. As a rule, this season begins from the moment of snow cover in the forest to the onset of stable rainy autumn weather or the formation of snow cover.

During the fire season, forest fires are distributed unevenly. There are periods of fire maxima, which are also called forest fire maxima, which are periods (period) of the fire season during which the number of fires exceeds their average monthly number. More often there is one fire maximum, less often – two. This period may include three to four months with a different number of fires.

There is a period of time that accounts for the largest number of fires, called the period of the fire (forest fire) peak. With one fire peak, it accounts for the largest number of fires during the fire season. The time of the forest fire peak depends on the geographical latitude of the area, the terrain of the territory, and, as a rule, it is stable.

Table 1. Distribution of forest lands by fire hazard classes.

No.	Name of the forest area	Area, ga						Middle class of natural fire hazard
		General	By natural fire hazard classes					
			I	II	III	IV	V	
1	2	3	4	5	6	7	8	9
1	Arkharinsky forest district	904748	72289	242376	309551	237376	43156	3.0
2	Belogorsky forestry	803600	11455	8218	726844	57083	–	3.0
3	Blagoveshchensk forestry	73540	73540	–	–	–	–	1.0
4	Bureysky forestry	495118	78639	67937	336327.0	11459.0	756.0	2.6
5	Zavitinsky forestry	114527	39004	23019	52504			2.2
6	Zeysky forestry	8793205	312637	1354091	5745137	1214831	166509	3.0
7	Magdagachinsky forestry	1318106	92925	363590	827423	34168	–	2.6
8	Mazanovsky forestry	2298184	124986	128942	1655221	382497	6538	3.1
9	Norsk forestry	4012232	486465	268290	2685034	572443	–	2.9
10	Svobodnensky forestry	349575	75610	204687	58650	8556	2072	2.0
11	Tynda forest district	7857877	1207556	1997653	4097516	525722	29430	2.7
12	Urushinsky forestry	2484272	490042	1307100	610741	76389	–	2.0
13	Shimanovsky forestry	1248253	276777	632075	304426	34416	559	2.2

(*continued*)

Table 1. *(continued)*

No.	Name of the forest area	Area, ga						Middle class of natural fire hazard
		General	By natural fire hazard classes					
			I	II	III	IV	V	
Total for the Ministry of Forestry and Fire Safety of the region		30753237	3341925	6597978	17409374	3154940	249020	2.5
14	Federal State Budgetary Institution "Khingansky State Nature Reserve"	97073	45504	21442	15158	14969	–	1.6
15	Federal State Budgetary Institution "Norsky State Nature Reserve"	211168	–	5036	72651	124858	8623	3.6
16	Federal State Budgetary Institution "Zeysky State Nature Reserve"	100128	392	3885	13080	47374	35397	4.1
Total for state nature reserves		408369	45896	30363	100889	187201	44020	3.1
17	Military Forestry of the Russian Defense Ministry	86248	16714	53929	11932	3673	–	2
Total for Military forestry		86248	16714	53929	11932	3673	–	2
Total for the subject of the Russian Federation:		31247854	3404535	6682270	17522195	3345814	293040	2,5

From the analysis of the table of distribution of the number of forest fires by month in the context of forest districts for a ten-year period, it follows that for most forest districts of the Amur region, the fire maximum falls in April, May, and June, with a fire peak in May. For a number of forest districts (Verkhnezeyskoye, Dipkunskoye, Dugdinsky, Tyndinsky, Ekimchkanskoye), the fire peak is shifted to the months of June and July. The second forest fire maximum is observed in the forest areas of the Amur region in the autumn period (September, October). The average duration of the fire season in the Amur region: from April 01 to November 10. There are no peatlands on the territory of the Amur Region.

Taking into account the extent of the region's territory and the variety of natural and anthropogenic factors, as well as for the efficiency of obtaining data on wildfires, it is necessary to constantly monitor the quantitative and qualitative characteristics of forests. In such conditions, it is advisable to use remote sensing of the Earth as the main monitoring method [12, 13]. The satellite ob-servations come from sensors with various spectral characteristics but most frequently use infrared bands as they can detect fires through smoke [14].

2 Materials and Methods of Research

Space monitoring of fire hazards in 2019 was carried out using the ISDM-Rosleskhoz remote monitoring information system and the Kaskad automated information system.

If it is necessary to conduct an urgent satellite survey of the area of an emergency or a predicted emergency, the time to receive information on a telephone request is from the regional reception center from 30 min to 1.5 h, from other laboratories-at least a day.

Space monitoring of fire danger in 2020 was carried out with the help of ISDM-Rosleskhoz, automated information systems "Cascade", "Thermal Points", also with the help of two systems: "Forest Watch", which has 12 all-round view cameras and the "Forest Guard" system, which has 4 cameras with access to the operational duty service of the Crisis Management Center (ODS CCS).

The provision of information satellite products was carried out from the geostationary satellite of the earth by the Research Center for Space Hydrometeorology of the SIC "Planeta". Employees of the supervisory activity units took part in the work of 1816 patrol and control groups in 24 municipalities of the region.

In the course of the joint work, 12543 courtyard territories and residential buildings were surveyed and inspected. 21,814 people were instructed in fire safety measures, and 19656 campaign materials on fire-fighting topics were distributed. 543 citizens were brought to administrative responsibility for violations of fire safety requirements.

3 Results

In total, since the beginning of the fire-hazardous period, 587 cases have been registered in the Amur Region, on a total area of 172,189 ha. For the same period last year (APPG), 455 fires were recorded, on a total area of 271733 ha. Despite the increase in the number of registered fires, there is a decrease in the area of fires by 36.7% – 171089 ha in comparison with 270490 ha for the same period last year. The number of fires eliminated in the first day increased by 53.3% and amounted to 437 fires. The decrease in the area and number of fires eliminated in the first day indicates their timely detection. The positive results are due to the use of new fire detection systems, timely distribution of forces and resources in accordance with forecasts, as well as ongoing preventive measures.

Temporary fire-fighting posts played an important role in responding to wildfires and preventing their transfer to economic facilities and localities. According to the calculations of the specialists of the Amur Center for Hydrometeorology and Environmental Monitoring, at the end of March, the indicators of mountain weather conditions reached 1–2 classes in the southern regions. According to the data of the space sensing of the Far-Eastern Center of the Federal State Budgetary Institution "Research Center of Space Hydrometeorology "Planeta" of the Federal Hydrometeorological Service," hot spots were recorded in the southern regions of the region at the end of March. In the third decade of March, mass preventive annealing of dry vegetation began in the southern regions that were freed from snow cover.

Since March 30, the fire season and special fire protection regime have been in effect in the territories of the Arkharinsky, Bureysky, Zavitinsky, Blagoveshchensk, Konstantinovsky, Mikhailovsky and Romnensky districts (Decree of the Government of the Amur Region No. 147 of 27.03.2020). Since 03.04.2020, the fire protection season and fire protection regime have been opened in a number of other districts of the region (in 2019-from March 15). In the last days of March, there were isolated forest fires in the region.

The predominance of warm weather during April contributed to the intense melting of snow. By the end of the first decade, the snow cover completely disappeared in the central and southern regions. Throughout the region, there was an increase in the indicators of forest burnability according to weather conditions, and by the end of the second decade, in most of the central and southern regions, as well as in a number of northern (Skovorodinsky, Magdagachinsky) regions, the fire danger reached class IV (high).

By April 16, a special fire-fighting regime and a fire-dangerous season have been established throughout the region. From April 17 to 19, by the order of the head of the Magdagachinsky district No. 40 of 14.04.2020, the "Emergency situation in the forests" regime was in effect on the territory of the district.

Periods of dry weather and high temperatures increased the degree of fire danger of forests according to various indicators. From 5 to 13 May, the IV (high) class of gorimost was noted in Skovorodinsky and Magdagachinsky districts, from 19 to 24 May – in most areas of the south of the region, at the end of the month – in Skovorodinsky, in some places in Zeysky and Magdagachinsky districts. A special fire-fighting regime was maintained throughout the entire territory of the region until the end of May.

The weather conditions in June did not contribute to the increase in the fire danger of the forest according to meteorological indicators, classes 1–3 were mainly noted. A special fire-hazardous regime was maintained in 8 municipalities – Zeysky, Mazanovsky, Magdagachinsky, Tyndinsky, Skovorodinsky and Selemdzhinsky districts, the cities of Tynda, Zey.

The hot and relatively dry weather of the first two decades of July contributed to the increase in fire danger according to weather indicators. In some periods, in some places in the Selemdzhinsky, Zeysky, Tyndinsky, Skovorodinsky, Magdagachinsky, Shimanovsky, Blagoveshchensky districts, the 4th (high) class was observed.

The rainy weather in August did not contribute to the increase in fire danger according to weather indicators, classes 1–2 prevailed, in some places-Class 3.

During September, fire hazard classes 1–3 prevailed according to weather indicators. Only in some periods in the north, there was an increase to the 4 (high) class in the Tynda, Skovorodinsky, Magdagachinsky districts. Meteorological conditions in October contributed to the occurrence of wildfires in the central and southern regions of the region (Table 2 and Table 3).

A preliminary forecast of emergency situations for the forest fire period of 2021 in the Amur region has been compiled (according to the average long-term indicators).

Table 2. Number of fires in the Amur region in 2020 compared to 2012–2019.

2012	2013	2014	2015	2016	2017	2018	2019	2020
578	209	639	523	411	332	433	455	587
Total area of fires	Total area of fires	Total area of fires	Total area of fires	Total area of fires	Total area of fires	Total area of fires	Total area of fires	Total area of fires
42377 ga	47191 ga	384443 ga	234928 ga	698810 ga	221985 ga	2286176 ga	271733 ga	172189 ga

Based on the analysis of the average long-term data on fires, climatic data of the region and taking into account man-made factors, it was determined that the occurrence of wildfires can be expected in the first decade of April. The occurrence of wildfires in the third decade of March is also possible (due to the establishment of daytime positive temperatures, early snow cover, windy weather and annealing of dry vegetation).

In a favorable scenario, the expected number of wildfires will not exceed 30–50 per month, with elimination within the first day after detection. Exceeding the average long-term parameters of the fire situation is unlikely.

In an unfavorable scenario, the expected number of wildfires will exceed 100 or more per month (April, May), with the preservation of free burning time for several days and spread to large areas.Gorenje, the main source of the fire, is the fire of the Russian Federation.

The greatest risk of wildfires is predicted in the Arkharinsky, Zeysky, Mag-Dagachinsky, Svobodnensky, Shimanovsky districts.

Under the condition of a sharp early warming, the absence of significant precipitation, the preservation of windy weather in the years when temperatures are set above the average long-term indicators, a sharp deterioration of the forest and fire situation is possible from the first decade of April to the first decade of June, similar situations are most frequent. The increase in the number of fires during this period is due to the careless handling of fire and the conduct of annealing of dry grass by the population, as well as the prolonged absence of precipitation.

It is also possible to maintain a high fire risk in the summer period (June–July) when hot and windy weather is established.

Analyzing the data of previous years, the largest number of wildfires is predicted in April–June and the first decade of July, both in forest and non-forest areas, including in protected areas and specially protected areas of the Amur region.

Since the second decade of July and in August, due to the systematic precipitation associated with active cyclonic activity, the number of wildfires, as a rule, decreases.

In the autumn months in September–October, wildfires may again occur due to a decrease in precipitation and the population conducting annealing of dry grass. In October, there is a possibility of forest fires in the northern regions of the region, where earlier there was a decrease in average daily temperatures and the grass cover dries up. According to the results of long-term observations, the number of natural fires in the autumn period decreases in comparison with the spring-summer period by 3 or more times.

Table 3. Dynamics of wildfires in the districts of the region for the period from 2004 to 2020.

District	2004	2005	2006	2007	2008	2009	2010	2011	2012	2013	2014	2015	2016	2017	2018	2019	2020	Average value
1	2	3	4	5	6	7	8	9	10	11	12	13	14	15	16	17	18	19
Areas with rich forest resources																		
Archarinsky	33	32	49	21	32	50	32	54	29	20	87	47	28	32	32	49	78	41
Belogorsky	3	0	15	7	14	6	1	15	1	2	12	4	1	0	1	0	1	5
Blagoveshchensk	20	40	40	19	19	23	2	1	9	16	33	30	36	28	34	19	23	23
Bureysky	7	7	29	24	35	33	41	43	42	16	52	27	24	30	31	20	49	30
Zavitinsky	6	–	15	12	10	7	8	16	10	13	31	23	14	9	20	16	20	14
Zeysky	36	52	60	32	80	63	32	88	116	34	108	79	63	70	53	75	87	66
Magdagachinsky	5	33	33	34	32	30	32	61	81	16	43	57	24	39	51	35	65	39
Mazanovsky	7	6	17	30	8	12	10	29	18	18	40	34	40	12	29	26	16	21
Selemdzhinsky	4	5	12	29	29	10	13	47	69	21	25	23	20	14	30	8	14	22
Svobodnensky	12	59	91	34	62	56	54	29	33	20	40	36	54	33	30	41	37	42
Skovorocinsky	10	24	29	5	18	7	8	31	111	11	48	26	21	19	28	40	73	30
Tyndyn	40	36	21	22	63	5	3	17	24	5	32	31	24	37	18	89	79	32
Szymanowski	8	42	64	48	85	51	35	32	28	16	55	65	38	6	50	21	27	39

(*continued*)

Table 3. *(continued)*

District	2004	2005	2006	2007	2008	2009	2010	2011	2012	2013	2014	2015	2016	2017	2018	2019	2020	Average value
Districts with no forest funds																		
Ivanovsky								0	0	0	0	0	0	0	1	0	0	0
Mikhailovsky								0	0	1	0	0	0	0	0	0	1	0
Konstantinovsky								0	0	0	0	0	0	0	0	0	0	0
Octobersky								1	3	0	15	20	9	3	9	3	8	7
Romnensky								5	2	0	12	9	12	0	7	7	4	6
Seryshevsky								4	2	0	5	8	2	0	9	4	1	4
Tambovsky								0	0	0	0	0	0	0	0	0	0	0
GO Blagoveshchensk								0	0	0	0	0	0	0	0	2	0	0
Svobodny city								0	0	0	0	0	0	0	0	0	4	2
Lands of the Ministry of Defense								0	0	0	1	4	1	0	0	0	0	1
TOTAL:	191	336	475	317	487	353	271	473	578	209	639	523	411	332	433	455	587	416

The greatest risk of fire spread from wildfires is predicted for settlements close to the forest zone, according to the list (122 populated areas). In the event of an unfavorable forest fire situation, there may be a threat to the destruction of the housing stock, outbuildings and economic facilities.

Also, the occurrence of fires in woodlands entails the spread of smoke, which in turn negatively affects the life of the population.

In the spring (April–May) and autumn (September–October) periods, there is a threat of fire moving from agricultural fires to settlements and horticultural areas. The highest probability is in areas with extensive uncultivated agricultural land: Blagoveshchensk, Belogorsky, Tambov, Ivanovo, Konstantinovsky, Mikhailovsky, Oktyabrsky municipal districts.

The main reasons for the spread of forest fires: natural and climatic conditions, low population density, untimely detection and untimely elimination of fire centers in hard-to-reach areas. Possible transition of forest fires from the Trans-Baikal and Khabarovsk territories, the Republic of Sakha (Yakutia).

According to a preliminary assessment in 2021:

there will be an increased probability of fire transmission from natural fires, including dry vegetation fires, to settlements and economic objects April–May);

there is an increased probability of deterioration of the environmental situation and smoke in settlements in the region (May–August).

there will be an increased risk of smoke in the border areas of the region (the reason: fires and grass fires, both in the territory of neighboring subjects, and the PRC), the greatest risk is April–May and September–October.

In connection with the implementation of preventive measures on the territory of the region, aimed at protecting settlements from wildfires, preventive work with the population and organizations, the occurrence of an emergency is predicted, not higher than a municipal nature, with a probability of 0.1.

References

1. Senchik, A.V., Kukharenko, N.S., Sosnovsky, I.E.: Characteristics of winter habitat of Siberian roe deer (capreolus pygargus pall) on zeysa-bureinskaya plain. Izvestiya Nizhnevolzhskogo agrouniversitetskogo kompleksa: Nauka i vysshego professional education **1**(53), 189–196 (2019). https://doi.org/10.32786/2071-9485-2019-01-25/
2. Romanova, N.A., Zhirnov, A.B., Yust, N.A., Fucheng, X.: Influence of forest growth conditions on the density of wood in the Amur region. Cent. Eur. For. J. **65**(1), 41–50 (2019). https://doi.org/10.2478/forj-2019-0005
3. Andela, N., et al.: A human-driven decline in global burned area. Science **356**, 1356–1362 (2017). https://doi.org/10.1126/science.aal4108/
4. Doerr, S.H., Santín, C.: Global trends in wildfire and its impacts: perceptions versus realities in a changing world. Philos. Trans. R. Soc. B **371**, 20150345 (2016). https://doi.org/10.1098/rstb.2015.0345
5. Soja, A.J., Tchebakova, N.M., French, N.H., et al.: Climate-induced boreal forest change: predictions versus current observations. Global Planet Change **56**(3–4), 274–296 (2007). https://doi.org/10.1016/j.gloplacha.2006.07.028
6. Ponomarev, E.I., Kharuk, V.I., Ranson, K.J.: Wildfires dynamics in Siberian larch forests. Forests **7**, 125 (2016). https://doi.org/10.3390/f7060125

7. Turetsky, M.R., Kane, E.S., Harden, J.W., et al.: Recent acceleration of biomass burning and carbon losses in Alaskan forests and peatlands. Nat. Geosci. **4**, 27–31 (2011). https://doi.org/10.1038/ngeo1027
8. Flannigan, M.D., Cantina, A.S., de Groot, W.J., Wotton, M., Newbery, A., Gowman, L.M.: Global wildland fire season. severity in the 21st century. For. Ecol. Manage. **294**, 54–61 (2013). https://doi.org/10.1016/j.foreco.2012.10.022
9. De Groot, W.J., Flannigan, M.D., Cantin, A.S.: Forest Ecol and Manag **294**, 35–44 (2013). https://doi.org/10.1016/j.foreco.2012.09.027
10. Shvidenko, A.Z., Schepaschenko, D.G.: Climate change and wildfires in Russia. Contemp. Probl. Ecol. **6**(7), 683–692 (2013). https://doi.org/10.1134/S199542551307010X
11. Ruffault, J., Curt, T., Martin-StPaul, N.K., Moron, V., Trigo, R.M.: Extreme wildfire events are linked to global-change-type droughts in the northern Mediterranean. Nat. Hazards Earth Syst. Sci. **18**, 847–856 (2018). https://doi.org/10.5194/nhess-18-847-2018
12. Kotelnikov, R.V., Lupyan, E.A., Bartalev, S.A., Ershov, D.V.: Space monitoring of forest fires: history of the creation and development of ISDM-Rosleskhoz. Contemp. Probl. Ecol. **13**, 795–802 (2020). https://doi.org/10.1134/S1995425520070045
13. Lupyan, E.A., Milekhin, O.E., Antonov, V.N., et al.: System of operation of joint information resources based on satellite data in the Planeta Research Centers for Space Hydrometeorology. Russ. Meteorol. Hydrol. **39**(12), 847–853 (2014). https://doi.org/10.3103/S1068373914120103
14. Janice, L., Coen, W., Schroeder, S., Conway, L.: Computational modeling of extreme wildland fire events: a synthesis of scientific understanding with applications to forecasting, land management, and firefighter safety. J. Comput. Sci. **45**, 101–152 (2020). https://doi.org/10.1016/j.jocs.2020.101226

Assessment of the Condition of Coniferous Trees in Urban and Suburban Forests of the Krasnoyarsk City

Nina Pakharkova(✉), Olga Shabalina, Galina Sorokina, Irina Bezkorovainaia, and Ekaterina Tegnerenko

Siberian Federal University, pr. Svobodniy, 79, 660041 Krasnoyarsk, Russia
npakharkova@sfu-kras.ru

Abstract. The urboecosystems of large cities differ significantly from natural ecosystems that form in similar climatic conditions. This research is devoted to the assessment of the state of woody plants in urban and suburban forests of Krasnoyarsk in order to optimize the assortment lists for landscaping areas. In the work, both traditional morphometric methods were used, as well as physiological and biochemical methods (for example, registration of chlorophyll fluorescence parameters), which make it possible to register at the early stages "hidden" disturbances that do not yet manifest themselves externally. Having ranked the studied species according to their resistance to urban conditions, the following row was obtained from the most resistant *Larix sibirica* → *Pinus sylvestris* → *Picea obovata*. In areas with a high and critical level of pollution, a decrease in the decorativeness of evergreen conifers was observed due to defoliation of needles starting from the second year of life.

Keywords: Air pollution · *Pinus* · *Picea* · *Larix* · Chlorophyll fluorescence

1 Introduction

Coniferous species are highly vulnerable to air pollution [1–3], yet they are very common in urban forests and parks in Northern Eurasia and North America. Although there is a large literature base on their ecology in urbanised environments [4–7], some aspects of their ecology remain underexplored in the literature, such as relationships between environmental pollution, suppressed photosynthetic activity, premature termination of winter dormancy and subsequent defoliation, especially in continental climates.

This research aims towards a better understanding of the responses of trees to air pollution stress in urban and semi-urban conditions of Krasnoyarsk (Southern Siberia, Russia). We research the role of urban air pollution on the depth of winter dormancy and its effect on the health of *Larix sibirica, Pinus sylvestris* and *Picea obovata.* The species were selected due to their broad occurrence in Siberia, but also for their difference in ecology and vulnerability.

A. Muratov and S. Ignateva (Eds.): AFE 2021, LNNS 353, pp. 553–562, 2022.
https://doi.org/10.1007/978-3-030-91402-8_62

Since the evolution of plants took place in conditions of sufficiently clean atmospheric air, modern plant species, including woody ones, do not possess specific adaptability to the action of toxic gases, which can lead to significant changes in adaptive responses to various stressors, including cold and frost. Therefore, it can be assumed that air pollution leads to disruption of physiological and biochemical processes in needles, which are expressed over time in changes in the morphological parameters of trees.

An important role in the survival of plants is played by the seasonal phenomena which cause complex adaptive response having rhythmic character. The seasonal manifestations of vital activity of organisms are clearly consistent with the most favorable periods of environmental conditions.

Transition to dormancy in the period of low temperatures plays a special role in maintaining the vitality of plants in temperate and high latitudes.

It is well-established that during the transition from active vegetation into winter dormancy the photosynthesizing cells of needle and bark undergo a number of changes. These changes include: the amount and the proportions of pigments, an increase in the fraction of unsaturated fatty acids in membranes, in the concentration of saccharides and cryoprotective proteins proteins, changes in the structure of the photosynthetic apparatus [8]. The zero level of chlorophyll fluorescence in condition of gradual (2,5° min-1) heating from 25–30 to 75–80 °C is shown by the chlorophyll fluorescence temperature curve (CFTC). This curve, starting with the study of U. Shreiber with colleagues [9, 10] is used for determination of the stability of PS-2 and for evaluation of the structural arrangement of chloroplasts in vegetating plants [11, 12]. Also, based on the changes in the shape of the CFTC it can be deduced whether the plant is in the state of winter dormancy or is vegetating. Low-temperature maximum of fluorescence at 49–52 °C in summer points to the lowering of the effectiveness of energy capturing by the reaction centers of PS-2. The high-temperature fluorescence peak around 60 °C originates from PS-1 in stroma and from PS-2 under specific conditions. The ratio of the amplitudes of low-/high-temperature peaks ranges typically from 2 to 3. The CFTC for needles or bark during winter dormancy is drastically different that during active vegetation: the low-temperature peak is absent, whereas the high-temperature peak rises and shifts to 70–75 °C. In intermediate periods (spring or autumn) there were gradual changes in recording CFTC. The absence of the lowtemperature peak is explained by a special state of the photosynthetic apparatus during the winter period due to reactions that protect the chlorophyll-protein complexes against a photodestruction. Fluorescence around 75 °C originates from chlorophyll-protein complexes after enhance interaction with the lipid phase of thylakoid membranes [12]. For the period of winter dormancy the ratio of the low-/high-temperature peaks is around 0.08, increasing in bark to around 1.7 during the transition of plants to active metabolism. The calculated the ratio may be used as an indicator of the degree of the depth of winter dormancy.

2 Materials and Methods

The study area is the city of Krasnoyarsk and its neighbourhood. The city is located in Eastern Siberia (56°0′N, 92°55′E) in the geographical region of forested steppes. The annual average air temperature is +1.6 °C, the warmest month is July (the average is +18.7 °C), the coldest month is January (−15.5 °C), the highest recorded temperature is +36.5 °C, and the lowest is −52.8 °C. The period with stable daily temperatures above 0 °C typically lasts from April 22 to October 2. With its population over 1 million people, Krasnoyarsk is a major industrial and logistical centre with high level of air pollution. Due to high concentrations of benzpyrene, formaldehyde, airborne particles, nitrogen dioxide and ammonia, the city is one of the Russia's most polluted municipal areas.

As a control, a sample plot (SP1) was used, located on the outskirts of the city of Krasnoyarsk from the western (windward) side, in the absence of technogenic pollution (Fig. 1A). The first experimental sample plot (SP2) is located in the industrial area of the city of Krasnoyarsk and is exposed to air pollution from both industrial enterprises and road transport. The second experimental sample plot (SP3) is located in the industrial area of the city of Krasnoyarsk, exposed to air pollution from industrial enterprises and is located in the temperature anomaly zone (Fig. 1B). The temperature in the places of temperature anomalies is 5°–8° higher than the average surface temperature of the city.

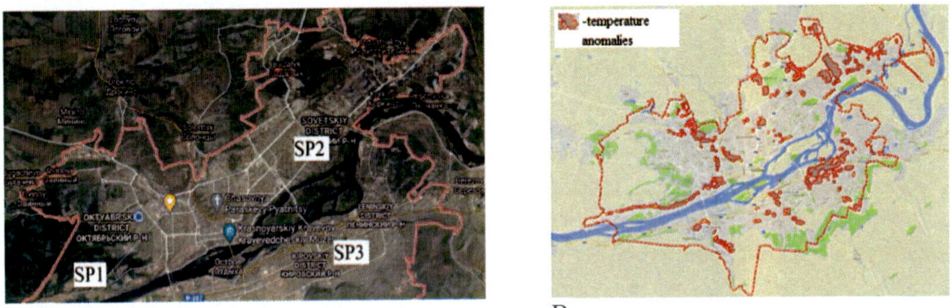

A B

Fig. 1. Krasnoyarsk map (A – sampling plots, B – distribution of maximum temperature values [13]).

As objects of research, we used branches of free-standing trees of Scots pine (*Pinus sylvestris* L.), Siberian spruce (*Picea obovata* Ledeb.), Siberian larch (*Larix sibirica* Ledeb.) at the age of 30–40 years. All measurements were carried out for branch, gymnosperm needles and phelloderm of angiosperm woody plants. Branches Samples were collected from the middle part of the crown at a height of ~2.5–3 m. Two-year old healthy-looking needles of Scots pine and Siberian spruce, the phelloderm of brunches of Siberian spruce and Siberian larch were collected during September 2020 through April 2021.

For our study we employed 10 trees for each species. The samples were collected from the same trees at the different sampling campaigns. Four sub-samples were taken from four perpendicular directions around the tree and were mixed subsequently to make one sample per tree. After taking needles from each tree, we mixed needles from all ten trees to make a combined sample. Five samples for measurements were taken from the combined sample (that represented all 10 trees).

The analysis of morphological parameters included the measurement of the brunch length, the length of the needles, the number of needles per unit of brunch length, the percentage of chlorosis and necrosis.

The chlorophyll fluorescence temperature curve was recorded with needle segments using the pulse-shaped modulated chlorophyll fluorimeter Junior-PAM (Walz, Germany). The results were processed using the WinControl full-featured software. The ratio of the intensity of the zero-level fluorescence at 50 °C and at 70 °C (coefficient R) was used as an indicator of the depth of the dormancy state (Fig. 2).

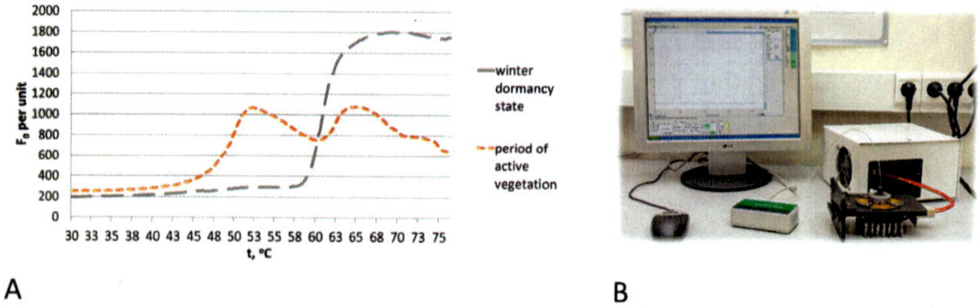

Fig. 2. The curves of temperature-induced changes in the zero-level of chlorophyll fluorescence (A) and fluorimeter Junior-PAM (B).

The needles were linearly heated from 25 to 75 °C at the rate of 2.5 °C using a computer-controlled heating device. The height of the peak is measured relative to the lowest value of the baseline, which reflects the F_0 value at room temperature. Values of the coefficient R below unity indicate that the woody plant is in the state of winter dormancy, whereas values above unity indicate that the woody plant is actively vegetating [14].

To determine inorganic anions on needle surfaces, pine and spruce needle leachate was analysed using liquid chromatograph LC-20 Prominence equipped with chromatographic column Shodex IC SI-90 4E (4 × 250 mm). The analytical accuracy of the chromatographer is n × 10^{-3} mg/l, and in the range of concentrations from 0.1 to 1000 mg/l it identifies up to 10 ions per sample in 30 min. The chromatograph was calibrated using national standards based on ISO 10304-1-92. We used mixture of sodium bicarbonate and hydrocarbonate (1,8 mM Na_2CO_3 + 1,7 mM $NaHCO_3$) as eluent additive. Needle samples for the analysis were collected in February to make sure that water-soluble agents precipitated on the needles during the winter had not leached with melting water in spring.

The statistical treatment of the data was carried out using the Microsoft Excel software. We used ANOVA (two-factor with replication).

3 Results

The chemical composition of needle leachate clearly demonstrates differences in air pollution statuses of the two sampling areas. Table 1 shows that concentrations of sulphur in the leachate from pine needles collected in SP2 and SP3 are higher than in SP1 by a factor of about 18; concentrations of chlorine and nitrogen are higher by factors of 7 and 2. In case of the spruce, this difference is even more significant, and on average it is a factor of 19 for sulphur, 7 for chlorine, 3 for nitrogen and 1.7 for total phosphorus.

Table 1. Anion concentrations in needle samples (n = 2, P = 0.95).

Sampling plot (tree species)	$C \pm \Delta C$, mg/kg			
	Cl^-	PO_4^{3-}	NO_3^-	SO_4^{2-}
SP1 (*Pinus sylvestris*)	2.9 ± 0.4	Not detected	28.0 ± 4.0	33.0 ± 6.0
SP1 (*Picea obovata*)	13.0 ± 2.0	3.2 ± 0.5	30.0 ± 5.0	27.0 ± 4.0
SP2 (*Pinus sylvestris*)	22.0 ± 3.0	Not detected	54.0 ± 8.1	270.0 ± 40.3
SP2 (*Picea obovata*)	92.0 ± 14.0	5.5 ± 0.8	84.0 ± 12.0	530.0 ± 80.0
SP3 (*Pinus sylvestris*)	24.2 ± 2.7	Not detected	56.0 ± 8.5	292.0 ± 39.0
SP3 (*Picea obovata*)	101.9 ± 11.7	4.8 ± 0.7	78.0 ± 11.1	540.0 ± 84.5

Studies have shown that different species of conifers during the winter are characterized by different depths of dormancy. As follows from the data presented in Fig. 3, Scots pine has a greater depth of dormancy compared to Siberian spruce. In March, pine brunches come out of dormancy on the fourth day (Fig. 3A), while spruce brunches in the contaminated areas – for one day (Fig. 3B), this trend persists in the other months. Natural emergence from dormancy in natural conditions also occurs earlier in the spruce – in early April, while in the pine – in early May.

Recording of thermoinduced curves allows to show difference between species of tree plants. Phelloderm of *Picea obovata* do not show winter dormancy period even in extremely cold weather because low temperature peak is visible on the curve in any season. It may be connected with energy function of phelloderm.

According to existing theoretical concept transition to winter dormancy leads to the lower ratio of low- and high-temperature maximums. The coefficient R for winter dormancy period is normally bellow 0,3. In the periods of transition to winter dormancy and back in the areas with higher level of pollution the values of R-coefficient are higher. It tells that process of transition is incomplete.

Our research confirms that the depth of winter dormancy of *Pinus sylvestris* and *Picea obovata* is highly sensitive to air pollution stress. The measurements photosynthetic activity and coefficient R (Figs. 3, 4) further demonstrate that regardless their age, under the air pollution stress pine and spruce needles are often unable to enter deep dormancy (or it can be substantially delayed and/or prematurely terminated). In polluted environments, in late winter and early spring the final release from dormancy is controlled only by low temperatures, and it can be easily triggered by the first daytime thawing. In such a situation, phothosynthetic and transpiration functions would fully recover, and frost conditions returning on one of the following nights, cause injuries of needles and even their desiccation. The registration of chlorophyll fluorescence parameters makes it possible to register at the early stages "hidden" disturbances that do not yet manifest themselves externally.

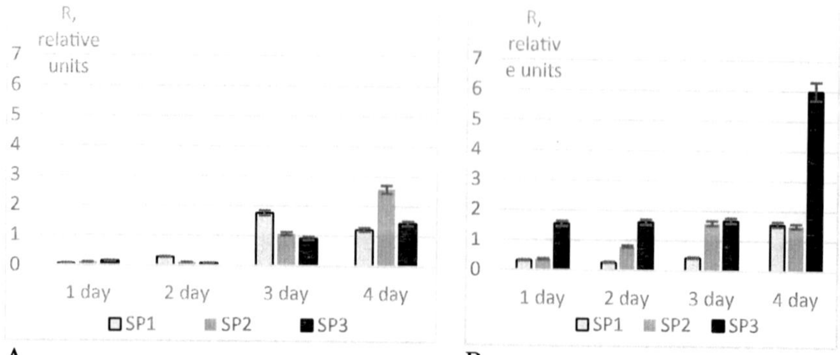

A B

Fig. 3. Coefficient R (the ratio of the intensity of the zero-level fluorescence at 50 °C and at 70 °C as an indicator of the depth of the dormancy state) change in needle A – *Pinus sylvestris*, B – *Picea obovata* (under laboratory conditions, March 2021).

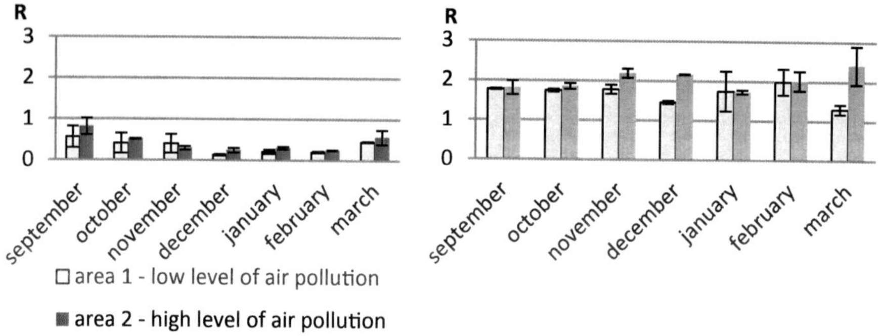

A B

Fig. 4. Coefficient R (the ratio of the intensity of the zero-level fluorescence at 50 °C and at 70 °C as an indicator of the depth of the dormancy state) change A – in the needle of *Picea obovate*, B – in the phelloderm of *Larix sibirica* (under nature conditions).

Chronic disturbances in the photosynthetic activity of needles lead to morphological changes in the brunch (see Table 2).

Under air pollution stress, coniferous species accumulate toxic compounds in the needles, which either prematurely drop, or stay on the shoot continuing to accumulate toxic substances that also can be indicated by necrosis and chlorosis. Our research confirms that *Pinus sylvestris* uses the first strategy, while *Picea obovata* employs the second one (Tables 1, 2). As a result, the pine has more ecological plasticity and less vulnerable to environmental pollution than the spruce.

Table 2. Morphological parameters of brunch.

Species	SP	Needles age, years	Brunch length, cm	Number of needles per unit of brunch length	Length of the needles, cm	% chlorosis and necrosis of needles
Picea obovata	SP1	1	4.02 ± 1.20	21.30 ± 1.41	1.80 ± 0.12	0.92 ± 0.04
		2	5.10 ± 0.80	19.02 ± 0.95	1.77 ± 0.14	1.05 ± 0.03
	SP2	1	2.90 ± 0.60	20.51 ± 1.12	1.46 ± 0.09	1.65 ± 0.09
		2	4.03 ± 0.55	17.36 ± 0.87	1.83 ± 0.13	8.59 ± 0.43
	SP3	1	2.2 ± 0.09	20.6 ± 0.74	1.79 ± 0.15	2.77 ± 0.19
		2	3.4 ± 0.19	17.9 ± 0.71	1.67 ± 0.11	13.56 ± 0.92
Pinus sylvestris	SP1	1	7.32 ± 1.30	12.14 ± 0.80	5.95 ± 1.93	0.72 ± 0.04
		2	6.90 ± 0.91	11.51 ± 0.58	6.73 ± 1.52	1.06 ± 0.05
	SP2	1	5.71 ± 0.62	12.52 ± 0.71	5.17 ± 1.55	1.63 ± 0.10
		2	5.96 ± 0.80	7.06 ± 0.33	5.14 ± 1.62	2.08 ± 0.12
	SP3	1	4.99 ± 0.61	11.4 ± 0.51	6.83 ± 1.59	2.41 ± 0.11
		2	5.08 ± 0.63	6.2 ± 0.17	6.38 ± 1.32	5.18 ± 0.43
Larix sibirica	SP1	1	10.6 ± 0.54	12.3 ± 0.69	2.4 ± 0.12	0.2 ± 0.05*
	SP2	1	11.0 ± 0.57	7.1 ± 0.68	3.6 ± 0.10	1.0 ± 0.18*
	SP3	1	11.4 ± 0.67	13.1 ± 0.69	2.9 ± 0.08	0.4 ± 0.08*

*The number of dead brachyblasts per 10 cm of the shoot, pieces.

We have got a statistically significant difference between SP1, SP2 and SP3 on the parameters: number of needles per unit of brunch length, % chlorosis and necrosis of needles (Table 3).

Table 3. Two-way ANOVA results on the influence of air pollution (SP1, SP2 and SP3) on morphological parameters of brunch of *Picea obovata, Pinus sylvestris* and *Larix sibirica.*

Parameter	Factor (source of variation)	SS[1]	df	MS	F	P-value	F critical
Brunch length	Species	6.845	1	6.845	0.5188	0.4852	4.7472
	SP	55.6811	2	27.8405	2.1100	0.1639	3.8853
	Species x SP	201.5033	2	100.7517	7.6359	0.0072	3.8853
	Error	158.3333	12	13.19444			

(*continued*)

Table 3. *(continued)*

Parameter	Factor (source of variation)	SS[1]	df	MS	F	P-value	F critical
Number of needles per unit of brunch length	Species	1.9143	1	1.9143	5.2738	0.6103*	4.7472
	SP	132.5339	2	66.2669	9.4768	0.0034*	3.8853
	Species x SP	209.6139	2	104.8069	14.9884	0.0005	3.8853
	Error	83.9106	12	6.9926			
% chlorosis and necrosis of needles (for *Pinus sylvestris, Picea obovate)*	Species	3.4409	1	3.4409	1.5877	0.2316	4.7472
	SP	199.3145	2	99.6572	45.9830	0.00001*	3.8853
	Species x SP	71.5159	2	35.7579	16.4991	0.0004	3.8853
	Error	26.0071	12	2.1673			

*indicates significant difference at $\alpha = 0.05$; [1]SS = within-group dispersion, MS = between-group dispersion, df = degree of freedom, F = value of ANOVA test; [a]notation "x" denoted interaction of factors.

4 Discussion

Analysis of the data obtained (Tables 2, 3) showed that there are no statistically significant differences in the length of the annual growth in the stands of *Larix sibirica, Pinus sylvestris* and *Picea obovata.* At the same time, in the SP2 plantation growing under conditions of high traffic load, the number of living brachyblasts is significantly lower and the number of dead brachyblasts are significantly higher. Thus, under conditions of a high level of traffic pollution in larch, a decrease in brunch coverage was noted due to increased death of brachyblasts. An increase in the average length of needles may act as a compensatory mechanism, which in the SP2 plantation was 3.6 ± 0.10 cm, while in the SP1 and SP3 plantations its length does not exceed 2.4–2.9 cm on average. For *Pinus sylvestris* and *Picea obovata,* statistically significant differences were obtained in the number of needles on a branch and the percentage of needles damaged by chlorosis and necrosis.

Phelloderm of *Larix sibirica* (Fig. 4) does not show the differences in the winter dormancy period even in low-temperature conditions. The transition into the dormancy state of the other woody plants (*Pinus sylvestris* and *Picea obovata*) growing in polluted regions happens later and the waking from this state is sooner. The depth of dormancy for these plants is less during the whole winter period.

Thus, we can conclude that in areas with a low level of environmental pollution, was noted a decrease in the rate of electron transport of primary reactions of photosynthesis in *Pinus sylvestris* and *Picea obovata.* In areas with a mid-level level of pollution, temporary shifts in the passage of phenophases in coniferous and deciduous trees were observed, in particular, a decrease in the duration and depth of winter dormancy. In trees, an increase in the proportion of needles with chlorosis and a slight decrease in the weight of needles, a decrease in the linear growth of brunches were noted. In areas with a high and critical level of pollution as a result of the action of acid gases (such as nitrogen oxide and dioxide, sulfur dioxide, included in the list of priority pollutants for Krasnoyarsk), a process of rapid necrotization of needles is taking place – necrosis is formed at the site of chlorosis. Also, as a result of the premature emergence of trees from the state of winter dormancy during short-term winter-spring thaws, physiological drying of the needles was noted.

Our research illustrated the strategies used by *Pinus sylvestris* and *Picea obovata* to deal with air pollution in urban and periurban environments. In response to high levels of air pollution, *Picea obovata* accelerates needle ageing, with necrosis and chlorosis abundantly developing on old needles. In contrast, *Pinus sylvestris* accumulates toxic substances only in some needles (of unidentified age group), which are quickly yellowing and shedding, while other needles maintain the photosynthetic function.

In climate conditions of Southern Siberia, disturbance of winter dormancy under air pollution stress represents a major threat to the health status of *Pinus sylvestris* and *Picea obovata.* Our data demonstrate that regardles the age of needles, the depth of winter dormancy of both species clearly correlates with air pollution levels, and the trees growing in industrial areas are easier to release from dormancy and to be affected by late winter or spring frost. In urban environments the risk of frost injuries is even higher due to early spring warm spells associated with heat island effect. *Pinus sylvestris* is less susceptible to air pollution and temperature fluctuations than *Picea obovata*, and therefore represents a better choice for urban forestry projects.

5 Conclusions

In the research, changes in the morphological parameters of branch and physiological parameters of the needles were revealed and phelloderm. The depth of dormancy for coniferous plants can be judged from the speed with which the brunches waking from the dormancy state. This phenomenon can be used as an informative bioindicative parameter of the degree of atmospheric pollution around industrial factories and in the city environment.

Having ranked the studied species according to their resistance to urban conditions, the following row was obtained from the most resistant *Larix sibirica* → *Pinus sylvestris* → *Picea obovata.* In areas with a high and critical level of pollution, a decrease in the decorativeness of evergreen conifers was observed due to defoliation of needles starting from the second year of life.

Acknowledgements. This work was supported by a grant from the Regional Science Foundation No. 2020061906506 "Principles of conservation and restoration of urban green areas that perform landscape-forming, recreational and health-improving functions within the city of Krasnoyarsk".

References

1. Pridacha, V.B., Sazonova, T.A., Talanova, T.Y., Ol'chev, A.V.: Russ. J. Ecol. **42**(1), 22–29 (2011). https://doi.org/10.1134/S1067413611010073
2. Rodriguez, J.H., Wannaz, E.D., Pignata, M.L., Fangmeier, A., Franzaring, J.: CLEAN – Soil Air Water **40**, 1315–1319 (2012). https://doi.org/10.1002/clen.201100584

3. Rucandio, M.I., Petit-Domínguez, M.D., Fidalgo-Hijano, C., García-Giménez, R.: Environ. Sci. Pollut. Res. **18**(1), 51–63 (2011). https://doi.org/10.1007/s11356-010-0350-y
4. Swidrak, I., Gruber, A., Kofler, W., Oberhuber, W.: Tree Physiol. **31**(5), 483–493 (2011). https://doi.org/10.1093/treephys/tpr034
5. Acquaviva, R., et al.: Environ. Sci. Pollut. Res. **19**(9), 3850–3858 (2012). https://doi.org/10.1007/s11356-012-1030-x
6. Wuytack, T., Samson, R., Wuyts, K., Adriaenssens, S., Kardel, F., Verheyen, K.: Water Air Soil Pollut. **224**(8), 1635–1649 (2013). https://doi.org/10.1007/s11270-013-1635-9
7. Liu, N., Guan, L.L., Sun, F.F., Wen, D.Z.: J. Plant Res. **127**(4), 491–501 (2014). https://doi.org/10.1007/s10265-014-0638-y
8. Oguist, G., Huner, N.P.A.: Annu. Rev. Plant Biol. **54**, 329–355 (2003)
9. Schreiber, U., Berry, P.A.: Annu. Rev. Plant Biol., 341–346 (1977)
10. Schreiber, U., Armond, P.A.: Biochim. Biophys. Acta **502**(1), 138–151 (1978)
11. Downton, W.J.S., Berry, J.A.: Biochim. Biophys. Acta **679**(3), 474–478 (1982)
12. Gaevskiy, N.A., Sorokina, G.A., Gold, V.M., Mirolyubskaya, I.V.: Plant Physiol. **38**(4), 685–692 (1991)
13. Yakubailik, O.E., Kadochnikov, A.A., Tokarevm, A.V.: Optoelectron. Instrum. Data Process. **54**(3), 243–249 (2018). https://doi.org/10.3103/S8756699018030056
14. Pakharkova, N.V., et al.: Revista Brasileira de Botanica **39**(4), 1005–1014 (2016). https://doi.org/10.1007/s40415-016-0298-3

Study of Crocosmia × Crocosmiiflora (Lemoine) N.E.Br. in the Agrometeorological Conditions of the Southern Zone of the Amur Region for the Prospects of Use in Landscaping

Irina Kurkova(✉) and Svetlana Stokoz

Far Eastern State Agrarian University, ul. Politehnicheskaja, 86, Blagoveshchensk, Amur Oblast 675005, Russia

Abstract. The article presents the data of a field experiment conducted in 2018–2020 at the Far Eastern State University for the study of *Crocosmia* × *crocosmiiflora* (Lemoine) N.E.Br The aim of the research was to study the growth and development features of *Crocosmia* × *crocosmiiflora* (L.) N.E.Br. in the agrometeorological conditions of the southern zone of the Amur region for the prospects of use in landscaping. The research objectives included: to study the phenological development of *Crocosmia* × *crocosmiiflora* depending on the weather conditions of the warm season; to evaluate the decorative features of *Crocosmia* × *crocosmiiflora* (L.) N.E.Br. for the possibility of entering into the list of plants used for landscaping the territories of the southern zone of the Amur region. The conditions of 2018 were favorable for C. crocosmiiflora development. That year, the flowering phase came earlier and was the longest. Adverse weather conditions in 2019 and 2020 negatively affected the development of plants, reducing the timing of the onset of phases and the duration of flowering. The duration of flowering of *Crocosmia* × *crocosmiiflora* (L) in 2020 was 29 days less than in 2018, and in 2019 the opening phase of the flowers was not observed, but the buds were quite well formed. The correlation dependence of the duration of flowering on the prevailing weather conditions of the growing seasons was calculated for three years of research. The calculation was carried out based on the average values of temperatures and precipitation. A strong correlation r = 0.96 on temperature and a strong negative dependence on precipitation r = −0.75 were found.

Keywords: Family Iridaceae Juss. · Crocosmia × crocosmiiflora (Lemoine) N.E.Br. · Vegetation period · Phenological observations · Correlation dependence · Decorative effect

1 Introduction

The family Iridaceae includes 65 genera with more than 1,800 species in the monocotyledonous series Asparagales [1, 2]. Plants of this family are distributed in tropical and temperate regions of the world, with the greatest diversity in sub-Saharan Africa, South America, Europe and temperate regions of Asia [3, 4]. Significant components

A. Muratov and S. Ignateva (Eds.): AFE 2021, LNNS 353, pp. 563–572, 2022.
https://doi.org/10.1007/978-3-030-91402-8_63

of the floral diversity of Eurasia and North America of this family are the genus Crocus and Iris, and the main genera of flora common in sub-Saharan Africa are Gladiolus and Morea.

Plants belonging to this family are perennial grasses or semi-shrubs, easily recognizable among monocotyledons by the presence of curved equilateral leaves that intersect transversely with each other, forming a flat fan-shaped arrangement, which is a unique feature of this family. The leaves have parallel venation, as in other monocotyledonous species. The flowers have three colored petal-shaped sepals and three petals that differ in size, shape and color, three stamens and one pistil with a lower ovary consisting of three fused carpels, which are additional characteristic features of the flowers. The fruits are loculicidal pods, hard to cartilaginous and sometimes woody, rarely indistinguishable. Iridaceae is characterized by fully developed underground storage organs, such as rhizomes, bulbs, or corms [5]. Iridaceae is considered one of the most beautiful families in horticulture [6]. Many species have the potential for further study and use in ornamental horticulture and floristry [7]. *Crocosmia* is not an exception. The plant has bright orange-red flowers on a branching inflorescence that appear in their natural habitat from January to June, after which a fleshy seed capsule containing purple-black seeds develops. The plant is widely distributed in the eastern parts of South Africa, and is found mainly in humid areas from the coast to 2000 m above sea level [8]. Under natural conditions, it reproduces vegetatively through tubers or rhizomes, and less often generatively due to its poor ability to set seeds [7, 9]. Many breeders around the world have conducted studies on the genera Watsonia species, Iris L, *Crocosmia* × *crocosmiiflora*, and *Crocosmia aurea* [10].

Traditional breeding has led to a significant improvement in genera within the Iridaceae family and a large number of flower plants have entered the domestic and global market as a floral product. Currently, the timing of the development of new varieties and hybrids is accelerated due to micro-reproduction, which is a valuable tool for breeders of modern science [7]. Additionally, production of storage organs (corms, tubers or bulbs) during micropropagation can reduce the losses incurred during acclimatization and decrease the time to flowering.The first record of microcosmic micro-reproduction was made by Ascough G. D. and other authors. In floriculture, the most common is the crocosmia *crocosmiiflora* (*Crocosmia* × *crocosmiiflora* (Lemoine). This is a garden hybrid of *C. aurea* × *C. pottsii*, obtained by the French breeder Lemoine in 1980 as a result of crossing. It has been used in culture since the mid-19th century.

The Iridaceae family, in addition to its decorative qualities, has a wide range of applications in the production of perfumes, soaps or powders. Thus, the feathery stigmas of Crocus sativa are used as a flavoring and food coloring agent, as well as as a medicinal ingredient called saffron.

Crocosmia × *crocosmiiflora* (Lemoine) N. E. Br. (montbretia; family Iridaceae, in the monocotyledonous Asparagales series), is used as an antitumor agent in Japanese traditional medicine. Extracts of its corms exhibit antitumor activity, the active substance is a mixture of saponins. Montbrezia extracts have antioxidant activity. Saponins and flavonoid glycosides are the main secondary metabolites of this plant [11].

The industry of ornamental flower plants is characterized by a great variety, today more species are cultivated than agricultural and garden crops combined. A significant share is accounted for by the production of flower plants for cutting. Consumer demand for cut flowers leads to the development of new plant species with different colors and new spectacular features, which is currently an important task [8, 12].

Due to the inherent decorative, attractive and simplicity of growing, *C. crocosmiiflora* is an interesting material that requires further study for the prospect of use in ornamental gardening. Therefore, the aim of our research was to study the growth and development of *Crocosmia* × *crocosmiiflora* (L.) N.E.Br. in the agrometeorological conditions of the southern zone of the Amur region for the prospects of use in landscaping.

The research objectives included: to study the phenological development of *Crocosmia* × *crocosmiiflora* depending on the weather conditions of the warm season; to evaluate the decorative features of *Crocosmia* × *crocosmiiflora* (L.) N.E.Br. for the possibility of entering into the list of plants used for landscaping the territories of the southern zone of the Amur region.

2 Materials and Methods

Study of crocosmia × crocosmiiflora (Lemoine) N.E.Br. it was carried out in accordance with the methodology. The studies included the fixation of phenological phases (budding, flowering: the beginning of flowering, mass flowering and end of flowering) according to the method of phenological observations by I. N. Beideman, biometric measurements, assessment of decorativeness during the growing season, the influence of weather conditions of the year on the decorativeness and development of plants.

Crocosmia × crocosmiiflora, a plant up to 100 cm tall. The stem is straight, thin, and branched. The leaves are light green, broadly linear or narrowly xiphoid, up to 2,5 cm wide and 40–60 cm long, erect, in dense bundles. The flowers are not large-3–5 cm in diameter, funnel-shaped, orange-red or yellow, collected in a small apical paniculate inflorescence. Flowering is abundant and long-lasting, summer-autumn-from July to September. During flowering, up to 10 flowers can be opened at the same time. From 1 corm grows 3–4 flower stalks. C. crocosmiiflora is an unpretentious plant that requires minimal care. Prefers open sunny places or partial shade. It can grow on any permeable soil, except for very poor ones. Crocosmia forms the babies that bloom the next year. Requires digging for the winter in regions with cold winters.

The climatic conditions of the Amur region are formed under the influence of the general circulation of the atmosphere, continental polar, marine polar, continental tropical and marine tropical air. Their interaction at different times of the year forms the features of the climate of this territory, which is defined as monsoon by the nature of formation and continental by temperature conditions.

Winter is severe, long-lasting with little precipitation, little snow cover, high insolation and low temperatures. The temperature of the coldest month of January, ranges from –24 °C to −46 °C, with an absolute minimum of −56 °C. The duration of the cold period is 130–160 days.

Spring is late, prolonged, often arid with strong winds, which often leads to loss of moisture in the upper layer of the soil and its desiccation. In March-May, 7–14% of annual precipitation falls, soil moisture is mainly determined by summer-autumn reserves, usually they are 60–70 mm of productive moisture in a layer of 0–20 cm. Warming of the soil to +5 °C to a depth of 10 cm in the southern zone occurs at the beginning of the second decade of April. In May and the first decade of June, there are large fluctuations in air temperature during the day from +25 °C during the day to 1 °C at night.

Summers are usually warm, with moderate dry weather in the first half and wet in the second. The greatest amount of precipitation in years with excessive moisture falls in July and August (up to 150–160 mm), in years with a lack of moisture, its deficit is most felt in June. At the beginning of growth and development, plants lack moisture, and in the second half of the growing season they suffer from its excess. The average temperature of July, the warmest month, ranges from 20 to 21 °C.

The first half of autumn is wet, warm, and the second half is dry. The average monthly positive temperatures remain until the end of the second or third decade of October. Precipitation falls 15–20% of the annual norm, mainly in September, at this time the amount of precipitation often exceeds 100 mm and can reach the level of 170–195 mm. According to the temperature regime, the fluctuations over the years are less significant.

The sum of active air temperatures above +10 °C in the southern zone is 2160–2300 °C. The duration of the frost-free period is up to 144 days. The duration of the frost-free period and the period of active vegetation are close to each other. Thus, sufficient heat, precipitation, and plenty of light during the warmer months favor the cultivation of flowering plants. At the same time, the lack of moisture in spring and early summer, slow warming of the soil, a large amount of precipitation in July and August, leading to waterlogging and compaction of the soil, a relatively short growing season have a negative impact on the formation of plants and the formation of full-fledged seeds.. The peculiarities of the agro-climatic conditions indicate that in the Amur region it is possible to grow crops, mainly with a short growing season. By year, meteorological indicators may differ significantly from the average annual data [14].

Weather conditions in the years of the research were characterized by the following features. Spring 2018 came two weeks earlier than usual, was unstable with sharp temperature changes and a lack of precipitation, low relative humidity on some days up to 30% and below. A steady temperature transition of +5 °C, which means the beginning of the growing season, occurred on April 10–17, which is two weeks earlier than the long-term dates. A steady transition of the average daily air temperature through +10 °C, which means the beginning of active vegetation, occurred in the first decade from 3 to 9 May, this is 1–4 days earlier than the average long-term dates. Over +15 °C, the temperatures passed in the third decade – May 27–28, which slightly deviated or was within the normal range, the timing allowed for planting the corm of the studied plant in the ground. The average temperature in the last decade was higher than the long-term values by 1–2 °C and was 15–16 °C, with an absolute maximum on some days up to 27–33 °C. The minimum temperatures dropped to −1–2 °C. Frosts in the air during the spring period of this year in most areas of the zone were late with a deviation from the usual dates by

3–16 days with an intensity of up to-1–3. The corms were planted in the ground when the threat of frost passed. Rain in May was not enough, they were distributed unevenly, had different intensity, sometimes with strong winds, thunderstorms and hail. In some areas of the southern zone, there were short-term rains for 4–5 days. Days with rain of 1 mm or more per month were from 4 to 7. In total, precipitation for the month was 12–41 mm, which is from 27 to 77% of the long-term norm. The duration of the sunshine was 85–135 h. The wind prevailed moderate 6–14 m/s, on some days with an increase to 15–23 m/s. Agrometeorological conditions at the end of May for the planting of C. crocosmiiflora were mostly favorable.

The summer period of 2018 was characterized by an unstable temperature regime, frequent rains, sometimes heavy, reaching the criterion of a dangerous phenomenon, and high relative humidity. A characteristic feature of this summer is the presence of dangerous phenomena with damage to agriculture – "waterlogging of the soil" during the periods of plant growth and harvesting, the fall of "large hail", flooding with water reaching the floodplain and exceeding the mark of an unfavorable hydrological phenomenon.

The average air temperature in the summer season was within the usual values and amounted to 19–20 °C.

During the three summer months during the formation of the vegetative mass and active growth of C. crocosmiiflora, the average daily air temperatures were below the average annual values in June, July, and August within the normal range, or higher by 1 °C. On some days, hot days with a maximum air temperature of 30 °C and above were observed for 1–7 days. The low temperatures in June were accompanied by intense precipitation, which was 1.3–2.2 times more than normal. In July and August, the amount of precipitation was also more than 121–239% of the norm. The rains were intense, with 25–38 rainy days recorded during the three summer months, with a large proportion occurring in June–July. So, in the second decade of June, an average of 6 days with precipitation were recorded in the zone. The total precipitation averaged 420 mm. The duration of the summer period was 93–101 days, the heat supply for three months was 1739–1959 °C, which is more than the long-term one by 170–190 degrees. As a result of the prevailing weather conditions of the growing season, the plants managed to increase their vegetative mass, form flower stalks, and go through the flowering phase.

Weather conditions in the warm period of 2019 were not typical for the southern zone of the Amur region. The weather was characterized by inconstancy and deviation in all indicators from long-term values. A steady temperature transition over +15 °C occurred 2–4 days later than usual. The duration of the warm period within the long-term values is 96–99 days. The heat supply was insufficient, the sum of the temperatures deviated in a smaller direction from the norm by 63–124 °C. The absolute maximum air temperature was 28–33 °C. Minimum temperatures on some days dropped to 1–9 °C. Precipitation fell frequently in the warm season, was of varying intensity, often had a local stormwater character, sometimes with hail. For three summer months, 29–38 days with rain were recorded, they fell 104–152% of the norm. July was particularly rainy, with 1,7–2,3 precipitation rates. August was also rainy, 8–16 days were recorded with precipitation, on some days very intense, during the month in the areas of the southern zone of the Amur region, there was 126–191 mm, which is 1,2–1,5 times higher than normal. The relative humidity was high at 69–90%. Agrometeorological conditions were extremely severe, in July and August, a dangerous agrometeorological phenomenon "waterlogging of the soil" was recorded, some territories were flooded.

Spring in 2020 was early and moderately warm, quite humid, with a lot of precipitation in May, there was for the season 41–131 mm. The agrometeorological conditions for planting were mostly favorable.

Summer has arrived in the southern regions of the region, 2–5 days earlier than the long-term dates. The average daily air temperature steadily passed through +15 °C on May 27. The weather was unstable, June and August were cold and rainy, July was hot and dry. Periods of cold spells were followed by intense warming. The second decade of July was unusually hot, the air warmed up to 32–38 °C during the day, and temperature records were recorded at many meteorological stations. A significant cooling with a decrease in the average daily air temperatures relative to the climatic norm by 3–5° was observed on August 19, 20. As a result, the average air temperature for three summer months was within the long-term norm and amounted to 16–20 °C. The duration of the summer period was 106–113 days, the heat supply was 1873 °C, which is slightly more than the norm by 47 °C.

In autumn, the steady transition of the average daily air temperature through +15 °C in the southern regions was carried out later than the long-term dates from 2 to 11 days – September 10–17. Thus, the duration of the summer period in the southern regions was longer by 4–13 days, the duration was 97–113 days. The average values of the sum of active temperatures +10 °C were 2478 °C, which is higher than the norm by 153 °C.

In general, the agrometeorological conditions in 2020 were not quite favorable for ornamental plants, which affected the timing and duration of flowering.

3 Results

Table 1. Flowering dates and duration of the Crocosmia × crocosmiiflora (Lemoine) phase N.E.Br.

Year of study	Flowering			Duration of flowering, days
	Beginning	Mass	End	
2018	24.07	30.07	12.10	81
2019	buds	buds	buds	45 buds
2020	20.08	31.08	10.10	52

Table 2. Biometric measurements.

Indicators	Year of research		
	2018	2019	2020
Plant height, cm	71 ± 5.1	57.5 ± 3.1	68 ± 3.3
Length of lower leaves, cm	29.6 ± 0.7	31 ± 0.7	14.5 ± 0.4
Width of the lower leaves, cm	1.5 ± 0.3	1.6 ± 0.2	0.3 ± 0.2
Flower stalk length, cm	54.4 ± 1.1	31 ± 0.7	21 ± 0.4
The length of the spike from the lower (bud) flower, cm	36.4 ± 0.9	12 ± 0.7	22 ± 0.7
Number (buds) of flowers in a spike, pcs	15 ± 2	15 ± 1	15 ± 1
Number of simultaneously opened colors, pcs	4 ± 0	–	4 ± 0
Flower diameter, cm	4.1 ± 0.2	–	4.4 ± 0.1

4 Discussion

In a harsh continental climate with a short growing season, ornamental plants of the "spring – summer-autumn" phenorhythmotype are of particular value. In the conditions of the south of the Amur region, this issue remains relevant and is implemented through the study of a different range of plants. Phenological studies are a prerequisite for achieving high productivity, quality, and decorative properties of plants [13, 15].

Crocosmia × crocosmiiflora corms (L.) N.E.Br. annually planted at the same time with a difference of one or two days. In 2018 and 2019, corms were planted in the open ground on May 30 and 31, respectively, in 2020-on June 1.

In 2018, the beginning of budding phase was recorded on July 10, mass-on July 16. In 2019, the beginning of the phase was celebrated on August 20, and the mass phase was celebrated on August 30. In 2020, July 25 and August 5, respectively. That is, the terms of development differed by year, and were shifted in 2019 by 40–50, in 2020 by 15–20 days compared to 2018.

During the research period, the beginning of flowering was recorded in 2018 on July 24, in 2020-on August 20, in 2019, the formed buds did not open until the end of the growing season. The end of flowering was recorded at the beginning of October in all terms of research, the dates coincide with the transition of temperatures through 5 °C. The duration of flowering was 81 days in 2018, 52 days in 2020, and in 2019, the unopened buds were kept on the plant for 45 days. Noting the phases of plant development, it is necessary to compare them with meteorological data, this gives an idea of the life cycle of the plant in different environmental conditions. The phenological development of plants was significantly influenced by the prevailing weather conditions. The conditions of 2018 were favorable for the development of Crocosmia × crocosmiiflora. This year, the flowering phase came earlier and was the longest. Adverse weather conditions in 2019 and 2020 negatively affected the development of plants, reducing the timing of the onset of phases and flowering. The duration of flowering of Crocosmia × crocosmiiflora (L) in 2020 was 29 days less than in 2018, and in 2019 the opening phase of the flowers was not observed, but the buds were quite well formed (Table 1).

The correlation dependence of the duration of flowering on the prevailing weather conditions of the growing seasons was calculated for three years of research. The calculation was carried out based on the average values of temperatures and precipitation. A strong correlation $r = 0.96$ on temperature and a strong negative dependence on precipitation $r = -0.75$ were found.

Crocosmia × crocosmiiflora plants in the conditions of the southern zone of the Amur region undergo a full cycle of development, except for the formation of seeds, but they form full-fledged corms, which are dug up and laid for storage when temperatures pass through 5 °C.

On the presentation of the completeness of the research and the description of the characteristics of decorativeness, biometric measurements were made of the height of plants; the length of the flower stalks from its exit from the leaf; the length of the flower stalk from the lower flower; the length and width of the leaf blade; the number of flowers in the spike, the number of simultaneously open flowers, the diameter of the flower.

Biometric measurements carried out during the research years showed that the plants reached the maximum height in a more favorable weather conditions in 2018-71 ± 5,1 cm, in 2019 and 2020 this indicator was on average 13,5 and 3 cm lower. Larger leaf blades formed on plants in 2019, which is shown when measuring the length and width of the leaves. The average indicators of the length of the flower stalk and the length of the spike from the lower (bud) flower were greater in 2018, in other years – less. The number (buds) of flowers in the spike in all years averaged 15 pieces, simultaneously open flowers-4, the diameter of the flower deviated slightly from the average value. In general, all the signs noted indicate that the plants are sufficiently decorative (Table 2).

Due to the rather long flowering time, the Crocosmia × crocosmiiflora plant can be used in landscaping areas for various purposes in the southern zone of the Amur region, in creating flowerbeds, mixed borders, rabattes. C. crocosmiiflora looks good in groups as a plant of the middle plan of landscape-style flower beds. Cut flower stalks during flowering can be stored for a long time, gradually drying out, preserving the decorative effect, which makes it possible to use C. crocosmiiflora in floristics not only in the warm, but also in the cold period of the year.

The assessment of the decorative value of plants is carried out during the period of mass flowering according to the stobal system with coefficients. The assessment is carried out differentially according to the most important decorative features: color of the inflorescence; flower stalk; size and shape of the inflorescence; abundance of flowering; bush (shape, decorativeness); resistance of inflorescences to adverse meteorological conditions; originality; condition of plants.

When assessing the decorative value of a plant, depending on the significance of the trait, each species has its own conversion factor. Each feature of decorativeness is evaluated within a five-point scale. In the future, the points (for each trait separately) are multiplied by the conversion factor (the degree of significance of the trait) and the result, which is the final assessment of the trait, is entered in the corresponding column of the decorative assessment card (Table 3).

Table 3. Assessment of the decorativeness of Crocosmia × crocosmiiflora (Lemoine) N.E.Br.

Attribute name	Evaluation of the attribute on a five-point system	Conversion factor depending on the significance of the attribute	Evaluation of a feature based on the one-hundred-point system
Decorative features			
Flower color and its stability (strength, brightness, spot)	5	3	15
Number of petals (density)	5	1	5
Flower size (large, medium, small)	5	2	10
Inflorescence (the number of flowers is dense, loose)	3	2	6
The number of simultaneously opened flowers	3	2	6
The quality of the flower stalk (strength, straightness)	5	2	10
Abundance of flowering	4	1	4
Plant condition	5	1	5
The originality of the view	5	2	10
Overall rating			71
Economically valuable features			
Duration of flowering	4	2	8
Drought resistance	3	1	3
Winter hardiness	0	1	0
Resistance to diseases and pests	5	1	5
Resistance to adverse weather conditions	4	2	8
Intensity of vegetative reproduction	5	1	5
Overall rating			29
Total			100

5 Conclusions

Thus, the primary study of Crocosmia × crocosmiiflora (L) allows us to conclude that the plant under the conditions of the southern zone of the Amur region under typical weather conditions, goes through a full development cycle, forms full-fledged corms, does not form seeds; weather factors of a sharply continental climate affect the external characteristics of plants; Crocosmia × crocosmiiflora (L) can be attributed to promising plants for use in various types of landscaping in the conditions of the region.

References

1. Chaveau, O., Eggers, L., Souza-Chies, T.T., Nadot, S.: Oil-producing flowers within the Iridoideae (Iridaceae): evolutionary trends in the flower of the New World genera. Ann. Bot. (Oxford) **110**, 713–729 (2012). https://doi.org/10.1093/aob/mcs134
2. Ascough, G.D., Erwin, J.E., van Staden, J.: Micropropagation of Iridaceae – a review. Plant Cell Tissue Organ Cult. **97**, 1–19 (2009). https://doi.org/10.1016/j.sajb.2010.08.001
3. Bahali, D.: Habitat of iridaceae in India. Bull. Natl. Inst. Ecol. **17** (2006)
4. Singab, A.N.B., et al.: Shedding the light on Iridaceae: ethnobotany, phytochemistry andbiological activity. Ind. Crops Prod. **92**, 308–335 (2016). https://doi.org/10.1016/j.indcrop.2016.07.040
5. Goldblatt, P., Rodriguez, A., Powell, M.P.: Iridaceae "out of Australasia"? Phylogeny, biogeography, and divergence time based on plastid DNA sequences. Syst. Bot. **33**, 495–508 (2008). https://doi.org/10.1600/036364408785679806
6. Goldblatt, P., Manning, J.C.: Systematics of the Southern African genus Ixia (Iridaceae). 1. The I. rapunculoides complex. Bothalia Afr. Biodivers. Conserv. **38**(1), 1–22 (2008). https://doi.org/10.4102/abc.v38i1.257
7. Ascough, G.D., Swart, P.A., Finnie, J.F., van Staden, J.: Micropropagation of *Romulea minutiflora, Sisyrinchium laxum* and *Tritonia gladiolaris* – Iridaceae with ornamental potential. South Afric J of Botany (2011). https://doi.org/10.1016/j.sajb.2010.08.001
8. van Kleunen, M., Johnson, S.D., Fischer, M.: Predicting naturalisation of southern African Iridaceae in other regions. J. Appl. Ecol. **44**, 594–603 (2007). https://doi.org/10.11111/j.1365-2664,2007.01304.x
9. Krupa-MałKiewicz, M., Żurawik, P.: In vitrorooting of Crocosmia × crocosmiiflora 'Lucifer.' Folia Hortic. **29**(1), 3–10 (2017). https://doi.org/10.1515/fhort-2017-0001
10. Krupa-MałKiewicz, M., Żurawik, P., Piórkowska, P.: Micropropagation of Crocosmia × crocosmiiflora 'Lucifer.' Folia Pomer. Univ. Technol. Stetin. **305**, 29–34 (2013)
11. Ascough, G.D., Erwin, J.E., van Staden, J.: Effectiveness of colchicine and oryzalin at inducing polyploidy in Watsonia lepida. Hortic Sci. **43**, 2248–2251 (2008). https://doi.org/10.21273/HORTSCI.43.7.2248
12. Lecomte, C., Alabouvette, C., Edel-Hermann, V., Robert, F., Steinberg, C.: Biological control of ornamental plant diseases caused by *Fusarium oxysporum*: a review. Biol. Control **10**, 17–30 (2016). https://doi.org/10.1016/j.biocontrol.2016.06.004
13. Cleland, E.E., Chiariello, N.R., Loarie, S.R., Mooney, H.A., Field, C.B.: Diverse responses of phenology to global changes in a grassland ecosystem. Proc. Natl. Acad. Sci. USA **103**(2006), 13740–13744 (2009). https://doi.org/10.1073/pnas.0600815103
14. Stokoz, S.V.: Nitrates in vegetables and potatoes of the southern zone of the Amur region.: dis. ... cand. biol. sci.:03.02.08: protected 26.10.2013: approved 17.02.14, p. 188 (2013)
15. Kim, Y.-S., et al.: Phenological growth stages of Korean ginseng (Panax ginseng) according to the extended BBCH scale. J. Ginseng Res. (2020). https://doi.org/10.1016/j.jgr.2020.12.006

Climate Warming and Changes in the Reproductive Capacity of Woody Plants

Kirill Tkachenko(✉), Gennady Firsov, and Alexandra Volchanskaya

Komarov Botanical Institute of RAS, Professor Popov str., 2, 197376 St. Petersburg, Russia
ktkachenko@binran.ru

Abstract. Critical or abnormally severe winters, which are the main factor in selection, have a great influence on the condition and life expectancy of plants. If the consequences of warm winters affect, first of all, the flowering and fruiting of plants, then after very cold winters there was a massive death and/or severe damage to many species of plants. The introduction of trees and shrubs in the Peter the Great Botanical Garden of Komarov Botanical Institute RAS took place against the background of much harsher winter and summer temperatures than at present. Changes in the reproductive sphere are also closely related to the freezing of plants (kidneys, shoots, sometimes and the entire aboveground part). The abundance of flowering and, accordingly, the yield of fruits and the quality of seeds can sharply decrease. At the beginning of the XXI century in St. Petersburg, the number of species of trees and shrubs that have reached reproductive status increased markedly. Against the background of ongoing climate warming, self-settling of a number of species in which it was not previously observed was found (*Carpinus betulus* L., *Chamaecyparis pisifera* (Siebold & Zucc.) Endl., *Cerasus maximowiczii* (Rupr.) Kom. & Aliss., *Spiraea betulifolia* Pall.), Which is an important indicator of adaptation and a sign of possible future naturalization of the species in this territory. At the same time, many species can become and become potentially invasive, which raises questions about their careful monitoring and culture in controlled conditions.

Keywords: Introduction · Botanical garden · Arboreal · Shrubs · Adaptation · Quality of fruits · Seeds · X-ray of reproductive organs

1 Introduction

St. Petersburg is located in a transition zone between regions with a temperate continental and maritime climate. This region is characterized by a high frequency of air masses of Atlantic origin, which determines the marine features of its climate. Active cyclonic activity and frequent changes in air masses determine the unstable nature of the weather in all seasons of the year, and cyclonic processes intensify in the cold half of the year. One of the features of the St. Petersburg climate is its high humidity. The relative humidity averages 80% per year; our city has long been called the "wet corner" of the country. Cloudy weather prevails throughout the year. As a result of cyclonic activity and the transfer of air masses from the Atlantic, cloudiness in this region is significant

A. Muratov and S. Ignateva (Eds.): AFE 2021, LNNS 353, pp. 573–580, 2022.
https://doi.org/10.1007/978-3-030-91402-8_64

throughout the year, and especially in the autumn-winter period. The absolute minimum air temperature of −35.6 °C was noted on January 17, 1940. The well-known "warming of the Arctic", which began at the end of the 19th century, peaked at the end of the 1930s and captured St. Petersburg. The climate of the second half of the XX century after the warming of the 1930s began to be regarded as "modern", with an average annual air temperature of 4.3 °C (Table 1). Then a conclusion was made about the stability of the probabilistic characteristics of the climate [1, 2].

Table 1. Changes in air temperature in St. Petersburg.

Month	Average monthly air temperature according to the St. Petersburg meteorological station, T (°C)			
	Climate rate in the XX century	Climate norm in 1980–2009	Climate norm in 2001–2009	Climate norm in 2011–2016
I	−7.7	−5.4 ± 0.7	−5.3	−6.3
II	−7.9	−5.8 ± 0.7	−5.9	−4.1
III	−4.2	−1.3 ± 0.5	−1.1	−0.6
IV	3.0	5.10.3	5.5	5.5
V	9.6	11.1 ± 0.3	11.8	12.9
VI	14.8	15.5 ± 0.4	15.4	16.7
VII	17.8	18.5 ± 0.3	20.1	19.7
VIII	16.0	16.8 ± 0.3	17.9	17.8
IX	10.8	11.5 ± 0.3	12.7	13.1
X	4.8	6.2 ± 0.3	6.2	6.2
XI	−0.5	0.0 ± 0.5	0.9	2.2
XII	−5.1	−3.6 ± 0.6	−3.4	−0.9
Year	4.3	5.8 ± 0.2	6.2	6.9

2 Object and Methods of Research

The collection of the arboretum of living plants of the Peter the Great Botanical Garden of the Komarov Botanical Institute of RAS is the main object of study. Additionally, we carried out observations of the plants of the Botanical Garden of the Forestry Academy. The work uses classic works on the study of phenology, rhythms of growth and development.

3 Research Results and Discussion

Since the beginning of scientific introductory activities in St. Petersburg, it has become obvious that the main restriction for the introduction of woody plants in open ground is their insufficient resistance to climate features in winter conditions. The experience of breeding wood exotic in these conditions for three centuries shows that critical or abnormally severe winters, which are the main factor in selection, have a great influence on the condition and life expectancy of plants. If the effects of warm, or "provocative," winters affect, first of all, the flowering and fruiting of plants, then after very cold winters there was a massive death and/or severe damage to many species of woody and shrub plants. The effects of severe winters have been observed for a number of years. In the second half of the 20th century, the climate of which was considered for a long time the "norm of the modern climate," the introduction of trees and shrubs in the Peter the Great Botanical Garden of Komarov Botanical Institute RAS took place against the background of much harsher winter and summer temperatures than at present. Many introduces plants were considered unpromising or completely unsuitable for cultivation in open ground.

For example, intensive introductory activities and work to restore the gardens and parks of Leningrad after the end of World War II were carried out against the background of much harsher and more unfavorable winter and summer temperatures for plants than now. This was the period of publication of the fundamental publication "Trees and Shrubs of the USSR" (1949–1962) and other dendrological works, on the basis of which winter resistance indicators were established for many years to come and the possibilities of culture here of exotic were determined.

Changes in the reproductive sphere are also closely related to the freezing of plants (kidneys, shoots, sometimes the entire aboveground part). Trees and shrubs cease to bloom and bear fruit, sometimes for many years and decades. The abundance of flowering and, accordingly, the yield of fruits and the quality of seeds can sharply decrease [3].

The famous "warming of the Arctic," which began at the end of the 19th century, reached a maximum in the late 1930s and captured St. Petersburg. The climate of the second half of the XX century after the warming of the 1930s began to be considered as "modern," with an average annual air temperature of 4.3 °C (Fig. 1).

The noticeable warming of the climate in St. Petersburg began in 1989, which became the warmest (7.6 °C) in history at that time (later it was surpassed by the year 2015: 7.7 °C), and the year 2020: 8.3 °C.

In the 21st century, climate warming has intensified noticeably after 2006. The winter of 2006/07 was a record short and lasted only 41 day, but autumn lasted almost 5 months.

It was very warm in the summer of 2010, at record high temperatures in July (24.4 °C). In the second decade of the XXI century, before the end of the calendar year, winter does not come in individual years – there is no snow cover, and some plants continue to grow.

There is an obvious tendency to increase heat availability when comparing the average annual air temperature. It increased by 2.5° the climate norm of the 20th century, which can be considered a very significant warming. The year 2020 became the warmest in the period of observations from the middle of the XVIII century: 8.3 °C. The minimum air temperature is noticeably warmer.

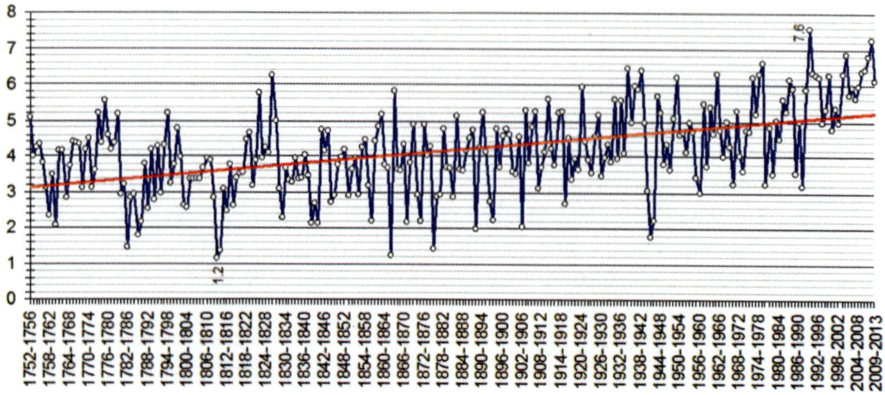

Fig. 1. The well-known "warming of the Arctic", which began in the late 19th century, peaked in the late 1930s and captured St. Petersburg. The climate of the second half of the XX century after the warming of the 1930s began to be regarded as "modern", with an average annual air temperature of 4.3 °C.

At the beginning of the 21st century, the number of species of trees and shrubs that have reached a reproductive state has noticeably increased in St. Petersburg. For a number of species, for the first time in a long period of introduction, seed offspring were obtained: *Acer japonicum* Thunb.; *Picea glehnii* (F. Schmidt) Mast.; *Fraxinus oxycarpa* M. Bieb. ex Willd.; *Fraxinus pojarkoviana* V. Vassil.; *Cydonia oblonga* Mill.; *Carpinus orientalis* Mill.; *Acer capillipes* Maxim.; *Pyrus zangezura* Maleev; *Aristolochia macrophylla* Lam. Some species show their first flowering, such as *Decaisnea fargesii* Franch. (synonym for *Decaisnea insignis* (Griff.) Hook.f. & Thomson).

Climate change towards warming makes it possible to grow more species from seeds, even local reproduction, and, ultimately, contribute to their acclimatization and the introduction of urban greening into practice [4].

Against the background of the ongoing warming of the climate, self-seeding of a number of species was found in which it had not previously been observed (*Carpinus betulus* L., *Chamaecyparis pisifera* (Siebold & Zucc.) Endl., *Cerasus maximowiczii* (Rupr.) Kom. & Aliss. [Synonym *Prunus maximowiczii* Rupr.], *Spiraea betulifolia* Pall. and many others), which is an important indicator of adaptation and a sign of possible future naturalization of the species in this area. At the same time, many species can become and are becoming potentially invasive, which raises questions about their careful monitoring and culture under controlled conditions [5].

Why there is such "surge" in reproductive abilities of plants?

– It is promoted by mitigation of winter temperatures. Increase in the minimum air temperature already led to the shift of limits of winter stability of wood plants, and borders of zones are more and stronger displaced to the North and the East. Increase in winter temperatures is especially important for those species of trees and bushes at which flower kidneys put in the summer of the year preceding blossoming. In the past at such types in cold winters they just froze.

- Considerable lengthening of a vegetative season promotes ripening of vegetative escapes and also fruits and seeds. In the past in adverse years, seeds even at species of local flora didn't grow ripe (as at *Acer platanoides* L. in 1976) [6].
- Increase in the frost-free period, especially is very important for early-flowering types, such as *Acer saccharinum* L. In the XIX and first half of the 20th century at this look only blossoming was noted, flowers were damaged by frosts, the first fructification is noted in 1980.
- For the first time in the conditions of St. Petersburg some began to fructify wood (including coniferous) breeds. But it turned out that the percent of the executed seeds doesn't exceed 5%. The executed seeds were viable.
- *Abies gracilis* Kom. synonym of *Abies sachalinensis* var. gracilis (Kom.) Farjon.

During the analysis of seeds, it turned out that part of them are struck with larvae of wreckers. Many executed seeds (4 and 5 classes) were struck by wreckers.

Larix gmelinii (Rupr.) Kuzen. (=*Larix dahurica* Turcz. ex Trautv.). The percent of the executed seeds isn't higher than 10%. Generally there were 0 and 1 classes, seeds of 3 and 4 classes there were no more than 5%. The executed seeds for 50% are struck by wreckers (Fig. 2).

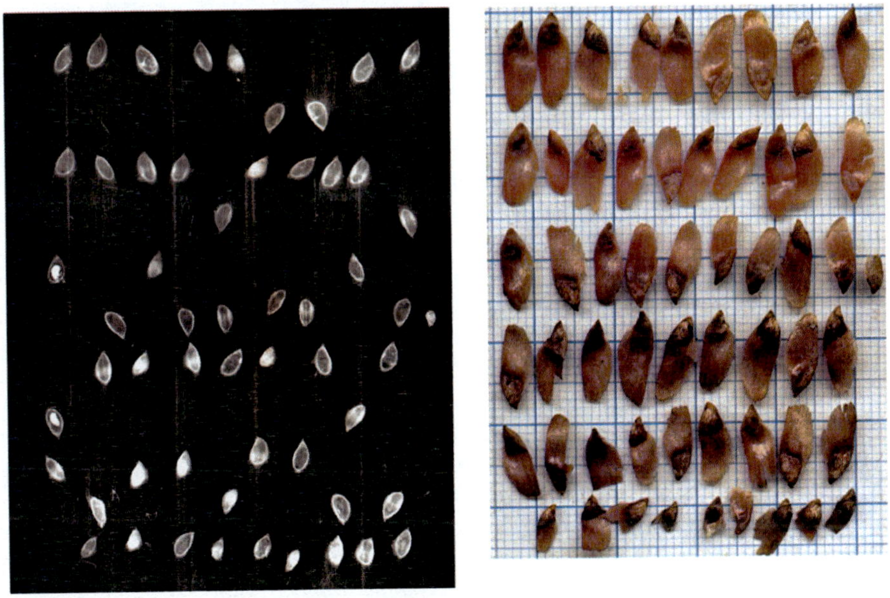

Fig. 2. Seeds of *Larix gmelinii* (Rupr.) Kuzen. (=*Larix dahurica* Turcz. Ex Trautv.).

The percentage of completed seeds is not more than 10%.

At some types the first blossoming as, for example, at *Decaisnea fargesii* Franch is noted Fig. 3). These are also *Acer capillipes* Maxim.; *Pyrus zangezura* Maleev and others.

It is possible that for insect-pollinated species higher air temperatures during flowering have a favorable effect on pollinating insects and on the possibility of pollination.

Fig. 3. Some species show their first flowering, such as *Decaisnea fargesii* Franch. Lardizabalaceae – dicotyledons, included in the order of Ranunculales.

This "surge" is also facilitated by a general increase in the species of cultivated dendroflora. If after the Great Patriotic War and in the 1980s the collection of the arboretum of the Peter the Great Botanic Garden totaled 600–800 species and forms, now it exceeds 1200 taxa. Climate warming significantly increases the ability to attract many heat-loving species of more southern latitudes to the culture.

Studying the features of the latent period, assessing the quality of reproductive diasporas for different plant species is important for the conservation of biodiversity and the development of methods for its conservation in botanical gardens [7].

As practice has shown, one of the most promising and absolutely non-destructive method is the method of x-ray of seeds. The use of the domestic X-ray apparatus "Elektronika-25" in the 60s–90s of the 20th century made it possible to make significant progress in the study of fruits and seeds of a number of plant species. The effectiveness of the use of fluoroscopy for assessing the quality of large fruits and seeds with dense covers (*Amygdalus*, *Cerasus*, *Hordeum*, *Malus*, *Phaseolus*, *Pinus*, *Prunus*, *Quercus*, *Triticum*, as well as *Helianthus annuus* L., *Cucumis sativus* L., *Cucurbita pepo* L.). The use of this method makes it possible to identify and select for sowing only full-fledged completed seeds (IV and V classes) with a normally developed embryo (reject empty and feeble seeds; 0 and I classes) without pest damage and annually monitor the quality of the resulting seeds [8].

With the further development and development of various modifications of microfocus X-ray devices, the transition to image digitization immediately after shooting objects, it became possible to analyze the quality of not only fruits and seeds, but also to evaluate the accretion of grafts. Using the example of the use of modernized X-ray analysis of seeds, it is shown that at present it is possible to quickly check the quality of reproductive diasporas (fruits and seeds) of both introduced and wild plants, to assess the degree of their fulfillment, to identify pest infestations and select from a batch of seeds only completed, full-fledged ones, which can be used for sowing or laying for storage (Fig. 4 a, b).

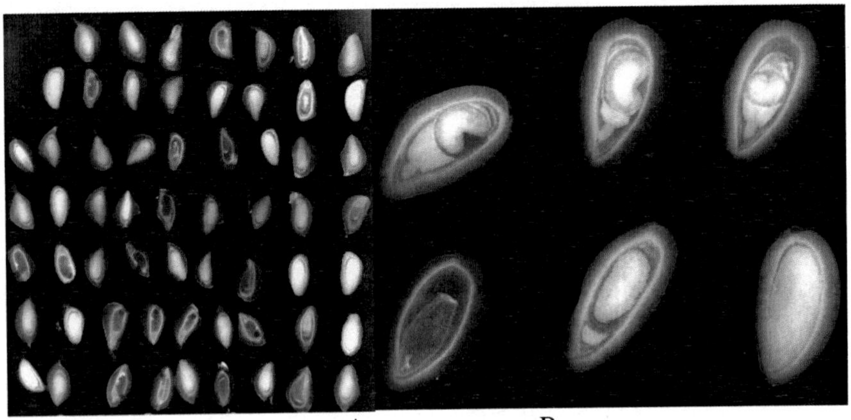
A B

Fig. 4. Nuts (seeds) *Rosa rugosa* Thunb. the larvae of insect pests are strongly affected (A). Pest larvae of varying degrees of development (B).

4 Conclusion

While we are talking about a small period of time. The most important of the surrounding factors acting on plant populations and leading to genetic adaptation will be climatic factor. Even minor climate changes lead to the formation of new populations. A special place is occupied by adaptation changes of plants in global changes in climatic conditions, covering completely or partially the territory occupied by the species. Shifts in climatic conditions do not directly lead to genetic changes in plants. Changes in the environment act through selection and appear only when generations change in populations.

There is a noticeable increase in the number of species and forms suitable for dendrological collections, the winter resistance of which allows them to grow in open ground (*Cryptomeria japonica, Podocarpus nivalis, Aucuba japonica*).

Climate change towards warming makes it possible to grow more species from seeds, even local reproductions, and ultimately contribute to their acclimatization and the introduction of urban landscaping.

In recent years, the Peter the Great Botanic Garden has significantly contributed to the replenishment of urban green spaces. Many new heat-loving species grown at the Garden nursery were transferred, for example, to the Dubki Park.

Acknowledgements. Work is performed within a state task on a planned subject "Collection of living plants of Botanical institute of V. L. Komarov (history, the current state, the prospects of use)", number AAAA-A18-118032890141 – 4.

References

1. Chernyavskikh, V.I.: Biological resources of Urtica dioica L.: areas of study and prospects of use. Biolog. Field J. **1**(3), 131–149 (2019). https://doi.org/10.18413/2658-3453-2019-1-3-131-149
2. Schiappacasse, F., Szigeti, J.C., Manzano, E., Kamenetsky, R.: Eremurus as a new cut flower crop in Aysen, Chile: Introduction from the Northern Hemisphere. Acta Hortic. **1002**, 115–122 (2013). https://doi.org/10.17660/ActaHortic.2013.1002.13
3. de Oliveira Gentil, D.F., do Nascimento Ferreira, S.A., Rebouças, E.R.: Germination of Psidium friedrichsthalianum (O. Berg) Nied. seeds under different temperature and storage conditions. J. Seed Sci. **40**(3), 246–252 (2018). https://doi.org/10.1590/2317-1545v40n3179617
4. Kerdyashkin, A.V., Kerdyashkina, G.V.: Classification of Schrenk spruce seeds by the degree of development of embryos and endosperms. Bull. St Petersburg For. Acad. **227**, 57–67 (2019). https://doi.org/10.21266/2079-4304.2019.227.57-67
5. Chernyavskikh, V.I., Dumacheva, E.V., Degtyar, O.V., Degtyar, A.V., Borodaeva, Z.: Analysis of the potential productivity of herbaceous vegetation of the gully-ravine complexes of the Belgorod region. Biolog. Field J. **1**, 55–63 (2019). https://doi.org/10.18413/2658-3453-2019-1-1-55-63
6. Flores, P.C., Poggi, D., García, S.M., Gariglio, N.F.: Topographic tetrazolium testing of Black Walnut (Juglans nigra L.). Seed Sci. Technol. **39**, 230–235 (2011). https://doi.org/10.15258/sst.2011.39.1.23
7. Pechenitsyn, V.P., Abdullaev, D.A., Akhmedzhanov, I.G.: Intrapopulation diversity of seeds of species of the genus Eremurus (Xanthorroeaceae) of the flora of Uzbekistan. Plant Res. **55**(4), 1–11 (2019). https://doi.org/10.1134/S0033994619040083
8. Tkachenko, K., Staroverov, N., Gryaznov, A.: X-ray quality control of fruits and seeds. Hortus Bot. (2018). https://doi.org/10.15393/j4.art.2018.5022

The History of the Introduction of Species of the Genus *Rosa* to St. Petersburg, Russia

Kirill Tkachenko(✉) and Alla Kapelian

Komarov Botanical Institute of RAS, ul Prof. Popova 2, 197376 St. Petersburg, Russia

Abstract. Species, forms and varieties of the genus Rosa (rose or wild rose) are known ornamental plants that are used as essential oil and medicinal plants. Interest in this group of plants began many centuries ago. The beginning of their rapid introduction into gardens and parks began in the 18th century. Over the past 100–200 years, the number of varieties that are beginning to be used in urban landscaping has been constantly increasing. Rosaries are created - as a special technique in landscape gardening art. In the new, XXI century, there is a new rise in interest in the revival of mass plantings of roses (and species of rose hips) both in parks and for landscape solutions for decorative design of urban spaces. The species that are of the greatest practical interest (decorative during the flowering period and autumn coloring of leaves, winter-, drought-, pollutant resistant) – *Rosa amblyotis* C.A. Mey, *R. altaica* Willd., *R. corymbifera* Borkh., *R. davurica* Pall., *R. gallica* L., *R. sweginzowii* Koehne. and, of course, *R. rugosa* Thunb., even though in recent years this species has been classified as invasive in a number of European countries.

Keywords: Gardens · Parks · Botanical gardens · Landscaping · History of gardening

1 Introduction

Roses as cult, ornamental and medicinal plants from ancient times accompany man. The history of their cultivation and introduction into culture dates back to the heyday of Assyrian, Egyptian, Greek and Roman cultures. The genus Rosa L. in the understanding of various authors has from 200 to 400 species, and a very large number of varieties and hybrids (according to some authors, there are at least about 18,000 of them) [1–9].

2 Materials and Methods

Collection of species and varieties of the genus Rose, grown in the Botanical Garden. Analysis of a significant number of published works on the introduction of species of this genus in the North-West of Russia, as well as available historical and archival materials.

A. Muratov and S. Ignateva (Eds.): AFE 2021, LNNS 353, pp. 581–588, 2022.
https://doi.org/10.1007/978-3-030-91402-8_65

3 Results

As a park rose, this is perhaps one of the main plants, starting from the 16th century, but it is quite possible earlier. The most famous roses are park plants in France. First of all, these are French (or Old French roses, varieties and hybrids from the Gallica or French rose R. gallica - Hybrid Gallica) and the rose variety Centifolia. Speaking of the first two main species, one cannot fail to mention such widely known and popular groups of old garden roses (Old Garden Roses), which include bourbon (Hybrid Bourbon), Damask (Damask), moss (Moos), and repairable (Hybrid Perpetual) appeared in the 19 centuries. From modern garden roses (Modern Roses), which have been introduced to decorate the park, it is necessary to highlight braided large-flowered (Large flowered Clamber), semi-flightless (Shrub) and polyantha (Polyantha). But the current rosaries are simply not conceivable without the inclusion of tea-hybrid, floribunda-roses (Floribunda), grandiflora (Grandiflora), miniature (Miniature), and real braided roses, or ramblers (Rambler). The modern variety assortment for all groups of roses has at least 1500–2500 (but at the same time only in the catalogs of the American society of rose gardens of the order of 3500 varieties!!!) Names that can be easily ordered and purchased in different farms of Russia and Europe, America and China.

Rosary (by Rosary means a special exposition of the most valuable and stable species and varieties isolated from collection plantations at the previous stage of introductory study, and designed taking into account the characteristic features of garden and park architecture. The task of the Rosary is to promote the best zoned varieties for this climatic zone, and at the same time demonstrate the techniques and methods of using roses in garden and park construction and individual gardening) – as an independent unit of the park was formed for several centuries (mainly in the period from the 16th to the 18th, but the heyday falls at the end of the 17th and beginning of the 18th centuries, when the rosary itself turned into a special dedicated garden in the park, as is done in the Malmaison garden in Paris; or "Italian Rose Garden" in Mainau, Germany; or rosary in the Boulogne Forest, as well as in Lyon and Orleans; in Germany in Baden-Baden, Frankfurt, Dortmund, Hamburg; in England in St. Albans [where about 450 varieties are located on an area of almost 3 hectares]), simultaneously with the transformations of both the gardens themselves and the parks, as such, into garden and park complexes. At this time (XVIII century), trellis, pergolas, gazebos dotted with roses appear in the parks. Stem roses, bosquets from cut species and hybrid R. gallica L., R. × damascena Herrm., R. rugosa Thunb., R. majalis Herrm. and their varieties with different colors of petals and a degree of rusticity. It is important to immediately emphasize here - in the conditions of North-West Russia, for better flowering, these plants need to be sheltered for the winter, to preserve the kidneys on young annual shoots, on which flower buds are formed that open for the next growing season.

It is important to note that varieties and hybrids of Chinese rose (R. chinensis Jacq.; giving the name of the group "tea-hybrid," which came from the English pronunciation of the word China [tʃaɪ.nə] – China, and not from the "invented" that "the smell of these roses resembles tea"), brought to Europe only in 1789. Groups of modern garden roses are tea-hybrid (Hybrid Tea), floribunda (Floribunda), miniature (Miniature) until 1948 in the conditions of St. Petersburg were grown only in closed soil, they were preserved in cold greenhouses, contained in tall and narrow ceramic pots (vases), which for the summer (only in the 50s of the twentieth century did roses begin to be grown in open ground, partially maintaining the practice of growing roses in pots with the preservation of cold greenhouses in a non-freezing basement).

Parks of the XVII–XVIII centuries included in the range used more often species of roses, which are commonly called rosehips, or widely known and popular Gallic or French and/or Proven and centifolia roses that have a single flowering. The magnificence of these varieties is reflected in drawings, on numerous figurines, vases, plates, services of the Saxon or Meissen (Germany), Sevres (France), St. Petersburg (Russia) and other porcelain factories of different countries.

For the conditions of St. Petersburg, wrinkled rosehips (R. rugosa) are very suitable as the most resistant to the climatic features of the region, having several flowering waves, quite diverse varieties and hybrids with simple, semi-and terry flowers of white, pink, red flowers.

Rose Damascus R. damascena and Rosa French R. gallica in the climate of St. Petersburg can be significantly damaged in winter by severe frosts (not covered shoots), which leads to a loss of their decorativeness for several years (until they grow and form new shoots with flower buds).

Other types of rosehips of the flora of Russia can be used in parks. This is cinnamon rosehips R. cinnamomea L. (according to the modern classification, has the synonymous name R. majalis), R. majalis, R. dahurica Pall., R. multiflora Thunb. and some other species. These plants can be planted in groups, as well as in solitary (single specimen set). But the most spectacular roses look precisely in Rosarium, where everything is focused on the beauty of these plants, in order to enjoy their charming aroma during their flowering on a warm sunny day.

Building the new Russian capital St. Petersburg, Peter I was also engaged in designing a beautiful garden, taking the most active part in the selection of the assortment. Having inherited an interest in gardening from his father, he had previously set up gardens at the palace in Preobrazhensky, where foreign flower crops grew – tulips, lilies, narcissists and others. Here, in an honorable place, the "serroborin color" grew – rosehips. It was from the Moscow Region that plants were originally sent for the new capital. Later, Peter I began to plant trees, shrubs and flowers from other countries. The nobles followed the king's example and surrounded their palaces with magnificent gardens.

Fashion played a great role in shaping the assortment. In the first half of the 18th century, Buxus sp. and roses (rosans) were in vogue. They had to be protected from winter frosts. The Dutchman gardener Fok began to cover them with a spruce lapel, horns, felt. Roses bred in Europe were brought to St. Petersburg. These were varieties of Western European selection, which are now referred to as old garden roses (Old Garden Roses) belonging to the groups of French, or Gallic (Gallica), Damascena (Damascena), Centifolia (Centifolia), moss (Moos – Rosa centifolia f. muscosa (Aiton) C.K.Schneid.), alba (Rosa×alba L.), or white (Alba) roses.

In addition to vernacular gardens, house gardens and palisades began to appear in St. Petersburg. There were several trees, shrubs, flower beds with roses, a gazebo with curly plants.

In the 19th century, with the involvement of Chinese tea roses (Tea) in the selection, new garden groups of roses appeared in Europe, which also began to be grown in the gardens of St. Petersburg. These were ancient garden roses belonging to the Bourbon, Portland, repairman (Hybrid Perpetual) groups. Many of them have the ability to flower. These roses required shelter for the winter. Those who did not need shelter were from the white rose group of Maiden's Blush variety, from the yellow rose group Rosa foetida f. persiana hort. = Persian Yellow.

Public gardens appeared in St. Petersburg in the second half of the 19th century, and with them the need for unpretentious roses that do not require much care and winter shelter, in the so-called park roses, i.e., roses that can winter in natural conditions [10].

And first of all, wild roses can be attributed to them – rosehips, their forms and hybrids. In the second half of the 19th century, the following were grown in parks: Rosa acicularis Lindl. (needle rose), Rosa cinnamomea (cinnamon rose) and its varieties having terry flowers, R. pimpinellifolia L. (R. spinosissima L.), as well as a fruit plant – R. pomifera Herrm.

The center of introduction of new species has always been botanical gardens. The Imperial Botanical Garden, which grew out of the Pharmacy Garden by 1823, played a large role in introducing new types of roses into the "green outfit" of the city. He introduced R. albertii Regel, flowering in white flowers, and which has become the main species in the landscaping of the city of R. rugosa.

Wrinkled rosehips (R. rugosa), unlike other species, blooms on the shoots of the current year, which makes it suitable for haircuts without losing flowering, while such a feature leads to the duration of flowering until autumn.

Another important biological feature of R. rugosa has now been discovered. In addition to air pollution in St. Petersburg, plants are influenced by salt used to clean sidewalks and roadways from snow and ice. This dirty snow-salt mixture is dumped on shrubs along the roads. The wrinkled rose withstands such salinization. Obviously, it has resistance to salinization of the soil, since it grows in its homeland along the sea coast. It is noted that this species is pollutant-resistant, and cleanses urban air and land, from heavy metal pollution [11, 12].

In the second half of the 19th century, the Pomological Garden, organized in 1861 by E. Regel, director of the Imperial Botanical Garden, and J. Kesselring, became important for the introduction and further distribution of plants in St. Petersburg. The garden was engaged in the cultivation and introduction into the culture of fruit, ornamental and perennial plants, new varieties and valuable species. In the Pomological Garden and acclimatization nurseries, plants were grown not only for introductory purposes, but also for sale.

Of the seeds sent by Karl Maksimovich from Japan, in the nursery of Regel and Kesselring, the Rusa rugosa purpurea plena (Fig. 1), called E. Regel "Queen of the North," appeared. It turned out that R. rugosa is able to cross with other species and varieties of roses. Subsequently, magnificent varieties were obtained from it, forming the Hybrid Rugosa group, which have terry flowers of different colors, and new winter-resistant groups of garden roses – R. × Kordessii and Canadian roses.

Thanks to the invention and the introduction into practice of air-dry shelter of heat-loving groups of plants to protect them from frost, made prof. N.I. Kichunov, roses appeared in the parks and urban landings of Leningrad (St-Petersburg). But in the future, due to the laboriousness of work, a change in funding, almost all rose planting in the city

Fig. 1. Rusa rugosa Thunb. forma purpurea plena, cv. “Queen of the North”.

Fig. 2. Thickets of R. rugosa Thunb. of the Gulf of Finland.

was removed. However, in recent years, in a number of areas of the city, rose planting begins to revive (restore) (at the expense of the district budget, and with the active participation of the district offices of the green building).

Currently, interest in garden roses is being revived as an element of landscaping of the city. For mass planting, the most winter-resistant groups of modern garden roses are used - soil-blooded, Canadian, less often floribunda.

Huge areas along roads, avenues and routes, again occupy planting R. rugosa. To create bright color spots, it is used in different forms, having flowers of various colors from white to purple. But along the shores of the Gulf of Finland, this species has become invasive (Fig. 2 and 3) [13–17].

Fig. 3. Thickets of R. rugosa Thunb. of the Gulf of Finland. Fruiting time.

Despite the long experience of introducing roses in St. Petersburg in the parks and gardens of the city, the assortment of rosehips is not large, mainly these are those species that have long been used in the landscaping of R. canina L., R. glauca Pourr., R. rugosa, Rosa spinosissima L.

In the historical parks of Pavlovsk and Peterhof, collections of ancient garden roses have been restored, where they occupy their worthy place.

4 Conclusion

However, there are species that have been growing in the Botanical Garden collection for more than 30 years, and according to the complex of signs they can be recommended for wide use in landscaping: Rosa amblyotis C.A. Mey, R. altaica Wild., R. corymbifera Borkh., R. davurica Pall., R. sweginzowii Koehne.

French and Damascus roses, in the conditions of St. Petersburg, are less wintering. The main limiting reason for the introduction of a wider range of roses in the conditions of St. Petersburg is cold and raw springtime and summers, interspersed late-spring and early autumn frosts, winter-spring provocative thaws, leading to the development of bacterial and fungal diseases, significant freezing of the kidneys after their awakening, and damage by insect pests.

Based on many years of experience in growing roses (as well as rosehips) in the Botanical Garden of Peter the Great BIN RAS, park roses in St. Petersburg, in addition to species rose-rosehips and their hybrids, can be used for urban landscaping. Currently, a promising range can be made up of garden roses of the alba group, centifolia, moss, Portland. And, of course, planting these plants in the city will greatly decorate the modern urban landscape.

Acknowledgements. Work is performed within a state task on a planned subject "Collection of living plants of Botanical institute of V.L. Komarov (history, the current state, the prospects of use)", number AAAA-A18-118032890141 – 4.

References

1. Basim, E., Basim, H.: Antibacterial activity of *Rosa damascena* essential oil. Fitoterapia **74**, 394–396 (2003)
2. Cairns, T.: Rose collections and trials I rose collections and herbaria. In Encyclopedia of Rose Science, pp. 664–681 (2003). https://doi.org/10.1016/b0-12-227620-5/00086-0
3. Joyaux, F.: History of roses in cultivation I European (Pre-1800). Encyclopedia of Rose Science, pp. 395–402 (2003). https://doi.org/10.1016/b0-12-227620-5/00047-1
4. Marriott, M.: History of roses in cultivation I Morden (Post -1800). In: Encyclopedia of Rose Science, pp. 395–402 (2003). https://doi.org/10.1016/b0-12-227620-5/00047-1
5. Van de Pol, P.A.: History of roses in cultivation I history of the perfume industry. In: Encyclopedia of Rose Science, pp. 410–414. https://doi.org/10.1016/b0-12-227620-5/00116-6
6. Kovacheva, N., Rusanov, K., Atanassov, I.: Industrial cultivation of oil bearing rose and rose oil production in Bulgaria during 21st century, directions and challenges. Biotechnol. Biotechnol. Equip. **24**, 1793–1798 (2010)
7. Baydar, N.G., Baydar, H.: Phenolic compounds, antiradical activity and antioxidant capacity of oil-bearing rose (*Rosa damascena* Mill.) extracts. Ind. Crop. Prod. **41**, 375–380 (2013)
8. Olsson, A., Nybom, H., Prentice, H.C.: Relationships between nordic dogroses (*Rosa* L. Sect. Caninae, Rosaceae) assessed by RAPDs and elliptic Fourier analysis of leaflet shape. Syst. Bot. **25**(3), 511 (2000). https://doi.org/10.2307/2666693
9. Shabbir, F., Hanif, M.A., Ayub, M.A., Jilani, M.I., Rahman, S.: Damask rose. Med. Plants South Asia, 217–230 (2020). https://doi.org/10.1016/b978-0-08-102659-5.00017-3
10. Calzoni, G.L., Antognoni, F., Pari, E., Fonti, P., Gnes, A., Speranza, A.: Active biomonitoring of heavy metal pollution using *Rosa rugosa* plants. Env Poll **149**(2), 239–245 (2007). https://doi.org/10.1016/j.envpol.2006.12.023
11. Ahsan, M., et al.: Heavy metal accumulation imparts structural differences in fragrant *Rosa* species irrigated with marginal quality water. Ecotox and Envir Saf **159**, 240–248 (2018). https://doi.org/10.1016/j.ecoenv.2018.05.003
12. Bruun, H.H.: *Rosa rugosa* Thunb. ex Murray. J. Ecol. **93**(2), 441–470 (2005). https://doi.org/10.1111/j.1365-2745.2005.01002.x
13. Isermann, M.: Expansion of *Rosa rugosa* and *Hippophaë rhamnoides* in coastal grey dunes: Effects at different spatial scales. Flora Morph. Distr. Funct. Ecol. Plants **203**(4), 273–280 (2008). https://doi.org/10.1016/j.flora.2007.03.009
14. Jørgensen, R.H., Kollmann, J.: Invasion of coastal dunes by the alien shrub *Rosa rugosa* is associated with roads, tracks and houses. Flora Morph. Distr. Funct. Ecol. Plants **204**(4), 289–297 (2009). https://doi.org/10.1016/j.flora.2008.03.002

15. Muñoz-Vallés, S., Cambrollé, J.: The threat of native-invasive plant species to biodiversity conservation in coastal dunes. Ecol. Eng. **79**, 32–34 (2015). https://doi.org/10.1016/j.ecoleng.2015.03.002
16. Stefanowicz, A.M., Zubek, S., Stanek, M., et al.: Invasion of *Rosa rugosa* induced changes in soil nutrients and microbial communities of coastal sand dunes. Sci. Total Env. (2019). https://doi.org/10.1016/j.scitotenv.2019.04.408
17. Petersburg gardens. History and modernity. SPb: SPb MEDIA (2011)

Comparative Analysis of Dendroflora of Parks in Blagoveschensk

Natalia Yust(✉), Natalia Timchenko, and Olesia Shcherbakova

Far Eastern State Agrarian University, St. Polytechnic 86, Blagoveshchensk, Amur Region, Russia

Abstract. Since the dendroflora is a living organism that is in a constant life process under the influence of the technogenic load of cities, it is necessary to conduct a systematic examination of its condition in order to identify defects in development and their timely elimination, because after death, the process of restoration of dendroflora takes decades, which can adversely affect health of the city population. It is important to carry out ecological and biological observations of woody and shrub plants in a timely manner and track the dynamics of change, for further competent selection of measures for the conservation and restoration of trees and shrubs in parks. Four parks were investigated on the territory of Blagoveshchensk: City Park of Culture and Leisure, Pervomaisky Park, Park Druzhby and 40 Years of the Komsomol Park, a list of dendroflora of parks in Blagoveshchensk was compiled, a taxonomic analysis of the dendroflora was carried out, the ecological-cenotic belonging and ecological-geographical confinement of the identified species, as well as 6 species (13%) included in the Red Book of the Amur Region. Thus, the use of species in the Red Data Book in landscaping parks is a fairly effective way to preserve rare and endangered species.

Keywords: Landscaping · Tree and shrub vegetation · Park · State · Green zone

1 Introduction

Historically, dendrological parks and botanical gardens represent one of the oldest scientific and practical information networks and are one of the most important, unique repository of the planet's genetic resources [1–5].

Research carried out to study the dendroflora of city parks is of interest to scientists, since the issues of improving the quality of life of the population of cities through the ecologization and eco-restoration of the habitat by improving the greening system of cities are relevant. The study of large objects of urban greening was carried out by Russian and foreign scientists [6–8].

Purpose of the work: to analyze the dendroflora of the parks of the city of Blagoveshchensk.

A. Muratov and S. Ignateva (Eds.): AFE 2021, LNNS 353, pp. 589–601, 2022.
https://doi.org/10.1007/978-3-030-91402-8_66

Objectives of the work: – to compile a list of dendroflora of parks in Blagoveshchensk, – to carry out a taxonomic analysis of dendroflora in parks in Blagoveshchensk, – to determine the ecological-cenotic affiliation and ecological-geographical confinement of the identified species; to identify rare and in need of protection plant species.

In the late 1980s, Blagoveshchensk became a large trade, economic, administrative and cultural center of the Amur region. The peculiarity of Blagoveshchensk is that it is buried in greenery. Blagoveshchensk is located in the southern part of the Amur Region and covers an area of 321 km^2. The city is located on the banks of the Amur River, at the confluence of the Zeya River, on the border with China.

The forests of the region are floristically diverse. The main forest-forming species are Mongolian oak, white birch, black birch, Scots pine and David's aspen. The rest of the species (Amur velvet, ash-leaved maple, small-leaved elm) are not significant in the composition of the plantations.

Of the total area covered by forest vegetation (as of 01.01.2018), coniferous stands account for 64.1%, soft – leaved stands for 24.2%, hard – leaved stands for 2.0%, and shrubs for 9.7% [8–10].

2 Materials and Methods of Research

Ecological and biological observations of woody, shrub plants were carried out during the growing seasons in the city of Blagoveshchensk in the period from 2018–2020. The assessment of the vital state of tree plantations was determined by two indicators according to the weighted average score of the plantation state and by the index of the vital state of the plantation using a 5-point scale E.H. Andreeva. Parks were used as test plots, where a detailed inventory of trees was carried out by the method of continuous enumeration and the characteristics of each plant. The route method was used to identify the species composition of plants [11–13].

Routes within the park were chosen relative to the location of the road and path network. The identification of species was carried out on the basis of samples of the collected photoherbarium according to reference guides. Ecological-cenotic groups and geographical elements are indicated according to the works of G. A. Peshkova, V. M. Starchenko. In floristic complexes and ecological-coenotic groups are combined species that gravitate in their ecological-coenotic characteristics and the nature of distribution to homogeneous in the botanical-geographical relation natural-territorial complexes or their individual elements. Life forms are given according to the summary of A. B. Bezdelev and T. A. Bezdeleva [14, 15].

3 Results

Within the city of Blagoveshchensk, four parks were investigated: the City Park of Culture and Leisure, Pervomaisky Park, Park Druzhby and 40 Years of the Komsomol Park.

Table 1. Species composition of trees and shrubs.

No.	Plant name	Life form	Ecological-cenotic group	Geographic element
	2	3	4	5
1	*Acer ginnala* Maxim. – Riverine maple	12	FO-NM	EA
2	*A. negundo* L. – Ash-leaved maple	3	HM, AD-RU	C, AD
3	*Betula davurica* Pall. – Daurian birch	3	FO-NM	EA
4	*B. platyphylla* Sukacz. – Flat-leaved birch	3	FO-NM	CA
5	*Lonicera edulis* Turcz. ex Freyn – Edible honeysuckle	13	FO-LC, NM	EA, NA
6	*Viburnum sargentii* Koehne – Kalina Sargent	12	FO-NM	EA
7	*Swida alba* (L.) Opiz – White svidina	12	FO-NM	EuA
8	*Caragana arborescens* Lam. – Tree caragana	12, 13, 4	FO-CT	C, AD
9	*Quercus mongolica* Fisch. ex Ledeb. – Mongolian oak	3	FO-NM	EA
10	*Ribes* s. p. – Currant	13	FO-NM	EA
11	*Juglans mandshurica* Maxim. – Manchurian walnut	3	FO-NM	EA
12	*Syringa amurensis* Rupr. – Amur lilac	4	FO-NM	EA
13	*S. vulgaris* L. – Common lilac	12	FO, FO-CT	C, AD
14	*Picea obovata* Ledeb. – Siberian pine	1	FO-DC	NA
15	*Pinus sylvestris* L. – Scots pine	1	FO-LC	EuA
16	*Cerasus glandulosa* (Thunb.) Loisel.– Glandular cherry	13, 22	FO-NM	EA
17	*Crataegus dahurica* Koehneet Schneid. – Hawthorn daurian	12, 4	FO-NM	EA
18	*C. pinnatifida* Bunge – Hawthorn pinnate	12, 4	FO-NM	EA
19	*Malus baccata* (L.) Borkh. – Berry apple	4, 12	FO-NM	EA, SS
20	*Padus asiatica* Kom. – Asian bird cherry	3	FO-NM	EA
21	*Dasiphora fruticosa* (L.) Rydb. – Kuril shrub tea	13, 17	AM-MM	CA

(*continued*)

Table 1. *(continued)*

No.	Plant name	Life form	Ecological-cenotic group	Geographic element
22	*Physocarpus amurensis* (Maxim.) Maxim. – Amur bubble	13	FO-ST	EA
23	*Rosa davurica* Pall. – Daurian rosehip	13	FO	NA, EA
24	*Sorbaria sorbifolia* (L.) A.Br. – Rowan-leaved fieldberry	12, 13, 22	FO-NM	NA, EA
25	*Sorbus amurensis* Koehne – Rowan Amur	4	FO-NM	EA
26	*Populus balsamifera* L. – Balsam poplar	3	FO	C, AD
27	*Ulmus japonica (Rehd.) Sarg.* – Ilm dolinny	3	FO-NM	EA
28	*Ulmus pumila* L. – Small-leaved Ilm	3	FO-ST	CA, EA
29	*Vitis amurensis* Rupr. – Amur grapes	21	FO-NM	EA

1. Part of the city at the intersection of Komsomolskaya and Lenin streets with an exit to the Amur embankment.

Initially, aboriginal plant species were used for landscaping the park, now, as a result of research, it has been determined that the species composition of the tree and shrub vegetation of the park includes 29 species, of which 25 species belong to the representatives of the native flora and 4 adventive species (Table 1).

Ecological-cenotic groups: FO – species of the forest complex without a clearly expressed confinement to a certain belt; FO-LC – light coniferous forest; FO-DC – dark coniferous forest; FO-NM – nemoral (preboreal); FO-ST – forest-steppe; AD-RU – adventive-ruderal.

Geographic element: EuA – Eurasian extratropical; EA – East Asian; NA – North Asian; CA, EA – species with an area covering North and East Asia; CA – common Asian; SS – South Siberian; CA, EA – species with the main range in East and Central Asia; AD – adventive; C – cultivated plants.

The leading family in terms of the number of species is Rosaceae (10 species), which is explained by the widespread use of representatives of this family in the landscaping of settlements (Fig. 1).

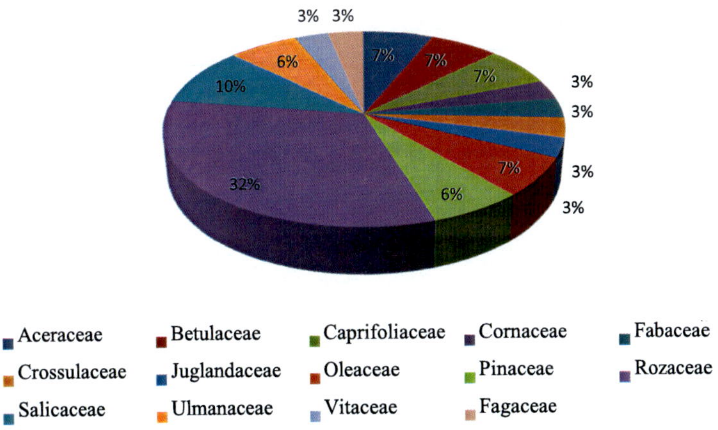

Fig. 1. Composition of tree and shrub plantations in the city park of culture and recreation.

However, the tree plantations of the park are dominated by a representative of the Salicaceae family, Populus balsamifera. In addition, the park's tree and shrub species includes 2 species included in the Red Data Book of the Amur Region: Juglans mandshurica and Vitis amurensis.

Summer-green shrubs (48.2%) and summer-green trees (41.04%) account for the largest share of tree and shrub vegetation, which is associated with the southern location of the city itself and favorable climatic conditions. The largest number of species is represented by forest species – 86.2%. Among the adventive species, the forest complex includes A. negundo L., Caragana arborescens, S. vulgaris L., Populus balsamifera L.

Geographic analysis of tree and shrub vegetation showed that the composition includes 9 geographical elements. The largest number of species belongs to the East Asian element (51.7%).

2. Pervomaisky Park of Culture and Leisure is located at st. Krasnoflotskaya, 2. Pervomaisky Park was first mentioned at the beginning of the 20th century. The area occupied by Pervomaisky Park is 7.6 ha. The species composition of the tree and shrub vegetation of the park includes 35 species, of which 31 species belong to the representatives of the native flora and 4 adventive species (Table 2).

Table 2. Species composition of trees and shrubs.

No.	Plant name	Life form	Ecological-cenotic group	Geographic element
1	2	3	4	5
1	*Maackia amurensis* – Maakia Amurskaya	3	FO-NM	EA
2	*Lespedeza bicolor* – Lespedeza Bicolor	3	FO-NM	EA

(*continued*)

Table 2. (*continued*)

No.	Plant name	Life form	Ecological-cenotic group	Geographic element
3	*Acer ginnala* Maxim. – Riverine maple	12	FO-NM	EA
4	*A. negundo L.* – Ash-leaved maple	3	FO, AD-RU	C, AD
5	*Betula davurica* Pall. – Daurian birch	3	FO-NM	EA
6	*B. platyphylla* Sukacz. – Flat-leaved birch	3	FO-NM	CA
7	*Alnus hirsuta* – Alder fluffy	3	FO-LC	NA, EA
8	*Corylus heterophylla* – Hazel-leaved	3	FO-NM	EA
9	*Lonicera tatarica L.* – Honeysuckle Tatarskaya	13	FO	AD
10	*Lonicera chrysantha* – Honeysuckle Golden	13	FO-NM	EA
11	*Sambucus nigra L.* – Elderberry Black	13	FO-EuA	C, AD
12	*Phellodendron amurense* – Velvet Amur	3	FO-NM	EA
13	*Quercus mongolica Fisch. ex Ledeb.* – Mongolian	3	FO-NM	EA
14	*Berberis amurensis* – Barberry Amur	13	FO-NM	EA
15	*Juglans mandshurica Maxim.* – Manchurian walnut	3	FO-NM	EA
16	*S. vulgaris L.* – Common lilac	12	FO, FO-ST	C, AD
17	*Fraxinus mandshurica* – Manchurian ash	3	FO-NM	EA
28	*Pinus koraiensis L.* – Korean pine	1	FO-DC	EA
19	*Pinus sylvestris L.* – Scots pine	1	FO-LC	EuA
20	*Larix gmelinii* – Daurian larch	3	FO-LC	EuA

(*continued*)

Table 2. (*continued*)

No.	Plant name	Life form	Ecological-cenotic group	Geographic element
21	*C. pinnatifida Bunge* – Hawthorn pinnate	12, 4	FO-NM	EA
22	*Malus baccata (L.) Borkh.* – Berry apple	4, 12	FO-NM	EA, SS
23	*Pyrus ussuriensis* – Ussuri pear	4, 12	FO-NM	EA
24	*Padus asiatica Kom.* – Asian bird cherry	3	FO-NM	EA
25	*Prunus maackii* – Bird cherry Maak	3	FO-NM	EA
26	*Rosa davurica Pall.* – Daurian rosehip	13	FO	EA, NA
28	*Tilia amurensis* – Linden Amurskaya	3	FO-NM	EA
28	*Populus suaveolens* – Poplar fragrant	3	FO-NM	NA, EA
29	*Populus maximowiczii* – Poplar Maximovich	3	FO-NM	EA
30	*Populus tremula L.* – Poplar Trembling	3	FO-NM	EuA
31	*Salix caprea L.* – Willow goat	3	FO-NM	LC, EuA
32	*Salix rorida* – Willow Rosistaya	3	FO-NM	CA
33	*Ulmus japonica (Rehd.) Sarg.* – Ilm dolinny	3	FO-NM	EA
34	*Ulmus pumila L.* – Small-leaved Ilm	3	FO-ST	CA, EA
35	*Philadelphus tenuifolius* – Chubushnik Thin-leaved	21	FO-ST	CA, EA

Ecological-cenotic groups: FO – species of the forest complex without a clearly expressed confinement to a certain belt; FO-LC – light coniferous forest; FO-DC – dark coniferous forest; FO-NM – nemoral (preboreal); FO-ST – forest-steppe; AD-RU – adventive-ruderal.

Geographic element: EuA – Eurasian extratropical; EA – East Asian; NA – North Asian; NA, EA – species with an area covering North and East Asia; CA – common

Asian; SS – South Siberian; CA, EA – species with the main range in East and Central Asia; AD – adventive; C – cultivated plants.

In the dendroflora of the park, the leading family in terms of the number of species is Rosaceae (6 species), which is explained by the widespread use of representatives of this family in landscaping settlements.

However, the tree plantations of the park are dominated by a representative of the Fagaceae family – Quercus mongolica Fisch. ex Ledeb. In addition, the tree and shrub species of the park include 7 species included in the Red Data Book of the Amur Region: Juglans mandshurica, Maackia amurensis, Pyrus ussuriensis, Phellodendron amurense, Fraxinus mandshurica, Philadelphus tenuifolius, Pinus koraiensis L.

Arboreal and shrub vegetation is represented by a variety of life forms from evergreen trees to summer-green dwarf shrubs. The largest share falls on summer-green shrubs (22.6%) and summer-green trees (68.6%), which is associated with the southern location of the city itself and favorable climatic conditions.

Ecological-cenotic analysis showed that the largest number of species is represented by forest species – 68.6%. Among the adventive species, the forest complex includes S. vulgaris L., Sambucus nigra L., Lonicera tatarica L., A. negundo L.

Geographic analysis of tree and shrub vegetation showed that the composition includes 9 geographical elements. The largest number of species belongs to the East Asian element (65.7%).

3. Park Druzhby of Blagoveshchensk is located in the second microdistrict between the street. Kalinin-Dyachenko and Novotroitskoe highway. Park Druzhby was created in the 90s, at the time of the revival of Russian-Chinese relations.

The species composition of the tree and shrub vegetation of the park includes 20 species, of which 17 species belong to the representatives of the native flora and 3 adventive species (Table 3).

Table 3. Species composition of trees and shrubs.

No.	Plant name	Life form	Ecological-ecenotic group	Geographic element
	2	3	4	5
1	*Acer ginnala* Maxim. –Riverine maple	12	FO-NM	EA
2	*Salix schwerinii* – Iva Schwerina	3	FO-NM	CA
3	*Populus balsamifera L.* – Balsam poplar	3	FO-NA	C, AD
4	*B. platyphylla* Sukacz. – Flat-leaved birch	3	FO-NM	CA

(*continued*)

Table 3. (*continued*)

No.	Plant name	Life form	Ecological-ecenotic group	Geographic element
4	*Betula davurica* Pall. – Daurian birch	3	FO-NM	EA
6	*Viburnum sargentii* Koehne – Kalina Sargent	12	FO-NM	EA
7	*Swida alba* (L.) Opiz – White svidina	12	FO-NM	EuA
8	*Caragana arborescens Lam.* – Tree caragana	12, 13, 4	FO-ST	C, AD
9	*Juglans mandshurica* Maxim. – Manchurian walnut	3	FO-NM	EA
10	*Syringa amurensis* Rupr. – Amur lilac	4	FO-NM	EA
11	*S. vulgaris* L. – Common lilac	12	FO, FO-ST	C, AD
12	*Picea obovata* Ledeb. – Siberian pine	1	FO-LC	NA
13	*Pinus sylvestris* L. – Scots pine	1	FO-LC	EuA
14	*Cerasus glandulosa* (Thunb.) Loisel.– Glandular cherry	13, 22	FO-NM	EA
15	*Malus baccata* (L.) Borkh. – Berry apple	4, 12	FO-NM	EA, SS
16	*Padus asiatica* Kom. – Asian bird cherry	3	FO-NM	EA
17	*Prunus Ps* –Almond	3, 14	FO-ST	CA, EA
18	*Sorbus amurensis* Koehne – Rowan Amur	4	FO-NM	EA
19	*Ulmus japonica* (Rehd.) Sarg. – Ilm dolinny	3	FO-NM	EA
20	*Ulmus pumila* L. – Small-leaved Ilm	3	FO-ST	CA, EA

Ecological-cenotic groups: FO – species of the forest complex without a clearly expressed confinement to a certain belt; FO-LC – light coniferous forest; FO-DC – dark coniferous forest; FO-NM – nemoral (preboreal); FO-ST – forest-steppe; AD-RU – adventive-ruderal.

Geographic element: EuA – Eurasian extratropical; EA – East Asian; NA – North Asian; CA, EA – species with an area covering North and East Asia; CA – common Asian; SS – South Siberian; CA, EA – species with the main range in East and Central Asia; AD – adventive; C – cultivated plants.

In the dendroflora of the park, the leading family in terms of the number of species is Rosaceae (4 species), which is explained by the widespread use of representatives of this family in landscaping settlements.

However, the tree plantations of the park are dominated by a representative of the Salicaceae family, Populus balsamifera. In addition, the tree and shrub species of the park includes 1 species included in the Red Data Book of the Amur Region: Juglans mandshurica.

Vegetation is represented by a variety of life forms from evergreen trees to summer-green dwarf shrubs. The largest share falls on summer-green shrubs (45%) and summer-green trees (55%), which is due to the southern location of the city itself and favorable climatic conditions.The largest number of species is represented by forest species – 65%. Among the adventive species, the forest complex includes Caragana arborescens Lam., S. vulgaris L., Populus balsamifera L.

The composition includes 9 geographic elements. The largest number of species belongs to the East Asian element (55%).

4. Park 40 years of the Komsomol. The species composition of tree and shrub vegetation in the park includes 13 species, of which 12 species are representatives of the native flora and 1 is an adventive species (Table 4).

Table 4. Species composition of trees and shrubs.

No.	Plant name	Life form	Ecological-cenotic group	Geographic element
1	2	3	4	5
1	*Acer ginnala* Maxim. – Riverine maple	12	FO-NM	EA
2	*Betula davurica* Pall. – Daurian birch	3	FO-NM	EA
3	*B. platyphylla* Sukacz. – Flat-leaved birch	3	FO-NM	CA
4	*Swida alba* (L.) Opiz – White svidina	12	FO-NM	EuA
5	*Pinus sylvestris* L. – Scots pine	1	FO-LC	EuA
6	*Cerasus glandulosa* (Thunb.) Loisel.– Glandular cherry	13, 22	FO-NM	EA

(*continued*)

Table 4. (*continued*)

No.	Plant name	Life form	Ecological-cenotic group	Geographic element
7	*Malus baccata* (L.) Borkh. – Berry apple	4, 12	FO-NM	EA, SS
8	*Padus asiatica* Kom. – Asian bird cherry	3	FO-NM	EA
9	*Rosa davurica* Pall. – Daurian rosehip	13	FO	EA, NA
10	*Populus balsamifera* L. – Balsam poplar	3	FO	C, AD
11	*Salix schwerinii* – Iva Schwerina	3	FO–NM	NA
12	*Ulmus japonica* (Rehd.) Sarg. – Ilm dolinny	3	FO-NM	EA
13	*Ulmus pumila* L. – Small-leaved Ilm	3	FO-ST	CA,EA

Ecological-cenotic groups: FO – species of the forest complex without a clearly expressed confinement to a certain belt; FO-LC – light coniferous forest; FO-DC – dark coniferous forest; FO-NM – nemoral (preboreal); FO-ST – forest-steppe; AD-RU – adventive-ruderal.

Geographic element: EuA – Eurasian extratropical; EA – East Asian; NA – North Asian; CA, EA – species with an area covering North and East Asia; CA – common Asian; SS – South Siberian; CA, EA – species with the main range in East and Central Asia; AD – adventive; C – cultivated plants.

The leading family in terms of the number of species is Rosaceae (4 species), which is explained by the widespread use of representatives of this family in the landscaping of settlements.

4 Discussion

According to the research results, the dendroflora of parks in Blagoveshchensk has 45 species and consists of representatives of the native flora (40) and adventive species (5).

In the dendroflora of parks in Blagoveshchensk, the first three places are occupied by the families Rosovye (13 species), Honeysuckle (5), Birch (4) and Sosnovye (4). The adventive dendroflora of parks contains only 4 families: two species of Honeysuckle and one species each Maple, Olive and Leguminous.

The dendroflora of parks in Blagoveshchensk contains a variety of life forms from evergreen trees to summer-green dwarf shrubs. Summer-green trees were most often noted (16 species/35.6%). Such a composition of life forms is determined by the southern position of the city and the dominant forest nemoral type of vegetation.

The dendroflora of the parks of the city of Blagoveshchensk presents types of forest, steppe, mountain complexes. Ecological-cenotic analysis shows the overwhelming superiority of forest (31 species/68.9%), primarily forest nemoral species.

This ratio confirms that before human development, nemoral forests were widespread on the territory of Blagoveshchensk. The presence of steppe species is associated with climatic conditions (low precipitation) and the proximity of the Manchu steppes on the border territory of the PRC southeast of Blagoveshchensk.

Ecological-cenotic analysis of adventive species of dendroflora demonstrates the predominance of forest species, for which the conditions of Blagoveshchensk create rather comfortable growing conditions.

5 Conclusions

As a result of research in the dendroflora of parks, 6 species (13%), included in the Red Book of the Amur Region, were noted. There are five types of declining (vulnerable) status:

- Amur velvet (Phellodendron amurense Rupr.), A species with declining numbers. The only corknose in Russia of industrial importance, an ornamental, medicinal, melliferous plant (2);
- Manchurian walnut (Juglans mandshurica Maxim.), A species whose numbers are declining as a result of changes in conditions of existence or destruction of habitats. Valuable food, medicinal, ornamental tree (2a);
- Manchurian Ash (Fraxinus mandshurica Rupr.), A species that is declining in numbers as a result of overuse by humans and can be stabilized by special conservation measures. Decorative view with valuable wood (2b);
- Amur grape (Vitis amurensis Rupr.), A species that is declining in numbers as a result of overuse by humans and can be stabilized by special conservation measures. Food, decorative and medicinal plant valuable for breeding (2c);
- Amur Maakia (Maackia amurensis Rupr. Et Maxim.), A species whose numbers are decreasing as a result of changes in conditions of existence or destruction of habitats. The only native tree of the legume family. Ornamental, melliferous, medicinal plant.

The rare status has one type:

- Ussuri pear (Pyrus ussuriensis Maxim.), A rare species in the northwestern limit of distribution. A very valuable plant for breeding cultivars of pears.

Thus, the use of species from the Red Data Book in the landscaping of city parks is undoubtedly an effective and promising way to preserve rare and endangered species.

References

1. Dorofeeva, V.D., Degtyareva, S.I., Komarova, O.V., Shipilova, V.F.: Generative and vegetative reproduction of pseudotsuga menziesii (mirb.) franco in the central chernozem region. In: IOP Conference Series: Earth and Environmental Science. International Scientific and Practical Conference. Forest ecosystems as global resource of the biosphere: calls, threats, solutions (Forestry-2019), p. 012044 (2019)
2. Popova, A., Popova, V.: Change of Quercus robur L. Acorns seedlings cytogenetic parameters under anthropogenic impact. In: IOP Conference Series: Earth and Environmental Science. Cep. International Forestry Forum Forest Ecosystems as Global Resource of the Biosphere: Calls, Threats, Solutions, p. 012047 (2020)
3. Sjöman, H.: Busse A (2010) Nielsen selecting trees for urban paved sites in Scandinavia – a review of information on stress tolerance and its relation to the requirements of tree planners. Urban For. Urban Greening **9**(4), 281–293 (2010). https://doi.org/10.1016/j.ufug.2010.04.001
4. Deutscher, J., Kupec, P., Futter, M.: Ecohydrological consequences of tree removal in an urban park evaluated using open data, free software and a minimalist measuring campaign. Sci. Total Environ. **655**, 1495–1504 (2019). https://doi.org/10.1016/j.scitotenv.2018.11.277
5. Li, Y., Wang, S., Chen, Q.: Potential of thirteen urban greening plants to capture particulate matter on leaf surfaces across three levels of ambient atmospheric pollution. Int. J. Environ. Res. Public Health **16**(3), E402 (2019). https://doi.org/10.3390/ijerph16030402
6. Dyderski, M.K., Pawlik, L.: Spatial distribution of tree species in mountain national parks depends on geomorphology and climate. Forest Ecol. Manage. **474**, 118366 (2020). https://doi.org/10.1016/j.foreco.2020.118366
7. Dyderski, M.K., Pawlik, L.: Drivers of forest aboveground biomass and its increments in the Tatra Mountains after 15 years. CATENA **205**, 105468 (2021). https://doi.org/10.1016/j.catena.2021.105468
8. Romanova, N.A., Zhirnov, A.B., Yust, N.A., Fucheng, X.: Influence of forest growth conditions on the density of wood in the Amur region. Cent. Eur. For. J. **65**(1), 41–50 (2019). https://doi.org/10.2478/forj-2019-0005
9. Senchik, A.V., Kukharenko, N.S., Sosnovsky, I.E.: Characteristics of winter stations of Siberian roe deer (capreolus pygargus pall) on the Zeysko-Bureinskaya plain. Izvestiya Nizhnevolzhskogo grouniversitetskogo kompleksa: Nauka i vysshego professional education **1**(53), 189–196 (2019). https://doi.org/10.32786/2071-9485-2019-01-25/
10. Romanova, N.A., Zhirnov, A.B., Yust, N.A., Fucheng, X.: Impact of forest growth conditions on the wood density: the case of Amur region. Folia Forestalia Pol. Ser. A **60**(4), 292–298 (2018)
11. Talovskaya (Kolegova), E.B.: Thymus baicalensis (Lamiaceae) morphological transformation under different environmental conditions. Contemp. Probl. Ecol. **8**(5), 607–613 (2015)
12. Turner, T.: Open Space Planning in London, pp. 366–386. Strategic Planning Advice for London, London (1988)
13. Ibriyev, D.M.-E.: Groznenskiy dendrariy – vchera, segodnya, zavtra. Vestnik nauki i obrazovaniya **7–1**(85), 28–34 (2020). https://doi.org/10.24411/2312-8089-2020-10703
14. Kalaev, V.N., Popova, A.A.: Cytogenetic polymorphism of english oak (Quercus robur L.). Seedlings from areas with different levels of anthropogenic pollution. Silvae Genet. **63**(6), 245–252 (2014)
15. Galychyn, O., Üstundağ, K.: Organic Urbanism: Human-oriented Design for Metropolises. Procedia Environ. Sci. **37**, 396–407 (2017)

Fortified Sour-Milk Beverages with the Use of the Far Eastern Region's Wild Berries

Elena Zakipnaya(✉) and Svetlana Parfyonova

Far Eastern State Agrarian University, 86, Politeknicheskaya Str., Blagoveschensk, Russia

Abstract. The article is devoted to the enrichment of sour-milk beverages, using wild-growing raw materials from the Far Eastern region, for nutrition of various age groups of the population. The possibility and feasibility of introducing wrinkled rosehip syrup into the milk base of a fermented milk drink has been established. The possibility of preparing a fermented milk drink with a predetermined optimal dose of the herbal component has been confirmed. The influence of the plant component on the organoleptic, physicochemical indicators, as well as the improvement of the functional and technological properties of the fermented milk drink has been revealed. The production of sour-milk beverages using wild-growing rose hips, which have bioenergetic and healing properties, will maintain a high level of the body's work, increase the body's resistance to harmful environmental factors.

Keywords: Rosehip · Sour-milk · Organoleptic assessment · Food products

1 The Introducing Part

One of the most important ways to develop the range of food products can be considered the creation of balanced products in nutritional and biological value, which can provide interest in nutrients and energy.

Human activity depends on consuming good nutrition. The nutrition is not the last social problem at all times. There is a whole system of measures that are aimed at protecting people from the effects of nutritional deficiencies. The government's scientific and technical nutrition policy has the aim to strengthen the health of the people [1]. The importance of functional nutrition is becoming more and more serious, the use of food products that saturate the physiological needs for nutrients and energy, which have a beneficial effect on the system of the whole body, have a beneficial effect on human health, on his well-being in general, and also helps to reduce the risk of development diseases [2].

A functional product is a food product that improves and maintains health, reduces the risk of developing diseases connected with nutrition, because of having physiologically functional food ingredients in its composition, and it is intended for use by all age groups of the population. One of the conditions for the effectiveness of functional nutrition is their systematic use in the composition of food rations and the scientifically grounded and proven properties aimed at reducing the risk of developing diseases associated with

A. Muratov and S. Ignateva (Eds.): AFE 2021, LNNS 353, pp. 602–610, 2022.
https://doi.org/10.1007/978-3-030-91402-8_67

nutrition. The ability of functional nutrition to influence the functions of organs and body systems is due to the presence of functional ingredients in them, the purpose of which is to prevent or replenish nutritional deficiencies [3].

Despite many of activities carried out in this direction, the Russian market for functional food products is rather weak, and the market for ingredients for product enrichment is even less developed. In recent years, the production of fermented milk products has been increasing at the fastest pace in the dairy industry. The dairy industry plays an important role in solving the problem of improving the health of the nation, occupying a special place among other sectors of the food industry.

Nutrition should help to protect the human's health from adverse conditions of environment. Sour-milk beveragesplay an important role in dietary and medical nutrition, as they contain nutrients and biologically active substances. It is important that sour-milk beveragesare considered to be a valuable product by their nature, they have excellent taste and nutritional properties, and are also medicinal. Therefore, if you add such drinks to your diet, you get proteins, carbohydrates and vitamins, improve digestion, and increase immunity.

The beneficial properties can make the basis for a strategy for promoting a product in the food industry market and distinguishing it from competing brands. The functional dairy and beverages market is a dynamic segment where innovation is the key to success. The problem of optimal provision of the population with enriched products in modern conditions turns out to be insoluble by traditional methods. It requires qualitatively new approaches. A reliable way that guarantees an effective solution to this problem is adding foods enriched with substances that are necessary for the physiological needs of the human body to the diet.

Currently, in the development and production of the beneficial properties with a high biological value, the naturalness of the added components, in particular berries, is of great importance. Many technologies are known with the addition of Far Eastern plant raw materials, such as: sea buckthorn, cranberries, Manchurian aralia, Chinese magnolia vine, etc., rose hips. Other Far Eastern plants are used in enriched products, but there is very little information of its use in sour-milk beverages.

Rosehip is considered to be a very good multivitamin and mineral supplement in the diet, the content of vitamin C in it is 17 mg/100 g of the product, it is 10 times more than black currant, lemons by 40 or more, and apples 100–200 times. In terms of chemical composition and nutritional value, rose hips surpass all wild fruits and berries up to 7.1 mg/100 g. Rose hips are among those unique berries that have a beneficial effect on the body. [4, 5]. The aim of the study is to enrich sour-milk beverages using wild berries from the Far East region, and to assess their functional and technological properties.

In accordance with theobjective, the following tasks were solved in the course of the study: study of the effect of the amount of herbal components added to the formulation and substantiation of the optimal amount for use insour-milk beverages; assessment of organoleptic indicators of the quality of finishedsour-milk beverages.

2 The Materials and the Methods

The main raw materials for the production of the fortificated sour-milk beverages were:

- The milk taken in accordance with GOST 31449-2013 raw cow's milk. Technical conditions.
- Vivo yoghurt leaven. The ingredients: freeze-dried strains: Streptococcusthermophilus, Lactobacillusdelbrueckiissp. bulgaricus, Lactobacillusacidophilus, Bifidobacteriumlactis, lactose.The number of live bacteria in 1.0 g of the product is not less than 1 × 109 CFU.Manufacturer: LLC "VIVO Industry", 121087, Russia, Moscow, Beregovoyproezd, house No. 7, building 1, office 1.
- flavoring filler – wrinkled rosehip syrup, which is a viscous liquid of dark brown color, has a pleasant, sweet and sour taste. Nutritional value 100 ml. product: carbohydrates – 84 g; energy value (caloric content) – 1410 kJ/336 kcal.

Doing the work, the generally accepted, standard researching methods of physical, chemical and microbiological indicators of raw materials and finished products were used:

Sampling for laboratory studies of samples of fermented milk products was carried out according to a single method in accordance with the requirements of GOST R 52969.

- determination of the mass fraction of fat – according to GOST R ISO 2446-2011;
- the mass fraction of protein in the product is determined according to GOST 25179-2014;
- determination of the mass fraction of dry substances – according to GOST R 54761-2011.
- Recommended methods for sensoryassessment.The increase in acidity in the process of obtaining a fermented milk clot was determined according to GOST 3624-92 "Milk and dairy products. Titrimetric methods for determining acidity"by titration using the phenolphthalein indicator.

The organoleptic assessment of finished products was carried out on the basis of GOST R ISO 22935-2-2011. "Milk and dairy products. organoleptic analysis. Part 2." The following indicators were investigated: smell, taste, texture, appearance and color. These characteristics must comply with the requirements of the standard.

Methods for the quantitative determination of the chemical composition of substances based on measuring their physical properties are called physicochemical methods of analysis. All these measurements are associated with the use of appropriate instruments and therefore they are often called instrumental methods of analysis.

To determine the temperature, a TU U 33.2-24667973-001-2001 thermometer was used. To determine the active acidity pH, a portable pH meter HI 8314 was used, manufactured by HANNA instrument for the Russian Federation (State Register SN No. 14300-99). The viscosity was determined using an SV-10 viscometer.

The energy value of the product was calculated from the data of the chemical composition using the coefficients for proteins — 4.1; lipids — 9; carbohydrates — 3.75.

3 The Results of the Study

The main goal of the research stage is to study the complex of indicators of model products and determine the maximum doses for adding rosehip syrup to the sour-milk beverages' recipe. The objects of research were samples of sour-milk beverages, developed with the addition of sourdough; samples of sour-milk beverages with different percentages of rosehip syrup. In Russia, in the Far East, eleven species of rose hips have been identified, but only five of them can reasonably be considered sources of vitamin C. In the Amur Region, three types of rose hips are most common:

1. prickly rose
2. rosarugosa
3. rosadavurica

For the study, the fruits of two types of rose hips were selected: the fruits of the prickly rose (Rosaacicularis) and the fruits of the rosarugosa (Rosarugosa). For the study of physicochemical indicators, including biologically active substances, the fruits of the prickly rose and rosarugosawere collected at the stage of their technological maturity. When using the fruit, they were given the following characteristic:

- the fruits of prickly rose are narrowly oblong, ovate or pear-shaped, bright red, drooping, with a constriction at the top, fruit length about 2 cm (up to 3), up to 1.5–2.5 cm in diameter. The fruit's walls are solid,the outer surface is sparkle, smooth.
- the fruits of rosarugosa are oblate-spherical, up to 4 cm in diameter, fleshy, bright orange-red color. Unlike other common types of rose hips, its fruits are not oblong, but like a ball flattened from the poles. The color of ripe fruits is from light orange to bright red; sepals are preserved on ripest fruits. The walls of the fruit are solid, the outer surface is sparkle, less frequently matte, smooth. There are long, very stiff bristly hairs inside the fruit.

a) b) c)

Fig. 1. Rosehip of the Far East region, general view. a) rosarugosa, b) prickly rose, c) rosadavurica.

As a result of the comparative characteristics of rose hips, rosarugosahas valuable vitamins – the fruits are fleshy, large, high yield of about 3–4 kg per bush (Fig. 1).

Fig. 2. Rosehip of the Far East region, sectional view. a) rosarugosa, b) prickly rose, c) rosadavurica.

As a result of the research, it was found that the vitamin C in the rosehip berry contains 650 mg per 100 g, which includes the daily food consumption for an adult by 7.2 times, about 10 times more than in black currant berries and 50 times more than in lemon.

At the first stage, for the enrichment of the sour-milk beverages, the raw material of the Far Eastern region of wild berries – with bioenergetic and healing properties, rosarugosa syrup was used as an enriching component. The basis of the sour-milk beverages will be milk prepared with the use of starter microorganisms. It will be free of preservatives and dyes, and you can enrich its taste with natural berries or syrups. As an enriching component, wild-growing rosarugosa's berries were used, which have bioenergetic and healing properties. Dry bacterial starter cultures of the VIVO brand are one of the best starter cultures for the preparation of sour-milk beverages and they are very popular now in Russia.

VIVO starter cultures are living bacteria for making homemade sour-milk beverages that have probiotic properties. For the research, models of the base product were compiled with a regulating factor – the type and amount of ingredients. Research step $\Delta = 5$ wt%. The developed sour-milk beverages have an increased content of carbohydrates due to a decrease in the content of proteins and fats. Due to the adding of rosehip syrup, there was an increase in carbohydrates in the product.

Base product models are shown in the Table 1.

Table 1. Base product models.

Variants	Milk-base	Rosehipsyrup
Control	100	–
Sample 1	90	10
Sample 2	85	15
Sample 3	80	20

Table 2 shows the base product models. As an additive to the milk base, we add rosehip syrup, which has undergone a special heat treatment. Three samples of the finished milk base were prepared: the first sample was added 10% rosehip syrup, the second 15% and the third 20% rosehip syrup.

When developing a new product, the organoleptic properties of the product were studied: flavor, color, smell, consistency, which are shown in Table 2.

Table 2. Organoleptic characteristics of the base product with rosehip syrup.

Variants	Organolepticindicators			
	Color	Flavor	Smell	Consistency
Control	Milky white, uniform throughout the mass	Pure fermented milk, without tastes and odors	Fermentedmilk	Moderately homogeneous viscous liquid, with an undisturbed clot
Sample 1	Has a light beige tint	Sweet, sour	Pleasantlyacid	Homogeneous, a bit viscous
Sample 2	Richbeige	Sweet flavor, aroma of added rosehip component, pleasantly acid	Pleasantly acidis absent, strong smell of the added component	Homogeneous, moderatelyviscous
Sample 3	darkbeigeз	Verysweetflavor	Sour-sweet	Moderately viscous, the presence of particles of the added component

The analysis of Table 2 has shown that taking into account the increase in the dosage of rosehip syrup from 10 to 20% by weight, the sour-milk beverages had a color from white to dark beige, the taste and smell changed from milky to milky-sweet and from milky to sweet-sour, respectively. Sample 1 tastes milky with a pleasant sourness of rosehip. Sample 2 has an added rosehip aroma, with a slight sourness. Sample 3 is sweet.

Taste panel has determined the most optimal in terms of taste indicators of rosehip syrup in the production of the sour-milk beverages in test sample 1 and 2, which have pleasant flavor and can be used in production. The chemical properties of fermented milk products are characterized by the values of titratable and active acidity. Titratable acidity is the content of all acidic compounds in the sour-milk beverages, active is the concentration of active hydrogen ions.

There are discrepancies between them, with an increase in titrated activity, active acidity, due to the presence of buffer systems, remains at a constant level for a long time [7]. The results of studies of titratable activity in the sour-milk beverages are presented in Fig. 2.

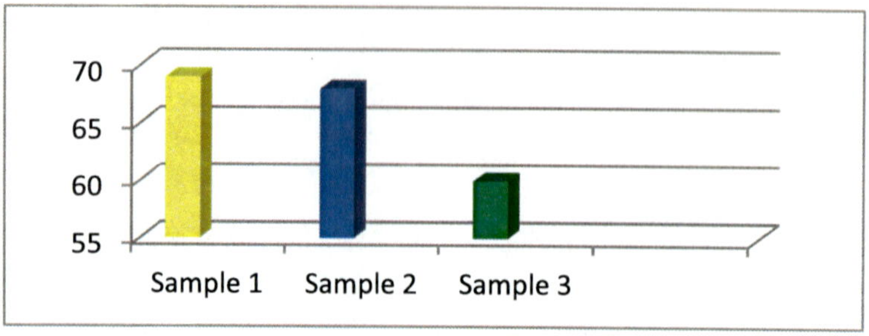

Fig. 3. Titratable acidity of samples of fermented milk drinks with rosehip syrup.

The results of studies to determine the titratable acidity in the studied fermented milk drinks show that in the test samples, with an increase in the concentration of rosehip syrup, the acidity decreases. The decrease in acidity occurs due to the added dosage of rosehip syrup, which is confirmed by experimental samples (Fig. 3).

Active acidity is shown in Fig. 4.

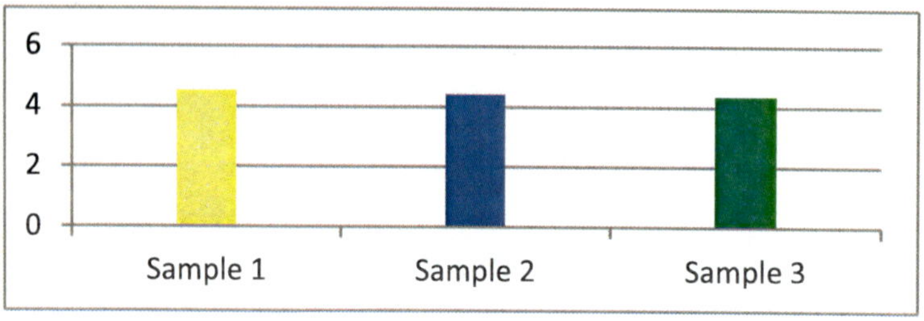

Fig. 4. Active acidity of samples of fermented milk drinks with rosehip syrup.

The results of studies on the determination of active acidity have shown that in some reactions, only free hydrogen ions, which are present in the solution under these conditions, take part. With an increase in the concentration of rosehip syrup from 5 to 20% and a decrease in the content of the milk base, the active acidity in the test samples naturally decreased in the drink. The concentration of hydrogen ions and the pH value characterize precisely this value and express the active acidity of the solution [6, 7].

The ratio between active and total acidity depends only on the degree of acid dissociation in a given solution. The lower the degree of dissociation, the lower the proportion of the total acidity is active acidity. On the contrary, at a high degree of dissociation, the difference between them is insignificant, and in the limit when the degree of dissociation is equal to one (a = 1), the active acidity is equal to the total acidity. From the experiments we see that the degree of dissociation is equal to one, since the titratable and active acidity naturally decreased.

Table 3 shows the results of measurements of the dynamic viscosity of the prototypes carried out on the SV-10 viscometer.

From the figures shown in Table 3, we can see that with an increase in the dose of added syrup from 5 to 20%, the viscosity of the sour-milk beverages naturally decreased.

Table 3. Viscosity of fermented milk drinks with rosehip syrup.

Samples	Viscosity, mPa·S
Sample 1	4.51
Sample 2	4.44
Sample 3	4.36

Based on the research results, the formulation of a model product with a percentage of rosehip syrup was compiled: milk base –85%, rosehip syrup –15% [10]. The developed drink has a natural fermented milk taste, with the aroma of the introduced component, light, non-nutritive, homogeneous [6]. The chemical composition of this product determines not only the usefulness, but also the taste, aroma and color of the drink. Thanks to the nutritional value, you can calculate the ratio of calorie content, digestibility and quality of food. The chemical composition of the developed sour-milk beverages with rosehip syrup is presented in Table 4.

Table 4. Chemical composition of sour-milk beverages with rosehip syrup.

Variant	Proteins	Fats	Carbohydrates	Energyvalue, kcals
Control	16.4	22.5	22.42	61.32
Sample 1	14.84	20.25	46.97	82.06
Sample 2	14.08	19.17	59.28	92.53
Sample 3	13.28	18.0	71.52	102.8

The developed sour-milk beverages have an increased content of carbohydrates due to a decrease in the content of proteins and fats. The increase in carbohydrates in the product was due to the addition of rosehip syrup, which contains sugar, and their calorie content was also studied in the products. The energy value of sample 3 was 102.8 kcal or was 1.43% higher than in sample 1; in relation to the control sample - 1.68%. Sample 2 was respectively 1.29%; 1.51%. Sample 1 was respectively 1.14%; 1.34%. Energy value also depends on the increase of carbohydrates in the developed dairy products.

4 Discussion

In the selected topic, a model product was formulated with a percentage of rosarugosa syrup. Thanks to the research carried out, the possibilities of expedient use as a component of rosarugosa syrup for enrichment of a sour-milk beverages have been studied, which will give a new understanding of the reasonable quantitative use of this plant raw material of the Far Eastern region in the dairy industry.

The creation of a technology for the production of a qualitatively new food product will provide:

- a product of mass consumption for various age groups of the population, including children of various ages and the elderly, pregnant and lactating women, workers of industrial enterprises of various professional groups;
- products for therapeutic and prophylactic purposes;
- products for the prevention of various diseases and strengthening the protective functions of the body, helping to reduce the risk of exposure to harmful substances, including for the population of zones ecologically unfavorable for various types of pollution;
- food products for military people and certain groups of the population in extreme conditions.

5 Conclusions

Based on the performed theoretical and experimental studies, formulations of sour-milk beverages enriched with berry components have been developed. In particular, rosehip syrup, for the purpose of its further use in the technology of sour-milk beverages. The chemical composition and nutritional value of wild rose berries, useful properties have been studied. Due to the rich chemical composition of wild rose berries, they can serve as an excellent raw material for the production of beverages. In the course of the studies, the optimal dose of the rosehip syrup for the quality of the sour-milk beverageswas established, which should be 15% of the product mass. Organoleptic, physicochemical, functional and technological indicators, energy and nutritional value of sour-milk beverages have been determined.

The expediency of using wild berries of the Far Eastern region for further enrichment of sour-milk beverageshasbeen substantiated and confirmed.

References

1. Akimov, M., Bessonov, V.V., Kodentsova, V.M., Eller, K.I., et al.: Biological value of russian-made fruits and berries. Nutr. Issues **89**(4), 220–232 (2020). https://doi.org/10.24411/0042-8833-2020-10055
2. Gavrilova, N., Chernopolskaya, N., Rebezov, M., et al.: Advanced biotechnology of specialized fermented milk products. Int. J. Recent Technol. Eng. **8**(2), 2718–2722 (2019). https://doi.org/10.35940/ijrte.B3158.078219
3. Zaushintsena, A., et al.: Extracts of Rhodiola Rosea l. and Scutellaria Galericulata l. in functional dairy products. Foods Raw Mater. **8**(1), 163–170 (2020). https://doi.org/10.21603/2308-4057-2020-1-163-170
4. Zakharova, L.M.: Development and introduction of new dairy technologies. Foods Raw Mater. **2**(2), 68–74 (2014). https://doi.org/10.12737/5462
5. Musina, O., Lisin, P.: An approach to the choice of alternatives of the optimized formulations. Foods Raw Mater. **3**(2), 65–73 (2015). https://doi.org/10.12737/13120
6. Reshetnik, E.I., Utochkina, E.A.: Healthy food products with probiotic and prebiotic properties. Foods Raw Mater. **1**(1), 88–94 (2013). https://doi.org/10.12737/1562
7. Reshetnik, E., Derzhapolskaya, Y., Gribanova, S.: Study of starter cultures in biotechnology of medical and preventive nutrition products. E3S Web Conf. **203**, 04002 (2020). https://doi.org/10.1051/e3sconf/202020304002

Use of Pine Nut Resources in Food Technology as One of the Steps of Sustainable Forestry

Yulia Derzhapolskaya, Ekaterina Reshetnik, and Svetlana Gribanova(✉)

Far Eastern State Agrarian University, 86, Politeknicheskaya Str., Blagoveschensk, Russia

Abstract. The aim of the work is the scientific substantiation and practical implementation of complex processing of Siberian cedar pine nuts for the production of food products for general and functional purposes. As a result of the work, studies were carried out to study the moisture-binding, water-retaining, fat-retaining and emulsifying properties of cedar cake. The nutraceutical potential of pine nut cake has been studied. Data on the content of macrocomponents of the chemical com-position of pine nut cake are used to substantiate the list of functional and techno-logical properties and determine the regulated values of these indicators. Studies have shown that cedar nut cake has a high moisture-binding capacity, which al-lows it to be recommended for use in formulations of dairy products with a high content of fatty components. The results obtained indicate the advisability of us-ing pine nut cake as a new food source of protein in view of its value in terms of the ratio and amount of nutrients, their properties.

Keywords: Pine nut cake · Nutrients · Functional and technological properties

1 Introduction

The Russian Federation has unique forest reserves and favorable opportunities for the development of the timber industry complex. It accounts for more than 20% of the world's forested area, 23% of the world's timber reserves, including 50% of coniferous trees, which is 35 times more than in the United States and 3 times more than in Canada.

The enormous wealth of the Far Eastern region in view of the poor exploration of resources and the absence of a scientific concept of continuous and non-depleting nature management does not effectively use the renewable bioresource potential. Sustainable forest management contributes to food security, poverty reduction, economic development and sustainable land management in the broader context of sustainable development.

Siberian cedar pine (Pinus Sibirica Du Tour) is a valuable forest-forming species that has no analogues in terms of a variety of useful functions: ecological, climate-regulating, water-saving; forest products, including the irreplaceable pine nut, widely used in the food and pharmaceutical industries.

Cedar oil cake obtained from the extraction of oil from pine nuts belongs to secondary raw materials. However, it is of great importance as an additional source of complete protein, easily digestible carbohydrates, vitamins and minerals. Research on the chemical

A. Muratov and S. Ignateva (Eds.): AFE 2021, LNNS 353, pp. 611–619, 2022.
https://doi.org/10.1007/978-3-030-91402-8_68

composition of pine nut cake [1–3] confirms the promise of its complex processing with the possibility of using it in food technology.

Also, this hypothesis is confirmed by studies [4, 5] in which the use of pine nut cake in the technology of liver pate is proposed, the author [6] carried out data on the possibility of enrichment with this component of semi-finished products. Developed [7] and conducted research on concentrates of second meals based on pine nut cake. The work [8] is devoted to the use of cedar nut cake in the technology of confectionery products. The study of the possibility of using pine nut cake in the technology of dairy products are devoted to the works [9]. The purpose of the presented work is the scientific substantiation and practical implementation of complex processing of Siberian cedar pine nuts for the production of food products for general and functional purposes.

2 Materials and Methods

When conducting analytical studies, we used conventional and special chemical, physicochemical and organoleptic methods for studying the properties of raw materials. Functional and technological characteristics were determined according to the following methods [10].

To determine the moisture binding capacity (MBC) of dry cedar oil cake, samples weighing 1 g were hydrated with distilled water in various ratios with a step of 1 cm^3 per 1 g of oil cake: 1:1, 1:2, 1:3, 1:4 and up to 1:10. Then the hydrated cake was transferred into centrifuge tubes with a volume of 10 cm^3 and centrifuged for 10 min at a speed of 100 c^{-1} to separate unbound moisture. The amount of bound moisture as a percentage of the total volume of water introduced during hydration was taken as the value of the moisture-binding capacity.

$$MBC = \frac{a}{b} \times 100 \tag{1}$$

where a – the amount of moisture bound, cm^3, b – total volume of water added during hydration, cm^3.

Determination of the water-holding capacity was carried out after heat treatment of the hydrated samples. Samples weighing 1 g were hydrated with distilled water in different ratios with a step of 1 cm^3 per 1 g: 1:1, 1:2, 1:3, 1:4, 1:5. The hydration time is 10 min.

The hydrated samples were placed on the inner surface of the wide part of the milk butyrometer. The butyrometer was placed in a boiling water bath for 15 min. The calculation of the water-holding capacity (WHC) was carried out according to the recommendations.

$$\text{WHC} = W - MRA \tag{2}$$

where W – total moisture content in the sample, %; MRA – moisture releasing ability;

$$MRA = anm^{-1} \times 100, \tag{3}$$

where a – division price of butyrometer, a = 0,01 sm^3; n – number of divisions on the butyrometer; m – sample weight, g.

When determining fat-absorption capacity (FAC), 1 g of the preparation was placed in centrifuge tubes with a volume of 30 cm^3 and 1 to 5 g of vegetable oil was added with an interval of 1 g. The contents of the tubes were stirred with glass rods for 10 min, after which the tubes with suspensions of the preparation were kept for 15 min at a temperature of 74–76 °C in a thermostat. After incubation, the tubes were cooled with cold water to room temperature and centrifuged for 15 min at 30 c^{-1}.

The maximum amount of added oil, at which no separation of the oil phase is observed during the test, in terms of 1 g of the preparation, was taken as the value of the FAC. FAC was expressed in grams of oil per 1 g of the preparation.

$$\text{FAC} = \frac{a - b}{c} \times 100, \quad (4)$$

where a – weight of the test tube with protein and associated oil, g; b – weight of the test tube with protein, g; c – protein sample, g.

When determining the fat-emulsifying ability (FEA) and stability of the emulsion, a sample of the product is suspended in 100 cm^3 of water in a homogenizer (or mixer) at a speed of 66.6 c^{-1} for 60 c. Then add 100 cm^3 of refined sunflower oil and the mixture is emulsified in a homogenizer or mixer at a speed of 1500 s^{-1} for 5 min. After that, the emulsion is poured into 4 calibrated centrifuge tubes with a capacity of 50 cm^3 and centrifuged at 500 c^{-1} for 10 min. Next, the volume of the emulsified oil is determined.

Fat-emulsifying ability (%):

$$FEA = \frac{V1}{V} \times 100, \quad (5)$$

where V1 – volume of emulsified oil, cm^3, V – total volume of emulsion, cm^3.

3 Results

The functional and technological properties of processed products, first of all, depend on the chemical composition of the raw material from which it is obtained.

Cedar cake, formed after the extraction of oil from nuts, is a dry light mass in the form of flakes with a pleasant smell and sweetish, with a slight bitterness taste. For use in formulations of dairy products, it is advisable to use pre-ground cake to the state of a fine powder (Fig. 1).

Cedar cake should be considered as a concentrate of biologically valuable nutrients – proteins, lipids, carbohydrates. For example, the mass fraction of proteins in it is, on average, 34.0%. Carbohydrates are of great nutritional and technological importance, which determine the organoleptic value and digestibility of products. A distinctive feature of the carbohydrate composition of pine nut cake is a high content of polysaccharides, which account for 65% of their total amount, including dietary fiber (cellulose and pentosans).

Fig. 1. Appearance of cedar nut cake, natural (left) and crushed (right).

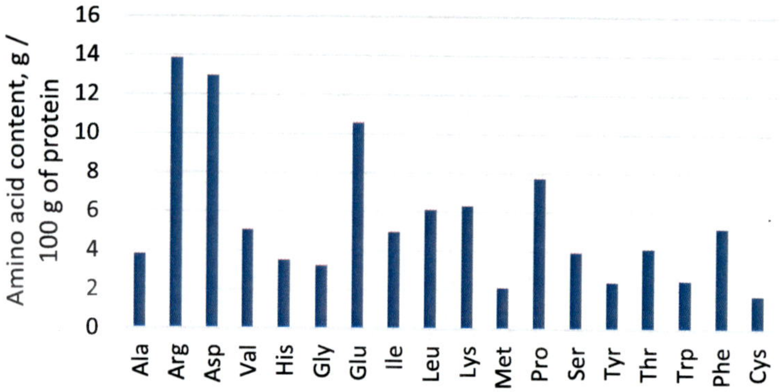

Fig. 2. Amino acid composition of cedar oil cake.

It has been confirmed that cedar cake contains a large amount of all essential amino acids (Fig. 2), and the content of the most deficient ones (tryptophan, lysine and methionine) is 2.5, respectively; 6.29 and 3.86 g/100 g protein (higher than ideal protein).

The study of the mineral composition of cedar cake showed that it contains important macro- and microelements for the human body, including calcium, phosphorus, potassium, sodium, magnesium, iron, manganese and zinc. A feature of the mineral composition of pine nut cake is a low calcium content (46.8 g/kg of the product) and a pronounced predominance of potassium over sodium (Table 1).

Table 1. Mineral composition of cedar oil cake

Mineral name	Content in cake cedar nut, mg/100 g
Macronutrients	
K	1206.3
Ca	48.6
Mg	601.9
Na	19.7
P	2273.9
Trace elements	
Mn	16.1
Fe	13.2
Cu	0.18
Zn	21.3

Vitamins are a necessary component of enzymes, an animal organism in some cases, without the introduction of vitamins from the outside, is not able to synthesize enzymes, which are the main engines of chemical transformations of a substance in an animal organism. For this purpose, an analysis of the vitamin composition of pine nut cake was carried out (Fig. 3).

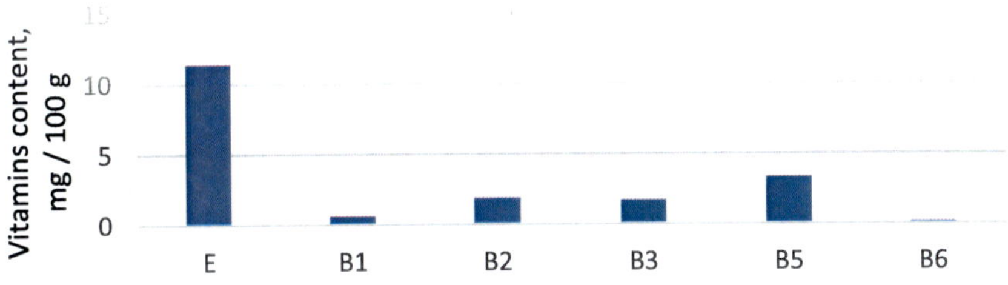

Fig. 3. Vitamin composition of cedar oil cake.

The vitamin composition of pine nut cake is represented mainly by water-soluble B-vitamins.

Cedar oil cake contains riboflavin (B2) in the amount of 1.86 mg/100 g, pantothenic acid (B3) 1.72 mg/100 g, niacin (B5) – 3.24 mg/100 g. Consumption of pine nut oil cake in the amount of 100 g will satisfy the daily requirement for these vitamins by 103.3%; 34.4%; 7.0% respectively. Tocopherol (vitamin E) in pine nut cake contains 11.4 mg/100 g, which, when consumed, will satisfy the daily requirement of the human body by 76.2%.

The functional and technological properties of substances must be clarified taking into account the technological methods and modes of their processing into food products, since any change in the environment around protein molecules, causing a change in their conformation, which may entail a modification of the functional properties of these proteins [11].

The most important functional properties of proteins include water-holding, water-binding and fat-holding, fat-emulsifying properties. The moisture-binding and water-holding capacity of the cake is characterized by the ability to bind and retain moisture as a result of technological processing. These indicators depend on the structure of the molecules and the chemical composition of the investigated product. Water-holding ability is the property of protein preparations to absorb and retain water due to the presence of hydrophilic groups in one polymer chain.

In the course of experimental studies, the ability of pine nut cake to hydrate was studied.

The data presented in Figs. 4 and 5 indicate the ability of cedar oil cake to absorb and retain moisture. Moreover, with an increase in the hydromodule in the mixture, the moisture-binding and water-holding capacity decreases in direct proportional relationship. The higher the moisture content of the mixture, the lower the moisture absorption and water holding capacity.

Also, studies were carried out to study the effect of the temperature of the hydronic module and the time of its contact with the cake. According to the results of this experiment, it was revealed that the most intense moisture absorption is observed in the first 10–15 min, regardless of the amount of hydromodule, and the most optimal temperature of the hydromodule is 80 °C.

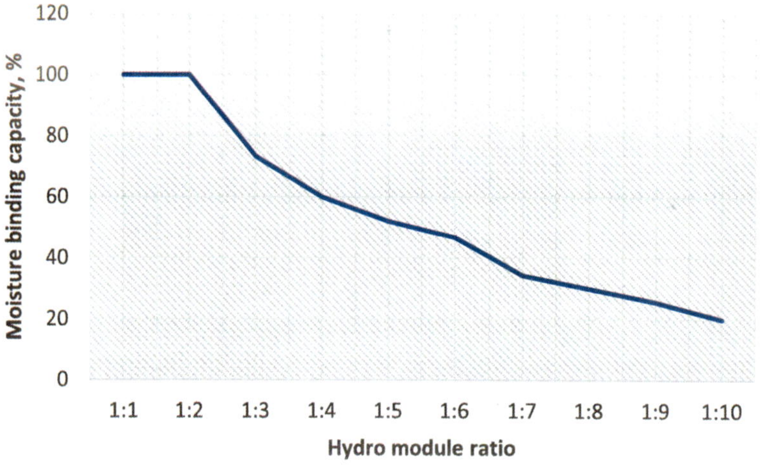

Fig. 4. Change in moisture binding capacity.

The ability of pine nut cake to interact with fat is characterized by fat emulsifying ability, the determination of which was carried out in the concentration range from 0.05 to 0.4 g of product per 1 g of fat.

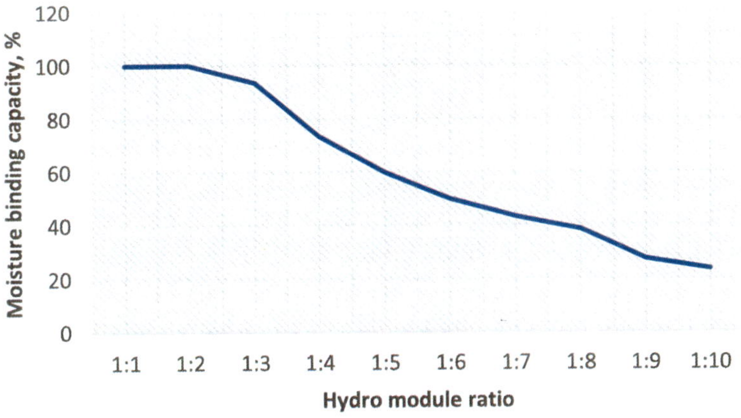

Fig. 5. Changing the water-holding capacity of the cake.

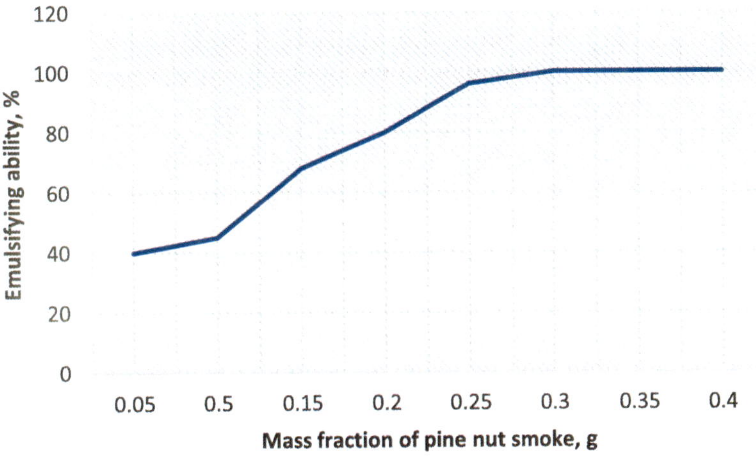

Fig. 6. Dependence of fat emulsifying ability on the mass fraction of pine nut cake.

The results of the study presented in the Fig. 6 indicate a direct dependence of the fat emulsifying ability of pine nut cake on its concentration per 1 g of fat. So with an increase in the concentration of cake 0.05–0.4 g/g, the ability to emulsify fat increased from 40 to 100%.

As can be seen from the Fig. 7 the fat-absorption capacity of cedar nut cake increases with increasing temperature of mixture processing and reaches its maximum of 91% at 80 °C. A further increase in temperature has no effect on the change in the studied index.

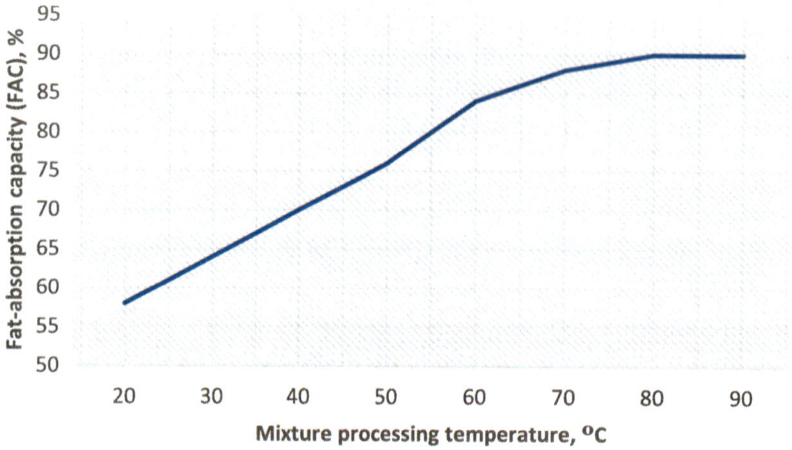

Fig. 7. Dependence of the fat-absorption capacity of pine nut cake on the processing temperature of the mixture.

4 Discussion

The studied indicators of the functional and technological properties of pine nut cake are of practical interest from the point of view of the possibility of its use in the technology of functional milk-based products. The most promising is the combination of vegetable and animal proteins, which will make it possible to create a protein product with an amino acid composition that is closest to the ideal protein.

5 Conclusions

Studies have shown that cedar nut cake has a high moisture-binding capacity, which allows it to be recommended for use in formulations of dairy products with a high content of fatty components. The results obtained indicate the advisability of using pine nut cake as a new food source of protein in view of its value in terms of the ratio and amount of nutrients, their properties.

The use of cedar nut cake in food formulations will make it possible to more efficiently use the resource potential of the pine nut, which will lead to an improvement in the welfare of the population and will allow obtaining additional revenues to the regional budgets.

References

1. Baker, E.J., Miles, E.A., Calder, P.: A review of the functional effects of pine nut oil, pinolenic acid and its derivative eicosatrienoic acid and their potential health benefits. Prog. Lipid Res. **82**, 101097 (2021). https://doi.org/10.1016/j.plipres.2021.101097
2. Kemerli-Kalbaran, T., Ozdemir, M.: Multi-response optimization of oil extraction from pine nut (Pinus pinea L.) by response surface methodology: Extraction efficiency, physicochemical properties and antioxidant activity. LWT **103**, 34–43 (2019). https://doi.org/10.1016/j.lwt.2018.12.067

3. Khamagaeva, I.S., Zambalova, N.A., Tsyzhipova, A.V., Bubeev, A.T.: The development of a biologically active additive to reduce the blood cholesterol level. E3S Web Conf. **161**, 01093 (2020). https://doi.org/10.1051/e3sconf/202016101093
4. Lisin, P.A., Musina, O.N.: An approach to the choice of alternatives of the optimized formulations. Foods Raw Mater. **3**(2), 65–73 (2015). https://doi.org/10.12737/13120
5. Xie, K., Miles, E.A., Caldera, P.C.: A review of the potential health benefits of pine nut oil and its characteristic fatty acid pinolenic acid. J. Funct. Foods **23**, 464–473 (2016). https://doi.org/10.1016/j.jff.2016.03.003
6. Destaillats, F., Cruz-Hernandez, C., Giuffrida, F., Dionisi, F.: Identification of the botanical origin of pine nuts found in food products by gas–liquid chromatography analysis of fatty acid profile. J. Agric. Food Chem. **58**(4), 2082–2087 (2010). https://doi.org/10.1021/jf9041722
7. Zadernowski, R., Naczk, M., Czaplicki, S.: Chemical composition of Pinus Sibirica nut oils. Eur. J. Lipid Sci. Technol. **111**(7), 698–704 (2009). https://doi.org/10.1002/ejlt.200800221
8. Dyshluk, L., et al.: Prospects for using pine nut products in the dairy industry. Foods Raw Mater. **6**(2), 264–280 (2018). https://doi.org/10.21603/2308-4057-2018-2-264-280
9. Prosekov, A.Y., Dyshlyuk, L.S., Milent'eva, I.S., Pavsky, V.A., Ivanova, S.A., Garmashov, S.Y.: Study of the biofunctional properties of cedar pine oil with the use of testing cultures. Foods Raw Mater. **6**(1), 136–143 (2018). https://doi.org/10.21603/2308-4057-2018-1-136-143
10. Lin, S., Yang, R., Cheng, S., Wang, K., Qin, L.: Moisture absorption and dynamic flavor changes in hydrolysed and freeze-dried pine nut (Pinus koraiensis) by-products during storage. Food Res. Int. **103**, 243–252 (2018). https://doi.org/10.1016/j.foodres.2017.10.016
11. Reshetnik, E., Derzhapolskaya, Y., Gribanova, S.: Study of starter cultures in biotechnology of medical and preventive nutrition products. E3S Web Conf. **203**(6), 04020 (2020). https://doi.org/10.1051/e3sconf/202020304002

Development Technology of Functional Drink

Anna Ermolaeva and Elena Gartovannaya(✉)

Far Eastern State Agrarian University, ul. Politehnicheskaja 86, 675005 Blagoveshchensk, Amur Oblast, Russia

Abstract. The Far Eastern region of Russia is a unique territory where a huge number of plants grow. According to the latest data, the list of wild edible berry plants in the Russian Far East (RFE) includes 160 species from 44 genera and 21 families. Of these, actually wild (native to the Russian Far East) - 143 species (89.4%) and adventive (naturalized in the Russian Far East) - 17 (0.6%); definitely edible - 125 species (78.1%) and conditionally edible - 35 (21.9%). The development of technology for non-alcoholic functional drinks with the additional of natural ingredients into the composition will allow saturating the market with useful products for various categories of consumers. The addition of black currant berries, containing pectin substances in their composition, to soy-based drinks makes it possible to exclude the additional of stabilizers. When black currant is added in a ratio of 1: 0.3 (soy base - black currant), delamination occurs only after 48 h. In addition to the ratio 1: 0.3, ratios of 1: 0.1 and 1: 0.5 (soy base - black currant) were considered. The sample (1: 0.3) turned out to be the best option for all organoleptic and technological indicators. Also, the addition of black currant berries allows to enrich the developed drink with vitamin C and make this indicator one of the functional directions of the drink.

Keywords: Drink · Currant berries · Soy grain · Soy base · Questionnaire · Technology · Recipe

1 Introduction

Functional foods are food products that have additional properties in addition to the traditional nutritional value due to the addition (fortification) of additional ingredients, new or existing. When creating a recipe for a functional drink, it was necessary to solve a number of problems: to ensure good taste of the product with the correct ratio of proteins, fats and carbohydrates; enrich the product with components that increase its nutritional and biological value; improve quality indicators [1, 2].

Soy raw materials are a source of unique complete proteins, the composition of indispensable amino acids of which almost corresponds to the composition of animal origin proteins, has a perfect ratio of fatty acids, and the absence of cholesterol and the presence of lecithin makes it possible to recommend a soy-based product for people with lactose deficiency and diabetes mellitus [3].

A. Muratov and S. Ignateva (Eds.): AFE 2021, LNNS 353, pp. 620–627, 2022.
https://doi.org/10.1007/978-3-030-91402-8_69

The choice of currant berries (Ribesnigrum L.) as a functional component for the enrichment of soft drinks is due to the high content of biologically active substances and environmental friendliness. Currant berries stand out significantly among other berry raw materials, they contain more than 700 biologically active substances, including vitamin C-96 mg%, vitamin B-0.06 mg%, vitamin P-1.1–1.5 mg%, carotene - 0.7 mg%, various sugars - from 4.5 to 16.8 (mainly glucose and fructose), organic acids - 2.5–4.5% (citric, malic), proteins - 1%, pectin substances - 0.2–0.8%, tannins - 0.39–0.43%, nitrogenous substances, essential oil, flavonoids (5-methylquercitin, quercitrin), oxycinnamic acids (caffeic, n-coumaric), anthocyanins (cyanidin-3-glucoside, cyanidin-3-rhamnoglucoside, delphinidin-3-rhamnoglucoside, delphinidin-3-glucoside) [4, 5]. They are rich in vitamin E (tocopherol), its content is about 0.72 mg/100 g, surpassing almost all fruits and berries, with the exception of sea buckthorn, cloudberries, rose hips and black chokeberry. It protects vitamin A from oxidation and improves its absorption [6]. 100 g of fresh berries contain 0.86 mg of vitamin K - phylloquinone. The daily requirement for this vitamin is 0.2–0.3 mg, so a small handful of black currants per day can fully satisfy the need for this vitamin.

Given the presence of a diverse and affordable raw material base in the Russian Far East, as well as the high concentration of processing industries in this region, equipped with modern high-performance equipment, the research focuse on developing the technology of functional drinks based on soy and berry raw materials is relevant.

Objects and methods of research. The objects of research were: soy processing products - soy base, soybean grain of the "Batya" variety, which meets the requirements of GOST 17109–88 "Soy. Requirements for harvesting and supply"; black currant species (Ribesnigrum L), model samples of developed drinks. During performing the research, were used the generally accepted, standard research methods, the energy value of raw materials and finished products was calculated using Rubner's coefficients. For the organoleptic assessment of the quality of the developed products, was used the method of point scales, as well as the profile method (T.M. Safronova). To determine the physicochemical parameters and nutritional value of raw materials, intermediate products and prepared functional drinks were used modern generally accepted research methods, in accordance with the current regulatory and technical documentation. The studies were carried out on a Kapel-105 device. Statistical processing of the obtained research results was carried out using the computer program Statistica 6.0 for Windows.

2 Results

The develoiping blackcurrant soy drink is a prime example of a functional drink. Black currant in the drink, enriched with vitamin C - allows it to be attributed to the group of preventive action. And soy milk, as a high-protein product, is included in the special-purpose group. The comparative chemical composition of currant berries is presented in Table 1.

There are certain categories of consumers whom can be manifested intolerances to animal products such as milk and dairy products, because of this intolerance, such prod ucts cannot be used in the daily diet, or at least their consumption should be significantly reduced. A well-known alternative to the use of milk of animal origin is the use of milk of plant origin, which once again proves the relevance of this topic [7].

As part of this work, was carried out research of the preferences of residents of Blagoveshchensk regarding to soy drinks, which are a high-protein product and belong to the group of special-purpose products.

Table 1. Content of vitamins, bioflavonoids, carbohydrates and organic acids in 100 g of berries.

Berry	Drysubstances, g	Vitamins		Bioflavonoids, g	Carbohydrates			Organicacids, g
		C, mg	P, mg		Polysaccharides		Mono anddisaccharides, g	
					Fibre, g	Pectin, g		
Honeysuckle	30.20	124.80	778.91	2.15	3.58	0.78	12.45	3.37
Wildstrawberry	18.56	85.95	191.00	0.92	2.30	1.35	6.78	1.32
Seabuckthorn	23.47	197.13	723.87	4.20	2.53	0.80	4.95	2.75
Blackcurrant	20.52	455.57	886.20	1.40	3.62	1.88	8.98	2.21

To solve the set tasks, the questionnaire method was used. According to picture 1, the total number of respondents was 117 persons, and the bulk (78 people, 66%) of the respondents are people aged 18 to 34 years. Picture 2 shows that soy drinks are preferred by only a small part of the respondents (16%). Half of the respondents are informed about the dietary properties of soy products (absence of animal fat, cholesterol, etc.), 50% of those respondents have ever tried soy milk or soy drinks. There is a tendency - half of the respondents have heard about the benefits of soy milk and also half have tasted soy milk. Picture 3 informs that 50% of the respondents know at least one manufacturer of soy milk or soy drinks, and 42 people (36%) do not know a single one. Of no small importance for people is the availability of the products under consideration in terms of pricing policy, the majority of consumers (76%) consider the optimal price from 50 to 100 rubles (Figs. 1, 2 and 3).

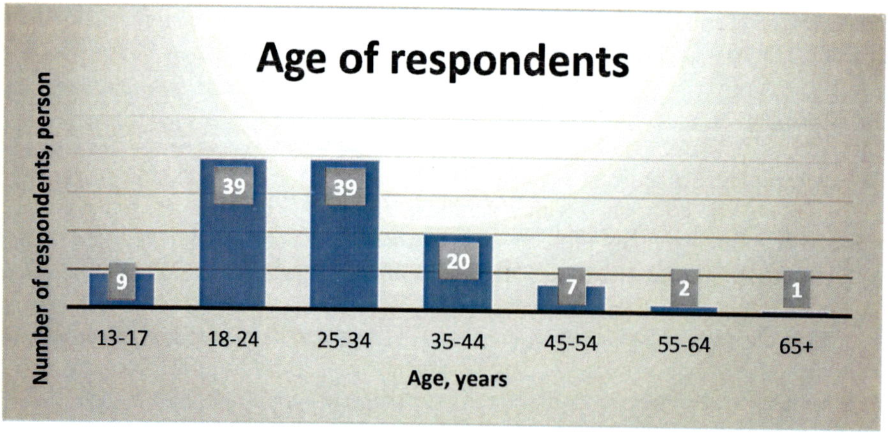

Fig. 1. Age of the respondents.

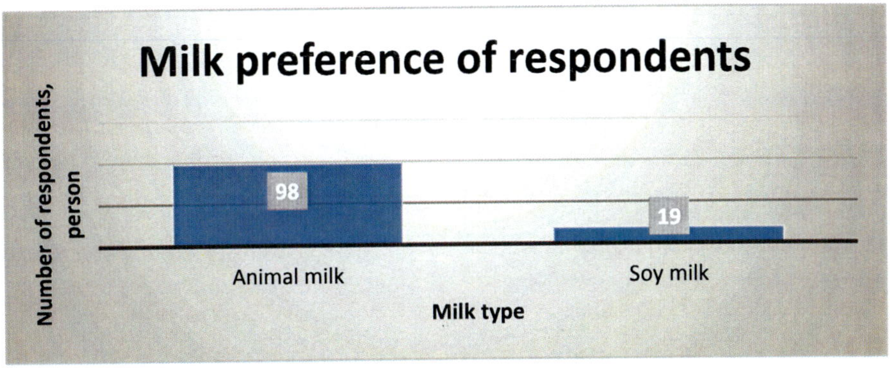

Fig. 2. Preferences of the respondents in the choice of milk.

Fig. 3. Result of the public awareness regarding knowledge of brands of soy milk or soy drinks.

The survey of consumer preferences showed that the consumer (for the most part) is not aware of the benefits of soy drinks, has not tasted it, but ready to taste. Summing up, we can conclude that the development and implementation of functional drinks on the consumer market of this region using plant raw materials will satisfy the population's need for this type of product and expand the range of the market segment.

One of the problems in the development of soy-based drinks is the spontaneous stratification of the product, which occurs due to the instability of the protein complex when a slightly acidic berry part is added into the soy base. The pectin-containing substances of currant berries made it possible to exclude this problem [8].

To obtain the base of the drink, the soybean variety "Batya" was used. The soy base was prepared according to the following technology: the required amount of soybeans was sorted out, cleaned of foreign matter, washed with water at a temperature of 30–35 °C for 30 min, then soaked in water at a temperature of 20 °C for 12 h. After 12 h, the swollen seeds were washed with running water to remove the seed coat. Then, in a ratio of 1: 9 (soy: water), a soy suspension was obtained using a grinder-extractor SC. The soy suspension was filtered and additionally subjected to heat treatment. The resulting soy base was later used for the production of drinks.

3 Discussion

One of the stages of this work was to determine the dose of adding fresh black currant berries to the soybean base. During the development of the recipe, pilot batches were produced with different percentages between the soy base and the applied berry. To determine the structure of the plant composition, 3 samples were prepared, in which the ratio of soy base and fresh currant berries was varied (picture 4). The classic 19-point system was used for organoleptic evaluation of the model drinks. A tasting commission of 10 people carried out an organoleptic assessment.

The results of the organoleptic evaluation of the samples are presented in Table 2.

Table 2. Organoleptic assessment of the quality of a drink with a different ratio of soy base and fresh black currant berries.

Indicators	SampleNo. 1 (1: 0.1)	SampleNo. 2 (1: 0.3)	SampleNo. 3 (1: 0.5)
Appearance	Opaqueliquidwithsuspensions	Opaque, moderately thick liquid, with suspensions	Opaque, thick liquid with suspensions
Color	Lightpink	Crimson	Ruby
Scent	Faint aroma corresponding to the aroma of the raw materials. Soybaseflavorprevails	Characteristic, expressive aroma corresponding the aroma of raw materials	An obsessive scent that corresponding the scent of raw materials
Taste	Weakly expressed, the taste of the soy base prevails in the taste	Sweet and sour, typical of black currants	Sour, pronounced

The analysis of the results of the organoleptic evaluation showed that sample No. 2 was the best option, the drink had a pleasant, sour taste, rich color and aroma, the taste of soybeans was minimized. It was also noted that with an increase in the dosage of black currant, the organoleptic assessment of the drink worsened in all respects.

To improve the organoleptic characteristics of a functional drink, it is proposed to add honey, which is a natural sweetener (Fig. 4).

Fig. 4. Prototypes.

Having studied the chemical composition and energy value of the developed drink, the following results were obtained: one portion of the drink is 200 ml, with the protein content - 3.97 g, fat - 2.10 g, carbohydrates - 18.0 g, the calorie content was 112.3 kcal (Table 3).

Table 3. Chemical composition of the drink.

Indicator	Contentin 100g /, g	Indicator	Contentin 100g /, g
Chemicalcomposition, g		Mineralsubstances, mg	
Water	90.0	Calcium	66.4
Protein	1.95	Phosphorus	85.0
Fats	1.05	Magnesium	56.0
Carbohydrates	9.1	Iron	1.94
Fiber	5.4	Potassium	328.0
Vitamins, mg			
Vitamin A	10.2	Vitamin PP	0.9
Vitamin B1 (thiamine)	0.09	Vitamin C	30.0
Vitamin B2 (riboflavin)	0.11	Vitamin B4 (choline)	30.98
Energyvalue 100g, kcal			
Energyvalue		56.15	

The content of vitamin C in drinks will provide more than 50% of the daily requirement of the human body for ascorbic acid (daily requirement 90 mg) with a single consumption of 200 cm3 of the drink.

During determining the amino acid composition, the results were obtained (Table 4).

Table 4. The content of essential amino acids in a functional drink.

Aminoacid	Content, g	Percentage of the daily consumption rate,%	Aminoacid	Content, g	Percentage of the daily consumption rate,%
Valine	0.19	3.8	Leucine	0.31	4.43
Phenylalanine	0.16	2.67	Lysine	0.21	3.82
Arginine	0.27	6.75	Methionine	0.05	1.43
Histidine	0.15	7.5	Threonine	0.2	5.0
Isoleucine	0.18	4.5	Tryptophan	0.07	7.0

4 Conclusions

Black currant berries are rich in sugars 8.98%. The total acidity is 2.21%. Great value is the high content of ascorbic acid - 456mg/100g. Its amount depends on the variety and conditions of the year. Currants have a very high amount of rutin - 886 mg/100g and an acceptable amount of flavonoids (1.4%). According to the obtained data, black currant berries contain 20.5% of dry matter. The berries contain 3.62% fiber and 1.88% pectins. The total acidity of the currant is 2.21%. Thus, as follows from the literature data and the results of the author's research on the content of biologically active substances, currants are one of the most valuable berry crops. The recipe of the resulting drink was characterized by the absence of unnatural food additives. Natural pigments included in the composition of plants were used as dyes.

During use soybean processing products, a certain economic effect can be obtained, since soybeans are cheaper and more accessible raw materials, and the use of berry raw materials will improve the organoleptic and physicochemical characteristics of the developed drink.

References

1. Litvinenko, O.V., Korneva, N.Yu.: Prospects for using new soybean varieties breeding by All-Russian SRI of Soybean in the production of a soy-chocolate drink. Bull. Moscow State Tech. Univ. Proc. Murmansk State Tech. Univ. **22**(3), 413–432 (2019). https://doi.org/10.21443
2. Pantelidis, G.E., Vasilakakis, M., Manganaris, G.A.: Antioxidant capacity, phenol, anthocyanin and ascorbic acid contents in raspberries, blackberries, red currants, gooseberries and Cornelian cherries. Food Chem. 777–783 (2007). https://doi.org/10.1016/j.foodchem.2006.06.021
3. Statsenko, E.S., Litvinenko, O.V.: Assessment of technological properties of soybean grain of the FSBSI ARSRI of soybean and its processing products to determine their suitability for use in food production. Bull South Ural State Univ. Ser. Food Biotech. **7**(3), 31–40 (2019). https://doi.org/10.14529/food190304
4. Frolova, N.A., ReznichenkoI, Yu.: Investigation of the chemical composition of fruit and berry raw materials of the Far Eastern Region as a perspective source of nutrients and bioactive compounds. Probl of Nutr **288**(2), 83–90 (2019). https://doi.org/10.24411/0042-88332019-10021

5. Galstyan, A.G., Aksyonova, L.M., Lisitsyn, A.B., Oganesyants, L.A., Petrov, A.N.: Modern approaches to storage and effective processing of agricultural products for obtaining high-quality food products. Bull. Russ. Acad. Sci. **89**(5), 539–542 (2019). https://doi.org/10.31857/S0869-5873895539-542
6. Sulaiman, M., Tijani, H.I., Abubakar, B.M., et al.: An overview of natural plant antioxidants: analysis and evaluation. Afr. J. Microb. Res. **1**(4), 64–72 (2013). https://doi.org/10.11648/j.ab.20130104.12
7. VelichkoNA, R.E.A.: Himicheskijsostavvodnyhjekstraktovpiceaabovatairazrabotkare-cepturybezalkogol'nogonapitkanaegoosnove. Vestnik Kras GAU **3**, 143–146 (2018). https://doi.org/10.36718/1819-4036-2021-3-138-143-10418
8. Tyupkina, G.I.: Studying of consumer preference of norilsk population concerning soft drinks. Dostizheniyanaukiitekhniki APK **32**(4), 74–76 (2018). https://doi.org/10.24411/0235-2451-2018

The Use of the Far East Wild Plants in the Production of Functional Alimentary Products

Elena Osipenko(✉) and Svetlana Kostrykina

Far Eastern State Agrarian University, ul. Politehnicheskaja 86, 675005 Blagoveshchensk, Amur Oblast, Russia

Abstract. The main priority aim of the Russian Federation government's policy, as well as of a number of other states, is to provide the population with food products enriched with a variety of physiologically significant ingredients, of high quality and affordable. The suboptimal structure of nutrition when consuming food products with low consumer characteristics is the reason for a decreased quality of life and the development of a number of diseases of the population, also due to the unreasonably high caloric content of food products, reduced nutritional value, excessive consumption of saturated fats, deficiency of micronutrients and dietary fiber. The optimal amount of added functional additives has been determined. The authors have carried out a comprehensive assessment of the quality of the finished product. The practical interest is the development and usage of technologies, which allow using the traditionally demanded products, taking into account the principles of safety, functionality and environmental friendliness. The objects of the study were the products of processing of wild berry raw materials of the red bilberry kind, a food additive from the wood of Daurian larch "Arabinogalactan". Prescription compositions for the production of flour confectionery products for different categories of the population were scientifically substantiated. Based on the study of the chemical composition, quality indicators and biological value of the data obtained, the expediency of using Far Eastern raw materials has been established.

Keywords: Food additive · Arabinogalactan · Berries of the lingonberry family · Confectionery products · Functional food products

1 The Introducing Part

The strategy for improving the quality of food products in the Russian Federation until 2030 (Order of the Government of the Russian Federation dated June 29, 2016 No. 1364-r) is focused on ensuring adequate nutrition, preventing diseases, increasing the duration and improving the quality of life of the population of the Russian Federation. At present, Russians, like residents of other countries of the world, receive insufficient amounts of micro- and macronutrients, the deficiency of vitamins in the diet is up to 30%.

A. Muratov and S. Ignateva (Eds.): AFE 2021, LNNS 353, pp. 628–636, 2022.
https://doi.org/10.1007/978-3-030-91402-8_70

Violation of the balanced diet's rules leads to a decrease in the body's resistance to unfavorable environmental factors, an increase in the risk of developing diseases depending on the composition of the diet.

The use of food products with functional properties allows you to optimize nutrition due to its correction for deficient macro- and micronutrients. The development of functional products is aimed at preventing and replenishing the deficit with the required amount of effective components that improve the functioning of organs and body systems.

Fortification of basic, daily consumed food products such as flour, confectionery, pasta, vegetable oils, seasonings is compulsory for many developed countries, such as the USA, Canada and developing countries in Asia, Latin America, Africa, this is legally regulated [1–3].

Recent years have been characterized by the intensive development of industrial technologies for functional food products intended for regular use as a part of food rations by different age groups of the healthy population.

The main principle of creating functional products is to achieve the highest possible level of usefulness and guaranteed food safety of the product.

The improvement of modern food production technologies is directly related to the use of natural food components containing minor biologically active substances.

Wild berries are the leaders in the content of the most important dietary biologically active components created by nature.

Wild edible berry plants of the Russian Far East cover a very significant taxonomic range. On the territory of the Amur region, 78 species of berry plants grow. Among the wild berries, berries of the lingonberry kind are of particular interest (cranberries, blueberries, blueberries, redberries, cranberries, blueberries).

The nutritional value of fruit and berry raw materials and products of its processing is the main quality characteristic. The defining properties of the processed products of the common lingonberry berries, as well as the whole lingonberry family kind, are low energy and high nutritional value, original organoleptic properties that form stable consumer preferences [4, 5].

The chemical composition of wild berries and lingonberry fruits is very diverse. It includes organic acids, sugars, pectin and tannins, anthocyanins, leukoanthocyanins and catechins, vitamins, nitrogenous compounds, macro- and microelements. Berries of the lingonberry's kind are rich in vitamins A, B, especially vitamin C. Vitamin C is an antioxidant, the value of which for the body can hardly be overestimated. Ascorbic acid takes part in the redox processes continuously occurring in a living cell, and it is considered as a powerful stimulating factor for strengthening the immune system, increasing the body's resistance to aggressive environmental influences [6–9].

In recent years, active work has been carried out on the development of new stabilizers and structure-forming agents, which make it possible to improve the organoleptic indicators of the quality of products and stabilize the quality indicators during storage [10].

The food supplement "Arabinogalactan" demonstrates a wide range of functional properties. Arabinogalactan is a polysaccharide of plant origin, obtained from the wood of Dahurian larch, a biologically active substance with a wide spectrum of biological activity: gastroprotective, immunobiological properties.

Arabinogalactan can be used as the main component of dietary supplements intended for the prevention of cardiovascular, bronchopulmonary, infectious diseases; normalization of the gastrointestinal tract; increasing immunity.

Possessing the properties of a stabilizer, a prebiotic, a source of fiber and soluble dietary fiber, arabinogalactan can be used to prolong the period of storing and increase the shelf life of food products for the production of functional food products.

Based on the foregoing, it becomes necessary to further study the possibility of using natural Far Eastern raw materials in the technology of functional food products of systemic action.

2 The Aim of the Study

The aim of the study is the development of new types of confectionery products of a functional orientation using raw materials from the Far Eastern region - wild berries of the lingonberry's kind, a food additive from Daurian larch wood "Arabinogalactan" and assessment of quality indicators.

3 Research Objectives

Research objectives are to study the possibility of using powder from berries (pulp) of common lingonberry (Vaccinium vitis-ideal) and the using of arabinogalactan in confectionery technology in order to increase the nutritional and biological value of finished products and conduct a comprehensive assessment of the quality of finished products.

4 Materials and Methods

The experimental part of the work was carried out in the laboratories of the Far Eastern State Agrarian University, the Faculty of Technology, chair of product technology and organization of public catering, chemistry' chair.

The objects of the study were the berries of common lingonberry (Vaccinium vitis-ideal), powder from berries (pulp) of common lingonberry, food additive "Arabinogalactan" produced from Daurian (Siberian) larch, flour confectionery product - sugar cookies, produced according to a standard recipe (control samples), sugar cookies "Taiga", enriched with biologically active components (prototypes).

While performing experimental studies, a complex of generally accepted, standard and modified research methods was used: physicochemical, microbiological and organoleptic in accordance with GOST 53159 - 2008 "Organoleptic analysis. Methodology", GOST 24901 – 2014 "Cookies. General technical conditions". Technological operations were carried out according to the traditional scheme for experimental and control samples [11].

Powder from dried berries (pulp) of common lingonberry (Vaccinium vitis-ideal) was added to the dough at the stage of its preparation in an amount of 3.5%, 5.5%, 7.5%, arabinogalactan in an amount of 3% to the mass of standard recipe flour.

5 The Results

To obtain a powder from common lingonberry, berries growing in the Far East under the conditions of the Amur Region were used. Lingonberry powder was obtained according to the following technological scheme: inspection and sorting of raw materials → preparation of raw materials → drying at t not higher than 40 °C for 7–10 days → grinding of raw materials → packaging → storage at a temperature not higher than 20 °C. The appearance of the common lingonberry powder is shown in Fig. 1. The pure Arabinogalactan food additive was provided by Ametis LLC, the city of Blagoveshchensk (Fig. 2).

Fig. 1. Lingonberry pulp powder.

Fig. 2. Food supplement "Arabinogalactan".

The technological process for the production of Taezhnoye sugar cookies consists of the following operations: preparation of raw materials for production → preparation of a recipe mix (combining powdered sugar with liquid components, stirring for 10 min, adding a chemical baking powder dissolved in water at a temperature of 15–20 °C, softened at a temperature of 40 °C margarine, food essence, stirring until a homogeneous consistency for 15–20 min, the temperature of the recipe mixture should not exceed 30 °C) → dough preparation (adding 5.5% lingonberry powder based on the total flour weight and 3% Arabinogalactan food additive to the premium wheat flour, mixing with the recipe mixture, stirring) → molding → baking at a temperature of 200–220 °C for 6 min (duration and baking modes may vary depending on the type of oven, the degree of filling) → cooling → finishing.

Fig. 3. Control sample.

Fig. 4. Sugar cookies "Taiga".

In laboratory conditions, we have tested the technology for preparing sugar cookies "Taezhnoye" (Fig. 4), the control sample is shown in Fig. 3.

The quality indicators based on the principles of HACCP (Hazard Analysis and Critical Control Points) were determined (Table 2), the nutritional and energy value of the experimental sample was calculated.

The characteristics of organoleptic indicators of the quality of sugar cookies "Taezhnoe" are presented in Table 1.

Table 1. Organoleptic indicators of the quality of sugar cookies "Taezhnoe".

The appearance of the product	Complies with the norm, the shape is correct, the surface is smooth without cracks and blisters
Taste and flavor of the product	Pleasant, sweet, typical for a baked product, with a taste and flavors typical for lingonberries
The color	The surface is brown, at the break is soft pink, uniform
Consistency	Brittle, crumbly

It has been established that the addition of powder from dried pulp of common lingonberry in an amount of 5.5% to the weight of flour in sugar biscuits "Taiga" leads to a change in organoleptic characteristics. The flour product acquires the taste and flavor, which is usual for of lingonberry berries and for pink-purple hue.

The results of the organoleptic assessment have showed that the sample with a dosage of powder from dried pulp of common lingonberry in an amount of 5.5% by weight of flour had the best parameters and the highest number of points (Fig. 5). The tasting analysis was carried out according to the point assessment of organoleptic indicators developed by the authors.

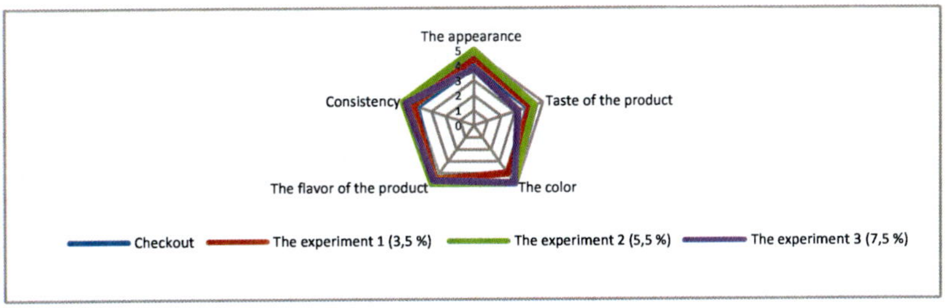

Fig. 5. Tasting analysis of organoleptic indicators of sugar cookies "Tayozhnoe".

The next stage of the research included the identification of dangerous factors in the production of prototypes from the moment the raw material was received to the final product, including all stages of the life cycle of the production of Tayozhnoye sugar cookies. The identification of critical control points in the production of prototype cookies was carried out in order to eliminate or minimize the risk or the possibility of its manifestation (Table 2).

Table 2. Analysis of dangerous factors (production of sugar cookies "Tayozhnoe").

The name of the operation	Considered hazard	Controlled signs	Preventive action
Acceptance and storage of raw materials	Increased moisture content of dry products, the presence of foreign impurities, increased acidity (dairy products), the presence of unacceptable defects	Moisture (flour, granulated sugar, food additive "Arabinogalactan", lingonberry pulp powder), change in organoleptic parameters	Control of goods supporting documentation, entrance control, compliance with storage conditions (relative air humidity, temperature)
Mixing of components (recipe mixture, flour, lingonberry powder, food additive "Arabinogalactan")	Non-compliance of the specific value of the proportion of the constituents	The specific value of the proportion of the constituents according to the recipe	Strict adherence to recipes' standards
Mixing the components	Inhomogeneity of consistency	Appearance, consistency, mixing time	Compliance with the technological regime (the duration of the operation is 15–20 min)
Molding	Inhomogeneity of consistency,violation of the size of the dough's thickness of semi-finished product	Appearance, consistency, product thickness	Compliance with the technological regime
Baking	Burnt odor, discoloration of the product	Color, consistency, duration of the operation	Compliance with the technological regime (baking at t 200–220 °C, within 6 min)

The data list 2 show that for each potential hazardous factor, a risk analysis was carried out taking into account the likelihood of the hazardous factor and the severity of its consequences. Critical control points at various stages of the production process were identified.

When determining the physicochemical indicators of the quality of sugar cookies "Taezhnoe" in comparison with the control sample, changes were observed (Table 3).

Table 3. Physical and chemical indicators of products.

The name of indicators	Name of samples		Standard
	Control	Cookies"Taezhnoe"	
Alkalinity, degrees	1.7	1.4	2.0
Humidity,%	6.2	7.6	10.0
Wetness,%, not less	167	178	180

In the course of the research, it was found that when adding experimental additives to the recipe for sugar cookies, the wetness changes insignificantly, the moisture content increases in the Taezhnoye cookies to 7.6% in the experimental sample, versus 6.2% in the control sample. The addition of arabinogalactan promotes the formation of heat-resistant protein-polysaccharide complexes, increases the plasticity of the dough. It leads to an increase in the proportion of tightly bound moisture in finished products, which reduces shrinkage and increases the yield of finished products. The alkalinity index of the experimental samples decreases slightly.

The main aim of the research is to obtain a product of increased nutritional and biological value. The chemical composition of the obtained Tayozhnoye biscuits makes it possible to judge its functional orientation. The content of nutrients in the finished product with the addition of lingonberry powder and arabinogalactan, in 100 g of the product, is 4.7 g of protein; fat - 20.5; carbohydrates - 43.4. Energy value is 361.8 kcal (prototype 372 kcal). Vitamins, mg - B1 - 0.03; B2 - 0.1; P - 0.46; E - 0.24; C - 0.68. Minerals, mg: Na - 43.6; Ca 28.2; Mg - 19.2.

The inclusion of arabinogalactan in the recipe prolongs the freshness of the finished product. The investigated control and experimental samples were stored in a package at room temperature. After ten, twenty and thirty days of storage, the indicators of the control sample worsened - the hardness, flavor of the product decreased, and crumbling appeared. The prototype had a lingonberry's flavor. The microbiological parameters of the products remained within the normal range. The use of experimental additives in the production of a prototype helps to increase the organoleptic properties of the product, preserve freshness, and increase absorption. Enrich the product with vitamins, minerals, dietary fiber.

6 Discussion

In the production of food, the primary goal is to choose the raw materials, justifying economically the way it is processed. Raw materials should be affordable, cheap and technologically advanced. When using regional raw materials, the cost of production is reduced. Taking into account these factors, it is necessary to study possible uses of Far Eastern raw materials for food purposes.

The choice of wild fruits of common lingonberry (Vaccinium vitis-ideal) as a functional ingredient is due to the high content of biologically active substances, a fairly large raw material base and ecological cleanliness characteristic of the Far Eastern region of Russia. Today, many experts working on healthy eating and nutrition, note that in a few years, specialized food products, including functional ones, will make up more than half of the daily diet, which corresponds to the main directions of food policy, not only of the Government of Russia, but also of other states. Improving modern technologies and developing formulations of fortified food products, it is necessary to use natural ingredients with biological activity, excluding artificial additives. The Far East of Russia possesses quite diverse and poorly studied raw materials for obtaining food products of a functional orientation Due to the seasonality of lingonberry harvesting, the remoteness of the place of collection and processing, the high cost of fresh biologically valuable plant raw materials, many food manufacturers prefer to work with processed products. Lingonberries are harvested in large quantities in autumn. The traditional storage method is freezing, which requires production facilities and refrigerated storage facilities. Processing of lingonberry berries at the place of collection with separation into juice and pulp with subsequent drying is the best option for both the supplier and the producer. In addition, the biological potential of raw materials can be used more fully.

The use of biologically active additives in the form of a powder is one of the modern trends that have a number of advantages in terms of industrial production. It is less costly to transport and store additives in powder form. The technology for producing the powder makes it possible to obtain non-oxidized, finely dispersed, retaining color, smell and taste inherent in lingonberries, characterized by relatively high quality and low cost.

Water holding capacity of arabinogalactan is especially important in the production of flour confectionery products for dietary purposes. Possessing the properties of a prebiotic, arabinogalactan maintains the normal balance of microflora in the gastrointestinal tract. It is a source of dietary fiber, improves nutrition and nutrient absorption.

The additives used for prototyping are recommended as functional food ingredients for daily nutrition.

The developed prototype will make it possible to more fully use raw materials, expand the range of flour confectionery products, and obtain a product of increased nutritional and biological value.

7 Conclusions

The entire world community's efforts are aimed at finding environmentally friendly, technologically advanced, renewable raw materials that contain essential nutrients for humans.

The chemical composition of raw materials, its biological and nutritional value have been studied. The expediency of using the products of processing of wild berry raw materials and Dahurian larch in food production has been experimentally substantiated. The optimal dosages of the added additives have been established: arabinogalactan in the amount of 3%, powder from the fruits of common lingonberry 5.5% to the mass of flour. It has been proven that with the help of functional additives obtained from natural wild-growing raw materials, it is possible to adjust the amount of dietary fiber, vitamins, micro- and macroelements in a manufactured product, as well as to influence an increase in shelf life, improve taste and aroma, and reduce energy value.

The use of the potential of raw materials from the Far Eastern region will expand the range and obtain affordable food products for the consumers.

References

1. Luthringer, C.L., Rowe, L.A., Vossenaar, M., Garrett, G.S.: Regulatory monitoring of fortified foods: identifying barriers and good practices. Glob. Health SciPract. **3**(3), 446–461 (2015). https://doi.org/10.9745/GHSP-D-15-00171
2. Tulchinsky, T.H.: The key role of government in addressing the pandemic of micronutrient deficiency conditions in Southeast Asia. Nutrients **7**(4), 2518–2523 (2015). https://doi.org/10.3390/nu7042518
3. Sharma, S., Sheehy, T., Kolonel, L.N.: Ethnic differences in grains consumption and their contribution to intake of B-vitamins: results of the Multiethnic Cohort Study. Nutr. J. **12**(65) (2013). https://doi.org/10.1186/1475-2891-12-65
4. Shkolnikova, M.N.: About the use of pectin from wild berries pomace in the marmalade recipes. Food Ind. **4**(1), 81–86 (2019). https://doi.org/10.29141/2500-1922-2019-4-1-9
5. Szajdek, A., Borowska, E.J.: Bioactive compounds and health-promoting properties of berry fruits: a review. Plant Foods Hum. Nutr. **63**, 147–156 (2008). https://doi.org/10.1007/s11130-008-0097-5
6. Klavins, L., Kviesis, J., Nakurte, I., Klavins, M.: Berry press residues as a valuable source of polyphenolics: extractionoptimisation and analysis. LWT - Food Sci. Tech. **93**, 583–591 (2018). https://doi.org/10.1016/j.lwt.2018.04.021
7. Kitrytė, V., Kavaliauskaitė, A., Tamkutė, L., Pukalskienė, M., Syrpas, M., Venskutonis, P.: Zero waste biorefining of lingonberry (Vaccinium vitis-idaea L.) pomace into functional ingredients by consecutive high pressure and enzyme assisted extractions with green solvents. Food Chem. **322**, 126767 (2020). https://doi.org/10.1016/j.foodchem.2020.126767
8. Lyutikova, M.N.: Chemical composition and practical application of lingonberry and cranberry berries. Chem. Veget Raw Mater. **2**, 49–53 (2015). https://doi.org/10.14258/jcprm.201502429
9. Gartovannaya, E.A., Ermolaeva, A.V., Kostrykina, S.A., Ivanova, C.S.: Chemical composition of natural raw materials in producing innovative food products. Eur. J. Anal. Chem. **13**(5), 67 (2018). https://doi.org/10.29333/ejac/97823
10. Trivedi, P., Karppinen, K., Klavins, L., et al.: Compositional and morphological analyzes of wax in northern wild berry species. Food Chem. **295**, 441–448 (2019). https://doi.org/10.1016/j.foodchem.2019.05.134
11. Osipenko, E., Denisovich, Y., Gavrilova, G.A., Vodolagina, E.: The use of bioactive components of plant raw materials from the far eastern region for flour confectionery production. AIMS Aquic Food **4**(1), 73–87 (2019). https://doi.org/10.3934/agrfood.2019.1.73

Efficiency of the Application of Growth Stimulants Ribav-Extra and Zircon in the Germination of Manchurian Fir (*Abies Holophylla* Maxim.) Seeds

Valentina Ostroshenko[1] and Lyudmila Ostroshenko[2(✉)]

[1] Far Eastern Branch of the Russian Academy of Sciences, Federal Scientific Center of the East Asia Terrestrial Biodiversity, Prosp. 100-let, 159, 690022 Vladivostok, Russia

[2] Primorye State Agricultural Academy, Prosp. Blyuhera, 44, 692510 Ussuriysk, Russia

Abstract. The stimulating effect of the aqueous solutions of growth stimulants Ribav-Extra and Zircon of different concentrations have been studied and the doses activating germinative energy, laboratory germination of seeds and the growth of biometric parameters of sprouts of Manchurian fir (*Abies holophylla* Maxim.) have been identified. It was found, that the stimulator Ribav-Extra was effective for the germinative energy of seeds of Manchurian fir when using all the concentrations, comprising 13.3–21.3% (excess compared to the control – 33–113%); for laboratory germination – at the concentrations of $1{\cdot}3{\cdot}10^{-3}-1{\cdot}6{\cdot}10^{-3}$ ml/l, amounting to 43.1–53%, where the indicators exceeded the control by 12.2–38%. Zircon, respectively, had a more active effect at the concentrations of $1{\cdot}2{\cdot}10^{-3}-1{\cdot}7{\cdot}10^{-3}$ and $1{\cdot}4{\cdot}10^{-3}-1{\cdot}6{\cdot}10^{-3}$ ml/l, comprising 11–16 и 42–47%, exceeding the control by 10–60 и 9.4–22.4%. Seed quality classes had increased: from the third to the second and the first. For the growth of sprouts along the length Ribav-Extra had a more positive effect when using all solution concentrations (excess to the control – 7.1–33.3%); Zircon – when using the concentrations of $1{\cdot}2{\cdot}10^{-3}-1{\cdot}5{\cdot}10^{-3}$ ml/l, exceeding the control by 7.1–25%. For the weight of sprouts both drugs were effective at the concentrations of $1{\cdot}3{\cdot}10^{-3}-1{\cdot}5{\cdot}10^{-3}$ ml/l, exceeding the control values by 2.7–28.8 and 1.3–25.4%.

Keywords: Manchurian fir · Seeds · Ribav-Extra · Zircon · Germinative energy · Laboratory germination · Length and weight of sprouts

1 Introduction

Forests of the Primorsky Territory are spread over the large area. Coniferous tree species occupy 51.9% of their territory [1]. They are: Scots pine (Pinus sylvestris L.), Amur larch (Larix amurensis B. Kolesn.), Khingan fir (Abies nephrolepis (Trautv.) Maxim.) and Manchurian fir (A. holophylla Maxim.).

A. Muratov and S. Ignateva (Eds.): AFE 2021, LNNS 353, pp. 637–647, 2022.
https://doi.org/10.1007/978-3-030-91402-8_71

Manchurian fir is found only in the southern part of Primorye, not moving north Partizansky and Ussuriysky districts. This is the most powerful tree species that grows in the south of the Russian Far East [2, 3].

Manchurian fir is the main forest-forming species in the southern part of the Far East [2, 3]. Together with Korean pine it makes up complex black-fir-broad-leaved forests different of their composition [3]. It is a rock-fortifying species due to its powerful root system [3].

This species is characterized by high polymorphism, which confirms its antiquity and unpretentiousness to an unfavorable climate [3].

Artificial breeding can be complicated by the exceptional thermophilicity of fir and the difficulty of harvesting its seeds.

Recently, areas and stocks of Manchurian fir have greatly reduced due to long-term anthropogenic influence, destruction of undergrowth by forest fires, increased logging and felling by the local population for «New Year trees». Today, the area of black-fir-broad-leaved forests comprises 23 thousands hectares [2, 3]. Manchurian fir is included in the document «On approval of the List of species of trees and shrubs, the logging of which is prohibited», approved by the order of Federal Forestry Agency dated 05.12.2011 г. No. 513 [2, 3].

Ecological and biological characteristics, small area of distribution, high national economic value of this tree species cause the need of its protection and active reproduction. This is possible through the use of growth stimulants [2, 3].

Growth stimulants are the substances that stimulate or inhibit growth and development of plants. In forestry their study is carried out on an experimental basis. Studies carried out in different regions of Russia and abroad indicate on the prospects of their use. So, when using growth stimulants: Baikal, brassinosteroid, DNA, Ecogel, Ekopin, Epin-Extra, Ethamon, Gibberellin, Humate + 7, Kornevin, Krezacin, Obereg, Ribav-Extra, PRG, Zircon, etc. the increase in the sowing qualities of seeds (germinative energy, laboratory and soil germination) and growth enhancement of coniferous and deciduous tree species: Brazilian pine (Araucaria angustifolia (Bertol.) Kuntze), Scots pine (Pinus sylvestris L.), omatsu (P. densiflora Siebold et Zucc.), Siberian larch (Larix sibirica Ledeb.), Douglas fir (Pseudotsuga menziesii (Mirb.) Franco), Siberian spruce (Picea obovata Ledeb.), Norvay spruce (P. abies (L.) H. Karst.), European spruce (P. abies (L.) H. Karst.), English oak (Quercus robur L.), etc. were observed [1–10].

In this research the study of the influence of the growth stimulants Ribav-Extra and Zircon on the germination of far eastern tree species Manchurian fir was carried out for the first time.

2 The Purpose of the Research

The purpose of the research is to study the effect of aqueous solutions of the growth regulators Ribav-Extra and Zircon on the sowing qualities of seeds (germinative energy, laboratory germination) and the growth dynamics of biometric parameters (length and weight) of Manchurian fir sprouts.

3 Research Objectives

- soaking of Manchurian fir seeds in aqueous solutions of the growth stimulants Ribav-Extra and Zircon of different concentrations;
- germination of seed material in laboratory conditions;
- analysis of the effect of these stimulants on the germinative energy, laboratory germination of seeds and biometric parameters (length and weight) of sprouts.

4 Materials and Methods

The object of the research is Manchurian fir seeds. The place of collection of seeds is the territory of the GTS – a branch of the Federal Research Center for Biodiversity of Terrestrial Biota of the Far East Branch of the Russian Academy of Sciences. The climate of the GTS territory – is monsoon, however, it is worth noting a significant continentality, which manifests itself in large seasonal amplitudes in temperature and humidity of air and soil. On the territory of the research object three groups of soils are distinguished: mountain-forest burozems, podzolic soils and soddy soils.

The study of the stimulating effect of the growth regulators Ribav-Extra and Zircon on germinative energy and laboratory germination was carried out in accordance to current GOST 13056.6–97 and OST 56–27-77.

The experiments included seven variants (the concentrations of drug solutions included: $1 \cdot 10^{-3}$ – $1 \cdot 7 \cdot 10^{-3}$ ml/l) and control – seeds, kept in distilled water. For germination, seeds were used that were externally intact and they were soaked in the solution drugs for 18–20 h in the indicated solution concentrations. The ratio of the volume of seeds and the solution of the drug was 1:5. A fourfold replication was used.

The seeds prepared for the experiments were washed and placed in Petri dishes. For germination, a thermostat TS-80 – «KZMA» was used. The germination bed, which consisted of filter paper, folded in four layers, was periodically moistened by wetting it with distilled water. The temperature was set within 25–27 °C. Seed sprouts were counted on days 7, 10, 15, 20, 25. The germinative energy of seeds was determined on the 7th and 10th days.

On the days of counting, for every replication, the amount of germinated and not germinated seeds were counted; the indicators of sprout length were determined using an electronic caliper, with an accuracy of 0.1 mm. The weight was determined using an electronic balance BLKT-500, with an accuracy of 0.01 g. On the last day of laboratory germination record the number of seeds left on the bed was determined: not germinated, disease-free, decayed, steamed, empty, abnormally germinated and pest-infested seeds. The data was recorded in special card. The results of germination were processed by the methods of the mathematical statistics in the Microsoft Excel program. The basic of the mean, the reliability of the calculations and the accuracy of the experiment were calculated. The significance of the differences between the mean values and the control was determined by the Student's t-test.

5 Results

As the result of the experiments it was found, that the growth stimulator Ribav-Extra had a positive effect on the indicators seed germinative energy when using all concentrations, comprising 13.3–21.3% (excess compared to the control – 33–113%) (Table 1).

When using Zircon the most positive effect on seeds germinative energy was noted when they were treated by the solution concentrations of $1{\cdot}2{\cdot}10^{-3}–1{\cdot}7{\cdot}10^{-3}$ ml/l, where the average indicators comprised 11–16%, exceeding the control by 10–60%. The concentration of $1{\cdot}10^{-3}$ ml/l had not positive effect. The decrease to the control comprised 37% (Table 2).

On laboratory seed germination the active effect of the growth regulator Ribav-Extra was noted at the concentrations of $1{\cdot}3{\cdot}10^{-3}–1{\cdot}6{\cdot}10^{-3}$ ml/l, comprising 43.1–53%. These indicators exceeded the values of the control group by 12.2–38%. There was an increase in seed quality classes: from the third to the second and the first. The significance of differences with the control was noted when using the concentrations of $1{\cdot}4{\cdot}10^{-3}–1{\cdot}6{\cdot}10^{-3}$ ml/l: $t_{0.05} = 3.4 > t_{st} = 2.45$, $t_{0.01} = 4.8 > t_{st} = 3.71$ and $t_{0.05} = 2.6 > t_{st} = 2.45$. The solution concentrations of $1{\cdot}2{\cdot}10^{-3}$ and $1{\cdot}7{\cdot}10^{-3}$ ml/l were less effective, exceeding the indicators of the control values by 1.6–2.6%. The concentration of $1{\cdot}10^{-3}$ ml/l led to a decrease in laboratory seed germination indicators by 3.6% in comparison with the control values (Table 1).

Table 1. The influence of the growth stimulator Ribav-Extra on germinative energy and laboratory germination of Manchurian fir seeds.

Concentrations of drug solutions, ml/l								
Seeds accounting with duration germination	Control	$1{\cdot}10^{-3}$	$1{\cdot}2{\cdot}10^{-3}$	$1{\cdot}3{\cdot}10^{-3}$	$1{\cdot}4{\cdot}10^{-3}$	$1{\cdot}5{\cdot}10^{-3}$	$1{\cdot}6{\cdot}10^{-3}$	$1{\cdot}7{\cdot}10^{-3}$
1	2	3	4	5	6	7	8	9
Number of germinated seeds, %								
7th day	2.5 ± 0.9	4.5 ± 0.7	4.8 ± 0.6	4.3 ± 0.6	8.0 ± 1.3	6.8 ± 0.9	6.0 ± 0.9	5.3 ± 1.4
Percentage to control		+ 80.0	+ 92.0	+ 72.0	+ 220.0	+ 172.2	+ 140.0	+ 112.0
Reliability of calculations, t_m	2.9	6.9	7.6	6.8	6.2	7.9	6.5	3.8
Accuracy of experiment (P),%	34.8	14.4	13.1	14.7	16.1	12.6	15.3	26.0
10th day	7.3 ± 1.1	8.8 ± 0.5	8.5 ± 0.7	9.0 ± 1.3	13.3 ± 1.3	12.8 ± 0.8	11.8 ± 1.9	11.8 ± 0.8
Percentage to control		+ 20.5	+ 16.4	+ 23.3	+ 82.2	+ 75.3	+ 61.6	+ 61.6
Reliability of calculations, t_m	6.6	18.3	13.1	7.0	10.1	17.1	6.2	15.7

(continued)

Table 1. (*continued*)

Concentrations of drug solutions, ml/l								
Accuracy of experiment (P), %	15.2	5.5	7.6	14.3	9.9	5.9	16.0	6.4
15th day	15.3 ± 1.6	9.3 ± 1.4	12.0 ± 1.1	14.0 ± 1.8	11.3 ± 1.1	12.3 ± 0.9	12.5 ± 0.7	10.0 ± 1.1
Percentage to control		-39.2	-21.6	-8.5	-26.1	-19.6	-18.3	-34.6
Reliability of calculations. t_m	9.9	6.5	11.1	7.9	10.2	14.3	19.2	9.3
Accuracy of experiment (P), %	10.1	15.5	9.0	12.7	9.8	7.0	5.2	10.8
20th day	9.3 ± 0.6	9.5 ± 0.7	10.0 ± 0.4	12.8 ± 1.4	10.0 ± 1.1	13.3 ± 0.9	9.5 ± 0.7	9.5 ± 1.0
Percentage to control		+ 2.2	+ 7.5	+ 37.6	+ 7.5	+ 43.0	+ 2.2	+ 2.2
Reliability of calculations, t_m	14.8	14.6	24.4	9.3	9.3	15.5	14.6	9.1
Accuracy of experiment (P), %	6.8	6.8	4.1	10.8	10.8	6.5	6.8	10.9
25th day	4.0 ± 0.9	4.8 ± 0.5	3.3 ± 0.5	3.0 ± 0.4	6.0 ± 1.1	7.5 ± 0.7	4.5 ± 0.3	2.8 ± 0.3
Percentage to control		+ 20.0	-17.5	-25.0	+ 50.0	+ 87.5	+ 12.5	-30.0
Reliability of calculations, t_m	4.3	10.0	6.9	7.3	5.6	11.5	15.5	11.2
Accuracy of experiment (P), %	23.0	10.0	14.5	13.7	18.0	8.7	6.4	8.9
Seed germinative energy, %	10.0	13.3	13.3	13.3	21.3	20.0	18.0	17.1
Seed laboratory germination, %	38.4	37.0	39.0	43.1	49.0*	53.0*	44.3*	39.4
Significance of differences, t_{fact}		0.8	0.1	2.0	3.4	4.8	2.6	0.5
Number of not germinated seeds, %	62.0	63.0	61.0	57.0	51.0	47.0	56.0	61.0
Among them:								
disease-free	17.0	20.0	18.0	15.0	16.0	16.0	16.0	17.0
decayed	5.0	3.0	4.0	4.0	2.0	2.0	3.0	2.0
steamed	5.0	2.0	3.0	3.0	1.0	1.0	2.0	2.0
empty	11.0	16.0	17.0	15.0	14.0	11.0	18.0	20.0
abnormally germinated	20.0	22.0	18.0	17.0	17.0	15.0	16.0	18.0
pest-infested	4.0	-	1.0	3.0	1.0	2.0	1.0	2.0

* The differences are significant

Table 2. The influence of the growth stimulator Zircon on germinative energy and laboratory germination of Manchurian fir seeds.

Seeds accounting with duration germination	Control	Concentrations of drug solutions. ml/l						
		$1{\cdot}10^{-3}$	$1{\cdot}2{\cdot}10^{-3}$	$1{\cdot}3{\cdot}10^{-3}$	$1{\cdot}4{\cdot}10^{-3}$	$1{\cdot}5{\cdot}10^{-3}$	$1{\cdot}6{\cdot}10^{-3}$	$1{\cdot}7{\cdot}10^{-3}$
1	2	3	4	5	6	7	8	9
	Number of germinated seeds. %							
7th day	2.5 ± 0.9	2.0 ± 0.4	4.0 ± 0.7	5.5 ± 0.9	7.0 ± 0.6	7.0 ± 1.1	6.0 ± 0.4	5.0 ± 0.4
Percentage to control		-20.0	+ 60.0	+ 120.0	+ 180.0	+ 180.0	+ 140.0	+ 100.0
Reliability of calculations, t_m	2.9	4.9	5.6	6.3	18.1	6.5	14.6	12.2
Accuracy of experiment (P), %	34.8	20.5	17.8	15.8	8.3	15.4	6.8	8.2
10th day	7.3 ± 1.1	4.3 ± 0.6	6.8 ± 0.5	6.0 ± 0.9	8.0 ± 0.9	8.5 ± 0.3	7.0 ± 0.4	6.0 ± 0.4
Percentage to control		-41.1	-6.8	-17.8	+ 9.6	+ 16.4	-4.1	-17.8
Reliability of calculations, t_m	6.6	6.8	14.2	6.5	8.7	29.3	17.1	14.6
Accuracy of experiment (P), %	15.2	14.7	7.1	15.3	11.5	3.4	5.9	6.8
15th day	15.3 ± 1.6	9.8 ± 1.0	9.8 ± 0.5	6.3 ± 0.5	12.3 ± 1.1	15.0 ± 2.7	9.0 ± 1.1	9.0 ± 1.8
Percentage to control		-35.9	-35.9	-58.8	-19.6	-2.0	-41.2	-41.2
Reliability of calculations, t_m	9.9	9.5	20.4	13.1	11.1	5.5	8.3	4.9
Accuracy of experiment (P), %	10.1	10.5	4.9	7.6	9.0	18.3	12.0	20.3
20th day	9.3 ± 0.6	11.5 ± 2.5	9.0 ± 0.7	11.0 ± 1.2	11.0 ± 1.8	11.0 ± 0.9	11.0 ± 0.9	10.0 ± 1.8
Percentage to control		+ 23.7	-3.2	+ 18.3	+ 18.3	+ 18.3	+ 18.3	+ 7.5
Reliability of calculations, t_m	14.8	4.5	12.7	8.9	6.2	12.0	12.0	5.5
Accuracy of experiment (P), %	6.8	22.1	7.9	11.2	16.2	8.4	8.4	18.3
25th day	4.0 ± 0.9	7.3 ± 0.3	6.5 ± 1.0	6.3 ± 0.6	4.8 ± 1.0	5.3 ± 1.4	9.0 ± 0.4	7.0 ± 0.9
Percentage to control		+ 82.5	+ 62.5	+ 57.5	+ 20.0	+ 32.5	+ 125.0	+ 75.0
Reliability of calculations, t_m	4.3	29.2	6.3	10.0	4.7	3.8	6.3	7.6
Accuracy of experiment (P), %	23.0	3.4	16.0	10.0	21.5	26.0	15.8	13.1
Seed germinative energy, %	10.0	6.3	11.0	12.0	15.0	16.0	13.0	11.0
Seed laboratory germination, %	38.4	35.0	36.1	35.1	43.1	47.0*	42.0	37.0

(*continued*)

Table 2. *(continued)*

Seeds accounting with duration germination	Control	Concentrations of drug solutions. ml/l						
		$1{\cdot}10^{-3}$	$1{\cdot}2{\cdot}10^{-3}$	$1{\cdot}3{\cdot}10^{-3}$	$1{\cdot}4{\cdot}10^{-3}$	$1{\cdot}5{\cdot}10^{-3}$	$1{\cdot}6{\cdot}10^{-3}$	$1{\cdot}7{\cdot}10^{-3}$
Significance of differences, t_{fact}		1.1	0.7	1.2	1.6	3.2	1.6	0.3
Number of not germinated seeds, %	62.0	65.0	64.0	65.0	57.0	53.0	58.0	63.0
Among them:								
disease-free	17.0	18.0	20.0	19.0	18.0	18.0	21.0	21.0
decayed	5.0	8.0	4.0	5.0	3.0	-	4.0	4.0
steamed	5.0	3.0	1.0	2.0	2.0	3.0	1.0	4.0
empty	11.0	14.0	17.0	15.0	14.0	11.0	12.0	14.0
abnormally germinated	20.0	20.0	19.0	20.0	18.0	19.0	19.0	19.0
pest-infested	4.0	2.0	3.0	4.0	2.0	2.0	1.0	1.0

* The differences are significant

The positive effect of Zircon on the growth of laboratory seed germination indicators was found at the concentrations of $1{\cdot}4{\cdot}10^{-3}-1{\cdot}6{\cdot}10^{-3}$ ml/l, comprising 42–47% (excess to the control – 9.4–22.4%). An increase in seed quality classes was noted from the third to the second and the first. The significance of differences with the control values was noted at the concentration of $1{\cdot}5{\cdot}10^{-3}$ ml/l: $t_{0.05} = 3.2 > t_{st} = 2.45$. The laboratory germination decreased relative to the control by 6–8.9% at the concentrations of $1{\cdot}10^{-3}$-$1{\cdot}3{\cdot}10^{-3}$ ml/l. The concentration of $1{\cdot}7{\cdot}10^{-3}$ ml/l also had an inhibitory effect. The decrease to the control was 3.6% (Table 2).

On the growth of biometric parameters of sprouts in length when using the stimulator Ribav-Extra all concentrations were effective and exceeded the control by 7,1–33,3%. The differences with the control group were significant at P = 0.05% and P = 0.01% (Fig. 1).

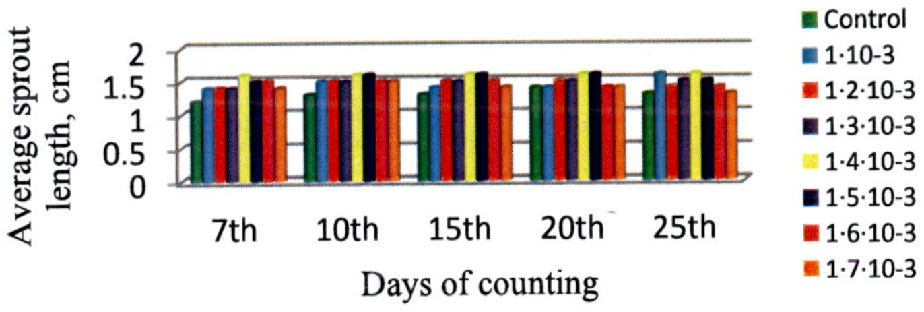

Fig. 1. The influence of the growth stimulator Ribav-Extra on the average length of sprouts when germinating Manchurian fir seeds.

Zircon showed an active growth in the length of sprouts when using the concentrations of $1{\cdot}2{\cdot}10^{-3}-1{\cdot}5{\cdot}10^{-3}$ ml/l. Excess to the control – 7.1–25%. There was a significant difference with the control indicators: $t_{fact} > t_{table}$ at P = 0.01%, P = 0.05%. The use of high ($1 \cdot 10^{-3}$ ml/l) and low ($1 \cdot 6 \cdot 10^{-3} - 1 \cdot 7 \cdot 10^{-3}$ ml/l) concentrations had an insignificant effect on the growth of sprouts (Fig. 2).

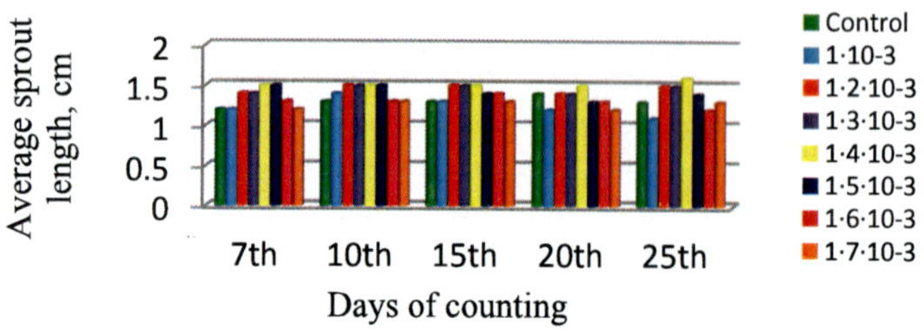

Fig. 2. The influence of the growth stimulator Zircon on the average length of sprouts when germinating Manchurian fir seeds.

The concentrations of 1·3·10–3–1·5·10–3 ml/l when using the drug Ribav-Extra accelerated the growth of sprout weight, exceeding the control values by 2.7–28.8%. At the concentration of 1·3·10–3 ml/l on the 25th day of counting the differences with the control values were significant: t0.05 = 2.8 > tst = 2.45. The concentrations of 1·10–3 and 1·7·10–3 ml/l had an inhibitory effect, decreasing the sprout weight relative to the control by 4.1–24.1%. The concentrations of 1·2·10–3 and 1·6·10–3 ml/l also had an inhibitory effect. The indicators decreased in relation to the control by 1.3–1.4%. The exceptions were the use of the concentration 1·2·10–3 ml/l on the 7th and 25th days of counting as well as the concentration of 1·6·10–3 ml/l on the 25th day. The indicators exceeded the control by 1.4–20.3% (Fig. 3).

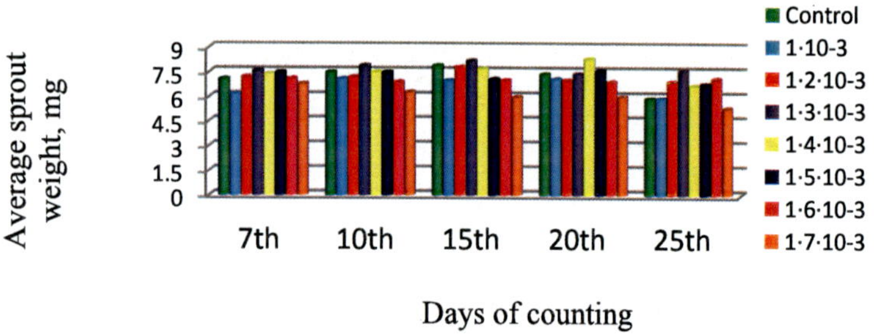

Fig. 3. The influence of the growth stimulator Ribav-Extra on the average weight of sprouts when germinating Manchurian fir seeds.

The use of the solution concentrations of $1 \cdot 3 \cdot 10^{-3} - 1 \cdot 5 \cdot 10^{-3}$ ml/l of the drug Zircon also had an inhibitory effect on the increase in weight indicators. The indicators exceeded the control by 1,3–25,4%. The reliability of the differences between the average values with the control was noted at the solution concentration of $1 \cdot 3 \cdot 10^{-3}$ ml/l on the 25th day of accounting: $t_{0,05} = 3{,}3 > t_{st} = 2{,}45$. The concentration of $1 \cdot 6 \cdot 10^{-3}$ ml/l had a weak effect. The concentrations of $1{\cdot}10^{-3} - 1{\cdot}2{\cdot}10^{-3}$ and $1{\cdot}7{\cdot}10^{-3}$ ml/l did not have a positive influence. There was a decrease in relation to the control by 8,1–24,1% (Fig. 4).

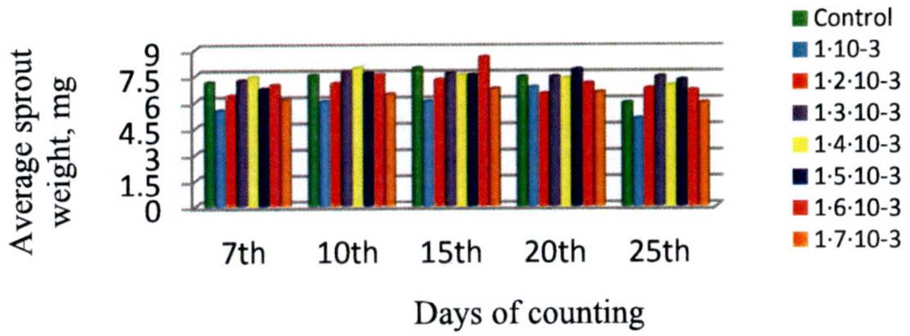

Fig. 4. The influence of the growth stimulator Zircon on the average weight of sprouts when germinating Manchurian fir seeds.

6 Discussion

Thus, the authors found, that the drugs Ribav-Extra and Zircon had a positive effect on the germinative energy, laboratory germination of seeds and the growth of sprout roots in the length and weight of Manchurian fir.

At the same time, the authors also conducted experiments to study the influence of these growth stimulants on the sowing qualities of seeds and biometric indicators of other economically valuable far eastern tree species: Scots pine (Pinus sylvestris L.), omatsu (P. densiflora Siebold et Zucc.) and Khingan fir (Abies nephrolepis (Trautv.) Maxim.). The high efficiency of the tested drugs was also noted.

So, when germinating Scots pine seeds the most positive effect of the growth stimulator Ribav-Extra on the germinative energy was noted at the solution concentrations of 1·3·10–3 and 1·5·10–3 ml/l, comprising 76–78% and on the indicators of laboratory germination when use the solution concentrations of 1·4·10–3 and 1·5·10–3 ml/l, where it comprised 90,1–93%. Excess to the control, respectively: 11,8–14,7 and 15,1–18,8%. Zircon showed a high efficiency on germinative energy at the concentration of 1·6·10–3 ml/l (up to 74,1%) and on laboratory germination at the concentrations of 1·3·10–3-1·6·10–3 ml/l (up to 86,0–90,1%). Excess to the control, respectively: 9 and 9,8–15,1%.

In omatsu, Ribav-Extra did not have a significant influence on germinative energy. However, its positive effect on laboratory germination was observed at the solution concentrations of 1·4·10–3–1·6·10–3 ml/l, at which it comprised 92,3–95,3%, exceeding the control by 6,1–9,5%. The stimulator Zircon was more effective at the solution concentrations of 1·10–3–1·3·10–3 ml/l, at which germinative energy and laboratory germination comprised, respectively: 82–87 and 94–96,3% (excess to the control – 9,3–16 и 8–10,7%).

In Khingan fir, a more active growth of sowing qualities of seeds was noted. So, Ribav-Extra had a positive effect on germinative energy at the solution concentrations of 1·2·10–3-1·7·10–3 ml/l; on laboratory germination – at the concentrations of 1·3·10–3—1·6·10–3 ml/l, comprising 4–16% (excess to the control – 21,2–384,8%) and 21,4–33,1%) (excess to the control – 127,7–252,1%). Zircon was more effective at the solution concentrations of 1·3·10–3-1·6·10–3 ml/l, at which germinative energy and laboratory germination comprised 7–10 and 15–27%, exceeding the control by 112,1–203% and, respectively, by 59,6–187,2%.

The positive effect of the stimulants can be explained both by the biological peculiarities of the tested seeds and by the positive influence of the growth stimulants used in the experiments. The growth stimulants also had an active effect on the increase of sprouts in length and weight.

Thus, in Scots pine, with an increase in the biometric parameters of sprouts along the length, the growth stimulator Ribav-Extra showed a high activity at the solution concentrations of 1·2·10–3—1·5·10–3 ml/l, exceeding the control by 5,6–53,3%; for the increase in the weight of sprouts – at the concentrations of 1·3·10–3—1·5·10–3 ml/l, exceeding the control indicators by 6,3–14,5%. Zircon had a positive effect on the increase in the length of sprouts at the concentrations of 1·4·10–3—1·6·10–3 ml/l (excess to the control –6,3–46,7%); on the increase in the weight of sprouts – at the concentrations of 1·5·10–3—1·6·10–3 ml/l, exceeding the control by 11,6–28,4%.

In omatsu, the stimulator Ribav-Extra had a positive effect on the increase of sprouts in length and weight at the concentrations of 1·2·10–3—1·6·10–3 ml/l, exceeding the control by 5,9–93,8 and 1,4–31,2%. Zircon showed high activity on the growth of sprouts along the length at the concentrations of 1·2·10–3—1·7·10–3 ml/l (excess to the control – 5,9–37,5%). However, the positive effect of the simulator on the growth of sprout weight was noted at the concentrations of 1·10–3—1·3·10–3 ml/l only on the 10th day of counting, exceeding the control by 16,4–23,9%.

In Khingan fir, the accelerated increase in the length of sprouts was observed at the concentrations of 1·10–3—1·6·10–3 ml/l (excess to the control – 7,7–50%). For the growth of sprout weight the drug Ribav-Extra was effective at all concentrations, exceeding the control values by 5,1–79,5%. The growth stimulator Zircon had an active effect on the growth of sprout length at the concentrations of 1·4·10–3-1·5·10–3 ml/l (excess to the control – 8,3–25%); on the increase in the sprout weight when using the concentrations of 1·2·10–3—1·7·10–3 ml/l (excess to the control – 6,1–46,5%).

In the future, it is planned to study the influence of these growth stimulants on the increase in germinating energy, laboratory germination and the growth of biometric sprout indicators of other coniferous tree species.

7 Conclusions

Thus, for the indicators of the germinative energy of Manchurian fir seeds the growth stimulator Ribav-Extra was effective when using all concentrations, for laboratory germination – at the concentrations of $1{\cdot}3{\cdot}10^{-3}$—$1{\cdot}6{\cdot}10^{-3}$ ml/l. Zircon, respectively, at the concentrations of $1{\cdot}2{\cdot}10^{-3}$-$1{\cdot}7{\cdot}10^{-3}$ and $1{\cdot}4{\cdot}10^{-3}$—$1{\cdot}6{\cdot}10^{-3}$ ml/l. The increase in seed quality classes was noted – from the third to the second and the first. For the growth of sprout indicators along the length Ri-bav-Extra had a positive effect at all solution concentrations, Zircon – when using the concentrations of $1{\cdot}2{\cdot}10^{-3}$—$1{\cdot}5{\cdot}10^{-3}$ ml/l.

Both the drugs are effective for the growth of sprout weight at the concentrations of $1{\cdot}3{\cdot}10^{-3}-1{\cdot}5{\cdot}10^{-3}$ ml/l.

It is necessary to study the influence of these growth stimulants on sowing qualities of seeds and biometric sprout indicators of other coniferous tree species.

References

1. Ostroshenko, V.Y.: Efficiency of use of the growth stimulator Ecopin when germinating omatsu (*Pinus densiflora* Siebold et Zucc.) seeds. General question of world science. Collection of scientific papers on materials of the international scientific-practical conference 30.03.2019. Ed. SIC Science Russia, pp. 95–104 (2019). https://doi.org/10.18411/gq-30-03-2019-23
2. Ostroshenko, V.Y., Ostroshenko, L.Y.: The effectiveness of the application of growth stimulants in growing Manchurian fir (*Abies holophylla* Maxim.) seedlings. Adv. Curr. Nat. Sci. **4**, 41–47 (2020). https://doi.org/10.17513/use.37360
3. Ostroshenko, V.Y., Ostroshenko, L.Y.: Effect of the growth stimulator Ecopin on sowing qualities of seeds and biometric parameters of Manchurian fir (*Abies holophylla* Maxim.) sprouts. Eurasion Union Sci. **2**(71), 17–23 (2020). https://doi.org/10.31618/ESU.2413-9335.2020.6.71.611
4. Egorova, A.V., Chernibrovkina, N.P., Robonen, E.V.: Effect of a coniferous drug on the growth and elemental composition of *Pins sylvestris* L. seedlings in a forest nursery. Chem. Plant Raw Mater **2**,171–180 (2017). https://doi.org/10.14258/jcprm.2017021720
5. Fraga, H.P.F., Vieira, L.N., Heringer, A.S., Puttkammer, C.C., Silveira, V., Guerra, M.P.: DNA methylation and proteome profiles of *Araucaria angustifolia* (Bertol.) Kuntze embryogenic cultures as affected by plant growth regulators supplementation. Plant Cell Tiss. Organ Cult. **125**(2), 353–374 (2016). https://doi.org/10.1007/s11240-016-0956-y
6. Kabanova, S.A., Danchenko, M.A., Kochegarov, I.S., Kabanov, A.N.: The experience of intensive cultivation of one-yar-old seedlings of *Pinus sylvestris* L. in Pavlodar region of the republic of Kazakhstan. Rus. J. **6**, 104–117 (2019). https://doi.org/10.17238/issn0536-1036.2019.6.104
7. Kirienko, M.A., Goncharova, I.A.: The prolonged influence of growth stimulants on morphometric indicators of three-year seedlings of main forest forming species of Central Siberia. Sib For. J. **1**, 65–70 (2018). https://doi.org/10.15372/SJFS20180107
8. Kuneš, I., Baláš, M., Linda, R., Gallo, J., Nováková, O.: Effects of bras- sinosteroid application on seed germination of Norway spruce, scots pine, douglas fir and English oak. iForest **10**(1), 121–127 (2016). https://doi.org/10.3832/ifor1578-009
9. Muhametshina, A.R., Petrova, G.A., Shaihraziev, S.H., Gibadullin, N.F., Rusakova, A.S.: Effectiveness of growth stimulants in European spruce cultivation under cover. For. Bull **24**(3):81–86. (2020). https://doi.org/10.18698/2542-1468-2020-3-81-86
10. Shchukin, R.A., Bogdanov, O.E., Zavoloka, I.P., Ryazanov, G.S., Kruglov, N.M.: Biotechnological basis for application of growth regulators for rooting of green cuttings of trees and shrubs in a greenhouse with a misting system. In: BIO Web of Conferences, vol. 23 (2020). https://doi.org/10.1051/bioconf/20202301009

Changes in the Areas of the Predominant Tree-And-Shrub Species of the Transbaikal Territory

Larisa Pak(✉)

Tver State Agricultural Academy, Marshal Vasilevsky Street (Sakharovo), 7, Tver 170100, Russia

Abstract. The results of analyzing the changes in the species composition of the forest fund in the territory of the Transbaikal territory have been provided for the period between 1956 and 2019 with the aim of identifying trends in the spatiotemporal variability. For obtaining information about resources of the forests in the region, materials of the State Forest Inventories and State Forest Registers have been retrospectively studied, which contain information about the main forest-forming tree species with division into coniferous, hardwood and softwood plants, as well as about other tree species and shrubs. The total of 23 dominant tree and shrub categories involved in the forests formation of the Transbaikal territory have been allocated, of which 11 categories have remained unchanged. As a result, it has been found that over the past 63 years, significant growth has been observed in the diversity of tree and shrub species, as well as the displacement of the zonal categories of trees and shrubs; the process of "green desertification" has been observed. Geographically, this process occurs unevenly, causing spatial diversity of changes in the areas. The most prosperous of all forest areas in the Transbaikal territoryis the Baikal mountain-forest district. The territory passed by the Trans-Siberian Railway – the Trans-Baikal mountain forest district – was significantly developed.

Keywords: Transbaikal territory · Biodiversity · Trees and shrubs · Forest fund · Forest-covered area

1 Introduction

Of the numerous issues that have been widely discussed lately, one of the priorities is the problem of studying global warming for predicting the response of the forest cover to climatic variability. Numerous studies predict further reduction of forests' sustainability on large areas and performing a variety of functions. Despite the fact that there are many factors and several uncertainties in assessing the effect of climate variability on the forests, there is also quite a reasonable assumption that forest transformation has started and boundaries of the forest vegetation are shifting [1–13].

Transbaikal territory is one of the entities of the Russian Federation rich in forests. Its share in the category of the Russia's "wooded land" is 3.6% [14, 15]. Over the period

A. Muratov and S. Ignateva (Eds.): AFE 2021, LNNS 353, pp. 648–657, 2022.
https://doi.org/10.1007/978-3-030-91402-8_72

from 1950 until present, intense climatic variability has been observed in the territory of the Krai: the surface air temperature has increased by 1.9 °C, and in the changes in the annual amount of precipitation, clearly visible cyclicity has been observed: the period from 1965 to 1981 was a period of reduced humidity, from 1982 to 1998 – a period of increased humidity, from 1999 to 2009 – a period of reduced humidity, and from 2010 until present – again a period of increased humidity. An assumption has been made that if the temperature keeps rising, hard climatic conditions will be observed in the next period of low humidity. This will naturally affect the forest cover in the region. One of the real ways of obtaining regional information about forests is the forest inventory data from forest management. This paper was aimed at studying the trends of spatiotemporal variability of the species composition in the region over the period between 1956 and 2019.

2 Materials

The data of the State Inventories of Forest Resources (SIFR) and State Forest Registers (SFR) are generalized information for obtaining the information about forest resources in the Transbaikal territory. The very first data about SIFR in the region were published in 1956 in the Statistical Handbook on Forest Fund of Russia. Afterward, such data were published until 2003 in approximately five years' intervals. According to the Forest Code, the term "State Forest Fund" was replaced with "State Forest Register". Since 2008, SFR of the region has been updated annually.

This article is based on the information obtained from the databases of SIFR and SFR for the Zabaykalsky Krai as of first of January in years 1956, 1961, 1966, 1973, 1978, 1983, 1988, 1993, 1998, 2003, 2008 – 2019, and State reports for years 1995, 1997, 1999, 2002 – 2019.

SIFR and SFR provide information about the main forest-forming tree species with division into coniferous, hardwood and softwood plants, as well as about other tree and shrub species. Moreover, according to the data available, hardwood species were for the first time allocated in the SIFR dated 01.01.1998. The exact number of species in the forests of the region is very difficult to be determined since the forest inventory in the process of forest management indicates only the generic name in the plantations composition. According to the materials of SIFR and SFR, over the 63 year period, the total of 23 dominant tree and shrub categories involved in the formation of forests of the Transbaikal territory have been allocated. These include: "larch", "birch", "pine", "cedar", "aspen", "spruce", "poplar", "fir", "gray alder", "black alder", "willow tree", "elm", "apricot", "bird cherry", "apple tree", "shrub birch", "dwarf pine", "purple willow", "shrub willow", "shrub alder", "sea-buckthorn", "other tree species", "other shrubs". Given the fact that some categories have not been always allocated over this long period due to their scarcity, it turned necessary to join "poplar", "gray alder", "black alder", "elm and other elms", "apricot", "bird cherry", "apple tree", "shrub alder", "sea-buckthorn", and "other shrubs" into category "other tree species". Category "purple willow", which was allocated in the SIFR until 1966, has been called "shrub willows" from 1973 until the present, and category "dwarf birch" has been renamed to "birch shrubs". Category "elm" includes all elms in the region. As a result, an ordered series of data have been obtained about the areas of tree and shrub species prevailing in the region. In the end, 11 categories remained unchanged, which had been considered based on the List of Forest Zones of the Russian Federation and the List of Forest Areas of the Russian Federation for the period between 2005 and 2019 (the available period).

3 Results

Over the 63-year period, the area covered with forest vegetation has increased significantly (by 20.3%). This increase was uneven in terms of the predominant species. The area covered by coniferous species (larch) has reduced, and the area covered by softwood species (birch), shrubs (shrub birch) has increased (Table 1). For the remaining categories, minor fluctuations have been noted.

Table 1. Dynamics of the areas with the species predominant in SIFR and SFR for Eastern Transbaikalia.

Years of accounting	Category name. % of the area covered with forest vegetation											
	larch	birch	pine	cedar	aspen	spruce	fir	shrub birch	dwarf pine	shrub willow	other tree species	Total area covered with forest vegetation
1956	81.0	3.1	9.3	1.5	0.1	0.003	0.007	0.4	4.2	0.002	0.01	100
1961	75.8	6.1	10.1	2.9	0.2	0.04	0.01	0.2	4.3	0.0004	0.05	100
1966	68.3	12.2	9.6	2.9	0.8	0.03	0.01	1.8	3.8	0.03	0.09	100
1973	63.8	14.0	10.2	3.5	1.2	0.03	0.02	3.07	3.8	0.03	0.1	100
1978	61.0	14.5	9.8	3.4	1.2	0.03	0.02	4.7	4.7	0.03	0.1	100
1983	58.9	14.8	9.7	3.1	1.3	0.04	0.02	5.8	5.8	0.03	0.1	100
1988	57.8	15.2	9.5	3.5	1.5	0.04	0.02	5.8	5.8	0.03	0.3	100
1993	56.9	15.5	9.1	3.4	1.5	0.04	0.01	5.8	5.8	0.03	0.4	100
1998	55.2	16.0	8.6	3.5	1.6	0.04	0.01	5.8	5.8	0.03	0.6	100
2003	57.0	16.3	8.8	3.6	1.7	0.04	0.01	5.8	5.8	0.03	0.6	100
2008	56.8	16.6	8.6	3.6	1.7	0.04	0.01	5.8	5.8	0.02	0.6	100
2009	55.6	18.0	9.1	3.4	1.8	0.03	0.01	5.5	5.5	0.03	0.6	100
2010	55.6	18.1	9.0	3.0	1.8	0.03	0.01	5.6	5.5	0.03	0.6	100
2011	55.6	18.0	9.0	3.4	1.8	0.03	0.01	5.5	5.5	0.02	0.6	100
2012	55.6	18.0	9.0	3.4	1.8	0.03	0.01	5.5	5.5	0.03	0.6	100
2013	55.8	18.0	9.0	3.4	1.8	0.03	0.01	5.5	5.5	0.03	0.6	100
2014	55.8	17.9	9.0	3.4	1.8	0.03	0.01	5.5	5.5	0.03	0.6	100
2015	56.1	18.0	9.0	3.4	1.8	0.03	0.01	5.5	5.5	0.03	0.6	100
2016	55.8	18.0	8.9	3.4	1.8	0.03	0.01	5.5	5.5	0.03	0.6	100
2017	55.7	18.0	9.0	3.4	1.8	0.03	0.01	5.5	5.5	0.03	0.6	100
2018	55.7	18.0	9.0	3.4	1.8	0.03	0.01	5.5	5.5	0.03	0.6	100
2019	55.7	18.0	9.0	3.4	1.8	0.03	0.01	5.5	5.5	0.03	0.6	100
Changes Over the 1956 – 2019 period. %	−25.3	+14.9	−0.3	+1.9	+1.7	+0.027	+0.003	+5.1	+1.3	+0.028	+0.59	

Such a situation indicates the territorial diversity of changes in the areas and leads to doubts about the climatogenicity of their causes.

Identification of the dependencies between changes in the acreage of the main species and the climate has revealed the presence of statistically significant correlations for several selected categories. A positive correlation (0.25 – 0.5) was observed for birch, cedar, shrub birch and shrub willow, dwarf pine, and other tree species, while negative correlation (0.1 – 0.4) was observed for pine, larch, aspen, spruce, and fir.

Analysis of the dynamics of the areas covered by tree and shrub species across forest zones and regions of the Krai has revealed spatial variability in the number of species in the forests of the region. Over the 14-year period, 45% of its area has undergone changes in the composition of the tree-and-shrub vegetation (Fig. 1, Table 2, Table 3).

In the Steppe zone that includes the forest-steppe Zabaykalsky region, slight growth was observed (by 0.015%) in the diversity of shrub plantings in three categories: "shrub birch", "dwarf pine", and "other wood species", as well as decreased shares of larch, birch, and spruce. A significant reduction in the diversity of tree and shrub species was noted in the Argun (by 0.3%) and Aginsk (by 0.08%) forest districts of the zone.

Of all studied categories in the South Siberian mountain zone, the increased shares of larch, birch, pine, cedar, spruce, fir, shrub birch, dwarf pine, and other tree species were found (by 44.3%). Among these, softwood species (birch) and shrubs (shrub birch) were dominating. For other categories, no changes have been noted in these zones.

In the forest areas within the South Iberian mountain zone, mixed trends in the dynamics of the forest cover have been noted.

Thus, in the Baikal mountain forest district, increased (by 1.16%) areas covered with larch, pine, cedar, spruce, fir were observed, as well as decreased areas covered with birch, shrub birch, and dwarf pine. In this forest area, larch, spruce, and pine were dominating. A significant reduction in the diversity of tree and shrub species was observed in Khiloksky (by 0.45%) and Krasnochikoysky (by 0.13%) forest areas of the zone.

In the Trans-Baikal mountain-permafrost area, positive changes (43.3%) were observed for the following categories: "larch", "birch", "cedar", "spruce", "shrub birch", "dwarf pine" and "other tree species". Among these, the shares of birch, shrub birch, and dwarf pine were the greatest. A negative trend was noted for the "fir" category. There was no reduction in the diversity of tree and shrub species in the forest areas of the zone.

In the Trans-Baikal mountain-forest area, the areas of birch, spruce, fir, and dwarf pine increased slightly. These were dominated by category "birch". The areas covered with larch, pine and birch shrub, reduced. Moreover, these were dominated by "shrub birch".

The Trans-Baikal mountain-forest district is the only district with a negative trend (0.14%) in the species diversity of the tree-and-shrub vegetation. A significant reduction in the diversity of tree and shrub species was noted in Ingodinsky (by 2.75%), Kyrinsky (by 0.52%), Chernyshevsky (by 0.38%), Sretensky (by 0.31%), Verkhne-Chitinsky (by 0.23%), Karymsky (by 0.21%), Beklemishevsky (by 0.14%), part of Argunsky (by 0.3%), Olenguysky (by 0.2%), Nerchinsky (by 0.03%), forest areas of the zone.

Border of forest zones
Border of forest regins
I – Southem-Siberian mountain zone
II – Forest-steppe zone
1 – The Baikal mountain forest district
2 – The Transbaikalian mountain-permafrost-affected district
3 – The Transbaikalian mountain forest district
4 – The Transbaikalian forest-steppe district

Fig. 1. Forest zones and forest areas layout in the territory of the Transbaikal territory.

Table 2. Dynamics of the areas covered with the species predominant in SIFR and SFR for the Transbaikal territory.

Forest division name	Changes over the period of 2005 – 2019. %										
	larch	birch	pine	cedar	aspen	spruce	fir	shrub birch	dwarf pine	shrub willow	other tree species
Aginskoye	0	0.02	0	0	0	-0.11	0.01	0	0	0	0
Akshinskoe	0.02	– 0.08	0	0	0	0.06	0	0.03	0.01	0	0
Aleksandrovsko–Zavodskoye	0.005	0.03	0	0	0	0.11	0	0	0	0	0.02
Argunskoye	– 0.01	– 0.13	0	0	0	– 0.18	– 0.01	0.03	0	0	0
Badinskoye	0.5	0.33	0.15	0	0	0.13	0.05	0	0	0	0
Baleyskoye	0.02	0.28	0	0	0	0.46	0	0.01	0	0	0
Beklemishevskoye	– 0.01	– 0.09	0	0	0	– 0.02	0	– 0.02	0	0	0
Verkhne-Chitinskoye	– 0.04	– 0.15	0	0	0	– 0.03	0	– 0.01	0	0	0
Gazimuro-Zavodskoye	– 0.03	2.32	0	0	0	1.02	0.14	0.12	0	0	0
Duldurginskoye	0.02	0.12	0	0	0	0.06	– 0.01	0	0	0	0
Ingodinskoye	– 0.09	– 0.32	– 0.07	0	0	– 0.07	– 0.01	– 2.22	0.03	0	0
Karymskoe	0	– 0.17	0	0	0	– 0.02	0	– 0.02	0	0	0
Krasnochikoyskoe	0.1	– 0.54	0.42	0.01	0	– 0.04	– 0.01	– 0.06	– 0.01	0	0

(continued)

Table 2. *(continued)*

Forest division name	Changes over the period of 2005 – 2019. %										
	larch	birch	pine	cedar	aspen	spruce	fir	shrub birch	dwarf pine	shrub willow	other tree species
Kyrinskoye	– 0.08	– 0.33	– 0.04	0	0	0	– 0.02	– 0.03	– 0.02	0	0
Mogochinskoye	1.11	9.36	0	0	0	1.62	0.01	10.2	1.93	0	0
Nerchinskoye	– 0.02	– 0.05	0	0	0	0.04	0	0	0	0	0
Olenguyskoe	– 0.06	– 0.09	0	0	0	– 0.04	– 0.01	0	0	0	0
Ononskoye	– 0.06	0	0	0	0	0.05	0.01	0	0	0	0
Petrovsk-Zabaykalskoye	0.23	0.19	0.07	0	0	0.05	0	0.04	0	0	0
Sretenskoye	– 0.04	0.1	0	0	0	– 0.42	0.02	0.03	0	0	0
Tungokochenskoye	– 0.03	9.2	0	0	0	– 0.021	– 0.02	– 0.17	– 0.07	0	0.01
Khilokskoye	– 0.07	– 0.23	– 0.04	0	0	– 0.07	– 0.01	– 0.03	0	0	0
Charskoye	0.14	6.4	0	0.01	0	0.19	0	– 0.06	2.88	0	0.67
Chernyshevskoye	– 0.01	– 0.24	0	0	0	– 0.1	0.02	– 0.05	0	0	0
Chitinskoye	0.04	0.03	0	0	0	– 0.02	– 0.01	0	0	0	0
Shilkinskoye	0	– 0.05	0	0	0	0.12	0.48	0	0	0	0

Table 3. Dynamics of the areas covered with the species predominant in forest zones and forest regions for the Transbaikal territory.

Forest zones and forest region name	Changes over the period of 2005 – 2019. %										
	larch	birch	pine	cedar	aspen	spruce	fir	shrub birch	dwarf pine	shrub willow	other tree species
Southern-Siberian Mountain zone											
Baikal Mountain Forest district	0.76	-0.25	0.6	0.01	0	0.07	0.03	-0.05	-0.01	0	0
Zabaykalsky mountain-permafrost region	1.22	24.96	0	0.01	0	1.789	-0.01	9.97	4.74	0	0.68
Zabaykalsky mountain forest district	-0.33	1.11	-0.11	0	0	0.74	0.6	-2.16	0.01	0	0
Forest-steppe zone											
Zabaikalsky forest-steppe region	-0.025	-0.04	0	0	0	-0.01	0	0.06	0.01	0	0.02

4 Discussion

The positive dynamics in the variation of the area of the lands in the Forest Fund territory covered with forest vegetation were associated with the handover of the lands unfit for agricultural use and abandoned lands, which have been gradually covered by coniferous, small-leaved species and shrubs. The decrease in the volume of forest harvesting, self-overgrowth in the areas of "idle" (burning, dead stands, cuttings, clearings) and areas occupied by natural woodlands also led to an increase in the area covered by forest.

Regional trends in larch, pine, spruce and fir indicate a low degree of involvement of climatic processes. Of the coniferous species, a significant decrease in the share of larch, of pine and an insignificant one in that of spruce, fir have been related to numerous forest fires, industrial logging, and forest drying due to the spread of pests, diseases, and other reasons. Significant areas of larch forests of the region are located in remote areas on permafrost. This suggests that the fight against forest fires in these areas is not carried out, only the consequences of fires are recorded. The reduction in the area of larch and pine forests in the Trans-Baikal mountain forest region is due to the excessive development of the forest territory through which the Trans-Siberian Railway passes. There is a greater decrease in the diversity of tree and shrub vegetation than in any other area.

According to a number of authors, the reasons for the weakening and drying of Siberian spruce forests are an increase in the temperature of the surface air layer, leading to the activation of root rot [16]. The reason for the negative correlation of temperature growth with changes in the areas of spruce forests in the Trans-Baikal forest-steppe region is found.

A decrease in the area in relation to 2005 was noted in the Zabaykalsky mountain-permafrost region for fir (by 0.01%). There is also no statistically significant correlation with the temperature increase.

The ongoing reforestation at forest departments, along with the good natural renewal of the coniferous species over the last 20 years have failed to rectify the trend of reducing the area of coniferous forests (larch, pine). Measures for the restoration of coniferous species were carried out only in relation to pine. But they were not enough. More than half of the forest crops were written off as a result of their death. For larch forests, no measures have been taken or not enough has been done.

The transition from the liquidation of the old management system - the State Forest Service of the Transbaikal territory to the formation of a new one – the Ministry of Natural Resources of the Transbaikal territory with the inclusion of two departments (the Department of Forest Management and the Department of State Supervision and Protection of Forests) led to a reduction in the area of forest nurseries. Preference began to be given to a cheaper and easier method of reforestation – the implementation of measures to promote natural forest regeneration and self-regrowth. Self-regrowth of the lands of the reforestation fund, as a rule, takes place naturally, with the change of coniferous species to deciduous ones. Therefore, the area of soft-leaved stands has increased and exceeded the share of coniferous forests. Economically valuable coniferous plantings were replaced by less valuable, small-leaved ones.

Birch is characterized by a significant increase in area, often associated with an increase in temperature. Given that birch is an early-successional species, it is necessary to recognize the increased probability of climatogenicity of this trend. Small-leaved species are usually considered as evidence of anthropogenic transformations of coniferous forests as a result of fires and violations of forest management. Probably, therefore, in the Trans-Baikal mountain-permafrost region, there is a significant increase in the areas occupied by birch and shrub birch. Within this forest area, in the Mogochinsky district, was located LLC "Pulp and Industrial Plant "Polar" (Amazar pulp and Paper Mill), the laying of which began in 2013. To provide the plant with raw materials, huge areas of forests were allocated in the Mogochinsky district of Transbaikalia. The Amazar Pulp

Mill project had more negative consequences than positive ones. Therefore, in 2019, the investment project of the Amazar Pulp and Paper Mill was deprived of the priority status in the field of forest development.

A significant reduction in the area of birch over a 14-year period was observed in the Baikal mountain forest region (by 0.25%) in the Khiloksky and Krasnochikoysky districts of the region. Here, the increase in the share of coniferous species was reflected in a decrease in the diversity of soft-leaved species and shrubs. This is because conifers tend to form tree-stands with few other tree species. It is the most prosperous forest district in the entire territory of the Transbaikal territory, into which species that are unusual for this area do not penetrate.

The growth of shrubs (bush birch) over a 63-year period also indicates a catastrophic transformation of the forest environment, the death of soft-leaved species due to the passage of fires in these areas or their drying out due to changes in the hydrological regime, the replacement of the area of soft-leaved species with shrubs, the appearance of forest-suitable areas for a long time. This situation is particularly evident in the Trans-Baikal forest-steppe region. In other words, there is a shift in the zonal categories of tree and shrub vegetation, the species diversity changes with the displacement and replacement of some species with others that are most adapted to the current conditions, that is, there is a process of "green desertification" of the region. Geographically, this process occurs unevenly, causing spatial diversity of changes in the areas (Fig. 2).

In artificial reforestation and afforestation, it is necessary to strive to recreate the possible local (natural) fullness of biocenoses for a given forest zone or area. This is what will prevent the emergence of free ecological niches, where species that are not

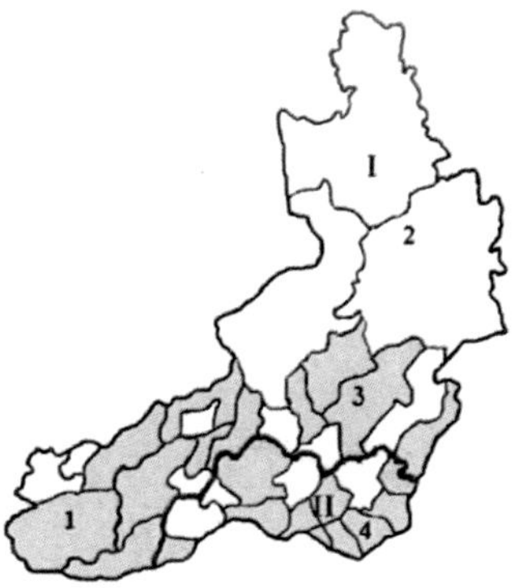

Border of forest zones
Border of forest regins
I – Southem-Siberian mountain zone
II – Forest-steppe zone
1 – The Baikal mountain forest district
2 – The Transbaikalian mountain-permafrost-affected district
3 – The Transbaikalian mountain forest district
4 – The Transbaikalian forest-steppe district

Fig. 2. Forest areas where the diversity of tree and shrub species is reduced in Transbaikal territory.

typical for this community could penetrate, and will allow maintaining the stability of the natural ecosystem and species diversity of this region.

5 Conclusion

Summing up the aforesaid, it should be noted that a significant increase in the diversity of tree and shrub species has been observed in the Transbaikal territory over the past 63 years. This fact shows not only the influence of climatogenic factors but also the effect of the anthropogenic factor on the reduction of the indigenous coniferous biocoenoses and on the appearance of free ecological niches that are occupied by species unnatural for this area.

References

1. Grigorieva, S.O., Konstantinov, A.V., Shkolnik, I.M.: The effect of climate changes on the composition of tree stands, their stability, and the habitats of major forest tree species. Proc Petersburg Res. Instit. For. **3**, 4–21 (2016). https://doi.org/10.21178/2079-6080.2016.3.4
2. Nazimova, D.I., Tsaregorodtsev, V.G., Andreeva, N.M.: Forest vegetation zones of southern Siberia and current climate change. Geogr. Nat. Res. **31**(2), 124–131 (2010). https://doi.org/10.1016/j.gnr.2010.06.006
3. Zamolodchikov, D.G.: Predicting the growth of global temperature in the twenty-first century based on a simple statistical model. Comp. Res. Mod. **8**(2), 379–390 (2016). https://doi.org/10.20537/2076-7633-2016-8-2-379-390
4. Zamolodchikov, D.G.: Natural and anthropogenic components of the modern changes of global temperature. Dokl. Earth Sci. **456**(1), 631–634 (2014). https://doi.org/10.1134/S1028334X14050390
5. Semenov, M.A., Vysotsky, A.A., Pashchenko, V.I.: Forecast of adaptive adaptations in forestry in connection with possible climatic changes. Izvestiya vysshikh uchebnykh zavedeniy. For. Mag. **5**(371), 57–69 (2019)
6. Golubyatnikov, L.L., Mokhov, I.I., Eliseev, A.V.: Nitrogen cycle in the Earth's climate system and its modeling. Izvestiya. Atmosph. Ocean Phys. **49**(3), 229–243 (2013). https://doi.org/10.7868/S0002351513030073
7. Crookston, N.L., Rehfeldt, G.E., Dixon, G.E., Weiskittel, A.R.: Addressing climate change in the forest vegetation simulator to assess impacts on landscape forest dinamics. J. For. Ecol. Manag. **7**, 1198–1211 (2010). https://doi.org/10.1016/j.foreco.2010.07.013
8. Eliseev, A.V., Mokhov, I.I.: Uncertainty of climate response to natural and anthropogenic forcings due to different land use scenarios. Adv. Atmos. Sci. **28**(5), 1215–1232 (2011). https://doi.org/10.1007/s00376-010-0054-8
9. Zaehle, S., Dalmonech, D.: Carbon nitrogen interactions on land at global scales: current understanding in modelling climate biosphere feedbacks. Curr. Opin. Environ. Sustain. **3**, 311–320 (2011). https://doi.org/10.5194/bg-2019-298
10. Gerber, S., Hedin, L.O., Oppenheimer, M., et al.: Nitrogen cycling and feedbacks in a global dynamic land model. Glob. Biogeochem. Cycles **24**(GB1001) (2010). https://doi.org/10.1029/2008GB003336
11. Klimenko, V.V., Tereshin, A.G., Mikushina, O.V.: An impact of atmospheric and climate changes on the energy potential of Russian forest recources. Rep. Acad. Sci. **488**(6), 612–618 (2019). https://doi.org/10.31857/S0869-56524886612-618

12. Filipchuk, A.N., Malysheva, N.V., Zolina, T.A., Yugov, A.N.: Boreal forests of Russia: opportunities for climate change mitigation. For. Inform. **1**, 92–114 (2020). https://doi.org/10.24419/LHI.2304-3083.2019.4.10
13. Sergienko, V.G., Konstantinov, A.V.: Forecast of influence of climate change on ecosystems and natural diversity species of Russian flora and fauna biotic complexes. Proc. Petersburg Sci. Res. Inst. For. **2**, 29–44 (2016). https://doi.org/10.21178/2079-6080.2016.2.27
14. Pak, L.N.: Dynamics and state of the forest fund of the Trans-Baikal territory for the period from 1998 to 2016. Ecol. Environ. Conserv. J. Pap. **23**(2), 1056–1062 (2017)
15. Pak, L.N.: Variability of quality indicators of Pinus sylvestris L. seeds in the territory of the Zabaikalsky krai. For. Magaz **6**, 262–269 (2019). https://doi.org/10.17238/issn0536-1036.2019.6.262
16. Nikolaeva, M.A., Faizulin, D.K.H., Potokin, A.P., Jamaleev, O.A.: Comparative evaluation of preservation and growth of spruce climatypes based on long-term provenance trials in Russia. Folia For. Pol. **56**(1), 56–67 (2014). https://doi.org/10.2478/ffp-2014-0006

Ecological and Chemical Assessment of the Habitats of Cranes in the Khingan State Nature Reserve, Russia

Antonina Pakusina[1(✉)], Tatyana Platonova[2], Tatyana Parilova[3], Mikhail Parilov[3], Nikita Malinovsky[1], and Irina Balan[3]

[1] Far Eastern State Agrarian University, ul. Politehnicheskaja, 86, Blagoveshchensk, Amur Oblast 675005, Russia
[2] Amur State University, 21, Ignatievskoe St, Blagoveshchensk 675028, Russia
[3] Khingan State Nature Reserve Arkhara, Amur region, Arkhara, Russia

Abstract. The article discusses the results of studies of hydrochemical indicators, the content of heavy metals in ecosystem components (water, fish, macrophytes, eggshells of birds) of the Khingan State Nature Reserve, where rare and endangered bird species such as White-naped crane, Red-crowned crane and others. At the Station for the reintroduction of rare species of birds of the Khingan State Nature Reserve, cranes are raised and released into the wild. In the course of the study, hydrochemical indicators were determined in the Borzya river, Kleshenskoye Lake and Dolgoe Lake. In summer, the water of reservoirs is saturated with oxygen, characterized by a high value of permanganate oxidizability. In terms of phosphate content, the water bodies are eutrophic. The content of zinc and copperin water exceeded the fishery standard. The content of lead and cadmium in the water of the lakes did not exceed the fishery standard. The high content of manganese and total iron in water is due to a natural factor. Macrophytes (*Potamogeton maackianus*, *Myriophyllum spicatum*, *Trapa maximowiczii*) showed a high concentration of lead (26.9 mg/kg, 60.7 mg/kg and 29.3 mg/kg respectively). High concentrations of lead and cadmium were found in the muscle tissue of fish. In the eggshells of the White-naped crane and Red-crowned crane, the content of heavy metals decreases in the order Fe > Cu > Pb > Zn > Ni > Cd.

Keywords: Wetlands · Heavy metals · Macrophytes · Bird habitat

1 Introduction

The Khingan State Natural Reserve is located in the southeast of the Amur Region within the Khingano-Arkharinskaya Lowland. It is listed as wetlands of international importance in the frame of Ramsar convention. Its wetlands are the breeding place of many rare and endangered birds species such as White-naped crane Grus vipio (Pallas, 1811), Red-crowned crane Grus japonensis (S. Müller 1776) and others. In Khingan nature reserve the Station for the reintroduction of rare species of birds is established

A. Muratov and S. Ignateva (Eds.): AFE 2021, LNNS 353, pp. 658–666, 2022.
https://doi.org/10.1007/978-3-030-91402-8_73

where cranes are raised and released into the wild. Recently much attention is paid on ecological condition of territories that have high conservation status [1, 2].

Wetlands in the Middle stream Amur river basin are transforming because of agricultural activity [3], climate change [4] and changing rivers water regime [5–7]. Identification of heavy metal contamination in different environmental components is the one of main environmental exploration aspect. Heavy metals are environmental pollutants that can make negative influence on any species and can have anthropogenic or natural origin. Macrophytes and fish accumulate heavy metals, which enter water bodies as a result of runoff from the surface soil layers during severe floods. Macrophytes are sensitive indicators of climate change and anthropogenic pollution [8]. Birds feed on macrophytes, birds live in thickets of aquatic plants and hide from danger. Environmental pollution with heavy metals has a negative impact on birds, contributes to the disease and death of birds [9]. The accumulation of heavy metals in the habitat leads to a decrease in the growth rate of chicks and a decrease in the number of birds [10]. The limiting factors for birds are natural and climatic conditions that affect the survival of clutches and young stock. The decline in the number of birds occurs as a result of forest fires and spring agricultural burns. At the same time, birds' nests burn down and chicks perish [11, 12]. Floods, that are common for Amur River floodplain, can affect the transport of chemical pollutants [13].

Bottom sediments, macrophytes and hydrobionts, including fish, are widespread objects for studying the accumulation of heavy metals in wetland ecosystems [8]. Taking into account the fact that the accumulation of ecotoxicants increases as it moves along the trophic chains, it can be assumed that birds, whose food is based on hydrobionts of lakes and wetlands (including Red-crowned and White naped Cranes), could accumulate toxic elements and suffer from their negative impact [9, 10].

Territory of Antonovskoe forestry of Khingan nature reserve, the place of our investigation, was excluded from any economic use in 1978 and nowadays has minimum anthropogenic press from individual farms located on the periphery of the reserve. Because of that, this territory could be used as a reference for determining the background concentrations of heavy metals in various components of ecosystems.

The aim of this work is to study the ecological state of lakes and small rivers of the Khingan State Nature Reserve by hydrochemical indicators and the content of heavy metals in water, in macrophytes, fish and eggshells of Red-crowned and White-naped cranes kept at the reintroduction station for the of rare bird species of the Khingan nature reserve.

2 Object and Methods of Research

The investigations were carry out in Antonovskoe forestry of Khingan State nature Reserve. Water samples, bottom sediments, macrophytes and fish for analysis were collected in Kleshenskoye Lake (N 49°24′3″, E 129°43′37″), Dolgoe Lake (N 49°22′55, E 129°44′22″) and Borzya River (N 49°24′5″, E 129°44′12″). Eggshells of Red-crowned and White-naped cranes were collected form birds, kept in Khingansky Reserve Reintroduction Station. In spring and summer period the birds are kept in enclosures on the bank of Kleshenskoe Lake. Birds are fed by fish from this lake and they get drinking water from this lake as well.

Water samples were taken during the dry season in spring and summer 2016–2017. Measurements to determine the hydrochemical parameters were performed on a PD-303S digital spectrophotometer (Apel, Japan). Macrophytes, fish were sampled in summer, washed with bidistilled water, and dried in an oven at 50 °C. Sample preparation of samples of macrophytes, bottom sediments, fish, eggshells was carried out using the system of microwave mineralization of samples at high pressure "Minotavr" (manufactured by "Lumex"). Heavy metals in water, macrophytes, fish, eggshells of birds were determined on an AAnalyst 400 atomic absorption spectrometer with flame atomization of samples.

3 Research Results and Discussion

3.1 Hydrochemical Indicators of the Water of the Borzya River, Kleshenskoe and Dolgoe Lakes

The water temperature in the reservoirs of the Khingan Nature Reserve corresponded to the hydrological season. The acid-base properties (pH) of the waters of Borzya river were within 6.26 – 6.63, Kleshenskoe Lake – 5.96 – 6.5, Dolgoe Lake – 6.63 – 6.69, which is below the norm. Low values of specific electrical conductivity of water in reservoirs (63 – 103 μS/cm) indicate low mineralization. The lowest values of specific electrical conductivity were observed in spring, and maximum in summer. The total hardness of water in reservoirs did not exceed 1.35 mg-eq/L. Water in reservoirs has a high color. The color of water in Kleshenskoye Lake ranged from 201° of color in spring to 663 degrees of color (Cr-Co) in summer. In Lake Dolgoe and in the Borzya River, the chromaticity reached 273 degrees of color. The color of water depends on the type of soil, the content of iron and manganese compounds in the water.

The oxygen saturation of water in the summer period in the Kleshenskoe and Dolgoe lakes is 95–101%. In the Borzya River, the oxygen content did not exceed 3.6 mgO_2/dm^3. Water in reservoirs had high values of permanganate oxidizability (PO), which characterizes a high content of organic matter in water. The BOD5 value did not exceed 3.9 mgO_2/dm^3 (Table 1).

Table 1. Oxygen indices of water in reservoirs of the Khingan nature reserve.

Name	Dissolved oxygen, mgO_2/dm^3	BOD_5, mgO_2/dm^3	PO, mgO/dm^3
Borzya River	3.1 – 3.6	0.2 – 3.9	13.5 – 18.8
Kleshenskoe Lake	2.2 – 9.5	0.9 – 3.7	14.1 – 17.92
Dolgoe Lake	4.8 – 8.8	0.4 – 3.6	10.7 – 15.33

Nitrate nitrogen predominated among nitrogen compounds in summer. In Borzya River, the content of nitrate nitrogen was 4.07 mg/dm^3, in Kleshenskoye Lake 10.23 mg/dm^3. In spring, the content of nitrates was minimal and did not exceed 0.34 mg/dm^3. However, in spring, the content of ammonium nitrogen increased to

0.51 mg/dm^3 in the Borzya River and to 0.37 mg/dm^3 in Kleshenskoye Lake. Nitrite in the waters of reservoirs was found in small amounts 0.002–0.005 mg/dm^3. Seasonal dynamics is observed in the content of nitrogen compounds: in the fall, their content is minimal in water, since nitrates are assimilated by aquatic plants. The forms of nitrogen in water are interrelated with the processes of nitrification and denitrification that occur in rivers and lakes, with physicochemical parameters at different times of the year [14].

The waters of the Borzya River, Kleshenskoye and Dolgoe Lakes are characterized by a high content of phosphates. For example, in the summer of 2016 in the Borzya River and Kleshenskoye Lake, the phosphate content was 0.71 mg/dm^3 and 0.92 mg/dm^3, respectively. In spring 2017, in Kleshenskoe and Dolgoe Lakes, the content of total phosphorus did not exceed 0.071 mg/dm^3. A large amount of total phosphorus in the Borzya River and Kleshenskoye Lake is the cause of eutrophication.

The content of pollutants in the water of the Kleshenskoe and Dolgoe Lakes and the Borzya river was within the normal range: oil products - 0.01 mg/dm^3, phenols –0.0008 mg/dm^3, anionic surfactants less than 0.1 mg/dm^3.

The cadmium content in the water of the Kleshenskoe and Dolgoe Lakes and the Borzya River during the study period did not exceed the norm. In the summer of 2016, the lead content in the Borzya River reached 6.00 μg/dm^3, during the subsequent study period, the lead content in the Borzya River water corresponded to the norm. The content of zinc and copper in the water of the Borzya River and Kleshenskoe and Dolgoe Lakes exceeded the fishery standard (Table 2). The highest concentrations of heavy metals in water were observed in summer. In the spring of 2017, the zinc content in Dolgoe Lake did not exceed the standard and amounted to 3.67 μg/dm^3, in other water bodies - no more than 15.9 μg/dm^3. The lead content in the water of the reserve's reservoirs ranged from 1.2 μg/dm^3 to 3.42 μg/dm^3, cadmium not more than 0.06 μg/dm^3. The formation of the chemical composition of natural waters is influenced by soil, rocks, relief features, climatic conditions, and floods. The high iron content in the watercourses of the Amur basin is due to the presence of iron-bearing minerals in soils and rocks of taiga territories. It is known that high levels of manganese and iron are considered as the regional geochemical feature of Amur River basin natural landscapes [1, 16].

Table 2. The content of heavy metals in the water of reservoirs of the Khingan nature reserve in July 2016, μg/dm^3.

Name	Pb	Cd	Cu	Zn
Borzya river	5.95 ± 1; 0.5	0.19 ± 0.02	11.99 ± 1.2	31.1 ± 3.2
Kleshenskoe lake	3.68 ± 1; 0.4	0.18 ± 0.05	14.69 ± 1.4	30.1 ± 3.2
Dolgoe lake	4.10 ± 1; 0.4	0.73 ± 0.1	3.37 ± 0.3	20.14 ± 2.0
$MPC_{f\text{-}r}$	6	5	1	10
$MPC_{h.\text{-}d.c.\text{-}p}$	10	1	1000	1000

Note: $MPC_{f\text{-}r}$ is the maximum permissible concentration of fisheries, $MPC_{h.\text{-}d.c.\text{-}p}$ is the maximum allowable concentration of household and drinking and cultural purposes.

3.2 Content of Heavy Metals in Macrophytes of the Borzya River and the Kleshenskoe, Dolgoe Lakes

Potamogeton maackianus and *Myriophyllum spicatum* are hydrophytes capable of accumulating heavy metals from water. Calla palustris grows along the banks of the Borzya River and is a poisonous plant for many animals. *Trapa maximowiczii* is found in the Far East in the floodplain water bodies of the Amur River basin; it is included in the Red Data Books of various regions of Russia. In the macrophytes of the reservoirs of the Khingan Nature Reserve *Potamogeton maackianus* and *Trapa maximowiczii*, high concentrations of lead (>30 mg/kg) and manganese (>300 mg/kg) were found (Table 3).

Table 3. The average level of heavy metals in aquatic plants of the small Borzya River, Dolgoe and Kleshenskoe Lakes, mg/kg (n = 5).

Macrophyte	Habitat	Pb	Cu	Zn	Fe	Mn
Calla palustris (rhizomes) *Calla palustris* (shoot)	Borzya River	57.9 ± 5.8 13.8 ± 1.4	24.7 ± 2.4 13.7 ± 1.3	126.7 ± 12.7 180.0 ± 18.1	7711 ± 771 2616 ± 262	98 ± 10 242 ± 24
Myriophyllum spicatum L	Borzya River	26.9 ± 2.7	14.9 ± 1.5	64.2 ± 6.5	4019 ± 402	548 ± 55
Potamogeton maackianus A. Benn	Kleshenskoe Lake	60.7 ± 6.1	13.1 ± 1.8	106.0 ± 10.6	2315 ± 235	205 ± 21
Trapa maximowiczii (nut fruit) *Trapa maximowiczii* Korsch. (leaf)	Dolgoe Lake	25.5 ± 3.0 29.3 ± 4.0	0.5 ± 0.01 69 ± 7.0	27.1 ± 3.0 94.3 ± 9.5	2440 ± 244 3917 ± 391	716 ± 72 771 ± 77

In general, the levels of accumulation of lead, manganese and zinc in the studied plants are comparable with the data obtained by Chinese ecologists when studying the content of heavy metals in some types of macrophytes under conditions of anthropogenic pollution [15]. Under the conditions of agricultural pollution in Zeya-Bureya plain, a high concentration of lead (13.5 mg/kg) was found in macrophytes Najas marina, Ceratophyllum oryzetorum - a high concentration of cadmium (2.39 mg/kg), but both plants have a deficient concentration of zinc (less than 20 mg/kg) [16]. In macrophytes Nymphoides peltata and Myriophyllum spicatum L., the Pb content was 30.5 ± 3.5 mg/kg [17]. Hydrophytes submerged in water are capable of accumulating heavy metals in large quantities. The content of zinc, copper, lead, and manganese in macrophytes depends on the content of these elements in water and bottom sediments [16, 17].

3.3 Content of Heavy Metals in Fish from Kleshenskoye Lake

At the Station for the reintroduction of rare species of birds of the Khingan State Nature Reserve, Red-crowned and White-naped cranes, which are listed in the Red Book, are grown and released into the wild. The scientific station of the reserve and the summer camp of the Station for the reintroduction of rare bird species are located on the bank of Kleshenskoye Lake. By this way, pollutants get in and accumulate in cranes bodies.

We studied the content of heavy metals in the muscle tissue of Silver Carp Carassius gibelio (Bloch, 1782) weighing from 33 g to 89 g, Lake Minnow Rhynchocypris percnurus (Pallas, 1814) weighing from 33 g to 50 g, Amur Sleeper Perccottus glenii (Dybowski, 1877) weighing from 35 g to 53 g. The content of iron, zinc and manganese in the muscle tissue of the fish was much lower than the MPL (maximum permissible levels). In the muscle tissue of Carassius gibelio, the copper content ranged from 6.5 mg/kg to 17.7 mg/kg, which exceeds the MPL. In the muscle tissue of the fish of Kleshenskoye Lake, a higher content of nickel, cadmium and lead was found (Table 4).

Table 4. Average content of heavy metals in muscle tissue of fish from Kleshenskoye Lake, mg/kg (n = 9).

Fish	Weight, g	Cu	Ni	Fe	Zn	Cd	Mn	Pb
Carassius gibelio (Bloch, 1782)	59.3 ± 10.5	11.7 ± 1.5	0.8 ± 0.2	8.0 ± 1.1	20.7 ± 2.3	0.26 ± 0.09	0.27 ± 0.01	2.0 ± 0.6
Rhynchocypris percnurus (Pallas, 1814)	39.3 ± 5.3	6.6 ± 1.1	0.9 ± 0.2	7.2 ± 0.4	5.3 ± 0.2	0.23 ± 0.06	0.30 ± 0.01	1.2 ± 0.6
Perccottus glenii (Dybowski, 1877)	43.3 ± 3.3	3.3 ± 0.1	1.2 ± 0.1	7.4 ± 0.1	3.3 ± 0.1	0.33 ± 0.33	0.26 ± 0.01	2.2 ± 0.6
MPL*		10	0.5	30	40	0.2	10	1

* *Hygienic requirements for safety and nutritional value of food products. SanPiN 2.3.2.1078–01*

Amur Sleeper muscle tissue contains more toxic elements such as nickel, lead and cadmium than Minnow and Carp. Amur Sleeper as ichthyophage, absorbs heavy metals in a bioavailable form in higher concentrations than Carp and Minnow, that feed on plankton, algae, and insect larvae.

Scientists, characterizing the habitat of cranes in the wetlands of northeastern China, indicated that the content of zinc and copper in cyprinids Cyprinidae is higher than in Odontobutidae [18].

3.4 Content of Heavy Metals in Eggshells of Cranes

The accumulation of heavy metals in the tissues and organs of birds is being actively studied [20]. Determination of heavy metals in eggshells and feathers of birds is a non-invasive way of obtaining information [18, 20]. Chicks and adult birds are used to register pollutants in the environment. If these are migratory birds, then it is better to study birds of the first year of life [21]. Relevant works could be the studies to determine pollutants in the habitat of birds of wetlands, endangered [1–3, 18], waterfowl and hunting bird species [17].

In the eggshells of White-naped and Red-crowned cranes of Khingan Reserve, iron and copper were present in the highest quantities, and cadmium in the smallest quantities (Table 5). The content of manganese and mercury was less than the detection limit. The presence and concentration of heavy metals in eggshell of cranes, inhabited in Khingan Reserve, in our opinion, could be recognized as natural geochemical background of this territory.

Table 5. Heavy metal concentration in eggshells of birds, mg/kg (n = 3).

Name	Pb	Cd	Cu	Zn	Fe	Ni
White-naped crane *Grus vipio* (Pallas, 1811)	7.9 ± 0.8	0.5	8.3 ± 0.8	2.7 ± 0.4	8.8 ± 0.8	1.5 ± 0.2
Red-crowned crane *Grus japonensis* (S. Müller 1776)	6.1 ± 0.6	0.4	8.4 ± 0.8	3.8 ± 0.4	13.2 ± 1.2	3.0 ± 0.3

4 Conclusions

The obtained research results were the result of monitoring studies in 2016 and 2017 of the habitats of cranes in the Khingan State Nature Reserve in terms of hydrochemical indicators and the content of heavy metals in ecosystem components (water, bottom sediments, fish, macrophytes, eggshells of birds). The waters of the Kleshenskoe and Dolgoe lakes are characterized by high oxygen saturation. However, in the Borzya River, the oxygen content did not exceed 3.6 mgO_2/dm^3. In spring and summer, high values of permanganate oxidizability (PO) were noted in the water bodies, which characterizes the high content of organic matter in the water. A high content of phosphates and nitrates was found in summer samples, and in spring ammonium nitrogen was dominant in the water. The lowest concentrations of nitrogen and phosphorus compounds were observed after the autumn flood. In terms of phosphate content, the water bodies are eutrophic. The content of zinc (17.4–31.1 $\mu g/dm^3$) and copper (3.4–14.7 $\mu g/dm^3$) in the water of the reserve's reservoirs exceeded the fishery standard. The content of lead in water bodies was 3.4–4.1 $\mu g/dm^3$, cadmium –0.20–0.73 $\mu g/dm^3$.

Macrophytes (*Potamogeton maackianus, Myriophyllum spicatum, Trapa maximowiczii*) Showed a high concentration of lead (26.9 ± 3.5 mg/kg, 60.7 ± 6.2 mg/kg and 29.3 ± 3.2 mg/kg respectively). High concentrations of lead and cadmium were found in the muscle tissue of fish (Carassius gibelio, Perccottus glenii, Rhynchocypris percnurus). In the eggshells of the White-naped crane and Red-crowned crane, the content of heavy metals decreases in the order Fe > Cu > Pb > Zn > Ni > Cd. The content of heavy metals in the eggshell of cranes indicates a natural geochemical background of this territory.

References

1. Pakusina, A.P., Platonova, T.P., Lobarev, S.A., Smirenski, S.M.: Chemical and ecological characteristics of lakes located in the Muraviovka park. Asian J. Water Envir. Poll **15**(4), 27–34 (2018). https://doi.org/10.3233/AJW-18005410
2. Xu, P., et al.: Use of aquaculture ponds by globally endangered redcrowned crane (*Grus japonensis*) during the wintering period in the Yancheng National Nature Reserve, a Ramsar wetland. Glob. Ecol. Conserv. **23**, e01123 (2020). https://doi.org/10.1016/j.gecco.2020.e01123
3. Zhang, W., Ma, J.: Waterbirds as bioindicators of wetland heavy metal pollution. Proc. Environ. Sci. **10**, 2769–2774 (2011). https://doi.org/10.1016/j.proenv.2011.09.429

4. Gong, M., et al.: The path forward: conservation of climate change-affected breeding habitat of red-crowned cranes near Zhalong reserve. China Pak. J. Zoo **53**(2), 733–742 (2021)
5. Nikitina, O.I., Dubinina, V.G., Bolgov, M.V., Parilov, M.P., Parilova, T.A.: Environmental flom releasesfor wetland biodiversity conservation in the amur river basin. Water **12**, 2812 (2020). https://doi.org/10.3390/w12102812
6. Vlasova, S.N., Voropayeva, I.S., Dugina, I.O., et al.: Giltchin river watershed. history. wetlands. water resources. In: Smirenski, S.M., Smirenski, E.M. (ed.) Dalnauka: Vladivostok, Russia, p. 202 (2016)
7. Akhtyamov, M.K.H., Morozova, G.Y., Boldovski, N.B., Baburin, A.A.: Muraviovka Park. Natural conditions and vegetation. Vladivostok EFB RAS (2002)
8. Manasypov, R.M., et al.: Biogeochemistry of macrophytes, sediments and porewaters in thermokarst lakes of permafrost peatlands, western Siberia. Sci. Total Environ. **763**, 144201 (2021). https://doi.org/10.1016/j.scitotenv.2020.144201
9. Jaramillo, E., Lagos, N.A., Labra, F., et al.: Recovery of black-necked swans, macrophytes and water quality in a Ramsar wetland of southern Chile: assessing resilience following sudden anthropogenic disturbances. Sci. Total Environ. **628–629**, 291–301 (2018). https://doi.org/10.1016/j.scitotenv.2018.01.333
10. Wenzhi, J.D., Wang, Y.S., Zhang, H., et al.: Effects of environmental metal pollution on reproduction of a free-living resident songbird, the tree sparrow (Passer montanus). Sci. Total Environ. **721**, 137674 (2020). https://doi.org/10.1016/j.scitotenv.2020.137674
11. Toushkin, A., Toushkina, A., Matweeva, O., Senchik, A.: Resources of the manchurian pheasant (Phasianus colchicus Pallasi) in the Amur region. In: E3S Web of Conference, vol. 203, p. 01001 (2020). https://doi.org/10.1051/e3sconf/202020301001
12. Toushkin, A.A., Toushkina, A.F., Matweeva, O.A.: Numbers changes and economic use of the manchurian pheasant (Phasianus colchicus Pallasi) in the Amur region. Far. East Agri. Bull **1**, 84–90 (2020). https://doi.org/10.24411/1999-6837-2020-11011
13. Makhinov, A.N., et al.: Impact of large floods on water resources and infrastructure of urban areas in the Amur river region. In: IOP Conference Series: Earth and Environmental Science, vol. 666, p. 032019 (2021).https://doi.org/10.1088/1755-1315/666/3/032019
14. Bu, H., Meng, W., Zhang, Y.: Nitrogen pollution and source identification in the Haicheng River basin in Northeast China. Sci. Total. Environ. **409**, 3394–3402 (2011). https://doi.org/10.1016/j.scitotenv.2011.05.030
15. Peng, K., Luo, C., Lou, L., Li, X., Shen, Z.: Bioaccumulation of heavy metals by the aquatic plants *Potamogeton pectinatus* L. and *Potamogeton malaianus* Miq. and their potential use for contamination indicators and in wastewater treatment. Sci. Total Environ. **392**, 22–29 (2008). https://doi.org/10.1016/j.scitotenv.2007.11.032
16. Pakusina, A.P., Platonova, T.P., Demidenok, Z.H.A.: Environmental and chemical characteristics of the Zeya-Bureya plain small river Filinovka. In: IOP Conference Series Earth Environment Science, vol. 548, 062015p (2020). https://doi.org/10.1088/1755-1315/548/6/062015
17. Pakusina, A.P., Platonova, T.P.: Ecological and chemical characteristics of small river Arguzikha (Zeya-Bureya plain, Russia). In: E3S Web of Conference, vol. 203, p. 03001 (2020). https://doi.org/10.1051/e3sconf/202020303001
18. Luo, J., Ye, Y., Gao, Z., Wang, Y., Wang, W.: Characterization of heavy metal contamination in the habitat of red-crowned crane (*Grus japonensis*) in Zhalong Wetland, Northeastern China. Bull. Environ. Contam. Toxicol. **93**(3), 327–333 (2014). https://doi.org/10.1007/s00128-014-1331-5
19. Mukhtar, H., Chan, C.-Y., Lin, Y.-P., Lin, C.-M.: Assessing the association and predictability of heavy metals in avian organs, feathers, and bones using crowdsourced samples. Chemosphere **252**, 126583 (2020). https://doi.org/10.1016/j.chemosphere.2020.126583

20. Espin, S., Garcia-Fernandez, A.J., Herzke, D., et al.: Tracking pan-continental trends in environmental contamination using sentinel raptors-what types of samples should we use? Ecotoxicol **25**(4), 777–801 (2016). https://doi.org/10.1007/s10646-016-1636-8
21. Borghesi, F., Dinelli, E., Migani, F., et al.: Assessing environmental pollution in birds: a new methodological approach for interpreting bioaccumulation of trace elements in feather shafts using geochemical sediment data. Meth. Ecol. Evol. **8**, 96–108 (2017). https://doi.org/10.1111/2041-210X.12644

Distribution of Hippophae Rhamnoides L. In Primorsky Krai and Its Invasive Potential

Nina Kolyada[1](✉), Tatyana Moskaliuk[1], and Aleksandr Kolyada[2]

[1] Far Eastern Branch of the Russian Academy of Sciences, Federal Scientific Center of the East Asia Terrestrial Biodiversity, pr-t 100-let Vladivostoka, 159, Vladivostok 690022, Russia
Kolyada18@rambler.ru

[2] Primorye State Agricultural Academy, pr-t Bluckera, 44, Ussuriysk 692500, Russia

Abstract. *Hippophae rhamnoides* L. (Elaeagnaceae Juss.) is a valuable multipurpose plant with vast secondary area which covers many regions of Europe, Asia and North America. When introduced in new places, it often exhibit invasive properties. In Primorsky Krai (Russian Federation) sea buckthorn is detected in 139 settlements where it grows mainly in household plots and is used as medicinal and food plant. Easily forming numerous root suckers, sea buckthorn often create more or less dense thickets. Six places were found in Primorsky Krai where *Hippophae rhamnoides* actively reproduces vegetatively. These places are disturbed areas and sea buckthorn cultivation plots. The plant don't penetrate into natural phytocoenoses which allows regarded it as epecophyte.

Keywords: Sea buckthorn · *Hippophae rhamnoides* · Economic value · Distribution · Russian Federation · Primorsky Krai · Cultivation · Vegetative propagation · Epecophyte

1 Introduction

Hippophae rhamnoides L. (sea buckthorn) is a dioecious woody plant of the family Elaeagnaceae Juss., which has a vast area in the temperate zone of Eurasia [1].

It's a very valuable and multipurpose plant. Its medicinal properties (antioxidant, immunomodulatory, antitumoral, gepatoprotective) are most known [2, 3]. Fruits are a significant source of flavonoids, carotenoids, vitamins (especially A, E and K). Medicinal parts are also twigs and leaves. Due to the high content of antioxidants in fruits, sea buckthorn preparations are effective in the treatment of diabetes, cardiovascular diseases, osteoporosis, skin aging, brain diseases, allergies, ulcerative gastritis, cancer, as well as atrophy and erosion of the mucous tissues of the stomach and genitourinary system, and many other diseases. Vital microelements: Cr, Fe, Mn, Mo and Co, in sea buckthorn are in an easily assimilable biogenic form. It accumulates these and other elements more intensively than other plants and does not absorb toxic ones: lead and arsenic. Not only fruits – almost all sea buckthorn organs contain valuable bioactive elements. The leaves lack caffeine, they have antimicrobial and antiviral activity and serve as raw materials

A. Muratov and S. Ignateva (Eds.): AFE 2021, LNNS 353, pp. 667–675, 2022.
https://doi.org/10.1007/978-3-030-91402-8_74

for vitamin tea. The bark is a valuable raw material for the production of serotonin, a hormone that plays an important role in the body of animals and plants.

Often Hippophae rhamnoides is cultivated in Russia, Europe, United States and other countries as a food plant and source of enrichment of products with biologically active constituents – sugars, organic acids, vitamins, tannins etc. [4]. The most consumed part of the plant are berries [5]. Methods of using sea buckthorn powder in the food industry are being developed including cryochemical processing.

The species is used in cosmetics [6], animal breeding [7], as decorative in urban landscapes. It possesses high anti-erosion properties and used in the reclamation of disturbed areas [8], have nitrogen-fixing ability [9] which allows using it in soil and water conservation [10]. The anti-erosion function is realized due to its intensive sucking formation activity [11]. The root systems of sea buckthorn are superficial. Cord-like, fleshy roots with underdeveloped mechanical tissue, unlike shoots, branch weakly. They move in different directions from the stems far beyond the crown – by 10–12 m, exceeding the diameter of the latter by 1.5 (tall forms of sea buckthorn) – 2.5 (3) times (depressed individuals). The depth of root penetration into the soil is 10–30 cm. Sea buckthorn roots penetrate the upper layers of disturbed areas, holding the loose soil, and the growing offspring prevent the drift and washout of organic matter. The unevenness of the surface between the roots is smoothed out by litter and undifferentiated mass, preventing soil erosion and contributing to the emergence of other plants. In the conditions of the mountainous terrain, these properties of sea buckthorn are especially important for nature protection.

Wide distribution of Hippophae rhamnoides is connected with its ecological properties. It's a very heliophilous, hardy, drought-tolerant, salt-tolerant species which successfully adapts to different extreme conditions. Hippophae rhamnoides can grow on poor soils preferring aerated substrates. Nitrogen-fixing bacteria develop on its roots, so it often settles on substrates poor in organic matter, withstands some soil salinization, but does not tolerate waterlogging. It grows and bears fruit best of all in areas with a high level of groundwater and flowing moisture (along rivers, on the slopes of ravines) on sandy loam and light loamy soils with a slightly acidic and neutral reaction and a high calcium content.

The plant reproduces by seeds and vegetatively [12]. It easily forms numerous root suckers which may create thickets.

The economic use and ecological characteristics have resulted in the formation of an extensive secondary area of Hippophae rhamnoides. Currently, it is found in culture in many countries in Europe, East Asia [13] as well as in North America [14].

The ability of the plant to form root suckers resulted in invasive properties in various places of cultivation.

In Russia Hippophae rhamnoides is included in the Black Book of Flora of Central Russia. It exhibits potentially invasive properties in the Pskov region, Siberia.

It is important that together with Hippophae rhamnoides its phytophagous insects also penetrate.

In the Russian Far East, Hippophae rhamnoides first appeared in the middle of the 20th century. Now it is used mostly as food and medicinal plant, less often as ornamental plant in the Primorsky Krai, Amur Oblast', Khabarovsky Krai. In places of culture the plant often reproduces vegetatively forming root suckers.

In the Amur Oblast', sea buckthorn appeared in the late 1940s as a fruit plant and for protective afforestation. At present, it is found in most settlements of the region, spreading both by root suckers and by seeds carried by rodents and birds.

In the vicinity of Blagoveshchensk (Amur Oblast') it naturalizes most actively in disturbed areas, which are characterized by weakened interspecies competition, high insolation, increased soil salinity and the presence of various water basins.

In the Khabarovskii Krai, sea buckthorn has been found in culture since the second half of the 20th century. Natural spreading has been especially active since the end of the 20th century. On coastal sands, artificially washed sandy hills, it forms thickets. Occurs along railway embankments, weedy places. Thickets can reach an area of tens of square kilometers.

It shows invasive properties in the Khabarovsky Krai, Amur Oblast' (for the valley of the river Bureya it regarded as ergaziophyte, agriophyte, euneophyte).

In the Primorsky Krai, the first plantings of sea buckthorn appeared in the middle of the last century. Back in the 1950s. it was rare enough. It was brought to the Gornotezhnaya Station from Siberia and showed itself as an exceptionally winter-hardy, unpretentious and resistant to diseases and pests plant. Due to its good survival rate and the ability to give massive root shoots, sea buckthorn quickly develops new habitats and becomes one of the most widespread berry crops.

2 Materials and Methods

The distribution of Hippophae rhamnoides in the Primorsky Krai (Russian Far East) was studied. By the traditional route-reconnaissance method more than 200 settlements of the Primorsky Krai were surveyed in the period of 2016–2020.

3 Results and Discussion

In the Primorsky Krai, Hippophae rhamnoides is widely cultivated mainly as a food and medicinal plant. It represented by clone systems, single shrubs or small trees, reaching significant sizes (up to 5 m in height and 28 cm in diameter). According to the results of route studies, Hippophae rhamnoides was found in 139 settlements of the region (Fig. 1).

It is found mainly in household plots, near residential buildings next to highways.

In places of planting, it often creates root suckers. This happens most intensively in the territories of abandoned or destroyed residential buildings or industrial facilities.

We found a number of places in the Primorsky Krai, where Hippophae rhamnoides more or less actively reproduces vegetatively.

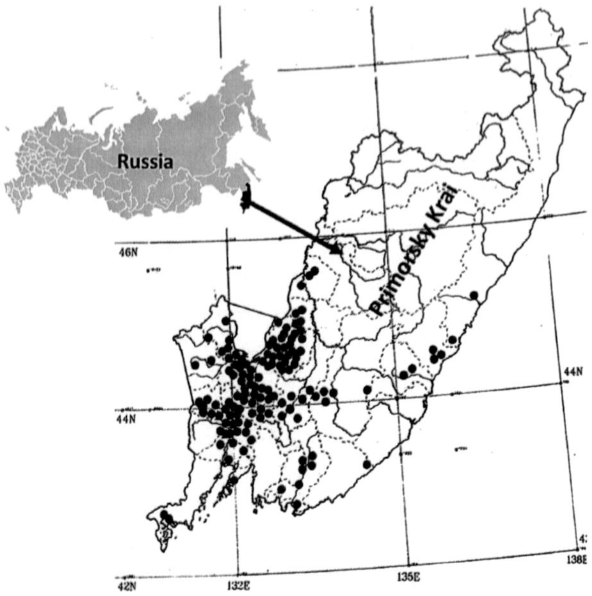

Fig. 1. Distribution of sea buckthorn in settlements in Primorsky Krai.

3.1 Ussuriysk City District

Ussuriysk city (43°48′ N, 131°57′ E). One of the places is the territory of an unfinished combined heat and power plant in the valley of the river Komarovka. Here, before the construction of the power plant begun, there were a dacha settlement and a biological station of the Ussuriysk State Pedagogical Institute. Hippophae rhamnoides were also cultivated on personal plots. This territory was freed from buildings, and the fertile layer was taken out, only small fragments of the original plantings remained.

At the end of the 20th century, colonization of this territory and the adjacent slope by Hippophae rhamnoides fell began. Probably, first diasporas were introduced here by birds, and then vegetative dispersal took place for 10–15 years. In the early 2000s the plant formed a number of clone systems of different sizes. In 2006, the maximum age of maternal individuals was 15 years. Probably, the invasion of sea buckthorn began at the end of the last century. This is quite logical, given that this time coincides with the restoration of summer cottage gardening and the completion of the acclimatization of sea buckthorn in cultural plantings in the region. The variety of colors and shapes of sea buckthorn fruits in the thickets indicates that they are formed by clones of different varieties, which once again emphasizes the fact of introduction in the region.

In 2020, there were 10 clone systems here, each of them includes 5 to 20–30 suckers. The largest one consists of 30 shrub-like root suckers (tree-like forms are absent) up to 2–3 m in height and with a trunk diameter of 1.2 to 2.5 cm. There are many dried up plants.

In twenty years, Hippophae rhamnoides from the valley climbed the slope of the adjacent hill. At present, it has survived only on bare, eroded, heavily indented sections of the slope of the hill, where the fire did not reach. In the areas below, at the base of the hill, there were also quite extensive thickets of sea buckthorn, but because of the fires, now there are only dried burnt individuals and single living ones. It should be noted that fires are a possible reason for the low preservation of sea buckthorn when using it to fix disturbed lands on the dumps of the Luchegorsk coal mine.

The largest of the surviving thickets has an area of 50 × 15 m and has about 150 suckers up to 1.5 (2) m in height and up to 3.5 cm in diameter (Fig. 2). Above, Mongolian oak (Quercus mongolica Fisch. ex Ledeb.) forest begins, into which Hippophae rhamnoides does not enter. Below there are shrub-herbaceous thickets formed by Lespedeza bicolor Turcz., wormwood species (Artemisia L.) and various grasses (Poaceae Barnh.).

In Ussuriysk we also found small clone system of Hippophae rhamnoides on the railway gravel embankment in the southeastern part of the city. It consists of 13 suckers 0.3–1 m high and with stems up to 2 cm in diameter.

Fig. 2. Hippophae *rhamnoides* in Ussuriysk city vicinity.

3.2 Kavalerovsky District

Vicinity of the village Vysokogorsk (44°24′15″ N, 135°24′06″ E). Hippophae rhamnoides formed here a thicket of 20 suckers from 0.7 to 2.5 m in height on an area of 6 × 12 m, with a stem diameter of 1.5–2.5 cm.

3.3 Dalnegorsk City District

Dalnegorsk city district is characterized by wide distribution of Hippophae rhamnoides, both in seliteb areas and vicinity of settlements. In Dalnegorsk city (44°34′ N, 135°37′ E), Rudnaya Pristan' settlement, Monomakhovo settlement etc. it is found in summer cottages, on the hill slopes near the highway, in the valley of the river Rudnaya and seashore dunes. Along the highway from Rudnaya Pristan' to Dalnegorsk we found about 18 clone systems, in which Hippophae rhamnoides often forms a continuous canopy (Fig. 3).

- In Rudnaya Pristan' (44°21′ N, 135°48′ E), the largest clone system consists of 30 specimens from 0.5 to 1.8 cm in height, with trunk diameter 0.8–1.5 cm.
- In Monomakhovo (44°23′53″ N, 135°46′05″ E), a clone system with an area of 10 × 22 m was found, consisting of 35 fruit-bearing suckers, with a trunk diameter of 5–15 cm. One thicket was found on the bank of the river Rudnaya (probably in the place of a previously standing building). It consists of 13 suckers up to 5.5 m in height and with a trunk diameter of up to 10 cm.

Fig. 3. Hippophae *rhamnoides* near Rudnaya Pristan' settlement.

3.4 Nakhodka City District

Nakhodka city (42°49′ N, 132°53′ E). Hippophae rhamnoides is often found here in the landscaping of seliteb areas. Wild individuals were noted on Severny Prospekt]. In 2020, the authors on Severny Avenue (Fig. 4) on a soil and gravel embankment along the road recorded 5 clone systems of sea buckthorn, numbering from 5 to 23 individuals. The largest clone system had an area of 3 × 15 m with 23 individuals 0.5–6 m high and a trunk diameter at the base of up to 15 cm; in addition, there were numerous (about 40) suckers from 0.25 m in height. The largest ones were fruiting.

Fig. 4. Hippophae *rhamnoides* in Nakhodka city.

3.5 Chernigowsky District

In Orekhovo settlement (45°32′49″ N, 134°22′00″ E) Hippophae rhamnoides abundantly grew at the site of the destroyed building on an area of about 100 m2 (Fig. 5). The total number of aboveground stems was 205 specimens 0.3–3 m high, which also sprouted between the remaining foundation slabs.

Fig. 5. Hippophae *rhamnoides* in Orekhovo settlement.

According to our observations, in the Primorsky Krai, Hippophae rhamnoides does not penetrate into natural phytocenoses, settling in disturbed areas. With the appearance of aboriginal trees in the areas and further formation of a closed tree layer above the thickets of Hippophae rhamnoides, it dies from shading. This behavior is typical for sea buckthorn growing on sand dunes in Northern Ireland where Hippophae rhamnoides is well-known dune invader [15].

According to results obtained we regarded sea buckthorn as epecophyte.

4 Conclusions

At present Hippophae rhamnoides in the Primorsky Krai is found in cultivation in most settlements and used as food and medicinal plant. In the absence of planting care, it can run wild and more or less intensively reproduce vegetatively and by seed, in some cases forming dense thickets. The following factors contribute to the appearance of sea buckthorn in disturbed territories: destruction of the soil and vegetation cover, which ensured the absence of competitors; availability of seed sources; the presence of birds – distributers of sea buckthorn fruits.

Today, the introduction of Hippophae rhamnoides into natural phytocenoses is not observed, which allows regarded the species as epecophyte according to the degree of naturalization. However, such factors as ability of the plant for vegetative dispersal and anthropogenic transformation of the vegetation cover of the region can promote penetration of Hippophae rhamnoides into natural communities. So, control over the dispersal of this species is necessary. Considering the high anti-erosion value of Hippophae rhamnoides due to active vegetative mobility, and high adaptation to droughts and frost, it should be recommended for the reclamation of completely disturbed landscapes in the Primorsky Krai.

References

1. Kalia, R.K., Singh, R., Rai, M.K., et al.: Biotechnological interventions in sea buckthorn (Hippophae L.): current status and future prospects. Trees **25**, 559–575 (2011)
2. Suryakumar, G., Gupta, A.: Medicinal and therapeutic potential of Sea buckthorn (*Hippophae rhamnoides* L.). J. Ethnopharmac. **138**, 268–278 (2011). https://doi.org/10.1016/j.jep.2011.09.024
3. Pundira, S., Gargb, P., Dviwedi, A., et al.: Ethnomedicinal uses, phytochemistry and dermatological effects of Hippophae rhamnoides L.: a review. J. Ethnopharmac. **266**, 113434 (2021). https://doi.org/10.1016/j.jep.2020.113434
4. Ciesarova, Z., Murkovic, M., Cejpek, K., et al.: Why is sea buckthorn (Hippophae rhamnoides L.) so exceptional? a review. Food Res. Internat **133**, 109–170 (2020). https://doi.org/10.1016/j.foodres.2020.109170
5. Vilas-Franquesa, A., Saldo, J., Juan, B.: Potential of sea buckthorn-based ingredients for the food and feed industry – a review. Food Prod. Process Nutr. **2**, 17 (2020)
6. Bal, M.L., Meda, V., Naik, N.S., Satya, S.: Sea buckthorn: a potential source of valuable nutrients for nutraceuticals and cosmeceuticals. Food Res. Int. **44**, 1718–1727 (2011)
7. Krejcarova, J., Strakova, E., Suchy, P., et al.: Sea buckthorn (Hippophae rhamnoides L.) as a potential source of nutraceutics and its therapeutic possibilities – a review. Acta Vet. Brno **84**, 257–268 (2015). https://doi.org/10.2754/avb201584030257
8. Enescu, C.M.: Sea-buckthorn: a species with a variety of uses, especially in land reclamation. Dendrobiology **72**, 41–46 (2014)

9. Shan, S.R.P.: Root system of Seabuckthorn (Hippophae rhamnoides L.). Morphology, Metabolism and Gene Expression. Thesis for PhD. Uppsala, p. 65 (2015). https://doi.org/10.13140/RG.2.2.27701.73441
10. Zho, X., Tian, L., Zhang, J., Ma, L., Li, X., Tian, C.H.: Rhizospheric fungi and their link with the nitrogen-fixing Frankia harbored in host plant Hippophae rhamnoides L. Basic Microbiol. **57**(12), 1055–1064 (2017). https://doi.org/10.1002/jobm.201700312
11. Dumitrescu, C.G.: Research on the sea-buckthorn (Hippophae rhamnoides L.) eco-restoration value in avoiding erosion and supporting soil fixation. Bull UASVM Agricul. **73**(2), 343–344 (2016)
12. Li Thomas, S.C., Beveridge Thomas, H.J.: Sea Buckthorn (Hippophae rhamnoides L.): Production and Utilization. Ottawa: NRC Research Press, p. 133 (2013)
13. Nawaz, M.A., et al.: Morphological and genetic diversity of sea buckthorn (hippophae rhamnoides l.) in the Karakoram mountains of Northern Pakistan. Diversity **10**(3), 76 (2018). https://doi.org/10.3390/d10030076
14. Letchamo, W., Ozturk, M., Altay, V., Musayev, M., Mamedov, N.A., Hakeem, K.R.: An alternative potential natural genetic resource: sea buckthorn [Elaeagnus rhamnoides (syn.: Hippophae rhamnoides)]. In: Ozturk, M., Hakeem, K.R., Ashraf, M., Ahmad, M.S.A. (eds.) Global Perspectives on Underutilized Crops, pp. 25–82. Springer, Cham (2018). https://doi.org/10.1007/978-3-319-77776-4_2
15. Richards, E., Burningham, H.: Hippophae rhamnoides on a coastal dune system: a thorny issue? J. Coast Conserv. **15**, 73–85 (2011). https://doi.org/10.1007/s11852-010-0122-3

Control Over the State of Specially Protected Natural Areas and Objects of Regional Significance of the Amur Region

Tatiana Zolotukhina and Galina Stekolnikova(✉)

Far Eastern State Agrarian University, Polytechnical street, 86, Blagoveshchensk, Russia

Abstract. An analysis of the current state and history of the creation of specially protected natural areas and objects of regional significance in the Amur region was carried out. The concept of specially protected natural areas and objects was disclosed; the legislative basis for the creation, functioning and protection of specially protected natural areas and objects was studied. The main categories of specially protected natural areas were considered. The main directions of control over the state of organized specially protected natural areas of regional significance in the Amur region were identified. As a result, it was established that work on the development of a system of specially protected natural areas in the Amur region in recent years has been carried out at a fairly high rate, both in relation to the creation of new protected areas and the improvement of the functioning of already existing reserves and natural monuments. Recently, more and more cases of direct violation of the regime of special protection of specially protected natural areas have been recorded, the most serious of which are the illegal allotment of sites within specially protected natural areas for construction, laying of roads and engineering communications. Often the reason that prevents the prosecution of those responsible is the lack of clearly defined boundaries of specially protected natural areas, which allows developers to dispute the fact that buildings are located in the protected area.

Keywords: Specially protected natural areas · Objects of regional significance · Control over the state of specially protected natural areas · History of the creation of specially protected natural areas · Amur region

1 Introduction

The preservation of unique landscapes, habitats of rare species of flora and fauna, as well as the preservation of unique monuments of cultural heritage is one of the most pressing problems today both in the world, in the state, and in its individual territories. It is for this reason that land legislation provides for a category of specially protected natural territories and objects, which, depending on the decision of the authorities, can be withdrawn partially or completely from circulation and economic use. In addition, a special legal regime is established for such territories [1, 2].

A. Muratov and S. Ignateva (Eds.): AFE 2021, LNNS 353, pp. 676–686, 2022.
https://doi.org/10.1007/978-3-030-91402-8_75

As cities grow, increasing the need for residents in the natural environment; this need becomes the more acute, the more rapidly urbanization proceeds. An increasing number of people consider the natural environment to be the most important foundation for their quality of life. The formation of specially protected natural areas, the composition of objects related to them and the content of nature conservation measures are regulated by law.

Lands of specially protected natural areas and objects are lands of particular importance. At the same time, the significance of these lands is determined by substantive and formal characteristics in accordance with the current legislation, in the scientific, environmental, historical and cultural spheres. When establishing the value of such lands, take into account their aesthetic, therapeutic, recreational or other purpose.

One of the main tasks in the field of preserving and restoring the natural environment is the creation and development of specially protected natural areas of different levels and regimes, primarily specially protected natural areas of federal significance, the formation on their basis, as well as on the basis of other territories, with a predominance of natural processes, natural reserve fund of Russia, preservation of unique natural complexes [3, 4].

In accordance with the law, specially protected natural areas are understood as areas of land, water surface and air space above them, on which natural complexes and objects of special environmental, scientific, cultural, aesthetic, recreational and health-improving significance are located, which are completely or partially excluded from economic use and for which a special protection regime has been established [5, 6].

The scale, configuration and territorial boundaries of specially protected natural areas are established depending on the whole set of natural, climatic and socio-economic conditions, as well as the position and significance of natural, historical and cultural complexes and objects within the boundaries of the planned specially protected natural area.

Taking into account the peculiarities of the regime of specially protected natural areas, the following categories of specially protected natural areas are distinguished:

1. State nature reserves;
2. National parks;
3. Natural parks;
4. State nature protected areas;
5. Natural monuments;
6. Dendrological parks and botanical gardens.

In addition, specially protected natural areas are distinguished by their significance:

- federal significance;
- regional significance;
- local significance.

Specially protected natural areas of federal significance belong to the property of the Russian Federation and are under the jurisdiction of federal government bodies; regional significance refers to the property of the constituent entities of the Russian

Federation; specially protected natural areas of local significance belong to the property of municipalities and are under the jurisdiction of local governments [7].

The allocation of specially protected natural areas in the Constitution of the Russian Federation as an independent object and state interest means that the state attaches considerable independent significance to it. Accordingly, the state should take targeted actions to develop a system of specially protected natural areas and create an adequate, scientifically grounded legal regime to preserve their favorable state.

The basis for initiating the process of creating regional specially protected natural areas is the available data on the position and significance of natural, historical and cultural complexes and objects. The subjects initiating the creation of specially protected natural areas may be state and municipal authorities, public and other organizations, as well as citizens of the Russian Federation [8].

2 Materials and Methods

The history of the creation of specially protected natural areas in the Amur region began on April 11, 1958 by the Resolution of the Council of Ministers of the RSFSR on the creation of the "Khingano-Arkharinsky" reserve of republican significance. The territory chosen for it was really unique - the largest arrays of cedar-deciduous forests on the left bank of the then pristine Arkhara river. Therefore, 1958 is the starting point for specially protected natural areas in the Amur region. The next year, the first reserves of regional significance were established - "Orlovsky" and "Tolbuzinsky".

The most significant for the development of specially protected natural areas of the Amur region was 1963, when on the same day, October 3, two reserves were established - "Zeysky" and "Khingansky". In July of the same year, by the Decree of the Amur Regional Executive Committee No. 304, 6 new zoological reserves of the Hunting department were organized. In 1967, 5 more regional reserves appeared. The next expansion of specially protected natural areas occurred in 1975–76 and by 1986 in the region were 21 regional protected areas, 2 republican protected areas and 2 reserves on a total area of 1.7 million hectares, which amounted to 4.6% of the territory of the Amur region.

The Amur region is distinguished by a fairly high level of biological diversity, which is associated with the peculiarities of the historical development of the territory, its location at the junction of several large floral and faunal zones. Representatives of the northern, eastern Siberian, Okhotsk and southern Dauro-Mongolian and Manchurian floras and faunas meet here, complementing and enriching the natural complexes of the territory. In this regard, the environmental, scientific, economic, recreational value of this territory is very high, creating the potential for sustainable development of the Amur region [9].

At the same time, active economic development of the region's territory, associated primarily with the implementation of large industrial projects for exploration and production of minerals, oil transportation, development of hydropower resources of the territory, requires taking measures to preserve the potential of natural complexes. The greatest efficiency in this direction is demonstrated by the creation of specially protected natural areas, which make it possible to preserve both individual components of natural complexes and the community as a whole. Currently, in the Amur region are 3 federal reserves, 2 federal protected areas, 32 regional protected areas, 117 natural monuments of regional significance. The total area of specially protected natural areas is 3.69 million hectares - 10.17% of the area of the Amur region. This corresponds to the standards adopted by the international community for the share of specially protected natural areas in a particular region [12].

At the same time, in terms of the ratio of the shares of specially protected natural areas in the total area of the subjects of the Far Eastern Federal District, the Amur region occupies only the fifth place, behind the Kamchatka Territory (35.2%), the Republic of Sakha (Yakutia) (29.4%), the Jewish Autonomous region (15.3%), Primorsky Territory (14.4%).

In the Amur region, there is the following ratio of the areas of specially protected natural areas of various types and ranks (Table 1) [21].

Table 1. The ratio of the areas of protected areas of the Amur region.

Type of protected area	Area, ha	In % of the total area of the Amur region
Reserves	407 671	1.13
Protected areas of federal significance	170 256	0.47
Protected areas of regional significance	2 996 004	8.28
Total	407 671	1.13

Specially protected natural areas are unevenly distributed throughout the Amur region. So, in the Arkharinsky district, specially protected natural areas of various ranks occupy 20.7% of the area of the district, in the Skovorodinsky district, 4.4% of the territory, in the Zeysky district, 4.5% of the district's territory. Thus, the most vulnerable natural complexes of the northern territories are protected in the region to the least extent. At the same time, it is here that in recent years the most ambitious industrial development projects have been carried out - the construction of the ESPO oil pipeline, the Ulak-Elga railway, the development of the Garinsky iron ore deposit, the creation of the Nizhne-Zeya hydroelectric power station. All this raises the question of the development and expansion of the system of specially protected natural areas for the government of the Amur region [9].

3 Results

Since mid-2009, in the Amur region, in cooperation with the World Wildlife Fund, work has begun on the creation of new specially protected natural areas of regional significance. First of all, it was carried out in relation to the creation of specially protected natural areas in the northern regions of the region - Skovorodinsky, Zeysky and Mazanovsky.

Based on data on the distribution and number of animals, species diversity of the flora, existing threats to natural complexes, using geoinformation systems, a spatial analysis of the region's territory was carried out, key points for biodiversity conservation were identified, the boundaries of planned specially protected natural areas were determined [11].

Three territories were identified as priorities for creation: the "Verkhne-Amursky" protected area in the Skovorodinsky district, "Tokinsky" in the Zeysky district, and "Nizhne-Norsky" in the Mazanovsky district. In addition, it was decided to expand the territory of the "Andreevsky" protected area in the Arkharinsky district by including in its territory relict arrays of cedar-deciduous forests in the basins of the Maly and Bolshoy Tigan rivers [9].

State natural complex protected area of regional significance "Tokinsky" named after G.A. Fedoseev in the Zeysky district, area of 25,100 hectares. The protected area is located on the southern slopes and foothills of the Stanovoy ridge in the area of the lake Toko, the upper reaches of the river Okonon. The purpose of creating the protected area is to preserve the high-mountain complexes of the Stanovoy ridge, which include elements of the Okhotsk-Kamchatka and East Siberian flora and serve as habitats for the bighorn sheep listed in the Red Data Books of the Khabarovsk Territory and the Amur region. The territory is the place where the traditional use of natural resources is carried out by the representatives of the indigenous peoples of the north of the Amur region.

State natural complex protected area "Verkhne-Amursky" in the Skovorodinsky district, area of 50,700 hectares. The protected area is located in the interfluve of the Oldy and Bolshoi Never rivers, including the Cherpelskie Krivuny tract, unique for the upper reaches of the Amur in terms of floristic composition. The protected area serves as a compensator for the negative impact on the natural complexes of the upper Amur by the construction of the Skovorodino-Dzhalinda-Mohe oil pipeline and the potential construction of a cascade of the Amazar and Dzhalinden hydroelectric power station.

State natural zoological protected area of regional significance "Nizhne-Norsky" Mazanovsky district, area 30278 hectares. A seasonal protected area - from March 1 to November 1. Located on the right bank of the lower reaches of the river Nora. The purpose of creation is to protect summer habitats and migration routes of the Norsk subpopulation of Siberian roe deer, which forms the basis of the herd of this animal in the Amur region. Territory for the prospective discovery of nesting sites for a number of rare bird species, in particular, the fish owl and the black and Far Eastern stork. The territory serves as a buffer zone for the Norsk State Nature Reserve [9].

State complex protected area "Andreevsky" Arkharinsky district, the area expected to be included in the protected area is 15740 hectares. The expansion is supposed to include relict intact arrays of cedar-deciduous forests in the basins of the Maly Tigan and Bolshoi Tigan rivers into the territory. The age of individual Korean pine trees (Korean cedar) here is 600–700 years old, there is an intensive renewal of Korean pine. In the last five years, the presence of the Amur tiger, listed in the Red Data Books of the Russian Federation and the International Union for Conservation of Nature, has been noted on the area.

It should be noted that the need to create the first two territories has been considered since the beginning of the 90th year of the last century, however, several attempts to create them over the past 15 years have not been brought to their logical conclusion.

Comprehensive surveys were conducted in respect of all these territories, as a result of which packages of documents were prepared justifying the creation of new specially protected natural territories-justification for the creation, environmental impact assessment. Public discussions of projects for the creation and expansion of nature protected areas were held, the approval of interested authorities and users carrying out activities in the specially protected natural territories planned for the creation was obtained. A positive conclusion of the state environmental expertise was received for all the protected areas, now they are active [9].

In parallel with the work on preparing the creation of new reserves, in 2009 and 2010 an analysis of the existing system of specially protected natural areas was carried out, on the basis of which a scheme for the development and placement of specially protected natural areas of regional significance in the Amur region was developed, adopted by the decree of the Governor of the Amur region No. 194 dated 25.05 .2010. According to this scheme, work will be carried out to optimize the scheme of specially protected natural areas of regional significance for the coming years. The scheme reflects the creation of the above-mentioned nature reserves, as well as protected zones of the State Nature Reserves "Voskresensky", "Zhelundinsky", "Ust-Tygdinsky", which should compensate for the negative anthropogenic impact on the specified specially protected natural areas.

In addition, one of the main tasks defined by the scheme is the study of the current state of natural monuments of regional and local significance, located on the territory of the region. Currently, these specially protected natural areas, created mainly in the 70–80s of the last century, do not have boundaries and regimes established by the relevant decrees, the state of most of them does not correspond to the environmental requirements established by law [9].

Based on the results of the study, decisions will be made to optimize the system of natural monuments, to enact the provisions on them in accordance with the modern requirements of legislation on specially protected natural areas.

Another program document defining the development of specially protected natural areas was the regional target program "Improving the functioning conditions of the system of specially protected natural areas of the Amur region and the system of protection of wildlife objects of the Amur region for 2009–2013", approved by the Decree of the Government of the Amur region No. 264 of 26.08 2009.

The program identified a number of measures that made it possible to increase the efficiency of the existence of a system of specially protected natural areas of regional significance - the creation of a network of cordons, the improvement of the technical equipment of the reserves, the introduction of methods of aviation registration of animals. The program defines the amount of funding for these activities, as well as the amount of costs for an inventory of natural monuments, the creation of new specially protected natural areas, the development of tourism infrastructure, etc. The total amount of funding for this program was slightly less than 170 million rubles.

Thus, the information of the characteristics of the landscape, its structure, natural and climatic conditions, the relationship of all phenomena, the natural complex and its change under the influence of human activities form the basis for the development of all measures for the protection and improvement of nature. In modern conditions, the problem of optimizing the qualitative and quantitative composition of the entire network of the region is extremely relevant, since the functionality and development prospects of each specially protected natural area and the entire network as a whole depend on these indicators [10].

4 Discussion

The natural complexes of the Amur region are characterized by a low capacity for self-recovery, especially in the northern territories of the region, which makes the issue of preserving slightly disturbed territories even more urgent. The situation is aggravated by the implementation of a number of large industrial projects in hydropower, oil transportation, development of mineral resources and other areas in the region. Strengthening the anthropogenic impact on natural objects poses the task for the government authorities and environmental protection structures to minimize the negative consequences of economic activity for unique and valuable natural complexes, to preserve and develop specially protected natural areas that compensate for this impact [11].

In 2019, in the Amur region, work was carried out to set the boundaries of specially protected natural areas for state cadastral registration. All specially protected natural areas that exist now are put on the state cadastral register. The result of this work was the availability of information about the boundaries and regime of specially protected natural areas for land users, which will allow avoiding unintentional violation of the regime of special protection and the boundaries of specially protected natural areas during any economic and other work in the region.

The overwhelming majority of the existing specially protected natural areas of regional significance in the Amur region were organized in the 1970–1980s of the last century, and their passports and regulations contained extremely incomplete data about the object: there was no comprehensive description of natural features; objects of special protection of specially protected natural areas were not identified; in some cases, it was impossible to clearly define the boundaries of the object from the available description. The areas of specially protected natural areas indicated in the documents often differed significantly from their actual areas.

The work on the state cadastral registration of the boundaries of specially protected natural areas has significantly strengthened the system of specially protected natural areas in the Amur region. All materials are approved by state authorities and specially protected natural areas are now reflected on the Public Cadastral Map. In the future, it is necessary to strengthen monitoring over the observance of the special protection regime for specially protected natural areas in order to timely prevent violations of the current legislation.

The legal regime of specially protected natural areas is regulated by acts of environmental legislation of a general nature, laws on the legal regime of certain types of natural resources, as well as by special legislation on specially protected natural areas - the Federal Laws "On Specially Protected Natural Areas" 1995, "On Natural Medicinal Resources, medical and health-improving areas and resorts "1995, the Resolution of the Government of the Russian Federation" On the procedure for maintaining the state cadastre of specially protected natural areas" dated 19.10.1996 No. 1249, provisions on specific types of specially protected natural areas (reserves, protected areas, national parks, etc.).

The legal regime of specially protected natural areas is established with the aim of:

- preservation of typical and unique landscapes;
- maintaining genetic biodiversity;
- protection of objects of animate and inanimate nature;
- conservation of cultural heritage objects [12].

Control and management in the sphere of functioning and organization of specially protected natural areas shall be carried out by executive bodies of state power or local self-government bodies that have created a specially protected natural area.

In order to take into account specially protected natural areas, a state cadastre of specially protected natural areas is maintained, which includes information about the status of these areas, about their geographical position and boundaries, about the regime of special protection of these areas, nature users, about environmental education, scientific, economic, historical and cultural value. The state cadastre of specially protected natural areas is maintained in order to assess the state of the natural reserve fund, determine the prospects for the development of the network of these territories, increase the efficiency of state control over compliance with the relevant regime, as well as take into account these territories when planning the socio-economic development of regions. The procedure for maintaining the state cadastre of specially protected natural areas is established by Order of the Ministry of Natural Resources of Russia dated 19.03.2012 No. 69 "On approval of the Procedure for maintaining the state cadastre of specially protected natural areas" [13].

For the protection of natural complexes and objects in the territories of state natural reserves and national parks, special state inspectorates are created for the protection of the territories of state natural reserves and national parks, whose employees are included in the staff of the relevant nature conservation institutions. Executive authorities of the constituent entities of the Russian Federation and local self-government bodies can protect specially protected natural areas of regional and local significance through structural units specially created for this purpose, endowed with appropriate powers [14].

The primary normative legal act governing relations in the field of organization, protection and use of specially protected natural areas in the Russian Federation is Federal Law No. 33-FL of March 14, 1995 "On Specially Protected Natural Areas". In addition, the regulatory framework in the field of environmental protection and nature constitute the laws and other legal acts of the Russian Federation and its subjects, as well as specific legislation criminal, civil and administrative branches of law.

The creation and operation of regional specially protected natural areas is governed by the legislation of the respective constituent entities of the Russian Federation. Analysis of the legislation of a number of constituent entities of the Russian Federation revealed that, in contrast to specially protected natural areas of federal significance, the procedures for organizing regional specially protected natural areas are not clearly defined, there are no regulations on the procedure for their creation, the list of documents required for preparing a decision of the executive authority of the constituent entity Of the Russian Federation on the organization of specially protected natural areas is not defined. In addition, the constantly changing federal legislation introduces additional difficulties in resolving these issues. Nevertheless, a number of subjects of Russia have accumulated some experience in the creation of regional specially protected natural areas. The main subject on the part of the state, exercising management and control in the field of organization and functioning of specially protected natural areas of regional significance, are regional power structures [15].

If we consider from the point of view of the content of the management mechanism of specially protected natural areas, then it is a methodological, organizational, economic and legal activity for the formation and use of tools for the development of specially protected natural areas. Management mechanisms for specially protected natural areas of regional significance can be conditionally divided into two groups:

1. Monotype of management, corresponding to the level of public law education (federal, regional, local). This type of mechanism is inherent in the presence of only one status within the boundaries of the territory of objects;
2. Taking into account the combination of statuses located within the boundaries of the administrative-territorial unit, this group combines the following combined types of management mechanisms:

- regional and federal;
- municipal and federal;
- municipal and regional;
- municipal-regional-federal types [16].

Recently, more and more cases of direct violation of the regime of special protection of specially protected natural areas have been recorded, the most serious of which are illegal allotment of sites within specially protected natural areas for construction, laying of roads and engineering communications. Often the reason that prevents the prosecution of those responsible is the lack of clearly defined boundaries of specially protected natural areas, which allows developers to dispute the fact that buildings are located in the protected area.

5 Conclusion

In general, it can be noted that work on the development of a system of specially protected natural areas in the Amur region in recent years has been carried out at a fairly high rate, both in relation to the creation of new protected areas and the improvement of the functioning of already existing reserves and natural monuments. Moreover, a significant role in the activation of this work was played by the interaction of the government of the Amur region with the World Wildlife Fund, which initiated the creation of new specially protected natural areas.

The existence of specially protected natural areas is one of the most important measures to preserve the natural complexes of the Amur region. The totality of the categories of specially protected natural areas available in the region makes it possible to take under protection not only individual components of the environment, but also their complex as a whole, which significantly increases the efficiency of environmental protection measures.

Completion of work on the creation of all reserves and protected areas specified in the scheme, an inventory of natural monuments, the implementation of measures to improve the functioning of the system of specially protected natural areas, will form a sufficiently flexible and stable system of specially protected areas in the Amur region. It will become the ecological framework of the region's territory, which will make it possible to conduct sustainable and non-depleting nature management on the territory of the Amur region in the interests of all residents of the Amur region.

References

1. Dudley, N., Hockings, M., Stolton, S., et al.: Priorities for protected area research. Parks **24**(1), 35–50 (2018). https://doi.org/10.2305/IUCN.CH.2018.PARKS-24-1ND.en
2. Alekseenko, N.A., Koshkarev, A.V., Kuramagomedov, B.M., Medvedev, A.A.: Geoportals of Russian specially protected natural areas. Geodesy Cartography **80**(5), 34–46 (2019). https://doi.org/10.22389/0016-7126-2019-947-5-34-46
3. Lipka, O.N.: Methodological approaches to climate change vulnerability assessment of protected areas. Nat. Conserv. Res. **2**(3), 68–79 (2017). https://doi.org/10.24189/ncr.2017.036
4. Hole, D.G., et al.: Toward a management framework for networks of protected areas in the face of climate change. Conserv. Biol. **25**(2), 305–315 (2011). https://doi.org/10.1111/j.1523-1739.2010.01633.x
5. Elbakidze, M., Angelstam, P., Sobolev, N., et al.: Protected area as an indicator of ecological sustainability? a century of development in Europe's boreal forest. Ambio **42**(2), 201–214 (2013). https://doi.org/10.1007/s13280-012-0375-1
6. Juvonen, S.K., Kuhmonen, A., Opdahl, T., et al.: Evaluation of the Protected Area Network in the Barents Region. Helsinki. Edita Prima Ltd 314 (2013). https://doi.org/10.15838/esc.2019.3.62.7
7. Semin, P.O.: Some patterns of establishing regional categories of specially protected natural areas. Econ. Soc. **21**(2), 11–38 (2020). https://doi.org/10.17323/1726-3247-2020-2-11-38
8. Mueller, L., Sheudshen, A.K., Eulenstein, F.: Novel Methods for Monitoring and Managing Land and Water Resources in Siberia. Series Title: Springer Water, p. 760 (2016).https://doi.org/10.1134/S2079096120020134
9. Alekseenko, N.A.: Topical issues of cartographic support of specially protected natural areas of Russia. Geodesy Cartography **80**(1), 13–23 (2019). https://doi.org/10.22389/0016-7126-2019-943-1-13-23
10. Degteva, S.V., Ponomarev, V.I., Eisenman, S.W., Dushenkov, V.: Striking the balance: challenges and perspectives for the protected areas network in northeastern European Russia. Ambio **44**(6), 473–490 (2015). https://doi.org/10.1007/s13280-015-0636-x
11. Manakov, Y.A., Kupriyanov, O.A.: The system of specially protected natural areas of the Kemerovo region as a factor in mitigating the impact of coal mining on biodiversity. Coal **7**(1120), 89–94 (2019). https://doi.org/10.18796/0041-5790-2019-7-89-94
12. Tikhonova, T.V., Shchenyavsky, V.A.: Accounting ecosystem services in assessing the resource efficiency of specially protected natural areas of the Komi Republic. Econ. Soc. Changes: Facts Trends Forecast **14**(1), 110–124 (2021). https://doi.org/10.15838/esc.2021.1.73.8
13. Smirennikova, E.V., Ukhanova, A.V., Voronina, L.V.: Conflicts in specially protected natural areas of the Arctic region: identification, analysis and solutions. Econ. Soc. Changes: Facts Trends Forecast **12**(3), 107–123 (2019). https://doi.org/10.15838/esc.2019.3.62.7

14. Andreeva, I.V., Tsilikina, S.V.: Transformational processes in environmental systems (on the example of the regions of Western Siberia). Geogr. Nat. **2**, 50–58 (2019). https://doi.org/10.21782/GIPR0206-1619-2019-2(50-58)
15. Wendland, K.J., Baumann, M., Lewis, D.J., Sieber, A., Radeloff, V.C.: Protected area effectiveness in European Russia: a post-matching panel data analysis. Land Econ. **91**(1), 149–168 (2015). https://doi.org/10.3368/le.91.1.149
16. Kuhmonen, A., Mikkola, J., Storrank, B., Lindholm, T., (eds.).: Protected areas and high conservation value forests in the Barents Euro-Arctic Region – Sweden, Finland and Russia. (2017). https://helda.helsinki.fi/handle/10138/229432. DOI: https://doi.org/10.15838/esc.2019.3.62.7

Author Index

A
Abramova, Irina, 88, 339
Amakova, Tatyana, 131

B
Balabanov, Viktor, 397
Balan, Irina, 658
Baranov, Alexander, 479, 505
Barashkova, Anastasia, 70, 96
Barsukova, Elena, 80, 174
Begun, Stepan, 147
Belyaeva, Galina, 105
Berkal, Irina, 487
Bezkorovainaia, Irina, 553
Bodnar, Olena, 54
Bogdan, Polina, 386
Boginskaya, Natalia, 434
Boiarskaia, Anastasiia, 255
Boiarskii, Boris, 255
Bolsheshapova, Nadezhda, 88
Boyarkin, Evgeny, 88, 131
Burkovska, Alla, 54
Burlov, Sergey, 88
Burmaga, Andrey, 36
Butovets, Ekaterina, 219, 275
Bykovskaya, Natalia, 404

C
Chaikina, Elena, 80, 174
Chausov, Nikolai, 264
Chausova, Lyudmila, 264
Chekusov, Maxim, 164
Chikachev, Roman, 495
Chugaeva, Natalia, 404

D
Danilenko, Irina, 275
Darwish, Fadi, 122
Derzhapolskaya, Yulia, 611
Dubkov, Alexandr, 469
Dyadchenko, Olga, 505

E
Emelyanov, Alexey, 44
Epifantsev, Viktor, 19
Ermolaeva, Anna, 357, 620

F
Fedoreeva, Olga, 285, 376
Filatov, Alexander, 202
Firsov, Gennady, 573
Fisenko, Petr, 1, 434
Fokin, Sergey, 246
Fokina, Evgeniya, 105

G
Gainatulina, Vera, 329
Ganzei, Kirill, 514
Gartovannaya, Elena, 357, 620
Goncharuk, Aleksey, 28
Gorbacheva, Natalia, 542
Grebenshchikova, Elena, 542
Gribanova, Svetlana, 611
Guchenko, Svetlana, 9

H
Hasegawa, Hideo, 255

A. Muratov and S. Ignateva (Eds.): AFE 2021, LNNS 353, pp. 687–689, 2022.
https://doi.org/10.1007/978-3-030-91402-8

I
Ilyushko, Marina, 9
Imtosimi, Olga, 293
Inshakov, Sergey, 397
Ivanova, Elena, 303
Ivanova, Natalia, 311
Ivanova, Natalya, 505
Ivashchenko, Anna, 321

K
Kalenik, Tatiana, 122
Kapelian, Alla, 581
Kem, Aleksander, 164
Khasbiullin, Roman, 329
Khasbiullina, Olga, 329
Khovanets, Evgeniy, 542
Kim, Irina, 1, 113, 434, 443
Kiselyova, Alyona, 514
Klimenko, Natalia, 339
Klykov, Aleksey, 113, 443
Klykov, Alexey, 80, 174, 386, 469
Klyuchnikov, Michael, 347
Klyuchnikova, Elena, 347
Klyuchnikova, Natalia, 347
Kolyada, Aleksandr, 667
Kolyada, Nina, 667
Konovalova, Inna, 386
Kostrykina, Svetlana, 628
Kozlova, Lyubov, 70
Kraskovskaya, Natalia, 275
Kryuchkova, Nadezhda, 386
Kuhareva, Olga, 62
Kuliakina, Nadezhda, 193
Kurkova, Irina, 563
Kushaeva, Elena, 469
Kushnarev, Alexey, 28
Kuzmitskaya, Galina, 193
Kuznetsov, Evgeny, 28, 36
Kuznetsova, Alexandra, 366
Kuznetsova, Elena, 339
Kvashko, Ludmila, 185

L
Lapshakova, Lyudmila, 532
Latkin, Alexander, 285, 376
Li, Natalia, 122
Lonkina, Ekaterina, 522
Lontseva, Irina, 415
Loskin, Mikhail, 202
Lukyanchuk, Ludmila, 219
Lysa, Olena, 62

M
Makannikova, Marina, 532
Malinovsky, Nikita, 658
Mamai, Oksana, 185
Marchuk, Lyudmila, 469
Matsishina, Natalia, 443
Matsishina, Nathalia, 1, 434
Melnikova, Tatiana, 238
Mikhailichenko, Oleg, 211
Mishustina, Tetiana, 62
Mitrokhina, Olesya, 28
Mokhan, Oksana, 44
Molchanova, Tatiana, 542
Moskaliuk, Tatyana, 667
Motkina, Elena, 122
Mukhina, Dina, 285, 376
Muratov, Alexey, 156
Murugova, Galina, 174, 386

N
Nemykin, Aleksej, 451
Nianko, Viktoriia, 54, 62

O
Ogorodnikova, Natalia, 264
Osipenko, Elena, 487, 628
Osipov, Yakov, 19
Ostroshenko, Lyudmila, 637
Ostroshenko, Valentina, 637

P
Pak, Larisa, 648
Pakharkova, Nina, 553
Pakusina, Antonina, 658
Panasiuk, Alexander, 19
Parfyonova, Svetlana, 602
Parilov, Mikhail, 658
Parilova, Tatyana, 658
Pashina, Lubov, 311
Piletskaya, Olga, 246
Platonova, Tatyana, 658
Potenko, Tatiana, 185
Prisyazhnaya, Irina, 230
Prisyazhnaya, Serafima, 230
Pshenichnikova, Nina, 514

R
Rafalskaya, Olga, 238
Rafalsky, Sergey, 238
Razgonova, Maya, 122
Redkokashin, Aleksander, 397
Redkokashina, Anna, 397
Reshetnik, Ekaterina, 611
Reshetnikov, Alexander, 70, 138
Reshetnikov, Alexandr, 202
Rodnikova, Ilona, 514
Romashova, Marina, 9
Rubtsova, Tamara, 522

S
Saveleva, Ekaterina, 285, 376
Schmidt, Andrey, 164
Schmidt, Yuliya, 424
Sennikov, Vyacheslav, 415
Senotrusova, Tamara, 122
Sergeeva, Victoria, 479
Shabalina, Olga, 553
Sharipova, Tatiana, 36
Sharov, Maxim, 460
Shchegorets, Olga, 36
Shcherbakova, Olesia, 589
Shchitov, Sergey, 28, 36
Shelkovkina, Natalia, 542
Sherstyukova, Tamara, 321
Shishlov, Sergey, 404
Shishlova, Marina, 404
Shmidt, Yurij, 424
Shuravin, Alexandr, 28
Sinegovskii, Mikhail, 255
Sinegovskiy, Mikhail, 230
Sinyegovskaya, Valentina, 230
Sirenko, Natalia, 54
Slepenkov, Alexander, 36
Sobko, Olga, 1, 434
Soldatova, Lyubov, 424
Solodun, Vladimir, 131
Sorokina, Arina, 147
Sorokina, Galina, 553
Stekolnikova, Galina, 676
Stokoz, Svetlana, 563

T
Tegnerenko, Ekaterina, 553
Telichko, Olga, 44
Tikhomirova, Evgeniya, 211
Tikhonchuk, Pavel, 156
Timchenko, Natalia, 589
Timchenko, Natalya, 505
Timoshinov, Roman, 469
Timoshinova, Oksana, 174
Tishkova, Anna, 293
Tkachenko, Kirill, 573, 581
Tokareva, Olga, 211
Tuaeva, Eugenia, 156

U
Ukhvatkina, Olga, 514

V
Vaitekhovich, Yuriy, 19
Vasina, Evgenia, 219
Vlasenko, Galina, 321
Volchanskaya, Alexandra, 573
Volkov, Dmitriy, 1, 113
Volkov, Dmitry, 443
Vozniuk, Valentina, 113

Y
Yakimenko, Mariya, 147
Yayuk, Lyubov, 303
Yekimov, Sergey, 54, 62
Yudaeva, Natalia, 211
Yust, Natalia, 487, 589

Z
Zaitsev, Alexsander, 131
Zakharova, Elena, 451
Zakipnaya, Elena, 602
Zhupley, Irina, 185, 424
Zolotukhina, Tatiana, 676